AF443203

The **DAWN** of the **LHC ERA**

TASI 2008

The **DAWN** of the **LHC ERA**

TASI 2008

Editor

Tao Han

University of Wisconsin, USA

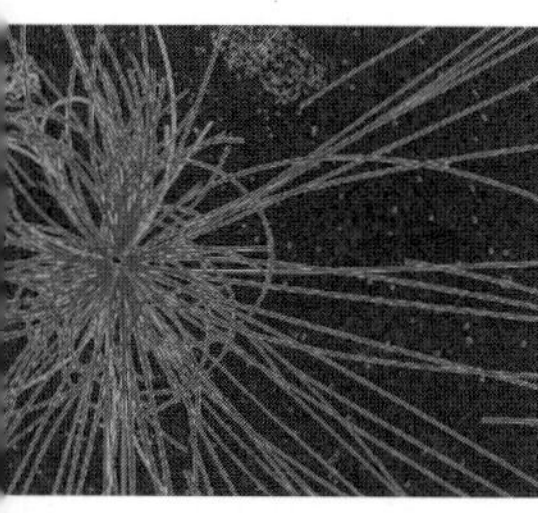

Proceedings of the
2008 Theoretical Advanced
Study Institute in
Elementary Particle Physics

Boulder, Colorado, USA
2 – 27 June 2008

NEW JERSEY · LONDON · SINGAPORE · BEIJING · SHANGHAI · HONG KONG · TAIPEI · CHENNAI

Published by

World Scientific Publishing Co. Pte. Ltd.

5 Toh Tuck Link, Singapore 596224

USA office: 27 Warren Street, Suite 401-402, Hackensack, NJ 07601

UK office: 57 Shelton Street, Covent Garden, London WC2H 9HE

British Library Cataloguing-in-Publication Data
A catalogue record for this book is available from the British Library.

THE DAWN OF THE LHC ERA
TASI 2008
Proceedings of the 2008 Theoretical Advanced Study Institute in Elementary Particle Physics

ISBN-13 978-981-283-835-3
ISBN-10 981-283-835-X

Printed in Singapore by B & JO Enterprise

Preface

When this volume of the Proceedings is published, the CERN Large Hadron Collider (LHC) will have started taking data, and a new era in high energy physics, as well as in science overall, has since begun. The depth of the LHC program in understanding the nature at such short space-time scales, its extent in exploring fundamental questions, the scope of the involvement in the community world-wide are all unprecedented. Revolutionary scientific progress is highly anticipated from the LHC experiments. The LHC will fully explore the Tera-scale physics and possibly beyond: the mechanism of the electroweak symmetry breaking, fermion masses, their flavor mixing and CP violation, new forces and strongly interacting dynamics, the nature of dark matter, larger unification of fundamental interactions, extended symmetries such as Supersymmetry and extra spatial dimensions, to name a few exciting possibilities.

In the recent years, we have also witnessed the major discoveries in the fields of cosmology and neutrino physics. The nature of dark energy, the origin of dark matter, and the cause of the inflation are all mysteries to uncover. The tiny neutrino masses, their large flavor mixing, and the nature of Dirac or Majorana type are all pressing issues to address.

The design of the TASI-2008 lectures reflects these upcoming developments in the field. There are four parts of the lectures:

- The Standard Model And LHC Phenomenology
- LHC Experimentation
- Advanced Theoretical Topics
- Neutrino Physics, Astroparticle Physics, And Cosmology

I feel deeply impressed by the balance of the contents, the coverage of the materials, the pedagogical nature of the presentation and the writing of the lecture notes, for which I am grateful to those dedicated lecturers. I believe that the students in TASI 08 must have learnt a lot from them. The Proceedings will also benefit other researchers in the related fields.

In closing, I would like to thank the TASI general Director, Prof. K.T. Mahanthappa for the enjoyable collaboration during TASI 08; Susan Spika and Elizabeth Price for their efficient secretarial help in making TASI 08 run smoothly; Tom DeGrand for organizing the mountain hikes that challenged our limitations and imagination; and my co-scientific Director Robin Erbacker for helping me putting the program together. TASI Schools thank the National Science Foundation, the Department of Energy and the University of Colorado for financial and material support. I would also like to thank the TASI Scientific Advisor Board to have invited me as one of the Scientific Program Directors, that gave me the opportunity to have worked with these leading physicists as lecturers, and also gave me the pleasure to have interacted with these motivated and talented young students, who are our future in taking high energy physics to a brand new stage.

Go LHC!

Tao Han
University of Wisconsin, Madison

Contents

PART 1

The Standard Model and LHC Phenomenology

Chapter 1

Introduction to the Standard Model and Electroweak Physics

Paul Langacker

School of Natural Sciences, Institute for Advanced Study
Princeton, NJ 08540, USA
E-mail: pgl@ias.edu

A concise introduction is given to the standard model, including the structure of the QCD and electroweak Lagrangians, spontaneous symmetry breaking, experimental tests, and problems.

1.1. The Standard Model Lagrangian

1.1.1. *QCD*

The standard model (SM) is a gauge theory[1,2] of the microscopic interactions. The strong interaction part, quantum chromodynamics (QCD)* is an $SU(3)$ gauge theory described by the Lagrangian density

$$\mathcal{L}_{SU(3)} = -\frac{1}{4} F^i_{\mu\nu} F^{i\mu\nu} + \sum_r \bar{q}_{r\alpha} i \, \not{D}^\alpha_\beta \, q^\beta_r, \tag{1.1}$$

where g_s is the QCD gauge coupling constant,

$$F^i_{\mu\nu} = \partial_\mu G^i_\nu - \partial_\nu G^i_\mu - g_s f_{ijk} \, G^j_\mu \, G^k_\nu \tag{1.2}$$

is the field strength tensor for the gluon fields G^i_μ, $i = 1, \cdots, 8$, and the structure constants f_{ijk} $(i, j, k = 1, \cdots, 8)$ are defined by

$$[\lambda^i, \lambda^j] = 2i f_{ijk} \lambda^k, \tag{1.3}$$

where the $SU(3)$ λ matrices are defined in Table 1.1. The λ's are normalized by $\mathrm{Tr}\, \lambda^i \lambda^j = 2\delta^{ij}$, so that $\mathrm{Tr}\, [\lambda^i, \lambda^j] \lambda^k = 4i f_{ijk}$.

---

*See Ref. [3] for a historical overview. Some recent reviews include Ref. [4] and the QCD review in Ref. [5].

3

Table 1.1. The $SU(3)$ matrices.

$$\lambda^i = \begin{pmatrix} \tau^i & 0 \\ 0 & 0 \end{pmatrix}, \qquad i = 1,2,3$$

$$\lambda^4 = \begin{pmatrix} 0 & 0 & 1 \\ 0 & 0 & 0 \\ 1 & 0 & 0 \end{pmatrix} \qquad \lambda^5 = \begin{pmatrix} 0 & 0 & -i \\ 0 & 0 & 0 \\ i & 0 & 0 \end{pmatrix}$$

$$\lambda^6 = \begin{pmatrix} 0 & 0 & 0 \\ 0 & 0 & 1 \\ 0 & 1 & 0 \end{pmatrix} \qquad \lambda^7 = \begin{pmatrix} 0 & 0 & 0 \\ 0 & 0 & -i \\ 0 & i & 0 \end{pmatrix}$$

$$\lambda^8 = \frac{1}{\sqrt{3}} \begin{pmatrix} 1 & 0 & 0 \\ 0 & 1 & 0 \\ 0 & 0 & -2 \end{pmatrix}$$

The F^2 term leads to three and four-point gluon self-interactions, shown schematically in Figure 1.1. The second term in $\mathcal{L}_{SU(3)}$ is the gauge covariant derivative for the quarks: q_r is the rth quark flavor, $\alpha, \beta = 1, 2, 3$ are color indices, and

$$D^\alpha_{\mu\beta} = (D_\mu)_{\alpha\beta} = \partial_\mu \delta_{\alpha\beta} + i g_s \, G^i_\mu \, L^i_{\alpha\beta}, \tag{1.4}$$

where the quarks transform according to the triplet representation matrices $L^i = \lambda^i/2$. The color interactions are diagonal in the flavor indices, but in general change the quark colors. They are purely vector (parity conserving). There are no bare mass terms for the quarks in (1.1). These would be allowed by QCD alone, but are forbidden by the chiral symmetry of the electroweak part of the theory. The quark masses will be generated later by spontaneous symmetry breaking. There are in addition effective ghost and gauge-fixing terms which enter into the quantization of both the $SU(3)$ and electroweak Lagrangians, and there is the possibility of adding an (unwanted) term which violates CP invariance.

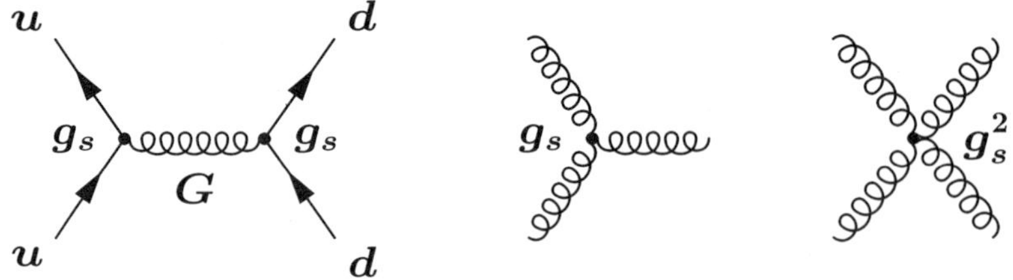

Fig. 1.1.   Interactions in QCD.

QCD has the property of asymptotic freedom,[6,7] i.e., the running coupling becomes weak at high energies or short distances. It has been extensively tested in this regime, as is illustrated in Figure 1.2. At low energies or large distances it becomes strongly coupled (infrared slavery),[8] presumably leading to the confinement of quarks and gluons. QCD incorporates the observed global symmetries of the strong interactions, especially the spontaneously broken global $SU(3) \times SU(3)$ (see, e.g., Ref. [9]).

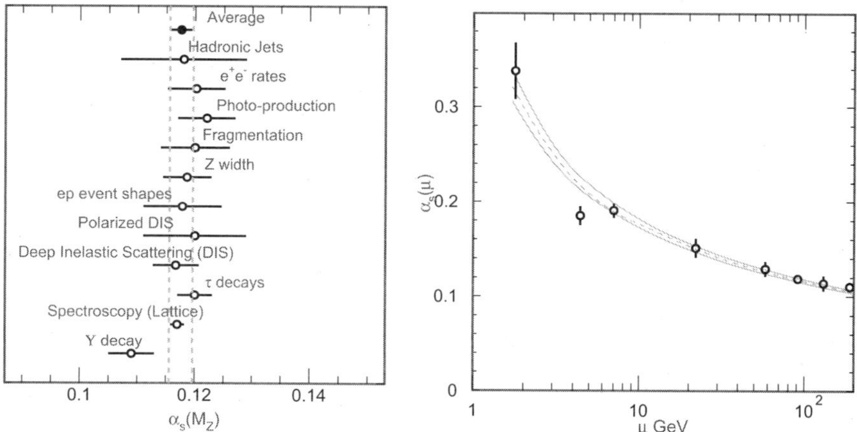

Fig. 1.2. Running of the QCD coupling $\alpha_s(\mu) = g_s(\mu)^2/4\pi$. Left: various experimental determinations extrapolated to $\mu = M_Z$ using QCD. Right: experimental values plotted at the μ at which they are measured. The band is the best fit QCD prediction. Plot courtesy of the Particle Data Group,[5] http://pdg.lbl.gov/.

1.1.2. *The Electroweak Theory*

The electroweak theory[10–12] is based on the $SU(2) \times U(1)$ Lagrangian[†]

$$\mathcal{L}_{SU(2) \times U(1)} = \mathcal{L}_{gauge} + \mathcal{L}_\phi + \mathcal{L}_f + \mathcal{L}_{Yuk}. \tag{1.5}$$

The gauge part is

$$\mathcal{L}_{gauge} = -\frac{1}{4} W_{\mu\nu}^i W^{\mu\nu i} - \frac{1}{4} B_{\mu\nu} B^{\mu\nu}, \tag{1.6}$$

where W_μ^i, $i = 1$, 2, 3 and B_μ are respectively the $SU(2)$ and $U(1)$ gauge fields, with field strength tensors

$$\begin{aligned} B_{\mu\nu} &= \partial_\mu B_\nu - \partial_\nu B_\mu \\ W_{\mu\nu}^i &= \partial_\mu W_\nu^i - \partial_\nu W_\mu^i - g\epsilon_{ijk} W_\mu^j W_\nu^k, \end{aligned} \tag{1.7}$$

[†]For a recent discussion, see the electroweak review in Ref. [5].

where $g(g')$ is the $SU(2)$ $(U(1))$ gauge coupling and ϵ_{ijk} is the totally antisymmetric symbol. The $SU(2)$ fields have three and four-point self-interactions. B is a $U(1)$ field associated with the weak hypercharge $Y = Q - T^3$, where Q and T^3 are respectively the electric charge operator and the third component of weak $SU(2)$. (Their eigenvalues will be denoted by y, q, and t^3, respectively.) It has no self-interactions. The B and W_3 fields will eventually mix to form the photon and Z boson.

The scalar part of the Lagrangian is

$$\mathcal{L}_\phi = (D^\mu \phi)^\dagger D_\mu \phi - V(\phi), \tag{1.8}$$

where $\phi = \begin{pmatrix} \phi^+ \\ \phi^0 \end{pmatrix}$ is a complex Higgs scalar, which is a doublet under $SU(2)$ with $U(1)$ charge $y_\phi = +\frac{1}{2}$. The gauge covariant derivative is

$$D_\mu \phi = \left(\partial_\mu + ig\frac{\tau^i}{2}W^i_\mu + \frac{ig'}{2}B_\mu \right) \phi, \tag{1.9}$$

where the τ^i are the Pauli matrices. The square of the covariant derivative leads to three and four-point interactions between the gauge and scalar fields.

$V(\phi)$ is the Higgs potential. The combination of $SU(2) \times U(1)$ invariance and renormalizability restricts V to the form

$$V(\phi) = +\mu^2 \phi^\dagger \phi + \lambda(\phi^\dagger \phi)^2. \tag{1.10}$$

For $\mu^2 < 0$ there will be spontaneous symmetry breaking. The λ term describes a quartic self-interaction between the scalar fields. Vacuum stability requires $\lambda > 0$.

The fermion term is

$$\mathcal{L}_f = \sum_{m=1}^{F} \left(\bar{q}^0_{mL} i\,\slashed{D}q^0_{mL} + \bar{l}^0_{mL} i\,\slashed{D}l^0_{mL} + \bar{u}^0_{mR} i\,\slashed{D}u^0_{mR} \right. \tag{1.11}$$
$$\left. + \bar{d}^0_{mR} i\,\slashed{D}d^0_{mR} + \bar{e}^0_{mR} i\,\slashed{D}e^0_{mR} + \bar{\nu}^0_{mR} i\,\slashed{D}\nu^0_{mR} \right).$$

In (1.11) m is the family index, $F \geq 3$ is the number of families, and $L(R)$ refer to the left (right) chiral projections $\psi_{L(R)} \equiv (1 \mp \gamma_5)\psi/2$. The left-handed quarks and leptons

$$q^0_{mL} = \begin{pmatrix} u^0_m \\ d^0_m \end{pmatrix}_L \qquad l^0_{mL} = \begin{pmatrix} \nu^0_m \\ e^{-0}_m \end{pmatrix}_L \tag{1.12}$$

transform as $SU(2)$ doublets, while the right-handed fields u^0_{mR}, d^0_{mR}, e^{-0}_{mR}, and ν^0_{mR} are singlets. Their $U(1)$ charges are $y_{q_L} = \frac{1}{6}$, $y_{l_L} = -\frac{1}{2}$, $y_{\psi_R} = q_\psi$.

The superscript 0 refers to the weak eigenstates, i.e., fields transforming according to definite $SU(2)$ representations. They may be mixtures of mass eigenstates (flavors). The quark color indices $\alpha = r$, g, b have been suppressed. The gauge covariant derivatives are

$$D_\mu q^0_{mL} = \left(\partial_\mu + \tfrac{ig}{2}\vec{\tau}\cdot\vec{W}_\mu + \tfrac{ig'}{6}B_\mu\right)q^0_{mL} \qquad D_\mu u^0_{mR} = \left(\partial_\mu + \tfrac{2ig'}{3}B_\mu\right)u^0_{mR}$$

$$D_\mu l^0_{mL} = \left(\partial_\mu + \tfrac{ig}{2}\vec{\tau}\cdot\vec{W}_\mu - \tfrac{ig'}{2}B_\mu\right)l^0_{mL} \qquad D_\mu d^0_{mR} = \left(\partial_\mu - \tfrac{ig'}{3}B_\mu\right)d^0_{mR}$$

$$D_\mu e^0_{mR} = \left(\partial_\mu - ig'B_\mu\right)e^0_{mR}$$

$$D_\mu \nu^0_{mR} = \partial_\mu \nu^0_{mR},$$

$$(1.13)$$

from which one can read off the gauge interactions between the W and B and the fermion fields. The different transformations of the L and R fields (i.e., the symmetry is chiral) is the origin of parity violation in the electroweak sector. The chiral symmetry also forbids any bare mass terms for the fermions. We have tentatively included $SU(2)$-singlet right-handed neutrinos ν^0_{mR} in (1.11), because they are required in many models for neutrino mass. However, they are not necessary for the consistency of the theory or for some models of neutrino mass, and it is not certain whether they exist or are part of the low-energy theory.

The standard model is anomaly free for the assumed fermion content. There are no $SU(3)^3$ anomalies because the quark assignment is non-chiral, and no $SU(2)^3$ anomalies because the representations are real. The $SU(2)^2Y$ and Y^3 anomalies cancel between the quarks and leptons in each family, by what appears to be an accident. The $SU(3)^2Y$ and Y anomalies cancel between the L and R fields, ultimately because the hypercharge assignments are made in such a way that $U(1)_Q$ will be non-chiral.

The last term in (1.5) is

$$\mathcal{L}_{Yuk} = -\sum_{m,n=1}^{F}\left[\Gamma^u_{mn}\bar{q}^0_{mL}\tilde{\phi}u^0_{nR} + \Gamma^d_{mn}\bar{q}^0_{mL}\phi d^0_{nR}\right.$$

$$\left.+ \Gamma^e_{mn}\bar{l}^0_{mn}\phi e^0_{nR} + \Gamma^\nu_{mn}\bar{l}^0_{mL}\tilde{\phi}\nu^0_{nR}\right] + h.c.,$$

$$(1.14)$$

where the matrices Γ_{mn} describe the Yukawa couplings between the single Higgs doublet, ϕ, and the various flavors m and n of quarks and leptons. One needs representations of Higgs fields with $y = +\tfrac{1}{2}$ and $-\tfrac{1}{2}$ to give masses to the down quarks and electrons $(+\tfrac{1}{2})$, and to the up quarks and neutrinos $(-\tfrac{1}{2})$. The representation $\phi^\dagger$ has $y = -\tfrac{1}{2}$, but transforms as the 2^* rather than the 2. However, in $SU(2)$ the 2^* representation is related to the 2 by a similarity transformation, and $\tilde{\phi} \equiv i\tau^2 \phi^\dagger = \begin{pmatrix} \phi^{0\dagger} \\ -\phi^- \end{pmatrix}$ transforms as a 2 with $y_{\tilde{\phi}} = -\frac{1}{2}$. All of the masses can therefore be generated with a single Higgs doublet if one makes use of both ϕ and $\tilde{\phi}$. The fact that the fundamental and its conjugate are equivalent does not generalize to higher unitary groups. Furthermore, in supersymmetric extensions of the standard model the supersymmetry forbids the use of a single Higgs doublet in both ways in the Lagrangian, and one must add a second Higgs doublet. Similar statements apply to most theories with an additional $U(1)'$ gauge factor, i.e., a heavy Z' boson.

1.2. Spontaneous Symmetry Breaking

Gauge invariance (and therefore renormalizability) does not allow mass terms in the Lagrangian for the gauge bosons or for chiral fermions. Massless gauge bosons are not acceptable for the weak interactions, which are known to be short-ranged. Hence, the gauge invariance must be broken spontaneously,[13–18] which preserves the renormalizability.[19–22] The idea is that the lowest energy (vacuum) state does not respect the gauge symmetry and induces effective masses for particles propagating through it.

Let us introduce the complex vector

$$v = \langle 0|\phi|0 \rangle = \text{ constant}, \tag{1.15}$$

which has components that are the vacuum expectation values of the various complex scalar fields. v is determined by rewriting the Higgs potential as a function of v, $V(\phi) \to V(v)$, and choosing v such that V is minimized. That is, we interpret v as the lowest energy solution of the classical equation of motion[‡]. The quantum theory is obtained by considering fluctuations around this classical minimum, $\phi = v + \phi'$.

The single complex Higgs doublet in the standard model can be rewritten in a Hermitian basis as

$$\phi = \begin{pmatrix} \phi^+ \\ \phi^0 \end{pmatrix} = \begin{pmatrix} \frac{1}{\sqrt{2}}(\phi_1 - i\phi_2) \\ \frac{1}{\sqrt{2}}(\phi_3 - i\phi_4) \end{pmatrix}, \tag{1.16}$$

[‡]It suffices to consider constant v because any space or time dependence $\partial_\mu v$ would increase the energy of the solution. Also, one can take $\langle 0|A_\mu|0 \rangle = 0$, because any non-zero vacuum value for a higher-spin field would violate Lorentz invariance. However, these extensions are involved in higher energy classical solutions (topological defects), such as monopoles, strings, domain walls, and textures.[23,24]

where $\phi_i = \phi_i^\dagger$ represent four Hermitian fields. In this new basis the Higgs potential becomes

$$V(\phi) = \frac{1}{2}\mu^2\left(\sum_{i=1}^{4}\phi_i^2\right) + \frac{1}{4}\lambda\left(\sum_{i=1}^{4}\phi_i^2\right)^2, \tag{1.17}$$

which is clearly $O(4)$ invariant. Without loss of generality we can choose the axis in this four-dimensional space so that $\langle 0|\phi_i|0\rangle = 0$, $i = 1, 2, 4$ and $\langle 0|\phi_3|0\rangle = \nu$. Thus,

$$V(\phi) \to V(v) = \frac{1}{2}\mu^2\nu^2 + \frac{1}{4}\lambda\nu^4, \tag{1.18}$$

which must be minimized with respect to ν. Two important cases are illustrated in Figure 1.3. For $\mu^2 > 0$ the minimum occurs at $\nu = 0$. That is, the vacuum is empty space and $SU(2) \times U(1)$ is unbroken at the minimum. On the other hand, for $\mu^2 < 0$ the $\nu = 0$ symmetric point is unstable, and the minimum occurs at some nonzero value of ν which breaks the $SU(2) \times U(1)$ symmetry. The point is found by requiring

$$V'(\nu) = \nu(\mu^2 + \lambda\nu^2) = 0, \tag{1.19}$$

which has the solution $\nu = \left(-\mu^2/\lambda\right)^{1/2}$ at the minimum. (The solution for $-\nu$ can also be transformed into this standard form by an appropriate $O(4)$ transformation.) The dividing point $\mu^2 = 0$ cannot be treated classically. It is necessary to consider the one loop corrections to the potential, in which case it is found that the symmetry is again spontaneously broken.[25]

We are interested in the case $\mu^2 < 0$, for which the Higgs doublet is replaced, in first approximation, by its classical value $\phi \to \frac{1}{\sqrt{2}}\begin{pmatrix}0\\\nu\end{pmatrix} \equiv v$. The generators L^1, L^2, and $L^3 - Y$ are spontaneously broken (e.g., $L^1 v \neq 0$). On the other hand, the vacuum carries no electric charge ($Qv = (L^3 + Y)v = 0$), so the $U(1)_Q$ of electromagnetism is not broken. Thus, the electroweak $SU(2) \times U(1)$ group is spontaneously broken to the $U(1)_Q$ subgroup, $SU(2) \times U(1)_Y \to U(1)_Q$.

To quantize around the classical vacuum, write $\phi = v + \phi'$, where ϕ' are quantum fields with zero vacuum expectation value. To display the physical particle content it is useful to rewrite the four Hermitian components of ϕ' in terms of a new set of variables using the Kibble transformation:[26]

$$\phi = \frac{1}{\sqrt{2}}e^{i\sum \xi^i L^i}\begin{pmatrix}0\\\nu + H\end{pmatrix}. \tag{1.20}$$

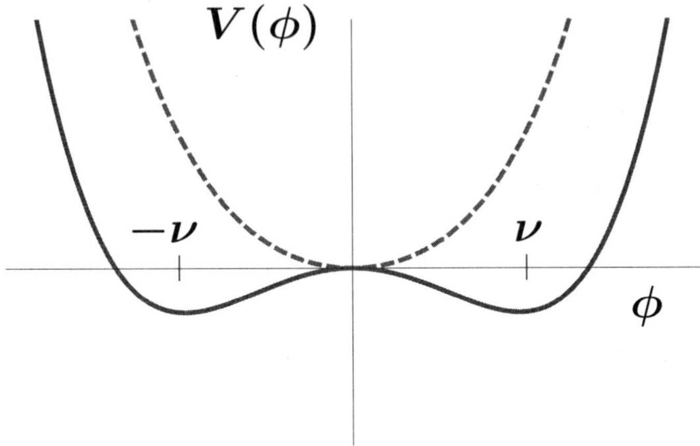

Fig. 1.3.   The Higgs potential $V(\phi)$ for $\mu^2 > 0$ (dashed line) and $\mu^2 < 0$ (solid line).

H is a Hermitian field which will turn out to be the physical Higgs scalar. If we had been dealing with a spontaneously broken global symmetry the three Hermitian fields ξ^i would be the massless pseudoscalar Nambu-Goldstone bosons[27–30] that are necessarily associated with broken symmetry generators. However, in a gauge theory they disappear from the physical spectrum. To see this it is useful to go to the unitary gauge

$$\phi \to \phi' = e^{-i \sum \xi^i L^i} \phi = \frac{1}{\sqrt{2}} \begin{pmatrix} 0 \\ \nu + H \end{pmatrix}, \qquad (1.21)$$

in which the Goldstone bosons disappear. In this gauge, the scalar covariant kinetic energy term takes the simple form

$$(D_\mu \phi)^\dagger D^\mu \phi = \frac{1}{2}(0 \; \nu) \left[\frac{g}{2} \tau^i W_\mu^i + \frac{g'}{2} B_\mu \right]^2 \begin{pmatrix} 0 \\ \nu \end{pmatrix} + H \text{ terms}$$

$$\to M_W^2 W^{+\mu} W_\mu^- + \frac{M_Z^2}{2} Z^\mu Z_\mu + H \text{ terms}, \qquad (1.22)$$

where the kinetic energy and gauge interaction terms of the physical H particle have been omitted. Thus, spontaneous symmetry breaking generates mass terms for the W and Z gauge bosons

$$W^\pm = \frac{1}{\sqrt{2}}(W^1 \mp i W^2)$$

$$Z = -\sin\theta_W B + \cos\theta_W W^3. \qquad (1.23)$$

The photon field

$$A = \cos\theta_W B + \sin\theta_W W^3 \tag{1.24}$$

remains massless. The masses are

$$M_W = \frac{g\nu}{2} \tag{1.25}$$

and

$$M_Z = \sqrt{g^2 + g'^2}\,\frac{\nu}{2} = \frac{M_W}{\cos\theta_W}, \tag{1.26}$$

where the weak angle is defined by

$$\tan\theta_W \equiv \frac{g'}{g} \;\Rightarrow\; \sin^2\theta_W = 1 - \frac{M_W^2}{M_Z^2}. \tag{1.27}$$

One can think of the generation of masses as due to the fact that the W and Z interact constantly with the condensate of scalar fields and therefore acquire masses, in analogy with a photon propagating through a plasma. The Goldstone boson has disappeared from the theory but has reemerged as the longitudinal degree of freedom of a massive vector particle.

It will be seen below that $G_F/\sqrt{2} \sim g^2/8M_W^2$, where $G_F = 1.16637(5) \times 10^{-5}$ GeV^{-2} is the Fermi constant determined by the muon lifetime. The weak scale ν is therefore

$$\nu = 2M_W/g \simeq (\sqrt{2}G_F)^{-1/2} \simeq 246 \text{ GeV}. \tag{1.28}$$

Similarly, $g = e/\sin\theta_W$, where e is the electric charge of the positron. Hence, to lowest order

$$M_W = M_Z \cos\theta_W \sim \frac{(\pi\alpha/\sqrt{2}G_F)^{1/2}}{\sin\theta_W}, \tag{1.29}$$

where $\alpha \sim 1/137.036$ is the fine structure constant. Using $\sin^2\theta_W \sim 0.23$ from neutral current scattering, one expects $M_W \sim 78$ GeV, and $M_Z \sim 89$ GeV. (These predictions are increased by $\sim (2-3)$ GeV by loop corrections.) The W and Z were discovered at CERN by the UA1[31] and UA2[32] groups in 1983. Subsequent measurements of their masses and other properties have been in excellent agreement with the standard model expectations (including the higher-order corrections).[5] The current values are

$$M_W = 80.398 \pm 0.025 \text{ GeV}, \qquad M_Z = 91.1876 \pm 0.0021 \text{ GeV}. \tag{1.30}$$

1.3. The Higgs and Yukawa Interactions

The full Higgs part of $\mathcal{L}$ is

$$
\begin{aligned}
\mathcal{L}_\phi &= (D^\mu \phi)^\dagger D_\mu \phi - V(\phi) \\
&= M_W^2 W^{\mu+} W_\mu^- \left(1 + \frac{H}{\nu}\right)^2 + \frac{1}{2} M_Z^2 Z^\mu Z_\mu \left(1 + \frac{H}{\nu}\right)^2 \\
&\quad + \frac{1}{2} \left(\partial_\mu H\right)^2 - V(\phi).
\end{aligned}
\tag{1.31}
$$

The second line includes the W and Z mass terms and also the ZZH^2, $W^+W^-H^2$ and the induced ZZH and W^+W^-H interactions, as shown in Table 1.2 and Figure 1.4. The last line includes the canonical Higgs kinetic energy term and the potential.

Table 1.2. Feynman rules for the gauge and Higgs interactions after SSB, taking combinatoric factors into account. The momenta and quantum numbers flow into the vertex. Note the dependence on M/ν or M^2/ν.

$W_\mu^+ W_\nu^- H$:	$\frac{1}{2} i g_{\mu\nu} g^2 \nu = 2 i g_{\mu\nu} \frac{M_W^2}{\nu}$	$W_\mu^+ W_\nu^- H^2$:	$\frac{1}{2} i g_{\mu\nu} g^2 = 2 i g_{\mu\nu} \frac{M_W^2}{\nu^2}$
$Z_\mu Z_\nu H$:	$\frac{i g_{\mu\nu} g^2 \nu}{2\cos^2\theta_W} = 2 i g_{\mu\nu} \frac{M_Z^2}{\nu}$	$Z_\mu Z_\nu H^2$:	$\frac{i g_{\mu\nu} g^2}{2\cos^2\theta_W} = 2 i g_{\mu\nu} \frac{M_Z^2}{\nu^2}$
H^3:	$-6 i \lambda \nu = -3 i \frac{M_H^2}{\nu}$	H^4:	$-6 i \lambda = -3 i \frac{M_H^2}{\nu^2}$
$H \bar{f} f$:	$-i h_f = -i \frac{m_f}{\nu}$		

$$
\begin{aligned}
&W_\mu^+(p)\gamma_\nu(q)W_\sigma^-(r) && ie\, \mathcal{C}_{\mu\nu\sigma}(p,q,r) \\
&W_\mu^+(p)Z_\nu(q)W_\sigma^-(r) && i\frac{e}{\tan\theta_W}\mathcal{C}_{\mu\nu\sigma}(p,q,r) \\
&W_\mu^+ W_\nu^+ W_\sigma^- W_\rho^- && i\frac{e^2}{\sin^2\theta_W}\mathcal{Q}_{\mu\nu\rho\sigma} \\
&W_\mu^+ Z_\nu \gamma_\sigma W_\rho^- && -i\frac{e^2}{\tan\theta_W}\mathcal{Q}_{\mu\rho\nu\sigma} \\
&W_\mu^+ Z_\nu Z_\sigma W_\rho^- && -i\frac{e^2}{\tan^2\theta_W}\mathcal{Q}_{\mu\rho\nu\sigma} \\
&W_\mu^+ \gamma_\nu \gamma_\sigma W_\rho^- && -ie^2 \mathcal{Q}_{\mu\rho\nu\sigma} \\
&\mathcal{C}_{\mu\nu\sigma}(p,q,r) \equiv g_{\mu\nu}(q-p)_\sigma + g_{\mu\sigma}(p-r)_\nu + g_{\nu\sigma}(r-q)_\mu \\
&\mathcal{Q}_{\mu\nu\rho\sigma} \equiv 2 g_{\mu\nu} g_{\rho\sigma} - g_{\mu\rho} g_{\nu\sigma} - g_{\mu\sigma} g_{\nu\rho}
\end{aligned}
$$

After symmetry breaking the Higgs potential in unitary gauge becomes

$$
V(\phi) = -\frac{\mu^4}{4\lambda} - \mu^2 H^2 + \lambda \nu H^3 + \frac{\lambda}{4} H^4.
\tag{1.32}
$$

The first term in the Higgs potential V is a constant, $\langle 0|V(\nu)|0\rangle = -\mu^4/4\lambda$. It reflects the fact that V was defined so that $V(0) = 0$, and therefore $V < 0$ at the minimum. Such a constant term is irrelevant to physics in

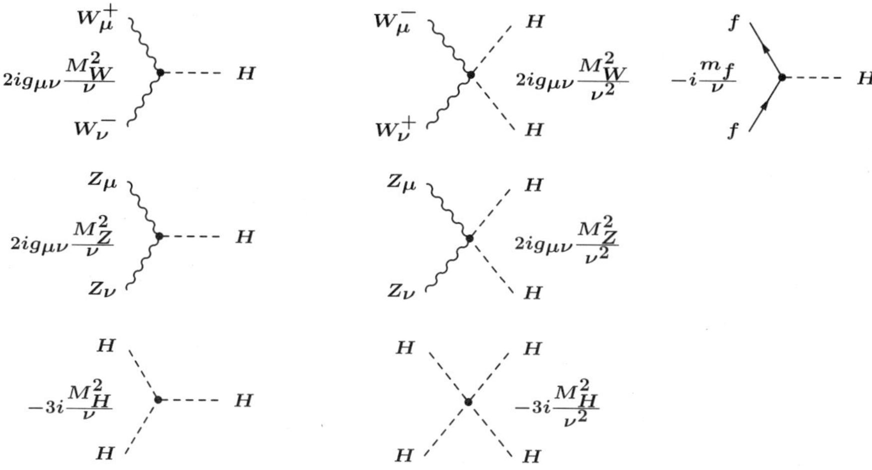

Fig. 1.4. Higgs interaction vertices in the standard model.

the absence of gravity, but will be seen in Section 1.5 to be one of the most serious problems of the SM when gravity is incorporated because it acts like a cosmological constant much larger (and of opposite sign) than is allowed by observations. The third and fourth terms in V represent the induced cubic and quartic interactions of the Higgs scalar, shown in Table 1.2 and Figure 1.4.

The second term in V represents a (tree-level) mass

$$M_H = \sqrt{-2\mu^2} = \sqrt{2\lambda}\nu, \tag{1.33}$$

for the Higgs boson. The weak scale is given in (1.28), but the quartic Higgs coupling λ is unknown, so M_H is not predicted. A priori, λ could be anywhere in the range $0 \le \lambda < \infty$. There is an experimental lower limit $M_H \gtrsim 114.4$ GeV at 95% cl from LEP.[33] Otherwise, the decay $Z \to Z^*H$ would have been observed.

There are also plausible theoretical limits. If $\lambda > \mathcal{O}(1)$ the theory becomes strongly coupled ($M_H > \mathcal{O}(1\text{ TeV})$). There is not really anything wrong with strong coupling a priori. However, there are fairly convincing triviality limits, which basically say that the running quartic coupling would become infinite within the domain of validity of the theory if λ and therefore M_H is too large. If one requires the theory to make sense to infinite energy, one runs into problems[§] for any λ. However, one only needs for the theory to hold up to the next mass scale Λ, at which point the standard model breaks down. In that case,[34-36]

$$M_H < \begin{cases} \mathcal{O}(180)\ GeV, & \Lambda \sim M_P \\ O(700)\ GeV, & \Lambda \sim 2M_H. \end{cases} \qquad (1.34)$$

The more stringent limit of $O(180)$ GeV obtains for Λ of order of the Planck scale $M_P = G_N^{-1/2} \sim 10^{19}$ GeV. If one makes the less restrictive assumption that the scale Λ of new physics can be small, one obtains a weaker limit. Nevertheless, for the concept of an elementary Higgs field to make sense one should require that the theory be valid up to something of order of $2M_H$, which implies that $M_H < O(700)$ GeV. These estimates rely on perturbation theory, which breaks down for large λ. However, they can be justified by nonperturbative lattice calculations,[37-39] which suggest an absolute upper limit of $650 - 700$ GeV. There are also comparable upper bounds from the validity of unitarity at the tree level,[40] and *lower* limits from vacuum stability.[34,41-43] The latter again depends on the scale Λ, and requires $M_H \gtrsim 130$ GeV for $\Lambda = M_P$ (lowered to ~ 115 GeV if one allows a sufficiently long-lived metastable vacuum[42,43]), with a weaker constraint for lower Λ.

The Yukawa interaction in the unitary gauge becomes

$$-\mathcal{L}_{Yuk} \rightarrow \sum_{m,n=1}^{F} \bar{u}_{mL}^{0} \Gamma_{mn}^{u} \left(\frac{\nu + H}{\sqrt{2}} \right) u_{mR}^{0} + (d,e,\nu)\ \text{terms} + \ h.c.$$

$$= \bar{u}_L^0 \left(M^u + h^u H \right) u_R^0 + (d,e,\nu)\ \text{terms} + h.c., \qquad (1.35)$$

where in the second form $u_L^0 = \left(u_{1L}^0 u_{2L}^0 \cdots u_{FL}^0 \right)^T$ is an F-component column vector, with a similar definition for u_R^0. M^u is an $F \times F$ fermion mass matrix $M_{mn}^u = \Gamma_{mn}^u \nu/\sqrt{2}$ induced by spontaneous symmetry breaking, and $h^u = M^u/\nu = gM^u/2M_W$ is the Yukawa coupling matrix.

In general M is not diagonal, Hermitian, or symmetric. To identify the physical particle content, it is necessary to diagonalize M by separate unitary transformations A_L and A_R on the left- and right-handed fermion fields. (In the special case that M^u is Hermitian, one can take $A_L = A_R$).

[§]This is true for a pure λH^4 theory. The presence of other interactions may eliminate the problems for small λ.

Then,

$$A_L^{u\dagger} M^u A_R^u = M_D^u = \begin{pmatrix} m_u & 0 & 0 \\ 0 & m_c & 0 \\ 0 & 0 & m_t \end{pmatrix} \qquad (1.36)$$

is a diagonal matrix with eigenvalues equal to the physical masses of the charge $\frac{2}{3}$ quarks[¶]. Similarly, one diagonalizes the down quark, charged lepton, and neutrino mass matrices by

$$\begin{aligned} A_L^{d\dagger} M^d A_R^d &= M_D^d \\ A_L^{e\dagger} M^e A_R^e &= M_D^e \\ A_L^{\nu\dagger} M^\nu A_R^\nu &= M_D^\nu. \end{aligned} \qquad (1.37)$$

In terms of these unitary matrices we can define mass eigenstate fields $u_L = A_L^{u\dagger} u_L^0 = (u_L \ c_L \ t_L)^T$, with analogous definitions for $u_R = A_R^{u\dagger} u_R^0$, $d_{L,R} = A_{L,R}^{d\dagger} d_{L,R}^0$, $e_{L,R} = A_{L,R}^{e\dagger} e_{L,R}^0$, and $\nu_{L,R} = A_{L,R}^{\nu\dagger} \nu_{L,R}^0$. Typical estimates of the quark masses are[5,9] $m_u \sim 1.5 - 3$ MeV, $m_d \sim 3 - 7$ MeV, $m_s \sim 70 - 120$ MeV, $m_c \sim 1.5 - 1.8$ GeV, $m_b \sim 4.7 - 5.0$ GeV, and $m_t = 170.9 \pm 1.8$ GeV. These are the current masses: for QCD their effects are identical to bare masses in the QCD Lagrangian. They should not be confused with the constituent masses of order 300 MeV generated by the spontaneous breaking of chiral symmetry in the strong interactions. Including QCD renormalizations, the u, d, and s masses are running masses evaluated at 2 GeV^2, while $m_{c,b,t}$ are pole masses.

So far we have only allowed for ordinary Dirac mass terms of the form $\bar{\nu}_{mL}^0 \nu_{nR}^0$ for the neutrinos, which can be generated by the ordinary Higgs mechanism. Another possibility are lepton number violating Majorana masses, which require an extended Higgs sector or higher-dimensional operators. It is not clear yet whether Nature utilizes Dirac masses, Majorana masses, or both[‖]. What is known, is that the neutrino mass eigenvalues are tiny compared to the other masses, $\lesssim \mathcal{O}(0.1)$ eV, and most experiments are insensitive to them. In describing such processes, one can ignore Γ^ν, and the ν_R effectively decouple. Since $M^\nu \sim 0$ the three mass eigenstates are effectively degenerate with eigenvalues 0, and the eigenstates are arbitrary. That is, there is nothing to distinguish them except their weak interactions,

---

[¶]From (1.36) and its conjugate one has $\hat{A}_L^{u\dagger} M^u M^{u\dagger} \hat{A}_L^u = \hat{A}_R^{u\dagger} M^{u\dagger} M^u \hat{A}_R^u = M_D^{u2}$. But $MM^\dagger$ and $M^\dagger M$ are Hermitian, so $A_{L,R}$ can then be constructed by elementary techniques, up to overall phases that can be chosen to make the mass eigenvalues real and positive, and to remove unobservable phases from the weak charged current.

[‖]For reviews, see Refs. [44–47].

so we can simply define ν_e, ν_μ, ν_τ as the weak interaction partners of the e, μ, and τ, which is equivalent to choosing $A_L^\nu \equiv A_L^e$ so that $\nu_L = A_L^{e\dagger} \nu_L^0$. Of course, this is not appropriate for physical processes, such as oscillation experiments, that *are* sensitive to the masses or mass differences.

In terms of the mass eigenstate fermions,

$$-\mathcal{L}_{Yuk} = \sum_i m_i \bar{\psi}_i \psi_i \left(1 + \frac{g}{2M_W} H\right) = \sum_i m_i \bar{\psi}_i \psi_i \left(1 + \frac{H}{\nu}\right). \quad (1.38)$$

The coupling of the physical Higgs boson to the ith fermion is $gm_i/2M_W$, which is very small except for the top quark. The coupling is flavor-diagonal in the minimal model: there is just one Yukawa matrix for each type of fermion, so the mass and Yukawa matrices are diagonalized by the same transformations. In generalizations in which more than one Higgs doublet couples to each type of fermion there will in general be flavor-changing Yukawa interactions involving the physical neutral Higgs fields.[48] There are stringent limits on such couplings; for example, the $K_L - K_S$ mass difference implies $h/M_H < 10^{-6} \, \mathrm{GeV}^{-1}$, where h is the $\bar{d}s$ Yukawa coupling.[49–51]

1.4. The Gauge Interactions

The major quantitative tests of the electroweak standard model involve the gauge interactions of fermions and the properties of the gauge bosons. The charged current weak interactions of the Fermi theory and its extension to the intermediate vector boson theory** are incorporated into the standard model, as is quantum electrodynamics. The theory successfully predicted the existence and properties of the weak neutral current. In this section I summarize the structure of the gauge interactions of fermions.

1.4.1. *The Charged Current*

The interaction of the W bosons to fermions is given by

$$\mathcal{L} = -\frac{g}{2\sqrt{2}} \left(J_W^\mu W_\mu^- + J_W^{\mu\dagger} W_\mu^+\right), \quad (1.39)$$

**For a historical sketch, see Ref. [50].

where the weak charge-raising current is

$$J_W^{\mu\dagger} = \sum_{m=1}^{F} \left[\bar{\nu}_m^0 \gamma^\mu (1 - \gamma^5) e_m^0 + \bar{u}_m^0 \gamma^\mu (1 - \gamma^5) d_m^0 \right]$$

$$= (\bar{\nu}_e \bar{\nu}_\mu \bar{\nu}_\tau) \gamma^\mu (1 - \gamma^5) V_\ell \begin{pmatrix} e^- \\ \mu^- \\ \tau^- \end{pmatrix} + (\bar{u}\ \bar{c}\ \bar{t}) \gamma^\mu (1 - \gamma^5) V_q \begin{pmatrix} d \\ s \\ b \end{pmatrix}. \tag{1.40}$$

$J_W^{\mu\dagger}$ has a $V - A$ form, i.e., it violates parity and charge conjugation maximally. The fermion gauge vertices are shown in Figure 1.5.

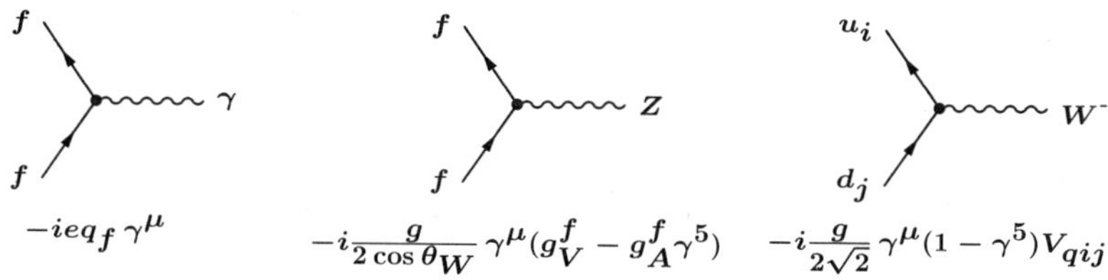

Fig. 1.5. The fermion gauge interaction vertices in the standard electroweak model. $g_V^f \equiv t_{fL}^3 - 2\sin^2\theta_W q_f$ and $g_A^f \equiv t_{fL}^3$, where $t_{uL}^3 = t_{\nu L}^3 = +\frac{1}{2}$, while $t_{dL}^3 = t_{eL}^3 = -\frac{1}{2}$. The $\bar{d}_j u_i W^-$ vertex is the same as for $\bar{u}_i d_j W^+$ except $V_{qij} \to \left(V_q^\dagger\right)_{ji} = V_{qij}^*$. The lepton-$W^\pm$ vertices are obtained from the quark ones by $u_i \to \nu_i$, $d_j \to e_j^-$, and $V_q \to V_\ell$.

The mismatch between the unitary transformations relating the weak and mass eigenstates for the up and down-type quarks leads to the presence of the $F \times F$ unitary matrix $V_q \equiv A_L^{u\dagger} A_L^d$ in the current. This is the Cabibbo-Kobayashi-Maskawa (CKM) matrix,[52,53] which is ultimately due to the mismatch between the weak and Yukawa interactions. For $F = 2$ families V_q takes the familiar form[††]

$$V_{Cabibbo} = \begin{pmatrix} \cos\theta_c & \sin\theta_c \\ -\sin\theta_c & \cos\theta_c \end{pmatrix}, \tag{1.41}$$

where $\sin\theta_c \simeq 0.22$ is the Cabibbo angle. This form gives a good zero[th]-order approximation to the weak interactions of the u, d, s and c quarks; their coupling to the third family, though non-zero, is very small. Including

---

[††]An arbitrary $F \times F$ unitary matrix involves F^2 real parameters. In this case $2F - 1$ of them are unobservable relative phases in the fermion mass eigenstate fields, leaving $F(F-1)/2$ rotation angles and $(F-1)(F-2)/2$ observable CP-violating phases. There are an additional $F - 1$ Majorana phases in V_ℓ for Majorana neutrinos.

P. Langacker

these couplings, the 3-family CKM matrix is

$$V_{CKM} = \begin{pmatrix} V_{ud} & V_{us} & V_{ub} \\ V_{cd} & V_{cs} & V_{cb} \\ V_{td} & V_{ts} & V_{tb} \end{pmatrix} \sim \begin{pmatrix} 1 & \lambda & \lambda^3 \\ \lambda & 1 & \lambda^2 \\ \lambda^3 & \lambda^2 & 1 \end{pmatrix}, \tag{1.42}$$

where the V_{ij} may involve a CP-violating phase. The second form, with $\lambda = \sin\theta_c$, is an easy to remember approximation to the observed magnitude of each element,[54] which displays a suggestive but not well understood hierarchical structure. These are order of magnitude only; each element may be multiplied by a phase and a coefficient of $\mathcal{O}(1)$.

$V_\ell \equiv A_L^{\nu\dagger} A_L^e$ in (1.40) is the analogous leptonic mixing matrix. It is critical for describing neutrino oscillations and other processes sensitive to neutrino masses. However, for processes for which the neutrino masses are negligible we can effectively set $V_\ell = I$ (more precisely, V_ℓ will only enter such processes in the combination $V_\ell^\dagger V_\ell = I$, so it can be ignored).

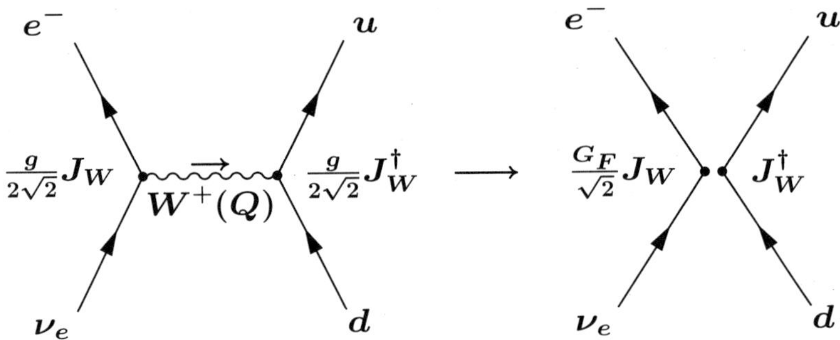

Fig. 1.6. A weak interaction mediated by the exchange of a W and the effective four-fermi interaction that it generates if the four-momentum transfer Q is sufficiently small.

The interaction between fermions mediated by the exchange of a W is illustrated in Figure 1.6. In the limit $|Q^2| \ll M_W^2$ the momentum term in the W propagator can be neglected, leading to an effective zero-range (four-Fermi) interaction

$$-\mathcal{L}^{cc}_{eff} = \frac{G_F}{\sqrt{2}} J_W^\mu J_{W\mu}^\dagger, \tag{1.43}$$

where the Fermi constant is identified as

$$\frac{G_F}{\sqrt{2}} \simeq \frac{g^2}{8M_W^2} = \frac{1}{2v^2}. \tag{1.44}$$

Thus, the Fermi theory is an approximation to the standard model valid in the limit of small momentum transfer. From the muon lifetime, $G_F = 1.16637(5) \times 10^{-5}$ GeV^{-2}, which implies that the weak interaction scale defined by the VEV of the Higgs field is $\nu = \sqrt{2}\langle 0|\phi^0|0\rangle \simeq 246$ GeV.

The charged current weak interaction as described by (1.43) has been successfully tested in a large variety of weak decays,[5,55–57] including β, K, hyperon, heavy quark, μ, and τ decays. In particular, high precision measurements of β, μ, and τ decays are a sensitive probe of extended gauge groups involving right-handed currents and other types of new physics, as is described in the chapters by Deutsch and Quin; Fetscher and Gerber; and Herczeg in Ref. [57]. Tests of the unitarity of the CKM matrix are important in searching for the presence of fourth family or exotic fermions and for new interactions.[58] The standard theory has also been successfully probed in neutrino scattering processes such as $\nu_\mu e \to \mu^- \nu_e, \nu_\mu n \to \mu^- p, \nu_\mu N \to \mu^- X$. It works so well that the charged current neutrino-hadron interactions are used more as a probe of the structure of the hadrons and QCD than as a test of the weak interactions.

Weak charged current effects have also been observed in higher orders, such as in $K^0 - \bar{K}^0$, $D^0 - \bar{D}^0$, and $B^0 - \bar{B}^0$ mixing, and in CP violation in K and B decays.[5] For these higher order processes the full theory must be used because large momenta occur within the loop integrals. An example of the consistency between theory and experiment is shown in Figure 1.7.

1.4.2. *QED*

The standard model incorporates all of the (spectacular) successes of quantum electrodynamics (QED), which is based on the $U(1)_Q$ subgroup that remains unbroken after spontaneous symmetry breaking. The relevant part of the Lagrangian density is

$$\mathcal{L} = -\frac{gg'}{\sqrt{g^2 + g'^2}} J_Q^\mu (\cos\theta_W B_\mu + \sin\theta_W W_\mu^3), \qquad (1.45)$$

where the linear combination of neutral gauge fields is just the photon field A_μ. This reproduces the QED interaction provided one identifies the combination of couplings

$$e = g\sin\theta_W \qquad (1.46)$$

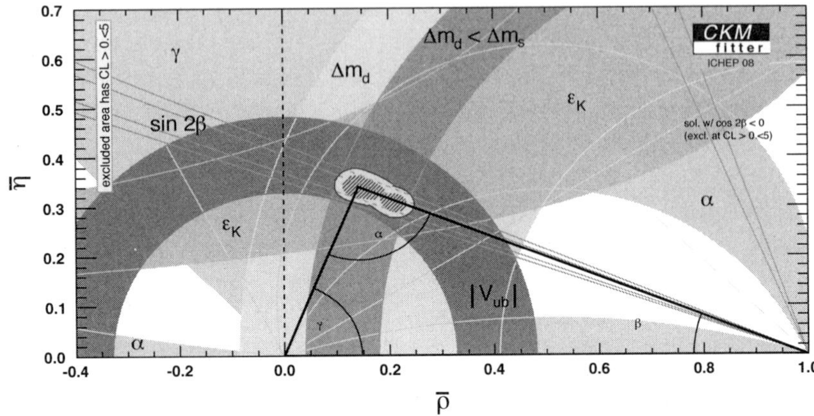

Fig. 1.7. The unitarity triangle, showing the consistency of various CP-conserving and CP-violating observables from the K and B systems. $\bar{\rho}$ and $\bar{\eta}$ are the same as ρ and η up to higher order corrections, where $\rho - i\eta = V_{ub}/(V_{cb}V_{us})$. Plot courtesy of the CKMfitter group,[59] http://ckmfitter.in2p3.fr.

as the electric charge of the positron, where $\tan\theta_W \equiv g'/g$. The electromagnetic current is given by

$$
\begin{aligned}
J_Q^\mu &= \sum_{m=1}^{F} \left[\frac{2}{3}\bar{u}_m^0 \gamma^\mu u_m^0 - \frac{1}{3}\bar{d}_m^0 \gamma^\mu d_m^0 - \bar{e}_m^0 \gamma^\mu e_m^0 \right] \\
&= \sum_{m=1}^{F} \left[\frac{2}{3}\bar{u}_m \gamma^\mu u_m - \frac{1}{3}\bar{d}_m \gamma^\mu d_m - \bar{e}_m \gamma^\mu e_m \right].
\end{aligned}
\tag{1.47}
$$

It takes the same form when written in terms of either weak or mass eigenstates because all fermions which mix with each other have the same electric charge. Thus, the electromagnetic current is automatically flavor-diagonal.

Quantum electrodynamics is the most successful theory in physics when judged in terms of the theoretical and experimental precision of its tests. A detailed review is given in Ref. [60]. The classical atomic tests of QED, such as the Lamb shift, atomic hyperfine splittings, muonium ($\mu^+ e^-$ bound states), and positronium ($e^+ e^-$ bound states) are reviewed in Ref. [61]. The most precise determinations of α and the other physical constants are surveyed in Ref. [62]. High energy tests are described in Refs. [63, 64]. The currently most precise measurements of α are compared in Table 1.3. The approximate agreement of these determinations, which involves the calculation of the electron anomalous magnetic moment $a_e = (g_e - 2)/2$ to high order, validates not only QED but the entire formalism of gauge invariance and renormalization theory. Other basic predictions of gauge invariance (assuming it is not spontaneously broken, which would lead to electric charge nonconservation), are that the photon mass m_γ and its charge q_γ (in units of e) should vanish. The current upper bounds are extremely impressive[5]

$$m_\gamma < 1 \times 10^{-18} \text{ eV} , \qquad q_\gamma < 5 \times 10^{-30}, \tag{1.48}$$

based on astrophysical effects (the survival of the Solar magnetic field and limits on the dispersion of light from pulsars).

There is a possibly significant discrepancy between the high precision measurement of the anomalous magnetic moment of the muon $a_\mu^{exp} = 11\ 659\ 208.0(5.4)(3.3) \times 10^{-10}$ by the Brookhaven 821 experiment,[65] and the theoretical expectation, for which the purely QED part has been calculated to 4 loops and the leading 5 loop contributions estimated (see the review by Höcker and Marciano in Ref. [5]). In addition to the QED part, there are weak interaction corrections (2 loop) and hadronic vacuum polarization and hadronic light by light scattering corrections. There is some theoretical uncertainty in the hadronic corrections. Using estimates of the hadronic vacuum polarization using the measured cross section for $e^+e^- \to$ hadrons in a dispersion relation, one finds

$$a_\mu^{SM} = 116\ 591\ 788(58) \times 10^{-11} \Rightarrow \Delta a_\mu = a_\mu^{exp} - a_\mu^{SM} = 292(86) \times 10^{-11}, \tag{1.49}$$

a 3.4σ discrepancy. However, using hadronic τ decay instead, the discrepancy is reduced to only 0.9σ. If real, the discrepancy could single the effects of new physics, such as the contributions of relatively light supersymmetric particles. For example, the central value of the discrepancy in (1.49) would be accounted for[66] if

$$m_{SUSY} \sim 67\sqrt{\tan\beta} \text{ GeV}, \tag{1.50}$$

where m_{SUSY} is the typical mass of the relevant sleptons, neutralinos, and charginos, and $\tan\beta$ is the ratio of the expectation values of the two Higgs doublets in the theory.

Table 1.3. Most precise determinations of the fine structure constant $\alpha = e^2/4\pi$. Δ_e is defined as $\left[\alpha^{-1} - \alpha^{-1}(a_e)\right] \times 10^6$. Detailed descriptions and references are given in Ref. [62].

Experiment	Value of α^{-1}	Precision	Δ_e
$a_e = (g_e - 2)/2$	137.035 999 683 (94)	$[6.9 \times 10^{-10}]$	–
h/m (Rb, Cs)	137.035 999 35 (69)	$[5.0 \times 10^{-9}]$	0.33 ± 0.69
Quantum Hall	137.036 003 0 (25)	$[1.8 \times 10^{-8}]$	-3.3 ± 2.5
h/m (neutron)	137.036 007 7 (28)	$[2.1 \times 10^{-8}]$	-8.0 ± 2.8
$\gamma_{p,^3He}$ (J. J.)	137.035 987 5 (43)	$[3.1 \times 10^{-8}]$	12.2 ± 4.3
μ^+e^- hyperfine	137.036 001 7 (80)	$[5.8 \times 10^{-8}]$	-2.0 ± 8.0

1.4.3. *The Neutral Current*

The third class of gauge interactions is the weak neutral current, which was predicted by the $SU(2) \times U(1)$ model. The relevant interaction is

$$\mathcal{L} = -\frac{\sqrt{g^2 + g'^2}}{2} J_Z^\mu \left(-\sin\theta_W B_\mu + \cos\theta_W W_\mu^3 \right) = -\frac{g}{2\cos\theta_W} J_Z^\mu Z_\mu, \quad (1.51)$$

where the combination of neutral fields is the massive Z boson field. The strength is conveniently rewritten as $g/(2\cos\theta_W)$, which follows from $\cos\theta_W = g/\sqrt{g^2 + g'^2}$.

The weak neutral current is given by

$$\begin{aligned}
J_Z^\mu &= \sum_m \left[\bar{u}_{mL}^0 \gamma^\mu u_{mL}^0 - \bar{d}_{mL}^0 \gamma^\mu d_{mL}^0 + \bar{\nu}_{mL}^0 \gamma^\mu \nu_{mL}^0 - \bar{e}_{mL}^0 \gamma^\mu e_{mL}^0 \right] \\
&\quad - 2\sin^2\theta_W J_Q^\mu \\
&= \sum_m \left[\bar{u}_{mL} \gamma^\mu u_{mL} - \bar{d}_{mL} \gamma^\mu d_{mL} + \bar{\nu}_{mL} \gamma^\mu \nu_{mL} - \bar{e}_{mL} \gamma^\mu e_{mL} \right] \\
&\quad - 2\sin^2\theta_W J_Q^\mu.
\end{aligned} \quad (1.52)$$

Like the electromagnetic current J_Z^μ is flavor-diagonal in the standard model; all fermions which have the same electric charge and chirality and therefore can mix with each other have the same $SU(2) \times U(1)$ assignments, so the form is not affected by the unitary transformations that relate the mass and weak bases. It was for this reason that the GIM mechanism[67] was introduced into the model, along with its prediction of the charm quark. Without it the d and s quarks would not have had the same $SU(2) \times U(1)$ assignments, and flavor-changing neutral currents would have resulted. The absence of such effects is a major restriction on many extensions of the standard model involving exotic fermions.[68] The neutral current has two contributions. The first only involves the left-chiral fields and is purely $V - A$. The second is proportional to the electromagnetic current with coefficient $\sin^2 \theta_W$ and is purely vector. Parity is therefore violated in the neutral current interaction, though not maximally.

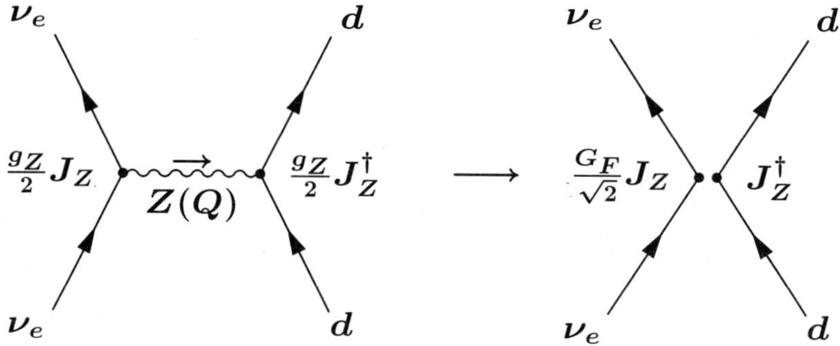

Fig. 1.8. Typical neutral current interaction mediated by the exchange of the Z, which reduces to an effective four-fermi interaction in the limit that the momentum transfer Q can be neglected. g_Z is defined as $\sqrt{g^2 + g'^2}$.

In an interaction between fermions in the limit that the momentum transfer is small compared to M_Z one can neglect the Q^2 term in the propagator, and the interaction reduces to an effective four-fermi interaction

$$-\mathcal{L}_{eff}^{NC} = \frac{G_F}{\sqrt{2}} J_Z^\mu J_{Z\mu}. \tag{1.53}$$

The coefficient is the same as in the charged case because

$$\frac{G_F}{\sqrt{2}} = \frac{g^2}{8M_W^2} = \frac{g^2 + g'^2}{8M_Z^2}. \tag{1.54}$$

That is, the difference in Z couplings compensates the difference in masses in the propagator.

The weak neutral current was discovered at CERN in 1973 by the Gargamelle bubble chamber collaboration[69] and by HPW at Fermilab[70] shortly thereafter, and since that time Z exchange and $\gamma - Z$ interference processes have been extensively studied in many interactions, including $\nu e \to \nu e$, $\nu N \to \nu N$, $\nu N \to \nu X$; polarized e^--hadron and μ-hadron scattering; atomic parity violation; and in $e^+ e^-$ and Z-pole reactions[‡‡]. Along

---

[‡‡]For reviews, see Refs. [57,71–75] and the Electroweak review in [5]. For a historical perspective, see Ref. [76].

with the properties of the W and Z they have been the primary quantitative test of the unification part of the standard electroweak model.

The results of these experiments have generally been in excellent agreement with the predictions of the SM, indicating that the basic structure is correct to first approximation and constraining the effects of possible new physics. One exception are the recent precise measurements of the ratios of neutral to charged current deep inelastic neutrino scattering by the NuTeV collaboration at Fermilab,[77] with a sign-selected beam which allowed them to minimize the effects of the c threshold in the charged current denominator. They obtained a value of $\sin^2 \theta_W = 1 - M_W^2/M_Z^2$ of $0.2277(16)$, which is 3.0σ above the global fit value of $0.2231(3)$, possibly indicating new physics. However, the effect is reduced to $\sim 2\sigma$ if one incorporates the effects of the difference between the strange and antistrange quark momentum distributions, $S^- \equiv \int_0^1 dxx[s(x) - \bar{s}(x)] = 0.00196 \pm 0.00135$, from dimuon events, recently reported by NuTeV.[78] Other possible effects that could contribute are large isospin violation in the nucleon sea, next to leading order QCD effects and electroweak corrections, and nuclear shadowing (for a review, see Ref. [5]).

1.4.4. *The Z-Pole and Above*

The cross section for e^+e^- annihilation is greatly enhanced near the Z-pole. This allowed high statistics studies of the properties of the Z at LEP (CERN) and SLC (SLAC) in $e^-e^+ \to Z \to \ell^-\ell^+$, $q\bar{q}$, and $\nu\bar{\nu}$ *. The four experiments ALEPH, DELPHI, L3, and OPAL at LEP collected some 1.7×10^7 events at or near the Z-pole during the period 1989-1995. The SLD collaboration at the SLC observed some 6×10^5 events during 1992-1998, with the lower statistics compensated by a highly polarized e^- beam with $P_{e^-} \gtrsim 75\%$.

The basic Z-pole observables relevant to the precision program are:

- The lineshape variables M_Z, Γ_Z, and σ_{peak}.
- The branching ratios for Z to decay into e^-e^+, $\mu^-\mu^+$, or $\tau^-\tau^+$; into $q\bar{q}$, $c\bar{c}$, or $b\bar{b}$; or into invisible channels such as $\nu\bar{\nu}$ (allowing a determination of the number $N_\nu = 2.985 \pm 0.009$ of neutrinos lighter than $M_Z/2$).
- Various asymmetries, including forward-backward (FB), hadronic FB charge, polarization (LR), mixed FB-LR, and the polarization

---

*For reviews, see Ref. [79] and the articles by D. Schaile and by A. Blondel in Ref. [57].

ata was

$$= 77^{+28}_{-22} \text{ GeV}, \qquad m_t = 171.1 \pm 1.9 \text{ GeV}$$

$$= 0.1217(17), \qquad \hat{\alpha}(M_Z^2)^{-1} = 127.909(19), \qquad \Delta\alpha^{(5)}_{\text{had}} = 0.02799(14)$$

$$= 0.23119(14), \qquad \bar{s}_\ell^2 = 0.23149(13), \qquad s_W^2 = 0.22308(30),$$

$$(1.55)$$

a good overall χ^2/df of 49.4/42. The three values of the weak angle s^2
to the values found using various renormalization prescriptions, viz.
$\overline{\text{IS}}$, effective Z-lepton vertex, and on-shell values, respectively. The
has a larger uncertainty because of a stronger dependence on the top
$\Delta\alpha^{(5)}_{\text{had}}(M_Z)$ is the hadronic contribution to the running of the fine
ure constant $\hat{\alpha}$ in the $\overline{\text{MS}}$ scheme to the Z-pole.
he data are sensitive to m_t, α_s (evaluated at M_Z), and M_H, which en-
e radiative corrections. The precision data alone yield $m_t = 174.7^{+10.0}_{-7.8}$
in impressive agreement with the direct Tevatron value 170.9 ± 1.9.
Z-pole data alone yield $\alpha_s = 0.1198(20)$, in good agreement with the
average of $0.1176(20)$, which includes other determinations at lower
. The higher value in (1.55) is due to the inclusion of data from
nic τ decays[†].
he prediction for the Higgs mass from indirect data[‡], $M_H = 77^{+28}_{-22}$ GeV,
d be compared with the direct LEP 2 limit $M_H \gtrsim 114.4\,(95\%)$ GeV.[33]
e is no direct conflict given the large uncertainty in the prediction,
he central value is in the excluded region, as can be seen in Figure
Including the direct LEP 2 exclusion results, one finds $M_H < 167$
at 95%. As of this writing CDF and D0 are becoming sensitive to
pper end of this range, and have a good chance of discovering or
ding the SM Higgs in the entire allowed region. We saw in Section
nat there is a theoretical range 115 GeV $< M_H < 180$ GeV in the
rovided it is valid up to the Planck scale, with a much wider allowed
otherwise. The experimental constraints on M_H are encouraging
persymmetric extensions of the SM, which involve more complicated
sectors. The quartic Higgs self-interaction λ in (1.10) is replaced by
couplings, leading to a theoretical upper limit $M_H \lesssim 130$ GeV in the
nal supersymmetric extension (MSSM), while M_H can be as high as
GeV in generalizations. In the decoupling limit in which the second
doublet is much heavier, the direct search lower limit is similar to the

---

cent reevaluation of the theoretical formula[82] lowers the τ value to $0.1187(16)$,
tent with the other determinations.
predicted value would decrease if new physics accounted for the value of $A^{(0b)}_{FB}$.[83]

of produced τ's.

The branching ratios and FB asymmetries could be
for e, μ, and τ, allowing tests of lepton family univers
LEP and SLC simultaneously carried out other pro
studies and tests of QCD, and heavy quark physics.
The second phase of LEP, LEP 2, ran at CERN 1
energies gradually increasing from ~ 140 to ~ 209 (
electroweak results were precise measurements of the
its width and branching ratios; a measurement of e^+
and single W, as a function of center of mass (CM)
the cancellations between diagrams that is character
able gauge field theory, or, equivalently, probes the 1
limits on anomalous quartic gauge vertices; measuren
sections and asymmetries for $e^+e^- \to f\bar{f}$ for $f = $
reasonable agreement with SM predictions; and a str
114.4 GeV on the Higgs mass, and even hints of an (
GeV. LEP2 also studied heavy quark properties, teste
for supersymmetric and other exotic particles.
The Tevatron $\bar{p}p$ collider at Fermilab has run from
energy of nearly 2 TeV. The CDF and D0 collaborati
the top quark in 1995, with a mass consistent with the
precision electroweak and B/K physics observations;
mass, the W mass and decay properties, and leptonic
out Higgs searches; observed $B_s - \bar{B}_s$ mixing and other
carried out extensive QCD tests; and searched for an
couplings, heavy W' and Z' gauge bosons, exotic fermi
and other types of new physics.[5] The HERA $e^{\pm}p$ collid
W propagator and Z exchange effects, searched for l
exotic interactions, and carried out a major progra
structure functions studies.[81]
The principal Z-pole, Tevatron, and weak neutral (
results are listed and compared with the SM best fit
and 1.5. The Z-pole observations are in excellent agr
expectations except for $A_{FB}^{0,b}$, which is the forward-bac
$e^-e^+ \to b\bar{b}$. This could be a fluctuation or a hint of
might be expected to couple most strongly to the t
November, 2007, the result of the Particle Data Grou

Table 1.4. Principal Z-pole observables, their experimental values, theoretical predictions using the SM parameters from the global best fit with M_H free (yielding $M_H = 77^{+28}_{-22}$ GeV), pull (difference from the prediction divided by the uncertainty), and Dev. (difference for fit with M_H fixed at 117 GeV, just above the direct search limit of 114.4 GeV), as of 11/07, from Ref. [5]. $\Gamma(\text{had})$, $\Gamma(\text{inv})$, and $\Gamma(\ell^+\ell^-)$ are not independent.

Quantity	Value	Standard Model	Pull	Dev.
M_Z [GeV]	91.1876 ± 0.0021	91.1874 ± 0.0021	0.1	-0.1
Γ_Z [GeV]	2.4952 ± 0.0023	2.4968 ± 0.0010	-0.7	-0.5
$\Gamma(had)$ [GeV]	1.7444 ± 0.0020	1.7434 ± 0.0010	—	—
$\Gamma(inv)$ [MeV]	499.0 ± 1.5	501.59 ± 0.08	—	—
$\Gamma(\ell^+\ell^-)$ [MeV]	83.984 ± 0.086	83.988 ± 0.016	—	—
σ_{had} [nb]	41.541 ± 0.037	41.466 ± 0.009	2.0	2.0
R_e	20.804 ± 0.050	20.758 ± 0.011	0.9	1.0
R_μ	20.785 ± 0.033	20.758 ± 0.011	0.8	0.9
R_τ	20.764 ± 0.045	20.803 ± 0.011	-0.9	-0.8
R_b	0.21629 ± 0.00066	0.21584 ± 0.00006	0.7	0.7
R_c	0.1721 ± 0.0030	0.17228 ± 0.00004	-0.1	-0.1
$A_{FB}^{0,e}$	0.0145 ± 0.0025	0.01627 ± 0.00023	-0.7	-0.6
$A_{FB}^{0,\mu}$	0.0169 ± 0.0013		0.5	0.7
$A_{FB}^{0,\tau}$	0.0188 ± 0.0017		1.5	1.6
$A_{FB}^{0,b}$	0.0992 ± 0.0016	0.1033 ± 0.0007	-2.5	-2.0
$A_{FB}^{0,c}$	0.0707 ± 0.0035	0.0738 ± 0.0006	-0.9	-0.7
$A_{FB}^{0,s}$	0.0976 ± 0.0114	0.1034 ± 0.0007	-0.5	-0.4
$\bar{s}_\ell^2(A_{FB}^{0,q})$ (LEP)	0.2324 ± 0.0012	0.23149 ± 0.00013	0.8	0.6
$\bar{s}_\ell^2(A_{FB}^{0,e})$ (CDF)	0.2238 ± 0.0050		-1.5	-1.6
A_e (*hadronic*)	0.15138 ± 0.00216	0.1473 ± 0.0011	1.9	2.4
(*leptonic*)	0.1544 ± 0.0060		1.2	1.4
(P_τ)	0.1498 ± 0.0049		0.5	0.7
A_μ	0.142 ± 0.015		-0.4	-0.3
A_τ (SLD)	0.136 ± 0.015		-0.8	-0.7
(P_τ)	0.1439 ± 0.0043		-0.8	-0.5
A_b	0.923 ± 0.020	0.9348 ± 0.0001	-0.6	-0.6
A_c	0.670 ± 0.027	0.6679 ± 0.0005	0.1	0.1
A_s	0.895 ± 0.091	0.9357 ± 0.0001	-0.4	-0.4

standard model. However, the direct limit is considerably lower in the non-decoupling region in which the new supersymmetric particles and second Higgs are relatively light.[33,84,85]

It is interesting to compare the Z boson couplings measured at different energy scales. The renormalized weak angle measured at different scales in the $\overline{\text{MS}}$ scheme is displayed in Figure 1.10.

The precision program has also been used to search for and constrain the

Table 1.5. Principal non-Z-pole observables, as of 11/07, from Ref. [5]. m_t is from the direct CDF and D0 measurements at the Tevatron; M_W is determined mainly by CDF, D0, and the LEP II collaborations; g_L^2, corrected for the $s - \bar{s}$ asymmetry, and g_R^2 are from NuTeV; $g_V^{\nu e}$ are dominated by the CHARM II experiment at CERN; A_{PV} is from the SLAC polarized Møller asymmetry; and the Q_W are from atomic parity violation.

Quantity	Value	Standard Model	Pull	Dev.
m_t [GeV]	$170.9 \pm 1.8 \pm 0.6$	171.1 ± 1.9	-0.1	-0.8
M_W ($\bar{p}p$)	80.428 ± 0.039	80.375 ± 0.015	1.4	1.7
M_W (LEP)	80.376 ± 0.033		0.0	0.5
g_L^2	0.3010 ± 0.0015	0.30386 ± 0.00018	-1.9	-1.8
g_R^2	0.0308 ± 0.0011	0.03001 ± 0.00003	0.7	0.7
$g_V^{\nu e}$	-0.040 ± 0.015	-0.0397 ± 0.0003	0.0	0.0
$g_A^{\nu e}$	-0.507 ± 0.014	-0.5064 ± 0.0001	0.0	0.0
$A_{PV} \times 10^7$	-1.31 ± 0.17	-1.54 ± 0.02	1.3	1.2
Q_W(Cs)	-72.62 ± 0.46	-73.16 ± 0.03	1.2	1.2
Q_W(Tl)	-116.4 ± 3.6	-116.76 ± 0.04	0.1	0.1

effects of possible new TeV scale physics*. This includes the effects of possible mixing between ordinary and exotic heavy fermions,[68] new W' or Z' gauge bosons,[88,89] leptoquarks,[80,90–92] Kaluza-Klein excitations in extra-dimensional theories,[5,93–95] and new four-fermion operators,[80,90,96,97] all of which can effect the observables at tree level. The oblique corrections,[98,99] which only affect the W and Z self energies, are also constrained. The latter may be generated, e.g., by heavy non-degenerate scalar or fermion multiplets and heavy chiral fermions,[5] such as are often found in models that replace the elementary Higgs by a dynamical mechanism.[100] A major implication of supersymmetry is through the small mass expected for the lightest Higgs boson. Other supersymmetric effects are small in the decoupling limit in which the superpartners and extra Higgs doublet are heavier than a few hundred GeV.[84,85,101,102] The precisely measured gauge couplings at the Z-pole are also important for testing the ideas of gauge coupling unification,[103] which works extremely well in the MSSM.[104–107]

---

*For reviews, see Refs. [5,74,75,87].

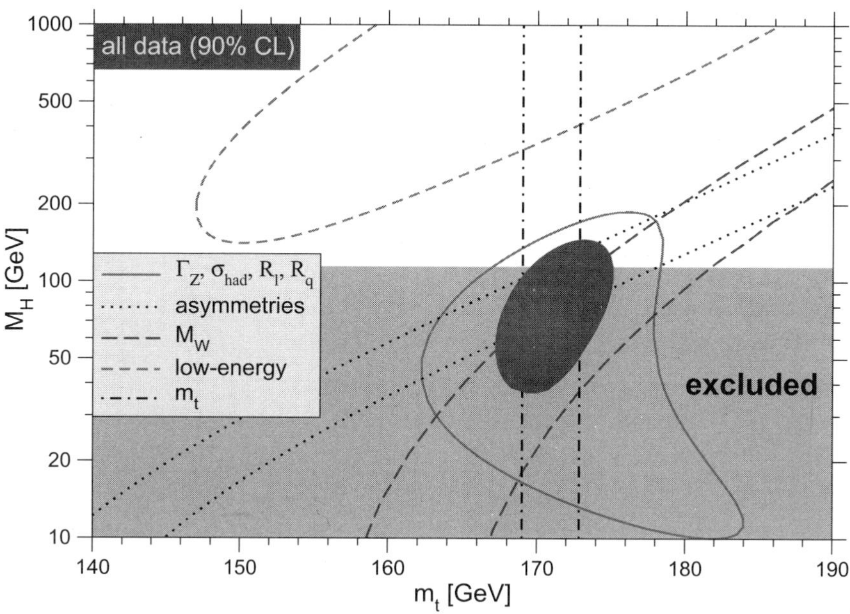

Fig. 1.9. 1σ allowed regions in M_H vs m_t and the 90% cl global fit region from precision data, compared with the direct exclusion limits from LEP 2. Plot courtesy of the Particle Data Group.[5]

1.4.5. *Gauge Self-interactions*

The $SU(2)$ gauge kinetic energy terms in (1.6) lead to 3 and 4-point gauge self-interactions for the W's,

$$
\begin{aligned}
\mathcal{L}_{W3} = &- ig\left(\partial_\rho W_\nu^3\right)W_\mu^+ W_\sigma^- \left[g^{\rho\mu}g^{\nu\sigma} - g^{\rho\sigma}g^{\nu\mu}\right] \\
&- ig\left(\partial_\rho W_\mu^+\right)W_\nu^3 W_\sigma^- \left[g^{\rho\sigma}g^{\mu\nu} - g^{\rho\nu}g^{\mu\sigma}\right] \\
&- ig\left(\partial_\rho W_\sigma^-\right)W_\nu^3 W_\mu^+ \left[g^{\rho\nu}g^{\mu\sigma} - g^{\rho\mu}g^{\nu\sigma}\right],
\end{aligned}
\tag{1.56}
$$

and

$$
\mathcal{L}_{W4} = \frac{g^2}{4}\left[W_\mu^+ W_\nu^+ W_\sigma^- W_\rho^- Q^{\mu\nu\rho\sigma} - 2W_\mu^+ W_\nu^3 W_\sigma^3 W_\rho^- Q^{\mu\rho\nu\sigma}\right],
\tag{1.57}
$$

where

$$
Q_{\mu\nu\rho\sigma} \equiv 2g_{\mu\nu}g_{\rho\sigma} - g_{\mu\rho}g_{\nu\sigma} - g_{\mu\sigma}g_{\nu\rho}.
\tag{1.58}
$$

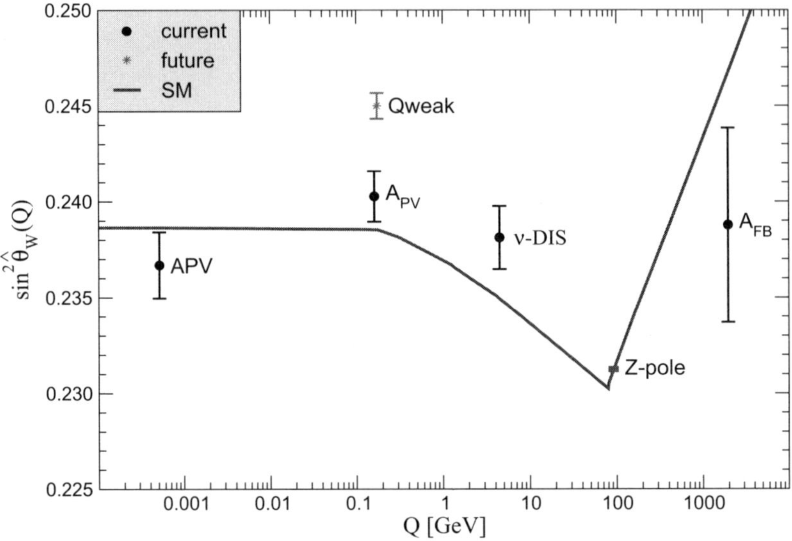

Fig. 1.10. Running $\hat{s}_Z^2(Q^2)$ measured at various scales, compared with the predictions of the SM.[86] The low energy points are from atomic parity violation (APV), the polarized Møller asymmetry (PV) and deep inelastic neutrino scattering (corrected for an $s - \bar{s}$ asymmetry). Q_{weak} shows the expected sensitivity of a future polarized e^- measurement at Jefferson Lab. Plot courtesy of the Particle Data Group.[5]

These carry over to the W, Z, and γ self-interactions provided we replace W^3 by $\cos\theta_W Z + \sin\theta_W A$ using (1.23) and (1.24) (the B has no self-interactions). The resulting vertices follow from the matrix element of $i\mathcal{L}$ after including identical particle factors and using $g = e/\sin\theta_W$. They are listed in Table 1.2 and shown in Fig. 1.11.

The gauge self-interactions are essential probes of the structure and consistency of a spontaneously-broken non-abelian gauge theory. Even tiny deviations in their form or value would destroy the delicate cancellations needed for renormalizability, and would signal the need either for compensating new physics (e.g., from mixing with other gauge bosons or new particles in loops), or of a more fundamental breakdown of the gauge principle, e.g., from some forms of compositeness. They have been constrained

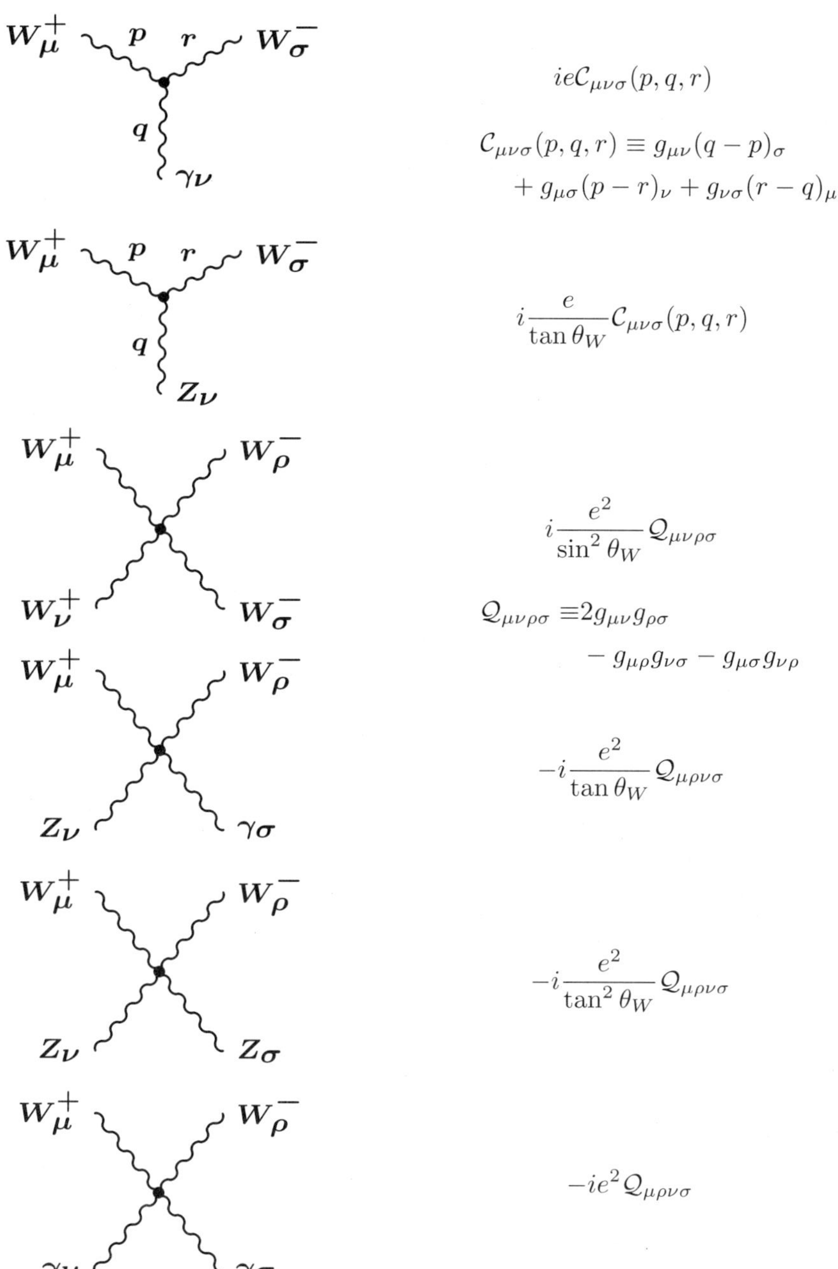

$$ie\mathcal{C}_{\mu\nu\sigma}(p,q,r)$$

$$\mathcal{C}_{\mu\nu\sigma}(p,q,r) \equiv g_{\mu\nu}(q-p)_\sigma$$
$$+ g_{\mu\sigma}(p-r)_\nu + g_{\nu\sigma}(r-q)_\mu$$

$$i\frac{e}{\tan\theta_W}\mathcal{C}_{\mu\nu\sigma}(p,q,r)$$

$$i\frac{e^2}{\sin^2\theta_W}\mathcal{Q}_{\mu\nu\rho\sigma}$$

$$\mathcal{Q}_{\mu\nu\rho\sigma} \equiv 2g_{\mu\nu}g_{\rho\sigma}$$
$$- g_{\mu\rho}g_{\nu\sigma} - g_{\mu\sigma}g_{\nu\rho}$$

$$-i\frac{e^2}{\tan\theta_W}\mathcal{Q}_{\mu\rho\nu\sigma}$$

$$-i\frac{e^2}{\tan^2\theta_W}\mathcal{Q}_{\mu\rho\nu\sigma}$$

$$-ie^2\mathcal{Q}_{\mu\rho\nu\sigma}$$

Fig. 1.11. The three and four point-self-interactions of gauge bosons in the standard electroweak model. The momenta and charges flow into the vertices.

by measuring the total cross section and various decay distributions for $e^-e^+ \to W^-W^+$ at LEP 2, and by observing $\bar{p}p \to W^+W^-, WZ$, and $W\gamma$ at the Tevatron. Possible anomalies in the predicted quartic vertices in Table 1.2, and the neutral cubic vertices for ZZZ, $ZZ\gamma$, and $Z\gamma\gamma$, which are absent in the SM, have also been constrained by LEP 2.[80]

The three tree-level diagrams for $e^-e^+ \to W^-W^+$ are shown in Figure 1.12. The cross section from any one or two of these rises rapidly with center of mass energy, but gauge invariance relates these three-point vertices to the couplings of the fermions in such a way that at high energies there is a cancellation. It is another manifestation of the cancellation in a gauge theory which brings higher-order loop integrals under control, leading to a renormalizable theory. It is seen in Figure 1.13 that the expected cancellations do occur.

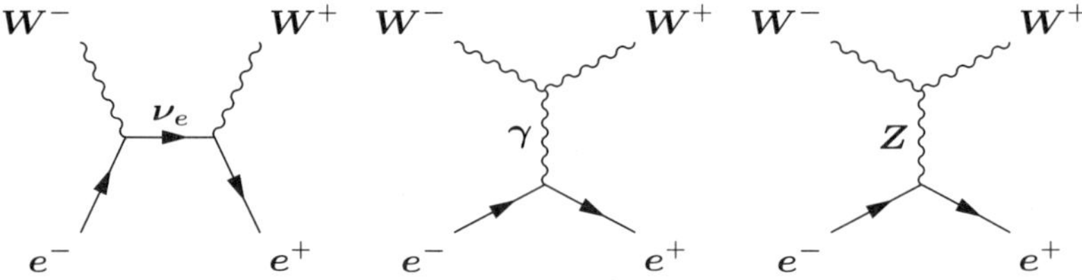

Fig. 1.12. Tree-level diagrams contributing to $e^+e^- \to W^+W^-$.

1.5. Problems with the Standard Model

For convenience we summarize the Lagrangian density after spontaneous symmetry breaking:

$$\mathcal{L} = \mathcal{L}_{gauge} + \mathcal{L}_\phi + \sum_r \bar{\psi}_r \left(i\,\slashed{\partial} - m_r - \frac{m_r H}{\nu} \right) \psi_r$$
$$- \frac{g}{2\sqrt{2}} \left(J_W^\mu W_\mu^- + J_W^{\mu\dagger} W_\mu^+ \right) - eJ_Q^\mu A_\mu - \frac{g}{2\cos\theta_W} J_Z^\mu Z_\mu, \tag{1.59}$$

where the self-interactions for the $W^\pm$, Z, and γ are given in (1.56) and (1.57), $\mathcal{L}_\phi$ is given in (1.31), and the fermion currents in (1.40), (1.47), and (1.52). For Majorana ν_L masses generated by a higher dimensional operator involving two factors of the Higgs doublet, as in the seesaw model, the ν

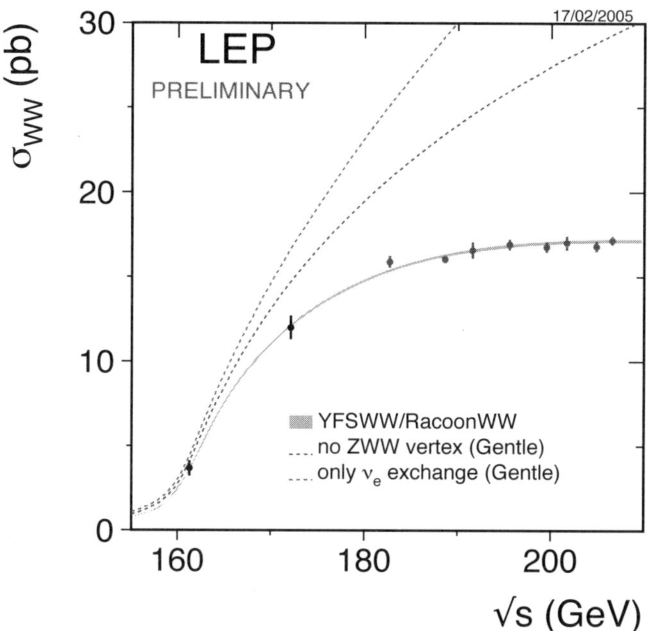

Fig. 1.13. Cross section for $e^-e^+ \rightarrow W^-W^+$ compared with the SM expectation. Also shown is the expectation from t channel ν_e exchange only, and for the ν_e and γ diagrams only. Plot courtesy of the LEP Electroweak Working Group,[80] `http://www.cern.ch/LEPEWWG/`.

term in (1.59) is replaced by

$$\mathcal{L} = \sum_r \bar{\nu}_{rL} i\,\not{\partial}\nu_{rL} - \frac{1}{2}m_{\nu r}\left(\bar{\nu}_{rL}\nu_{rR}^c + h.c.\right)\left(1 + \frac{H}{\nu}\right)^2, \qquad (1.60)$$

where ν_{rR}^c is the CP conjugate to ν_L (see, e.g., Ref. [45]).

The standard electroweak model is a mathematically-consistent renormalizable field theory which predicts or is consistent with all experimental facts. It successfully predicted the existence and form of the weak neutral current, the existence and masses of the W and Z bosons, and the charm quark, as necessitated by the GIM mechanism. The charged current weak interactions, as described by the generalized Fermi theory, were successfully incorporated, as was quantum electrodynamics. The consistency between theory and experiment indirectly tested the radiative corrections and ideas of renormalization and allowed the successful prediction of the top quark mass. Although the original formulation did not provide for massive neutrinos, they are easily incorporated by the addition of right-handed states ν_R (Dirac) or as higher-dimensional operators, perhaps generated by an underlying seesaw (Majorana). When combined with quantum chromodynamics for the strong interactions, the standard model is almost certainly the approximately correct description of the elementary particles and their interactions down to at least 10^{-16}cm, with the possible exception of the Higgs sector or new very weakly coupled particles. When combined with general relativity for classical gravity the SM accounts for most of the observed features of Nature (though not for the dark matter and energy).

However, the theory has far too much arbitrariness to be the final story. For example, the minimal version of the model has 20 free parameters for massless neutrinos and another 7 (9) for massive Dirac (Majorana) neutrinos*, not counting electric charge (i.e., hypercharge) assignments. Most physicists believe that this is just too much for the fundamental theory. The complications of the standard model can also be described in terms of a number of problems.

The Gauge Problem

The standard model is a complicated direct product of three subgroups, $SU(3) \times SU(2) \times U(1)$, with separate gauge couplings. There is no explanation for why only the electroweak part is chiral (parity-violating). Similarly, the standard model incorporates but does not explain another fundamental fact of nature: charge quantization, i.e., why all particles have charges which are multiples of $e/3$. This is important because it allows the electrical neutrality of atoms ($|q_p| = |q_e|$). The complicated gauge structure suggests the existence of some underlying unification of the interactions, such as one would expect in a superstring[108–110] or grand unified theory.[88,111–114] Charge quantization can also be explained in such theories, though the "wrong" values of charge emerge in some constructions due to different hypercharge embeddings or non-canonical values of Y (e.g., some string constructions lead to exotic particles with charges of $\pm e/2$). Charge quantization may also be explained, at least in part, by the existence of

---

*12 fermion masses (including the neutrinos), 6 mixing angles, 2 CP violation phases (+ 2 possible Majorana phases), 3 gauge couplings, M_H, ν, θ_{QCD}, M_P, Λ_{cosm}, minus one overall mass scale since only mass ratios are physical.

magnetic monopoles[115] or the absence of anomalies[†], but either of these is likely to find its origin in some kind of underlying unification.

The Fermion Problem

All matter under ordinary terrestrial conditions can be constructed out of the fermions (ν_e, e^-, u, d) of the first family. Yet we know from laboratory studies that there are ≥ 3 families: (ν_μ, μ^-, c, s) and (ν_τ, τ^-, t, b) are heavier copies of the first family with no obvious role in nature. The standard model gives no explanation for the existence of these heavier families and no prediction for their numbers. Furthermore, there is no explanation or prediction of the fermion masses, which are observed to occur in a hierarchical pattern which varies over 5 orders of magnitude between the t quark and the e^-, or of the quark and lepton mixings. Even more mysterious are the neutrinos, which are many orders of magnitude lighter still. It is not even certain whether the neutrino masses are Majorana or Dirac. A related difficulty is that while the CP violation observed in the laboratory is well accounted for by the phase in the CKM matrix, there is no SM source of CP breaking adequate to explain the baryon asymmetry of the universe.

There are many possible suggestions of new physics that might shed light on these questions. The existence of multiple families could be due to large representations of some string theory or grand unification, or they could be associated with different possibilities for localizing particles in some higher dimensional space. The latter could also be associated with string compactifications, or by some effective brane world scenario.[5,93–95] The hierarchies of masses and mixings could emerge from wave function overlap effects in such higher-dimensional spaces. Another interpretation, also possible in string theories, is that the hierarchies are because some of the mass terms are generated by higher dimensional operators and therefore suppressed by powers of $\langle 0|S|0\rangle/M_X$, where S is some standard model singlet field and M_X is some large scale such as M_P. The allowed operators could perhaps be enforced by some family symmetry.[116] Radiative hierarchies,[117] in which some of the masses are generated at the loop level, or some form of compositeness are other possibilities. Despite all of these ideas there is no compelling model and none of these yields detailed predictions. Grand unification by itself doesn't help very much, except for the prediction of m_b in terms of m_τ in the simplest versions.

[†]The absence of anomalies is not sufficient to determine all of the Y assignments without additional assumptions, such as family universality.

The small values for the neutrino masses suggest that they are associated with Planck or grand unification physics, as in the seesaw model, but there are other possibilities.[44–47]

Almost any type of new physics is likely to lead to new sources of CP violation.

The Higgs/Hierarchy Problem

In the standard model one introduces an elementary Higgs field to generate masses for the W, Z, and fermions. For the model to be consistent the Higgs mass should not be too different from the W mass. If M_H were to be larger than M_W by many orders of magnitude the Higgs self-interactions would be excessively strong. Theoretical arguments suggest that $M_H \lesssim 700$ GeV (see Section 1.3).

However, there is a complication. The tree-level (bare) Higgs mass receives quadratically-divergent corrections from the loop diagrams in Figure 1.14. One finds

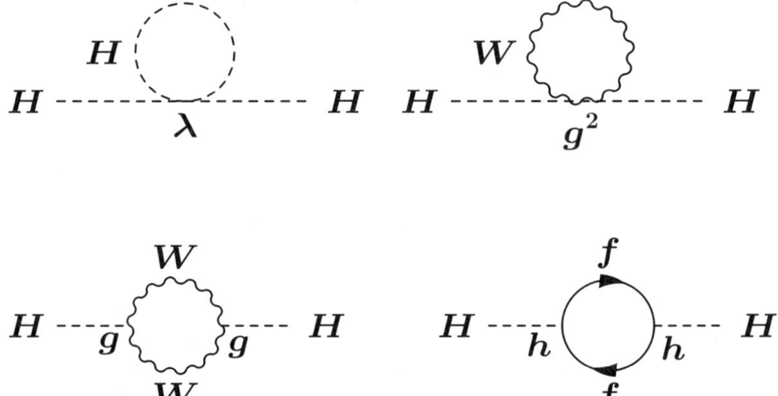

Fig. 1.14. Radiative corrections to the Higgs mass, including self-interactions, interactions with gauge bosons, and interactions with fermions.

$$M_H^2 = (M_H^2)_{bare} + \mathcal{O}(\lambda, g^2, h^2)\Lambda^2, \qquad (1.61)$$

where Λ is the next higher scale in the theory. If there were no higher scale one could simply interpret Λ as an ultraviolet cutoff and take the view that M_H is a measured parameter, with $(M_H)_{bare}$ not observable. However, the theory is presumably embedded in some larger theory that cuts off the momentum integral at the finite scale of the new physics*. For example, if the next scale is gravity Λ is the Planck scale $M_P = G_N^{-1/2} \sim 10^{19}$ GeV. In a grand unified theory, one would expect Λ to be of order the unification scale $M_X \sim 10^{14}$ GeV. Hence, the natural scale for M_H is $\mathcal{O}(\Lambda)$, which is much larger than the expected value. There must be a fine-tuned and apparently highly contrived cancellation between the bare value and the correction, to more than 30 decimal places in the case of gravity. If the cutoff is provided by a grand unified theory there is a separate hierarchy problem at the tree-level. The tree-level couplings between the Higgs field and the superheavy fields lead to the expectation that M_H is close to the unification scale unless unnatural fine-tunings are done, i.e., one does not understand why $(M_W/M_X)^2$ is so small in the first place.

One solution to this Higgs/hierarchy problem is TeV scale supersymmetry, in which the quadratically-divergent contributions of fermion and boson loops cancel, leaving only much smaller effects of the order of supersymmetry-breaking. (However, supersymmetric grand unified theories still suffer from the tree-level hierarchy problem.) There are also (non-supersymmetric) extended models in which the cancellations are between bosons or between fermions. This class includes Little Higgs models,[118,119] in which the Higgs is forced to be lighter than new TeV scale dynamics because it is a pseudo-Goldstone boson of an approximate underlying global symmetry, and Twin-Higgs models.[120]

Another possibility is to eliminate the elementary Higgs fields, replacing them with some dynamical symmetry breaking mechanism based on a new strong dynamics.[100] In technicolor, for example, the SSB is associated with the expectation value of a fermion bilinear, analogous to the breaking of chiral symmetry in QCD. Extended technicolor, top-color, and composite Higgs models all fall into this class.

Large and/or warped extra dimensions[121–123] can also resolve the difficulties, by altering the relation between M_P and a much lower fundamental scale, by providing a cutoff at the inverse of the extra dimension scale, or by using the boundary conditions in the extra dimensions to break the electroweak symmetry (Higgsless models[124]). Deconstruction models, in which no extra dimensions are explicity introduced,[125,126] are closely related.

Most of the models mentioned above have the potential to generate flavor changing neutral current and CP violation effects much larger than

---

*There is no analogous fine-tuning associated with logarithmic divergences, such as those encountered in QED, because $\alpha \ln(\Lambda/m_e) < \mathcal{O}(1)$ even for $\Lambda = M_P$.

observational limits. Pushing the mass scales high enough to avoid these problems may conflict with a natural solution to the hierarchy problem, i.e., one may reintroduce a little hierarchy problem. Many are also strongly constrained by precision electroweak physics. In some cases the new physics does not satisfy the decoupling theorem,[127] leading to large oblique corrections. In others new tree-level effects may again force the scale to be too high. The most successful from the precision electroweak point of view are those which have a discrete symmetry which prevents vertices involving just one heavy particle, such as R-parity in supersymmetry, T-parity in some little Higgs models,[128] and KK-parity in universal extra dimension models.[129]

A very different possibility is to accept the fine-tuning, i.e., to abandon the notion of naturalness for the weak scale, perhaps motivated by anthropic considerations.[130] (The anthropic idea will be considered below in the discussion of the gravity problem.) This could emerge, for example, in split supersymmetry.[131]

The Strong CP Problem

Another fine-tuning problem is the strong CP problem.[132–134] One can add an additional term $\frac{\theta_{QCD}}{32\pi^2} g_s^2 G\tilde{G}$ to the QCD Lagrangian density which breaks P, T and CP symmetry*. $\tilde{G}^i_{\mu\nu} = \epsilon_{\mu\nu\alpha\beta}G^{\alpha\beta i}/2$ is the dual field strength tensor. This term, if present, would induce an electric dipole moment d_N for the neutron. The rather stringent limits on the dipole moment lead to the upper bound $|\theta_{QCD}| < 10^{-11}$. The question is, therefore, why is θ_{QCD} so small? It is not sufficient to just say that it is zero (i.e., to impose CP invariance on QCD) because of the observed violation of CP by the weak interactions. As discussed in Sec. 1.4.1, this is believed to be associated with phases in the quark mass matrices. The quark phase redefinitions which remove them lead to a shift in θ_{QCD} by $\mathcal{O}(10^{-3})$ because of the anomaly in the vertex coupling the associated global current to two gluons. Therefore, an apparently contrived fine-tuning is needed to cancel this correction against the bare value. Solutions include the possibility that CP violation is not induced directly by phases in the Yukawa couplings, as is usually assumed in the standard model, but is somehow violated spontaneously. θ_{QCD} then would be a calculable parameter induced at loop level, and it is possible to make θ_{QCD} sufficiently small. However, such

---

*One could add an analogous term for the weak $SU(2)$ group, but it does not lead to observable consequences, at least within the SM.[133,135]

models lead to difficult phenomenological and cosmological problems[†]. Alternately, θ_{QCD} becomes unobservable (i.e., can be rotated away) if there is a massless u quark.[138] However, most phenomenological estimates[139] are not consistent with $m_u = 0$. Another possibility is the Peccei-Quinn mechanism,[140] in which an extra global $U(1)$ symmetry is imposed on the theory in such a way that θ_{QCD} becomes a dynamical variable which is zero at the minimum of the potential. The spontaneous breaking of the symmetry, along with explicit breaking associated with the anomaly and instanton effects, leads to a very light pseudo-Goldstone boson known as an axion.[141,142] Laboratory, astrophysical, and cosmological constraints suggest the range $10^9 - 10^{12}$ GeV for the scale at which the $U(1)$ symmetry is broken.

The Gravity Problem

Gravity is not fundamentally unified with the other interactions in the standard model, although it is possible to graft on classical general relativity by hand. However, general relativity is not a quantum theory, and there is no obvious way to generate one within the standard model context. Possible solutions include Kaluza-Klein[143] and supergravity[144–146] theories. These connect gravity with the other interactions in a more natural way, but do not yield renormalizable theories of quantum gravity. More promising are superstring theories (which may incorporate the above), which unify gravity and may yield *finite* theories of quantum gravity and all the other interactions. String theories are perhaps the most likely possibility for the underlying theory of particle physics and gravity, but at present there appear to be a nearly unlimited number of possible string vacua (the landscape), with no obvious selection principle. As of this writing the particle physics community is still trying to come to grips with the landscape and its implications. Superstring theories naturally imply some form of supersymmetry, but it could be broken at a high scale and have nothing to do with the Higgs/hierarachy problem (split supersymmetry is a compromise, keeping some aspects at the TeV scale).

In addition to the fact that gravity is not unified and not quantized there is another difficulty, namely the cosmological constant. The cosmological constant can be thought of as the energy of the vacuum. However, we saw in Sec. 1.3 that the spontaneous breaking of $SU(2) \times U(1)$ generates a value $\langle 0|V(\nu)|0\rangle = -\mu^4/4\lambda$ for the expectation value of the Higgs potential at the

---

[†]Models in which the CP breaking occurs near the Planck scale may be viable.[136,137]

minimum. This is a c-number which has no significance for the microscopic interactions. However, it assumes great importance when the theory is coupled to gravity, because it contributes to the cosmological constant. The cosmological constant becomes

$$\Lambda_{cosm} = \Lambda_{bare} + \Lambda_{SSB}, \tag{1.62}$$

where $\Lambda_{bare} = 8\pi G_N V(0)$ is the primordial cosmological constant, which can be thought of as the value of the energy of the vacuum in the absence of spontaneous symmetry breaking. (The definition of $V(\phi)$ in (1.10) implicitly assumed $\Lambda_{bare} = 0$.) Λ_{SSB} is the part generated by the Higgs mechanism:

$$|\Lambda_{SSB}| = 8\pi G_N |\langle 0|V|0\rangle| \sim 10^{56}\Lambda_{obs}. \tag{1.63}$$

It is some 10^{56} times larger in magnitude than the observed value $\Lambda_{obs} \sim (0.0024\ \text{eV})^4/8\pi G_N$ (assuming that the dark energy is due to a cosmological constant), and it is of the wrong sign.

This is clearly unacceptable. Technically, one can solve the problem by adding a constant $+\mu^4/4\lambda$ to V, so that V is equal to zero at the minimum (i.e., $\Lambda_{bare} = 2\pi G_N \mu^4/\lambda$). However, with our current understanding there is no reason for Λ_{bare} and Λ_{SSB} to be related. The need to invoke such an incredibly fine-tuned cancellation to 50 decimal places is probably the most unsatisfactory feature of the standard model. The problem becomes even worse in superstring theories, where one expects a vacuum energy of $\mathcal{O}(M_P^4)$ for a generic point in the landscape, leading to $\Lambda_{obs} \gtrsim 10^{123}|\Lambda_{obs}|$. The situation is almost as bad in grand unified theories.

So far no compelling solution to the cosmological constant problem has emerged. One intriguing possibility invokes the anthropic (environmental) principle,[147–149] i.e., that a much larger or smaller value of $|\Lambda_{cosm}|$ would not have allowed the possibility for life to have evolved because the Universe would have expanded or recollapsed too rapidly.[150] This would be a rather meaningless argument unless (a) Nature somehow allows a large variety of possibilities for $|\Lambda_{cosm}|$ (and possibly other parameters or principles) such as in different vacua, and (b) there is some mechanism to try all or many of them. In recent years it has been suggested that both of these needs may be met. There appear to be an enormous landscape of possible superstring vacua,[151–154] with no obvious physical principle to choose one over the other. Something like eternal inflation[155] could provide the means to sample them, so that only the environmentally suitable vacua lead to long-lived Universes suitable for life. These ideas are highly controversial and are currently being heatedly debated.

The New Ingredients

It is now clear that the standard model requires a number of new ingredients. These include

- **A mechanism for small neutrino masses.** The most popular possibility is the minimal seesaw model, implying Majorana masses, but there are other plausible mechanisms for either small Dirac or Majorana masses.[44–47]

- **A mechanism for the baryon asymmetry.** The standard model has neither the nonequilibrium condition nor sufficient CP violation to explain the observed asymmetry between baryons and antibaryons in the Universe[156–158]*. One possibility involves the out of equilibrium decays of superheavy Majorana right-handed neutrinos (leptogenesis[162,163]), as expected in the minimal seesaw model. Another involves a strongly first order electroweak phase transition (electroweak baryogenesis[164]). This is not expected in the standard model, but could possibly be associated with loop effects in the minimal supersymmetric extension (MSSM) if one of the scalar top quarks is sufficiently light.[165] However, it is most likely in extensions of the MSSM involving SM singlet Higgs fields that can generate a dynamical μ term, which can easily lead to strong first order transitions at tree-level.[166] Such extensions would likely yield signatures observable at the LHC. Both the seesaw models and the singlet extensions of the MSSM could also provide the needed new sources of CP violation. Other possibilities for the baryon asymmetry include the decay of a coherent scalar field, such as a scalar quark or lepton in supersymmetry (the Affleck-Dine mechanism[167]), or CPT violation.[168,169] Finally, one cannot totally dismiss the possibility that the asymmetry is simply due to an initial condition on the big bang. However, this possibility disappears if the universe underwent a period of rapid inflation.[170]

---

*The third necessary ingredient, baryon number nonconservation, is present in the SM because of non-perturbative vacuum tunnelling (instanton) effects.[159] These are negligible at zero temperature where they are exponentially suppressed, but important at high temperatures due to thermal fluctuations (sphaleron configurations), before or during the electroweak phase transition.[160,161]

- **What is the dark energy?** In recent years, a remarkable concordance of cosmological observations involving the cosmic microwave background radiation (CMB), acceleration of the Universe as determined by Type Ia supernova observations, large scale distribution of galaxies and clusters, and big bang nucleosynthesis has allowed precise determinations of the cosmological parameters:[5,171–173] the Universe is close to flat, with some form of dark energy making up about 74% of the energy density. Dark matter constitutes $\gtrsim 21\%$, while ordinary matter (mainly baryons) represents only about 4-5%. The mysterious dark energy,[174–176] which is the most important contribution to the energy density and leads to the acceleration of the expansion of the Universe, is not accounted for in the SM. It could be due to a cosmological constant that is incredibly tiny on the particle physics scale, or to a slowly time varying field (quintessence). Is the acceleration somehow related to an earlier and much more dramatic period of inflation[170]? If it is associated with a time-varying field, could it be connected with a possible time variation of coupling "constants"[177]?

- **What is the dark matter?** Similarly, the standard model has no explanation for the observed dark matter, which contributes much more to the matter in the Universe than the stuff we are made of. It is likely, though not certain, that the dark matter is associated with elementary particles. An attractive possibility is weakly interacting massive particles (WIMPs), which are typically particles in the $10^2 - 10^3$ GeV range with weak interaction strength couplings, and which lead naturally to the observed matter density. These could be associated with the lightest supersymmetric partner (usually a neutralino) in supersymmetric models with R-parity conservation, or analogous stable particles in Little Higgs or universal extra dimension models. There are a wide variety of variations on these themes, e.g., involving very light gravitinos or other supersymmetric particles. There are many searches for WIMPs going on, including direct searches for the recoil produced by scattering of Solar System WIMPs, indirect searches for WIMP annihilation products, and searches for WIMPs produced at accelerators.[178–180] Axions, perhaps associated with the strong CP problem or with string vacua,[181] are another possibility. Searches for axions produced in the Sun, in the laboratory, or from the early universe are currently underway.[134,182]

- **The suppression of flavor changing neutral currents, proton decay, and electric dipole moments.** The standard model has a number of accidental symmetries and features which forbid proton decay, preserve lepton number and lepton family number (at least for vanishing neutrino masses), suppress transitions such as $K^+ \to \pi^+ \nu\bar{\nu}$ at tree-level, and lead to highly suppressed electric dipole moments for the e^-, n, atoms, etc. However, most extensions of the SM have new interactions which violate such symmetries, leading to potentially serious problems with FCNC and EDMs. There seems to be a real conflict between attempts to deal with the Higgs/hierarchy problem and the prevention of such effects.

  Recently, there has been much discussion of minimal flavor violation, which is the hypothesis that all flavor violation, even that which is associated with new physics, is proportional to the standard model Yukawa matrices,[51,183] leading to a significant suppression of flavor changing effects.

Acknowledgments

I am grateful to Tao Han for inviting me to give these lectures. This work was supported by the organizers of TASI2008, the IBM Einstein Fellowship, and by NSF grant PHY-0503584.

References

1. H. Weyl, *Z. Phys.* **56**, 330 (1929).
2. C.-N. Yang and R. L. Mills, *Phys. Rev.* **96**, 191 (1954).
3. D. J. Gross, *Rev. Mod. Phys.* **77**, 837 (2005).
4. S. Bethke, *Prog. Part. Nucl. Phys.* **58**, 351 (2007).
5. C. Amsler *et al.*, *Phys. Lett.* **B667**, p. 1 (2008).
6. D. J. Gross and F. Wilczek, *Phys. Rev. Lett.* **30**, 1343 (1973).
7. H. D. Politzer, *Phys. Rev. Lett.* **30**, 1346 (1973).
8. H. Fritzsch, M. Gell-Mann and H. Leutwyler, *Phys. Lett.* **B47**, 365 (1973).
9. J. Gasser and H. Leutwyler, *Phys. Rept.* **87**, 77 (1982).
10. S. L. Glashow, *Nucl. Phys.* **22**, 579 (1961).
11. S. Weinberg, *Phys. Rev. Lett.* **19**, 1264 (1967).
12. A. Salam In *Elementary Particle Theory*, ed. N. Svartholm (Almquist and Wiksells, Stockholm, 1969), 367-377.
13. J. S. Schwinger, *Phys. Rev.* **125**, 397 (1962).
14. P. W. Anderson, *Phys. Rev.* **130**, 439 (1963).

15. P. W. Higgs, *Phys. Lett.* **12**, 132 (1964).

16. P. W. Higgs, *Phys. Rev.* **145**, 1156 (1966).

17. F. Englert and R. Brout, *Phys. Rev. Lett.* **13**, 321 (1964).

18. G. S. Guralnik, C. R. Hagen and T. W. B. Kibble, *Phys. Rev. Lett.* **13**, 585 (1964).

19. G. 't Hooft, *Nucl. Phys.* **B35**, 167 (1971).

20. G. 't Hooft and M. J. G. Veltman, *Nucl. Phys.* **B50**, 318 (1972).

21. B. W. Lee and J. Zinn-Justin, *Phys. Rev.* **D5**, 3121 (1972).

22. B. W. Lee and J. Zinn-Justin, *Phys. Rev.* **D7**, 1049 (1973).

23. S. R. Coleman, *Aspects of symmetry: selected Erice lectures* (Cambridge Univ. Press, Cambridge, 1985).

24. A. Vilenkin, *Phys. Rept.* **121**, p. 263 (1985).

25. S. R. Coleman and E. Weinberg, *Phys. Rev.* **D7**, 1888 (1973).

26. T. W. B. Kibble, *Phys. Rev.* **155**, 1554 (1967).

27. Y. Nambu, *Phys. Rev. Lett.* **4**, 380 (1960).

28. Y. Nambu and G. Jona-Lasinio, *Phys. Rev.* **122**, 345 (1961).

29. J. Goldstone, *Nuovo Cim.* **19**, 154 (1961).

30. J. Goldstone, A. Salam and S. Weinberg, *Phys. Rev.* **127**, 965 (1962).

31. G. Arnison *et al.*, *Phys. Lett.* **B166**, p. 484 (1986).

32. R. Ansari *et al.*, *Phys. Lett.* **B186**, p. 440 (1987).

33. R. Barate *et al.*, *Phys. Lett.* **B565**, 61 (2003).

34. N. Cabibbo, L. Maiani, G. Parisi and R. Petronzio, *Nucl. Phys.* **B158**, p. 295 (1979).

35. J. F. Gunion, S. Dawson, H. E. Haber and G. L. Kane, *The Higgs hunter's guide* (Westview, Boulder, CO, 1990).

36. T. Hambye and K. Riesselmann (1997), hep-ph/9708416.

37. A. Hasenfratz, K. Jansen, C. B. Lang, T. Neuhaus and H. Yoneyama, *Phys. Lett.* **B199**, p. 531 (1987).

38. J. Kuti, L. Lin and Y. Shen, *Phys. Rev. Lett.* **61**, p. 678 (1988).

39. M. Luscher and P. Weisz, *Nucl. Phys.* **B318**, p. 705 (1989).

40. B. W. Lee, C. Quigg and H. B. Thacker, *Phys. Rev.* **D16**, p. 1519 (1977).

41. G. Altarelli and G. Isidori, *Phys. Lett.* **B337**, 141 (1994).

42. J. A. Casas, J. R. Espinosa and M. Quiros, *Phys. Lett.* **B382**, 374 (1996).

43. G. Isidori, G. Ridolfi and A. Strumia, *Nucl. Phys.* **B609**, 387 (2001).

44. M. C. Gonzalez-Garcia and Y. Nir, *Rev. Mod. Phys.* **75**, 345 (2003).

45. P. Langacker, J. Erler and E. Peinado, *J. Phys. Conf. Ser.* **18**, 154 (2005).

46. R. N. Mohapatra *et al.*, *Rept. Prog. Phys.* **70**, 1757 (2007).

47. M. C. Gonzalez-Garcia and M. Maltoni, *Phys. Rept.* **460**, 1 (2008).

48. S. L. Glashow and S. Weinberg, *Phys. Rev.* **D15**, p. 1958 (1977).

49. M. K. Gaillard and B. W. Lee, *Phys. Rev.* **D10**, p. 897 (1974).

50. P. Langacker (1991), in *TeV Physics*, ed T. Huang et al., Gordon and Breach, N.Y., 1991.

51. Y. Nir (2007), 0708.1872.

52. N. Cabibbo, *Phys. Rev. Lett.* **10**, 531 (1963).

53. M. Kobayashi and T. Maskawa, *Prog. Theor. Phys.* **49**, 652 (1973).

54. L. Wolfenstein, *Phys. Rev. Lett.* **51**, p. 1945 (1983).

55. E. D. Commins and P. H. Bucksbaum, *Weak interactions of leptons and quarks* (Cambridge Univ. Press, Cambridge, 1983).

56. P. B. Renton, *Electroweak interactions* (Cambridge Univ. Press, Cambridge, 1990).

57. P. G. Langacker, *Precision tests of the standard electroweak model* (World Scientific, Singapore, 1995).

58. A. Czarnecki, W. J. Marciano and A. Sirlin, *Phys. Rev.* **D70**, p. 093006 (2004).

59. J. Charles *et al.*, *Eur. Phys. J.* **C41**, 1 (2005), with updated results and plots available at: http://ckmfitter.in2p3.fr.

60. T. Kinoshita, *Quantum electrodynamics* (World Scientific, Singapore, 1990).

61. S. G. Karshenboim, *Phys. Rept.* **422**, 1 (2005).

62. P. J. Mohr, B. N. Taylor and D. B. Newell, *Rev. Mod. Phys.* **80**, 633 (2008).

63. S. L. Wu, *Phys. Rept.* **107**, 59 (1984).

64. C. Kiesling, *Tests of the standard theory of electroweak interactions* (Springer, Berlin, 1988).

65. G. W. Bennett *et al.*, *Phys. Rev.* **D73**, p. 072003 (2006).

66. A. Czarnecki and W. J. Marciano, *Phys. Rev.* **D64**, p. 013014 (2001).

67. S. L. Glashow, J. Iliopoulos and L. Maiani, *Phys. Rev.* **D2**, 1285 (1970).

68. P. Langacker and D. London, *Phys. Rev.* **D38**, p. 886 (1988).

69. F. J. Hasert *et al.*, *Phys. Lett.* **B46**, 138 (1973).

70. A. C. Benvenuti *et al.*, *Phys. Rev. Lett.* **32**, 800 (1974).

71. J. E. Kim, P. Langacker, M. Levine and H. H. Williams, *Rev. Mod. Phys.* **53**, p. 211 (1981).

72. U. Amaldi *et al.*, *Phys. Rev.* **D36**, p. 1385 (1987).

73. G. Costa, J. R. Ellis, G. L. Fogli, D. V. Nanopoulos and F. Zwirner, *Nucl. Phys.* **B297**, p. 244 (1988).

74. P. Langacker, M.-x. Luo and A. K. Mann, *Rev. Mod. Phys.* **64**, 87 (1992).

75. J. Erler and P. Langacker (2008), 0807.3023.

76. P. Langacker (1993), hep-ph/9305255.

77. G. P. Zeller *et al.*, *Phys. Rev. Lett.* **88**, p. 091802 (2002).

78. D. Mason *et al.*, *Phys. Rev. Lett.* **99**, p. 192001 (2007).

79. S. Schael *et al.*, *Phys. Rept.* **427**, p. 257 (2006).

80. J. Alcaraz *et al.* (2006), hep-ex/0612034.

81. K. Wichmann (2007), 0707.2724.

82. K. Maltman and T. Yavin (2008), 0812.2457.

83. M. S. Chanowitz, *Phys. Rev.* **D66**, p. 073002 (2002).

84. S. Heinemeyer, W. Hollik and G. Weiglein, *Phys. Rept.* **425**, 265 (2006).

85. S. Heinemeyer, W. Hollik, A. M. Weber and G. Weiglein, *JHEP* **04**, p. 039 (2008).

86. A. Czarnecki and W. J. Marciano, *Int. J. Mod. Phys.* **A15**, 2365 (2000).

87. J. Erler, A. Kurylov and M. J. Ramsey-Musolf, *Phys. Rev.* **D68**, p. 016006 (2003).

88. J. L. Hewett and T. G. Rizzo, *Phys. Rept.* **183**, p. 193 (1989).

89. P. Langacker (2008), 0801.1345.

90. K.-m. Cheung, *Phys. Lett.* **B517**, 167 (2001).

91. M. Chemtob, *Prog. Part. Nucl. Phys.* **54**, 71 (2005).

92. R. Barbier *et al.*, *Phys. Rept.* **420**, 1 (2005).

93. J. L. Hewett and M. Spiropulu, *Ann. Rev. Nucl. Part. Sci.* **52**, 397 (2002).

94. C. Csaki (2004), hep-ph/0404096.

95. R. Sundrum (2005), hep-th/0508134.

96. G.-C. Cho, K. Hagiwara and S. Matsumoto, *Eur. Phys. J.* **C5**, 155 (1998).

97. Z. Han and W. Skiba, *Phys. Rev.* **D71**, p. 075009 (2005).

98. M. E. Peskin and T. Takeuchi, *Phys. Rev. Lett.* **65**, 964 (1990).

99. M. E. Peskin and T. Takeuchi, *Phys. Rev.* **D46**, 381 (1992).

100. C. T. Hill and E. H. Simmons, *Phys. Rept.* **381**, 235 (2003).

101. J. Erler and D. M. Pierce, *Nucl. Phys.* **B526**, 53 (1998).

102. J. R. Ellis, S. Heinemeyer, K. A. Olive, A. M. Weber and G. Weiglein, *JHEP* **08**, p. 083 (2007).

103. H. Georgi, H. R. Quinn and S. Weinberg, *Phys. Rev. Lett.* **33**, 451 (1974).

104. U. Amaldi, W. de Boer and H. Furstenau, *Phys. Lett.* **B260**, 447 (1991).

105. J. R. Ellis, S. Kelley and D. V. Nanopoulos, *Phys. Lett.* **B249**, 441 (1990).

106. C. Giunti, C. W. Kim and U. W. Lee, *Mod. Phys. Lett.* **A6**, 1745 (1991).

107. P. Langacker and M.-x. Luo, *Phys. Rev.* **D44**, 817 (1991).

108. M. B. Green, J. H. Schwarz and E. Witten, *Superstring theory* (Cambridge Univ. Press, New York, NY, 1987).

109. J. Polchinski, *String Theory* (Cambridge Univ. Press, Cambridge, 1998).

110. K. Becker, M. Becker and J. Schwarz, *String Theory and M-Theory: A Modern Introduction* (Cambridge Univ. Press, Cambridge, 2007).

111. H. Georgi and S. L. Glashow, *Phys. Rev. Lett.* **32**, 438 (1974).

112. P. Langacker, *Phys. Rept.* **72**, p. 185 (1981).

113. G. G. Ross, *Grand Unified Theories* (Westview Press, Reading, MA, 1985).

114. S. Raby (2008), 0807.4921.

115. J. Preskill, *Ann. Rev. Nucl. Part. Sci.* **34**, 461 (1984).

116. C. D. Froggatt and H. B. Nielsen, *Nucl. Phys.* **B147**, p. 277 (1979).

117. K. S. Babu and R. N. Mohapatra, *Phys. Rev. Lett.* **66**, 556 (1991).

118. N. Arkani-Hamed, A. G. Cohen, E. Katz and A. E. Nelson, *JHEP* **07**, p. 034 (2002).

119. M. Perelstein, *Prog. Part. Nucl. Phys.* **58**, 247 (2007).

120. Z. Chacko, H.-S. Goh and R. Harnik, *Phys. Rev. Lett.* **96**, p. 231802 (2006).

121. N. Arkani-Hamed, S. Dimopoulos and G. R. Dvali, *Phys. Lett.* **B429**, 263 (1998).

122. K. R. Dienes, E. Dudas and T. Gherghetta, *Nucl. Phys.* **B537**, 47 (1999).

123. L. Randall and R. Sundrum, *Phys. Rev. Lett.* **83**, 3370 (1999).

124. C. Csaki, C. Grojean, L. Pilo and J. Terning, *Phys. Rev. Lett.* **92**, p. 101802 (2004).

125. N. Arkani-Hamed, A. G. Cohen and H. Georgi, *Phys. Lett.* **B513**, 232 (2001).

126. C. T. Hill, S. Pokorski and J. Wang, *Phys. Rev.* **D64**, p. 105005 (2001).

127. T. Appelquist and J. Carazzone, *Phys. Rev.* **D11**, p. 2856 (1975).

128. H.-C. Cheng and I. Low, *JHEP* **09**, p. 051 (2003).

129. T. Appelquist, H.-C. Cheng and B. A. Dobrescu, *Phys. Rev.* **D64**, p. 035002

(2001).

130. V. Agrawal, S. M. Barr, J. F. Donoghue and D. Seckel, *Phys. Rev.* **D57**, 5480 (1998).

131. N. Arkani-Hamed and S. Dimopoulos, *JHEP* **06**, p. 073 (2005).

132. R. D. Peccei, *Adv. Ser. Direct. High Energy Phys.* **3**, 503 (1989).

133. M. Dine (2000), hep-ph/0011376.

134. J. E. Kim and G. Carosi (2008), 0807.3125.

135. A. A. Anselm and A. A. Johansen, *Nucl. Phys.* **B412**, 553 (1994).

136. A. E. Nelson, *Phys. Lett.* **B143**, p. 165 (1984).

137. S. M. Barr, *Phys. Rev.* **D30**, p. 1805 (1984).

138. D. B. Kaplan and A. V. Manohar, *Phys. Rev. Lett.* **56**, p. 2004 (1986).

139. D. R. Nelson, G. T. Fleming and G. W. Kilcup, *Phys. Rev. Lett.* **90**, p. 021601 (2003).

140. R. D. Peccei and H. R. Quinn, *Phys. Rev. Lett.* **38**, 1440 (1977).

141. S. Weinberg, *Phys. Rev. Lett.* **40**, 223 (1978).

142. F. Wilczek, *Phys. Rev. Lett.* **40**, 279 (1978).

143. A. Chodos, P. G. O. Freund and T. W. Appelquist, *Modern Kaluza-Klein theories* (Addison-Wesley, Merlo Park, CA, 1987).

144. H. P. Nilles, *Phys. Rept.* **110**, p. 1 (1984).

145. J. Wess and J. A. Bagger, *Supersymmetry and supergravity* (Princeton Univ. Press, Princeton, NJ, 1992).

146. J. Terning, *Modern Supersymmetry: Dynamics and Duality* (Clarendon Press, Oxford, 2006).

147. J. D. Barrow and F. J. Tipler, *The anthropic cosmological principle* (Clarendon Press, Oxford, 1986).

148. M. J. Rees (2004), astro-ph/0401424.

149. C. J. Hogan, *Rev. Mod. Phys.* **72**, 1149 (2000).

150. S. Weinberg, *Rev. Mod. Phys.* **61**, 1 (1989).

151. R. Bousso and J. Polchinski, *JHEP* **06**, p. 006 (2000).

152. S. Kachru, R. Kallosh, A. Linde and S. P. Trivedi, *Phys. Rev.* **D68**, p. 046005 (2003).

153. L. Susskind (2003), hep-th/0302219.

154. F. Denef and M. R. Douglas, *JHEP* **05**, p. 072 (2004).

155. A. D. Linde, *Mod. Phys. Lett.* **A1**, p. 81 (1986).

156. A. D. Sakharov, *JETP Lett.* **5**, 24 (1967).

157. W. Bernreuther, *Lect. Notes Phys.* **591**, 237 (2002).

158. M. Dine and A. Kusenko, *Rev. Mod. Phys.* **76**, p. 1 (2004).

159. G. 't Hooft, *Phys. Rev. Lett.* **37**, 8 (1976).

160. F. R. Klinkhamer and N. S. Manton, *Phys. Rev.* **D30**, p. 2212 (1984).

161. V. A. Kuzmin, V. A. Rubakov and M. E. Shaposhnikov, *Phys. Lett.* **B155**, p. 36 (1985).

162. M. Fukugita and T. Yanagida, *Phys. Lett.* **B174**, p. 45 (1986).

163. S. Davidson, E. Nardi and Y. Nir, *Phys. Rept.* **466**, 105 (2008).

164. M. Trodden, *Rev. Mod. Phys.* **71**, 1463 (1999).

165. M. Carena, G. Nardini, M. Quiros and C. E. M. Wagner, *JHEP* **10**, p. 062 (2008).

166. V. Barger, P. Langacker and G. Shaughnessy, *New J. Phys.* **9**, p. 333 (2007).
167. I. Affleck and M. Dine, *Nucl. Phys.* **B249**, p. 361 (1985).
168. A. G. Cohen and D. B. Kaplan, *Phys. Lett.* **B199**, p. 251 (1987).
169. H. Davoudiasl, R. Kitano, G. D. Kribs, H. Murayama and P. J. Steinhardt, *Phys. Rev. Lett.* **93**, p. 201301 (2004).
170. D. H. Lyth and A. Riotto, *Phys. Rept.* **314**, 1 (1999).
171. E. W. Kolb and M. S. Turner, *The early universe* (Addison-Wesley, Redwood City, CA, 1990).
172. P. J. E. Peebles, *Principles of physical cosmology* (Princeton Univ. Press, Princeton, NJ, 1993).
173. J. Dunkley *et al.* (2008), 0803.0586.
174. P. J. E. Peebles and B. Ratra, *Rev. Mod. Phys.* **75**, 559 (2003).
175. J. Frieman, M. Turner and D. Huterer (2008), 0803.0982.
176. J. Martin, *Mod. Phys. Lett.* **A23**, 1252 (2008).
177. J.-P. Uzan, *Rev. Mod. Phys.* **75**, p. 403 (2003).
178. G. Jungman, M. Kamionkowski and K. Griest, *Phys. Rept.* **267**, 195 (1996).
179. G. Bertone, D. Hooper and J. Silk, *Phys. Rept.* **405**, 279 (2005).
180. D. Hooper and E. A. Baltz, *Ann. Rev. Nucl. Part. Sci.* **58**, 293 (2008).
181. P. Svrcek and E. Witten, *JHEP* **06**, p. 051 (2006).
182. S. J. Asztalos, L. J. Rosenberg, K. van Bibber, P. Sikivie and K. Zioutas, *Ann. Rev. Nucl. Part. Sci.* **56**, 293 (2006).
183. G. D'Ambrosio, G. F. Giudice, G. Isidori and A. Strumia, *Nucl. Phys.* **B645**, 155 (2002).

Chapter 2

Topics in Flavor Physics

K.S. Babu

Department of Physics, Oklahoma State University
Stillwater, OK 74078, USA
babu@okstate.edu

In this review, I first summarize our present knowledge on the fundamental parameters of the flavor sector. Then I discuss various scenarios going beyond the standard model which attempt to explain aspects of the "flavor puzzle". Relating quark masses and mixing angles via flavor symmetry is explored. Explaining the mass hierarchy via the Froggatt–Nielsen mechanism is reviewed and illustrated. Grand unification ideas are pursued to seek a pattern in the observed masses and mixings of quarks and leptons. Generating light fermion masses as radiative corrections is explained and illustrated. The popular solutions to the strong CP problem are summarized. Finally, specific processes in B meson system where significant new flavor contributions can arise are discussed.

2.1. Overview

This set of lectures will focus on the flavor sector of the Standard Model (SM). As you know, most of the free parameters of the SM reside in this sector. In case you have not thought about it lately, let me remind you of the counting of parameters of the SM. Not including neutrino masses, there are 19 parameters in the SM. Five of these are flavor universal – the three gauge couplings (g_1, g_2, g_3), one Higgs quartic coupling λ, and one Higgs mass-squared μ^2, while the remaining fourteen are parameters associated with the flavor sector. Six quark masses, three charged lepton masses, four quark mixing angles (including one weak CP violating phase) make up thirteen, while the strong CP violating parameter $\bar{\theta}$, which is intimately related to the quark masses, is the fourteenth flavor parameter. If we include small neutrino masses and mixing angles into the SM, as needed to explain neutrino oscillation data from a variety of experiments, an additional nine parameters will have to be introduced (three neutrino masses, three neutrino mixing angles and three CP violating phases, in the case of Majorana neutrinos). You see that twenty three of the twenty eight parameters describe flavor physics in the SM.

While there is abundant information on the numerical values these parameters take, a fundamental understanding of the origin of these parameters is currently lacking. Why are there three families of quarks and leptons in the first place? Are the flavor parameters all arbitrary, or are they inter-connected? Why do the charged fermion masses exhibit a strong hierarchical structure spanning some six orders of magnitude? Why are the mixing angles in the quark sector hierarchical? Are the mixing parameters related to the mass ratios? Why is $\bar{\theta} < 10^{-9}$? Why are neutrino masses so much smaller than the charged fermion masses? What causes (at least two of) the neutrino mixing angles to be much larger than the corresponding quark mixing angles? What is the origin of CP violation? The lack of a fundamental understanding of such issues is often referred to as the "flavor puzzle".

Various solutions to this puzzle have been proposed, inevitably leading to physics beyond the Standard Model, for within the SM these parameters can only be accommodated, and not explained. Forthcoming experiments, especially at the LHC, have the potential to confirm or refute some, but not all, of these proposed non–standard scenarios. If the new flavor dynamics occurs near the TeV scale, it is potentially accessible to the LHC, but if it occurs at a much higher scale, then it will not be directly accessible. It should be mentioned at the outset that there is no compelling reason for the flavor dynamics to occur near the TeV scale, most puzzles can be explained even when the dynamics takes place near the Planck scale. This is because the small parameters of the flavor sector are quite stable under radiative corrections, owing to chiral symmetries. If the smallness of a certain parameter has an explanation from Planck scale physics, it is an equally good explanation at the low energy scale. Testing such high scale theories would be more difficult in general. In some cases, for example, with low energy supersymmetry, information from the high scale flavor dynamics will be carried by particles which survive to the TeV scale (the SUSY particles), in which case flavor physics may be tested at colliders. Processes such as lepton flavor violating $\mu \to e\gamma$ decay and $b \to s\gamma$ transition appear to be promising setups to test such scenarios.

The Higgs boson is waiting to be discovered at the LHC. Its production and decay rates can be significantly modified relative to the SM expectations in some of the flavor–extensions of the SM. I will describe explicit models in this category. Very little is known about the top quark properties currently. LHC will serve as a top quark factory where modifications in the top sector arising from flavor–extensions can be studied. These include flavor changing decays of the top and its possible anomalous couplings to the gauge bosons. We have learned a lot about the B meson system from the B factories lately, but there are still many open issues and some puzzles which will be probed at the LHC. These include precise determination of the CP violating parameters, rare processes allowed in the SM but not yet observed, and new physics processes in B decays that require modification of the SM structure.

In Sec. 2.2, we will take a tour of the flavor parameters of the SM and review how these are measured and interpreted. Various ideas attempting to understand aspects of the flavor puzzle will then be introduced and their experimental consequences outlined. In Sec. 2.3 we will seek inter–relations between quark masses and mixing angle. Sec. 2.4 will be devoted to an understanding of the fermion mass and mixing hierarchies based on the Frogatt–Nielsen mechanism. In Sec. 2.5 we will develop grand unification as a possible clue to the flavor puzzle. Sec. 2.6 discusses radiative fermion mass generation, Sec. 2.7 summarizes the suggested solutions to the strong CP problem, and in Sec. 2.8 we introduce specific beyond the SM scenarios for the flavor sector and study their experimental manifestations at the LHC.

2.2. Flavor structure of the Standard Model

Because of the chiral structure of weak interactions, bare fermion masses are not allowed in the Standard Model. Fermion masses arise via Yukawa interactions given by the Lagrangian

$$\mathcal{L}_{\text{Yukawa}} = Q^T Y_u u^c H - Q^T Y_d d^c \tilde{H} - L^T Y_\ell e^c \tilde{H} + h.c. \qquad (2.1)$$

Here I have used the standard notation for quark (Q, u^c, d^c) and lepton (L, e^c) fields. (Q, L) are $SU(2)_L$ doublets, as is the Higgs field H and its conjugate $\tilde{H} = i\tau_2 H^*$, while the (u^c, d^c, e^c) fields are $SU(2)_L$ singlets. All fermion fields are left–handed, a charge conjugation matrix C is understood to be sandwiched between all of the fermion bi-linears in Eq. (2.1). Contraction of the color indices is not displayed, but should be obvious. $Y_{u,d,\ell}$ are the Yukawa couplings matrices spanning generation space which are complex and non–Hermitian. $SU(2)_L$ contraction between the fermion doublet and Higgs doublet involves the matrix $i\tau_2$. Explicitly, we have (for a family labeled by index i)

$$Q_i = \begin{pmatrix} u_i \\ d_i \end{pmatrix} \; ; \; L_i = \begin{pmatrix} \nu_i \\ e_i \end{pmatrix} \; ; \; H = \begin{pmatrix} H^+ \\ H^0 \end{pmatrix} \; ; \; \tilde{H} = \begin{pmatrix} H^{0*} \\ -H^- \end{pmatrix} , \qquad (2.2)$$

so that Eq. (2.1) expands to

$$\mathcal{L}_{\text{Yukawa}} = (Y_u)_{ij}[u_i u_j^c H^0 - d_i u_j^c H^+] + (Y_d)_{ij}[u_i d_j^c H^- + d_i d_j^c H^{0*}]$$
$$+ (Y_\ell)_{ij}[\nu_i e_j^c H^- + e_i e_j^c H^{0*}] + h.c. \qquad (2.3)$$

The neutral component of H acquires a vacuum expectation value (VEV) $\langle H^0 \rangle = v$, spontaneously breaking the electroweak symmetry ($v \simeq 174$ GeV). The Higgs field can then be parametrized in the unitary gauge as $H^0 = (\frac{h}{\sqrt{2}} + v)$ where h is a real physical field (the Higgs boson). In this gauge $H^\pm$, which are eaten up by the $W^\pm$ gauge bosons, and the phase of H^0, which is eaten up by the Z^0 gauge boson, do not appear.

The VEV of H^0 generates the following fermion mass matrices:

$$M_u = Y_u v , \qquad M_d = Y_d v , \qquad M_\ell = Y_\ell v . \qquad (2.4)$$

The Yukawa coupling matrices contained in $(Y_u)_{ij}/\sqrt{2}(uu^c h)$, etc in each of the up, down and charged lepton sector becomes proportional to the corresponding mass matrix. Once the mass matrices are brought to diagonal forms, the Yukawa coupling matrices will be simultaneously diagonal. There is thus no tree–level flavor changing current mediated by the neutral Higgs boson in the Standard Model. This is a feature that is generally lost as we extend the SM to address the flavor issue (for example by introducing multiple Higgs doublets or extra fermions).

We make unitary rotations on the quark fields in family space. Unitarity of these rotations will ensure that the quark kinetic terms remain canonical. Specifically, we define mass eigenstates $(u^0, u^{c0}, d^0, d^{c0})$ via

$$u = V_u \, u^0 , \quad u^c = V_{u^c} \, u^{c0} ,$$
$$d = V_d \, d^0 , \quad d^c = V_{d^c} \, d^{c0} , \qquad (2.5)$$

and we choose the unitary matrices such that

$$V_u^T (Y_u v) V_{u^c} = \begin{pmatrix} m_u & & \\ & m_c & \\ & & m_t \end{pmatrix} , \quad V_d^T (Y_d v) V_{d^c} = \begin{pmatrix} m_d & & \\ & m_s & \\ & & m_b \end{pmatrix} . \quad (2.6)$$

We have assumed here that the number of families is three, but the procedure applies to any number of families. Bi-unitary transformations such as the ones in Eq. (2.6) can diagonalize non–Hermitian matrices. The same transformations should be applied to all interactions of the quarks. As already noted, these transformations will bring the Yukawa interactions of quarks with the Higgs boson into diagonal forms. The couplings of the Z^0 boson and the photon to quarks will have the original diagonal form even after this rotation. For example, $(\overline{u}\gamma_\mu I u)Z^\mu$ where I is the identity matrix acting on family space will transform to $(\overline{u^0}\gamma_\mu (V_u^\dagger I V_u)u^0)Z^\mu$, which is identical to $(\overline{u^0}\gamma_\mu I u^0)Z^\mu$. Similarly, $(\overline{u^c}\gamma_\mu I u^c)Z^\mu$ will transform to $(\overline{u^{c0}}\gamma_\mu I u^{c0})Z^\mu$. We see that there is no tree level flavor changing neutral current (FCNC) mediated by the Z^0 boson and the photon in the SM.

Most significantly, the transformations of Eq. (2.6) will bring the charged current quark interaction, which originally is of the form $\mathcal{L}_{cc} = g/\sqrt{2}(\overline{u}\gamma_\mu d)W^{+\mu} + h.c.$, into the form

$$\mathcal{L}_{cc} = \frac{g}{\sqrt{2}}\left[\overline{u^0}\gamma_\mu V d^0\right] W^{\mu+} + h.c. \tag{2.7}$$

where

$$V = V_u^\dagger V_d \tag{2.8}$$

is the quark mixing matrix, or the Cabibbo–Kobayashi–Maskawa (CKM) matrix.[1,2] In the SM, all the flavor violation is contained in V. Being product of unitary matrices, V is itself unitary. This feature has thus far withstood experimental scrutiny, with further scrutiny expected from LHC experiments.

Note that the right–handed rotation matrices V_{u^c} and V_{d^c} have completely disappeared, a result of the purely left–handed nature of charged weak current.

We can repeat this process in the leptonic sector. We define, in analogy with Eq. (2.5),

$$\nu = V_\nu\, \nu^0\,, \quad e = V_e\, e^0\,, \quad e^c = V_{e^c}\, e^{c0}\,. \tag{2.9}$$

We choose Y_e and Y_{e^c} such that

$$Y_e^T(Y_\ell v)Y_{e^c} = \begin{pmatrix} m_e & & \\ & m_\mu & \\ & & m_\tau \end{pmatrix}. \tag{2.10}$$

Note that there is no right–handed neutrino in the SM. If the Yukawa Lagrangian is as given in Eq. (2.1), there is no neutrino mass. In that case one can choose $V_\nu = V_e$, so that the charged current weak interactions will remain flavor diagonal. However, it is now well established that neutrinos have small masses. Additional terms must be added to Eq. (2.1) in order to accommodate them. The simplest possibility is to add a non–renormalizable term

$$\mathcal{L}_{\nu-\text{mass}} = \frac{(L^T Y_\nu L)HH}{2M_*} + h.c. \tag{2.11}$$

where the $SU(2)_L$ contraction between the H fields is in the triplet channel and Y_ν is a complex symmetric matrix in generation space. Here M_* is a mass scale much above the weak interaction scale. Eq. (2.11) can arise by integrating out some heavy fields with mass of order M_*. The most celebrated realization of this is the seesaw mechanism, where M_* corresponds to the mass of the right–handed neutrino.[3] The neutrino masses are suppressed, compared to the charged fermion masses, because of the inverse dependence on the heavy scale M_*. Right–handed neutrinos, if they exist, are complete singlets of the SM gauge symmetry, and can possess bare SM invariant mass terms, unlike any other fermion of the SM. This is an elegant explanation of why the neutrinos are much lighter than other fermions, relying only on symmetry principles and dimensional analysis. Eq. (2.11) leads to a light neutrino mass matrix given by

$$M_\nu = Y_\nu \frac{v^2}{M_*}. \tag{2.12}$$

Now we choose V_ν so that

$$V_\nu^T Y_\nu \frac{v^2}{M_*} V_\nu = \begin{pmatrix} m_1 & & \\ & m_2 & \\ & & m_3 \end{pmatrix}, \tag{2.13}$$

with $m_{1,2,3}$ being the tiny masses of the three light neutrinos. The leptonic charge current interaction now becomes

$$\mathcal{L}_{cc}^\ell = \frac{g}{\sqrt{2}} [\overline{e^0}\gamma_\mu U \nu^0] \, W^{-\mu} + h.c. \tag{2.14}$$

where

$$U = V_e^\dagger V_\nu \tag{2.15}$$

is the leptonic mixing matrix, or the Pontecorvo–Maki–Nakagawa–Sakata (PMNS) matrix.[4] As V, U is also unitary. Neutrino oscillations observed in experiments are attributed to the off–diagonal entries of the matrix U. We assumed here that the neutrino mass generation mechanism violated total lepton number by two units. While this is very attractive, it should be mentioned that neutrinos could acquire masses very much like the quarks.

That would require the right–handed ν^c states to be part of the low energy theory. M_ν will then be similar to M_ℓ of Eq. (2.10). Neutrino oscillation phenomenology will be identical to the case of L–violating neutrino masses. In this case, however, the neutrino Yukawa couplings will have to be extremely tiny to accommodate the observed masses. Furthermore, some global symmetries, such as total lepton number, will have to be assumed in order to forbid gauge invariant mass terms for the right–handed neutrinos.

The fermionic states (e_i^0) are simply the physical electron, the muon, and the tau lepton states. Similarly, the quark fields with a superscript ^{0} are the mass eigenstates. It is conventional to drop these superscripts, which we shall do from now on.

2.2.1. *Lepton masses*

Conceptually charged lepton masses are the easiest to explain. Leptons are propagating states, and their masses are simply the poles in the propagators. Experimental information on charged lepton masses is rather accurate:[5]

$$m_e = 0.510998902 \pm 0.000000021 \text{ MeV},$$

$$m_\mu = 105.658357 \pm 0.000005 \text{ MeV},$$

$$m_\tau = 1777.03^{+0.30}_{-0.26} \text{ MeV}. \tag{2.16}$$

The direct kinematic limits on the three neutrino masses are:[5]

$$m_{\nu_e} \leq 3 \text{ eV}, \ m_{\nu_\mu} \leq 0.19 \text{ MeV}, \ m_{\nu_\tau} \leq 18.2 \text{ MeV}. \tag{2.17}$$

Neutrino oscillation experiments have provided much more accurate determinations of the squared mass differences $\Delta m_{ij}^2 = m_i^2 - m_j^2$. Solar and atmospheric neutrino oscillation experiments, when combined with accelerator and reactor neutrino experiments, suggest the following allowed values (with 2σ error quoted):[6]

$$\Delta m_{21}^2 = (7.25 - 8.11) \times 10^{-5} \text{ eV}^2,$$

$$\Delta m_{31}^2 = \pm(2.18 - 2.64) \times 10^{-3} \text{ eV}^2. \tag{2.18}$$

While this still leaves some room for the absolute masses, when combined with the direct limit on $m_{\nu_e} \leq 3$ eV, the options become limited. Current data allow for two possible ordering of the mass hierarchies: (i) normal hierarchy where $m_1 \leq m_2 \ll m_3$, and (ii) inverted hierarchy where $m_1 \simeq m_2 \gg m_3$. More specifically, ν_e is mostly in the lightest eigenstate in the case of normal hierarchy, while it is mostly in the heavier eigenstate in the case of inverted hierarchy. The sign of Δm_{31}^2 is not known at the moment, which gives these two ordering options. On the other hand, the sign of Δm_{21}^2 is fixed from the condition that MSW resonance occurs inside the Sun.

2.2.2. *Leptonic mixing matrix*

The PMNS matrix U, being unitary, has N^2 independent components for N families of leptons. Out of these, $N(N-1)/2$ are Euler angles, while the remaining $N(N+1)/2$ are phases. Many of these phases can be absorbed into the fermionic fields and removed. If one writes $U = Q\hat{U}P$, where P and Q are diagonal phase matrices, then by redefining the phases of e fields as $e \to Qe$, the N phases in Q can be removed. P has only $N-1$ non–removable phases (an overall phase is irrelevant). For $N = 3$, $P = diag.(e^{i\alpha},\ e^{i\beta},\ 1)$. α, β are called the Majorana phases. (If the neutrino masses are of the Dirac type, these phases can also be removed by redefining the ν^c fields.) $\hat{U}$ will then have $N(N+1)/2 - (2N-1) = \frac{1}{2}(N-1)(N-2)$ phases. For $N = 3$, there is a single "Dirac" phase in U. This single phase will be relevant for neutrino oscillation phenomenology. The two Majorana phases (α, β) do not affect neutrino oscillations, but will be relevant for neutrino-less double beta decay.

In general, the PMNS matrix for three families of leptons can be written as

$$U = \begin{pmatrix} U_{e1} & U_{e2} & U_{e3} \\ U_{\mu 1} & U_{\mu 2} & U_{\mu 3} \\ U_{\tau 1} & U_{\tau 2} & U_{\tau 3} \end{pmatrix}. \tag{2.19}$$

To enforce the unitarity relations it is convenient to adopt specific parametrizations. The Euler angles, as you know, can be parametrized in many different ways. Furthermore, the Dirac phase can be chosen to appear in different ways (by field redefinitions). The "standard parametrization" that is now widely used[5] has $U_{PMNS} = U.P$ where

$$U = \begin{pmatrix} c_{12}c_{13} & s_{12}c_{13} & s_{13}e^{-i\delta} \\ -s_{12}c_{23} - c_{12}s_{23}s_{13}e^{i\delta} & c_{12}c_{23} - s_{12}s_{23}s_{13}e^{i\delta} & s_{23}c_{13} \\ s_{12}s_{23} - c_{12}c_{23}s_{13}e^{i\delta} & -c_{12}s_{23} - s_{12}c_{23}s_{13}e^{i\delta} & c_{23}c_{13} \end{pmatrix}. \tag{2.20}$$

Here $s_{ij} = \sin\theta_{ij}$, $c_{ij} = \cos\theta_{ij}$.

Our current understanding of these mixing angles arising from neutrino oscillations can be summarized as follows (2 σ error bars quoted):[6]

$$\sin^2 \theta_{12} = 0.27 - 0.35\,,$$
$$\sin^2 \theta_{23} = 0.39 - 0.63\,,$$
$$\sin^2 \theta_{13} \leq 0.040\,. \tag{2.21}$$

Here θ_{12} limit arises from solar neutrino data (when combined with Kam-Land reactor neutrino data), θ_{23} from atmospheric neutrinos (when combined with MINOS accelerator neutrino data), and θ_{13} from reactor neutrino data.

It is intriguing that the current understanding of leptonic mixing can be parametrized by the unitary matrix

$$U_{TB} = \begin{pmatrix} \sqrt{\frac{2}{3}} & \frac{1}{\sqrt{3}} & 0 \\ -\frac{1}{\sqrt{6}} & \frac{1}{\sqrt{3}} & -\frac{1}{\sqrt{2}} \\ -\frac{1}{\sqrt{6}} & \frac{1}{\sqrt{3}} & \frac{1}{\sqrt{2}} \end{pmatrix} P\,. \tag{2.22}$$

This mixing is known as tri-bimaximal mixing.[7] This nomenclature is based on the numerology $\sin^2 \theta_{12} = 1/3$, $\sin^2 \theta_{23} = 1/2$, $\sin^2 \theta_{13} = 0$ that follows from Eq. (2.22). As we will see, such a geometric structure is far from being similar to the quark mixing matrix. Note that currently θ_{13} is allowed to be zero, in which case the Dirac phase δ becomes irrelevant. We also have no information on the Majorana phases (α, β in P), which can only be tested in neutrino-less double beta decay experiments.

There have been considerable activity in the literature in trying the reproduce the tri-bimaximal mixing matrix of Eq. (2.22) based on symmetries. The most popular idea has been to adopt the non-Abelian flavor symmetry A_4, which is the symmetry group of a regular tetrahedron. It is also the group of even permutations of four letters. This finite group has twelve elements, which fall into one three–dimensional ($\mathbf{3}$) and three one–dimensional ($\mathbf{1} + \mathbf{1'} + \mathbf{1''}$) irreducible representations. A_4 is the simplest symmetry group with a triplet representation. Assigning the lepton doublets to the $\mathbf{3}$, and the three charged lepton singlets to the the ($\mathbf{1} + \mathbf{1'} + \mathbf{1''}$), it is possible, assuming a specific vacuum structure, to reproduce the "geometric" form of the leptonic mixing matrix.[8]

2.2.3. *Quark masses*

Unlike the leptons, quarks are not propagating particles. So their masses have to be inferred indirectly from properties of hadrons. There are various techniques to do this. Let me illustrate this for the light quark masses $(u,\ d,\ s)$ by the method of chiral perturbation theory.[9]

Consider the QCD Lagrangian at low energy scales. Electroweak symmetry has already been broken, and heavy quarks (t, b, c) have decoupled. The Lagrangian for the light quarks $(u,\ d,\ s)$ and the gluon fields takes the form

$$\mathcal{L} = \sum_{k=1}^{N_F=3} \bar{q}_k(i\slashed{D} - m_k)q_k - \frac{1}{4}G_{\mu\nu}G^{\mu nu}, \qquad (2.23)$$

where $G_{\mu\nu}$ is the gluon field strength and $\slashed{D}$ is the covariant derivative. m_k is the mass of the k-th quark and q_k denotes the quark field. This Lagrangian has a chiral symmetry in the limit where the quark masses vanish. The three left–handed quarks can be rotated into one another, and the three right–handed quarks can be rotated independently. The symmetry is $SU(3)_L \times SU(3)_R \times U(1)_V$, with the axial $U(1)_A$ (of the classical symmetry $U(3)_L \times U(3)_R$) explicitly broken by anomalies. The $U(1)_V$ is baryon number, which remains unbroken even after QCD dynamics. QCD dynamics breaks the $SU(3)_L \times SU(3)_R$ symmetry down to the diagonal subgroup $SU(3)_V$. In the limit of vanishing quark masses, there must be 8 Goldstone bosons corresponding to this symmetry breaking. These Goldstone bosons are identified as the pseudoscalar mesons, which are however, not exactly massless. The (small) quark masses actually break the chiral symmetry explicitly and thus generate small masses for the mesons.

Chiral perturbation theory is a systematic expansion in p/Λ_χ, where p is the particle momentum and $\Lambda_\chi \sim 1$ GeV is the chiral symmetry breaking scale. Since the masses of the light quarks $(u,\ d,\ s)$ are smaller than Λ_χ, we can treat them as small perturbations and apply chiral expansion. The explicit breaking of chiral symmetry occurs via the mass term

$$M = \begin{pmatrix} m_u & & \\ & m_d & \\ & & m_s \end{pmatrix}. \qquad (2.24)$$

M can be thought of as a spurion field which breaks the chiral symmetry spontaneously. Under $SU(3)_L \times SU(3)_R$ symmetry $q_L \to U_L\, q_L$, $q_R \to U_R\, q_R$, while $M \to U_L\, M\, U_R^\dagger$. That is, M transforms as a $(3, 3^*)$ of this group. Under the unbroken diagonal $SU(3)_V$ subgroup, both q_L and q_R transform as triplets, while M splits into a $\mathbf{1} + \mathbf{8}$. Thus M can be written as $M = M_1 + M_8$, where M_1 is a singlet of $SU(3)_V$, while M_8 is an octet:

$$M_1 = \frac{(m_u + m_d + m_s)}{3} \begin{pmatrix} 1 & & \\ & 1 & \\ & & 1 \end{pmatrix},$$

$$M_8 = \frac{(m_u - m_d)}{2} \begin{pmatrix} 1 & & \\ & -1 & \\ & & 0 \end{pmatrix} + \frac{(m_u - m_d - 2m_s)}{6} \begin{pmatrix} 1 & & \\ & 1 & \\ & & -2 \end{pmatrix}. \quad (2.25)$$

The octet (under $SU(3)_V$) of mesons can be written down as a (normalized) matrix

$$\Phi = \begin{pmatrix} \frac{\pi^0}{\sqrt{2}} + \frac{\eta^0}{\sqrt{6}} & \pi^+ & K^+ \\ \pi^- & -\frac{\pi^0}{\sqrt{2}} + \frac{\eta^0}{\sqrt{6}} & K^0 \\ K^- & \overline{K^0} & -\sqrt{\frac{2}{3}}\eta^0 \end{pmatrix}. \quad (2.26)$$

The lowest order invariants involving Φ bilinear and M are

$$A \,\mathrm{Tr}(\Phi^2)M_1 + B \,\mathrm{Tr}(\Phi^2 M_8). \quad (2.27)$$

Here A and B are arbitrary coefficients. Eq. (2.27) can be readily expanded, which will give relations for the masses of mesons. Now, in the limit of $m_u = 0, m_d = 0, m_s \neq 0$, the $SU(2)_L \times SU(2)_R$ chiral symmetry remains unbroken, and so the pion fields should be massless. Working out the mass terms, and demanding that the pion mass vanishes in this limit, one finds a relation $A = 2B$. Using this relation we can write down the pseudoscalar meson masses. In doing so, let us also recall that electromagnetic interactions will split the masses of the neutral and charged members. To lowest order, this splitting will be universal. Then we have

$$m_{\pi^0}^2 = B(m_u + m_d)$$
$$m_{\pi^\pm}^2 = B(m_u + m_d) + \Delta_{\mathrm{em}}$$
$$m_{K^0}^2 = m_{\overline{K^0}}^2 = B(m_d + m_s)$$
$$m_{K^\pm}^2 = B(m_u + m_s) + \Delta_{\mathrm{em}}$$
$$m_\eta^2 = \frac{1}{3}B(m_u + m_d + 4m_s). \quad (2.28)$$

Here small $\pi^0 - \eta^0$ mixing has been neglected, which vanishes in the limit $m_u - m_d$ vanishes.

Eliminating B and $\Delta_{\rm em}$ from Eq. (2.28) we obtain two relations for quark mass ratios:

$$\frac{m_u}{m_d} = \frac{2m_{\pi^0}^2 - m_{\pi^+}^2 + m_{K^+}^2 - m_{K^0}^2}{m_{K^0}^2 - m_{K^+}^2 + m_{\pi^+}^2} = 0.56$$

$$\frac{m_s}{m_d} = \frac{m_{K^0}^2 + m_{K^+}^2 - m_{\pi^+}^2}{m_{K^0}^2 - m_{K^+}^2 + m_{\pi^+}^2} = 20.1 \qquad (2.29)$$

This is the lowest order chiral perturbation theory result for the mass ratios. Second order chiral perturbation theory makes important corrections to these ratios as discussed in more detail in Ref. [10]. Note that the absolute masses cannot be determined in this way. Alternative techniques, such as QCD sum rules and lattice calculations which provide the most precise numbers have to be applied for this.

For heavy quarks (c and b), one can invoke another type of symmetry, the heavy quark effective theory (HQET).[11] When the mass of the quark is heavier than the typical momentum of the partons $\Lambda \sim m_p/3 = 330$ MeV, one can make another type of expansion. In analogy with atomic physics, where different isotopes exhibit similar chemical behavior, the behavior of charm hadrons and bottom hadrons will be similar. In fact, there will be an $SU(2)$ symmetry relating the two, to lowest order in HQET expansion. One consequence is that the mass splitting between the vector and scalar mesons in the b and c sector should be related. This leads to a relations $M_{B^*} - M_B = \Lambda^2/m_b$ and $M_{D^*} - M_D = \Lambda^2/m_c$, leading to the prediction

$$\frac{M_{B^*} - M_B}{M_{D^*} - M_D} = \frac{m_c}{m_b}, \qquad (2.30)$$

which is in good agreement with experiments.

The most reliable determination of light quark masses come from lattice QCD. The QCD Lagrangian of Eq. (2.23) has only very few parameters, the strong coupling constant, and the three light quark masses. All the hadron masses and decay constants should in principle be calculable in terms of these parameters. Since QCD coupling is strong at low energies, perturbation theory is not reliable. Lattice QCD is formulated on discrete space time lattice points, rather than in the continuum. When the lattice spacing takes small value, lattice QCD should reproduce continuum QCD. No approximation is made as regards the value of the strong coupling constant α_s. It is thus a non-perturbative technique which, upon matching certain measured quantities, can be used to calculate the light quark mass parameters. In the last five years there has been tremendous advances in lattice QCD, owing to improved lattice action, as well as increased computing power. Early results on light quark masses assumed "quenching", i.e., ignored fermions propagating inside loops, but now full three flavor unquenched calculation with dynamical fermions are available. There have been several independent evaluations of the light quark masses, which generally are in good agreement with one another. Conventionally these masses are presented as running masses at $q = 2$ GeV in the $\overline{\text{MS}}$ scheme.

The MILC collaboration,[12] which adopted a partially quenched approximation, finds for the light quark masses

$$m_u(2 \text{ GeV}) = 1.7 \pm 0.3 \text{ MeV},$$
$$m_d(2 \text{ GeV}) = 3.9 \pm 0.46 \text{ MeV},$$
$$m_s(2 \text{ GeV}) = 76 \pm 7.6 \text{ MeV}. \tag{2.31}$$

Here I have combined the various uncertainties (statistical, systematic, simulation, and electromagnetic) in quadrature. The ratios of light quark masses are thought to be more reliable, as many of the uncertainties cancel in the ratios. It is customary to define an average mass of up and down quarks $\hat{m} = (m_u + m_d)/2$. The results of MILC collaboration corresponds to the following mass ratios:

$$\frac{m_u}{m_d} = 0.43 \pm 0.08,$$
$$\frac{m_s}{\hat{m}} = 27.4 \pm 4.2. \tag{2.32}$$

The JLQCD collaboration,[13] which includes three flavors of dynamical quarks finds

$$\hat{m}(2 \text{ GeV}) = 3.55^{+0.65}_{-0.28} \text{ MeV},$$
$$m_s(2 \text{ GeV}) = 90.1^{+17.2}_{-6.1} \text{ MeV},$$
$$\frac{m_u}{m_d} = 0.577 \pm 0.025. \tag{2.33}$$

The RBC & UKQCD collaboration,[14] which includes $2+1$ dynamical domain wall quarks finds

$$\hat{m}(2 \text{ GeV}) = 3.72 \pm 0.41 \text{ MeV},$$
$$m_s(2 \text{ GeV}) = 107.3 \pm 11.7 \text{ MeV},$$
$$\hat{m} : m_s = 1 : 28.8 \pm 1.65. \tag{2.34}$$

And finally, the HPQCD collaboration finda[15]

$$m_u(2\text{ GeV}) = 1.9 \pm 0.24 \text{ MeV},$$
$$m_d(2\text{ GeV}) = 4.4 \pm 0.34 \text{ MeV},$$
$$m_s(2\text{ GeV}) = 87 \pm 5.7 \text{ MeV}$$
$$\hat{m}(2\text{ GeV}) = 3.2 \pm 0.89 \text{ MeV},$$
$$\frac{m_u}{m_d} = 0.43 \pm 0.08. \tag{2.35}$$

One sees that the lattice calculations are settling down, and have become quite reliable. It should be mentioned that the same lattice QCD calculations also provide several of the hadronic form factors which enter into the determination of the CKM mixing angles.

The masses of the c and b quarks can be determined in a variety of ways. Charmonium and Upsilon spectroscopy, in conjunction with lattice calculations seem to be the most reliable. We summarize the masses of these quarks thus obtained, along with the ranges for the light quark masses.[5]

$$m_u(2\text{ GeV}) = 1.5 \text{ to } 3.3 \text{ MeV},$$
$$m_d(2\text{ GeV}) = 3.5 \text{ to } 6.0 \text{ MeV},$$
$$m_s(2\text{ GeV}) = 105^{+25}_{-35} \text{ MeV},$$
$$\frac{m_u}{m_d} = 0.35 \text{ to } 0.60,$$
$$\frac{m_s}{m_d} = 17 \text{ to } 22,$$
$$\frac{m_s}{(m_u + m_d)/2} = 25 \text{ to } 30,$$
$$m_c(m_c) = 1.27^{+0.07}_{-0.11} \text{ GeV},$$
$$m_b(m_b) = 4.20^{+0.17}_{-0.07} \text{ GeV}. \tag{2.36}$$

Sometimes the light quark masses are quoted at $q = 1$ GeV, rather at $q = 2$ GeV. There are significant differences in these two sets of values due to the rapid running of the strong coupling in this regime. Typically one finds for example, $m_u(1\text{ GeV}) \simeq 1.35\, m_u(2\text{ GeV})$.

The top quark mass is more directly determined leading to the value[5]

$$m_t = 171.3 \pm 1.1 \pm 1.2 \text{ GeV}. \tag{2.37}$$

Any ambitious theory of flavor should aim to address these observed values of quark masses.

2.2.4. *Running quark and lepton masses*

In attempting to explain the observed masses of fermions, it will be convenient to compare their masses at a common momentum scale μ. Usually this scale is taken to be much heavier than the QCD scale of about 1 GeV, or even the weak scale of 246 GeV, since new flavor dynamics cannot happen at lower scales. The measured quark and lepton masses then have to be extrapolated to a common momentum scale μ. Below the weak scale, this extrapolation would require the renormalization group evolution of the mass parameters caused by QCD and QED loops. The beta functions and the gamma functions necessary to do this have been computed to three–loop (and in some cases four–loop) accuracy.[16] In the $\overline{MS}$ scheme, which is widely used, the contributions to the beta functions and gamma functions from a specific flavor of fermion will decouple for momenta μ less than the mass of the particle. Before discussing this evolution, it is necessary to remark on the differences between "pole mass" and "running mass" of a fermion. For heavy quarks (c, b, t) the pole mass M_q is the physical mass, which appears as the pole in the propagator. (For light quarks (u, d, s) pole mass is not defined because of the non–perturbative nature of strong interactions at their mass scales.) The running mass $m_q(M_q)$ includes corrections from QCD and QED loops. The two are related for quarks via

$$M_q = m_q(M_q) \left[1 + \frac{4}{3}\frac{\alpha_s(M_q)}{\pi} + \kappa_q^{(2)}\left(\frac{\alpha_s(M_q)}{\pi}\right)^2 + \kappa_q^{(3)}\left(\frac{\alpha_s(M_q)}{\pi}\right)^3 \right],$$

$$(2.38)$$

where terms of order α_s^4 and higher have been neglected. The two–loop and the three–loop QCD correction factors are $\{\kappa_c^{(2)}, \kappa_b^{(2)}, \kappa_t^{(2)}\} = \{11, 21, 10.17, 9.13\}$ and $\{\kappa_c^{(3)}, \kappa_b^{(3)}, \kappa_t^{(3)}\} = \{123.8, 101.5, 80.4\}$. There can be significant differences between M_q and $m_q(M_q)$. For example, using $\alpha_s(M_Z) = 0.1176$ and $M_t = 172.5$ GeV, one obtains, with QCD evolution of α_s from M_Z to M_t, $\alpha_s(M_t) = 0.108$, and then from Eq. (2.38) $m_t(M_t) = 162.8$ GeV. For c and b quarks the differences are even bigger.

The running masses of leptons can be defined analogously, but now the QCD corrections are replaced by QED corrections. Consequently the differences between the pole mass M_ℓ and running mass $m_\ell(M_\ell)$ are less significant. The two masses are related via

$$m_\ell(\mu) = M_\ell \left[1 - \frac{\alpha}{\pi}\left\{ 1 + \frac{3}{2}\ln\frac{\mu}{m_\ell(\mu)} \right\} \right]. \qquad (2.39)$$

For momentum scales higher than the electroweak symmetry breaking scale, one should evolve the Yukawa couplings of the fermions, rather then their masses. One can define the running mass in this momentum regime as

$$m_i(\mu) = Y_i(\mu)\, v\,. \tag{2.40}$$

Here $v = 174$ GeV is the VEV of the Higgs doublet evaluated at the weak scale. Since the VEV v also is a function of momentum (owing to wave function renormalization of the Higgs filed), one could in principle define the running mass as $m_i(\mu) = Y_i(\mu)v(\mu)$. But this is usually not necessary, and will not be adopted here. The renormalization group evolution equations for the Yukawa couplings of the SM have been worked out to two–loop accuracy.[16]

While extrapolating the Yukawa coupling above the weak scale one has to specify the theory valid in that regime. Often it will be assumed to be the minimal supersymmetric standard model (MSSM). In the fermion Yukawa sector there are significant differences between the MSSM and the SM. The main difference is that supersymmetry requires two Higgs doublets, H_u with $(Y/2) = +1/2$ and H_d with $(Y/2) = -1/2$. The extra doublet is needed for anomaly cancelation and also for generating all fermion masses. Recall that in the SM Yukawa interaction of Eq. (2.1) we used H for generating the up–type quark masses and its conjugate $\tilde{H}$ for the down–type quark and charged lepton masses. Supersymmetric Yukawa couplings must be derived from a superpotential W, which is required to be holomorphic. This means that if H appears in W, then H^* cannot appear. The MSSM Yukawa interactions arise from the following superpotential.

$$\mathcal{W}_{\mathrm{Yukawa}}^{\mathrm{MSSM}} = Q^T Y_u u^c H_u - Q^T Y_d d^c H_d - L^T Y_\ell e^c H_d\,. \tag{2.41}$$

If we denote the VEVs of H_u and H_d as v_u and v_d, then the mass matrices for the three charged fermion sectors are

$$M_u = Y_u v_u\,, \qquad M_d = Y_d v_d\,, \qquad M_\ell = Y_\ell v_d\,. \tag{2.42}$$

The diagonalization procedure follows as in the SM. Notably, there is no Higgs boson mediated flavor changing couplings at tree level, in spite of having two Higgs doublets. The constraints of supersymmetry is the reason for its absence. (Only a single Higgs doublet couples to each one of the three sectors.) A new parameter appears, which is the ratio of the two Higgs vacuum expectation values:

$$\tan\beta = \frac{v_u}{v_d}\,. \tag{2.43}$$

Table 2.1. The running masses of quarks and leptons as a function of momentum μ. The last two columns correspond to the running masses in MSSM at $\Lambda_{\rm GUT} = 2 \times 10^{16}$ GeV.

$m_i \backslash \mu$	$m_c(m_c)$	2 GeV	$m_b(m_b)$	$m_t(m_t)$	1 TeV	$\Lambda_{\rm GUT}^{\tan\beta=10}$	$\Lambda_{\rm GUT}^{\tan\beta=50}$
m_u(MeV)	2.57	**2.2**	1.86	1.22	1.10	0.49	0.48
m_d(MeV)	5.85	**5.0**	4.22	2.76	2.50	0.70	0.51
m_s(MeV)	111	**95**	80	52	47	13	10
m_c(GeV)	**1.25**	1.07	0.901	0.590	0.532	0.236	0.237
m_b(GeV)	5.99	5.05	**4.20**	2.75	2.43	0.79	0.61
m_t(GeV)	*384.8*	*318.4*	*259.8*	**162.9**	150.7	92.2	94.7
m_e(MeV)	0.4955	$\sim$	0.4931	0.4853	0.4959	0.2838	0.206
m_μ(MeV)	104.474	$\sim$	103.995	102.467	104.688	59.903	43.502
m_τ(MeV)	1774.90	$\sim$	1767.08	1742.15	1779.74	1021.95	773.44

This parameter will influence many physical processes. $\tan\beta$ plays an important role in the RGE evolution of the Yukawa couplings. The range of $\tan\beta$ preferred in the MSSM is $\tan\beta = (1.7 - 60)$. When $\tan\beta < 1.7$ the top quark Yukawa couplings blows up before the momentum scale $\mu = \Lambda_{\rm GUT} \approx 2 \times 10^{16}$ GeV. $\Lambda_{\rm GUT}$ is associated with the scale of grand unification, where the three gauge couplings of the SM appear to meet, if there is low energy supersymmetry. For $\tan\beta > 60$ the b–quark and τ–lepton Yukawa couplings become non-perturbative before reaching $\Lambda_{\rm GUT}$.

In Table 2.1 we list the running masses of quarks and leptons as a function of the momentum scale μ. We have adopted the numbers listed from Ref. [17], but our independent calculations show general agreement at the level of few per cent with Ref. [17]. The input values for (c, b, t) quarks are the running masses indicated in bold. For this Table we have used light quark masses at $\mu = 2$ GeV as indicated in bold. For the charged lepton, we have used as input the masses given in Eq. (2.16). The masses of all fermions are listed at momentum scale $\mu = m_t$ and $\mu = 1$ TeV assuming the validity of the SM up to 1 TeV. Also listed are the running masses at $\mu = \Lambda_{\rm GUT} = 2 \times 10^{16}$ GeV assuming MSSM spectrum, for two values of $\tan\beta$ (10 and 50). The following input values have been used. $\alpha_s(M_Z) = 0.1176$, $\alpha^{-1}(M_Z) = 127.918$, and $\sin^2\theta_W(M_Z) = 0.23122$.

There are various noteworthy features in Table 2.1. The light quark masses $(m_u, m_d, , m_s)$ decrease by about a factor of two in going from $\mu = 2$ GeV to $\mu = 1$ TeV. This decrease is a result of QCD corrections. The d and s–quark masses decrease by about another factor of 4 in going from $\mu = 1$ TeV to $\mu = \Lambda_{\rm GUT}$, while m_u decreases by a factor of 2.3.

The net change in the values of $(m_u, , m_d, m_s)$ in going from $\mu = 2$ GeV to $\mu = \Lambda_{\mathrm{GUT}}$ for the case of $\tan\beta = 10$ is a factor $(4.9, 7.9, 7.3)$. The value of b–quark mass decreases considerably, by a factor of 6.9, in going from $\mu = m_b$ to $\mu = \Lambda_{\mathrm{GUT}}$ for $\tan\beta = 10$. $m_b(\mu = \Lambda_{\mathrm{GUT}})$ is close to the τ–lepton mass $m_\tau(\mu = \Lambda_{\mathrm{GUT}})$ (to within about 20%). The lepton masses decrease by about a factor of 2 in going from low energies to Λ_{GUT}. This decrease occurs because of the $SU(2)_L \times U(1)_Y$ contributions to the beta functions of Y_ℓ. These features will be relevant when we discuss predictions for fermion masses from Grand Unified theories in Sec. 2.5.

Sometimes the light quark masses are quoted at $\mu = 1$ GeV. In going from $\mu = 2$ GeV down to $\mu = 1$ GeV, the masses increase by a factor of 1.31, if $\alpha_s(M_Z) = 0.1176$ is used. The running factor to go from $\mu = 2$ GeV down to $\mu = m_c$ is indicated in Table 2.1, while the additional running factor to go from $\mu = m_c$ to $\mu = 1$ GeV is found to be 1.12. Thus, $(m_u, m_d, m_s) = (2.2, 5, 95)$ MeV at $\mu = 2$ GeV correspond to $(m_u, m_d, m_s) = (2.88, 6.58, 124)$ MeV at $\mu = 1$ GeV.

In Table 2.1 we have also included the top quark mass at momentum scales below M_t (indicated in italics). These values, which are un-physical, since the top quark decouples at its mass, will be rarely used.

2.2.5. *Quark mixing and CP violation*

The unitary matrix V of Eq. (2.8) which appears in the charged current interactions of Eq. (2.7) enters in a variety of processes. A lot of information has been gained on the matrix elements of V. The general matrix can be written as

$$
V = \begin{pmatrix} V_{ud} & V_{us} & V_{ub} \\ V_{cd} & V_{cs} & V_{cb} \\ V_{td} & V_{ts} & V_{tb} \end{pmatrix} . \tag{2.44}
$$

The standard parametrization of V is as in Eq. (2.20), but now understood to be for the quark sector. V has a single un-removable phase for three families of quarks and leptons. (The phases (α, β) which appeared in the case of Majorana neutrinos can be removed by right–handed quark field redefinition.) The single un-removable phase in V allows for the violation of CP symmetry in the quark sector. Unlike in the leptonic sector, the quark mixing angles turn out to be small. This enables one to make a perturbative expansion of the mixing matrix a la Wolfenstein.[18] The small parameter is taken to be $\lambda = |V_{us}|$ in terms of which one has (neglecting terms of order λ^5):

$$V = \begin{pmatrix} 1 - \frac{1}{2}\lambda^2 - \frac{1}{8}\lambda^4 & \lambda & A\lambda^3(\rho - i\eta) \\ -\lambda & 1 - \frac{1}{2}\lambda^2 - \frac{1}{8}\lambda^4(1 + 4A^2) & A\lambda^2 \\ A\lambda^3(1 - \rho - i\eta) & -A\lambda^2 + \frac{1}{2}A\lambda^4\left(1 - 2(\rho + i\eta)\right) & 1 - \frac{1}{2}A^2\lambda^4 \end{pmatrix} \quad (2.45)$$

Here the exact correspondence with Eq. (2.20) is given by

$$s_{12} \equiv \lambda, \quad s_{23} \equiv A\lambda^2, \quad s_{13}e^{-i\delta} \equiv A\lambda^3(\rho - i\eta). \quad (2.46)$$

Matrix elements of V are determined usually via semileptonic decays of quarks. In Fig. 2.1 we have displayed the dominant processes enabling determination of these elements. Fig. 2.1 (a) is the diagram for nuclear beta decay, from which $|V_{ud}|$ has been extracted rather accurately:[19]

$$|V_{ud}| = 0.97377 \pm 0.00027. \quad (2.47)$$

Fig. 2.1 (b) shows semileptonic K decay from which the Cabibbo angle $|V_{us}|$ can be extracted. The decays $K_L^0 \to \pi\ell\nu$ and $K^\pm \to \pi^0\ell^\pm\nu$ ($\ell = e, \mu$) have been averaged to obtain for the product $|V_{us}|f_+(0) = 0.21668 \pm 0.00045$. Here $f_+(0)$ is the form factor associated with this semileptonic decay evaluated at $q^2 = 0$. Using $f_+(0) = 0.961 \pm 0.008$ (obtained from QCD calculations, which are in agreement with lattice QCD evaluations), one obtains

$$|V_{us}| = 0.2257 \pm 0.0021. \quad (2.48)$$

$|V_{cd}|$ is extracted from $D \to K\ell\nu$ and $D \to \pi\ell\nu$ decays with assistance from lattice QCD for the computation of the relevant form factors. V_{cs} is determined from semileptonic D decays and from leptonic D_s decay ($D_s^+ \to \mu^+\nu$), combined with lattice calculation of the decay form factor f_{D_s}. Both $|V_{cd}|$ and $|V_{cs}|$ have rather large errors currently:

$$|V_{cd}| = 0.230 \pm 0.011,$$
$$|V_{cs}| = 0.957 \pm 0.010. \quad (2.49)$$

$|V_{cb}|$ is determined from both inclusive and exclusive decays of B hadrons into charm, yielding a value

$$|V_{cb}| = (41.6 \pm 0.6) \times 10^{-3}. \quad (2.50)$$

$|V_{ub}|$ is determined from charmless B decays and gives

$$|V_{ub}| = (4.31 \pm 0.30) \times 10^{-3}. \quad (2.51)$$

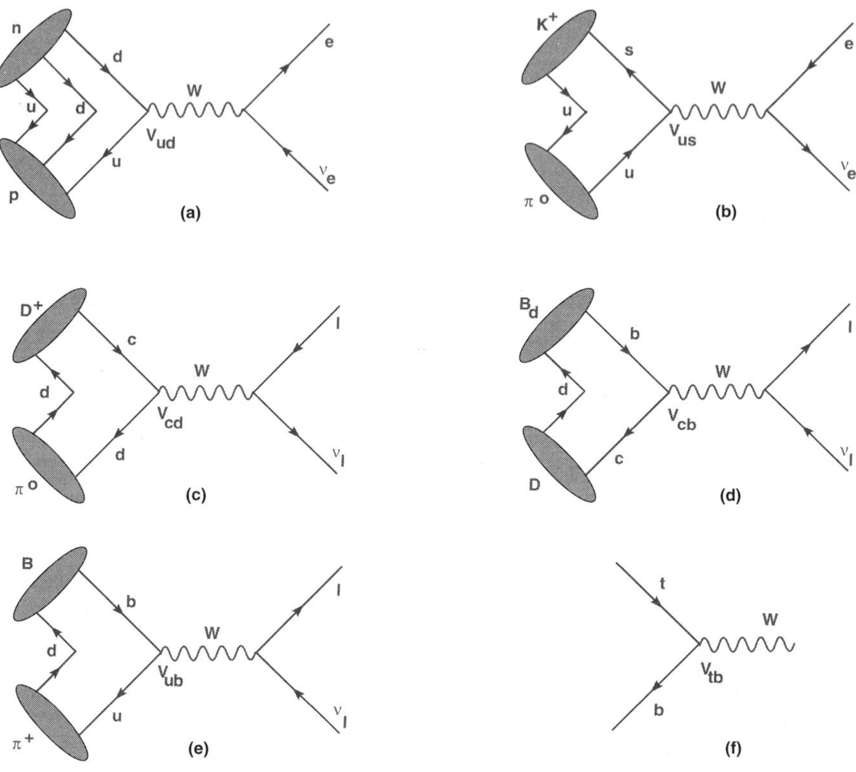

Fig. 2.1.   Processes determining $|V_{ij}|$.

Elements $|V_{td}|$ and $|V_{ts}|$ cannot be currently determined, for a lack of top quark events, but can be inferred from B meson mixings where these elements appear through the box diagram. The result is

$$|V_{td}| = (7.4 \pm 0.8) \times 10^{-3} \,,$$
$$\frac{|V_{td}|}{|V_{ts}|} = 0.208 \pm 0.008 \,. \tag{2.52}$$

Fig. 2.1 (f) depicts the decay of top quark into $W + b$. It can also decay into $W + q$ where q is d, s, b. By taking the ratio of branching ratios $R = B(t \to Wb)/\sum_q B(t \to Wq)$, CDF and D0 have arrived at a limit on $|V_{td}| > 0.74$.[5]

2.2.5.1. *Heavy quark symmetry*

Heavy quark symmetry plays an important role in the determination of $|V_{ub}|$ and $|V_{cb}|$. While a thorough discussion of HQET (Heavy Quark Effective Theory) is outside the scope of this write-up, it would be useful to get a feeling of some of the ideas involved. We refer the reader to Ref. [11] for a thorough review, and Ref. [20] for a pedagogical exposure.

Consider first the purely leptonic decay $B^- \to \ell^- \overline{\nu}_\ell$ for $\ell = e,\ \mu,\ \tau$. The transition amplitude for this decay is

$$T_{fi} = \frac{G_F}{\sqrt{2}} V_{ub} [\overline{u}_\ell \gamma_\mu (1 - \gamma_5) u_\nu] \left\langle 0 | \overline{u} \gamma^\mu (1 - \gamma_5) b | B^- \right\rangle . \tag{2.53}$$

Here G_F is the Fermi coupling. To compute the decay rate, the hadronic matrix element for the transition of B meson to vacuum needs to be evaluated. Note that the matrix element of vector current between pseudscalar meson and vacuum vanishes: $\langle 0 | \overline{u} \gamma^\mu b | B^- \rangle = 0$, while the axial vector current matrix element is parametrized as $\langle 0 | \overline{u} \gamma^\mu \gamma_5 b | B^- \rangle = i f_B q^\mu$, with f_B being the B meson decay constant and q^μ the B meson momentum. With this matrix element, the decay rate can be readily computed. One obtains

$$\Gamma(B^- \to \ell^- \overline{\nu}_\ell) = \frac{G_F^2}{8\pi} f_B^2 |V_{ub}|^2 M_B m_\ell^2 \left(1 - \frac{m_\ell^2}{M_B^2} \right)^2 . \tag{2.54}$$

Note the helicity suppression, which implies that the number of events in this channel will be small. Recently BELLE collaboration has observed the decay $B^- \to \tau^- \overline{\nu}$ with a 3.5 sigma statistical significance. Their results can be converted to a value for the product $|V_{ub}| f_B$ as

$$|V_{ub}| f_B = [10.1^{+1.6}_{-1.4}(\text{stat})^{+1.3}_{-1.4}(\text{syst})] \times 10^{-4} \, \text{GeV} . \tag{2.55}$$

Using lattice evaluations of f_B, one can obtain the value of $|V_{ub}|$ from Eq. (2.55). The accuracy of this determination, which is rather direct, suffers from the lack of events for this helicity suppressed decay.

Semileptonic decays do not suffer from the helicity suppression, and are therefore more promising. Unlike a single form factor that appears in the purely leptonic decay, now there will be two form factors. These two can be related via heavy quark symmetry, as we outline below. Consider the decay $\overline{B}_d^0 \to D^+ \ell \overline{\nu}_\ell$ which proceeds via Fig. 2.1 (d). The transition amplitude for this decay has the form

$$T_{fi} = \frac{G_F}{\sqrt{2}} V_{cb} [\overline{u}_\ell \gamma_\mu (1 - \gamma_5) u_\nu] \left\langle D^+ | \overline{c} \gamma^\mu (1 - \gamma_5) b | \overline{B}_d^0 \right\rangle . \tag{2.56}$$

A similar expression is obtained for the decay $\overline{B}_d^0 \to \pi^+ \ell \overline{\nu}_\ell$, with $|V_{cb}|$ replaced by $|V_{ub}|$ in Eq. (2.56). The matrix element of axial vector current between two pseudoscalar mesons vanishes: $\left\langle D^+ | \overline{u} \gamma^\mu \gamma_5 b | \overline{B}_d^0 \right\rangle = 0$. The vector current matrix element between two pseudoscalar mesons contains two form factors:

$$\left\langle D^+(k) | \overline{u} \gamma^\mu b | \overline{B}_d^0(p) \right\rangle = F_1(q^2) \left[(p+k)_\mu - \frac{M_B^2 - M_D^2}{q^2} q_\mu \right]$$
$$+ F_0(q^2) \frac{M_B^2 - M_D^2}{q^2} q_\mu , \tag{2.57}$$

where $q = p - k$.

To see how HQET can relate the two form factors $F_1(q^2)$ and $F_0(q^2)$, let me briefly review the crucial elements of HQET. In a hadron composed of one heavy (b) quark and one light anti-quark $\overline{u}$ or $\overline{d}$, the mass of b is much larger than the scale of QCD dynamics, $\Lambda_{\rm QCD}$. The b quark is then almost on-shell, moving with a velocity close to the hadron's four velocity. We write this as

$$p_Q^\mu = m_Q v^\mu + k^\mu , \tag{2.58}$$

where $k \ll m_Q$ is the residual momentum, and $v^2 = 1$. The b quark interacts with the light degrees of freedom, but such interactions can cause a change in the residual momentum by $\Delta k \sim \Lambda_{\rm QCD} \ll m_Q$. Thus $\Delta v \to 0$ as $\Lambda_{\rm QCD}/m_Q \to 0$.

In the heavy quark symmetry limit ($\Lambda_{\rm QCD}/m_Q \to 0$), the elastic scattering process $\overline{B}(v) \to \overline{B}(v')$ has the amplitude

$$\frac{1}{M_B} \left\langle \overline{B}(v') | \overline{b}(v') \gamma_\mu b(v) | \overline{B}(v) \right\rangle = \xi(v'.v)(v + v')_\mu . \tag{2.59}$$

A term of the type $(v - v')_\mu$ cannot appear on the right-hand side of Eq. (2.59) since $\slashed{v} b_v = b_v$ and $\overline{b}_{v'} \slashed{v}' = \overline{b}_{v'}$. The $1/M_B$ factor in Eq. (2.59) is associated with normalization of states, so the right-hand side of Eq. (2.59) has no dependence on the heavy quark flavor. Current conservation implies $\xi(v'.v = 1) = 1$, so that the function $\xi(v.v')$, the Isgur–Wise function,[21] is independent of the heavy quark flavor. Thus, in the heavy quark symmetry limit, we have

$$\frac{1}{\sqrt{M_D M_B}} \left\langle D(v') | \overline{c}_{v'} \gamma_\mu b_v | \overline{B}(v) \right\rangle = \xi(v.v')(v + v')_\mu . \tag{2.60}$$

This transition is now governed by a single form factor, $\xi(v'.v)$ with $\xi(1) = 1$. Comparing with Eq. (2.57), one finds

$$F_1(q^2) = \frac{M_D + M_B}{2\sqrt{M_D M_B}}\, \xi(w)$$

$$F_0(q^2) = \frac{2\sqrt{M_D M_B}}{M_D + M_B}\left(\frac{1+w}{2}\right)\xi(w) \tag{2.61}$$

where

$$w = v_D.v_B = \frac{M_D^2 + M_B^2 - q^2}{2M_D M_B}. \tag{2.62}$$

As an application of these ideas, consider the decay $\overline{B} \to D^*\ell\nu$. The differential decay rate for this process can be written as

$$\frac{d\Gamma}{dw} = G_F^2\, K\, F(w)^2 |V_{cb}|^2\,, \tag{2.63}$$

where K is a known kinematic function and $F(w)$ is related to the Isgur–Wise function (up to perturbative QCD corrections). It should obey the normalization

$$F(1) = \eta_A(\alpha_s)\left[1 + \frac{0}{m_c} + \frac{0}{m_b} + \mathcal{O}\left(\frac{\Lambda_{\text{QCD}}^2}{m_{b,c}^2}\right)\right]. \tag{2.64}$$

Here $\eta_A(\alpha_s)$ is a perturbatively calculable function. Note that $\mathcal{O}(\Lambda_{\text{QCD}}/m_{c,b})$ corrections vanish.[22] This decay distribution can be measured as a function of w, from which $F(w)|V_{cb}|$ can be extracted. Now, when extrapolated to zero recoil limit ($w = 1$), whence the decay rate vanishes), from Eq. (2.64), one obtains a value of $|V_{cb}|$.

2.2.5.2. *CP violation*

Charge conjugation (C) takes a particle to its antiparticle, Parity (spatial reflection) changes the helicity of the particle. Under CP, e_L^- will transform to e_R^+. Both C and P are broken symmetries in the SM, but the product CP is approximately conserved. Violation of CP has been seen only in weak interactions. The CKM mechanism predicts CP violation through a single complex phase that appears in the CKM matrix. Thus in the SM, various CP violating processes in K, B and other systems get correlated. So far such correlations have been consistent with CKM predictions, but more precise determinations in the B and D systems at the LHC may open up new physics possibilities.

                             K.S. Babu

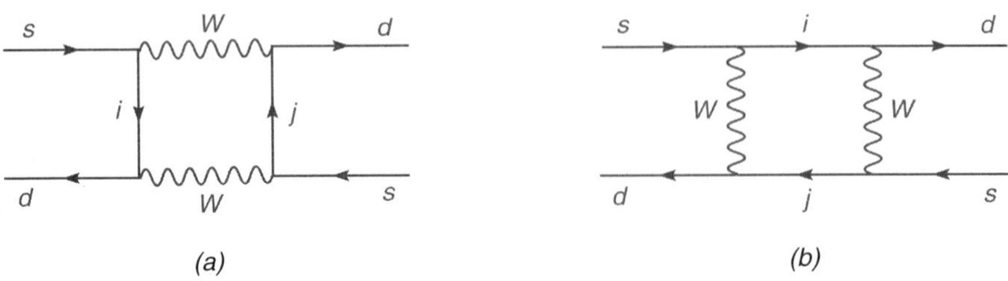

Fig. 2.2.   Box diagram inducing $K^0 - \overline{K^0}$ transition in the SM.

In the $K^0 - \overline{K^0}$ system, CP violation has been observed both in mixing and in direct decays. CP violation in mixing arises in the SM via the W–boson box diagram shown in Fig. 2.2. The CP asymmetry in mixing is parametrized by ϵ, which is a measure of the mixing between the CP even and CP odd states $K^0_{1,2} = (K^0 \pm \overline{K^0})/\sqrt{2}$. It has been measured to be

$$|\epsilon| = (2.229 \pm 0.010) \times 10^{-3} \,. \tag{2.65}$$

The measured value in in excellent agreement with expectations from the SM, and enables us to determine the single phase of the CKM matrix. The box diagram contribution to ϵ is given by

$$|\epsilon| = \frac{G_F^2 f_k^2 m_K m_W^2}{12\sqrt{2}\pi^2 \Delta m_K} \hat{B}_K \left\{ \eta_c S(x_c) \mathrm{Im}[(V_{cs} V_{cd}^*)^2] \right.$$
$$\left. + \eta_t S(x_t) \mathrm{Im}[(V_{ts} V_{td}^*)^2] + 2\eta_{ct} S(x_c, x_t) \mathrm{Im}[V_{cs} V_{cd}^* V_{ts} V_{td}^*] \right\} \,. \tag{2.66}$$

Here $S(x)$ and $S(x,y)$ are Inami–Lim functions[23] with $x_{c,t} = m_{c,t}^2/M_W^2$, and the η factors are QCD correction factors for the running of the effective $\Delta S = 2$ Hamiltonian from M_W to the hadron mass scale.

The direct CP violation parameter that leads to the decay $K \to \pi\pi$ has also been measured, leading to the value

$$Re(\epsilon'/\epsilon) = (1.65 \pm 0.26) \times 10^{-3} \,. \tag{2.67}$$

These decays occur via the penguin diagrams shown in Fig. 2.3. There are electromagnetic penguins and gluonic penguins, which tend to cancel each other. While the KM model predicts non-zero value of ϵ'/ϵ, estimating this value reliably has been difficult, partly because of this cancelation. Most estimates are in agreement with observations.

A wealth of information has been gained about CP violation from the B factories over the last decade. CP violation in B meson system is now

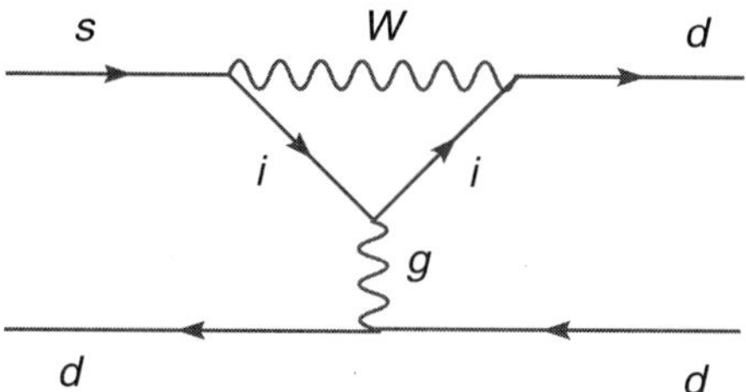

Fig. 2.3. One loop penguin diagram that generates CP violation in direct $K \to \pi\pi$ decay.

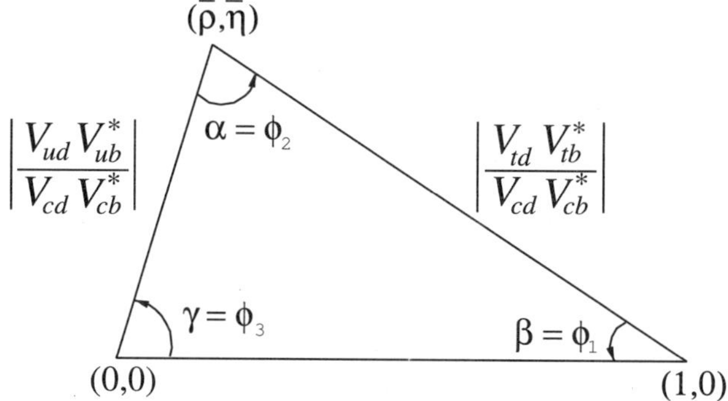

Fig. 2.4. Unitarity triangle in the CKM model.

well established. Several CP violating quantities have been measured in B_d meson system,[20] all of which show consistency with the CKM mixing matrix. Unitarity of the CKM matrix implies that $\sum_i V_{ij} V_{ik}^* = \delta_{jk}$ and $\sum_j V_{ij} V_{kj}^* = \delta_{ik}$. There are six vanishing combinations, which can be expressed as triangles in the complex plane. The areas of all of these triangles are the same. The most commonly used triangle arises from the relation

$$V_{ud}V_{ub}^* + V_{cd}V_{cb}^* + V_{td}V_{tb}^* = 0\,. \tag{2.68}$$

In the complex plane, the resulting triangle has sides of similar length (of order λ^3). This unitarity triangle relation is shown in Fig. 2.4. The three interior angles (α, β, γ), also referred to as (ϕ_2, ϕ_1, ϕ_3), can be written in the CKM model as

$$\alpha = \arg\left(\frac{-V_{td}V_{tb}^*}{V_{ud}V_{ub}^*}\right) \simeq \arg\left(-\frac{1-\rho-i\eta}{\rho+i\eta}\right),$$

$$\beta = \arg\left(\frac{-V_{cd}V_{cb}^*}{V_{td}V_{tb}^*}\right) \simeq \arg\left(\frac{1}{1-\rho-i\eta}\right),$$

$$\gamma = \arg\left(\frac{-V_{ud}V_{ub}^*}{V_{cd}V_{cb}^*}\right) \simeq \arg\left(\rho+i\eta\right). \tag{2.69}$$

One experimental test of the CKM mechanism is the measurement of $\alpha + \beta + \gamma = 180°$.

The angle β can be measured with the least theoretical uncertainty from the decay of $B_d \to J/\psi K_S$. It is found to be

$$\sin 2\beta = 0.68 \pm 0.03. \tag{2.70}$$

This value is in in good agreement with the CKM prediction.

The angle α is measured from decay modes where $b \to u\bar{u}d$ is dominant. Such decays includ $B \to \pi\pi$, $B \to \rho\rho$ and $B \to \pi\rho$. The value of α extracted is

$$\alpha = (88^{+6}_{-5})°. \tag{2.71}$$

The angle γ does not depend on the top quark, and can in principle be measured from tree–level decays of B meson. Strong interaction uncertainties are rather large in decays such as $B^\pm \to D^0 K^\pm$. The current value of the angle γ is

$$\gamma = (77^{+30}_{-32})°. \tag{2.72}$$

The current situation with the CKM mixing angles and CP violation phase is depicted in Fig. 2.5. The left panel is the result of a global analysis of flavor mixing and CP violation data by the UTFit[24] collaboration, while the right panel depicts the results from an independent CKMFitter[25] collaboration. The Wolfenstein parameters $\bar{\eta}$ is plotted against $\bar{\rho}$ in these figures. Here $\bar{\eta} = \eta(1-\lambda^2/2)$ and $\bar{\rho} = \rho(1-\lambda^2/2)$. A variety of input parameters have gone into these fits. Some of the constraints used are explicitly indicated in these figures. It is very non-trivial that the various constraint curves have a common intersection. This demonstrates the success of the CKM mechanism of flavor mixing and CP violation. The intersection of the various ellipses gives the best fit value for the Wolfenstein parameters $(\lambda,\ A,\ \rho,\ \eta)$, which are as follows:[5]

$$\lambda = 0.2272 \pm 0.0010, \quad A = 0.818^{+0.007}_{-0.017},$$

$$\rho = 0.221^{+0.064}_{-0.028}, \quad \eta = 0.340^{+0.017}_{-0.045}. \tag{2.73}$$

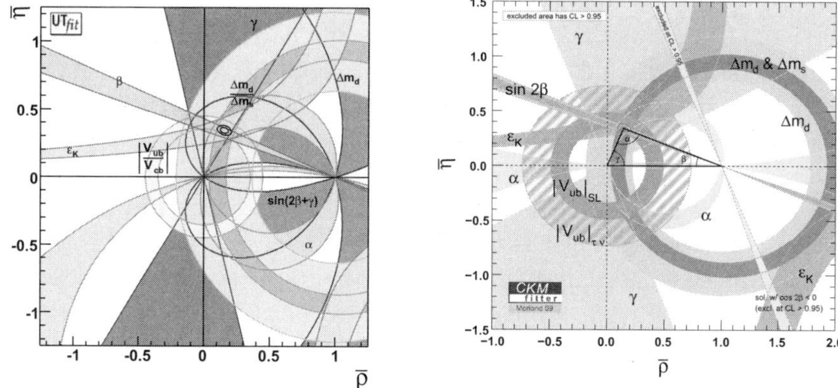

Fig. 2.5. Global fit to the mixing and CP violation data from the UTFit collaboration (left panel)[24] and the CKMFitter collaboration (right panel).[25]

Theories of flavor should provide an understanding of these fundamental parameters.

2.3. Relating quark mixings and mass ratios

Having reviewed the fundamental flavor parameters of the Standard Model, now we turn to attempts which explain some of the observed features. Necessarily one needs to invoke non–standard physics, which can be potentially tested at colliders.

We begin with a simple idea of relating quark masses and mixings by virtue of flavor symmetries. In the quark sector we have seen that the mass ratios such as m_d/m_s, m_u/m_c, etc are strongly hierarchical, while the mixing angles, such as V_{us} are also hierarchical, although the hierarchy here is not as strong. Can the quark mixing angle be computed in terms of the quark mass ratios? Clearly such attempts have to go beyond the SM. Here I give a simple two–family example which assumes a flavor $U(1)$ symmetry that distinguishes the two families.

2.3.1. *Prediction for Cabibbo angle in a two family model*

Consider the mass matrices for $(u,\ c)$ and $(d,\ s)$ quarks given by[26]

$$M_u = \begin{pmatrix} 0 & A_u \\ A_u^* & B_u \end{pmatrix}, \quad M_d = \begin{pmatrix} 0 & A_d \\ A_d^* & B_d \end{pmatrix}. \tag{2.74}$$

The crucial features of these matrices are (i) the zeros in the (1,1) entries, and (ii) their Hermiticity. Neither of these features can be realized within the SM. Recall that the SM symmetry would have arbitrary non–Hermitian matrices for M_u and M_d. The zero entries in Eq. (2.74) can be enforced by a flavor $U(1)$ symmetry, the Hermitian nature can be obtained if the gauge sector is left–right symmetric. Before constructing such a model, let us examine the consequences of Eq. (2.74). Matrices in Eq. (2.74) have factorizable phases. That is, $M_u = P_u \hat{M}_u P_u^*$, where $\hat{M}_u$ has the same form as M_u but with all entries real, and where $P_u = diag(e^{i\alpha_u}, 1)$ is a diagonal phase matrix. A similar factorization applies to M_d with a phase matrix $P_d = diag(e^{i\alpha_d}, 1)$. We can absorb these phase matrices into the quark fields, but since $\alpha_u \neq \alpha_d$, the matrix $P_u^* P_d = diag.(e^{i\psi}, 1)$ will appear in the charged current matrix ($\psi = \alpha_d - \alpha_u$). The matrices $\hat{M}_u$ and $\hat{M}_d$, which have all real entries, can be diagonalized readily, yielding for the mixing angles θ_u and θ_d

$$\tan^2 \theta_u = \frac{m_u}{m_c},$$

$$\tan^2 \theta_d = \frac{m_d}{m_s}. \tag{2.75}$$

This yields a prediction for the Cabibbo angle[26]

$$|\sin \theta_C| \simeq \left| \sqrt{\frac{m_d}{m_s}} - e^{i\psi} \sqrt{\frac{m_u}{m_c}} \right|. \tag{2.76}$$

This formula works rather well, especially since even without the second term, the Cabibbo angle is correctly reproduced. The phase ψ is a parameter, however, its effect is rather restricted. For example, since $\sqrt{m_d/m_s} \simeq 0.22$ and $\sqrt{m_u/m_c} \simeq 0.07$, $|\sin \theta_C|$ must lie between 0.15 and 0.29, independent of the value of ψ.

Now to a possible derivation of Eq. (2.74). Since SM interactions do not conserve Parity, it is useful to extend the gauge sector to the left-right symmetric group $G \equiv SU(3)_C \times SU(2)_L \times SU(2)_R \times U(1)_{B-L}$, wherein Parity invariance can be imposed.[27] The (1,2) and (2,1) elements of $M_{u,d}$ being complex conjugates of each other will then result. The left–handed and the right–handed quarks transform as $Q_{iL}(3,2,1,1/3) + Q_{iR}(3,1,2,1/3)$ under G. Under discrete parity operation $Q_{iL} \leftrightarrow Q_{iR}$. This symmetry can be consistently imposed, as $W_L \leftrightarrow W_R$ in the gauge sector under Parity. The leptons transform as $\psi_{iL}(1,2,1,-1) + \psi_{iR}(1,1,2,-1)$ under the gauge symmetry. Note that ψ_R, which is a doublet of $SU(2)_R$, contains the right–handed neutrino, as the partner of e_R. Thus there is a compelling reason for the existence of ν_R, unlike in the SM, where it is optional.

The Higgs field that couples to quarks should be $\Phi(1,2,2,0)$, and under Parity $\Phi \to \Phi^\dagger$. In matrix form Q_{iL}, Q_{iR}, Φ read as

$$Q_{iL} = \begin{pmatrix} u_i \\ d_i \end{pmatrix}_L , \quad Q_{iR} = \begin{pmatrix} u_i \\ d_i \end{pmatrix}_R , \quad \Phi = \begin{pmatrix} \phi_1^0 & \phi_2^+ \\ \phi_1^- & \phi_2^0 \end{pmatrix} , \qquad (2.77)$$

so that the Yukawa Lagrangian for quarks

$$\mathcal{L}_{\text{Yukawa}} = \overline{Q}_L \Phi Y Q_R + \overline{Q}_L \tilde{\Phi} \tilde{Y} Q_R + h.c. \qquad (2.78)$$

is gauge invariant. Here $\tilde{\Phi} \equiv \tau_2 \Phi^* \tau_2$. Imposing Parity, we see that the Yukawa matrices Y and $\tilde{Y}$ must be Hermitian, $Y = Y^\dagger$ and $\tilde{Y} = \tilde{Y}^\dagger$. This is the desired result for deriving Eq. (2.74). The VEVs $\langle \phi_1^0 \rangle$ and $\langle \phi_2^0 \rangle$ can be complex in general, but this will not affect the prediction for the Cabibbo angle of Eq. (2.76), since that only requires $|(M_{u,d})_{12}| = |(M_{u,d})_{21}|$. Additional Higgs fields, e.g., $\Delta_L(1,3,1,2) + \Delta_R(1,1,3,2)$, would be required for breaking the left–right symmetric gauge group down to the SM and for simultaneously generating large ν_R Majorana masses. However, these fields do not enter into the mass matrices of quarks.

To enforce zeros in the (1,1) entries of $M_{u,d}$ of Eq. (2.74), we can employ the following $U(1)$ flavor symmetry: $Q_{1L} : 2$, $Q_{1R} : -2$. $Q_{2L} : 1$, $Q_{2R} : -1$. $\Phi_1 : 2$, $\Phi_2 : 3$. Note that two Higgs bidoublet fields are needed. Φ_1 generates the (2,2) entries, while Φ_2 generates the (1,2) and (2,1) entries. There is no (1,1) entry generated, since there is no Higgs field with $U(1)$ charge of $+4$. Note also that the $\tilde{\Phi}_{1,2}$ fields, which have $U(1)$ charges $(-2, -3)$, do not couple to the quarks.

While we cannot determine the scale of flavor dynamics in this model, the $U(1)$ flavor symmetry and the left–right symmetry, which were crucial for the derivation of Eq. (2.76), could show up as new particles at the LHC. In general, one would also expect multiple Higgs bosons. We should note that the full theory is more elaborate compared to the minimal left–right symmetry without the flavor symmetry (two, instead of one bi-doublet Higgs fields are needed), but the effective theory is simpler, with the mass matrices being predictive.

2.3.2. *Three family generalization*

Eq. (2.74) can be generalized for the case of three families, a la Fritzsch.[28] The up and down quark mass matrices have Hermitian nearest neighbor interaction form:

$$M_{u,d} = \begin{pmatrix} 0 & A & 0 \\ A^* & 0 & B \\ 0 & B^* & C \end{pmatrix}_{u,d} . \tag{2.79}$$

Such matrices have factorizable phases, i.e., $M_{u,d} = P_{u,d}\hat{M}_{u,d}P^*_{u,d}$, where $\hat{M}_{u,d}$ are the same as in Eq. (2.79), but without any phases, and $P_{u,d}$ are diagonal phase matrices. Only two combinations of phases will enter into the CKM matrix, contained in the matrix $P = P^*_u P_d = \text{diag.}\{e^{i\alpha}, e^{i\beta}, 1\}$. The CKM matrix is then given by

$$V = O_u^T P O_d , \tag{2.80}$$

where $O_{u,d}$ are the orthogonal matrices that diagonalize $\hat{M}_{u,d}$ via

$$O_{u,d}^T \hat{M}_{u,d} \hat{M}_{u,d}^T O_{u,d} = \text{diag.}\{m^2_{u,d}, \ m^2_{c,s}, \ m^2_{t,b}\} . \tag{2.81}$$

In this model there are a total of eight parameters that describe quark masses, mixings and CP violation: six real parameters from $\hat{M}_{u,d}$ and the two phases (α, β). (Note that the six mixing angles that enter into O_u and O_d are determined in terms of the quark mass ratios.) These eight parameters must describe ten observables in the quark sector. There are thus two true predictions. Furthermore, since (α, β) are phases, they do not count as full parameters. One finds four relations between masses and mixings:[29]

$$|V_{us}| \simeq \left| \sqrt{\frac{m_d}{m_s}} - e^{i\psi}\sqrt{\frac{m_u}{m_c}} \right| ,$$

$$|V_{cb}| \simeq \left| \sqrt{\frac{m_s}{m_b}} - e^{i\phi}\sqrt{\frac{m_c}{m_t}} \right| ,$$

$$|V_{ub}| \simeq \left| \frac{m_s}{m_b}\sqrt{\frac{m_d}{m_b}} + e^{i\psi}\sqrt{\frac{m_u}{m_c}}\left(\sqrt{\frac{m_s}{m_b}} - e^{i\phi}\sqrt{\frac{m_c}{m_t}}\right) \right| ,$$

$$|V_{td}| \simeq \left| \frac{m_c}{m_t}\sqrt{\frac{m_u}{m_t}} + e^{i\psi}\sqrt{\frac{m_d}{m_s}}\left(\sqrt{\frac{m_c}{m_t}} - e^{i\phi}\sqrt{\frac{m_s}{m_b}}\right) \right| . \tag{2.82}$$

Here the two phases ψ and ϕ are related to the phases in the diagonal matrix P as $\psi = (\alpha - \beta)$ and $\phi = \beta$. Note that all independent elements of V are determined in this model in terms of the quark mass ratios and two phase parameters. In the expression for $|V_{ub}|$ in Eq. (2.82), the first term is numerically $\simeq 6 \times 10^{-4}$, which is about a factor of 8 less than the value of $|V_{ub}|$. Similarly, the first term in the expression for $|V_{td}|$ is $\sim 1 \times 10^{-5}$, which is negligible in relation to the value of $|V_{td}|$. (For these numerical estimates, I used the values of the running masses given in Table 2.1 evaluated at $\mu = 1$ TeV.) If these terms are neglected, one would have the following predictions:

$$\frac{|V_{ub}|}{|V_{cb}|} \simeq \sqrt{\frac{m_u}{m_c}}, \qquad \frac{|V_{td}|}{|V_{ts}|} \simeq \sqrt{\frac{m_d}{m_s}}. \tag{2.83}$$

These predictions are consistent with experimental data.

While the prediction for $|V_{us}|$ in the Fritzsch ansatz is the same as in the two family model of Eq. (2.76), which is successful, the relation for $|V_{cb}|$ will predict the mass of the top quark to be in the range $(40 - 80)$ GeV, which is now excluded by data.

There have been attempts to fix the problem of Fritzsch mass matrices by modifying its form slightly. If small (2,2) elements are allowed in $M_{u,d}$, the troublesome relation for $|V_{cb}|$ will be removed. However, adding (2,2) entries in M_u and M_d introduces two more complex parameters, and such a model will have no true prediction. The relation of Eq. (2.76) will however be maintained, provided that the (2,2) entries are not too large. Furthermore, the relations of Eq. (2.83), $|V_{ub}|/|V_{cb}| \simeq \sqrt{m_u/m_c}$ and $|V_{td}|/|V_{ts}| \simeq \sqrt{m_d/m_s}$, will be preserved,[30] if the new (2,2) entries are small perturbations.

A different alternative is to make the (2,3) and (3,2) entries of Eq. (2.79) different, while maintaining the relations between (1,2) and (2,1) entries. This can be achieved by non–Abelian discrete symmetries. Again, the number of parameters will increase by two compared to the original Fritszch ansatz. A special case where there is still a true prediction is worth mentioning.

Consider a non–Abelian discrete subgroup G of $SU(2)$ serving as a family symmetry. G is assumed to have pseudo–real doublet representations, just as $SU(2)$. Let the first two families of quarks be pseudo–real doublets of G, while the third family quarks are singlets of G. A *real* Higgs doublet which is a true singlet of G will generate the (3,3) entries of $M_{u,d}$ as well as (1,2) and (2,1) entries. Note that invariance under G will lead to the (1,2) entry being the negative of the (2,1) entry, a property of the original $SU(2)$ family symmetry. Now, if G is broken by a Higgs field transforming as a doublet of G, then unequal (2,3) and (3,2) entries in $M_{u,d}$ can be generated. This is a concrete modification of the Fritszch ansatz, with (1,2) and (2,1) entries having the same magnitude, but with the (2,3) and (3,2) entries unrelated.

A model of the type just described has been constructed in Ref. [31].

It is based on the dihedral group Q_6 which contains pseudo–real as well as real doublets. Most interestingly, if the origin of CP violation is taken to be spontaneous, then the phase matrix P appearing in Eq. (2.80) will have the form $P = diag.\{e^{-i\phi}, e^{i\phi}, 1\}$. (This happens since M_u and M_d each will have a single phase, appearing in the (2,3) and (3,2) entries, apart from irrelevant overall phases, if all the Yukawa couplings are assumed to be real by virtue of CP invariance.) Such a model will have one true prediction, since now there are nine parameters describing ten observables. It was found in Ref. [31] that this prediction, which relates $\bar{\eta}$ with $\bar{\rho}$, is fully consistent with data.

We shall return to mass matrix "textures" of the type described here when discussing fermion masses in the context of Grand Unification in Sec. 2.5. The nearest neighbor interaction, not necessarily symmetrical, will find useful applications.

2.4. Froggatt–Nielsen mechanism for mass hierarchy

The hierarchy in the masses and mixings of quarks and leptons can be understood by assuming a flavor $U(1)$ symmetry under which the fermions are distinguished. In this approach developed by Froggatt and Nielsen,[32] there is a "flavon" field S, which is a scalar, usually a SM singlet field, which acquires a VEV and breaks the $U(1)$ symmetry. This symmetry breaking is communicated to the fermions at different orders in a small parameter $\epsilon = \langle S \rangle / M_*$. Here M_* is the scale of flavor dynamics, and usually is associated with some heavy fermions which are integrated out. The nice feature of this approach is that the mass and mixing hierarchies will be explained as powers of the expansion parameter ϵ without assuming widely different Yukawa couplings. The effective theory below M_* is rather simple, while the full theory will have many heavy fermions, called Froggatt–Nielsen fields.

2.4.1. *A two family model*

Let me illustrate this idea with a two family example which is realistic when applied to the second and third families of quarks. Consider M_u and M_d for the $(c,\ t)$ and $(s,\ b)$ sectors given by

$$M_u = \begin{pmatrix} \epsilon^4 & \epsilon^2 \\ \epsilon^2 & 1 \end{pmatrix} v_u\,, \quad M_d = \begin{pmatrix} \epsilon^3 & \epsilon^3 \\ \epsilon & \epsilon \end{pmatrix} v_d\,. \tag{2.84}$$

Here $\epsilon \sim 0.2$ is a flavor symmetry breaking parameter. Every term in Eq. (2.84) has an order one coefficient which is not displayed. We obtain from Eq. (2.84) the following relations for quark masses and $|V_{cb}|$:

$$\frac{m_c}{m_t} \sim \epsilon^4 \,, \quad \frac{m_s}{m_b} \sim \epsilon^2 \,, \quad |V_{cb}| \sim \epsilon^2 \,. \tag{2.85}$$

All of these relations work well, for $\epsilon \sim 0.2$. Although precise predictions have not been made, one has a qualitative understanding of the hierarchies.

How do we arrive at Eq. (2.84)? We do it in two stages. First, let us look at the effective Yukawa couplings, which can be obtained from the Lagrangian:

$$\begin{aligned}
\mathcal{L}_{FN}^{\text{eff}} = & \left[Q_3 u_3^c H_u + Q_2 u_3^c H_u S^2 + Q_3 u_2^c H_u S^2 + Q_2 u_2^c H_u S^4 \right] \\
& + \left[Q_3 d_3^c H_d S + Q_3 d_2^c H_d S + Q_2 d_2^c H_d S^3 + Q_2 d_3^c H_d S^3 \right] \\
& + h.c.
\end{aligned} \tag{2.86}$$

Here I assumed supersymmetry, so that there are two Higgs doublets $H_{u,d}$. It is not necessary to assume SUSY, one can simply identify H_u as H of SM, and replace H_d by $\tilde{H}$. In Eq. (2.86) all couplings are taken to be of order one. The symmetry of Eq. (2.86) is a $U(1)$ with the following charge assignment.

$$\{Q_3, u_3^c\} : 0; \quad \{Q_2, u_2^c\} : 2; \quad \{d_2^c, d_3^c\} : 1; \quad \{H_u, H_d\} : 0; \quad S : -1 \,. \tag{2.87}$$

Now we wish to obtain Eq. (2.86) by integrating out certain Froggatt–Nielsen fields. This is depicted in Fig. 2.6 via a set of "spaghetti" diagrams. As you can see, there are a variety of fields denoted by $G_i, \overline{G}_i$ ($i = 1 - 4$) for the up–quark mass generation. G_i have the same gauge quantum numbers as the u^c quark of SM, while $\overline{G}_i$ have the conjugate quantum numbers. F_i have the quantum numbers of d^c quark, while $\overline{F}_i$ the conjugate quantum numbers.

You can readily read off the flavor $U(1)$ charges of the various F_i and G_i fields from the spaghetti diagrams. For example, the charge of G_1 is -2, while that of $\overline{G}_1$ is $+2$. The charges of G_2 is -1 and that of $\overline{G}_2$ is $+1$.

All flavor dynamics in this class of models could occur near the Planck scale. As long as the hierarchy between $\langle S \rangle$ and the masses of the Froggatt–Nielsen fields is not too strong, realistic fermion masses will be generated.

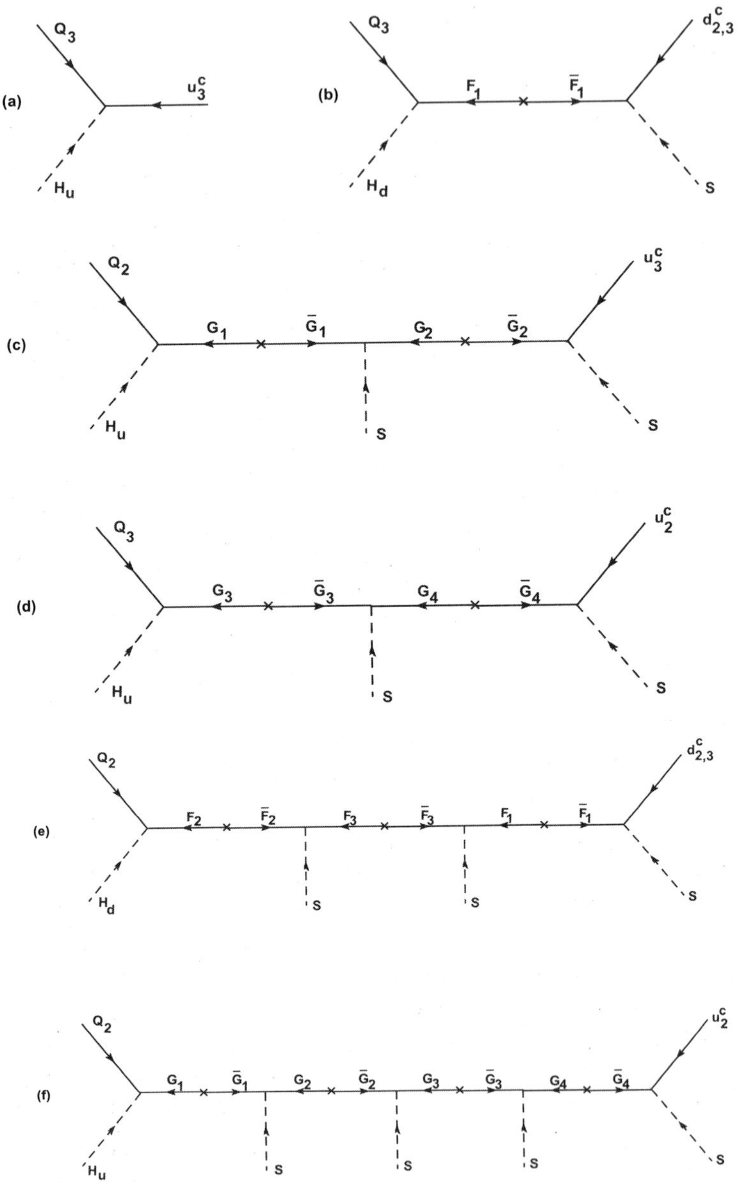

Fig. 2.6. Froggatt–Nielsen fields generating effective Yukawa couplings of Eq. (2.86).

Consider for example, Fig. 2.6 (b) which induces the b–quark mass. The effective interaction from this diagram goes as $\mathcal{L}_b^{\text{eff}} = Y_1 Y_2 \left(Q_3 d_3^c H_d\right)\left(S/M_{F_1}\right)$, where $Y_{1,2}$ are order one Yukawa couplings. If $\langle S\rangle/M_{F_1} \sim 0.2$ or so, realistic b–quark mass is obtained (with $\tan\beta \sim 10$). This allows for both $\langle S\rangle$ and M_{F_1} to be near the Planck scale. From Fig. 2.6 (f), one can read off the effective Lagrangian inducing the c–quark mass: $\mathcal{L}_c^{\text{eff}} = \Pi_{i=1}^5 Y_i' \left(Q_2 u_3^c H_u\right)\left(S^4/M_G^4\right)$. Here Y_i' are order one Yukawa couplings, and we assumed that all of G_i $(i=1-4)$ appearing in Fig. 2.6 (f) have a common mass M_G. With all couplings being order one, $m_c/m_t \sim 1/400$ can be reproduced, with $\epsilon \sim 0.2$. It should be emphasized that, although there are various Yukawa couplings, all of them can take order one values.

2.4.2. *A realistic three family Froggatt–Nielsen model*

Actually the flavor $U(1)$ that we used in the previous section is anomalous. String theory, when compactified to four dimension, generically gives an anomalous $U(1)_A$ with anomaly cancelation occuring by the Green–Schwartz mechanism.[33] In this case, we can get rid of the complicated Froggatt–Nielsen fields, and simply write down higher dimensional operators suppressed by the string scale. A bonus in this approach is that the small expansion parameter ϵ can be computed in specific models, where it tends to come out close to 0.2, of order the Cabibbo angle.

An explicit and complete anomalous $U(1)$ model that fits well all quark and lepton masses and mixings is constructed below. Consider the quark and lepton mass matrices of the following form:[34]

$$
M_u \sim \langle H_u\rangle \begin{pmatrix} \epsilon^8 & \epsilon^6 & \epsilon^4 \\ \epsilon^6 & \epsilon^4 & \epsilon^2 \\ \epsilon^4 & \epsilon^2 & 1 \end{pmatrix}, \qquad
M_d \sim \langle H_d\rangle \epsilon^p \begin{pmatrix} \epsilon^5 & \epsilon^4 & \epsilon^4 \\ \epsilon^3 & \epsilon^2 & \epsilon^2 \\ \epsilon & 1 & 1 \end{pmatrix},
$$

$$
M_e \sim \langle H_d\rangle \epsilon^p \begin{pmatrix} \epsilon^5 & \epsilon^3 & \epsilon \\ \epsilon^4 & \epsilon^2 & 1 \\ \epsilon^4 & \epsilon^2 & 1 \end{pmatrix}, \qquad
M_{\nu_D} \sim \langle H_u\rangle \epsilon^s \begin{pmatrix} \epsilon^2 & \epsilon & \epsilon \\ \epsilon & 1 & 1 \\ \epsilon & 1 & 1 \end{pmatrix},
$$

$$
M_{\nu^c} \sim M_R \begin{pmatrix} \epsilon^2 & \epsilon & \epsilon \\ \epsilon & 1 & 1 \\ \epsilon & 1 & 1 \end{pmatrix} \quad\Rightarrow\quad
M_\nu^{light} \sim \frac{\langle H_u\rangle^2}{M_R}\epsilon^{2s} \begin{pmatrix} \epsilon^2 & \epsilon & \epsilon \\ \epsilon & 1 & 1 \\ \epsilon & 1 & 1 \end{pmatrix}. \tag{2.88}
$$

Here we work with the MSSM gauge group with supersymmetry realized at the TeV scale. Each entry has an order one pre-factor in the matri-

Table 2.2. The flavor $U(1)_A$ charge assignment.

Field	$U(1)_A$ Charge	Charge notation
Q_1, Q_2, Q_3	4, 2, 0	q_i^Q
L_1, L_2, L_3	$1 + s, s, s$	q_i^L
u_1^c, u_2^c, u_3^c	4, 2, 0	q_i^u
d_1^c, d_2^c, d_3^c	$1 + p, p, p$	q_i^d
e_1^c, e_2^c, e_3^c	$4 + p - s, 2 + p - s, p - s$	q_i^e
$\nu_1^c, \nu_2^c, \nu_3^c$	1, 0, 0	q_i^ν
H_u, H_d, S	$0, 0, -1$	$(h, \bar{h}, q_s)$

ces of Eq. (2.88), which is not explicitly shown. These matrices can be obtained by the $U(1)$ charge assignment of Table 2.2. In Eq. (2.88), the integer p is allowed to take values 0, 1 or 2, corresponding to $\tan \beta$ taking large, medium or small values. The integer s only enters into neutrino masses. Green–Schwarz anomaly cancelation condition requires $s = p$ in the simplest scheme. With $s = p$, the charge assignment of Table 2.2 will be compatible with $SU(5)$ unification. That is to say that the $\{Q_i, u_i^c, e_i^c\}$ fields of a given generation all have the same $U(1)_A$ charge, and similarly, the $\{d_i^c, L_i\}$ fields of a given family have the same charge. As we discuss in Sec. 2.5, the former set of SM particles are grouped into a **10** of $SU(5)$, while the latter set forms a $\overline{\mathbf{5}}$.

In the last line of Eq. (2.88), M_{ν^c} stands for the heavy ν^c Majorana mass matrix. When the seesaw formula is applied one obtains the light neutrino mass matrix M_ν^{light}, shown also in the last line of Eq. (2.88).

All the qualitative features of quark and lepton masses and mixings are reproduced by these matrices. These include small quark mixings and large neutrino mixings. The mass ratios in the up–quark sector scale as $m_u : m_c : m_t \sim \epsilon^8 : \epsilon^4 : 1$, while those in the down quarks scale as $m_d : m_s : m_b \sim \epsilon^5 : \epsilon^2 : 1$ with an identical scaling for the charged lepton mass ratios. (See the diagonal entries of $M_{u,d,e}$ in Eq. (2.88).) These are all consistent with experimental data. The quark mixing angles scale roughly as the down quark mass ratios, which is also reasonable. In the charged lepton sector, the mixing angles are larger, compared to the quark sector. This arises because of the lopsided structure of M_d and M_e with

$M_d \sim M_e^T$. This is a feature of $SU(5)$ grand unification, where left–handed lepton doublets are paired with the conjugate of the right–handed down quarks. As a result, the left–handed leptonic mixing angles will be related to the right–handed down quark mixing angles, which are allowed to be large since they are unobservable in the SM.[35] Note also that the hierarchy between light neutrino masses is weaker, $(m_1 : m_2 : m_3) \sim (\epsilon^2 : 1 : 1)$, compared with the charged fermion mass hierarchy. This feature is also consistent with neutrino oscillation data.

A variety of models based on anomalous $U(1)$ flavor symmetry have been proposed in the literature. A cross section of these models can be found in Ref. [36–38].

2.4.2.1. *More about anomalous $U(1)$ flavor symmetry*

To see the consistency of the three family model described above, and to see how it may be subject to experimental scrutiny, let us explore the structure of anomalous $U(1)$ flavor symmetry and its applications a little further. This will also enable us to compute the small parameter ϵ in the model of Table 2.2.

In heterotic string theory the $U(1)_A$ anomalies are canceled by the Green–Schwarz mechanism[33] which requires

$$\frac{A_1}{k_1} = \frac{A_2}{k_2} = \frac{A_3}{k_3} = \frac{A_F}{3k_F} = \frac{A_{gravity}}{24}. \tag{2.89}$$

Here A_1, A_2, A_3, A_F and $A_{gravity}$ are the $U(1)_Y^2 \times U(1)_A$, $SU(2)_L^2 \times U(1)_A$, $SU(3)_C^2 \times U(1)_A$, $U(1)_A^3$ and $(Gravity)^2 \times U(1)_A$ anomaly coefficients. (The subscript F is used to indicate the anomalous $U(1)$ flavor symmetry group.) All other anomalies (such as $U(1)_A^2 \times U(1)_Y$) must vanish. k_i $(i = 1, 2, 3)$, k_F are the Kac-Moody levels. The non–Abelian levels k_2 and k_3 must be integers. The factor $1/3$ in front of the cubic anomaly A_F has a combinatorial origin owing to the three identical $U(1)_A$ gauge boson legs.

Even without a covering grand unified group, string theory predicts unification of all gauge couplings, including that of the $U(1)_A$ and g_F, at the fundamental scale M_{st}:[39,40]

$$k_i g_i^2 = k_F g_F^2 = 2g_{st}^2. \tag{2.90}$$

Here g_i are the $U(1)_Y$, $SU(2)_L$ and $SU(3)_C$ gauge couplings for $i = 1, 2, 3$.

With $k_2 = k_3 = 1$ we find from Table 2.2, $A_2 = (19 + 3s)/2$ and $A_3 = (19 + 3p)/2$. Equation (2.89) then requires $p = s$, i.e., a common exponent for the charged lepton and the neutrino Dirac Yukawa coupling matrices. With $p = s$, the condition $A_1/k_1 = A_2/k_2$ fixes k_1 to be 5/3, which is consistent with $SU(5)$ unification. Note also that the charges given in Table 2.2 become compatible with $SU(5)$ unification. Since $\text{Tr}(Y) = 0$ for the fermion multiplets of $SU(5)$, and since the Higgs doublets carry zero $U(1)_A$ charge, the anomaly coefficient $[U(1)_A]^2 \times U(1)_Y$ vanishes, as required. The last equality in Eq. (2.89) requires

$$A_{gravity} = \text{Tr}\,(q) = 12(19 + 3p).   \tag{2.91}$$

This cannot be satisfied with the MSSM fields alone, since $\text{Tr}(q)_{MSSM} = 5(13 + 3p)$, which does not match Eq. (2.91). We cancel this anomaly by introducing MSSM singlet fields X_k obeying $\text{Tr}\,(q)_X = A_{gravity} - \text{Tr}\,(q)_{MSSM} = 163 + 21p$. If all the X_k fields have the same charge equal to +1, they will acquire masses of order $M_{st}\epsilon^2$ through the coupling $X_k X_k S^2/M_{st}$ and will decouple from low energy theory. We will assume that these fields X_k have charge +1.

With the charges of all fields fixed, we are now in a position to determine the $U(1)_A$ charge normalization so that $g_F^2 = g_2^2 = g_3^2$ at the string scale, (We take $k_2 = k_3 = 1$.) This normalization factor, which we denote as $|q_s|$, is given by $|q_s| = 1/\sqrt{k_F}$. All the charges given in Table 2.2 are to be multiplied by $|q_s|$. From the Green–Schwarz anomaly cancelation condition $A_F/(3k_F) = A_2/k_2$, we have

$$\frac{\text{Tr}\,(q^3)}{3k_F} = \frac{19 + 3p}{2k_2},   \tag{2.92}$$

from which we find the normalization of the $U(1)_A$ charge $|q_s| = 1/\sqrt{k_F}$ to be

$$|q_s| = (0.179,\, 0.186,\, 0.181)\ \text{for } p = (0, 1, 2).   \tag{2.93}$$

The Fayet–Iliopoulos term for the anomalous $U(1)_A$, generated through the gravitational anomaly, is given by[41]

$$\xi = \frac{g_{st}^2 M_{st}^2}{192\pi^2}|q_s|A_{gravity}\,,   \tag{2.94}$$

where g_{st} is the unified gauge coupling at the string scale (see Eq. (2.90)). By minimizing the potential from the $U(1)_A$ D–term

$$V = \frac{|q_s|^2 g_F^2}{8}\left(\frac{\xi}{|q_s|} - |S|^2 + \sum_a q_a^f|\tilde{f}_a|^2 + \sum_k q_k^X|X_k|^2\right)^2,   \tag{2.95}$$

in such a way that supersymmetry remains unbroken ($\tilde{f}_a$ are the MSSM sfermions and X_k are the singlet fields, which do not acquire VEVs), one finds for the VEV of S

$$\epsilon = \langle S \rangle / M_{st} = \sqrt{g_{st}^2 A_{gravity} / 192\pi^2}. \tag{2.96}$$

For the fermion mass texture in Eq. (2.88), corresponding to the $U(1)_A$ charges given in Table 2.2, we find

$$\epsilon = (0.177, \, 0.191, \, 0.204) \text{ for } p = (0, 1, 2). \tag{2.97}$$

This shows that the small expansion parameter can indeed be calculated in string–inspired models. It should be noted that this is a bottom–up approach to model building, it would of course be desirable to start from string theory and arrive at the spectrum and charges listed in Table 2.2.

The masses of the $U(1)_A$ gauge boson and the corresponding gaugino are obtained from $M_F = |q_s| g_F \langle S \rangle / \sqrt{2}$ and found to be

$$M_F = \left(\frac{M_{st}}{54.5}, \, \frac{M_{st}}{52.5}, \, \frac{M_{st}}{53.9} \right) \text{ for } p = (0, 1, 2). \tag{2.98}$$

In the momentum range below M_{st} and above M_F, these gauge particles will be active and will induce flavor dependent corrections to the sfermion soft masses and the A–terms. Implications of these effects have been studied in Ref. [34,38], where it has been shown that the process $\mu \to e\gamma$ in this class of models is very close to the current experimental limits. Ongoing MEG experiment should be able to probe the entire allowed parameter space of these models, provided that the SUSY particles have masses not exceeding about 1 TeV.

2.4.3. *The SM Higgs boson as the flavon*

Can the SM Higgs field itself be the flavon field? Clearly, then new flavor dynamics must happen near the TeV scale. This is apparently possible with significant consequences for Higgs boson physics, as I shall now outline.[42,43]

Consider an expansion in $H^\dagger H / M^2$, which is a SM singlet that can play the role of S. Here H is the SM Higgs doublet and M is the scale of new physics. Immediately you may wonder how this is possible, since $H^\dagger H$ cannot carry any $U(1)$ quantum number. But think of SUSY at the TeV scale. SUSY has two Higgs doublets, H_u and H_d, in which case the combination $H_u H_d$ can carry $U(1)$ charge. When reduced to SM this expansion in terms of $H^\dagger H$ can be consistent.

Consider the following mass matrices for quarks in terms of the expansion parameter

$$\epsilon = \frac{v}{M}.$$ (2.99)

$$M_u = \begin{pmatrix} h_{11}^u \epsilon^6 & h_{12}^u \epsilon^4 & h_{13}^u \epsilon^4 \\ h_{21}^u \epsilon^4 & h_{22}^u \epsilon^2 & h_{23}^u \epsilon^2 \\ h_{31}^u \epsilon^4 & h_{32}^u \epsilon^2 & h_{33}^u \end{pmatrix} v, \quad M_d = \begin{pmatrix} h_{11}^d \epsilon^6 & h_{12}^d \epsilon^6 & h_{13}^d \epsilon^6 \\ h_{21}^d \epsilon^6 & h_{22}^d \epsilon^4 & h_{23}^d \epsilon^4 \\ h_{31}^d \epsilon^6 & h_{32}^d \epsilon^4 & h_{33}^d \epsilon^2 \end{pmatrix} v.$$ (2.100)

The charged lepton mass matrix is taken to have a form similar to M_d, with the couplings h_{ij}^d replaced by h_{ij}^ℓ. These matrices give good fit to masses and mixings, as in the case of anomalous $U(1)$ model with $\epsilon \sim 1/7$ and all the couplings $h_{ij}^{u,d}$ being of order one. The masses of the quarks and leptons can be read off from Eq. (2.100) in the approximation $\epsilon \ll 1$:

$$\{m_t, m_c, m_u\} \simeq \{|h_{33}^u|, \; |h_{22}^u|\epsilon^2, \; |h_{11}^u - h_{12}^u h_{21}^u / h_{22}^u|\epsilon^6\}v,$$

$$\{m_b, m_s, m_d\} \simeq \{|h_{33}^d|\epsilon^2, \; |h_{22}^d|\epsilon^4, \; |h_{11}^d|\epsilon^6\}v,$$

$$\{m_\tau, m_\mu, m_e\} \simeq \{|h_{33}^\ell|\epsilon^2, \; |h_{22}^\ell|\epsilon^4, \; |h_{11}^\ell|\epsilon^6\}v.$$ (2.101)

The quark mixing angles are found to be:

$$|V_{us}| \simeq \left| \frac{h_{12}^d}{h_{22}^d} - \frac{h_{12}^u}{h_{22}^u} \right| \epsilon^2,$$

$$|V_{cb}| \simeq \left| \frac{h_{23}^d}{h_{33}^d} - \frac{h_{23}^u}{h_{33}^u} \right| \epsilon^2,$$

$$|V_{ub}| \simeq \left| \frac{h_{13}^d}{h_{33}^d} - \frac{h_{12}^u h_{23}^d}{h_{22}^u h_{33}^d} - \frac{h_{13}^u}{h_{33}^u} \right| \epsilon^4.$$ (2.102)

With $\epsilon = 1/6.5$ and with all couplings $h_{ij}^{u,d}$ being of order one, excellent fits to the quark masses and CKM mixing angles can be found. As an example, take the couplings to be

$$\{|h_{33}^u|, |h_{22}^u|, |h_{11}^u - h_{12}^u h_{21}^u / h_{22}^u|\} \simeq \{0.96, 0.14, 0.95\},$$

$$\{|h_{33}^d|, |h_{22}^d|, |h_{11}^d|\} \simeq \{0.68, 0.77, 1.65\},$$

$$\{|h_{33}^\ell|, |h_{22}^\ell|, |h_{11}^\ell|\} \simeq \{0.42, 1.06, 0.21\}.$$ (2.103)

The corresponding quark masses at $\mu = m_t(m_t)$ are:

$$\{m_t, m_c, m_u\} \simeq \{166, \; 0.60, \; 2.2 \times 10^{-3}\} \text{ GeV},$$

$$\{m_b, m_s, m_d\} \simeq \{2.78, \; 7.5 \times 10^{-2}, \; 3.8 \times 10^{-3}\} \text{ GeV},$$

$$\{m_\tau, m_\mu, m_e\} \simeq \{1.75, \; 0.104, \; 5.01 \times 10^{-4}\} \text{ GeV}.$$ (2.104)

All these are in agreement with values quoted in Table 2.1. Furthermore, the CKM mixing angles are also reproduced correctly with this choice of couplings.

In this scheme, the Yukawa coupling matrices of the physical quark fields are no longer proportional to the corresponding mass matrices. We obtain for the Yukawa couplings,

$$Y_u = \begin{pmatrix} 7h^u_{11}\epsilon^6 & 5h^u_{12}\epsilon^4 & 5h^u_{13}\epsilon^4 \\ 5h^u_{21}\epsilon^4 & 3h^u_{22}\epsilon^2 & 3h^u_{23}\epsilon^2 \\ 5h^u_{31}\epsilon^4 & 3h^u_{32}\epsilon^2 & h^u_{33} \end{pmatrix} \,, Y_d = \begin{pmatrix} 7h^d_{11}\epsilon^6 & 7h^d_{12}\epsilon^6 & 7h^d_{13}\epsilon^6 \\ 7h^d_{21}\epsilon^6 & 5h^d_{22}\epsilon^4 & 5h^d_{23}\epsilon^4 \\ 7h^d_{31}\epsilon^6 & 5h^d_{32}\epsilon^4 & 3h^d_{33}\epsilon^2 \end{pmatrix} \quad (2.105)$$

Take for example, the (3,3) entry in M_d. It arises from the operator $h^d_{33}Q_3 d^c_3 \tilde{H}(H^\dagger H)/M^2$. The contribution to the mass matrix from this operator is $h^d_{33}v\epsilon^2$, while the contribution to the Yukawa coupling is $(h/\sqrt{2})h^d_{33}(3\epsilon^2)$. The flavor factors (3 in this example) are not the same for various entries, and would result in flavor violation in Higgs interactions.

There is a tree–level contribution mediated by the Higgs boson for $K^0 - \bar{K}^0$ mass difference in this scheme. The new contribution, $\Delta m_K^{\text{Higgs}}$, is given by

$$\Delta m_K^{\text{Higgs}} \simeq \frac{4}{3}\frac{f_K^2 m_K B_K}{m_{h^0}^2}\epsilon^{12} \left[\left\{\frac{1}{6}\frac{m_K^2}{(m_d+m_s)^2} + \frac{1}{6}\right\}\text{Re}\left[\left(\frac{h^d_{12}+h^{d*}_{21}}{\sqrt{2}}\right)^2\right] \right.$$

$$\left. - \left\{\frac{11}{6}\frac{m_K^2}{(m_d+m_s)^2} + \frac{1}{6}\right\}\text{Re}\left[\left(\frac{h^d_{21}-h^{d*}_{12}}{\sqrt{2}}\right)^2\right] \right] \,. \quad (2.106)$$

Here B_K is the bag parameter. Using $B_K = 0.75$, $f_K \simeq 160$ MeV, $\epsilon \simeq 1/6.5$ $m_s(1\text{ GeV}) = 175$ MeV, $m_d(1\text{ GeV}) = 8.9$ MeV, and with $h^d_{12} = 1, h^d_{21} = 0.5$, we obtain $\Delta m_K^{\text{Higgs}} \simeq 3.1 \times 10^{-17}$ GeV, for $m_{h^0} = 100$ GeV. This is two orders of magnitude below the experimental value. We see broad consistency with data, primarily because of the appearance of high powers of ϵ in processes involving the light generations. For heavy flavors, this suppression is not that strong. For example, the $\bar{t}ch^0$ vertex has a coefficient

$$\mathcal{L}^{t-c}_{\text{FCNC}} = \frac{2\epsilon^2 h^0}{\sqrt{2}}(h^u_{23}\,c\,t^c + h^u_{32}\,t\,c^c) + h.c. \quad (2.107)$$

This can lead to a branching ratio for $t \to ch^0$ at the level of $(0.1 - 1)\%$, depending on the actual value of the order one coupling h^u_{ij}. This decay may be observable at the LHC.

The most striking signature of this scenario is that the decay branching ratios of the Higgs boson will be modified considerably compared to the SM. Decays into light fermions are enhanced, while decay into W pair is not.

For a specific set of flavor quantum numbers, the decay branching ratios are shown in Fig. 2.7, adopted from Ref. [43]. The solid lines correspond to branching ratios in the present model, while the dashed lines are the corresponding ones in the SM. Note that the branching ratio for $h \to b\bar{b}$ is enhanced. While the $h \to WW^*$ decay rate becomes comparable to $h \to b\bar{b}$ in the SM for a Higgs boson mass of 135 GeV, this crossover occurs at $m_h = 175$ GeV in the present case. Branching ratio for $h \to \mu^+\mu^-$ has increased, while the branching ratio for $h \to \gamma\gamma$ has diminished. These predictions are readily testable at the LHC once the Higgs boson is detected.

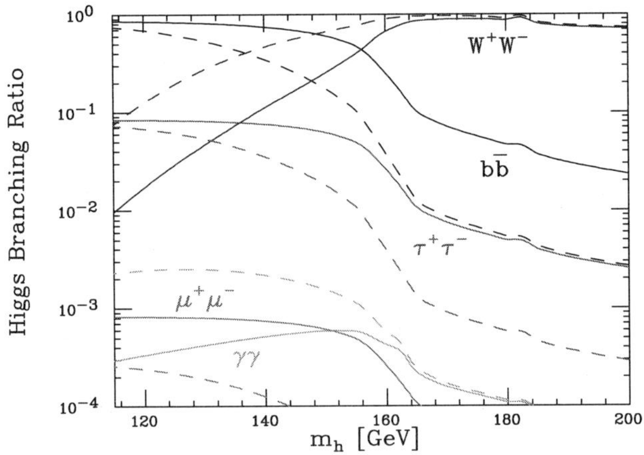

Fig. 2.7.   Higgs branching ratios with the SM Higgs as a flavon field.[43] The solid lines correspond to branching ratios with Higgs as a flavon, while the dashed lines are the corresponding SM branching fractions.

2.5. Grand Unification and the flavor puzzle

In this section we will develop ideas of Grand Unification which can provide significant insight into the flavor puzzle. When assisted by flavor symmetries, grand unified theories (GUTs) have great potential for addressing many of the puzzles.

Grand Unification is an ambitious program that attempts to unify the strong, weak and electromagnetic interactions.[44–46] It is strongly suggested by the unification of gauge couplings that happens in the minimal supersymmetric standard model. This is shown in Fig. 2.8, where the three gauge couplings of the standard model are extrapolated to high energies assuming weak scale supersymmetry. It is clear that data supports the merging of all three couplings to a common value. Besides its aesthetic appeal, in practical terms, grand unified theories reduce the number of parameters. For example, the three gauge couplings of the SM are unified into one at a very high energy scale $\Lambda_{\rm GUT} \simeq 2 \times 10^{16}$ GeV. The apparent differences in the strengths of the various forces is attributed to the spontaneous breakdown of the GUT symmetry to the MSSM and the resulting renormalization flow of the gauge couplings. SUSY GUTs are perhaps the best motivated extensions of the SM. They explain the quantization of electric charge, as well as the quantum numbers of quarks and leptons. They provide ideal settings for understanding the flavor puzzle, which will be the focus of this discussion.

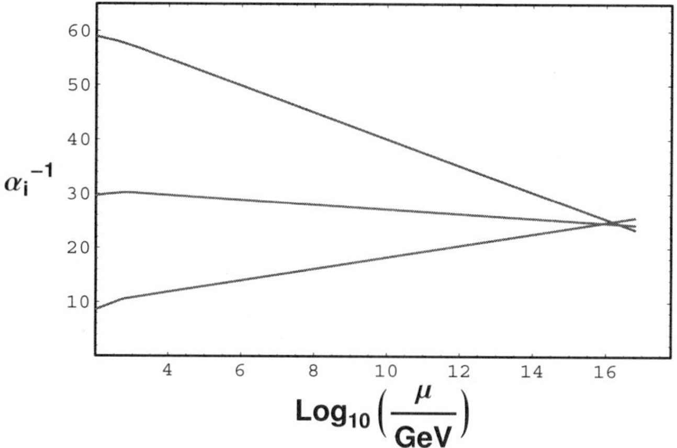

Fig. 2.8.   Evolution of the inverse gauge couplings (α_1^{-1}, α_2^{-1}, α_3^{-1}) (from top to bottom) in the MSSM as a function of momentum.

The simplest GUT model is based on $SU(5)$.[45] I will assume low energy supersymmetry, motivated by the gauge coupling unification and a solution to the hierarchy problem. For an understanding of quark–lepton masses and mixings SUSY is not crucial, but within the context of SUSY there will be many interesting flavor violating processes. In $SU(5)$, the fifteen components of one family of quarks and leptons are organized into two multiplets: A **10**–plet and a $\overline{\mathbf{5}}$–plet. The $\overline{\mathbf{5}}$ is of course the anti–fundamental representation of $SU(5)$, while the **10** is the anti-symmetric second rank tensor. These are represented by the following matrices:

$$
\psi(\mathbf{10}) : \frac{1}{\sqrt{2}}
\begin{pmatrix}
0 & u_3^c & -u_2^c & u_1 & d_1 \\
-u_3^c & 0 & u_1^c & u_2 & d_2 \\
u_2^c & -u_1^c & 0 & u_3 & d_3 \\
-u_1 & -u_2 & -u_3 & 0 & e^c \\
-d_1 & -d_2 & -d_3 & -e^c & 0
\end{pmatrix}
, \quad
\chi(\mathbf{\bar{5}}) :
\begin{pmatrix}
d_1^c \\
d_2^c \\
d_3^c \\
e \\
-\nu_e
\end{pmatrix} .
\qquad (2.108)
$$

Each family of quarks and leptons is organized in a similar form. It is very nontrivial that this assignment of fermions under $SU(5)$ is anomaly free. The anomaly from the $\mathbf{\bar{5}}$–plet is canceled by the anomaly from the $\mathbf{10}$–plet. Note that quarks and leptons are unified into common multiplets. Furthermore, particles and antiparticles are also unified. These features imply that baryon number, which is a global symmetry of the SM, is violated, and that proton will decay. Because the unification scale is rather large, $\Lambda_{\mathrm{GUT}} \approx 2 \times 10^{16}$ GeV, the decay rate of the proton is very slow, with a lifetime of order 10^{35} years. This is consistent with, but not very far from current experimental limits. Note that there is no ν^c field in the simplest version of $SU(5)$, but it can be added as a gauge singlet, as in the SM.

The symmetry breaking sector consists of two types of Higgs fields. One is an adjoint $\mathbf{24}_H$–plet Σ, which acquires vacuum expectation value and breaks $SU(5)$ down to the SM gauge symmetry. The VEV of this traceless Hermitian matrix is chosen as

$$
\langle \Sigma \rangle = V.\mathrm{diag}\left\{ 1,\, 1,\, 1,\, -\frac{3}{2},\, -\frac{3}{2} \right\} . \qquad (2.109)
$$

Under $SU(5)$ gauge transformation $\Sigma \to U \Sigma U^\dagger$. It is then clear that the VEV structure of Eq. (2.109) will leave invariant an $SU(3) \times SU(2) \times U(1)$ subgroup, identified as the SM gauge symmetry. 12 of the 24 gauge bosons of $SU(5)$ will acquire mass of order $V \sim \Lambda_{\mathrm{GUT}} \approx 2 \times 10^{16}$ GeV, leaving the remaining 12 SM gauge bosons massless.

Σ cannot couple to the fermions. A pair of $\{\mathbf{5}_H + \mathbf{\bar{5}}_H\}$ Higgs fields, denoted as $(H + \overline{H})$, are used for generating fermion masses and for electroweak symmetry breaking. H contain the H_u field of MSSM, while $\overline{H}$ contains the H_d field. These $(H + \overline{H})$ fields also contain color–triplet components, which must acquire GUT–scale masses, since they mediate proton decay. In minimal SUSY $SU(5)$ this splitting of color triplets and weak doublets is done by a special arrangement, by precisely tuning the mass term $M_H H \overline{H}$ and the coupling $\lambda H \overline{H} \Sigma$ of the superpotential, so that the $SU(2)_L$ doublet components remain light, while their color–triplet partners acquire large masses. This is possible, since the VEV of Σ breaks the $SU(5)$ symmetry.

The Yukawa couplings of fermions and the $(H, \overline{H})$ fields are obtained from the superpotential

$$W_{\text{Yuk}} = \frac{(Y_u)_{ij}}{4} \psi_i^{\alpha\beta} \psi_j^{\gamma\delta} H^\rho \epsilon_{\alpha\beta\gamma\delta\rho} + \sqrt{2}\, (Y_d)_{ij} \psi_i^{\alpha\beta} \chi_{j\alpha} \overline{H}_\beta \,. \tag{2.110}$$

Here (i, j) are family indices, and $(\alpha, \beta...)$ are $SU(5)$ indices with ϵ being the completely antisymmetric Levi–Cevita tensor. The $\overline{H}$ field has components similar to χ of Eq. (2.108), so that its fifth component is neutral and acquires a VEV: $\langle \overline{H}_5 \rangle = v_d$. Similarly, the fifth component of H acquires a VEV: $\langle H_5 \rangle = v_u$. When these VEVs are inserted in Eq. (2.110), the following mass terms for quarks and leptons are generated:

$$\mathcal{L}_{\text{mass}} = \frac{1}{2}(Y_u)_{ij}\, v_u \,(u_i u_j^c + u_j u_i^c) + (Y_d)_{ij}\, v_d \,(d_i d_j^c + e_i^c e_j) + h.c. \tag{2.111}$$

This leads to the following fermion mass matrices:

$$M_u = Y_u\, v_u \,, \quad M_d = Y_d\, v_d \,, \quad M_\ell = Y_d^T\, v_d \,. \tag{2.112}$$

Note that M_u is a symmetric matrix in family space. Furthermore, there are only two Yukawa coupling matrices describing charged fermion masses, unlike the three matrices we have in the SM. The reason for this reduction of parameters is the higher symmetry and the unification of quarks with leptons. Specifically, we have the relation

$$M_d = M_\ell^T \,. \tag{2.113}$$

This identity leads to the asymptotic (valid at the GUT scale) relations for the mass eigenvalues

$$m_b^0 = m_\tau^0, \; m_s^0 = m_\mu^0, \; m_d^0 = m_e^0 \,, \tag{2.114}$$

where the superscript ^{0} is used to indicate that the relation holds at the GUT scale.

In order to test the validity of the prediction of minimal SUSY $SU(5)$, we have to extrapolate the masses from GUT scale to low energy scale where the masses are measured. This is done by the renormalization group equations. The evolution of the b–quark and τ–lepton Yukawa couplings (λ_b and λ_τ), which are proportional to b–quark and τ–lepton masses, is shown in Fig. 2.9 for two differen values of $\tan\beta = (1.7, 50)$. $\tan\beta = v_u/v_d$ is the ratio of the two Higgs VEVs in MSSM. Here we have extrapolated the Yukawa couplings derived from the observed masses from low scale to the

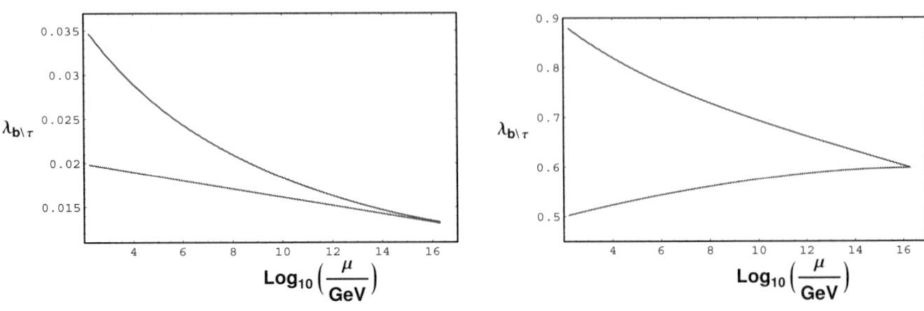

Fig. 2.9. Evolution of the b–quark and τ–lepton Yukawa couplings in the MSSM for $\tan\beta = 1.7$ (left panel) and 50 (right panel). $m_b(m_b) = 4.65$ GeV has been used here.

GUT scale. It is remarkable that unification of masses occurs in this simple context. The main effect on the evolution comes from QCD enhancement of b quark mass as it evolves from high energy to low energy scale, which is absent for the τ lepton.

Why these two specific values of $\tan\beta$? As it turns out, $b - \tau$ mass unification occurs only for specific values of $\tan\beta$, either for large values, or very small values. In Fig. 2.10 we plot the allowed values of $\tan\beta$ as a function of the strong coupling α_s.[47] From this figure, it is clear that intermediate values of $\tan\beta$ would lead to deviation from $m_b^0 = m_\tau^0$ by as much as 25%. For low and large values of $\tan\beta$, there is always good solution for $m_b(m_b)$, while for intermediate values there is no acceptable solution. It should be mentioned that there are significant finite corrections to the b–quark mass from loops involving the gluino, which is not included in the RGE analysis. These graphs, while loop suppressed, are enhanced by a factor of $\tan\beta$, and thus can be as large 30-40% for $m_b(m_b)$.[48] So even intermediate values of $\tan\beta$ are not totally excluded.

The last two relations of Eq. (2.114) turn out to be not acceptable when compared to low energy values of the masses. One can see this without going through the RGE evolution. Eq. (2.114) implies $m_s^0/m_d^0 = m_\mu^0/m_e^0$. These mass ratios are RGE independent, so one can compare them directly with observations. We have seen that $m_s/m_d \simeq 20$, while $m_\mu/m_e \simeq 200$. So this relation is off by an order of magnitude.

There is an elegant way of fixing the light fermion masses in $SU(5)$. Consider modifying Eq. (2.114) to the following relations:

$$m_b^0 = m_\tau^0, \quad m_s^0 = \frac{1}{3}\, m_\mu^0, \quad m_d^0 = 3\, m_e^0. \tag{2.115}$$

These relations were proposed by Georgi and Jarlskog and are known as the

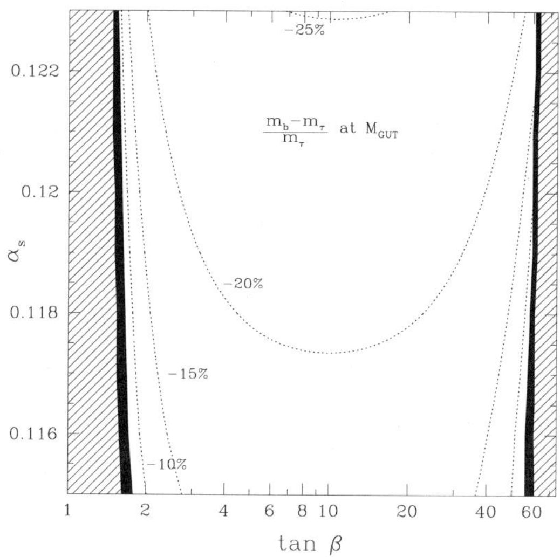

Fig. 2.10.   Deviation in the asymptotic relation $m_b^0 = m_\tau^0$ as a function of $\tan\beta$ and α_s.[47]

GJ relations.[49] The factors of 3 that appears in Eq. (2.115) have a simple group theoretic understanding in terms of $B-L$, under which lepton charges are (-3) times that of quark charges. The RGE independent quantity from Eq. (2.115) gives us

$$\frac{m_s}{m_d} = \frac{1}{9}\frac{m_\mu}{m_e},\qquad(2.116)$$

which is in good agreement with observations. There is one other prediction, which can be taken to be the value of $m_d(1\text{ GeV}) \simeq 8$ MeV, which is also is good agreement with data, although recent lattice calculations prefer somewhat smaller values of m_d.

2.5.1. *A predictive GUT framework for fermion masses*

How would one go about deriving the Georgi–Jarlskog mass relations? We invoke a flavor $U(1)$ symmetry as before. Consider the following mass matrices for up quarks, down quarks and charged leptons.[49–52]

$$M_u = \begin{pmatrix} 0 & a & 0 \\ a & 0 & b \\ 0 & b & c \end{pmatrix},\ M_d = \begin{pmatrix} 0 & de^{i\phi} & 0 \\ de^{-i\phi} & f & 0 \\ 0 & 0 & g \end{pmatrix},\ M_\ell = \begin{pmatrix} 0 & d & 0 \\ d & -3f & 0 \\ 0 & 0 & g \end{pmatrix}.\ (2.117)$$

The factor (-3) in charged lepton versus down quark mass matrix is attributed to the $B - L$ quantum number, and the zeros are enforced by a flavor symmetry. In $SU(5)$ GUT, the (1,2) and the (2,1) entries of M_d (and M_ℓ) are unrelated, but in $SO(10)$ GUT discussed in the next subsection, they can be related, as in Eq. (2.117). All parameters are complex to begin with, but after field redefinitions, only a single complex phase survives. There are 7 parameters in all to fit the 13 observables (9 masses, 3 mixing angles and one CP phase), thereby resulting in six predictions. Three of these predictions are the b, s and d–quark masses. We write them at the low energy scale by incorporating factors denoted as η which are the RGE evolution factors to go from the weak scale to the GUT scale. For light quark masses, there is a further evolution to go down from the weak scale to their respective mass (or hadron) scale. The predictions of the model for the quark masses are given by:

$$m_b = \eta_{b/\tau}^{-1} m_\tau; \quad \frac{m_d/m_s}{(1 - m_d/m_s)^2} = 9 \frac{m_e/m_\mu}{(1 - m_e/m_\mu)^2};$$

$$(m_s - m_d) = \frac{1}{3}\eta_{s/\mu}^{-1}(m_\mu - m_e). \tag{2.118}$$

The other three predictions are for the quark mixing angles and the CP phase J. J is the rephasing invariant CP violation parameter (Jarlskog invariant) which can be defined as

$$J = \mathrm{Im}(V_{us}V_{cb}V_{ub}^*V_{cs}^*) \tag{2.119}$$

and has a value of $J \simeq 2.8 \times 10^{-5}$. We have for the remaining three predictions[52]

$$|V_{cb}| = \eta_{KM}^{-1}\eta_{u/t}^{1/2}\sqrt{\frac{m_c}{m_t}}; \qquad \frac{|V_{ub}|}{|V_{cb}|} = \sqrt{\frac{m_u}{m_c}}; \tag{2.120}$$

$$J = \eta_{KM}^{-2}\eta_{u/t}\sqrt{\frac{m_d}{m_s}}\sqrt{\frac{m_c}{m_t}}\sqrt{\frac{m_u}{m_t}} \times$$

$$\left[1 - \frac{1}{4}\left(\sqrt{\frac{m_u}{m_c}}\sqrt{\frac{m_s}{m_d}} + \sqrt{\frac{m_c}{m_u}}\sqrt{\frac{m_d}{m_s}} - \sqrt{\frac{m_c}{m_u}}\sqrt{\frac{m_s}{m_d}}|V_{us}|^2\right)^2\right]^{\frac{1}{2}}.$$

Here $\eta_{ct} = [(m_c^0/m_t^0)/(m_c(m_t)/m_t(m_t))]$, $\eta_{KM} = |V_{cb}^0|/|V_{cb}|$, etc. One can write down semi–analytic results for the RGE factors, if the bottom–quark Yukawa coupling h_b is much smaller than the top Yukawa coupling h_t, (corresponding to $\tan\beta \lesssim 10$ or so). These RGE factors can be expressed then as

$$\eta_{KM} = \eta_{d/b} = \left(1 - \frac{Y_t}{Y_f}\right)^{\frac{1}{12}} ; \quad \eta_{s/\mu} = \left(\frac{\alpha_1}{\alpha_G}\right)^{-10/99} \left(\frac{\alpha_3}{\alpha_G}\right)^{-8/9} ; \quad (2.121)$$

$$\eta_{u/t} = \left(1 - \frac{Y_t}{Y_f}\right)^{\frac{1}{4}} ; \quad \eta_{b/\tau} = \left(\frac{\alpha_1}{\alpha_G}\right)^{-10/99} \left(\frac{\alpha_3}{\alpha_G}\right)^{-8/9} \left(1 - \frac{Y_t}{Y_f}\right)^{-1/12} .$$

Here α_G is the unified gauge coupling strength, $Y_t = h_t^2$ at the weak scale and Y_f is the fixed point value of Y_t. That is, Y_f is the largest value Y_t can take consistent with perturbation theory being valid upto the GUT scale. Numerically, $Y_f \simeq 1.2$. Y_t is of course obtained from $Y_t = [m_t(m_t)/v_u]^2$, which for $M_t = 172.5$ GeV is $Y_t \simeq 0.876$. Note that the CKM mixing parameters and the mass ratios in the same charge sector evolve only due to Yukawa couplings. The mass ratio m_s/m_μ does change with momentum proportional to the gauge interaction strength.

While five of the six predictions of this model agree well with experiments, the relation for $|V_{cb}|$ of Eq. (2.120) would imply that either the top quark mass is much higher than its observed value, or that the value of $|V_{cb}|$ is much larger than allowed. Indeed, if we use an acceptable value of $M_t = 172.5$ GeV, with $Y_f = 1.2$, Eq. (2.120) would lead to $|V_{cb}| \simeq 0.053$, which is more than 10 standard deviations away from its central value. If $|V_{cb}|$ is to be decreased down to any acceptable value, top quark mass will have to be very close to its perturbative upper limit, around 200 GeV, which is also excluded by experiments.

We conclude that, although very predictive and simple, the ansatz of Eq. (2.117) is excluded by data. It is interesting that while the original Fritszch ansatz of Eq. (2.79) was excluded since top quark mass was predicted to be too low, the present ansatz, which was very popular until a few years ago, is excluded for its prediction of top mass that is too large.

2.5.2. *Fermion masses in a predictive SO(10) model*

Now let us turn to an even more interesting class of GUTs, those based on the gauge symmetry $SO(10)$.[53] All members of a family are unified into a **16** dimensional spinor representation of $SO(10)$. This requires the existence of right–handed neutrino ν^c, leading naturally to the seesaw mechanism and small neutrino masses. $SU(5)$ has the option of having neutrino mass, but in that context there is no compelling argument for its existence. $SO(10)$ models are the canonical grand unified models, owing to the observed neutrino masses, and the fact that all members of a family are unified into a single **16**–dimensional spinor multiplet in $SO(10)$.

The spinor of $SO(10)$ breaks down under $SU(5)$ (which is one of its subgroups) as

$$16 = 10 + \overline{5} + 1 \,, \tag{2.122}$$

where the **1** is the ν^c field. The **10** and the $\overline{5}$ fields are identical to the case of $SU(5)$. We shall again assume low energy supersymmetry. Gauge symmetry breaking is accomplished in the SUSY limit by introducing Higgs fields in the adjoint $\mathbf{45_H}$, spinor $\{\mathbf{16_H} + \overline{\mathbf{16}}_\mathbf{h}\}$ and vector $\mathbf{10_H}$ representations. Because there is more symmetry in $SO(10)$, more scalars are needed to achieve symmetry breaking down to the SM. The spinor Higgs fields break $SO(10)$ down to $SU(5)$ changing the rank of the gauge group, while the adjoint $\mathbf{45_H}$–plet breaks this symmetry down to the SM. The vector $\mathbf{10_H}$–plet is used for fermion mass generation and for electroweak symmetry breaking. The MSSM Higgs doublets $H_{u,d}$ are contained partially in the $\mathbf{10_H}$ but can be partially also in the $\mathbf{16_H}$.

Let me work out a specific flavor model based on $SO(10)$ supplemented by a $U(1)$ symmetry.[54] While this model will not be as predictive as the ansatz that generated the GJ relations in the previous subsection, there are still a number of predictions, and these predictions are consistent with data. Several variations of the theme can be found in the literature,[55] but here I confine the discussions to the mass matrices of Ref. [54] and its slight generalization studied in Ref. [56].

The mass matrices for up and down quarks, and Dirac neutrino and charged leptons take the form:

$$M_u = \begin{bmatrix} 0 & \epsilon' & 0 \\ -\epsilon' & \zeta_{22}^u & \sigma + \epsilon \\ 0 & \sigma - \epsilon & 1 \end{bmatrix} \mathcal{M}_u^0; \qquad M_d = \begin{bmatrix} 0 & \eta' + \epsilon' & 0 \\ \eta' - \epsilon' & \zeta_{22}^d & \eta + \epsilon \\ 0 & \eta - \epsilon & 1 \end{bmatrix} \mathcal{M}_d^0 \tag{2.123}$$

$$M_\nu^D = \begin{bmatrix} 0 & -3\epsilon' & 0 \\ 3\epsilon' & \zeta_{22}^u & \sigma - 3\epsilon \\ 0 & \sigma + 3\epsilon & 1 \end{bmatrix} \mathcal{M}_u^0; \quad M_\ell = \begin{bmatrix} 0 & \eta' - 3\epsilon' & 0 \\ \eta' + 3\epsilon' & \zeta_{22}^d & \eta - 3\epsilon \\ 0 & \eta + 3\epsilon & 1 \end{bmatrix} \mathcal{M}_d^0 \,.$$

Here M_ν^D is the Dirac neutrino mass matrix.

Notice the various correlations in these matrices. The overall scale associated with M_u and M_ν^D are identical, while those for M_d and M_ℓ are the same. The "1" entry in all matrices have a common origin, arising from the operator $\mathbf{16_3 16_3 10_H}$. The ϵ entry appears with coefficient 1 in the up and down quark matrices, and with coefficient -3 in the leptonic mass matrices. This factor (-3) is the ratio of the $B - L$ charge of leptons versus quarks. Specifically, the ϵ entry arises from an operator $\mathbf{16}_2\mathbf{16}_3\,(\mathbf{10_H} \times \mathbf{45_H})/M$. Here the adjoint $\mathbf{45_H}$, which is a second rank antisymmetric tensor of $SO(10)$, acquires a VEV in a $B - L$ conserving direction:

$$\langle \mathbf{45_H} \rangle = i\tau_2 \times \text{diag.}(a,\, a,\, a,\, 0,\, 0) \,. \tag{2.124}$$

In the product $\mathbf{10_H} \times \mathbf{45_H}$, two fragments, an effective $\mathbf{10_H}$ and an effective $\mathbf{120_H}$, couple to the fermions. However, when the VEV of $\mathbf{45_H}$ from Eq. (2.124) is inserted, only the effective $\mathbf{120_H}$ is non-vanishing, leading to the relative factor of (-3) between leptons versus quarks. Note that the ϵ entry arises suppressed by $1/M$ so that $\epsilon \ll 1$, an idea familiar from the Froggatt–Nielsen mechanism. In an analogous fashion, the ϵ' entry arises from the operator $\mathbf{16}_1\mathbf{16}_2\,(\mathbf{10_H} \times \mathbf{45_H})\,S/M^2$, where S is an $SO(10)$ singlet flavon filed carrying a flavor $U(1)$ charge. This entry is then more suppressed compared to the ϵ entry. The σ entry originates from the operator $\mathbf{16}_2\mathbf{16}_3\,\mathbf{10_H}\,S/M$, and enters into all matrices with equal coefficient, just as the "1" entry. An operator $\mathbf{16}_2\mathbf{16}_3\mathbf{16_H}\mathbf{16_H}/M$ contributes equally to the down quark and charged lepton mass matrices, but not to M_u and M_ν^D, since $\mathbf{16_H}$ contains only an H_d–type field, and not an H_u–type field. The η entry in M_d and M_ℓ is the sum of the last two operators. The entry η' originates from $\mathbf{16}_1\mathbf{16}_2\mathbf{16_H}\mathbf{16_H}\,S^2/M^3$ operator.

These are precisely the operators one would obtain when the three families of fermions and the Higgs fields are assigned the following $U(1)$ charges:

$$\begin{array}{cccccccc} \mathbf{16}_3 & \mathbf{16}_2 & \mathbf{16}_1 & \mathbf{10}_H & \mathbf{16}_H & \overline{\mathbf{16}}_H & \mathbf{45}_H & \mathbf{S} \\ a & a+1 & a+2 & -2a & -a-1/2 & -a & 0 & -1 \end{array}. \tag{2.125}$$

In Ref. [54], where for simplicity, CP violation was ignored, the diagonal (2,2) entries were not introduced. In subsequent work these (2,2) entries, especially ζ_{22}^d, were used to accommodate CP violation. Here we present the predictions of the model as given in Ref. [54]. An acceptable fit to all mass and mixing parameters is obtained by the following choice of parameters at the GUT scale:

$$\sigma = -0.1096, \quad \eta = -0.1507, \quad \epsilon = 0.0954,$$
$$\epsilon' = 1.76 \times 10^{-4}, \quad \eta' = 4.14 \times 10^{-3}\,. \tag{2.126}$$

With these input, one obtains the following predictions:

$$m_b(m_b) = 4.9 \text{ GeV}, \quad m_s(1 \text{ GeV}) = 116 \,\text{MeV}, \quad m_d(1 \text{ GeV}) = 8 \text{ MeV}$$

$$\theta_C \simeq \left| \sqrt{\frac{m_d}{m_s}} - e^{i\phi} \sqrt{\frac{m_u}{m_c}} \right|, \quad \frac{|V_{ub}|}{|V_{cb}|} \simeq \sqrt{\frac{m_u}{m_c}} \simeq 0.07 \,. \qquad (2.127)$$

These predictions are in general agreement with data. When the (2,2) entries are included in the mass matrices, realistic CP violation phenomenology also follows.[56]

Light neutrino masses are generated in this scheme via the seesaw mechanism. Note that the Dirac neutrino mass matrix elements are completely fixed, because of $SO(10)$ symmetry, from the charged fermion sectors. The mechanism that generates heavy Majorana neutrino masses for the ν^c fields should be specified. The model already contains operators that do this, as given by

$$W_{\text{Maj}} = \mathbf{16}_i \mathbf{16}_j \, (\overline{\mathbf{16}}_{\mathbf{H}} \overline{\mathbf{16}}_{\mathbf{H}})/M \,. \qquad (2.128)$$

The natural scale of the cut–off M is $M = M_{\text{Planck}} = 2 \times 10^{18}$ GeV. Then with order one couplings in Eq. (2.128), one would obtain, for the (third family) right–handed Majorana mass, $M_{\nu_3^c} \sim \Lambda_{\text{GUT}}^2/M_{\text{Planck}} \sim 10^{14}$ GeV. This in turn leads to the light neutrino mass $m_\nu \sim m_t^2/M_{\nu_3^c} \simeq 0.05$ eV, nicely consistent with the value desired for atmospheric neutrino oscillation data.

In Ref. [54], it was shown, with a specific choice of the flavor structure of M_{ν^c}, that large neutrino oscillation angles arise naturally, while preserving the smallness of quark mixing angles. Specifically, while $|V_{cb}| \simeq 0.041$, $\sin^2 2\theta_{23} \simeq (0.9 - 0.99)$ was obtained, as a function of the light neutrino mass ratio m_2/m_3.

2.5.2.1. *Flavor violation in SUSY GUTs*

How do we go about testing ideas of grand unification in the flavor sector? Since the GUT scale is below the Planck scale, even though the flavor symmetry is broken near the GUT scale, soft SUSY breaking parameters can remember flavor violating interaction due to their running between the Planck scale and the GUT scale. Such running is expected in supergravity models, where the messengers of SUSY breaking have masses at the Planck scale. The most significant flavor violation in the model of Ref. [54] arises due to the splitting of the third family sfermions from those of the first two families. This is seen by the solution to the RGE equations for these

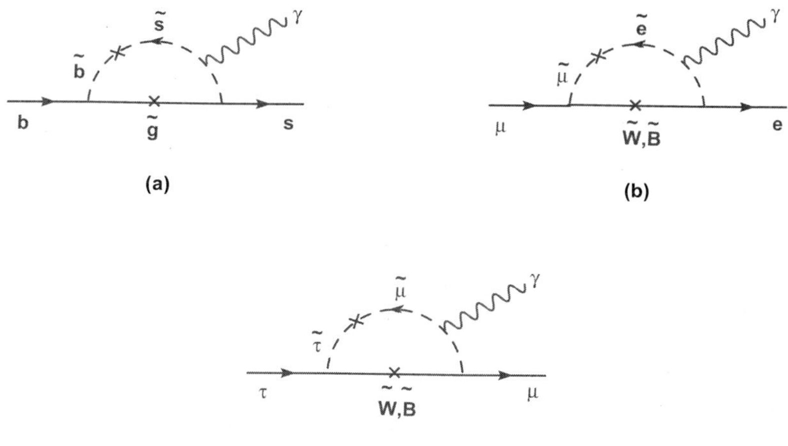

Fig. 2.11.   Rare decays induced by penguin diagrams via the exchange of SUSY particles. The flavor mixing occurs during the RGE flow between M_{GUT} and M_*.

masses.[57]

$$\Delta\hat{m}^2_{\tilde{b}_L} = \Delta\hat{m}^2_{\tilde{b}_R} = \Delta\hat{m}^2_{\tilde{\tau}_L} = \Delta\hat{m}^2_{\tilde{\tau}_R} \equiv \Delta$$

$$\approx -\left(\frac{30m_o^2}{16\pi^2}\right)h_t^2\, \log(M^*/M_{GUT})\,. \tag{2.129}$$

Here M^* is the fundamental scale where SUSY breaking messengers reside, with $M^* > M_{\mathrm{GUT}}$. h_t is the top quark Yukawa coupling. Note that leptons also feel the effect of top Yukawa, because leptons and quarks are unified. In Eq. (2.129) m_0 is the universal SUSY breaking scalar mass parameter. One sees that, because of the GUT threshold, universality is not preserved in this type of models. In going from gauge basis to the mass eigenbasis for the fermions, Eq. (2.129) would imply that there will be flavor changing scalar interactions. Because SUSY particles have masses of order TeV, these flavor violation can manifest in the MSSM sector via SUSY loops.

The most constraining FCNC process in the present model turns out to be $\mu \to e\gamma$. The diagrams inducing such processes in SUSY GUT models are shown in Fig. 2.11. In the present case it turns out that the decay $\tau \to \mu\gamma$ is not very significant, while the new contributions to $b \to s\gamma$ is not negligible. Predictions for the branching ratio for the decay $\mu \to e\gamma$ are depicted in Fig. 2.12 as a function of slepton mass.[57] Part of the parameter space is already ruled out, so there is a good chance that this process will be discovered at the MEG experiment at PSI.

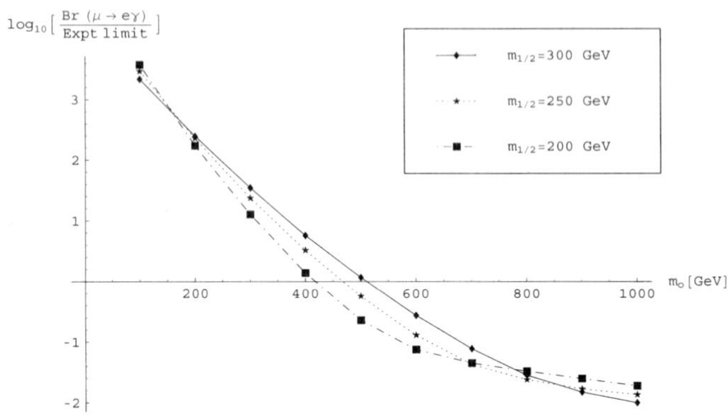

Fig. 2.12. Prediction for the branching ratio for $\mu \to e\gamma$ in the SUSY $SO(10)$ model as a function of slepton mass. The horizontal line indicates current experimental limit.[57]

There are other sources of flavor violation in SUSY GUTs. A widely discussed process is the $\ell_i \to \ell_j\gamma$ decay arising from neutrino mass physics.[58] The heavy right–handed neutrino mass is expected to be in the range $(10^{10} - 10^{14})$ GeV in SUSY GUTs. Even when the supergravity boundary conditions on the soft SUSY breaking parameters are valid at the GUT scale (and not the Planck scale), there is a momentum regime μ, $M_{\nu^c} \leq \mu \leq \Lambda_{\text{GUT}}$, where the ν^c fields are active. In this momentum regime the neutrino Dirac Yukawa couplings will affect the RGE evolution of the soft slepton mass parameters and generate lepton flavor violation. The FCNC effect in the slepton soft squared mass is given by

$$(\Delta m_{\tilde{L}}^2)_{ij} \simeq -\frac{\log(\Lambda_{\text{GUT}}/M_{\nu^c})}{8\pi^2}\left\{3\,m_0^2(Y_\nu^\dagger Y_\nu)_{ij} + (A_\nu^\dagger A_\nu)_{ij}\right\} . \quad (2.130)$$

Here Y_ν is the neutrino Dirac Yukawa coupling, while A_ν is the corresponding soft trilinear A–term.

In the MSSM, or in the SUSY $SU(5)$ model, the Yukawa coupling Y_ν cannot be determined from neutrino oscillation data. This is because the seesaw formula for light neutrinos goes as $m_\nu \sim Y_\nu^2 v^2/M_{\nu^c}$, and knowing m_ν does not determine Y_ν uniquely. However, if some of the entries of Y_ν are of order $(10^{-2} - 1)$, then the decay rate $\mu \to e\gamma$ will be within reach of ongoing experiments.

In SUSY $SO(10)$ there is a more crisp prediction for $\mu \to e\gamma$ arising from the neutrino sector. This happens because $SO(10)$ symmetry relates Y_ν

with the up–quark Yukawa couplings. Specifically, for the third family, we have $(Y_\nu)_{33} = Y_t$, the top quark Yukawa coupling. Since Y_t is of order one, the FCNC effects from the neutrino sector in SUSY $SO(10)$ are *predicted* to be significant. That is, they cannot be tuned to disappear, unlike in the SUSY $SU(5)$ model.

2.6. Radiative fermion mass generation

The hierarchical structure of the quark and lepton masses and the quark mixing angles can be elegantly understood by the mechanism of radiative mass generation. This is an alternative to the Froggatt–Nielsen mechanism. Here the idea is that only the heaviest fermions (eg. the third family quarks) acquire tree level masses. The next heaviest fermions (second family quarks) acquire masses as one loop radiative corrections, which are suppressed by a a typical loop factor $\sim 1/(16\pi^2) \sim 10^{-2}$ relative to the heaviest fermions. The lightest fermions (u and d quarks) acquire masses as two loop radiative corrections, which are then a factor $\sim [1/(16\pi^2)]^2 \sim 10^{-4}$ suppressed relative to the heaviest fermions. Thus, even without putting in small Yukawa couplings one understands the hierarchy in the mass spectrum of the fermions.

There is another appeal to this idea. If the electron mass is radiatively generated from the muon mass, then there must be no counter–term needed in the Lagrangian to absorb infinity associated with the electron mass. In other words, electron mass is "calculable", in terms of other parameters of the model. This idea was originally suggested by 'tHooft in his classic paper on the renormalizability of non–Abelian gauge theories.[59] This also implies that there must be some symmetry reason for the light fermions not to have tree level masses, otherwise the idea cannot be implemented consistently. Early attempts along this line were presented in Ref. [60]. More realistic models came along somewhat later.[61–65]

There is a resurgence of interest in this idea as the LHC turns on, since new particles with specific properties which may be seen at the LHC are predicted. There exist rather nice models of this type by Mohapatra and collaborators[61] from the late 80's. Recently Dobrescu and Fox have written a nice paper on the subject,[64] which I recommend to you. As in past examples, I will try to convey the main idea, with the understanding that implementation can vary considerably. I will discuss an implementation which I worked out with Mohapatra based on the permutation symmetry.[62]

Let us focus on the quark sector of the SM first. We wish to have a scenario where only the top quark and the bottom quark have tree level masses. In the same limit, there should be no CKM mixing induced. This can be realized if one has the following "democratic" mass matrices for up and down quarks.

$$M_{u,d} = \frac{m_{t,b}}{3} \begin{pmatrix} 1 & 1 & 1 \\ 1 & 1 & 1 \\ 1 & 1 & 1 \end{pmatrix} . \tag{2.131}$$

Of course, these matrices have rank 1, implying that only the top and the bottom acquire masses from here. A common unitary matrix will diagonalize M_u and M_d, so there is no CKM mixing induced at this stage.

How do we obtain democratic mass matrices of Eq. (2.131)? It turns out that the symmetry of these matrices is $S_{3L} \times S_{3R}$, where S_3 is the group of permutation of three letters. The Lagrangian that would generate Eq. (2.131) for M_u is of the form

$$\mathcal{L}_{\text{Yukawa}} = h_u(\overline{Q}_{1L} + \overline{Q}_{2L} + \overline{Q}_{3L})\tilde{H}(u_{1R} + u_{2R} + u_{3R}) \tag{2.132}$$

which is manifestly symmetric under separate permutations of the left–handed and the right–handed quark fields. So it is tempting to start with this symmetry group $S_{3L} \times S_{3R}$, but it is not necessary to have the S_{3R} group, since right-handed rotations are un-physical in the SM. So consider the following Lagrangian which only has the S_{3L} symmetry.

$$\begin{aligned} \mathcal{L}_{\text{Yukawa}} = {} & (\overline{Q}_{1L} + \overline{Q}_{2L} + \overline{Q}_{3L})\tilde{H}(h_1^u u_{1R} + h_2^u u_{2R} + h_3^u u_{3R}) \\ & + (\overline{Q}_{1L} + \overline{Q}_{2L} + \overline{Q}_{3L})H(h_1^d d_{1R} + h_2^d d_{2R} + h_3^d d_{3R}) . \end{aligned} \tag{2.133}$$

By right–handed rotations on u_R and d_R fields, we can bring Eq. (2.133) into the form of Eq. (2.132). Two combinations of the Q_{iL} and (u_{iR}, d_{iR}) fields orthogonal to Eq. (2.133) will be massless.

These massless Q_{iL} modes actually form the **2** dimensional representations of S_3. It is convenient to directly go to the irreducible representations of S_3. They are a true singlet **1**, an odd singlet **1′** and a doublet **2** = (x_1, x_2). The product of two **1′** gives a **1**, while the product of two **2** gives **1 + 1′ + 2**. The Clebsch–Gordon coefficients for this product (in a certain basis) are:[66]

$$\begin{pmatrix} x_1 \\ x_2 \end{pmatrix} \times \begin{pmatrix} y_1 \\ y_2 \end{pmatrix} = \mathbf{1} : (x_1 y_1 + x_2 y_2); \quad \mathbf{1'} : (x_1 y_2 - x_2 y_1);$$

$$\mathbf{2} : \begin{pmatrix} x_1 y_2 + x_2 y_1 \\ x_1 y_1 - x_2 y_2 \end{pmatrix} . \tag{2.134}$$

Now, consider the following assignment of quarks and scalars under S_3:

$$\begin{pmatrix} Q_{1L} \\ Q_{2L} \end{pmatrix} : \mathbf{2}; \quad Q_{3L} : \mathbf{1}; \quad u_{iR} : \mathbf{1} ,$$

$$H : \mathbf{1}, \quad \begin{pmatrix} \omega_1 \\ \omega_2 \end{pmatrix} : \mathbf{2}, \quad \omega_3 : \mathbf{1} . \tag{2.135}$$

Here the gauge structure is simply that of SM with H being the SM Higgs doublet. In order to radiatively generate light fermion masses, new ingredients are needed. The simplest possibility is to introduce scalar fields which have Yukawa couplings connecting the heavy (3rd generation) and the light fermions. We have assumed existence of $\omega_i(3, 1, -1/3)$ fields, which can have such Yukawa couplings, without inducing direct mass terms for the light fermions. Note that these ω_i fields are colored and charged, so they do not acquire vacuum expectation values.

The most general Yukawa couplings allowed in this SM $\times S_3$ model is given by

$$\begin{aligned}
\mathcal{L}_{\text{Yuk}} = {} & h_t \overline{Q}_{3L} t_R \tilde{H} + h_b \overline{Q}_{3L} b_R H + h_1 (Q_{1L}^T C Q_{3L} \omega_1 + Q_{2L}^T C Q_{3L} \omega_2) \\
& + h_2 (Q_{1L}^T C Q_{1L} + Q_{2L}^T C Q_{2L}) \omega_3 + h_3 Q_{3L}^T C Q_{3L} \omega_3 \\
& + h_4 \{ Q_{1L}^T C Q_{2L} + Q_{2L}^T C Q_{1L}) \omega_1 \\
& + (Q_{1L}^T C Q_{1L} - Q_{2L}^T C Q_{2L}) \omega_2 \} + h.c.
\end{aligned} \tag{2.136}$$

Here we have redefined the combination of u_R that couples to Q_{3L} as simply t_R (and similarly for b_R).

Clearly, from Eq. (2.136), only the top and bottom quarks acquire tree–level masses. There is no tree–level CKM mixing angle. So by symmetry reason, we have achieved the first stage of the program. Now, if S_3 is unbroken, none of the light fermions will acquire masses, even though they have Yukawa couplings via the ω_i fields. We can break S_3 spontaneously, or by soft bilinear terms in the Higgs potential:

$$V = \sum_{i,j=1}^{3} \mu_{ij}^2 \omega_i^* \omega_j + h.c. \tag{2.137}$$

With these soft breaking terms, light fermion masses will be induced. In Fig. 2.13 we have the one-loop and the two–loop mass generation diagrams.

The one–loop diagram of Fig. 2.13 only generates charm quark mass, and not the up quark mass. This can be understood as follows. At tree–level, among the down quarks, only b has a mass. There is a single linear combination of up quarks which couples to the b quark via the ω_i fields.

 K.S. Babu

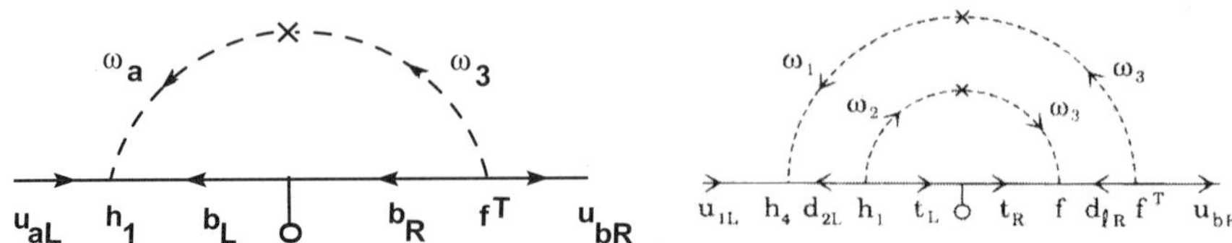

Fig. 2.13.   One–loop diagram inducing charm quark mass (left) and two–loop diagram inducing up quark mass (right).

It is this combination that picks up mass at one–loop. The orthogonal combination remains massless at this order. Now, the two–loop diagram connects up quarks to both b and s quarks. The inner loop of the two–loop diagram is the one–loop diagram that generates the s quark mass. As a result, u quark will acquire a mass proportional to the s quark mass at two–loop.

Including the one–loop diagram, the mass matrix for the (c, t) sector has the form

$$M_u^{1-\text{loop}} = \begin{pmatrix} \epsilon & a\epsilon \\ 0 & m_t^0 \end{pmatrix} \tag{2.138}$$

where a is of order one and the small parameter ϵ is found to be

$$\epsilon \simeq \left(\frac{h_1 f}{8\pi^2} \right) m_b \left(\frac{\mu_{a3}^2}{M_\omega^2} \right) \log \left(\frac{M_\omega^2}{m_b^2} \right) . \tag{2.139}$$

With the Yukawa couplings being order one, we can explain why the charm is much lighter than the top. The mixing angle V_{cb} is of order m_s/m_b, in agreement with observations. The two–loop diagrams which induce the up and down quark masses also induce the mixings of the first family. There is a natural hierarchy of mixing angles where $|V_{us}| \gg |V_{cb}| \gg |V_{ub}|$.

It is straightforward to extend the S_3 model to the leptonic sector. Consider the following assignment of leptons and ω_ℓ fields under S_3, where ω_ℓ are $(3^*, 1, -1/3)$ scalar fields. (These are not the conjugates of the ω_i fields from the quark sector, or else, there will be proton decay mediated by these scalars. We assume separate baryon number conservation, so the proton is stable.)

$$\begin{pmatrix} \psi_{1L} \\ \psi_{2L} \end{pmatrix} : \mathbf{2}, \quad \psi_{3L} : \mathbf{1}; \ e_{iR} : \mathbf{1}'$$

$$\omega_\ell : \mathbf{1}, \ \omega_\ell' : \mathbf{1}' \tag{2.140}$$

The general Yukawa coupling of leptons is given by

$$\mathcal{L}'_{\text{Yuk}} = f'_{ab} u^T_{aR} C e_{bR} \omega'_{\ell} + h'_1 Q^T_{3L} C \psi_{3L} \omega_{\ell} + h'_2 (Q^T_{1L} C \psi_{1L} + Q^T_{2L} C \psi_{2L}) \omega_{\ell} \\ + h'_3 (Q^T_{1L} C \psi_{2L} - Q^T_{2L} C \psi_{1L}) \omega'_{\ell} + h.c. \tag{2.141}$$

Note that all leptons are massless at the tree level. The one–loop diagram shown in Fig. 2.14 will induce the τ lepton mass, and is proportional to the top quark mass with a loop suppression. Only τ acquires a one–loop mass. The muon mass arises from the two–loop diagram of Fig. 14. The electron remains massless at this order, and acquires a mass only via a three–loop diagram.

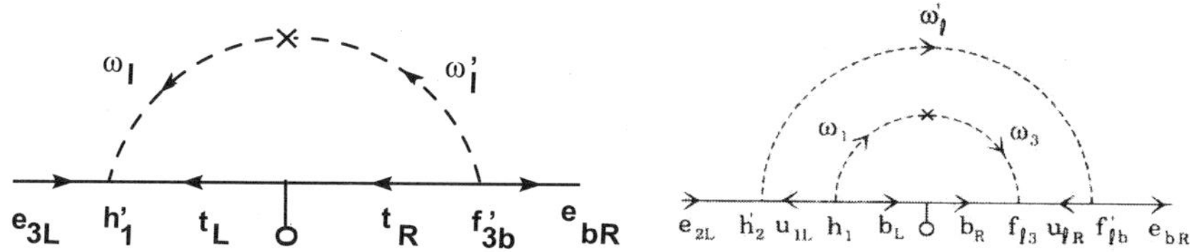

Fig. 2.14.   One-loop diagram inducing τ lepton mass (left) and two–loop diagram inducing the muon mass (right).

Note that we cannot constrain the masses of the ω fields from this process, since by taking the masses of the ω fields and the soft breaking μ^2 term to large values, the light fermion masses will remain unchanged.

However, in the supersymmetric version of the radiative mass generation mechanism, the new scalars should remain light, to about 1 TeV, since the superpotential is un-renormalized. That is to say that in a SUSY context, in the exact SUSY limit, the frmionic and bosonic loop diagrams add up to give zero. Once SUSY breaking terms are turned on, these diagrams will no longer cancel, and will generate finite quark and lepton masses. Thus, there is a prediction in this scenario. In addition to SUSY particles, LHC should discover these ω_i particles and their superpartners.

2.7. The strong CP problem and its resolution

There is no indication of CP violation in strong interactions. Yet, the QCD Lagrangian admits a term

$$\mathcal{L}_{QCD} = \frac{\theta\, g^2}{32\pi^2} G^a_{\mu\nu} \tilde{G}^{a\mu\nu} \tag{2.142}$$

which is P and T violating, and thus, owing to CPT invariance CP violating as well. In Eq. (2.142), $\tilde{G}^{a\mu\nu} = \frac{1}{2}\epsilon^{\mu\nu\rho\sigma}G^a_{\rho\sigma}$ is the dual field strength for the gluon. The Lagrangian in Eq. (2.142) is a total divergence, since $G^a_{\mu\nu}\tilde{G}^{a\mu\nu} = \partial_\mu K^\mu = \partial_\mu[\epsilon^{\mu\nu\rho\sigma}A^a_\nu(F^a_{\rho\sigma} - \frac{2}{3}\epsilon^{abc}A^b_\rho A^c_\sigma)]$. In a $U(1)$ gauge theory, the resulting surface term in the action would vanish for finite energy configurations. Thus a term analogous to Eq. (2.142) does not lead to P or T violation in QED. However, in QCD, the surface term gives rise to non-zero contributions, owing to finite energy "instanton" configurations, causing P and T violation.

It is not the parameter θ in Eq. (2.142) that is physical. Recall that the QCD Lagrangian also contains quark mass matrices M_u and M_d, which are generated after electroweak symmetry breaking. These matrices are complex, and generate the KM phase for CP violation in weak interactions. As discussed in Sec. 2.2, one makes bi–unitary transformations to bring these matrices into diagonal form: $U^{u\dagger}_L M_u U^u_R = \mathrm{diag}(m_u, m_c, m_t)$, and similarly for M_d. If U_L and U_R belong to the global $SU(N_f)_L \times SU(N_f)_R$ chiral symmetry (N_f is the number of quark flavors), which has no QCD anomaly, the diagonal quark masses cannot be made real. Specifically, $\mathrm{Det}(M_u) \to \mathrm{Det}(M_u)$ under such a special bi–unitary transformation. If the phases of the quark masses are denoted as $\theta_{u,c,t}$ and $\theta_{d,s,b}$, the combination

$$\theta_{\mathrm{QFD}} = \theta_u + \theta_c + \theta_t + \theta_d + \theta_s + \theta_b = \mathrm{Arg}[\mathrm{Det}(M_q)] \qquad (2.143)$$

cannot be removed by anomaly–free rotations. A chiral rotation on the quark fields is necessary in order to remove this phase. This however will generate an anomaly term in the Lagrangian, of the same form as in Eq. (2.142). The physical parameter is then

$$\overline{\theta} = \theta + \mathrm{Arg}[\mathrm{Det}M_q]\,. \qquad (2.144)$$

With $\overline{\theta}$ physical, there will be CP violation in strong interactions. However, there are stringent constraints on the value of $\overline{\theta}$ from experimental limits on the electric dipole moment (EDM) of the neutron: $\overline{\theta} < 10^{-10}$. This arises since in the presence of $\overline{\theta}$ neutron EDM can be shown to have a non-zero value given by

$$d_n \simeq \left[10^{-16} \times \overline{\theta}\right] \ \mathrm{e-cm}\,. \qquad (2.145)$$

From the experimental limit on neutron EDM, $d_n < 10^{-26}$ e-cm, one obtains the limit $\overline{\theta} < 10^{-10}$. Why is it that a fundamental dimensionless parameter of the Lagrangian, which should naturally be of order one, so small is the strong CP problem. If CP were a good symmetry of the entire Lagrangian, small $\bar{\theta}$ would have been quite natural. However, weak interactions do break CP invariance, which makes the strong CP problem acute.

There are various proposed solutions to the problem. At some point in time it was thought that the up quark mass may be zero. If true, that would solve the strong CP problem, since θ_u is then un-physical, and therefore $\bar{\theta}$ can be removed from the theory. But now we know, especially from lattice gauge theory results, that $m_u = 0$ is not an acceptable solution.

2.7.1. *Peccei–Quinn symmetry and the axion solution*

The most widely studied solution of the strong CP problem is the Peccei–Quinn (PQ) mechanism,[67] which yields a light pseudo–Goldstone boson, the axion.[68] Here the parameter $\bar{\theta}$ is promoted to a dynamical filed. This field acquires a non–perturbative potential induced by the QCD anomaly. Minimization of the potential yields the desired solution $\bar{\theta} = 0$, solving the strong CP problem.

In the presence of the $\bar{\theta}$ term in the Lagrangian, non–perturbative QCD effects will induce a vacuum energy given by

$$E_{\text{vac}} = \mu^4 \cos\bar{\theta}, \qquad (2.146)$$

where $\mu \sim \Lambda_{\text{QCD}} \sim 100$ MeV. This observation is crucially used in the PQ mechanism. What if $\bar{\theta}$ is a dynamical field? Then this non–perturbative potential will have to be minimized to locate the ground state (unlike the case where $\bar{\theta}$ is a constant in the Lagrangian). Minimization of this potential will yield $\bar{\theta} = 0$, as desired.

The essence of the PQ mechanism can be explained with a simple toy model.[69] Consider QCD with one quark flavor (q) and no weak interactions. Suppose there is a global $U(1)$ symmetry under which $q \to e^{-i\alpha\,\gamma_5/2}q$. Such a symmetry has a QCD anomaly, and can only be imposed at the classical level. A bare mass for q is then forbidden. Introduce now a complex color singlet scalar field ϕ which transforms under this $U(1)$ as $\phi \to e^{i\alpha}\phi$. The following Yukawa interaction is then allowed.

$$\mathcal{L}_{\text{Yuk}} = Y\bar{q}_L\phi q_R + Y^*\bar{q}_R\phi^* q_L. \qquad (2.147)$$

The potential for ϕ also respects the $U(1)$ symmetry, and is given by

$$V(\phi) = -m_\phi^2|\phi|^2 + \lambda|\phi|^4 \qquad (2.148)$$

With a negative sign for m_ϕ^2, the ϕ field will acquire a non-zero VEV, spontaneously breaking the $U(1)$. In this broken symmetric phase, we can parametrize ϕ as

$$\phi = \left[f_a + \tilde{\phi}(x^\mu) \right] e^{ia(x)/f_a} . \tag{2.149}$$

Here f_a is a real constant, while $\tilde{\phi}(x^\mu)$ and $a(x^\mu)$ are dynamical (real) fields. The quark q now acquires a mass, given by $M_q = Y f_a e^{ia(x)/f_a}$. Making the quark mass real by a field redefinition will induce a $\overline{\theta}$ given by

$$\overline{\theta}_{\text{eff}} = \theta + \text{Arg}[\text{Det}\, Y] + \frac{1}{f_a} a(x^\mu) . \tag{2.150}$$

The crucial point is that now $\overline{\theta}$ is a dynamical field, because of the presence of the a field, the axion. Without non–perturbative QCD effects, a will be massless, since it is the Goldstone boson associated with the spontaneous breaking of the global $U(1)$. The vacuum energy analog of Eq. (2.146) is now

$$E_{\text{vac}} = -\mu^4 \cos \overline{\theta}_{\text{eff}} . \tag{2.151}$$

Minimizing this potential with respect the dynamical a field would yield $\overline{\theta}_{\text{eff}} = 0$.

The field–dependent redefinition on q, $q(x^\mu) \rightarrow q(x^\mu)e^{-i(a(x^\mu)/f_a)(\gamma_5/2)}$ would remove the axion field from quark interactions except via derivatives, originating from the kinetic terms. The axion also will have couplings to the gluon field strength. These couplings are given by

$$\mathcal{L}_a = -\left(\frac{\partial_\mu a}{f_a} \right) \overline{q}\gamma_\mu\gamma_5 q + \frac{g^2}{32\pi^2} \left(\frac{a}{f_a} \right) G\tilde{G} . \tag{2.152}$$

It is the second term of Eq. (2.152) that actually induces the potential for the axion. Because of this potential, axion will have a mass of order $m_a \sim \Lambda_{\text{QCD}}^2/f_a$.

The essentials of realistic axion model are already present in this toy model. We need to turn on weak interactions, and we need to add three families of quarks. The straightforward implementation would involve the SM extended to have two Higgs doublets, one coupling to the up–type quarks, and the other coupling to the down–type quarks.[68] A global $U(1)$ can then be defined classically, which has a QCD anomaly. The axion will now be part of the Higgs doublet, with the axion decay constant $f_a \sim v \sim 10^2$ GeV. The couplings of the axion to quarks, Eq. (2.152), will now be rather strong. The decay $K^+ \rightarrow \pi^+ a$ will occur at an observable strength.

This process has been searched for, but has not been observed. Negative results in searches for this and other such processes have excluded the weak scale axion model.

Acceptable axion models of the "invisible" type[70,71] involving high scale PQ symmetry breaking are fully consistent. In the model of Ref. [70], in addition to the two Higgs doublets, a complex singlet Higgs scalar S is also introduced. The axion decay constant f_a is now the VEV of S, which can be much above the weak scale. The axion is primarily in S, with very weak couplings to the SM fermions. There are non–trivial constraints from astrophysics and cosmology on such a weakly interacting light particle. For example, axion can be produced inside supernovae. Once produced, they will escape freely, draining the supernova of its energy. Consistency with supernova observations requires that $f_a > 10^9$ GeV. Cosmological abundance of the axion requires that $f_a < 10^{12}$ GeV.

In the invisible axion model of Ref. [71], there is only a single Higgs doublet of the SM. A Higgs singlet R and a heavy quark Q, which has vectorial properties under the SM, are introduced. The PQ $U(1)$ symmetry acts on Q and the scalar R. Q acquires its mass only via its Yukawa coupling with R. (This example is essentially the same as the toy model described above.) The phase of R is the axion in this case, with phenomenology similar to, but somewhat different from, the axion model of Ref. [70].

It should be noted that axion is a leading candidate for the cosmological dark matter. For reviews of axion physics, astrophysics, cosmology, and detection techniques, see Ref. [72].

2.7.2. *Solving strong CP problem with Parity symmetry*

There is another class of solution to the strong CP problem. One can assume Parity[73,74] to set $\theta = 0$. If the fermion mass matrices have real determinant, then $\overline{\theta}$ can be zero at the tree level. Loop induced $\overline{\theta}$ needs to be small, but this is not difficult to realize.

Let me illustrate this idea with the left–right symmetric model which has Parity invariance. The Yukawa couplings are Hermitian in this setup. To make the mass matrices also Hermitian, we must ensure that the VEVs of scalars are real. This is easily done in the SUSY version, which is what I will describe.[74] In SUSY models, one should also take into account the contributions from the gluino to $\overline{\theta}$.

The model is the SUSY version of left–right symmetric model based on the gauge symmetry $SU(3)_C \times SU(2)_L \times SU(2)_R \times U(1)_{B-L}$ discussed in Sec. 2.3. Two bi-doublet scalars $\Phi_i(1, 2, 2, 0)$ $(i = 1, 2)$ are used to generate quark and lepton masses as well as CKM mixings. The relevant superpotential is given as

$$W = Y_u Q Q^c \Phi_u + Y_d Q Q^c \Phi_d \,. \tag{2.153}$$

The Yukawa coupling matrices Y_u and Y_d will be Hermitian, owing to Parity invariance. Parity also implies that the QCD Lagrangian parameter $\theta = 0$ and that the gluino mass is real. The soft SUSY breaking A–terms, analogous to W in Eq. (2.153) will also be Hermitian. We shall consider the case where these A terms are proportional to the respective Yukawa matrices. Furthermore, we assume universal masses for the squarks, as in minimal supergravity, or in gauge mediated SUSY breaking models.

The quark mass matrices $M_{u,d}$ are Hermitian at tree level since the VEVs of the bi-doublet scalars turn out to be real. Therefore $\bar{\theta} = 0$ at tree level. We wish to demonstrate that loop induced contributions to $\bar{\theta}$ are not excessive. Note that this setup has two Hermitian matrices Y_u and Y_d which are complex, with all other (flavor singlet) parameters being real.

Since parity is broken at a high scale (denoted as v_R), a nonzero value of $\bar{\theta}$ will be induced at the weak scale through renormalization group extrapolation below v_R. This is because the SM gauge symmetry does not permit the Yukawa couplings to remain Hermitian. The induced $\bar{\theta}$ will have the general structure given by

$$\delta\bar{\theta} = \mathrm{Im}\,\mathrm{Tr}[\Delta M_u M_u^{-1} + \Delta M_d M_d^{-1}] - 3\,\mathrm{Im}(\Delta M_{\tilde{g}} M_{\tilde{g}}^{-1}) \tag{2.154}$$

where $M_{u,d,\tilde{g}}$ denote the tree level contribution to the up–quark matrix, down–quark matrix and the gluino mass respectively, and $\Delta M_{u,d,\tilde{g}}$ are the loop corrections. To estimate the corrections from ΔM_u and ΔM_d, we note that the beta function for the evolution of Y_u below v_R is given by $\beta_{Y_u} = Y_u/(16\pi^2)(3Y_u^\dagger Y_u + Y_d^\dagger Y_d + G_u)$ with the corresponding one for Y_d obtained by the interchange $Y_u \leftrightarrow Y_d$ and $G_u \to G_d$. Here G_u is a family–independent contribution arising from gauge bosons and the $\mathrm{Tr}(Y_u^\dagger Y_u)$ term. The $3Y_u^\dagger Y_u$ term and the G_u term cannot induce non–Hermiticity in Y_u, given that Y_u is Hermitian at v_R. The interplay of Y_d with Y_u will however induce deviations from Hermiticity. Repeated iteration of the solution with $Y_u \propto Y_u Y_d^\dagger Y_d$ and $Y_d \propto Y_d Y_u^\dagger Y_u$ in these equations will generate the following structure:

$$\delta\bar{\theta} \simeq \left(\frac{\ln(M_U/M_W)}{16\pi^2}\right)^4 \left[c_1 \mathrm{Im}\,\mathrm{Tr}\left(Y_u^2 Y_d^4 Y_u^4 Y_d^2\right) + c_2 \mathrm{Im}\,\mathrm{Tr}\left(Y_d^2 Y_u^4 Y_d^4 Y_u^2\right)\right], \tag{2.155}$$

where M_U is the unification scale. Here c_1 and c_2 are order one coefficients which are not equal. To estimate the induced $\bar{\theta}$, we choose a basis where Y_u is diagonal, $Y_u = D$ and $Y_d = V D' V^\dagger$ where $D_u v_u = \mathrm{diag}(m_u, m_c, m_t)$, $D_d v_d = \mathrm{diag}(m_d, m_s, m_b)$ with V being the CKM matrix. The Trace of the first term in Eq. (2.155) is then $\mathrm{Im}(D_i^2 D_k^4 D_j'^4 D_l'^2 V_{ij} V_{kl} V_{il}^* V_{kj}^*)$. The leading contribution in this sum is $(m_t^4 m_c^2 m_b^4 m_s^2)/(v_u^6 v_d^6)\mathrm{Im}(V_{cb} V_{ts} V_{cs}^* V_{tb}^*)$. The second Trace in Eq.(2.155) is identical, except that it has an opposite sign. Numerically we find

$$\delta\bar{\theta} \sim 3 \times 10^{-27}(\tan\beta)^6(c_1 - c_2),\qquad(2.156)$$

which is well below the experimental limit of 10^{-10} from neutron EDM.

There are also finite corrections to the quark and gluino masses, which are not contained in the RG equations. Consider first the finite one loop corrections to the quark mass matrices. A typical diagram involving the exchange of squarks and gluino is shown in Fig. 2.15, where the crosses on the $\tilde{Q}$ and $\tilde{Q}^c$ lines represent (LL) and (RR) mass insertions that will be induced in the process of RGE evolution. From this figure we can estimate the form for $\Delta M_u = \frac{2\alpha_s}{3\pi} m_{\tilde{Q}}^2 A_u m_{\tilde{u}^c}^2$ where $\tilde{Q}$ is the squark doublet and $\tilde{u}^c$ is the right–handed singlet up squark. Without RGE effects, the trace of this term will be real, and will not contribute to $\bar{\theta}$. Looking at the RGE for $m_{\tilde{u}^c}^2$ upto two loop order, we see that for the case of proportionality of A_u and Y_u, $m_{\tilde{u}^c}^2$ gets corrections having the form $m_0^2 Y_u^2$ or $m_0^2 Y_u^4$ or $m_0^2 Y_u Y_d^2 Y_u$. Therefore in $\Delta M_u M_u^{-1}$, the M_u^{-1} always cancels and we are left with a product of matrices of the form $Y_u^n Y_d^m Y_u^p Y_d^q \cdots$. A similar comment applies when we look at the RGE corrections for $m_{\tilde{Q}}^2$ or A_u. If the product is Hermitian, then its trace is real. So to get a nonvanishing contribution to theta, we have to find the lowest order product of Y_u^2 and Y_d^2 that is non–Hermitian and we get

$$\delta\bar{\theta} = \frac{2\alpha_s}{3\pi}\left(\frac{\ln(M_U/M_W)}{16\pi^2}\right)^4 \left(k_1 \mathrm{Im}\mathrm{Tr}[Y_u^2 Y_d^4 Y_u^4 Y_d^2] + k_2 \mathrm{Im}\mathrm{Tr}[Y_d^2 Y_u^4 Y_d^4 Y_u^2]\right)$$

$$(2.157)$$

where $k_{1,2}$ are calculable constants. The numerical estimate of this contribution parallels that of the previous discussions, $\delta\bar{\theta} \sim (k_1 - k_2) \times 10^{-28}(\tan\beta)^6$. The contributions from the up–quark and down quark matrices tend to cancel, but since the $\tilde{d}^c$ and the $\tilde{u}^c$ squarks are not degenerate, $k_1 \neq k_2$ and the cancellation is incomplete.

In Fig. 2.15 we have also displayed the one–loop contribution to the gluino mass arising from the quark mass matrix. Here again one encounters the imaginary trace of two Hermitian matrices Y_u and Y_d, in the case

                         K.S. Babu

of universality and proportionality of SUSY breaking parameters. Our estimate for $\delta\bar{\theta}$ is similar to that of the quark mass matrix of Eq. (2.157).

This exercise shows that the strong CP problem can be consistently resolved with the imposition of parity symmetry.

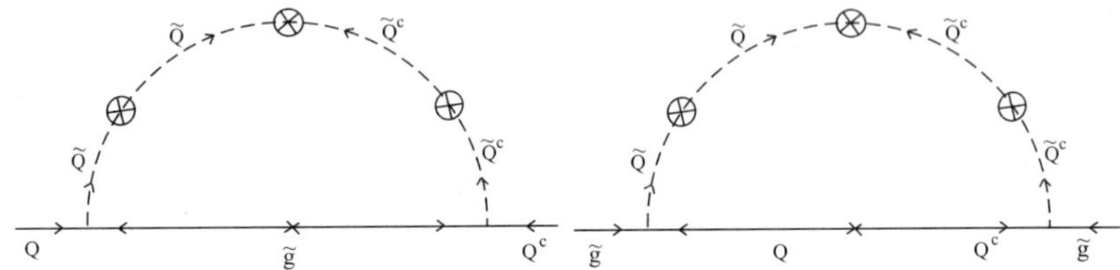

Fig. 2.15.   One-loop diagram inducing complex correction to the quark mass (left) and to the gluino mass (right).

2.7.3. *Solving the strong CP problem by CP symmetry*

The idea of Ref. [75] is to use CP as a spontaneously broken symmetry. The QCD θ is then zero. In order to generate KM CP violation in weak interactions, the mass matrices of the up and down quarks will have to be complex. This can be realized consistently, while keeping the determinant of the quark mass matrix real by breaking CP spontaneously. Then at tree–level $\bar{\theta}$ will be zero.

A model of this type can be readily constructed. Consider the addition of three vector–like $D + D^c$ quarks to the SM. These are $SU(2)$ singlets with $Y = \mp 2/3$, so that they can mix with the down–type quarks (d, d^c) of the SM. Suppose there is a discrete symmetry Z_2 under which the d^c quarks reflect sign. Several SM singlet Higgs scalar fields S_i with $i \geq 2$ are also needed. Under Z_2 these S_i fields are odd.

The Yukawa Lagrangian of this theory is given by

$$\mathcal{L}_{\text{Yuk}} = Y_u Q u^c H + Y_D Q D^c \tilde{H} + M_D D d^c + F_i D D^c S_i + h.c. \qquad (2.158)$$

CP invariance implies that all the coupling matrices (Y_u, Y_D, M_D, F_i) are real. Complex phases appear only in the VEVs of the S_i fields, which break CP spontaneously. The down–type quark mass matrix arising from Eq. (2.158) is given by

$$M_{d-D} = \begin{pmatrix} 0 & Y_D v \\ M_D & \sum_i F_i \langle S_i \rangle \end{pmatrix}. \qquad (2.159)$$

When the heavy D states are integrated out, the light 3×3 quark mass matrix for the down quarks will have a complex form, yielding weak CP violation. The determinant of M_{d-D} is real, owing to its structure (with all complex phases residing in the lower right–hand block). So $\bar{\theta} = 0$ in this model at tree level.

Loop corrections will induce non–zero $\bar{\theta}$ at the one loop level, which has a magnitude of order $\bar{\theta} \sim F^2/(16\pi^2)$. For $F \sim 10^{-4}$, this induced $\bar{\theta}$ will be within experimental limits.

2.8. Rare B meson decay and new physics

In this section we turn to specific processes where new physics may show up at colliders. It is quite likely that such processes will show up first in the heaviest fermion (t, b, τ) systems. Specifically, LHCb will be sensitive to such effects occurring in the B meson system. We focus on this system here.

New physics may show up at the LHC in decays of the B meson that are rare or forbidden in the SM. Low energy supersymmetry can provides such possibilities. Specifically, in the framework of SUSY with minimal flavor violation,[76] that is, flavor violation arising only via the MSSM Yukawa couplings, there are processes that are enhanced at large $\tan\beta$ which can be in the observable range.

One such example is the rare decay $B_{s,d} \to \mu^+\mu^-$ that has not been observed so far. In the SM, this process occurs via penguin and box diagrams. The branching ratio has been calculated to be[77]

$$Br(B_s \to \mu^+\mu^-) = (3.35 \pm 0.32) \times 10^{-9} \,,$$
$$Br(B_d \to \mu^+\mu^-) = (1.03 \pm 0.09) \times 10^{-10} \,. \qquad (2.160)$$

This prediction is to be compared with the current experimental limits from CDF and D0[5]

$$Br(B_s \to \mu^+\mu^-) < (5.8 \pm 0.32) \times 10^{-8} \,,$$
$$Br(B_d \to \mu^+\mu^-) < (1.8 \pm 0.09) \times 10^{-8} \,. \qquad (2.161)$$

There is a lot of room for new physics in these processes. At the LHC, sensitivity of the experiments will be better than the SM prediction.

2.8.1. $B_s \to \mu^+\mu^-$ in MSSM at large $\tan\beta$

Minimal supersymmetry at large $\tan\beta$ can significantly enhance the decay rate $B_s \to \mu^+\mu^-$. This occurs via exchange of Higgs bosons of MSSM.[78]

          K.S. Babu

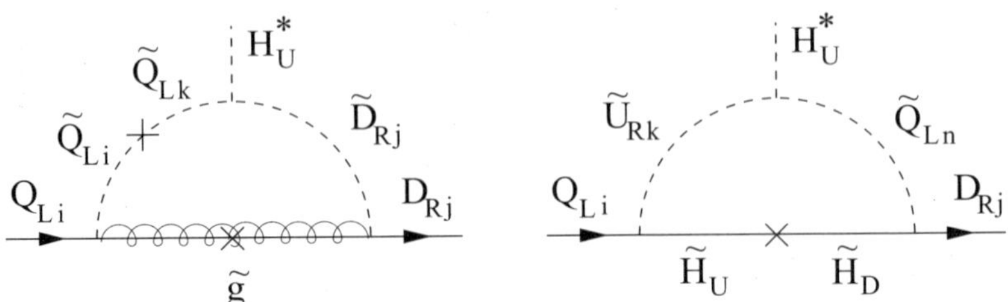

Fig. 2.16.   One-loop diagram inducing τ lepton mass (left) and two–loop diagram inducing the muon mass (right).

MSSM Yukawa couplings do preserve flavor at the tree level, see Eq. (2.41). That is, in the quark sector only H_u couples to the up–quarks, while only H_d couples to the down–quarks. There is no tree–level FCNC mediated by the Higgs boson. However, this situation changes once loop corrections to the Yukawa couplings are included.

To see this, let us begin by writing the effective Lagrangian for the interactions of the two Higgs doublets with the quarks in an arbitrary basis:

$$-\mathcal{L}_{eff} = \overline{D}_R \mathbf{Y_D} Q_L H_d + \overline{D}_R \mathbf{Y_D} \left[\epsilon_g + \epsilon_u \mathbf{Y_U^\dagger} \mathbf{Y_U} \right] Q_L H_u^* + h.c. \quad (2.162)$$

Here $\mathbf{Y_D}$ and $\mathbf{Y_U}$ are the 3×3 Yukawa matrices of the microscopic theory, while the $\epsilon_{g,u}$ are the finite, loop-generated non-holomorphic Yukawa coupling coefficients. The leading contributions to ϵ_g and ϵ_u are generated by the two diagrams in Fig. 2.16.

Consider the first diagram in Fig. 2.16. If all $\tilde{Q}_i$ masses are assumed degenerate at some scale M_{unif} then, at lowest order, $i = k$ and the diagram contributes only to ϵ_g:

$$\epsilon_g \simeq \frac{2\alpha_3}{3\pi} \mu^* M_3 f(M_3^2, m_{\tilde{Q}_L}^2, m_{\tilde{d}_R}^2) , \quad (2.163)$$

where

$$f(x,y,z) = -\frac{xy \log(x/y) + yz \log(y/z) + zx \log(z/x)}{(x-y)(y-z)(z-x)} . \quad (2.164)$$

Meanwhile, the second diagram of Fig. 2.16 contributes to ϵ_u:

$$\epsilon_u \simeq \frac{1}{16\pi^2} \mu^* A_U f(\mu^2, m_{\tilde{Q}_L}^2, m_{\tilde{u}_R}^2) . \quad (2.165)$$

(We assume that the trilinear A-terms can be written as some flavor-independent mass times $\mathbf{Y_U}$.) For typical inputs, one usually finds $|\epsilon_g|$ is about 4 times larger than $|\epsilon_u|$.

Owing to these loop corrections, the CKM mixing angles receive finite corrections. In particular,

$$V_{ub} \simeq V_{ub}^0 \left[\frac{1 + \epsilon_g \tan\beta}{1 + (\epsilon_g + \epsilon_u y_t^2)\tan\beta} \right]. \tag{2.166}$$

The same form also holds for the corrected V_{cb}, V_{td} and V_{ts}.

For $\epsilon_u \neq 0$, however, the rotation that diagonalized the mass matrix does not diagonalize the Yukawa couplings of the Higgs fields, leading to FCNC Higgs couplings given by

$$\mathcal{L}_{FCNC} = \frac{\bar{y}_b V_{tb}^*}{\sin\beta} \chi_{FC} \left[V_{td}\bar{b}_R d_L + V_{ts}\bar{b}_R s_L \right] \left(\cos\beta H_u^{0*} - \sin\beta H_d^0 \right) + h.c. \tag{2.167}$$

with the quark fields in the physical/mass eigenbasis, and defining

$$\chi_{FC} = \frac{-\epsilon_u y_t^2 \tan\beta}{(1 + \epsilon_g \tan\beta)[1 + (\epsilon_g + \epsilon_u y_t^2)\tan\beta]} \tag{2.168}$$

to parameterize the amount of flavor-changing induced.

We now consider the rare decay $B^0 \to \mu^+\mu^-$. This occurs via emission off the quark current of a single virtual Higgs boson which then decays leptonically. The amplitude for the process $B_{(d,s)}^0 \to \mu^+\mu^-$ is given by:

$$\mathcal{A} = \eta_{QCD} \frac{\bar{y}_b y_\mu V_{t(d,s)} V_{tb}^*}{2\sin\beta} \chi_{FC} \left\langle 0|\bar{b}_R d_L|B_{(d,s)}^0 \right\rangle \left[\bar{\mu} \left(a_1 + a_2 \gamma^5 \right) \mu \right] \tag{2.169}$$

where

$$a_1 = \frac{\sin(\beta - \alpha)\cos\alpha}{m_H^2} - \frac{\cos(\beta - \alpha)\sin\alpha}{m_h^2},$$

$$a_2 = -\frac{\sin\beta}{m_A^2}. \tag{2.170}$$

The partial width is then

$$\Gamma(B_{(d,s)}^0 \to \mu^+\mu^-) = \frac{\eta_{QCD}^2}{128\pi} m_B^3 f_B^2 \, \bar{y}_b^2 y_\mu^2 \, |V_{t(d,s)}^* V_{tb}|^2 \, \chi_{FC}^2 (a_1^2 + a_2^2). \tag{2.171}$$

For SUSY scalar masses of order 500 GeV, we can estimate the branching ratio to be near current experimental limit for $\tan\beta$ larger than about 30. The reason for this enhancement has to do with the dependence of this rate on $\tan\beta$. For large values of $\tan\beta$, the rate scales as $(\tan\beta)^6$. Two powers of $\tan\beta$ arise each from $\bar{y}_b^2$ and y_μ^2, while the remaining two powers arise from χ_{FC}^2.

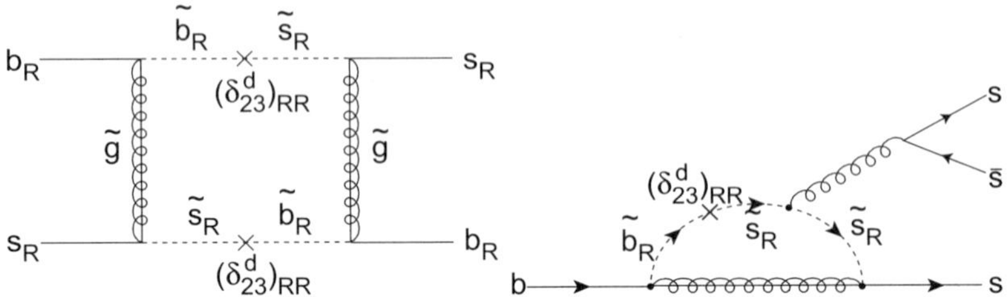

Fig. 2.17. New physics contributions to $B_s - \overline{B}_s$ mixing and $b \to s\bar{s}s$ in SUSY GUTs.

New physics contributions in B meson system can arise in SUSY GUTs.[79,80] Generically, these models predict large $\tilde{b}_R - \tilde{s}_R$ mixing, especially when large neutrino mixing angles are induced. As a result, there is a SUSY box diagram that contributes to $B_s - \overline{B}_s$ mixing, shown in Fig. 2.17. This contribution can be at the level of 30% of SM box diagram. Now, in the SM, CP violation arising from mixing in B_s is very small, but the new diagrams can significantly alter this scenario. There is also new contribution to direct B decays, which can also be comparable to the SM contribution. These ideas will therefore be tested in the near future at the LHC.

2.9. Conclusion and Outlook

Flavor physics is quite rich, in these lecture notes I have only scratched the surface of a subset of the various issues.

It is a great triumph for experiment and theory, that we know so much about the fundamental parameters of the flavor sector. Even a few years ago, it looked unlikely that so much would be learned with such high precision. On the experimental side, the two B factories, BABAR and BELLE, have contributed tremendously to the improved understanding. We have seen substantial progress on the theoretical understanding, especially from lattice gauge theory and Heavy Quark Effective Theory in the last decade. Both have played crucial roles in the precise determination of the fundamental parameters of the quark flavor sector, viz., quark masses, CKM mixing angles and CP violation. While we have learned a great deal about the fundamental parameters of the neutrino sector, in these lectures we focused primarily on the quark sector.

Knowing the fundamental parameters precisely is only the start. It is imperative that we seek explanations to these observations. Any such attempt will take us beyond the realms of the standard model. There is great hope that the LHC will actually test some of the new ideas introduced to explain some the puzzles in the flavor sector. It should, however, be cautioned that flavor dynamics could very well happen near the Planck scale, which would mask its direct effects. If there is low energy supersymmetry, there is a good chance that flavor physics, even if it occurs at a very highs scale, transmits information to the SUSY breaking sector, which may be observed. The prime candidate for these effects are rare decays of the type $\ell_i \to \ell_j \gamma$. Observing such decays will show the existence of new flavor physic, but it would be impossible, from these processes alone, to distinguish between various possibilities. We have seen that neutrino mass physics, GUT physics, and flavor physics related to anomalous $U(1)$, all lead to the prediction that $\mu \to e\gamma$ is in the observable range.

I have discussed at some length some, but not all, of the popular ideas that address the puzzles from the flavor sector. The mixing–mass sum rules in the quark sector appeared quite promising, but with more precise data, many of the models in this class have already been excluded. It has become increasingly difficult to find precise patterns in the masses and mixings that fit observations. Perhaps the best setting to address these issues is supersymmetric grand unification, supplemented by flavor symmetries. SUSY GUTs are well-motivated on independent grounds, they have the power to shed light on the flavor puzzle. Some recent ideas along this line are discussed in Sec. 2.5. I have also emphasized the close connection between the strong CP problem and the flavor puzzle. Axion solution to this problem is the most popular, but using P or CP symmetries seem to work equally well. These ideas may have collider signals, such as the discovery of right–handed $W_R^\pm$ gauge bosons.

With some luck, the path chosen by Nature may be revealed at the LHC in the coming years. Let us wait with hope.

Acknowledgments

I wish to thank Tao Han for inviting me to lecture at TASI and for his encouragement to write up these lecture notes. I also wish to acknowledge many enjoyable discussions with the participants at TASI 2008. It is a pleasure to thank K.T. Mahanthappa and the University of Colorado physics department for its warm hospitality. I have benefitted from discussions with Zurab Tavartkiladze. This work is supported in part by DOE Grant

Nos. DE-FG02-04ER41306 and DE-FG02-ER46140.

References

1. N. Cabibbo, Phys. Rev. Lett. **10**, 531 (1963).
2. M. Kobayashi and T. Maskawa, Prog. Theor. Phys. **49**, 652 (1973).
3. P. Minkowski, Phys. Lett. B **67**, 421 (1977);
   M. Gell-Mann, P. Ramond and R. Slansky, in *Supergravity* eds. P. van Nieuwenhuizen and D.Z. Freedman (North Holland, Amsterdam, 1979) p. 315;
   T. Yanagida, *In Proceedings of the Workshop on the Baryon Number of the Universe and Unified Theories, Tsukuba, Japan, 13-14 Feb 1979*;
   S. L. Glashow, NATO Adv. Study Inst. Ser. B Phys. **59**, 687 (1980);
   R. N. Mohapatra and G. Senjanovic, Phys. Rev. Lett. **44**, 912 (1980).
4. B. Pontecorvo, Sov. Phys. JETP **26**, 984 (1968) [Zh. Eksp. Teor. Fiz. **53**, 1717 (1967)];
   Z. Maki, M. Nakagawa and S. Sakata, Prog. Theor. Phys. **28**, 870 (1962).
5. C. Amsler *et al.* [Particle Data Group], Phys. Lett. B **667**, 1 (2008).
6. T. Schwetz, M. A. Tortola and J. W. F. Valle, New J. Phys. **10**, 113011 (2008).
7. P. F. Harrison, D. H. Perkins and W. G. Scott, Phys. Lett. B **530**, 167 (2002).
8. E. Ma and G. Rajasekaran, Phys. Rev. D **64**, 113012 (2001);
   K. S. Babu, E. Ma and J. W. F. Valle, Phys. Lett. B **552**, 207 (2003);
   G. Altarelli and F. Feruglio, Nucl. Phys. B **720**, 64 (2005);
   K. S. Babu and X. G. He, arXiv:hep-ph/0507217;
   W. Grimus and L. Lavoura, JHEP **0601**, 018 (2006);
   C. Luhn, S. Nasri and P. Ramond, Phys. Lett. B **652**, 27 (2007);
   I. de Medeiros Varzielas, S. F. King and G. G. Ross, Phys. Lett. B **644**, 153 (2007).
9. For a review see: J. Gasser and H. Leutwyler, Phys. Rept. **87**, 77 (1982).
10. A. V. Manohar and C. T. Sachrajda, "Quark Masses," in *Review of particle physics*, Phys. Lett. B **667**, 1 (2008).
11. For a review see: M. Neubert, Phys. Rept. **245**, 259 (1994).
12. C. Aubin *et al.* [MILC Collaboration], Phys. Rev. D **70**, 114501 (2004).
13. T. Ishikawa *et al.* [JLQCD Collaboration], Phys. Rev. D **78**, 011502 (2008).
14. C. Allton *et al.* [RBC-UKQCD Collaboration], Phys. Rev. D **78**, 114509 (2008).
15. Q. Mason, H. D. Trottier, R. Horgan, C. T. H. Davies and G. P. Lepage [HPQCD Collaboration], Phys. Rev. D **73**, 114501 (2006).
16. See for example the compilation in H. Arason, D. J. Castano, B. Keszthelyi, S. Mikaelian, E. J. Piard, P. Ramond and B. D. Wright, Phys. Rev. D **46**, 3945 (1992).
17. Z. z. Xing, H. Zhang and S. Zhou, Phys. Rev. D **77**, 113016 (2008).
18. L. Wolfenstein, Phys. Rev. Lett. **51**, 1945 (1983).

19. A. Ceccucci, Z. Ligeti and Y. Sakai, "The CKM quark-mixing matrix," in *Review of particle physics*, Phys. Lett. B **667**, 1 (2008).
20. For a review see: R. Fleischer, "Flavour Physics and CP Violation: Expecting the LHC," arXiv:0802.2882 [hep-ph].
21. N. Isgur and M. B. Wise, Phys. Lett. B **232**, 113 (1989); Phys. Lett. B **237**, 527 (1990).
22. M. E. Luke, Phys. Lett. B **252**, 447 (1990).
23. T. Inami and C. S. Lim, Prog. Theor. Phys. **65**, 297 (1981) [Erratum-ibid. **65**, 1772 (1981)].
24. M. Bona *et al.* [UTfit Collaboration], Nuovo Cim. **123B**, 666 (2008).
25. J. Charles *et al.* [CKMfitter Group], Eur. Phys. J. C **41**, 1 (2005). Updated fits from http://ckmfitter.in2p3.fr.
26. S. Weinberg, Trans. New York Acad. Sci. **38**, 185 (1977); F. Wilczek and A. Zee, Phys. Lett. B **70**, 418 (1977) [Erratum-ibid. **72B**, 504 (1978)]; H. Fritzsch, Phys. Lett. B **70**, 436 (1977).
27. J. C. Pati and A. Salam, Phys. Rev. D **8**, 1240 (1973); R. N. Mohapatra and J. C. Pati, Phys. Rev. D **11**, 566 (1975); G. Senjanovic and R. N. Mohapatra, Phys. Rev. D **12**, 1502 (1975).
28. H. Fritzsch, Phys. Lett. B **73**, 317 (1978).
29. See eg. K. S. Babu and Q. Shafi, Phys. Rev. D **47**, 5004 (1993).
30. L. J. Hall and A. Rasin, Phys. Lett. B **315**, 164 (1993).
31. K. S. Babu and J. Kubo, Phys. Rev. D **71**, 056006 (2005).
32. C. D. Froggatt and H. B. Nielsen, Nucl. Phys. B **147**, 277 (1979).
33. M.B. Green and J.H. Schwarz, Phys. Lett. **B149**, 117 (1984); Nucl. Phys. **B255**, 93 (1985); M.B. Green, J.H. Schwarz and P. West, Nucl. Phys. **B254**, 327 (1985).
34. K. S. Babu and T. Enkhbat, Nucl. Phys. B **708**, 511 (2005).
35. K. S. Babu and S. M. Barr, Phys. Lett. B **381**, 202 (1996); C. H. Albright, K. S. Babu and S. M. Barr, Phys. Rev. Lett. **81**, 1167 (1998); J. K. Elwood, N. Irges and P. Ramond, Phys. Rev. Lett. **81**, 5064 (1998); J. Sato and T. Yanagida, Phys. Lett. B **430**, 127 (1998).
36. L. E. Ibanez, G. G. Ross, Phys. Lett. **B332**, 100 (1994); P. Binetruy and P. Ramond, Phys. Lett. **B350**, 49 (1995); P. Binetruy, S. Lavignac and P. Ramond, Nucl. Phys. **B477**, 353 (1996).
37. T. Kobayashi, H. Nakano, H. Terao and K. Yoshioka, Prog. Theor. Phys. **110**, 247 (2003); K.S. Babu, I. Gogoladze and K. Wang, Nucl. Phys. **B660**, 322 (2003); H. K. Dreiner, H. Murayama and M. Thormeier, Nucl. Phys. B **729**, 278 (2005).
38. K.S. Babu, Ts. Enkhbat and I. Gogoladze, Nucl. Phys. **B678**, 233 (2004).
39. P. Ginsparg, Phys. Lett. **B197**, 139 (1987); V. S. Kaplunovsky, Nucl. Phys. **B307**, 145 (1988), Erratum-*ibid.* **B382**, 436 (1992).
40. M. Cvetic, L. L. Everett and J. Wang, Phys. Rev. **D59**, 107901 (1999).
41. M. Dine, N. Seiberg and E. Witten, Nucl. Phys. **B289**, 589 (1987);

J. Atick, L. Dixon and A. Sen, Nucl. Phys. **B292**, 109 (1987).

42. K. S. Babu and S. Nandi, Phys. Rev. D **62**, 033002 (2000).
43. G. F. Giudice and O. Lebedev, Phys. Lett. B **665**, 79 (2008).
44. J. C. Pati and A. Salam, Phys. Rev. D **10**, 275 (1974) [Erratum-ibid. D **11**, 703 (1975)].
45. H. Georgi and S. L. Glashow, Phys. Rev. Lett. **32**, 438 (1974).
46. H. Georgi, H. R. Quinn and S. Weinberg, Phys. Rev. Lett. **33**, 451 (1974).
47. K. S. Babu and C. F. Kolda, Phys. Lett. B **451**, 77 (1999).
48. L. J. Hall, R. Rattazzi and U. Sarid, Phys. Rev. D **50**, 7048 (1994).
49. H. Georgi and C. Jarlskog, Phys. Lett. B **86**, 297 (1979).
50. J. A. Harvey, D. B. Reiss and P. Ramond, Nucl. Phys. B **199**, 223 (1982).
51. S. Dimopoulos, L. J. Hall and S. Raby, Phys. Rev. Lett. **68**, 1984 (1992); Phys. Rev. D **45**, 4192 (1992); G. Anderson, S. Raby, S. Dimopoulos, L. J. Hall and G. D. Starkman, Phys. Rev. D **49**, 3660 (1994).
52. K. S. Babu and R. N. Mohapatra, Phys. Rev. Lett. **74**, 2418 (1995).
53. H. Georgi, in Particles and Fields, Ed. by C. Carlson (AIP, NY, 1975); H. Fritzsch and P. Minkowski, Annals Phys. **93**, 193 (1975).
54. K. S. Babu, J. C. Pati and F. Wilczek, Nucl. Phys. B **566**, 33 (2000).
55. C. H. Albright and S. M. Barr, Phys. Rev. D **58**, 013002 (1998); C. H. Albright, K. S. Babu and S. M. Barr, Phys. Rev. Lett. **81**, 1167 (1998); V. Lucas and S. Raby, Phys. Rev. D **55**, 6986 (1997); M. C. Chen and K. T. Mahanthappa, Int. J. Mod. Phys. A **18**, 5819 (2003).
56. K. S. Babu, J. C. Pati and P. Rastogi, Phys. Rev. D **71**, 015005 (2005).
57. K. S. Babu, J. C. Pati and P. Rastogi, Phys. Lett. B **621**, 160 (2005).
58. F. Borzumati and A. Masiero, Phys. Rev. Lett. **57**, 961 (1986); For a more recent analysis see: J. Hisano, T. Moroi, K. Tobe, M. Yamaguchi and T. Yanagida, Phys. Lett. B **357**, 579 (1995).
59. G. 't Hooft, Nucl. Phys. B **35**, 167 (1971).
60. S. Weinberg, Phys. Rev. Lett. **29**, 388 (1972); H. Georgi and S. L. Glashow, Phys. Rev. D **7**, 2457 (1973); S. M. Barr and A. Zee, Phys. Rev. D **15**, 2652 (1977); L. E. Ibanez, Phys. Lett. B **117**, 403 (1982).
61. B. S. Balakrishna, A. L. Kagan and R. N. Mohapatra, Phys. Lett. B **205**, 345 (1988); B. S. Balakrishna, Phys. Rev. Lett. **60**, 1602 (1988); K. S. Babu and E. Ma, Mod. Phys. Lett. A **4**, 1975 (1989); H. P. Nilles, M. Olechowski and S. Pokorski, Phys. Lett. B **248**, 378 (1990); R. Rattazzi, Z. Phys. C **52**, 575 (1991).
62. K. S. Babu and R. N. Mohapatra, Phys. Rev. Lett. **64**, 2747 (1990).
63. X. G. He, R. R. Volkas and D. D. Wu, Phys. Rev. D **41**, 1630 (1990).
64. B. A. Dobrescu and P. J. Fox, JHEP **0808**, 100 (2008).
65. S. M. Barr, Phys. Rev. D **76**, 105024 (2007); S. M. Barr and A. Khan, Phys. Rev. D **79**, 115005 (2009).
66. S. Pakvasa and H. Sugawara, Phys. Lett. B **73**, 61 (1978).
67. R. D. Peccei and H. R. Quinn, Phys. Rev. Lett. **38**, 1440 (1977); Phys. Rev.

D **16**, 1791 (1977).

68. S. Weinberg, Phys. Rev. Lett. **40**, 223 (1978);
F. Wilczek, Phys. Rev. Lett. **40**, 279 (1978).
69. S. Barr, in *CP violation and the limits of the standard model*, TASI 94 Proceedings, ed. J.F. Donoghue, World Scientific Publication (1995).
70. M. Dine, W. Fischler and M. Srednicki, Phys. Lett. B **104**, 199 (1981);
A. R. Zhitnitsky, Sov. J. Nucl. Phys. **31** (1980) 260 [Yad. Fiz. **31** (1980) 497].
71. J. E. Kim, Phys. Rev. Lett. **43**, 103 (1979);
M. A. Shifman, A. I. Vainshtein and V. I. Zakharov, Nucl. Phys. B **166**, 493 (1980).
72. J. E. Kim, Phys. Rept. **150**, 1 (1987);
P. Sikivie, Phys. Rev. Lett. **51**, 1415 (1983) [Erratum-ibid. **52**, 695 (1984)].
73. R. N. Mohapatra and G. Senjanovic, Phys. Lett. B **79**, 283 (1978);
R. N. Mohapatra and A. Rasin, Phys. Rev. Lett. **76**, 3490 (1996);
R. N. Mohapatra, A. Rasin and G. Senjanovic, Phys. Rev. Lett. **79**, 4744 (1997).
74. K. S. Babu, B. Dutta and R. N. Mohapatra, Phys. Rev. D **65**, 016005 (2002).
75. A. E. Nelson, Phys. Lett. B **136**, 387 (1984);
S. M. Barr, Phys. Rev. D **30**, 1805 (1984).
76. G. D'Ambrosio, G. F. Giudice, G. Isidori and A. Strumia, Nucl. Phys. B **645**, 155 (2002).
77. A. J. Buras, Phys. Lett. B **566**, 115 (2003);
C. Bobeth, M. Bona, A. J. Buras, T. Ewerth, M. Pierini, L. Silvestrini and A. Weiler, Nucl. Phys. B **726**, 252 (2005).
78. K. S. Babu and C. F. Kolda, Phys. Rev. Lett. **84**, 228 (2000).
79. D. Chang, A. Masiero and H. Murayama, Phys. Rev. D **67**, 075013 (2003).
80. M. Ciuchini, A. Masiero, P. Paradisi, L. Silvestrini, S. K. Vempati and O. Vives, Nucl. Phys. B **783**, 112 (2007).

Chapter 3

LHC Phenomenology for Physics Hunters

Tilman Plehn

SUPA, School of Physics and Astronomy, University of Edinburgh, Scotland

Welcome to the 2008 TASI lectures on the exciting topic of 'tools and technicalities' (original title). Technically, LHC physics is really all about perturbative QCD in signals or backgrounds. Whenever we look for interesting signatures at the LHC we get killed by QCD. Therefore, I will focus on QCD issues which arise for example in Higgs searches or exotics searches at the LHC, and ways to tackle them nowadays. In the last section you will find a few phenomenological discussions, for example on missing energy or helicity amplitudes.

3.1. LHC Phenomenology

When we think about signal or background processes at the LHC the first quantity we compute is the total number of events we would expect at the LHC in a given time interval. This number of events is the product of the hadronic (*i.e.* proton–proton) LHC luminosity measured in inverse femtobarns and the total production cross section measured in femtobarns. A typical year of LHC running could deliver around 10 inverse femtobarns per year in the first few years and three to ten times that later. People who build the actual collider do not use these kinds of units, but for phenomenologists they work better than something involving seconds and square meters, because what we typically need is a few interesting events corresponding to a few femtobarns of data. So here are a few key numbers and their orders of magnitude for typical signals:

$$N_{\text{events}} = \sigma_{\text{tot}} \cdot \mathcal{L} \qquad \mathcal{L} = 10 \cdots 300\,\text{fb}^{-1} \qquad \sigma_{\text{tot}} = 1 \cdots 10^4\,\text{fb} \quad (3.1)$$

Just in case my colleagues have not told you about it: there are two kinds of processes at the LHC. The first involves all particles which we know and love, like old-fashioned electrons or slightly more modern W and Z bosons or most recently top quarks. These processes we call *backgrounds* and find annoying. They are described by QCD, which means QCD is the theory of the evil. Top quarks have an interesting history, because when I was a graduate student they still belonged to the second class of processes, the *signals*. These typically involve particles we have not seen before. Such states are unfortunately mostly produced in QCD processes as well, so QCD is not entirely evil. If we see such signals, someone gets a call from Stockholm, shakes hands with the king of Sweden, and the corresponding processes instantly turn into backgrounds.

The main problem at any collider is that signals are much more rare that background, so we have to dig our signal events out of a much larger number of background events. This is what most of this lecture will be about. Just to give you a rough idea, have a look at Fig. 5.3: at the LHC the production cross section for two bottom quarks at the LHC is larger than 10^5 nb or 10^{11} fb and the typical production cross section for W or Z boson ranges around 200 nb or 2×10^8 fb. Looking at signals, the production cross sections for a pair of 500 GeV gluinos is 4×10^4 fb and the Higgs production cross section can be as big as 2×10^5 fb. When we want to extract such signals out of comparably huge backgrounds we need to describe these backgrounds with an incredible precision. Strictly speaking, this holds at least for those background events which populate the signal region in phase space. Such background event will always exist, so any LHC measurement will always be a statistics exercise. The high energy community has therefore agreed that we call a five sigma excess over the known backgrounds a signal:

$$\frac{S}{\sqrt{B}} = N_\sigma > 5 \qquad \text{(Gaussian limit)}$$

$$P_{\text{fluct}} < 5.8 \times 10^{-7} \qquad \text{(fluctuation probability)} \qquad (3.2)$$

Do not trust anybody who wants to sell you a three sigma evidence as a discovery, even I have seen a great number of those go away. People often have good personal reasons to advertize such effects, but all they are really saying is that their errors do not allow them to make a conclusive statement. This brings us to a well kept secret in the phenomenology community, which is the important impact of error bars when we search for exciting new physics. Since for theorists understanding LHC events and in particular background events means QCD, we need to understand where

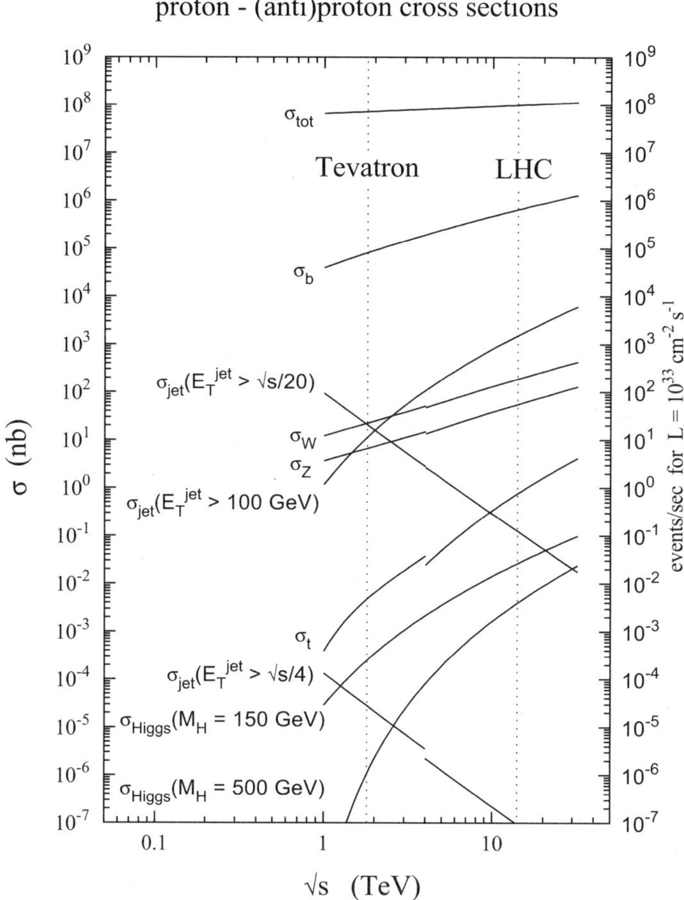

Fig. 3.1.   Production rates for different signal and background processes at hadron colliders. The discontinuity is due to the Tevatron being a proton–antiproton collider while the LHC is a proton–proton collider. The two colliders correspond to the x–axis values of 2 TeV and 14 TeV. Figure borrowed from CMS.

our predictions come from and what they assume, so here we go...

3.2. QCD and scales

Not all processes which involve QCD have to look incredibly complicated — let us start with a simple question: we know how to compute the production rate and distributions for Z production for example at LEP $e^+e^- \to Z$. To make all phase space integrals simple, we assume that the Z boson is on-shell, so we can simply add a decay matrix element and a decay phase space integration for example compute the process $e^+ e^- \to Z \to \mu^+ \mu^-$.

So here is the question: how do we compute the production of a Z boson at the LHC? This process is usually referred to as *Drell–Yan production*, even though we will most likely produce neither Drell nor Yan at the LHC. In our first attempts we explicitly do not care about additional jets, so if we assume the proton consists of quarks and gluons we simply compute the process $q\bar{q} \to Z$ under the assumption that the quarks are partons inside protons. Modulo the $SU(2)$ and $U(1)$ charges which describe the $Zf\bar{f}$ coupling

$$-i\gamma^\mu\left(\ell P_L + r P_R\right) \qquad \ell = \frac{e}{s_w c_w}\left(T_3 - Q s_w^2\right) \qquad r = \ell\Big|_{T_3=0} \qquad (3.3)$$

the matrix element and the squared matrix element for the partonic process $q\bar{q} \to Z$ will be the same as the corresponding matrix element squared for $e^+ e^- \to Z$, with an additional color factor. This color factor counts the number of $SU(3)$ states which can be combined to form a color singlet like the Z. This additional factor should come out of the color trace which is part of the Feynman rules, and it is N_c. On the other hand, we do not observe color in the initial state, and the color structure of the incoming $q\bar{q}$ pair has no impact on the Z–production matrix element, so we average over the color. This gives us another factor $1/N_c^2$ in the averaged matrix element (modulo factors two)

$$\overline{|\mathcal{M}|^2}(q\bar{q} \to Z) \sim \frac{1}{4N_c}\, m_Z^2\left(\ell^2 + r^2\right)\,. \qquad (3.4)$$

Notice that matrix elements we compute from our Feynman rules are not automatically numbers without a mass unit. Next, we add the phase space for a one-particle final state. In four space–time dimensions (this will become important later) we can compute a total cross section out of a matrix element squared as

$$s\,\frac{d\sigma}{dy} = \frac{\pi}{(4\pi)^2}\left(1 - \tau\right)\overline{|\mathcal{M}|^2}$$

The mass of the final state appears as $\tau = m_Z^2/s$ and can of course be m_W or the Higgs mass or the mass of a KK graviton (I know you smart-asses in the back row!). If we define s as the partonic invariant mass of the two quarks using the Mandelstam variable $s = (k_2 + k_2)^2 = 2(k_1 k_2)$, momentum conservation just means $s = m_Z^2$. This simple one-particle phase space has only one free parameter, the reduced polar angle $y = (1 + \cos\theta)/2 = 0\cdots 1$.

The azimuthal angle ϕ plays no role at colliders, unless you want to compute gravitational effects on Higgs production at Atlas and CMS. Any LHC Monte Carlo will either random-generate a reference angle ϕ for the partonic process or pick one and keep it fixed. The second option has at least once lead to considerable confusion and later amusement at the Tevatron, when people noticed that the behavior of gauge bosons was dominated by gravity, namely gauge bosons going up or down. So this is not as trivial a statement as you might think. At this point I remember that every teacher at every summer schools always feels the need to define their field of phenomenology — for example: phenomenologists are theorists who do useful things and know funny stories about experiment(alist)s.

Until now we have computed the same thing as Z production at LEP, leaving open the question how to describe quarks inside the proton. For a proper discussion I refer to any good QCD textbook and in particular the chapter on deep inelastic scattering. Instead, I will follow a pedagogical approach which will as fast as possible take us to the questions we really want to discuss.

If for now we are happy assuming that quarks move collinear with the surrounding proton, *i.e.* that at the LHC incoming partons have zero p_T, we can simply write a probability distribution for finding a parton with a certain fraction of the proton's momentum. For a momentum fraction $x = 0 \cdots 1$ this *parton density function* (pdf) is denoted as $f_i(x)$, where i describes the different partons in the proton, for our purposes u, d, c, s, g. All of these partons we assume to be massless. We can talk about heavy bottoms in the proton if you ask me about it later. Note that in contrast to structure functions a pdf is not an observable, it is simply a distribution in the mathematical sense, which means it has to produce reasonably results when integrated over as an integration kernel. These parton densities have very different behavior — for the valence quarks (uud) they peak somewhere around $x \lesssim 1/3$, while the gluon pdf is small at $x \sim 1$ and grows very rapidly towards small x. For some typical part of the relevant parameter space ($x = 10^{-3} \cdots 10^{-1}$) you can roughly think of it as $f_g(x) \propto x^{-2}$, towards x values it becomes even steeper. This steep gluon distribution was initially not expected and means that for small enough x LHC processes will dominantly be gluon fusion processes.

Given the correct definition and normalization of the pdf we can compute the *hadronic cross section* from its partonic counterpart as

$$\sigma_{\text{tot}} = \int_0^1 dx_1 \int_0^1 dx_2 \; f_i(x_1)\, f_j(x_2)\, \hat{\sigma}_{ij}(x_1 x_2 S) \tag{3.5}$$

where i, j are the incoming partons with the momentum factions $x_{i,j}$. The partonic energy of the scattering process is $s = x_1 x_2 S$ with the LHC proton energy $\sqrt{S} = 14$ TeV. The partonic cross section $\hat{\sigma}$ corresponds to the cross sections σ we already discussed. It has to include all the necessary Θ and δ functions for energy–momentum conservation. When we express a general n–particle cross section $\hat{\sigma}$ including the phase space integration, the x_i integrations and the phase space integrations can of course be swapped, but Jacobians will make your life hell when you attempt to get them right. Luckily, there are very efficient numerical phase space generators on the market which transform a hadronic n–particle phase space integration into a unit hypercube, so we do not have to worry in our every day life.

3.2.1. *UV divergences and the renormalization scale*

Renormalization, *i.e.* the proper treatment of ultraviolet divergences, is one of the most important aspects of field theories; if you are not comfortable with it you might want to attend a lecture on field theory. The one aspect of renormalization I would like to discuss is the appearance of the renormalization scale. In perturbation theory, scales arise from the regularization of infrared or ultraviolet divergences, as we can see writing down a simple loop integral corresponding to two virtual massive scalars with a momentum p flowing through the diagram:

$$B(p^2; m, m) \equiv \int \frac{d^4 q}{16\pi^2} \frac{1}{q^2 - m^2} \frac{1}{(q + p)^2 - m^2} \tag{3.6}$$

Such diagrams appear for example in the gluon self energy, with massless scalars for ghosts, with some Dirac trace in the numerator for quarks, and with massive scalars for supersymmetric scalar quarks. This integral is UV divergent, so we have to regularize it, express the divergence in some well-defined manner, and get rid of it by renormalization. One way is to introduce a cutoff into the momentum integral Λ, for example through the so-called Pauli–Villars regularization. Because the UV behavior of the integrand cannot depend on IR-relevant parameters, the UV divergence cannot involve the mass m or the external momentum p^2. This means that its divergence has to be proportional to $\log \Lambda/\mu^2$ with some scale μ^2 which is an artifact of the regularization of such a Feynman diagram.

This question is easier to answer in the more modern *dimensional regularization*. There, we shift the power of the momentum integration and use analytic continuation in the number of space–time dimensions to renormalize the theory

$$\int \frac{d^4 q}{16\pi^2} \cdots \to \mu^{2\epsilon} \int \frac{d^{4-2\epsilon} q}{16\pi^2} \cdots = \frac{i\mu^{2\epsilon}}{(4\pi)^2} \left[\frac{C_{-1}}{\epsilon} + C_0 + C_1\,\epsilon + \mathcal{O}(\epsilon^2) \right]$$

$$(3.7)$$

The constants C_i depend on the loop integral we are considering. The scale μ we have to introduce to ensure the matrix element and the observables, like cross sections, have the usual mass dimensions. To regularize the UV divergence we pick an $\epsilon > 0$, giving us mathematically well-defined poles $1/\epsilon$. If you compute the scalar loop integrals you will see that defining them with the integration measure $1/(i\pi^2)$ will make them come out as of the order $\mathcal{O}(1)$, in case you ever wondered about factors $1/(4\pi)^2 = \pi^2/(2\pi)^4$ which usually end up in front of the loop integrals.

The poles in ϵ will cancel with the counter terms, *i.e.* we renormalize the theory. Counter terms we include by shifting the renormalized parameter in the leading-order matrix element, *e.g.* $\overline{|\mathcal{M}|^2}(g) \to \overline{|\mathcal{M}|^2}(g + \delta g)$ with a coupling $\delta g \propto 1/\epsilon$, when computing $|\mathcal{M}_{\text{Born}} + \mathcal{M}_{\text{virt}}|^2$. If we use a physical renormalization condition there will not be any free scale μ in the definition of δg. As an example for a physical reference we can think of the electromagnetic coupling or charge e, which is usually defined in the Thomson limit of vanishing momentum flow through the diagram, *i.e.* $p^2 \to 0$. What is important about these counter terms is that they do not come with a factor $\mu^{2\epsilon}$ in front.

So while after renormalization the poles $1/\epsilon$ cancel just fine, the scale factor $\mu^{2\epsilon}$ will not be matched between the UV divergence and the counter term. We can keep track of it by writing a Taylor series in ϵ for the prefactor of the regularized but not yet renormalized integral:

$$\mu^{2\epsilon} \left[\frac{C_{-1}}{\epsilon} + C_0 + \mathcal{O}(\epsilon) \right] = e^{2\epsilon \log \mu} \left[\frac{C_{-1}}{\epsilon} + C_0 + \mathcal{O}(\epsilon) \right]$$

$$= \left[1 + 2\epsilon \log \mu + \mathcal{O}(\epsilon^2) \right] \left[\frac{C_{-1}}{\epsilon} + C_0 + \mathcal{O}(\epsilon) \right]$$

$$= \frac{C_{-1}}{\epsilon} + C_0 + 2 \log \mu\, C_{-1} + \mathcal{O}(\epsilon) \qquad (3.8)$$

We see that the pole C_{-1}/ϵ gives a finite contribution to the cross section, involving the *renormalization scale* $\mu_R \equiv \mu$.

Just a side remark for completeness: from Eq. (3.8) we see that we should not have just pulled out $\mu^{2\epsilon}$ out of the integral, because it leads to a logarithm of a number with a mass unit. On the other hand, from the way we split the original integral we know that the remaining $(4-2\epsilon)$-dimensional integral has to includes logarithms of the kind $\log m^2$ or $\log p^2$ which re-combine with the $\log \mu^2$ for example to a properly defined $\log \mu/m$. The only loop integral which has no intrinsic mass scale is the two-point function with zero mass in the loop and zero momentum flowing through the integral: $B(p^2 = 0; 0, 0)$. It appears for example as a self-energy correction of external quarks and gluons. Based on these dimensional arguments this integral has to be zero, but with a subtle cancellation of the UV and the IR divergences which we can schematically write as $1/\epsilon_{\rm IR} - 1/\epsilon_{\rm UV}$. Actually, I am thinking right now if following this argument this integral has to be zero or if it can still be a number, like $2376123/67523$, but it definitely has to be finite... And it is zero if you compute it.

Instead of discussing different renormalization schemes and their scale dependences, let us instead compute a simple renormalization scale dependent parameter, namely the *running strong coupling* $\alpha_s(\mu_R)$. It does not appear in our Drell–Yan process at leading order, but it does not hurt to know how it appears in QCD calculations. The simplest process we can look at is two-jet production at the LHC, where we remember that in some energy range we will be gluon dominated: $gg \to q\bar{q}$. The Feynman diagrams include an s–channel off-shell gluon with a momentum flow $p^2 \equiv s$. At next-to-leading order, this gluon propagator will be corrected by self-energy loops, where the gluon splits into two quarks or gluons and re-combines before it produces the two final-state partons.

The gluon self energy correction (or vacuum polarization, as propagator corrections to gauge bosons are often labelled) will be a scalar, *i.e.* fermion loops will be closed and the Dirac trace is closed inside the loop. In color space the self energy will (hopefully) be diagonal, just like the gluon propagator itself, so we can ignore the color indices for now. In Minkowski space the gluon propagator in unitary gauge is proportional to the transverse tensor $T^{\mu\nu} = g^{\mu\nu} - p^\nu p^\mu / p^2$. The same is true for the gluon self energy, which we write as $\Pi^{\mu\nu} \equiv \Pi\, T^{\mu\nu}$. The one useful thing to remember is the simple relation $T^{\mu\nu} T_\nu^\rho = T^{\mu\rho}$ and $T^{\mu\nu} g_\nu^\rho = T^{\mu\rho}$. Including the gluon, quark, and ghost loops the regularized gluon self energy with a momentum flow p^2

reads

$$\frac{1}{p^2} \Pi\left(\frac{\mu_R^2}{p^2}\right) = \frac{\alpha_s}{4\pi}\left(-\frac{1}{\epsilon} - \log\frac{\mu_R^2}{p^2}\right)\left(\frac{13}{6}N_c - \frac{2}{3}n_f\right) + \mathcal{O}(\log m_t^2)$$

$$\longrightarrow \frac{\alpha_s}{4\pi}\left(-\frac{1}{\epsilon} - \log\frac{\mu_R^2}{p^2}\right)\beta_g + \mathcal{O}(\log m_t^2)$$

$$\text{with} \quad \beta_g = \frac{11}{3}N_c - \frac{2}{3}n_f\,. \tag{3.9}$$

In the second step we have sneaked in additional contributions to the renormalization of the strong coupling from the other one-loop diagrams in the process. The number of fermions coupling to the gluons is n_f. We neglect the additional terms $\log(4\pi)$ and $\log\gamma_E$ which come with the poles in dimensional regularization. From the comments on the function $B(p^2; 0, 0)$ before we could have guessed that the loop integrals will only give a logarithm $\log p^2$ which then combines with the scale logarithm $\log\mu_R^2$. The finite top mass actually leads to an additional logarithms which we omit for now — this zero-mass limit of our field theory is actually special and referred to as its conformal limit.

Lacking a well-enough motivated reference point (in the Thomson limit the strong coupling is divergent, which means QCD is confined towards large distances and asymptotically free at small distances) we are tempted to renormalize α_s by also absorbing the scale into the counter term, which is called the $\overline{\text{MS}}$ scheme. It gives us a running coupling $\alpha_s(p)$. In other words, for a given momentum transfer p^2 we cancel the UV pole and at the same time shift the strong coupling, after including all relative $(-)$ signs, by

$$\alpha_s \longrightarrow \alpha_s(\mu_R^2)\left(1 - \frac{1}{p^2}\Pi\left(\frac{\mu_R^2}{p^2}\right)\right) = \alpha_s(\mu_R^2)\left(1 - \frac{\alpha_s}{4\pi}\beta_g\log\frac{p^2}{\mu_R^2}\right). \tag{3.10}$$

We can do even better: the problem with the correction to α_s is that while it is perturbatively suppressed by the usual factor $\alpha_s/(4\pi)$ it includes a logarithm which does not need to be small. Instead of simply including these gluon self-energy corrections at a given order in perturbation theory we can instead include all chains with Π appearing many times in the off-shell gluon propagator. Such a series means we replace the off-shell gluon propagator by (schematically written)

$$\frac{T^{\mu\nu}}{p^2} \longrightarrow \frac{T^{\mu\nu}}{p^2} + \left(\frac{T}{p^2} \cdot (-T\,\Pi) \cdot \frac{T}{p^2} \right)^{\mu\nu}$$

$$+ \left(\frac{T}{p^2} \cdot (-T\,\Pi) \cdot \frac{T}{p^2} \cdot (-T\,\Pi) \cdot \frac{T}{p^2} \right)^{\mu\nu} + \cdots$$

$$= \frac{T^{\mu\nu}}{p^2} \sum_{j=0}^{\infty} \left(-\frac{\Pi}{p^2} \right)^j = \frac{T^{\mu\nu}}{p^2} \frac{1}{1 + \Pi/p^2} \tag{3.11}$$

To avoid indices we abbreviate $T^{\mu\nu} T^{\rho}_{\nu} = T \cdot T$ which can be simplified using $(T \cdot T \cdot T)^{\mu\nu} = T^{\mu\rho} T^{\sigma}_{\rho} T^{\nu}_{\sigma} = T^{\mu\nu}$. This re-summation of the logarithm which occurs in the next-to-leading order corrections to α_s moves the finite shift in α_s shown in Eq. (3.10) into the denominator:

$$\alpha_s \longrightarrow \alpha_s(\mu_R^2) \left(1 + \frac{\alpha_s}{4\pi} \beta_g \, \log \frac{p^2}{\mu_R^2} \right)^{-1} \tag{3.12}$$

If we interpret the renormalization scale μ_R as one reference point p_0 and p as another, we can relate the values of α_s between two reference points with a *renormalization group equation* (RGE) which evolves physical parameters from one scale to another:

$$\alpha_s(p^2) = \alpha_s(p_0^2) \left(1 + \frac{\alpha_s(p_0^2)}{4\pi} \beta_g \, \log \frac{p^2}{p_0^2} \right)^{-1}$$

$$\frac{1}{\alpha_s(p^2)} = \frac{1}{\alpha_s(p_0^2)} \left(1 + \frac{\alpha_s(p_0^2)}{4\pi} \beta_g \, \log \frac{p^2}{p_0^2} \right) = \frac{1}{\alpha_s(p_0^2)} + \frac{1}{4\pi} \beta_g \, \log \frac{p^2}{p_0^2} \tag{3.13}$$

The factor α_s inside the parentheses can be evaluated at any of the two scales, the difference is going to be a higher-order effect. The interpretation of β_g is now obvious: when we differentiate the shifted $\alpha_s(p^2)$ with respect to the momentum transfer p^2 we find:

$$\frac{1}{\alpha_s} \frac{d\alpha_s}{d\log p^2} = -\frac{\alpha_s}{4\pi} \beta_g \qquad \text{or} \qquad \frac{1}{g_s} \frac{dg_s}{d\log p} = -\frac{\alpha_s}{4\pi} \beta_g = -g_s^2 \beta_g \tag{3.14}$$

This is the famous running of the strong coupling constant!

Before we move on, let us collect the logic of the argument given in this section: when we regularize an UV divergence we automatically introduce a reference scale. Naively, this could be a UV cutoff scale, but even the seemingly scale invariant dimensional regularization cannot avoid the introduction of a scale, even in the conformal limit of our theory. There are several ways of dealing with such a scale: first, we can renormalize our parameter at a reference point. Secondly, we can define a running parameter, *i.e.* absorb the scale logarithm into the $\overline{\text{MS}}$ counter term. This way, at each order in perturbation theory we can translate values for example of the strong coupling from one momentum scale to another momentum scale. If we are lucky, we can re-sum these logarithms to all orders in perturbation theory, which gives us more precise perturbative predictions even in the presence of large logarithms, *i.e.* large scale differences for our renormalized parameters. Such a (re–) summation is linked with the definition of scale dependent parameters.

3.2.2. *IR divergences and the factorization scale*

After this brief excursion into renormalization and UV divergences we can return to the original example, the Drell–Yan process at the LHC. In our last attempt we wrote down the hadronic cross sections in terms of parton distributions at leading order. These pdfs are only functions of the (collinear) momentum fraction of the partons in the proton.

The perturbative question we need to ask for this process is: what happens if we radiate additional jets which for one reason or another we do not observe in the detector. Throughout this writeup I will use the terms *jets and final state partons* synonymously, which is not really correct once we include jet algorithms and hadronization. On the other hand, in most cases a jet algorithms is designed to take us from some kind of energy deposition in the calorimeter to the parton radiated in the hard process. This is particularly true for modern developments like the so-called matrix element method to measure the top mass. Recently, people have looked into the question what kind of jets come from very fast collimated W or top decays and how such fat jets could be identified looking into the details of the jet algorithm. But let us face it, you can try to do such analyses after you really understand the QCD of hard processes, and you should not trust such analyses unless they come from groups which know a whole lot of QCD and preferable involve experimentalists who know their calorimeters very well.

So let us get back to the radiation of additional partons in the Drell–Yan process. These can for example be gluons radiated from the incoming quarks. This means we can start by compute the cross section for the partonic process $q\bar{q} \to Zg$. However, this partonic process involves renormalization as well as an avalanche of loop diagrams which have to be included before we can say anything reasonable, *i.e.* UV and IR finite. Instead, we can look at the crossed process $qg \to Zq$, which should behave similarly as a $2 \to 2$ process, except that it has a different incoming state than the leading-order Drell–Yan process and hence no virtual corrections. This means we do not have to deal with renormalization and UV divergences and can concentrate on parton or jet radiation from the initial state.

The amplitude for this $2 \to 2$ process is — modulo the charges and averaging factors, but including all Mandelstam variables

$$\overline{|\mathcal{M}|^2} \propto 8 \left[-\frac{t}{s} - \frac{s}{t} + \frac{2m_Z^2(s + t - m_Z^2)}{st} \right] \tag{3.15}$$

The new Mandelstam variables can be expressed in terms of the rescaled gluon-emission angle $y = (1 + \cos\theta)/2$ as $t = -s(1 - \tau)y$ and $u = -s(1 - \tau)(1 - y)$. As a sanity check we can confirm that $t + u = -s + m_Z^2$. The collinear limit when the gluon is radiated in the beam direction is given by $y \to 0$, which corresponds to $t \to 0$ with finite $u = -s + m_Z^2$. In that case the matrix element becomes

$$\overline{|\mathcal{M}|^2} \sim 8 \left[\frac{s^2 - 2sm_Z^2 + 2m_Z^4}{s(s - m_Z^2)} \frac{1}{y} - \frac{2m_Z^2}{s} + \mathcal{O}(y) \right] \tag{3.16}$$

This expression is divergent for collinear gluon radiation, *i.e.* for small angles y. We can translate this $1/y$ divergence for example into the transverse momentum of the gluon or Z according to

$$sp_T^2 = tu = s^2(1 - \tau)^2 \, y(1 - y) = (s - m_Z^2)^2 y + \mathcal{O}(y^2) \tag{3.17}$$

In the collinear limit our matrix element squared then becomes

$$\overline{|\mathcal{M}|^2} \sim 8 \left[\frac{s^2 - 2sm_Z^2 + 2m_Z^4}{s^2} \frac{(s - m_Z^2)}{p_T^2} + \mathcal{O}(p_T^0) \right] . \tag{3.18}$$

The matrix element for the tree-level process $qg \to Zq$ diverges like $1/p_T^2$. To compute the total cross section for this process we need to integrate it over the two-particle phase space. Without deriving this result we quote that this integration can be written in the transverse momentum of the outgoing particles, in which case the Jacobian for this integration introduces a factor p_T. Approximating the matrix element as C/p_T^2, we have to integrate

$$\int_{y^{\min}}^{y^{\max}} dy \frac{C}{y} = \int_{p_T^{\min}}^{p_T^{\max}} dp_T^2 \frac{C}{p_T^2} = 2 \int_{p_T^{\min}}^{p_T^{\max}} dp_T \, p_T \, \frac{C}{p_T^2}$$

$$\simeq 2C \int_{p_T^{\min}}^{p_T^{\max}} dp_T \frac{1}{p_T} = 2C \, \log \frac{p_T^{\max}}{p_T^{\min}} \quad (3.19)$$

The form C/p_T^2 for the matrix element is of course only valid in the collinear limit; in the remaining phase space C is not a constant. However, this formula describes well the collinear IR divergence arising from gluon radiation at the LHC (or photon radiation at e^+e^- colliders, for that matter).

We can follow the same strategy as for the UV divergence. First, we regularize the divergence using dimensional regularization, and then we find a well-defined way to get rid of it. Dimensional regularization now means we have to write the two-particle phase space in $n = 4 - 2\epsilon$ dimensions. Just for the fun, here is the complete formula in terms of y:

$$s \frac{d\sigma}{dy} = \frac{\pi(4\pi)^{-2+\epsilon}}{\Gamma(1-\epsilon)} \left(\frac{\mu^2}{m_Z^2}\right)^\epsilon \frac{\tau^\epsilon(1-\tau)^{1-2\epsilon}}{y^\epsilon(1-y)^\epsilon}\overline{|\mathcal{M}|^2} \sim \left(\frac{\mu^2}{m_Z^2}\right)^\epsilon \frac{\overline{|\mathcal{M}|^2}}{y^\epsilon(1-y)^\epsilon}.$$
$$(3.20)$$

In the second step we only keep the factors we are interested in. The additional factor $y^{-\epsilon}$ regularizes the integral at $y \to 0$, as long as $\epsilon < 0$, which just slightly increases the suppression of the integrand in the IR regime. After integrating the leading term $1/y^{1+\epsilon}$ we have a pole $1/(-\epsilon)$. Obviously, this regularization procedure is symmetric in $y \leftrightarrow (1-y)$. What is important to notice is again the appearance of a scale $\mu^{2\epsilon}$ with the n-dimensional integral. This scale arises from the IR regularization of the phase space integral and is referred to as *factorization scale* μ_F.

From our argument we can safely guess that the same divergence which we encounter for the process $qg \to Zq$ will also appear in the crossed process $q\bar{q} \to Zg$, after cancelling additional soft IR divergences between virtual and real gluon emission diagrams. We can write all these collinear divergences in a universal form, which is independent of the hard process (like Drell–Yan production). In the collinear limit, the probabilities of radiating additional partons or splitting into additional partons is given by universal *splitting functions*, which govern the collinear behavior of the parton-radiation cross section:

$$\frac{1}{\sigma_{\text{tot}}} \, d\sigma \sim \frac{\alpha_s}{2\pi} \frac{dy}{y} \, dx \, P_j(x) = \frac{\alpha_s}{2\pi} \frac{dp_T^2}{p_T^2} \, dx \, P_j(x) \quad (3.21)$$

The momentum fraction which the incoming parton transfers to the parton entering the hard process is given by x. The rescaled angle y is one way to integrate over the transverse-momentum space. The splitting kernels are different for different partons involved:

$$
\begin{aligned}
P_{q\leftarrow q}(x) &= C_F \, \frac{1+x^2}{1-x} \qquad\qquad P_{g\leftarrow q}(x) = C_F \, \frac{1+(1-x)^2}{x} \\
P_{q\leftarrow g}(x) &= T_R \left(x^2 + (1-x)^2\right) \\
P_{g\leftarrow g}(x) &= C_A \left(\frac{x}{1-x} + \frac{1-x}{x} + x(1-x)\right)
\end{aligned}
\tag{3.22}
$$

The underlying QCD vertices in these four collinear splittings are the qqg and ggg vertices. This means that a gluon can split independently into a pair of quarks and a pair of gluons. A quark can only radiate a gluon, which implies $P_{q\leftarrow q}(1-x) = P_{g\leftarrow q}(x)$, depending on which of the two final state partons we are interested in. For these formulas we have sneaked in the Casimir factors of $SU(N)$, which allow us to generalize our approach beyond QCD. For practical purposes we can insert the SU(3) values $C_F = (N_c^2 - 1)/(2N_c) = 4/3$, $C_A = N_c = 3$ and $T_R = 1/2$. Once more looking at the different splitting kernels we see that in the soft-daughter limit $x \to 0$ the daughter quarks $P_{q\leftarrow q}$ and $P_{q\leftarrow g}$ are well-defined, while the gluon daughters $P_{g\leftarrow q}$ and $P_{g\leftarrow g}$ are infrared divergent.

What we need for our partonic subprocess $qg \to Zq$ is the splitting of a gluon into two quarks, one of which then enters the hard Drell–Yan process. In the *collinear limit* this splitting is described by $P_{q\leftarrow g}$. We explicitly see that there is no additional soft singularity for vanishing quark energy, only the collinear singularity in y or p_T. This is good news, since in the absence of virtual corrections we would have no idea how to get rid of or cancel this soft divergence.

If we for example consider repeated collinear gluon emission off an incoming quark leg, we naively get a correction suppressed by powers of α_s, because of the strong coupling of the gluon. Such a chain of gluon emissions is illustrated in Fig. 3.2. On the other hand, the y integration over each new final state gluon combined with the $1/y$ or $1/p_T$ divergence in the matrix element squared leads to a possibly large logarithm which can be easiest written in terms of the upper and lower boundary of the p_T integration.

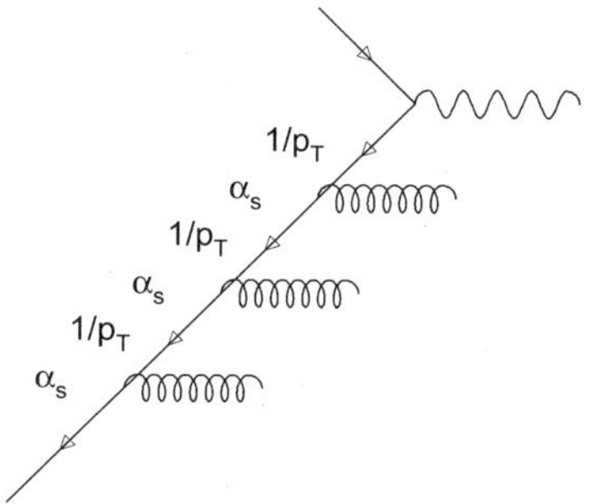

Fig. 3.2.   Feynman diagrams for the repeated emission of a gluon from the incoming leg of a Drell–Yan process. The labels indicate the appearance of α_s as well as the leading divergence of the phase space integration.

This means, at higher orders we expect corrections of the form

$$\sigma_{\text{tot}} \sim \sum_j C_j \left(\alpha_s \log \frac{p_T^{\max}}{p_T^{\min}} \right)^j \tag{3.23}$$

with some factors C_j. Because the splitting probability is universal, these fixed-order corrections can be re-summed to all orders, just like the gluon self energy. You notice how successful perturbation theory becomes every time we encounter a geometric series? And again, in complete analogy with the gluon self energy, this universal factor can be absorbed into another quantity, which are the parton densities.

However, there are three important differences to the running coupling:

First, we are now absorbing IR divergences into running parton densities. We are not renormalizing them, because renormalization is a well-defined procedure to absorb UV divergences into a redefined Lagrangian.

Secondly, the quarks and gluons split into each other, which means that the parton densities will form a set of coupled differential equations which describe their running instead of a simple differential equation with a beta function.

And third, the splitting kernels are not just functions to multiply the parton densities, but they are integration kernels, so we end up with a coupled set of integro-differential equations which describe the parton densities as a function of the factorization scale. These equation are called the

Dokshitzer–Gribov–Lipatov–Altarelli–Parisi or *DGLAP equations*

$$\frac{df_i(x,\mu_F)}{d\log\mu_F^2} = \frac{\alpha_s}{2\pi} \sum_j \int_x^1 \frac{dx'}{x'} \, P_{i\leftarrow j}\left(\frac{x}{x'}\right) f_j(x',\mu_F).$$
(3.24)

We can discuss this formula briefly: to compute the scale dependence of a parton density f_i we have to consider all partons j which can split into i. For each splitting process, we have to integrate over all momentum fractions x' which can lead to a momentum fraction x after splitting, which means we have to integrate z from x to 1. The relative momentum fraction in the splitting is then $x/z < 1$.

The DGLAP equation by construction resums collinear logarithms. There is another class of logarithms which can potentially become large, namely soft logarithms $\log x$, corresponding to the soft divergence of the diagonal splitting kernels. This reflects the fact that if you have for example a charged particle propagating there are two ways to radiate photons without any cost in probability, either collinear photons or soft photons. We know from QED that both of these effects lead to finite survival probabilities once we sum up these collinear and soft logarithms. Unfortunately, or fortunately, we have not seen any experimental evidence of these soft logarithms dominating the parton densities yet, so we can for now stick to DGLAP.

Going back to our original problem, we can now write the hadronic cross section production for Drell–Yan production or other LHC processes as:

$$\sigma_{\text{tot}}(\mu_F,\mu_R) = \int_0^1 dx_1 \int_0^1 dx_2 \, f_i(x_1,\mu_F) \, f_j(x_2,\mu_F) \, \hat{\sigma}_{ij}(x_1 x_2 S,\mu_R)$$
(3.25)

Since our particular Drell–Yan process at leading order only involves weak couplings, it does not include α_s at leading order. We will only see α_s and with it a renormalization scale μ_R appear at next-to-leading order, when we include an additional final state parton.

After this derivation, we can attempt a *physical interpretation* of the factorization scale. The collinear divergence we encounter for example in the $qg \to Zq$ process is absorbed into the parton densities using the universal collinear splitting kernels. In other words, as long as the p_T distribution of the matrix element follows Eq. (3.19), the radiation of any number of additional partons from the incoming partons is now included. These additional partons or jets we obviously cannot veto without getting into perturbative hell with QCD. This is why we should really write $pp \to Z + X$ when talking about factorization-scale dependent parton densities as defined in Eq. (3.25).

If we look at the $d\sigma/dp_T$ distribution of additional partons we can divide the entire phase space into two regions. The collinear region is defined by the leading $1/p_T$ behavior. At some point the p_T distribution will then start decreasing faster, for example because of phase space limitations. The transition scale should roughly be the factorization scale. In the DGLAP evolution we approximate all parton radiation as being collinear with the hadron, *i.e.* move them from the region $p_T < \mu_F$ onto the point $p_T = 0$. This kind of p_T spectrum can be nicely studied using bottom parton densities. They have the advantage that there is no intrinsic bottom content in the proton. Instead, all bottoms have to arise from gluon splitting, which we can compute using perturbative QCD. If we actually compute the bottom parton densities, the factorization scale is not an unphysical free parameter, but it should at least roughly come out of the calculation of the bottom parton densities. So we can for example compute the bottom-induced process $b\bar{b} \to H$ including resummed collinear logarithms using bottom densities or derive it from the fixed-order process $gg \to b\bar{b}H$. When comparing the $p_{T,b}$ spectra it turns out that the bottom factorization scale is indeed proportional to the Higgs mass (or hard scale), but including a relative factor of the order $1/4$. If we naively use $\mu_F = m_H$ we will create an inconsistency in the definition of the bottom parton densities which leads to large higher-order corrections.

Going back to the p_T spectrum of radiated partons or jets — when the transverse momentum of an additional parton becomes large enough that the matrix element does not behave like Eq. (3.19) anymore, this parton is not well-described by the collinear parton densities. We should definitely choose μ_F such that this high-p_T range is not governed by the DGLAP equation. We actually have to compute the hard and now finite matrix elements for $pp \to Z+$jets to predict the behavior of these jets. How to combine collinear jets as they are included in the parton densities and hard partonic jets is what the rest of this lecture will be about.

3.2.3. *Right or wrong scales*

Looking back at the last two sections we introduce the factorization and renormalization scales completely in parallel. First, computing perturbative higher-order contributions to scattering amplitudes we encounter divergences. Both of them we regularize, for example using dimensional regularization (remember that we had to choose $n = 4 - 2\epsilon < 4$ for UV and $n > 4$ for IR divergences). After absorbing the divergences into a re-definition of the respective parameters, referred to as renormalization for example of the strong coupling in the case of an UV divergence and as mass factorization absorbing IR divergences into the parton distributions we are left with a scale artifact. In both cases, this redefinition was not perturbative at fixed order, but involved summing possibly large logarithms. The evolution of these parameters from one renormalization/factorization scale to another is described either by a simple beta function in the case of renormalization and by the DGLAP equation in the case of mass factorization. There is one formal difference between these two otherwise very similar approaches. The fact that we can actually absorb UV divergences into process-independent universal counter terms is called renormalizability and has been proven to all orders for the kind of gauge theories we are dealing with. The universality of IR splitting kernels has not (yet) in general been proven, but on the other hand we have never seen an example where is failed. Actually, for a while we thought there might be a problem with factorization in supersymmetric theories using the supersymmetric version of the $\overline{\text{MS}}$ scheme, but this has since been resolved. A comparison of the two relevant scales for LHC physics is shown in Table 3.1.

Table 3.1. Comparison of renormalization and factorization scales appearing in LHC cross sections.

	renormalization scale μ_R	factorization scale μ_F
source	ultraviolet divergence	collinear (infrared) divergence
poles cancelled	counter terms (renormalization)	parton densities (mass factorization)
summation	resum self energy bubbles	resum collinear logarithms
parameter	running coupling $\alpha_s(\mu_R)$	parton density $f_j(x, \mu_F)$
evolution	RGE for α_s	DGLAP equation
large scales	typically decrease of σ_{tot}	typically increase of σ_{tot}
theory	renormalizability proven for gauge theories	factorization proven all order for DIS proven order-by-order DY...

The way I introduced factorization and renormalization scales clearly describes an artifact of perturbation theory and the way we have to treat divergences. What actually happens if we include all orders in perturbation theory? In that case for example the resummation of the self-energy bubbles is simply one class of diagrams which have to be included, either order-by-order or rearranged into a resummation. For example the two jet production rate will then not depend on arbitrarily chosen renormalization or factorization scales μ. Within the expression for the cross section, though, we know from the arguments above that we have to evaluate renormalized parameters at some scale. This scale dependence will cancel once we put together all its implicit and explicit appearances contributing to the total rate at all orders. In other words, whatever scale we evaluate the strong couplings at gets compensated by other scale logarithms in the complete expression. In the ideal case, these logarithms are small and do not spoil perturbation theory by inducing large logarithms. If we think of a process with one distinct external scale, like the Z mass, we know that all these logarithms have the form $\log \mu/m_Z$. This logarithm is truly an artifact, because it would not need to appear if we evaluated everything at the 'correct' external energy scale of the process, namely m_Z. In that sense we can even think of the running coupling as an *running observable*, which depends on the external energy of the process. This energy scale is not a perturbative artifact, but the cross section even to all orders really depends on the external energy scale. The only problem is that most processes after analysis cuts have more than one scale.

We can turn this argument around and estimate the minimum *theory error* on a prediction of a cross section to be given by the scale dependence in an interval around what we would consider a reasonable scale. Notice that this error estimate is not at all conservative; for example the renormalization scale dependence of the Drell–Yan production rate is zero, because α_s only enters are next-to-leading order. At the same time we know that the next-to-leading order correction to the cross section at the LHC is of the order of 30%, which far exceeds the factorization scale dependence.

Guessing the right scale choice for a process is also hard. For example in leading-order Drell–Yan production there is one scale, m_Z, so any scale logarithm (as described above) has to be $\log \mu/m_Z$. If we set $\mu = m_Z$ all scale logarithms will vanish. In reality, any observable at the LHC will include several different scales, which do not allow us to define just one 'correct' scale. On the other hand, there are definitely completely wrong scale choices. For example, using $1000 \times m_Z$ as a typical scale in the Drell–Yan process will if nothing else lead to logarithms of the size $\log 1000$ whenever a scale logarithm appears. These logarithms have to be cancelled to all orders in perturbation theory, introducing unreasonably large higher-order corrections.

When describing jet radiation, people usually introduce a phase-space dependent renormalization scale, evaluating $\alpha_s(p_{T,j})$. This choice gives the best kinematic distributions for the additional partons, but to compute a cross section it is the one scale choice which is forbidden by QCD and factorization: scales can only depend on exclusive observables, *i.e.* momenta which are given after integrating over the phase space. For the Drell–Yan process such a scale could be m_Z, or the mass of heavy new-physics states in their production process. Otherwise we double-count logarithms and spoil the collinear resummation. But as long as we are mostly concerned with distributions, we even use the transverse-momentum scale very successfully. To summarize this brief mess: while there is no such thing as the correct scale choice, there are more or less smart choices, and there are definitely very wrong choices, which lead to an unstable perturbative behavior.

Of course, these sections on divergences and scales cannot do the topic justice. They fall short left and right, hardly any of the factors are correct (they are not that important either), and I am omitting any formal derivation of this resummation technique for the parton densities. On the other hand, we can derive some general message from them: because we compute cross sections in perturbation theory, the absorption of ubiquitous UV and IR divergences automatically lead to the appearance of scales. These scales are actually useful because running parameters allow us to resum logarithms in perturbation theory, or in other words allow us to compute certain dominant effects to all orders in perturbation theory, in spite of only computing the hard processes at a given loop order. This means that any LHC observable we compute will depend on the factorization and renormalization scales, and we have to learn how to either get rid of the scale dependence by having the Germans compute higher and higher loop orders, or use the Californian/Italian approach to derive useful scale choices in a relaxed atmosphere, to make use of the resummed precision of our calculation.

3.3. Hard versus collinear jets

Jets are a major problem we are facing at the Tevatron and will be the most dangerous problem at the LHC. Let us face it, the LHC is not built do study QCD effects. To the contrary, if we wanted to study QCD, the Tevatron with its lower luminosity would be the better place to do so. Jets at the LHC by themselves are not interesting, they are a nuisance and they are the most serious threat to the success of the LHC program.

The main difference between QCD at the Tevatron and QCD at the LHC is the energy scale of the jets we encounter. *Collinear jets* or jets with a small transverse momentum, are well described by partons in the collinear approximation and simulated by a *parton shower*. This parton shower is the attempt to undo the approximation $p_T \to 0$ we need to make when we absorb collinear radiation in parton distributions using the DGLAP equation. Strictly speaking, the parton shower can and should only fill the phase space region $p_T = 0...\mu_F$ which is not covered by explicit additional parton radiation. Such so-called *hard jets* or jets with a large transverse momentum are described by hard matrix elements which we can compute using the QCD Feynman rules. Because of the logarithmic enhancement we have observed for collinear additional partons, there are much more collinear and soft jets than hard jets.

The problem at the LHC is the range of 'soft' or 'collinear' and 'hard'. As mentioned above, we can define these terms by the validity of the collinear approximation in Eq. (3.19). The maximum p_T of a collinear jet is the region for which the jet radiation cross section behaves like $1/p_T$. We know that for harder and harder jets we will at some point become limited by the partonic energy available at the LHC, which means the p_T distribution of additional jets will start dropping faster than $1/p_T$. At this point the logarithmic enhancement will cease to exist, and jets will be described by the regular matrix element squared without any resummation.

Quarks and gluons produced in association with gauge bosons at the Tevatron behave like collinear jets for $p_T \lesssim 20$ GeV, because the quarks at the Tevatron are limited in energy. At the LHC, jets produced in association with tops behave like collinear jets to $p_T \sim 150$ GeV, jets produced with 500 GeV gluinos behave like collinear jets to p_T scales larger than 300 GeV. This is not good news, because collinear jets means many jets, and many jets produce *combinatorical backgrounds* or ruin the missing momentum resolution of the detector. Maybe I should sketch the notion of combinatorical backgrounds: if you are looking for example for two jets to reconstruct an invariant mass you can simply plot all events as a function of this invariant mass and cut the background by requiring all event to sit around a peak in m_{jj}. However, if you have for example three jets in the event you have to decide which of the three jet-jet combinations should go into this distribution. If this seems not possible, you can alternatively consider two of the three combinations as uncorrelated 'background' events. In other words, you make three histogram entries out of your signal or background event and consider all background events plus two of the three signal combinations as background. This way the signal-to-background ratio decreases from N_S/N_B to $N_S/(3N_B + 2N_S)$, *i.e.* by at least a factor of three. You can guess that picking two particles out of four candidates with its six combinations has great potential to make your analysis a candidate for this circular folder under your desk. The most famous victim of such combinatorics might be the formerly promising Higgs discovery channel $pp \to t\bar{t}H$ with $H \to b\bar{b}$.

All this means for theorists that at the LHC we have to learn how to model collinear and hard jets reliably. This is what the remainder of the QCD lectures will be about. Achieving this understanding, I consider the most important development in QCD since I started working on physics. Discussing the different approaches we will see why such general–p_T jets are hard to understand and even harder to properly simulate.

3.3.1. *Sudakov factors*

Before we discuss any physics it makes sense to introduce the so-called Sudakov factors which will appear in the next sections. This technical term is used by QCD experts to ensure that other LHC physicists feel inferior and do not get on their nerves. But, really, Sudakov factors are nothing but simple survival probabilities. Let us start with an event which we would expect to occur p times, given its probability and given the number of shots. The probability of observing it n times is given by the Poisson distribution

$$\mathcal{P}(n;p) = \frac{p^n\, e^{-p}}{n!}\,. \tag{3.26}$$

This distribution will develop a mean at p, which means most of the time we will indeed see about the expected number of events. For large numbers it will become a Gaussian. In the opposite direction, using this distribution we can compute the probability of observing zero events, which is $\mathcal{P}(0;p) = e^{-p}$. This formula comes in handy when we want to know how likely it is that we do not see a parton splitting in a certain energy range.

According to the last section, the differential probability of a parton to split or emit another parton at a scale μ and with the daughter's momentum fraction x is given by the splitting kernel $P_{i \leftarrow j}(x)$ times dp_T^2/p_T^2. This energy measure is a little tricky because we compute the splitting kernels in the collinear approximation, so p_T^2 is the most inconvenient observable to use. We can approximately replace the transverse momentum by the *virtuality* Q, to get to the standard parameterization of parton splitting — I know I am just waving my hands at this stage, to understand the more fundamental role of the virtuality we would have to look into deep inelastic scattering and factorization. In terms of the virtuality, the splitting of one parton into two is given by the splitting kernel integrated over the proper range in the momentum fraction x

$$dP(x) = \frac{\alpha_s}{2\pi} \frac{dq^2}{q^2} \int dx\, P(x)$$

$$P(Q_{\min}, Q_{\max}) = \frac{\alpha_s}{2\pi} \int_{Q_{\min}}^{Q_{\max}} \frac{dq^2}{q^2} \int_{x_{\min}}^{x_{\max}} dx\, P(x) \qquad (3.27)$$

The splitting kernel we symbolically write as $P(x)$, avoiding indices and the sum over partons appearing in the DGLAP equation Eq. (3.24). The boundaries $x_{\min}$ and $x_{\max}$ we can compute for example in terms of an over-all minimum value Q_0 and the actual values q, so we drop them for now. Strictly speaking, the double integral over x and q^2 can lead to two overlapping IR divergences or logarithms, a soft logarithm arising from the x integration (which we will not discuss further) and the collinear logarithm arising from the virtuality integral. This is the logarithm we are interested in when talking about the parton shower.

In the expression above we compute the probability that a parton will split into another parton while moving from a virtuality $Q_{\max}$ down to $Q_{\min}$. This probability is given by QCD, as described earlier. Using it, we can ask what the probability is that we will not see a parton splitting from a parton starting at fixed $Q_{\max}$ to a variable scale Q, which is precisely the *Sudakov factor*

$$\Delta(Q, Q_{\max}) = e^{-P(Q, Q_{\max})}$$

$$= \exp\left[-\frac{\alpha_s}{2\pi} \int_Q^{Q_{\max}} \frac{dq^2}{q^2} \int_{x_{\min}}^{x_{\max}} dx\, P(x) \right] \sim e^{-\alpha_s \log^2 Q_{\max}/Q}$$

$$(3.28)$$

The last line omits all kinds of factors, but correctly identifies the logarithms involved, namely $\alpha_s^n \log^{2n} Q_{\max}/Q$.

3.3.2. *Jet algorithm*

Before discussing methods to describe jets at the LHC we should introduce one way to define jets in a detector, namely the k_T *jet algorithm.* Imagine we observe a large number of energy depositions in the calorimeter in the detector which we would like to combine into jets. We know that they come from a smaller number of partons which originate in the hard QCD process and which since have undergone a sizeable number of splittings. Can we try to reconstruct partons?

The answer is yes, in the sense that we can combine a large number of jets into smaller numbers, where unfortunately nothing tells us what the final number of jets should be. This makes sense, because in QCD we can produce an arbitrary number of hard jets in a hard matrix element and another arbitrary number via collinear radiation. The main difference between a hard jet and a jet from parton splitting is that the latter will have a partner which originated from the same soft or collinear splitting.

The basic idea of the k_T algorithm is to ask if a given jet has a soft or collinear partner. For this we have to define a collinearity measure, which will be something like the transverse momentum of one jet with respect to another one $y_{ij} \sim k_{T,ij}$. If one of the two jets is the beam direction, this measure simply becomes $y_{iB} \sim k_{T,i}$. We define two jets as collinear, if $y_{ij} < y_{\mathrm{cut}}$ where y_{cut} we have to give to the algorithm. The jet algorithm is simple:

    (1) for all final state jets find minimum $y^{\mathrm{min}} = \min_{ij}(y_{ij}, y_{iB})$
    (2a) if $y^{\mathrm{min}} = y_{ij} < y_{\mathrm{cut}}$ merge jets i and j, go back to (1)
    (2b) if $y^{\mathrm{min}} = y_{iB} < y_{\mathrm{cut}}$ remove jet i, go back to (1)
    (2c) if $y^{\mathrm{min}} > y_{\mathrm{cut}}$ keep all jets, done

The result of the algorithm will of course depend on the resolution y_{cut}. Alternatively, we can just give the algorithm the minimum number of jets and stop there. The only question is what 'combine jets' means in terms of the 4-momentum of the new jet. The simplest thing would be to just combine the momentum vectors $k_i + k_j \to k_i$, but we can still either combine the 3-momenta and give the new jet a zero invariant mass (which assumes it indeed was one parton) or we can add the 4-momenta and get a jet mass (which means they can come from a Z, for example). But these are details for most new-physics searches at the LHC. At this stage we run into a language issue: what do we really call a jet? I am avoiding this issue by saying that jet algorithms definitely start from calorimeter towers and not jets and then move more and more towards jets, where likely the last iterations could be described by combining jets into new jets.

From the QCD discussion above, it is obvious why theorists prefer a k_T algorithm over for other algorithms which define the distance between two jets in a more geometric manner: a jet algorithm combines the complicated energy deposition in the hadronic calorimeter, and we know that the showering probability or theoretically speaking the collinear splitting probability is best described in terms of virtuality or transverse momentum. A transverse-momentum distance between jets is from a theory point of view best suited to combine the right jets into the original parton from the hard interaction. Moreover, this k_T measure is intrinsically infrared safe, which means the radiation of an additional soft parton cannot affect the global structure of the reconstructed jets. For other algorithms we have to ensure this property explicitly, and you can find examples for this in QCD lectures by Mike Seymour.

One problem of the k_T algorithm is that noise and the underlying event can easiest be understood geometrically in the 4π detector. Basically, the low-energy jet activity is constant all over the detector, so the easiest thing to do is just subtract it from each event. How much energy deposit we have to subtract from a reconstructed jet depends on the actual area the jet covers in the detector. Therefore, it is a major step for the k_T algorithm that it can indeed compute an IR–safe geometric size of the jet. Even more, if this size is considerably smaller than the usual geometric measures, the k_T algorithm should at the end of the day turn out to be the best jet algorithm at the LHC.

3.4. Jet merging

So how does a traditional Monte Carlo treat the radiation of jets into the final state? It needs to reverse the summation of collinear jets done by the DGLAP equation, because jet radiation is not strictly collinear and does hit the detector. In other words, it computes probabilities for radiating collinear jets from other jets and simulates this radiation. Because it was the only thing we knew, Monte Carlos used to do this in the collinear approximation. However, from the brief introduction we know that at the LHC we should generally not use the collinear approximation, which is one of the reason why at the LHC we will use all-new Monte Carlos. Two ways how they work we will discuss here.

Apart from the collinear approximation for jet radiation, a second problem with Monte Carlo simulation is that they 'only do shapes'. In other words, the normalization of the event sample will always be perturbatively poorly defined. The simple reason is that collinear jet radiation starts from a hard process and its production cross section and from then on works with splitting probabilities, but never touches the total cross section it started from.

Historically, people use higher-order cross sections to normalize the total cross section in the Monte Carlo. This is what we call a *K factor*: $K = \sigma^{\text{improved}}/\sigma^{\text{MC}} = \sigma^{\text{improved}}/\sigma^{\text{LO}}$. It is crucial to remember that higher-order cross sections integrate over unobserved additional jets in the final state. So when we normalize the Monte Carlo we assume that we can first integrate over additional jets and obtain σ^{improved} and then just normalize the Monte Carlo which puts back these jets in the collinear approximation. Obviously, we should try to do better than that, and there are two ways to improve this traditional Monte Carlo approach.

3.4.1. *MC@NLO method*

When we compute the next-to-leading order correction to a cross section, for example to Drell–Yan production, we consider all contributions of the order $G_F \alpha_s$. There are three obvious sets of Feynman diagrams we have to square and multiply, namely the Born contribution $q\bar{q} \to Z$, the virtual gluon exchange for example between the incoming quarks, and the real gluon emission $q\bar{q} \to Zg$. Another set of diagrams we should not forget are the crossed channels $qg \to Zq$ and $\bar{q}g \to Z\bar{q}$. Only amplitudes with the same external particles can be squared, so we get the matrix-element-squared contributions

$$|\mathcal{M}_B|^2 \propto G_F \tag{3.29}$$

$$2\text{Re}\,\mathcal{M}_V^* \mathcal{M}_B \propto G_F \alpha_s \qquad |\mathcal{M}_{Zg}|^2 \propto G_F \alpha_s \qquad |\mathcal{M}_{Zq}|^2, |\mathcal{M}_{Z\bar{q}}|^2 \propto G_F \alpha_s$$

Strictly speaking, we should have included the counter terms, which are a modification of $|\mathcal{M}_B|^2$, shifted by counter terms of the order $\alpha_s(1/\epsilon + C)$. These counter terms we add to the interference of Born and virtual gluon diagrams to remove the UV divergences. Luckily, this is not the part of the contributions we want to discuss. IR poles can have two sources, soft and collinear divergences. The first kind is cancelled between virtual gluon exchange and real gluon emission. Again, we are not really interested in them.

What we are interested in are the collinear divergences. They arise from virtual gluon exchange as well as from gluon emission and from gluon splitting in the crossed channels. The collinear limit is described by the splitting kernels Eq. (3.22), and the divergences are absorbed in the redefinition of the parton densities (like an IR pseudo-renormalization).

To present the idea of MC@NLO Bryan Webber uses a nice toy model which I am going to follow in a shortened version. It describes simplified particle radiation off a hard process: the energy of the system before radiation is x_s and the energy of the outgoing particle (call it photon or gluon) is x, so $x < x_s < 1$. When we compute *next-to-leading order corrections* to a hard process, the different contributions (now neglecting crossed channels) are

$$\left. \frac{d\sigma}{dx} \right|_B = B\,\delta(x), \quad \left. \frac{d\sigma}{dx} \right|_V = \alpha_s \left(\frac{B}{2\epsilon} + V \right) \delta(x), \quad \left. \frac{d\sigma}{dx} \right|_R = \alpha_s \frac{R(x)}{x}. \quad (3.30)$$

The constant B describes the Born process and the assumed factorizing poles in the virtual contribution. The coupling constant α_s should be extended by factors 2 and π, or color factors. We immediately see that the integral over x in the real emission rate is logarithmically divergent in the soft limit, similar to the collinear divergences we now know and love. From factorization (*i.e.* implying universality of the splitting kernels) we know that in the collinear and soft limits the real emission part has to behave like the Born matrix element $\lim_{x \to 0} R(x) = B$.

The logarithmic IR divergence we extract in dimensional regularization, as we already did for the virtual corrections. The expectation value of any infrared safe observable over the entire phase space is then given by

$$\langle O \rangle = \mu_F^{2\epsilon} \int_0^1 dx \, \frac{O(x)}{x^{2\epsilon}} \left[\left. \frac{d\sigma}{dx} \right|_B + \left. \frac{d\sigma}{dx} \right|_V + \left. \frac{d\sigma}{dx} \right|_R \right]. \quad (3.31)$$

Dimensional regularization yields this additional factor $1/x^{2\epsilon}$, which is precisely the factor whose mass unit we cancel introducing the factorization scale $\mu_F^{2\epsilon}$. This renormalization scale factor we will casually drop in the following.

When we compute a distribution of for example the energy of one of the heavy particles in the process, we can extract a histogram from of the integral for $\langle O \rangle$ and obtain a normalized distribution. However, to compute such a histogram we have to numerically integrate over x, and the individual parts of the integrand are not actually integrable. To cure this problem, we can use the *subtraction method* to define integrable functions under the x integral. From the real emission contribution we subtract and then add a smartly chosen term:

$$\langle O \rangle_R = \int_0^1 dx \, \frac{O(x)}{x^{2\epsilon}} \left.\frac{d\sigma}{dx}\right|_R = \int_0^1 dx \, \frac{O(x)}{x^{2\epsilon}} \frac{\alpha_s R(x)}{x}$$

$$= \alpha_s \, B \, O(0) \int_0^1 dx \frac{1}{x^{1+2\epsilon}} + \int_0^1 dx \left(\frac{\alpha_s R(x) O(x)}{x^{1+2\epsilon}} - \frac{\alpha_s B O(0)}{x^{1+2\epsilon}} \right)$$

$$= \alpha_s \, B \, O(0) \int_0^1 dx \frac{1}{x^{1+2\epsilon}} + \alpha_s \int_0^1 dx \, \frac{R(x) O(x) - B O(0)}{x^{1+2\epsilon}}$$

$$= -\alpha_s \frac{B \, O(0)}{2\epsilon} + \alpha_s \int_0^1 dx \, \frac{R(x) O(x) - B O(0)}{x} \tag{3.32}$$

In the second integral we take the limit $\epsilon \to 0$ because the asymptotic behavior of $R(x \to 0)$ makes the numerator vanish and hence regularizes this integral without any dimensional regularization required. The first term precisely cancels the (soft) divergence from the virtual correction. We end up with a perfectly finite x integral for all three contributions

$$\langle O \rangle = B \, O(0) + \alpha_s V \, O(0) + \alpha_s \int_0^1 dx \, \frac{R(x) \, O(x) - B \, O(0)}{x}$$

$$= \int_0^1 dx \left[O(0) \left(B + \alpha_s V - \alpha_s \frac{B}{x} \right) + O(x) \, \alpha_s \frac{R(x)}{x} \right] \tag{3.33}$$

This procedure is one of the standard methods to compute next-to-leading order corrections involving one-loop virtual contributions and the emission of one additional parton. This formula is a little tricky: usually, the Born-type kinematics would come with an explicit factor $\delta(x)$, which in this special case we can omit because of the integration boundaries. We can re-write the same formula in terms of a derivative

$$\frac{d\sigma}{dO} = \int_0^1 dx \left[I(O)_{\text{LO}} \left(B + \alpha_s V - \alpha_s \frac{B}{x} \right) + I(O)_{\text{NLO}} \, \alpha_s \frac{R(x)}{x} \right] \tag{3.34}$$

The transfer function $I(O)$ is defined in a way that formally does precisely what we did before: at leading order we evaluate it using the Born kinematics $x = 0$ while allowing for a general $x = 0 \cdots 1$ for the real emission kinematics.

In this calculation we have integrated over the entire phase space of the additional parton. For a hard additional parton or jet everything looks well defined and finite. On the other hand, we cancel an IR divergence in the virtual corrections proportional to a Born-type momentum configuration $\delta(x)$ with another IR divergence which appears after integrating over small but finite values of $x \to 0$. In a histogram in x, where we encounter the real-emission divergence at small x, this divergence is cancelled by a negative delta distribution right at $x = 0$. Obviously, this will not give us a well-behaved distribution. What we would rather want is a way to smear out this pole such that it coincides with the in that range justified collinear approximation and cancels the real emission over the entire low-x range. At the same time it has to leave the hard emission intact and when integrated give the same result as the next-to-leading oder rate. Such a modification will use the emission probability or Sudakov factors. We can define an emission probability of a particle with an energy fraction z as $d\mathcal{P} = \alpha_s E(z)/z\, dz$. Note that we have avoided the complicated proper two–dimensional description in favor of this simpler picture just in terms of particle energy fractions.

Let us consider a perfectly fine observable, the radiated photon spectrum as a function of the (external) energy scale z. We know what this spectrum has to look like for the two kinematic configurations

$$\left.\frac{d\sigma}{dz}\right|_{\mathrm{LO}} = \alpha_s \frac{BE(z)}{z} \qquad\qquad \left.\frac{d\sigma}{dz}\right|_{\mathrm{NLO}} = \alpha_s \frac{R(z)}{z} \qquad (3.35)$$

The first term corresponds to parton shower radiation from the Born diagram (at order α_s), while the second term is the real emission defined above. The transfer functions we would have to include in Eq. (3.34) to arrive at this equation for the observable are

$$\left. I(z,1)\right|_{\mathrm{LO}} = \alpha_s \frac{E(z)}{z}$$

$$\left. I(z,x_M)\right|_{\mathrm{NLO}} = \delta(z-x) + \alpha_s \frac{E(z)}{z}\, \Theta(x_M(x) - z) \qquad (3.36)$$

The additional second term in the real-radiation transfer function arises from a parton shower acting on the real emission process. It explicitly requires that enough energy has to be available to radiate a photon with an energy z, where x_M is the energy available at the respective stage of showering, *i.e.* $z < x_M$.

These transfer functions we can include in Eq. (3.34), which becomes

$$
\begin{aligned}
\frac{d\sigma}{dz} &= \int_0^1 dx \left[I(z,1) \left(B + \alpha_s V - \alpha_s \frac{B}{x} \right) + I(z,x_M)\, \alpha_s \frac{R(x)}{x} \right] \\
&= \int_0^1 dx \left[\alpha_s \frac{E(z)}{z} \left(B + \alpha_s V - \alpha_s \frac{B}{x} \right) + (\delta(x-z) + \mathcal{O}(\alpha_s))\, \alpha_s \frac{R(x)}{x} \right] \\
&= \int_0^1 dx \left[\alpha_s \frac{BE(z)}{z} + \alpha_s \frac{R(z)}{z} \right] + \mathcal{O}(\alpha_s^2) \\
&= \alpha_s \frac{BE(z) + R(z)}{z} + \mathcal{O}(\alpha_s^2)
\end{aligned}
\tag{3.37}
$$

All Born–type contributions proportional to $\delta(z)$ have vanished by definition. This means we should be able to integrate the z distribution to the total cross section σ_{tot} with a $z_{\min}$ cutoff for consistency. However, the distribution we obtained above has an additional term which spoils this agreement, so we are still missing something.

On the other hand, we also knew we would fall short, because what we described in words about a subtraction term for finite x cancelling the real emission we have not yet included. This means, first we have to add a subtraction term to the real emission which cancels the fixed-order contributions for small x values. Because of factorization we know how to write such a subtraction term using the splitting function, called E in this example:

$$
\frac{R(x)}{x} \longrightarrow \frac{R(x) - BE(x)}{x}
\tag{3.38}
$$

To avoid double counting we have to add this parton shower to the Born-type contribution, now in the collinear limit, which leads us to a modified version of Eq. (3.34)

$$
\begin{aligned}
\frac{d\sigma}{dO} = \int_0^1 dx \Bigg[\; I(O,1) &\left(B + \alpha_s V - \frac{\alpha_s B}{x} + \frac{\alpha_s BE(x)}{x} \right) \\
&+ I(O,x_M)\, \alpha_s \frac{R(x) - BE(x)}{x} \Bigg]
\end{aligned}
\tag{3.39}
$$

When we again compute the z spectrum to order α_s there will be an additional contribution from the Born-type kinematics

$$\frac{d\sigma}{dz} = \int_0^1 dx \; \alpha_s \; \frac{BE(z) + R(z)}{z} + \mathcal{O}(\alpha_s^2)$$

$$\longrightarrow \int_0^1 dx \; \left[\alpha_s \; \frac{BE(z) + R(z)}{z} - \alpha_s \; \delta(x-z) \; \frac{BE(x)}{x} \right] + \mathcal{O}(\alpha_s^2)$$

$$= \int_0^1 dx \; \alpha_s \; \frac{BE(z) + R(z) - BE(z)}{z} + \mathcal{O}(\alpha_s^2)$$

$$= \alpha_s \; \frac{R(z)}{z} + \mathcal{O}(\alpha_s^2) \tag{3.40}$$

which gives us the distribution we expected, without any double counting.

In other words, this scheme implemented in the MC@NLO Monte Carlo describes the hard emission just like a next-to-leading order calculation, including the next-to-leading order normalization. On top of that, it simulates additional collinear particle emissions using the Sudakov factor. This is precisely what the parton shower does. Most importantly, it avoids double counting between the first hard emission and the collinear jets, which means it describes the entire p_T range of jet emission for the *first and hardest* radiated jet consistently. Additional jets, which do not appear in the next-to-leading order calculation are simply added by the parton shower, *i.e.* in the collinear approximation. What looked to easy in our toy example is of course much harder in the mean QCD reality, but the general idea is the same: to combine a fixed-order NLO calculation with a parton shower one can think of the parton shower as a contribution which cancels a properly defined subtraction term which we can include as part of the real emission contribution.

3.4.2. *CKKW method*

The one weakness of the MC@NLO method is that it only describes one hard jet properly and relies on a parton shower and its collinear approximation to simulate the remaining jets. Following the general rule that there is no such thing as a free lunch we can improve on the number of correctly described jets, which unfortunately will cost us the next-to-leading order normalization.

For simplicity, we will limit our discussion to final state radiation, for example in the inverse Drell–Yan process $e^+e^- \to q\bar{q}$. We know already that this final state is likely to evolve into more than two jets. First, we can radiate a gluon off one of the quark legs, which gives us a $q\bar{q}g$ final state, provided our k_T algorithm finds $y_{ij} > y_{\mathrm{cut}}$. Additional splittings can also give us any number of jets, and it is not clear how we can combine these different channels.

Each of these processes can be described either using matrix elements or using a parton shower, where 'describe' means for example compute the relative probability of different phase space configurations. The parton shower will do well for jets which are fairly collinear, $y_{ij} < y_{\mathrm{ini}}$. In contrast, if for our closest jets we find $y_{ij} > y_{\mathrm{ini}}$, we know that collinear logarithms did not play a major role, so we can and should use the hard matrix element. How do we combine these two approaches?

The CKKW scheme tackles this multi-jet problem. It first allows us to combine final states with a *different number of jets*, and then ensures that we can add a parton shower without any double counting. The only thing I will never understand is that they labelled the transition scale as 'ini'.

Using Sudakov factors we can first construct the *probabilities* of generating n–jet events from a hard two–jet production process. These probabilities make no assumptions on how we compute the actual kinematics of the jet radiation, *i.e.* if we model collinear jets with a parton shower or hard jets with a matrix element. This way we will also get a rough idea how Sudakov factors work in practice. For the two–jet and three–jet final states, we will see that we only have to consider the splitting probabilities for the different partons

$$\Gamma_q(Q_{\mathrm{out}}, Q_{\mathrm{in}}) \equiv \Gamma_{q\leftarrow q}(Q_{\mathrm{out}}, Q_{\mathrm{in}}) = \frac{2C_F}{\pi} \frac{\alpha_s(Q_{\mathrm{out}})}{Q_{\mathrm{out}}} \left(\log \frac{Q_{\mathrm{in}}}{Q_{\mathrm{out}}} - \frac{3}{4} \right)$$

$$\Gamma_g(Q_{\mathrm{out}}, Q_{\mathrm{in}}) \equiv \Gamma_{g\leftarrow q}(Q_{\mathrm{out}}, Q_{\mathrm{in}}) = \frac{2C_A}{\pi} \frac{\alpha_s(Q_{\mathrm{out}})}{Q_{\mathrm{out}}} \left(\log \frac{Q_{\mathrm{in}}}{Q_{\mathrm{out}}} - \frac{11}{12} \right)$$

$$(3.41)$$

The virtualities $Q_{\mathrm{in,out}}$ correspond to the incoming (mother) and outgoing (daughter) parton. Unfortunately, this formula is somewhat understandable from the argument before and from $P_{q\leftarrow q}$, but not quite. That has to do with the fact that these splittings are not only collinearly divergent, but also softly divergent, as we can see in the limits $x \to 0$ and $x \to 1$ in Eq. (3.22). These divergences we have to subtract first, so the formulas for the splitting probabilities $\Gamma_{q,g}$ look unfamiliar. In addition, we find finite terms arising from next-to-leading logarithms which spoil the limit

$Q_{\text{out}} \to Q_{\text{in}}$, where the probability of no splitting should go to unity. But at least we can see the leading (collinear) logarithm $\log Q_{\text{in}}/Q_{\text{out}}$. Technically, we can deal with the finite terms in the Sudakov factors by requiring them to be positive semi-definite, *i.e.* by replacing $\Gamma(Q_{\text{out}}, Q_{\text{in}}) < 0$ by zero.

Given the splitting probabilities we can write down the *Sudakov factor*, which is the probability of not radiating any hard and collinear gluon between the two virtualities:

$$\Delta_{q,g}(Q_{\text{out}}, Q_{\text{in}}) = \exp\left[- \int_{Q_{\text{out}}}^{Q_{\text{in}}} dq \, \Gamma_{q,g}(q, Q_{\text{in}}) \right] \qquad (3.42)$$

This integral boundaries are $Q_{\text{out}} < Q_{\text{in}}$. This description we can generalize for all splittings $P_{i \leftarrow j}$ we wrote down before.

First, we can compute the probability that we see exactly *two partons*, which means that none of the two quarks radiate a resolved gluon between the virtualities Q_2 and Q_1, where we assume that $Q_1 < Q_2$ gives the scale for this resolution. It is simply $[\Delta_q(Q_1, Q_2)]^2$, once for each quark, so that was easy.

Next, what is the probability that the two–jet final state evolves exactly into *three partons*? We know that it contains a factor $\Delta_q(Q_1, Q_2)$ for one untouched quark. If we label the point of splitting in the matrix element Q_q for the quark, there has to be a probability for the second quark to get from Q_2 to Q_q untouched, but we leave this to later. After splitting with the probability $\Gamma_q(Q_2, Q_q)$, this quark has to survive to Q_1, so we have a factor $\Delta_q(Q_1, Q_q)$. Let's call the virtuality of the radiated gluon after splitting Q_g, then we find the gluon's survival probability $\Delta_g(Q_1, Q_g)$. So what we have until now is

$$\Delta_q(Q_1, Q_2) \, \Gamma_q(Q_2, Q_q) \, \Delta_q(Q_1, Q_q) \, \Delta_g(Q_1, Q_g) \cdots \qquad (3.43)$$

That's all there is, with the exception of the intermediate quark. Naively, we would guess its survival probability between Q_2 and Q_q to be $\Delta_q(Q_q, Q_2)$, but that is not correct. That would imply no splittings resolved at Q_q, but what we really mean is no splitting resolved later at $Q_1 < Q_q$. Instead, we compute the probability of no splitting between Q_2 and Q_q from $\Delta_q(Q_1, Q_2)$ under the additional condition that splittings from Q_q down to Q_1 are now allowed. If no splitting occurs between Q_1 and Q_q this simply gives us $\Delta_q(Q_1, Q_2)$ for the Sudakov factor between Q_2 and Q_q. If one splitting happens after Q_q this is fine, but we need to add this combination to the

Sudakov between Q_2 and Q_q. Allowing an arbitrary number of possible splittings between Q_q and Q_1 gives us

$$\Delta_q(Q_1, Q_2) \left[1 + \int_{Q_q}^{Q_1} dq \, \Gamma_q(q, Q_1) + \cdots \right] =$$
$$= \Delta_q(Q_1, Q_2) \, \exp \left[\int_{Q_q}^{Q_1} dq \, \Gamma_q(q, Q_1) \right] = \frac{\Delta_q(Q_1, Q_2)}{\Delta_q(Q_1, Q_q)} .$$

$$(3.44)$$

So once again: the probability of nothing happening between Q_2 and Q_q we compute from the probability of nothing happening between Q_2 and Q_1 times possible splittings between Q_q and Q_1.

Collecting all these factors gives the combined probability that we find exactly three partons at a virtuality Q_1

$$\Delta_q(Q_1, Q_2) \, \Gamma_q(Q_2, Q_q) \, \Delta_q(Q_1, Q_q) \, \Delta_g(Q_1, Q_g) \, \frac{\Delta_q(Q_1, Q_2)}{\Delta_q(Q_1, Q_q)}$$
$$= \Gamma_q(Q_2, Q_q) \, [\Delta_q(Q_1, Q_2)]^2 \, \Delta_g(Q_1, Q_g) \qquad (3.45)$$

This result is pretty much what we would expected: both quarks go through untouched, just like in the two–parton case. But in addition we need exactly one splitting producing a gluon, and this gluon cannot split further. This example illustrates how it is fairly easy to compute these probabilities using Sudakov factors: adding a gluon corresponds to adding a splitting probability times the survival probability for this gluon, everything else magically drops out. At the end, we only integrate over the splitting point Q_q.

The first part of the CKKW scheme we illustrate is how to combine different n–parton channels in one framework. Knowing some of the basics we can write down the (simplified) *CKKW algorithm* for final state radiation. As a starting point, we compute all leading-order cross sections for n-jet production with a lower cutoff at y_{ini}. This cutoff ensures that all jets are hard and that all $\sigma_{n,i}$ are finite. The second index i describes different non-interfering parton configurations, like $q\bar{q}gg$ and $q\bar{q}q\bar{q}$ for $n = 4$. The purpose of the algorithm is to assign a weight (probability, matrix element squared,...) to a given phase space point, statistically picking the correct process and combining them properly.

(1) for each jet final state (n, i) compute the relative probability $P_{n,i} = \sigma_{n,i} / \sum \sigma_{k,j}$; select a final state with this probability $P_{n,i}$

(2) distribute the jet momenta to match the external particles in the matrix element and compute $\overline{|\mathcal{M}|^2}$

(3) use the k_T algorithm to compute the virtualities Q_j for each splitting in this matrix element

(4) for each internal line going from Q_j to Q_k compute the Sudakov factor $\Delta(Q_1, Q_j)/\Delta(Q_1, Q_k)$, where Q_1 is the final resolution of the evolution. For any final state line starting at Q_j apply $\Delta(Q_1, Q_j)$. All these factors combined give the combined survival probability described above.

The matrix element weight times the survival probability can be used to compute distributions from weighted events or to decide if to keep or discard an event when producing unweighted events. The line of Sudakov factors ensures that the relative weight of the different n–jet rates is identical to the probabilities we just computed. Their kinematics, however, are hard–jet configuration without any collinear assumption. There is one remaining subtlety in this procedure which I am skipping. This is the re-weighting of α_s, because the hard matrix element will be typically computed with a fixed hard renormalization scale, while the parton shower only works with a scale fixed by the virtuality of the respective splitting. But those are details, and there will be many more details in which different implementations of the CKKW scheme differ.

The second question is what we have to do to match the hard matrix element with the parton shower at a critical resolution point $y_{\text{ini}} = Q_1^2/Q_2^2$. From Q_1 to Q_0 we will use the parton shower, but above this the matrix elements will be the better description. For both regimes we already know how to combine different n–jet processes. On the other hand, we need to make sure that this last step does not lead to any double counting. From the discussion above, we know that Sudakovs which describe the evolution between scales but use a lower virtuality as the resolution point are going to be the problem. On the other hand, we also know how to describe this behavior using the additional splitting factors we used for the $Q_2 \cdots Q_q$ range. Carefully distinguishing the virtuality scale of the actual splitting and the scale of jet resolution is the key, which we have to combine with the fact that in the CKKW method starts each parton shower at the point where the parton first appears. It turns out that we can use this argument to keep the resolution ranges $y > y_{\text{ini}}$ and $y < y_{\text{ini}}$ separate, without any double counting. There is a simple way to check this, namely the question if the y_{ini} *dependence* drops out of the final combined probabilities. And the answer for final state radiation is yes, as proven in the original paper, including a hypothetical next-to-leading logarithm parton shower.

One widely used variant of CKKW is Michelangelo Mangano's *MLM scheme*, for example implemented in Alpgen or Madevent. Its main difference to the classical CKKW is that it avoids computing the corresponding survival properties using Sudakov form factors. Instead, it vetoes events which CKKW would have cut using the Sudakov rescaling. This way it avoids problems with splitting probabilities beyond the leading logarithms, for example the finite terms appearing in Eq. (3.41) which can otherwise lead to a mismatch between the actual shower evolution and the analytic expressions of the Sudakov factors. Its veto approach allows the MLM scheme to combine a set of n–parton events after they have been generated using hard matrix elements. Its parton shower is then not needed to compute a Sudakov reweighting. On the other hand, to combine a given sample of events the parton shower has to start from an external scale, which should be chosen as the hard(est) scale of the process.

Once the parton shower has defined the complete event, we need to decide if this event needs to be removed to avoid double counting due to an overlap of simulated collinear and hard radiation. After applying a jet algorithm (which in the case of Alpgen is a cone algorithm and in case of Madevent is a k_T algorithm) we can simply compare the hard event with the showered event by identifying each reconstructed showered jet with the partons we started from. If all jet–parton combinations match and there are not additional resolved jets apart from the highest-multiplicity sample we know that the showering has not altered the hard-jet structure of the event, otherwise the event has to go.

Unfortunately, the vetoing approach does not completely save the MLM scheme the backwards evolution of a generated event, since we still need to know the energy or virtuality scales at which partons split to fix the scale of the strong coupling. If we know the Feynman diagrams which lead to each event, we can check that a certain splitting is actually possible in its color structure.

In my non-expert user's mind, all merging schemes are conceptually similar enough that we should expect them to reproduce each others' results, and they largely do. But the devil is in the details, and we have to

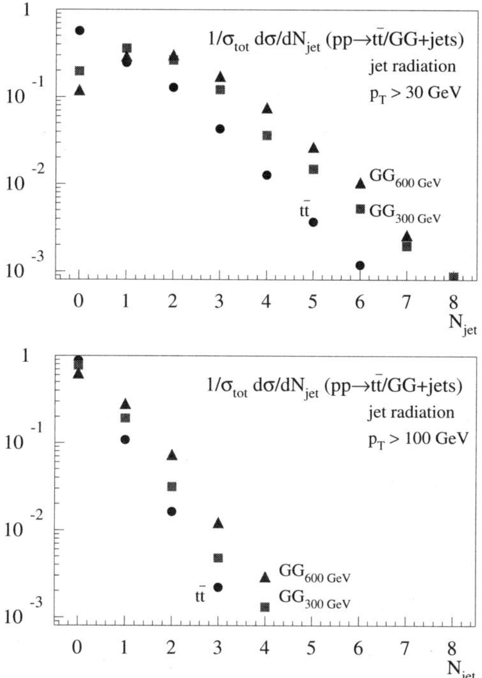

Fig. 3.3. Number of additional jets with a transverse momentum of at least 30 or 100 GeV radiated from top pair production and the production of heavy states at the LHC. As an example for such heavy states we use a pair of scalar gluons with a mass of 300 or 600 GeV, pair-produced in gluon fusion. The figures are from a forthcoming paper with Tim Tait (arXiv:0810:3919), produced with MadEvent using it's modified MLM algorithm — thanks to Johan Alwall.

watch out for example for threshold kinks in jet distributions which should not be there.

To summarize, we can use the CKKW or MLM schemes to combine n-jet events with variable n and at the same time combine matrix element and parton shower descriptions of the jet kinematics. In other words, we

Table 3.2. Comparison of the MC@NLO and CKKW schemes combining collinear and hard jets.

	MC@NLO (Herwig)	CKKW (Sherpa)
hard jets	first jet correct	all jets correct
collinear jets	all jets correct, tuned	all jets correct, tuned
normalization	correct to NLO	correct to LO plus real emission
variants	Powheg,...	MLM–Alpgen, MadEvent,...

can for example simulate $Z + n$ jets production at the LHC, where all we have to do is cut off the number of jets at some point where we cannot compute the matrix element anymore. This combination will describe all jets correctly over the entire collinear and hard phase space. In Fig. 3.3 we show the number of jets produced in association with a pair of top quarks and a pair of heavy new states at the LHC. The details of these heavy scalar gluons are secondary for the basic features of these distributions, the only parameter which matters is their mass, *i.e.* the hard scale of the process which sets the factorization scale and defines the upper limit of collinearly enhanced initial-state radiation. We see that heavy states tend to come with several jets radiated with transverse momenta up to 30 GeV, where most of these jets vanish once we require transverse momenta of at least 100 GeV. Looking at this figure you can immediately see that a suggested analysis which for example asks for a reconstruction of two W decay jets better give you a very good argument why it should not we swamped by combinatorics.

Looking at the individual columns in Fig. 3.3 there is one thing we have to keep in mind: each of the merged matrix elements combined into this sample is computed at leading order, the emission of real particles is included, while virtual corrections are not (completely) there. In other words, in contrast to MC@NLO this procedure gives us all jet distributions but leaves the normalization free, just like an old-fashioned Monte Carlo. The main features and shortcomings of the two merging schemes are summarized in Table 3.2. A careful study of the associated theory errors for example for Z+jets production and the associated rates and shapes I have not yet come across, but watch out for it.

As mentioned before — there is no such thing as a free lunch, and it is up to the competent user to pick the scheme which suits their problem best. If there is a well-defined hard scale in the process, the old-fashioned Monte Carlo with a tuned parton shower will be fine, and it is by far the fastest method. Sometimes we are only interested in one hard jet, so we can use MC@NLO and benefit from the correct normalization. And in other cases we really need a large number of jets correctly described, which means CKKW and some external normalization. This decision is not based on chemistry, philosophy or sports, it is based on QCD. What we LHC phenomenologists have to do is to get it right and know why we got it right.

On the other hand I am not getting tired of emphasizing that the conceptual progress in QCD describing jet radiation for all transverse-momentum scales is absolutely crucial for LHC analyses. If I were a string theorist I would definitely call this achievement a revolution or even two, like 1917 but with the trombones and cannons of Tchaikovsky's 1812. In contrast to a lot of progress in theoretical physics jet merging solves a very serious problem which would have limited our ability to understand LHC data, no matter what kind of Higgs or new physics we are looking for. And I am not sure if I got the message across — the QCD aspects behind it are not trivial at all. If you feel like looking at a tough problem, try to prove that CKKW and MLM work for initial-state and final-state radiation...

Before we move on, let me illustrate why in Higgs or exotics searches at the LHC we really care about this kind of progress in QCD. One way to look for heavy particles decaying into jets, leptons and missing energy is the variable

$$
\begin{aligned}
H_T &= \not{E}_T + \sum_j E_{T,j} + \sum_\ell E_{T,\ell} \\
&= \not{p}_T + \sum_j p_{T,j} + \sum_\ell p_{T,\ell} \qquad \text{(for massless quarks, leptons)} \quad (3.46)
\end{aligned}
$$

which for gluon-induced QCD processes should be as small as possible, while the signal's scale will be determined by the new particle masses. For the background process Z+jets, this distribution as well as the missing energy distribution using CKKW as well as a parton shower (both from Sherpa) are shown in Fig. 3.4. The two curves beautifully show that the naive parton shower is not a good description of QCD background processes to the production of heavy particles. We can probably use a chemistry approach and tune the parton shower to correctly describe the data even in this parameter region, but we would most likely violate basic concepts like factorization. How much you care about this violation is up to you, because we know that there is a steep gradient in theory standards from first-principle calculations of hard scattering all the way to hadronization string models...

3.5. Simulating LHC events

In the third main section I will try to cover a few topics of interest to LHC physicists, but which are not really theory problems. Because they are

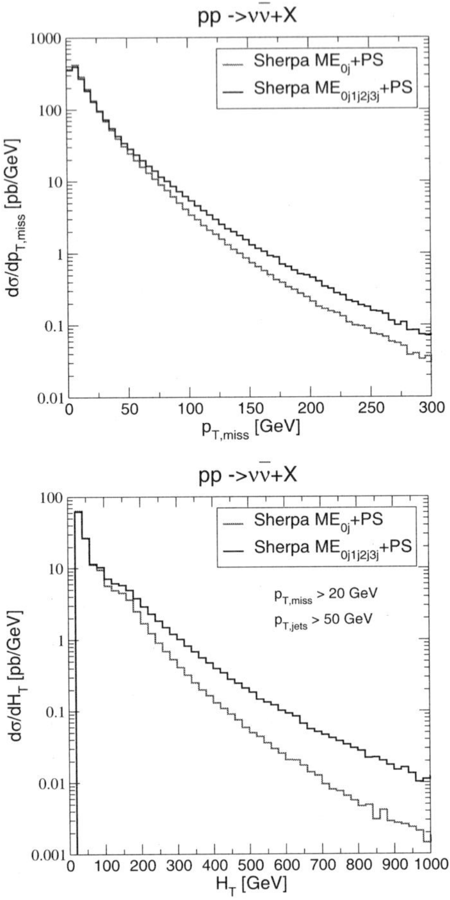

Fig. 3.4. Transverse momentum and H_T distributions for Z+jets production at the LHC. The two curves correspond to the Sherpa parton shower starting from Drell–Yan production and the fully merged sample including up to three hard jets. These distributions describe typical backgrounds for searches for jets plus missing energy, which could originate in supersymmetric squark and gluino production. Thank you to Steffen Schumann and Sherpa for providing these Figures.

crucial for our simulations of LHC signatures and can turn into sources of great embarrassment when we get them wrong in public.

3.5.1. *Missing energy*

Some of the most interesting signatures at the LHC involve dark matter particles. Typically, we would produce strongly interacting new particles

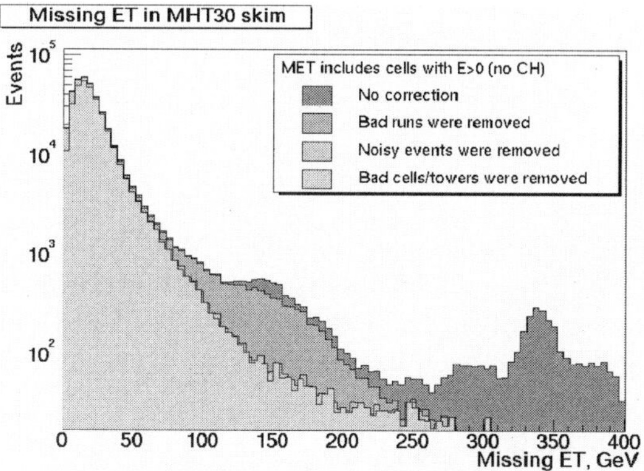

Fig. 3.5.   Missing energy distribution from the early running phase of the DZero experiment at the Tevatron. This figure I got from Beate Heinemann's lectures web site.

which then decay to the weakly interacting dark matter agent. On the way, the originally produced particles have to radiate quarks or gluons, to get rid of their color charge. If they also radiate leptons, those can be very useful to trigger on the events and reduce QCD backgrounds.

At the end of the last section we talked about the proper simulation of W+jets and Z+jets backgrounds to such signals. It turns out that jet merging predicts considerably larger missing transverse momentum from QCD sources, so theoretically we are on fairly safe ground. However, this is not the whole story of missing transverse momentum. I should say that I skipped most of this section, because Peter Wittich knows much more about it and covered it really nicely. But it might nevertheless be useful to include it in this writeup.

Figure 3.5 is a historic missing transverse energy distribution from DZero. It nicely illustrates that by just measuring missing energy, Tevatron would have discovered supersymmetry with two beautiful peaks in the missing-momentum distribution around 150 GeV and around 350 GeV. However, this distribution has nothing to do with physics, it is purely a detector effect.

The problem of missing energy can be illustrated with a simple number: to identify and measure a lepton we need around 500 out of 200000 calorimeter cells in an experiment like Atlas, while for missing energy we need all of them. Therefore, we need to understand our detectors really well to even cut on a variable like missing transverse momentum, and for this level of understanding we need time and luminosity. Unless something goes wrong with the machine, I would not expect us to find anything reasonable in early-running LHC data including a missing energy cut — really, we should not use the phrases 'missing energy' and 'early running' in the same sentences or papers.

There are three sources of missing energy which our experimental colleagues have to understand before we get to look at such distributions:

First, we have to subtract bad runs. This means that for a few hours parts of the detector might not have worked properly. We can identify such bad runs by looking at Standard Model physics, like gauge bosons, and remove them from the data sample.

Next, there is usually coherent noise in the calorimeter. Of 200000 cells we know that some of them will individually fail or produce noise. However, some sources of noise, like leaking voltage or other electronic noise can be correlated geometrically, *i.e.* coherent. Such noise will lead to beautiful missing momentum signals. In the same spirit, there might also be particles crossing our detector, but not coming from the interaction point. Such particles can be cosmic rays or errand beam radiation, and they will lead to unbalanced energy deposition in the calorimeter. The way to get rid of such noise is again looking for Standard Model candles and remove sets of events where such problems occur.

The third class of fake missing energy is failing calorimeter cells, like continuously hot cells or dead cells, which can be removed after we know the detector really well.

Once we understand all the source of fake missing momentum we can focus on real missing momentum. This missing transverse momentum is trivially computed from the momentum measurement of all tracks seen in the detector. This means that any uncertainty on these measurements, like the jet or lepton energy scale will smear the missing momentum. Moreover, we know that there is for example dead matter in the detector, so we have to compensate for this. This compensation is obviously a global correction to individual events, which means it will generally smear the missing energy distribution. So when we compute a realistic missing transverse momentum distribution at the LHC we have to smear all jet and lepton momenta, and in addition apply a Gaussian *smearing* of the order

$$\frac{\Delta \slashed{E}_T}{\text{GeV}} \sim \frac{1}{2} \sqrt{\frac{\sum E_T}{\text{GeV}}} \gtrsim 20 \qquad (3.47)$$

While this sounds like a trivial piece of information I cannot count the number of papers I get to referee where people forgot this smearing and discovered great channels to look for Higgs bosons or new physics at the LHC which completely fall apart when experimentalists take a careful look. Here comes another great piece of phenomenology wisdom: phenomenological studies are right or wrong based on the outcome if they can be reproduced by real experimentalists and real detectors — at least once we make sure our experimentalist friends did not screw it up again....

3.5.2. *Phase space integration*

At the very beginning of this lecture we discussed how to compute the total cross section for interesting processes. What we skipped is how to numerically compute such cross sections. Obviously, since the parton densities are not known in a closed analytical form, we will have to rely on numerical integration tools. Looking at a simple $2 \to 2$ process we can write the total cross section as

$$\sigma_{\text{tot}} = \int d\phi \int d\cos\theta \int dx_1 \int dx_2 \, F_{\text{PS}} \, |\mathcal{M}|^2 = \int_0^1 dy_1 \cdots dy_4 \, J_{\text{PS}}(\vec{y}) \, |\mathcal{M}|^2 \tag{3.48}$$

The different factors are shown in Eq. (3.20). In the second step we have rewritten the phase space integral as an integral over the four–dimensional unit cube, with the appropriate Jacobian. Like any integral we can numerically evaluate this phase space integral by binning the variable we integrate over:

$$\int_0^1 dy \, f(y) \quad \longrightarrow \quad \sum_j (\Delta y)_j f(y_j) \sim \Delta y \sum_j f(y_j) \tag{3.49}$$

Whenever we talk about numerical integration we can without any loss of generality assume that the integration boundaries are 0...1. The integration variable y we can divide into a discrete set of points y_j, for example defined as equi-distant on the y axis or by choosing some kind of random number $y_j \epsilon [0, 1]$. In the latter case we need to keep track of the bin widths $(\Delta y)_j$. In a minute, we will discuss how such a random number can be chosen in more or less smart ways; but before we discuss how to best evaluate such an integral numerically, let us first illustrate that this integral is much more useful than just providing the total cross section. If we are interested in a distribution of an observable, like for example the distribution of the transverse momentum of a muon in the Drell–Yan process, we need to compute $d\sigma(p_T)/dp_T$. This distribution is given by:

$$\sigma = \int dy_1 \cdots dy_N \; f(\vec{y}) = \int dy_1 \; \frac{d\sigma}{dy_1}$$

$$\left. \frac{d\sigma}{dy_1} \right|_{y_1^0} = \int dy_2 \cdots dy_N \; f(y_1^0) = \int dy_1 \cdots dy_N \; f(\vec{y}) \, \delta(y_1 - y_1^0) \qquad (3.50)$$

We can compute this distribution numerically in two ways. One way would be to numerically evaluate the $y_2 \cdots y_N$ integrations and just leave out the y_1 integration. The result will be a function of y_1 which we can evaluate at any point y_1^0. This method is what I for example used for Prospino, when I was a graduate student. The second and much smarter option corresponds to the last term in the equation above, with the delta distribution defined for discretized y_1. This is not hard to do: first, we define an array the size of the number of bins in the y_1 integration. Then, for each y_1 value of the complete $y_1 \cdots y_N$ integration we decide where it goes in this array and add $f(\vec{y})$ to this array. And finally, we print $f(y_1)$ to see the distribution. This array is referred to as a histogram and can be produced for example using the CernLib. This *histogram approach* does not look like much, but imagine you want to compute a distribution $d\sigma/dp_T$, where $p_T(\vec{y})$ is a complicated function of the integration variables, so you want to compute:

$$\frac{d\sigma}{dp_T} = \int dy_1 \cdots dy_N \; f(\vec{y}) \, \delta\left(p_T(\vec{y}) - p_T^0\right) \qquad (3.51)$$

Histograms mean that when we compute the total cross section entirely numerically we can trivially extract all distributions in the same process.

The procedure outlined above has an interesting interpretation. Imagine we do the entire phase space integrations numerically. Just like computing the interesting observables we can compute the momenta of all external particles. These momenta are not all independent, because of energy–momentum conservation, but this can be taken care of. The tool which translates the vector of integration variables $\vec{y}$ into the external momenta is called a *phase space generator*. Because the phase space is not uniquely defined in terms of the integration variables, the phase space generator also has to return the Jacobian J_{PS}, the phase space weight. If we think of the integration as an integration over the unit cube, this weight is combined with the matrix element squared $|\mathcal{M}|^2$. Once we compute the unique phase space configuration $(k_1, k_2, p_1 \cdots p_M)_j$ which corresponds to the vector $\vec{y}_j$ the combined weight $W = J_{\mathrm{PS}} |\mathcal{M}|^2$ is simply the probability that this configuration will appear at the LHC. Which means, we do not only integrate over the phase space, we really simulate events at the LHC. The only complication is that the probability of a certain configuration is not only given my the frequency with which it appears, but also by the additional explicit weight. So when we run our numerical integration through the phase space generator and histogram all the distributions we are interested in we really generate *weighted events*. These events, *i.e.* the momenta of all external particles and the weight W, we can for example store in a big file.

This simulation is not quite what experimentalists want — they want to represent the probability of a certain configuration appearing only by its frequency. This means we have to unweight the events and translate the weight into frequency. To achieve this we normalize all our event weights to the maximum weight W_{max}, *i.e.* compute the ratio $W_j/W_{\mathrm{max}}\epsilon[0,1]$, generate a flatly distributed random number $r\epsilon[0,1]$, and keep the event if $W_j/W_{\mathrm{max}} > r$. This guarantees that each event j survives with a probability W_j/W_{max}, which is exactly what we want — the higher the weight the more likely the event stays. The challenge in this translation is only that we will lose events, which means that our distributions will if anything become more ragged. So if it were not for the experimentalists we would never use *unweighted events*. I should add that experimentalists have a good reason to want such unweighted events, because they feed best through their detector simulations.

The last comment is that if the phase space configuration $(k_1, k_2, p_1 \cdots p_M)_j$ can be measured, its weight W_j better be positive. This is not trivial once we go beyond leading order. There, we need to add several contributions to produce a physical event, like for example different n–particle final states, and there is no need for all of them to be positive. All we have to guarantee is that after adding up all contributions and after integrating over any kind of unphysical degree of freedom we might have introduced, the probability of a physics configuration is positive. For example, negative values for parton densities are not problematic, as long as we always have a positive hadronic rate $d\sigma_{pp\to X} > 0$.

The numerical phase space integration for many particles faces two problems. First, the partonic phase space for M on-shell particles in the final state has $3(M+2) - 3$ dimensions. If we divide each of these directions in 100 bins, the number of phase space points we need to evaluate for a $2 \to 4$ process is $100^{15} = 10^{30}$, which is not realistic.

To integrate over a large number of dimensions we use *Monte Carlo integration.* In this approach we define a distribution $p_Y(y)$ such that for a one-dimensional integral we can replace the binned discretized integral in Eq. (3.49) with a discretized version based on a set of random numbers Y_j over the y integration space

$$\langle g(Y) \rangle = \int_0^1 dy\, p_Y(y)\, g(y) \quad \longrightarrow \quad \frac{1}{N} \sum_j g(Y_j) \qquad (3.52)$$

All we have to make sure is that the probability of returning Y_j is given by $p_Y(y)$ for $y < Y_j < y + dy$. This form has the advantage that we can naively generalize it to any number of n dimensions, just by organizing the random numbers Y_j in one large vector instead of an n-dimensional array.

Our n-dimensional phase space integral listed above we can rewrite the same way:

$$\int_0^1 d^n y\, f(y) = \int_0^1 d^n y\, \frac{f(y)}{p_Y(y)}\, p_Y(y) = \left\langle \frac{f(Y)}{p_Y(Y)} \right\rangle \quad \longrightarrow \quad \frac{1}{N} \sum_j \frac{f(Y_j)}{p_Y(Y_j)}$$

$$(3.53)$$

In other words, we have written the phase space integral in a discretized way which naively does not involve the number of dimensions any longer. All we have to do to compute the integral is average over N phase space values of f/p_Y. In the ideal case where we exactly know the form of the integrand and can map it into our random numbers, the error of the numerical integration will be zero. So what we have to find is a way to encode $f(Y_j)$ into $p_Y(Y_j)$. This task is called *importance sampling* and you will have to find some documentation for example on Vegas to look at the details.

Technically, you will find that Vegas will call the function which computes the weight $W = J_{\text{PS}}|\mathcal{M}|^2$ for a number of phase space points and average over these points, but including another weight factor W_{MC} representing the importance sampling. If you want to extract distributions via histograms you have to therefore add the total weight $W = W_{\text{MC}} J_{\text{PS}} |\mathcal{M}|^2$ to the columns.

The second numerical challenge is that the matrix elements for interesting processes are by no means flat, and we would like to help our adaptive (importance sampling) Monte Carlo by defining the integration variables such that the integrand is as flat as possible. Take for example the integration over the partonic momentum fraction, where the integrand is usually falling off at least as $1/x$. So we can substitute

$$\int_\delta dx \, \frac{C}{x} = \int_{\log \delta} d\log x \, \left(\frac{d\log x}{dx}\right)^{-1} \frac{C}{x} = \int_{\log \delta} d\log x \, C \qquad (3.54)$$

and improve our integration significantly. Moving on to a more relevant example: particularly painful are intermediate particles with Breit–Wigner propagators squared, which we need to integrate over the momentum $s = p^2$ flowing through:

$$P(s,m) = \frac{1}{(s-m^2)^2 + m^2\Gamma^2} \qquad (3.55)$$

For example the Standard-Model Higgs boson with a mass of 120 GeV has a width around 0.005 GeV, which means that the integration over the invariant mass of the Higgs decay products $\sqrt{s}$ requires a relative resolution of 10^{-5}. Since this is unlikely to be achievable, what we should really do is find a substitution which produces the inverse Breit–Wigner as a Jacobian and leads to a flat integrand — et voilá

$$\begin{aligned}
\int ds \, \frac{C}{(s-m^2)^2 + m^2\Gamma^2} &= \int dz \, \left(\frac{dz}{ds}\right)^{-1} \frac{C}{(s-m^2)^2 + m^2\Gamma^2} \\
&= \int dz \, \frac{(s-m^2)^2 + m^2\Gamma^2}{m\Gamma} \, \frac{C}{(s-m^2)^2 + m^2\Gamma^2} \\
&= \frac{1}{m\Gamma} \int dz \, C \qquad \text{with} \qquad \tan z = \frac{s-m^2}{m\Gamma}
\end{aligned}$$
$$(3.56)$$

This is the coolest *phase space mapping* I have seen, and it is incredibly useful. Of course, an adaptive Monte Carlo will eventually converge on such an integrand, but a well-chosen set of integration parameters will speed up our simulations significantly.

3.5.3. *Helicity amplitudes*

When we compute a transition amplitude, what we usually do is write down all spinors, polarization vectors, interaction vertices and propagators and square the amplitude analytically to get $|\mathcal{M}|^2$. Of course, nobody does gamma–matrix traces by hand anymore, instead we use powerful tools like Form. But we can do even better. As an example, let us consider the simple process $u\bar{u} \to \gamma^* \to \mu^+\mu^-$. The structure of the amplitude in the Dirac indices involves one vector current on each side ($\bar{u}_f \gamma_\mu u_f$). For each $\mu = 0 \cdots 3$ this object gives a c-number, even though the spinors have four components and each gamma matrix is a 4×4 matrix as well. The intermediate photon propagator has the form $g_{\mu\nu}/s$, which is a simple number as well and implies a sum over μ in both of the currents forming the matrix element.

Instead of squaring this amplitude symbolically we can first compute it numerically, just inserting the correct numerical values for each component of each spinor etc, without squaring it. MadGraph is a tool which automatically produces a Fortran routine which calls the appropriate functions from the Helas library, to do precisely that. For our toy process the MadGraph output looks roughly like:

```
      REAL*8 FUNCTION UUB_MUPMUM(P,NHEL)
C
C FUNCTION GENERATED BY MADGRAPH
C RETURNS AMPLITUDE SQUARED SUMMED/AVG OVER COLORS
C FOR PROCESS : u u~ -> mu+ mu-
C
      INTEGER    NGRAPHS,    NEIGEN,    NEXTERNAL
      PARAMETER (NGRAPHS=  1,NEIGEN=  1,NEXTERNAL=4)
      INTEGER    NWAVEFUNCS    , NCOLOR
      PARAMETER (NWAVEFUNCS=  5, NCOLOR=   1)

      REAL*8 P(0:3,NEXTERNAL)
      INTEGER NHEL(NEXTERNAL)

      INCLUDE 'coupl.inc'

      DATA Denom(1  )/            1/
      DATA (CF(i,1  ),i=1  ,1  ) /     3/

      CALL IXXXXX(P(0,1   ),ZERO ,NHEL(1   ),+1,W(1,1   ))
      CALL OXXXXX(P(0,2   ),ZERO ,NHEL(2   ),-1,W(1,2   ))
      CALL IXXXXX(P(0,3   ),ZERO ,NHEL(3   ),-1,W(1,3   ))
      CALL OXXXXX(P(0,4   ),ZERO ,NHEL(4   ),+1,W(1,4   ))
      CALL JIOXXX(W(1,1   ),W(1,2   ),GAU ,ZERO    ,ZERO    ,W(1,5   ))
      CALL IOVXXX(W(1,3   ),W(1,4   ),W(1,5   ),GAL ,AMP(1   ))
      JAMP(   1) = +AMP(    1)

      DO I = 1, NCOLOR
         DO J = 1, NCOLOR
            ZTEMP = ZTEMP + CF(J,I)*JAMP(J)
         ENDDO
         UUB_MUPMUM =UUB_MUPMUM+ZTEMP*DCONJG(JAMP(I))/DENOM(I)
      ENDDO
      END
```

The input to this function are the external momenta and the helicities of all fermions in the process. Remember that helicity and chirality are identical only for massless fermions. In general, chirality is defined as the eigenvalue of the projectors $(1 \pm \gamma_5)/2$, while helicity is defined as the projection of the spin onto the momentum direction, or as the left or right handedness. For each point in phase space and each helicity combination (± 1 for each external fermion) MadGraph computes the matrix element using Helas routines like for example:

- $\texttt{IXXXXX}(p, m, n_{\text{hel}}, n_{\text{sf}}, F)$ computes the wave function of a fermion with incoming fermion number, so either an incoming fermion or an outgoing anti-fermion. As input it requires the 4-momentum, the mass and the helicity of this fermion. Moreover, this particle with incoming fermion number can be a particle or an anti-particle. This means $n_{\text{fs}} = +1$ for the incoming u and $n_{\text{sf}} = -1$ for the outgoing μ^+, because the particles in MadGraph are defined as u and μ^-. The fermion wave function output is a complex array $F(1:6)$. Its first two entries are the left-chiral part of the fermionic spinor, *i.e.* $F(1:2) = (1-\gamma_5)/2\, u$ or $F(1:2) = (1-\gamma_5)/2\, v$ for $n_{\text{sf}} = \pm 1$. The entries $F(3:4)$ are the right-chiral spinor. These four numbers can be computed from the 4-momentum, if we know the helicity of the particles. Because for massless particles helicity and chirality are identical, our massless quarks and leptons will for example have only entries $F(1:2)$ for $n_{\text{hel}} = -1$ and $F(3:4)$ for $n_{\text{hel}} = +1$. The last two entries contain the 4-momentum in the direction of the fermion flow, namely $F(5) = n_{\text{sf}}(p(0) + ip(3))$ and $F(6) = n_{\text{sf}}(p(1) + ip(2))$. The first four entries in this spinor correspond to the size of each γ matrix, which is usually taken into account by computing the trace of the chain of gamma matrices.

- $\texttt{OXXXXX}(p, m, n_{\text{hel}}, n_{\text{sf}}, F)$ does the same for a fermion with outgoing fermion flow, *i.e.* our incoming $\bar{u}$ and our outgoing μ^-. The left-chiral and right-chiral components now read $F(1:2) = \bar{u}(1-\gamma_5)/2$ and $F(3:4) = \bar{u}(1+\gamma_5)/2$, and similarly for the spinor $\bar{v}$. The last two entries are $F(5) = n_{\text{sf}}(p(0)+ip(3))$ and $F(6) = n_{\text{sf}}(p(1)+ip(2))$.

- $\texttt{JIOXXX}(F_i, F_o, g, m, \Gamma, J_{io})$ computes the (off-shell) current for the vector boson attached to the two external fermions F_i and F_o. The coupling $g(1:2)$ is a complex array with the interaction of the left-chiral and right-chiral fermion in the upper and lower index. Obviously, we need to know the mass and the width of the intermediate vector boson. The output array J_{io} again has six components:

$$J_{io}(\mu+1) = -\frac{i}{q^2}\, F_o^T\, \gamma^\mu \left(g(1)\, \frac{1-\gamma_5}{2} + g(2)\, \frac{1+\gamma_5}{2} \right) F_i$$

$$J_{io}(5) = -F_i(5) + F_o(5) \sim -p_i(0) + p_o(0) + i\left(-p_i(3) - p_o(3)\right)$$

$$J_{io}(6) = -F_i(6) + F_o(6) \sim -p_i(1) + p_o(1) + i\left(-p_i(2) + p_o(2)\right)$$

$$\Rightarrow \quad q^\mu = (\mathrm{Re}J_{io}(5), \mathrm{Re}J_{io}(6), \mathrm{Im}J_{io}(6), \mathrm{Im}J_{io}(5)) \tag{3.57}$$

The last line illustrates why we need the fifth and sixth arguments of F_{io}. The first four entries in J_{io} correspond to the index μ in this vector current, while the index j of the spinors has been contracted between F_o^T and F_i.

- $\texttt{IOVXXX}(F_i, F_o, J, g, V)$ computes the amplitude of a fermion–fermion–vector coupling using the two external fermionic spinors F_i and F_o and an incoming vector current J. Again, the coupling $g(1:2)$ is a complex array, so we numerically compute

$$F_o^T \, \jmath \left(g(1) \, \frac{1 - \gamma_5}{2} + g(2) \, \frac{1 + \gamma_5}{2} \right) F_i \qquad (3.58)$$

We see that all indices j and μ of the three input arguments are contracted in the final result. Momentum conservation is not explicitly enforced by $\texttt{IOVXXX}$, so we have to take care of it beforehand.

Given the list above it is easy to see how MadGraph computes the amplitude for $u\bar{u} \to \gamma^* \to \mu^+\mu^-$. First, it always calls the wave functions for all external particles and puts them into the array $W(1:6, 1:4)$. The vectors $W(*, 1)$ and $W(*, 3)$ correspond to $F_i(u)$ and $F_i(\mu^+)$, while $W(*, 2)$ and $W(*, 4)$ mean $F_o(\bar{u})$ and $F_o(\mu^-)$. The first vertex we evaluate is the $\bar{u}\gamma u$ vertex, which given $F_i = W(*, 1)$ and $F_o = W(*, 2)$ uses $\texttt{JIOXXX}$ to compute the vector current for the massless photon in the s channel. Not much would change if we instead chose a massive Z boson, except for the arguments m and Γ in the $\texttt{JIOXXX}$ call. The $\texttt{JIOXXX}$ output is the photon current $J_{io} \equiv W(*, 5)$. The second step combines this current with the two outgoing muons in the $\mu^+\gamma\mu^-$ vertex. Since this number gives the final amplitude, it should return a c-number, no array. MadGraph calls $\texttt{IOVXXX}$ with $F_i = W(*, 3)$ and $F_o = W(*, 4)$, combined with the photon current $J = W(*, 5)$. The result $\texttt{AMP}$ is copied into $\texttt{JAMP}$ without an additional sign which could have come from the ordering of external fermions. The only remaining sum left to compute before we square $\texttt{JAMP}$ is the color structure, which in our simple case means one color structure with a color factor $N_c = 3$.

Of course, to calculate the transition amplitude MadGraph requires all masses and couplings. They are transferred through common blocks in the file coupl.inc and computed elsewhere. In general, MadGraph uses unitary gauge for massive vector bosons, because in the helicity amplitude approach it is easy to accommodate complicated tensors, in exchange for a large number of Feynman diagrams.

The function UUB_MUPMUM described above is not yet the full story. Remember that when we square $\mathcal{M}$ symbolically we need to sum over the spins of the outgoing states to transform a spinor product of the kind $u\bar{u}$ into the residue or numerator of a fermion propagator. To obtain the final result numerically we also need to sum over all possible helicity combinations of the external fermions, in our case $2^4 = 16$ combinations.

```fortran
      SUBROUTINE SUUB_MUPMUM(P1,ANS)
C
C FUNCTION GENERATED BY MADGRAPH
C RETURNS AMPLITUDE SQUARED SUMMED/AVG OVER COLORS
C AND HELICITIES FOR THE POINT IN PHASE SPACE P(0:3,NEXTERNAL)
C
C FOR PROCESS : u u~ -> mu+ mu-
C
      INTEGER     NEXTERNAL,    NCOMB,
      PARAMETER (NEXTERNAL=4, NCOMB= 16)
      INTEGER     THEL
      PARAMETER (THEL=NCOMB*1)

      REAL*8 P1(0:3,NEXTERNAL),ANS

      INTEGER NHEL(NEXTERNAL,NCOMB),NTRY
      REAL*8 T, UUB_MUPMUM
      INTEGER IHEL,IDEN,IC(NEXTERNAL)
      INTEGER IPROC,JC(NEXTERNAL)
      LOGICAL GOODHEL(NCOMB)

      DATA GOODHEL/THEL*.FALSE./
      DATA NTRY/0/

      DATA (NHEL(IHEL,  1),IHEL=1,4) / -1, -1, -1, -1/
      DATA (NHEL(IHEL,  2),IHEL=1,4) / -1, -1, -1,  1/
      DATA (NHEL(IHEL,  3),IHEL=1,4) / -1, -1,  1, -1/
      DATA (NHEL(IHEL,  4),IHEL=1,4) / -1, -1,  1,  1/
      DATA (NHEL(IHEL,  5),IHEL=1,4) / -1,  1, -1, -1/
      DATA (NHEL(IHEL,  6),IHEL=1,4) / -1,  1, -1,  1/
      DATA (NHEL(IHEL,  7),IHEL=1,4) / -1,  1,  1, -1/
      DATA (NHEL(IHEL,  8),IHEL=1,4) / -1,  1,  1,  1/
      DATA (NHEL(IHEL,  9),IHEL=1,4) /  1, -1, -1, -1/
      DATA (NHEL(IHEL, 10),IHEL=1,4) /  1, -1, -1,  1/
      DATA (NHEL(IHEL, 11),IHEL=1,4) /  1, -1,  1, -1/
      DATA (NHEL(IHEL, 12),IHEL=1,4) /  1, -1,  1,  1/
      DATA (NHEL(IHEL, 13),IHEL=1,4) /  1,  1, -1, -1/
      DATA (NHEL(IHEL, 14),IHEL=1,4) /  1,  1, -1,  1/
      DATA (NHEL(IHEL, 15),IHEL=1,4) /  1,  1,  1, -1/
      DATA (NHEL(IHEL, 16),IHEL=1,4) /  1,  1,  1,  1/
      DATA (  IC(IHEL,  1),IHEL=1,4) /  1,  2,  3,  4/
      DATA (IDEN(IHEL),IHEL=  1,  1) /  36/

      NTRY=NTRY+1

      DO IHEL=1,NEXTERNAL
         JC(IHEL) = +1
      ENDDO

      DO IHEL=1,NCOMB
         IF (GOODHEL(IHEL,IPROC) .OR. NTRY .LT. 2) THEN
            T = UUB_MUPMUM(P1,NHEL(1,IHEL),JC(1))
            ANS = ANS + T
```

```
      IF (T .GT. ODO .AND. .NOT. GOODHEL(IHEL,IPROC)) THEN
         GOODHEL(IHEL,IPROC)=.TRUE.
      ENDIF
    ENDIF
  ENDDO
  ANS = ANS/DBLE(IDEN)

  END
```

The important part of this subroutine is the list of possible helicity combinations stored in the array $\text{NHEL}(1:4, 1:16)$. Adding all different helicity combinations (of which some might well be zero) means a loop over the second argument and a call of UUB_MUPMUM with the respective helicity combination. The complete spin–color averaging factor is included as IDEN and given by $2 \times 2 \times N_c^2 = 36$. So MadGraph indeed provides us with a subroutine SUUB_MUPMUM which numerically computes $\overline{|\mathcal{M}|^2}$ for each phase space point, *i.e.* external momentum configuration. MadGraph also produces a file with all Feynman diagrams contributing to the given subprocess, in which the numbering of the external particles corresponds to the second argument of W and the argument of AMP is the numbering of the Feynman diagrams. After looking into the code very briefly we can also easily identify different intermediate results W which will only be computed once, even if they appear several times in the different Feynman diagrams.

The helicity method might not seem particularly appealing for a simple $2 \to 2$ process, but it makes it easily possible to compute processes with four and more particles in the final state and up to 10000 Feynman diagrams which we could never square symbolically, no matter how many graduate students' lives we turn into hell.

3.5.4. *Errors*

As argued in the very beginning of the lecture, LHC physics always means extracting signals from often large backgrounds. This means, a correct error estimate is crucial. For LHC calculations we are usually confronted with three types of errors.

The first and easiest one are the *statistical errors*. For small numbers of events these experimental errors are described by Poisson statistics, and for large numbers they converge to the Gaussian limit. And that is about the only complication we encounter for them.

The second set of errors are *systematic errors*, like for example the calibration of the jet and lepton energy scales, the measurements of the luminosity, or the efficiencies to identify a muon as a muon. Some of you might remember what happened last, when a bunch of theorists mistook a forward pion for an electron — that happened right around my TASI, and people had not only discovered supersymmetry, but also identified its breaking mechanism. Of course, our experimentalist CDF lecturer told us immediately that the whole thing was a joke. Naively, we would not assume that systematic are Gaussian, but remember that we determine these numbers largely from well-understood background processes. Such counting experiments in background channels like $Z \rightarrow$ leptons, however, do behave Gaussian. The only caveat is the shape of far-away tails, which can turn out to be bigger than the exponentially suppressed Gaussian shape.

The last source of errors are *theory errors*, and they are hardest to model, because they are dominated by higher-order QCD effects, fixed order or enhanced by large logarithms. If we could compute all remaining higher-order terms, we would do so, which means everything else is a wild guess. Moreover, higher-order effects are not any more likely to give a relative K factor of 1.0 than 0.9 or 1.1. In other words, theory errors cannot have a peak and they are definitely not Gaussian. There is a good reason to choose the Gaussian short cut, because we know that folding three Gaussian errors gives us another Gaussian error, which makes things so much easier. But this lazy approach assumes the we know much more about QCD than we actually do, so please stop lying. On the other hand, we also know that theory errors cannot be arbitrarily large. Unless there is a very good reason, a K factor for a total LHC cross section should not be larger than something like 3. If that were the case, we would conclude that perturbative QCD breaks down, and the proper description of error bars would be our smallest problem. In other words, the centrally flat theory probability distribution for an LHC observable has to go to zero for very large deviations from the currently best value.

A good solution to this problem is the so-called *Rfit scheme*, used for example by the CKMfitter or the SFitter collaborations. It starts from the assumption that for very large deviations there will always be tails from the experimental errors, so we can neglect the impact of the theory errors on this range. In the center of the distribution we simply cut open the experimental Gaussian-type distribution and insert a flat theory piece. We could also modify the transition region by changing for example the width of the experimental Gaussian error as an effect of a falling-off theory error, but in the simplest model we just use a log-likelihood $\chi^2 = -2 \log \mathcal{L}$ given a set of measurements $\vec{d}$ and in the presence of a general correlation matrix C

$$\chi^2 = \vec{\chi}_d^T \, C^{-1} \, \vec{\chi}_d$$

$$\chi_{d,i} = \begin{cases} 0 & |d_i - \bar{d}_i| < \sigma_i^{(\text{theo})} \\[2ex] \dfrac{d_i - \bar{d}_i + \sigma_i^{(\text{theo})}}{\sigma_i^{(\text{exp})}} & d_i - \bar{d}_i < -\sigma_i^{(\text{theo})} \\[2ex] \dfrac{d_i - \bar{d}_i - \sigma_i^{(\text{theo})}}{\sigma_i^{(\text{exp})}} & d_i - \bar{d}_i > \sigma_i^{(\text{theo})} \, . \end{cases} \tag{3.59}$$

And that is it, all three sources of LHC errors can be described correctly, and nothing stops us from computing likelihood maps to measure the top mass or identify new physics or just have some fun in life at the expense of the Grid.

Further reading, acknowledgments, etc.

This is the point where the week in beautiful Boulder is over and I should thank K.T. and his Boulder team as well as our two organizers for their kind invitation. I typed most of these notes in Boulder's many nice cafes and 11 years after I went here as a student TASI and Boulder still make the most enjoyable and most productive school in our field. Whoever might ever think about moving it away from Boulder cannot possibly have the success of the school in mind.

It has been great fun, even though QCD has a reputation of being a dry topic. I hope you enjoyed learning it as much as I enjoyed learning it while teaching it. Just like most of you I am really only a QCD user, but for an LHC phenomenologists there is no excuse for not knowing the relevant aspects of QCD. Have fun in the remaining lectures, write some nice theses, and I hope I will see as many of you as possible over the coming 20 years. LHC physics need all the help we can get, and it is great fun, so please come and join us!

Of course there are many people I need to thank for helping me write these notes: Fabio Maltoni, Johan Alwall and Steffen Schumann for having endured a great number of critical questions and for convincing me that jet merging is the future; Steffen Schumann, Ben Allanach and Tom DeGrand for their comments on this draft; Beate Heinemann for providing me with one of the most interesting plots from the Tevatron and for answering many stupid questions over the years — as did Dirk Zerwas and Kyle Cranmer.

You note that this writeup, just like the lectures, is more of an informal chat about LHC physics than a proper review paper. But if I had not cut as many corners we would never have made it to the fun topics. In the same spirit, there is no point in giving you a list of proper original references, so I would rather list a few books and review articles which might come in handy if you would like to know more:

- I started learning high–energy theory including QCD from Otto Nachmann's book. I still use his appendices to look up Feynman rules, because I have rarely seen another book with as few (if not zero) typos.[1] Similar, but maybe a little more modern is the primer by Cliff Burgess and Guy Moore.[2] At the end of it you will find more literature tips.
- For a more specialized book on QCD have a look at the pink book by Ellis, Stirling, Webber. It includes everything you ever wanted to know about QCD.[3] Maybe a little more phenomenology you can find in Günther Dissertori, Ian Knowles and Michael Schmelling's book on QCD and phenomenology.[4]
- If you would like to learn how to for example compute higher-order cross sections to Drell–Yan production, Rick Field works it all out in his book.[5]
- Unfortunately, there is comparably little literature on jet merging yet. The only review I know is by Michelangelo Mangano and Tim Stelzer.[6] There is a very concise discussion included with the comparison of the different models.[7] If you want to know more, you will have to consider the original literature or wait for the review article which Frank Krauss and Peter Richardson promised to write for Journal of Physics G.
- Recently, I ran across George Sterman's TASI lectures. They are comparably formal, but they are a great read if you know something about QCD already.[8]
- For MC@NLO there is nothing like the original papers. Have a look at Bryan Webber's and Stefano Frixione's work and you cannot but understand what it is about[9]!
- For CKKW, look at the original paper. It beautifully explains the general idea on a few pages, at least for final state radiation.[10]

– If you are using Madgraph to compute helicity amplitudes there is the original bright green documentation which describes every routine in detail. You might want to check the format of the arrays, if you use for example the updated version inside MadEvent.[11]

References

1. O. Nachtmann, "Elementary Particla Physics: Concepts and Phenomena", *Berlin, Germany: Springer (1990) 559 p.*
2. C. P. Burgess and G. D. Moore, "The Standard Model: A Primer", *Cambridge, UK: Cambridge Univ. Pr. (2007) 542 p.*
3. R. K. Ellis, W. J. Stirling and B. R. Webber, "QCD and collider physics", Camb. Monogr. Part. Phys. Nucl. Phys. Cosmol. **8**, 1 (1996).
4. G. Dissertori, I. G. Knowles and M. Schmelling, "QCD — High Energy Experiments and Theory", *Oxford, UK: Clarendon (2003) 538 p.*
5. R. D. Field, "Applications of Perturbative QCD", *Redwood City, USA: Addison-Wesley (1989) 366 p. (Frontiers in physics, 77).*
6. M. L. Mangano and T. J. Stelzer, "Tools For The Simulation Of Hard Hadronic Collisions", Ann. Rev. Nucl. Part. Sci. **55**, 555 (2005), CERN-PH-TH-2005-074.
7. J. Alwall *et al.*, "Comparative study of various algorithms for the merging of parton showers and matrix elements in hadronic collisions", Eur. Phys. J. C **53**, 473 (2008) [arXiv:0706.2569 [hep-ph]].
8. G. Sterman, "QCD and jets", arXiv:hep-ph/0412013.
9. S. Frixione and B. R. Webber, "Matching NLO QCD computations and parton shower simulations", JHEP **0206**, 029 (2002) [arXiv:hep-ph/0204244].
10. S. Catani, F. Krauss, R. Kuhn and B. R. Webber, "QCD matrix elements + parton showers", JHEP **0111**, 063 (2001) [arXiv:hep-ph/0109231].
11. H. Murayama, I. Watanabe and K. Hagiwara, "HELAS: HELicity amplitude subroutines for Feynman diagram evaluations".

Chapter 4

Collider Signal I : Resonance

Tim M.P. Tait

Argonne National Laboratory, Argonne, IL 60439
Northwestern University, 2145 Sheridan Road, Evanston, IL 60208,
tait@northwestern.edu

These TASI lectures were part of the summer school in 2008 and cover the collider signal associated with resonances in models of physics beyond the Standard Model. I begin with a review of the Z boson, one of the best-studied resonances in particle physics, and review how the Breit-Wigner form of the propagator emerges in perturbation theory and discuss the narrow width approximation. I review how the LEP and SLAC experiments could use the kinematics of Z events to learn about fermion couplings to the Z. I then make a brief survey of models of physics beyond the Standard Model which predict resonances, and discuss some of the LHC observables which we can use to discover and identify the nature of the BSM physics. I finish up with a discussion of the linear moose that one can use for an effective theory description of a massive color octet vector particle.

4.1. Introduction: The Z Boson

To begin with, let us look at the ordinary Z boson of the Standard Model. I am a big fan of using the Standard Model as a vehicle toward understanding new physics, and indeed, the Z boson is a perfect example of a vector resonance, one that illustrates almost any phenomena one could expect to encounter in the resonances of more exotic theories. We will discuss resonances with different spins when we turn to theories of physics beyond the Standard model further on. This review of the Z comes with two warnings: 1) it will by no means be complete (see the LEP EWWG report[1] for more details) and lacks any attempt at proper referencing, and 2) I made no real effort to match conventions with the rest of the world. Since I derived everything from scratch, it should be mostly self-consistent, but

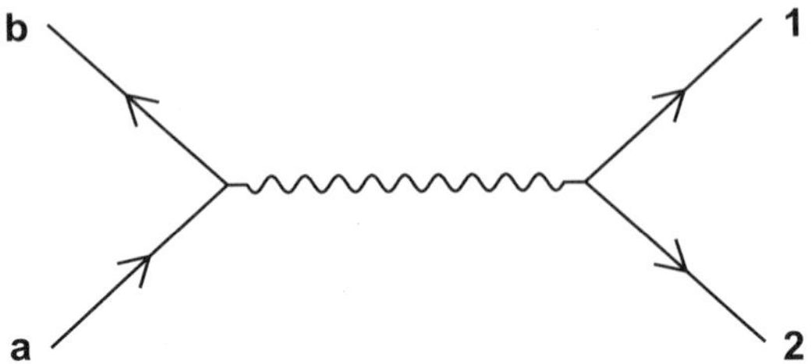

Fig. 4.1.   Feynman diagram showing $e^+e^- \to f\bar{f}$ through an intermediate photon or Z boson.

be careful when consulting another reference in tandem.

4.1.1. $e^+e^- \to f\bar{f}$

To begin with, we consider e^+e^- scattering into a pair of fermions, $f\bar{f}$. Some slight care is needed when the f's are themselves electrons, so we implicitly assume for now that $f \neq e$. In the Standard Model, this scattering is described at leading order in perturbation theory by two Feynman diagrams, one with an s-channel photon, and one with an s-channel Z. An example of one of these graphs is shown in Figure 4.1, where the labels indicate the incoming momenta p_a and p_b and outgoing momenta p_1 and p_2. The matrix element is given (in the unitary gauge) by,

$$\mathcal{M} = e\,[\bar{v}_b \gamma^\mu u_a]\,\frac{-g_{\mu\nu}}{s+i\epsilon}\,Q_f e\,[\bar{u}_1 \gamma^\nu v_2] + \tag{4.1}$$

$$[\bar{v}_b \gamma^\mu (g_R^e P_R + g_L^e P_L) u_a]\,\frac{-g_{\mu\nu} + p_\mu^Z p_\nu^Z / M_Z^2}{s - M_Z^2 + i\epsilon}\,\left[\bar{u}_1 \gamma^\nu (g_R^f P_R + g_L^f P_L) v_2\right]$$

where $P_{L/R}$ are chiral projectors, the labels on the four-spinors u and v remind us which momentum they take as their arguments, $s \equiv (p_a + p_b)^2$ is the usual Mandelstam variable, e is the electromagnetic coupling, Q_f the charge of f, and $g_{L/R}^i$ are the (chiral) Z boson couplings to fermion i,

$$g^i = \frac{e}{\sin\theta_W \cos\theta_W}\left(T_3^i - Q^i \sin^2\theta_W\right) \tag{4.2}$$

where T_3 is the third component of weak iso-spin and θ_W the weak mixing angle.

For now, let's consider unpolarized scattering, summing over the final state spins and averaging over the initial spins. For simplicity, I will assume $s \gg m_f$ and drop m_f in the calculation. For LEP, this was a good approximation for any fermion we had enough energy to produce anyway.

$$\overline{|\mathcal{M}|^2} = \frac{1}{4} \left\{ \frac{e^4 Q_f^2}{s^2} \text{Tr}\left[\not{p}_b \gamma^\mu \not{p}_a \gamma^\nu\right] \text{Tr}\left[\not{p}_1 \gamma^\mu \not{p}_2 \gamma^\nu\right] \right. \tag{4.3}$$

$$+ \frac{e^2 Q_f}{s(s - M_Z^2)} 2\text{Re}\,\text{Tr}\left[\not{p}_b \gamma^\mu \not{p}_a \gamma^\nu (g_R^e P_R + g_L^e P_L)\right] \text{Tr}\left[\not{p}_1 \gamma^\mu \not{p}_2 \gamma^\nu (g_R^f P_R + g_L^f P_L)\right]$$

$$\left. + \frac{1}{(s - M_Z^2)^2} \text{Tr}\left[\not{p}_b \gamma^\mu \not{p}_a \gamma^\nu (g_R^{e2} P_R + g_L^{e2} P_L)\right] \text{Tr}\left[\not{p}_1 \gamma^\mu \not{p}_2 \gamma^\nu (g_R^{f2} P_R + g_L^{f2} P_L)\right] \right\}$$

where I have dropped the $+i\epsilon$ terms in the Feynman propagator, which I do not need here. If f is a quark, I will also need to sum over their (mutual) colors, which will produce a factor of $N_c = 3$. Performing the traces leads to,

$$\overline{|\mathcal{M}|^2} = \frac{1}{4} \left\{ 4e^4 Q_f^2 (1 + \cos^2\theta) \right. \tag{4.4}$$

$$+ \frac{e^2 Q_f s}{(s - M_Z^2)} \left[(g_L^e + g_R^e)(g_L^f + g_R^f)(1 + \cos^2\theta) + 2(g_L^e - g_R^e)(g_L^f - g_R^f)\cos\theta \right]$$

$$\left. + \frac{s^2}{(s - M_Z^2)^2} \left[(g_L^{e2} + g_R^{e2})(g_L^{f2} + g_R^{f2})(1 + \cos^2\theta) + 2(g_L^{e2} - g_R^{e2})(g_L^{f2} - g_R^{f2})\cos\theta \right] \right\}$$

where θ is the scattering angle of f in the center-of-mass frame. We will come back to this expression in more detail after we fix up one important feature.

This expression for $\overline{|\mathcal{M}|^2}$ is very simply related to the differential cross section,

$$\frac{d\sigma}{d\cos\theta} = \frac{1}{64\pi s} \overline{|\mathcal{M}|^2} \ . \tag{4.5}$$

Since $\overline{|\mathcal{M}|^2}$ approaches a constant as $s \to 0$ or $s \to \infty$, we can see that the cross section is enhanced if we choose small center-of-mass energies (but note that this does not continue arbitrarily, since at some point we will not have enough energy to produce the $f\bar{f}$ final state and the process will switch off), and suppressed at very high energies.

The last term of Eq. (4.4) shows that the cross section is enhanced if we choose $s \simeq M_Z^2$. This is the resonant behavior we're looking for, but it also shows that the leading order to which we have been working is not cutting it. The cross section may be enhanced around the Z mass, but clearly this result, which says it should diverge, is unphysical*. To do better, we

---

*If you thought the Feynman $+i\epsilon$'s could save us, note that since $\epsilon \to 0$ they are just a temporary regulation of the problem.

must consider improving our calculation with some higher order effects in perturbation theory.

4.1.2. *Resummed Propagator*

To simplify the discussion, but without dropping any important details, in my discussion of higher order effects I will neglect the fact that the Z boson has spin, and treat it like a scalar. I will also restrict myself to the fermion sector, and ignore their spins too. So as far as the loop calculation goes, I am looking at the correction induced to a *scalar* Z boson by its coupling to a pair of *scalar* quarks or leptons. The full case, carrying around the vector and spin indices, is left as an exercise for the reader. It's not hard, just a little more messy. Seriously.

I actually need the first non-trivial correction to the propagator, the one that arises at one loop. However, I need to resum the important parts of it at all orders. Let us see how this works. The full correction to the propagator contains diagrams such as:

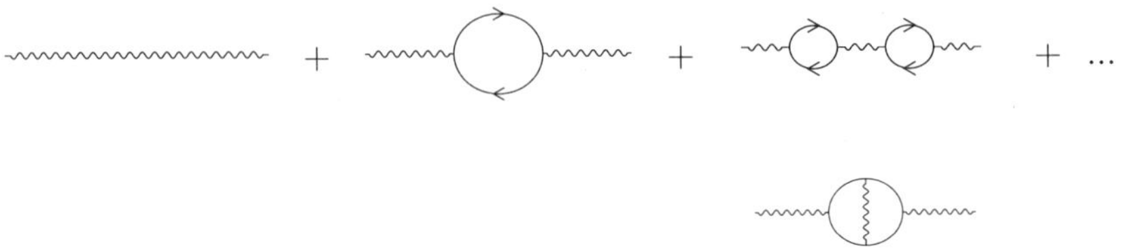

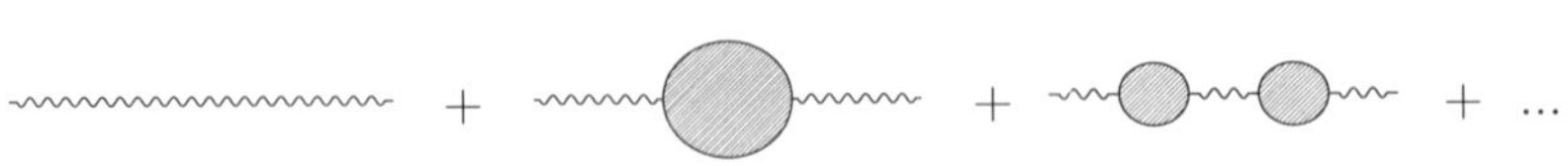

In addition to the genuine two loop and higher order corrections, we also have a two-loop term that is just the square of the one-loop correction itself. At every order n, there is a term that looks like the one-loop correction raised to the nth power. In fact, if I reorder my perturbative series so that instead of being organized by powers of the coupling, it is instead organized by the number of internal Z boson propagators, I can write this series as, where the blob represents the *one-particle irreducible* correction to the Z boson propagation:

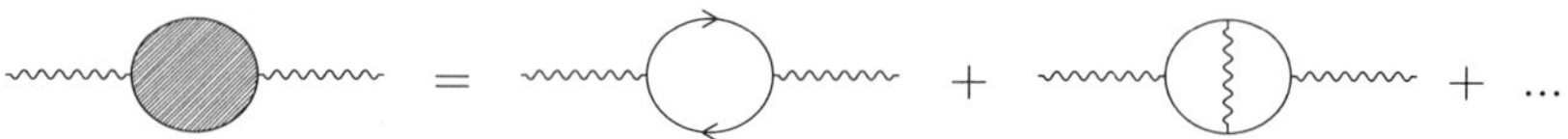

which I will call $i\Sigma(s)$ and will eventually compute, in our scalarized theory, to one loop in perturbation theory[†].

Organized this way, the propagator looks like a geometric series,

$$\frac{i}{s - M_Z^2} + \frac{i}{s - M_Z^2} i\Sigma(s) \frac{i}{s - M_Z^2}$$

$$+ \frac{i}{s - M_Z^2} i\Sigma(s) \frac{i}{s - M_Z^2} i\Sigma(s) \frac{i}{s - M_Z^2} + \ldots$$

$$= \frac{i}{s - M_Z^2} \left\{ 1 + i\Sigma(s) \frac{i}{s - M_Z^2} + \left(i\Sigma(s) \frac{i}{s - M_Z^2} \right)^2 + \ldots \right\}$$

$$= \frac{i}{s - M_Z^2} \left(\frac{1}{1 + \frac{\Sigma(s)}{s - M_Z^2}} \right) = \frac{i}{s - M_Z^2 + \Sigma(s)}$$

(4.6)

By computing $\Sigma(s)$ to whatever order in perturbation theory I like, and then using this modified propagator, I am capturing *some* effects at all orders in perturbation theory. This is great, but it does have some dangers, because while I get some effects at all orders, I am also missing some effects at *every* order[‡]. In general, I do need to be careful with this trick, because important features like gauge invariance hold order by order in perturbation theory. By keeping parts of a given order, I am potentially jeopardizing important cancellations and so forth. This issue will not bite us here, but it is good to keep the subtleties clearly in (the back of one's) mind even while not being rigorous.

We can write the most general expression for $\Sigma(s)$,

$$\Sigma(s) = Z^{-1}(s)s + B(s) + i\gamma(s) \tag{4.7}$$

[†] Note that Σ, as a Lorentz scalar, can at most be a function of s. It was precisely to avoid dealing with its tensor structure that I went to a scalarized toy example.
[‡] This is a general feature of resummations, and is i.e., part of the headache we get when we try to improve a parton shower at higher orders.

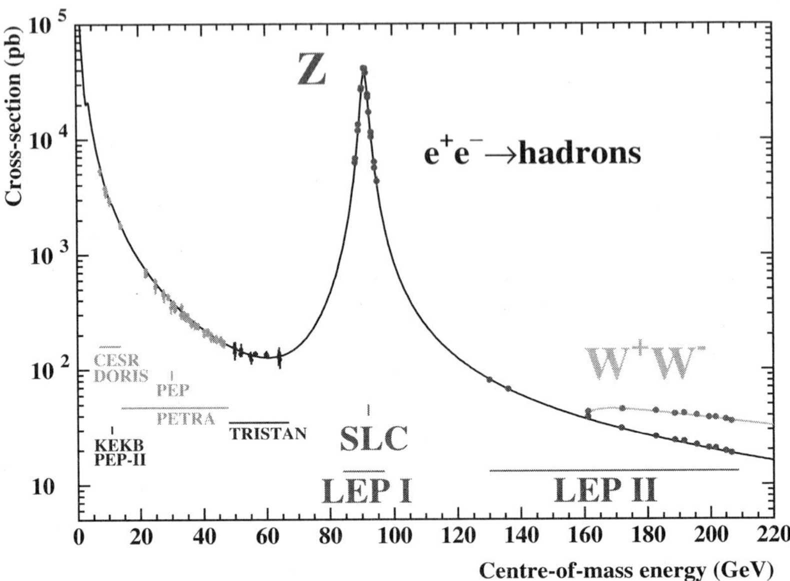

Fig. 4.2. The cross section as a function of E_{CM} after resumming the propagator (from the LEP EWWG[1]).

or in other words, there is a dimensionless complex coefficient $Z^{-1}(s)$ (the wave function renormalization) which multiplies s itself, and two real functions B and γ which have dimensions of mass^2 , but whose dimensions are not made up from s itself. In our problem, they both must be proportional to M_Z^2, since there is no other mass scale (having assumed $m_f \to 0$) at hand.

The functions $Z^{-1}(s)$ and $B(s)$ turn out to be divergent. They also are not going to help with our problem at $s \simeq M_Z^2$, because Z^{-1} just rescales the entire amplitude, and $B(s)$ just shifts the place where it occurs by some amount. The divergent functions need to be defined by a renormalization scheme; the obvious one for this problem is the *on-shell* scheme, for which,

$$Z^{-1}(M_Z^2) \equiv 1 \qquad B(M_Z^2) \equiv 0 \tag{4.8}$$

which allows us to ignore both $Z^{-1}(s)$ and $B(s)$ for $s \simeq M_Z^2$. These really just amount to requiring that the parameter M_Z in the Lagrangian is defined to be the center of the Z boson resonance, and that the field is canonically normalized there.

The $\gamma(s)$ term is the one we are looking for. It regulates the divergence, resulting in the most divergent term (the third term of Eq. (4.4)) becoming,

$$\overline{|\mathcal{M}|^2} \propto \frac{1}{\left(s - M_Z^2\right)^2 + \gamma^2} \tag{4.9}$$

As $s \to M_Z^2$, the cross section no longer goes to infinity, but instead is proportional to $1/\gamma^2$ (with a Breit-Wigner shape for s close, but not equal to M_Z^2). The parameter $\gamma(M_Z^2)$ controls both the magnitude of the peak in the distribution, and also the shape as the cross section falls off, slightly away from the peak position. In Figure 4.2, we show the result with the non-zero γ included, illustrating how γ regulates the behavior for $s \simeq M_Z^2$.

4.1.3. γ in the Toy Theory

Now let us compute γ in the scalarized theory. If you are very comfortable doing such calculations, I suggest you skip down to the next subsection, where we will interpret the results.

At leading order in perturbation theory, we need the graph,

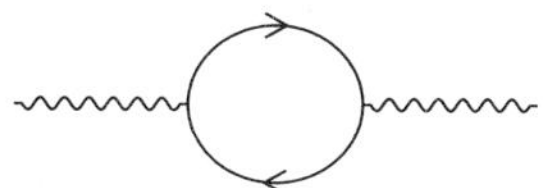

with (scalarized) leptons $\nu_e, \nu_\mu, \nu_\tau, e, \mu, \tau$ and quarks u, d, s, c, b, t running inside the loops. For a single one of these graphs, our loop correction is,

$$g^2 \int \frac{d^4\ell}{(2\pi)^4} \frac{1}{\ell^2 - m_f^2 + i\epsilon} \frac{1}{(\ell + p)^2 - m_f^2 + i\epsilon} \tag{4.10}$$

(remember that this loop is divergent, but that won't hurt us since ultimately we are after γ which is not). p_μ is the incoming/outgoing external momentum; p^2 is s in our previous discussion. We can rewrite this using the Feynman parameterization,

$$\frac{1}{AB} = \int_0^1 dx \frac{1}{[xA + (1-x)B]^2} \, , \tag{4.11}$$

which after some algebra results in,

$$g^2 \int \frac{d^4\ell}{(2\pi)^4} dx \left[\ell^2 + 2(1-x)\ell \cdot p + (1-x)p^2 - m_f^2 + i\epsilon\right]^{-2} \, . \tag{4.12}$$

Adding and subtracting $(1-x)^2 p^2$ inside the square brackets allows us to complete a square, and then changing integration variables to $\ell'_\mu \equiv \ell_\mu + (1-x)p_\mu$ yields,

$$g^2 \int \frac{d^4\ell'}{(2\pi)^4} dx \frac{1}{[\ell'^2 - \Delta + i\epsilon]^2} \qquad (4.13)$$

where

$$\Delta \equiv -x(1-x)p^2 + m_f^2 \ . \qquad (4.14)$$

To perform the integration over ℓ', we move into Euclidean space, taking $\ell_E^0 = i\ell'_0$, $\vec{\ell}_E = \vec{\ell}'$ (so $\ell'^2 \to -\ell_E^2$). The $+i\epsilon$'s allow us to deform the path integration back to the real ℓ_E^0 axis. Having kept them around long enough to keep us honest at this step, we can now afford to ignore them. The loop integral becomes,

$$g^2 \int \frac{d^4\ell_E}{(2\pi)^4} dx \frac{1}{[\ell_E^2 + \Delta]^2} = g^2 \frac{2\pi^2}{(2\pi)^4} \int_0^\infty d\ell_E^2 \int_0^1 dx \ \frac{\ell_E^2}{[\ell_E^2 + \Delta]^2} \ . \ (4.15)$$

We now shift ℓ_E^2 by Δ and cut the divergent integral off at some large scale Λ, resulting in,

$$\frac{g^2}{8\pi^2} \int_\Delta^{\Lambda^2} d\ell_E^2 \int_0^1 dx \left\{ \frac{1}{\ell_E^2} - \frac{\Delta}{\ell_E^4} \right\} \qquad (4.16)$$

$$= \frac{g^2}{8\pi^2} \int_0^1 dx \left\{ \log\left(\frac{\Lambda^2}{\Delta}\right) - 1 + \frac{\Delta}{\Lambda^2} \right\}$$

(where we note in passing that there is no term proportional to p^2, implying that Z^{-1} vanishes at the one loop order in our toy theory). The last term vanishes as we take the cut-off Λ to infinity. The middle term clearly does not contain an imaginary part. So to get a non-zero result for γ, we need the log to develop a branch cut, which will happen if $\Delta < 0$,

$$\log\left(\frac{\Lambda^2}{\Delta}\right) \to i\pi + \log\left(\frac{\Lambda^2}{|\Delta|}\right) \qquad (\Delta < 0) \qquad (4.17)$$

Provided this happens for some values of x in the range of its integration, the imaginary part of the self-energy correction is given by,

$$\gamma(p^2) = \frac{g^2}{8\pi} \int_{\Delta < 0} dx = \frac{g^2}{8\pi} \left(x^+ - x^- \right) \qquad (4.18)$$

where $x^\pm$ is the largest (smallest) value of x for which $\Delta < 0$. Note that the dependence on the regulator Λ has vanished for the imaginary part, justifying my initial statement that the imaginary part of Σ was not divergent.

From the definition of Δ, Eq. (4.14), and the fact that for $0 \leq x \leq 1$, $x(1-x)$ is maximal at $x = 1/2$, we see that to have a region with $\Delta < 0$, we need a large enough p^2 such that,

$$p^2 \geq 4m_f^2 \ . \tag{4.19}$$

This implies that we can ignore any particle in the loop whose mass is less than half of the center-of-mass energy – such particles induce contributions to $Z^{-1}(p^2)$ and $B(p^2)$, but they do not contribute to the imaginary part of the self-energy. At LEP energies, this means we didn't actually need the top quark in the loop.

Provided $p^2 \geq 4m_f^2$,

$$x^{\pm} = \frac{1}{2}\left(-1 \pm \sqrt{1 - \frac{4m_f^2}{p^2}}\right) \ , \tag{4.20}$$

for which,

$$\gamma(p^2) = \frac{g^2}{8\pi}\sqrt{1 - \frac{4m_f^2}{p^2}} \ . \tag{4.21}$$

4.1.4. *Back to* $e^+ e^- \to f\bar{f}$

For the region that sparked our interest in γ to begin with, $p^2 \simeq M_Z^2$, we can further recognize,

$$\gamma(M_Z^2) = M_Z\Gamma(Z \to f\bar{f}) \ . \tag{4.22}$$

We thus see at for energies close to the mass of the Z, the imaginary part of the one-loop contribution has produced the leading order decay width. Summing over all fermions for which $p^2 \geq 4m_f^2$, the imaginary parts of each of the one-loop amplitudes simply adds, and $\gamma(M_Z^2) = M_Z\Gamma_Z$ reproduces the inclusive decay width.

The fact that the width has appeared is not an accident – it is a general feature of quantum field theory, guaranteed by the optical theorem[2] as applied to the propagator. The optical theorem relates the imaginary part of the scattering $a \to b$ to the amplitudes for all possible final states $a \to F$ and $b \to F$,

$$2 \operatorname{Im} \mathcal{M}(a \to b) = \sum_F \int d\Pi_F \mathcal{M}^*(b \to F)\mathcal{M}(a \to F) \tag{4.23}$$

where $d\Pi_F$ is the integral over the phase space of F. Applied to the self-energy for the Z boson, the left-hand side of the equation (at one loop) becomes $2\,\mathrm{Im}\,\mathcal{M}(Z \to Z) = 2\gamma(M_Z^2)$ and we have,

$$\gamma(M_Z^2) = \frac{1}{2}\sum_F \int d\Pi_F |\mathcal{M}(Z \to F)|^2 \equiv M_Z \Gamma_Z \qquad (4.24)$$

Since nothing in our argument made explicit reference to the Z itself, it is clear that this will work for any particle which has decay amplitudes into on-shell states. In practice, this is how we compute γ in terms of the the width using only tree-level Feynman diagrams.

What we've learned is that the Z boson propagator has an imaginary part which receives contributions from every particle into which the Z can decay. At very low and very high energies, we don't notice the width because it is small, arising from higher order in perturbation theory than the zeroth order propagator. Around the Z mass, the width is crucial, because the zeroth order propagator vanishes, and the higher order effects are leading.

4.1.5. *The Narrow Width Approximation*

In the limit $\Gamma/M \to 0$, the Breit-Wigner becomes a δ-distribution:

$$\frac{1}{(s - M^2)^2 + M^2\Gamma^2} \;\to\; \frac{\pi}{M\Gamma}\delta(s - M^2) \qquad (\Gamma/M \to 0) \;(4.25)$$

In this limit we can only produce the Z boson on-shell. The δ-function puts the intermediate boson on-shell, and at leading order factorizes the production from the decay. The cross section becomes,

$$\sigma(e^+e^- \to f\bar{f}) \qquad\qquad\qquad\qquad\qquad\qquad (4.26)$$
$$\to \frac{1}{32\pi s}\int d\Pi_f d\Pi_{\bar{f}} |\mathcal{M}(e^+e^- \to Z|^2 \frac{\pi}{M\Gamma}\delta(s - M^2)|\mathcal{M}(Z \to f\bar{f}|^2$$
$$= \sigma(e^+e^- \to Z)\frac{\Gamma(Z \to f\bar{f})}{\Gamma} = \sigma(e^+e^- \to Z)\mathrm{BR}(Z \to f\bar{f})$$

the expected product of on-shell production times the branching ratio into $f\bar{f}$. For most weakly coupled resonances, this approximation holds very well.

In the narrow width limit, the unpolarized cross section for $e^+e^- \to f\bar{f}$

"on the pole" becomes,

$$\frac{d\sigma}{d\cos\theta} = \frac{1}{32}\frac{M}{\Gamma}\delta(s - M^2) \times \tag{4.27}$$

$$\left\{ 2(g_L^{e2} - g_R^{e2})(g_L^{f2} - g_R^{f2})\cos\theta + (g_L^{e2} + g_R^{e2})(g_L^{f2} + g_R^{f2})(1 + \cos^2\theta) \right\}$$

which reveals how we can start to learn about the couplings to the Z. When computing the inclusive cross section, integrated over all θ, the first term in the curly braces integrates to zero. So the inclusive cross section is proportional to the sum of the left- and right-handed couplings squared.

This is also a good place to remark on an interesting feature of the SM Z boson – because the weak mixing angle takes a particular value, $\sin^2\theta_W \simeq 1/4$, the electron couplings (see Eq. (4.2)) approximately satisfy $g_L^e \simeq -g_R^e$. This implies that there is an approximate cancellation in the coefficient of the first term in the curly braces for the SM Z. For the real Z boson, that is fine – it is the physics we have been handed. However, it does imply that taking the SM Z as the most generic possible model for a resonance may lead us to conclude that any effect from that term is suppressed, and this may not be a generic feature of new physics. So when you hear the words "Sequential Standard Model Z'" in an experimental search, do not be fooled.

Moving back to measuring the couplings, we saw that the production cross section is sensitive to the sum of the squares of the left- and right-handed couplings. What we would like to do next is to unravel the relative amounts of right- versus left-handed couplings. One way to do that is to follow the SLC route and polarize the incoming beams. For example, we can just count events when the electron is left-handed compared to the number of events when it is right-handed, and form the ratio,

$$A_{LR} \, (= A_e) \; \equiv \; \frac{N_L - N_R}{N_L + N_R} = \frac{g_L^{e2} - g_R^{e2}}{g_L^{e2} + g_R^{e2}} \tag{4.28}$$

the information about the final state cancels out in the ratio.

The other way to disentangle right- from left-handed couplings is to measure forward-backward asymmetries. From the differential cross section on the Z pole, it is clear that the way to do that is to pick up sensitivity to the $\cos\theta$ term in the differential cross section. θ is the angle between the incoming electron beam (which we know) and the out-going particle, f. Provided we can tell f from the anti-particle $\bar{f}$ in a given event[§], we can

---

[§]It's not always easy. For charged leptons it is moderately simple, and for heavy quarks with semi-leptonic decays, it can be done. For light quarks, it is probably hopeless.

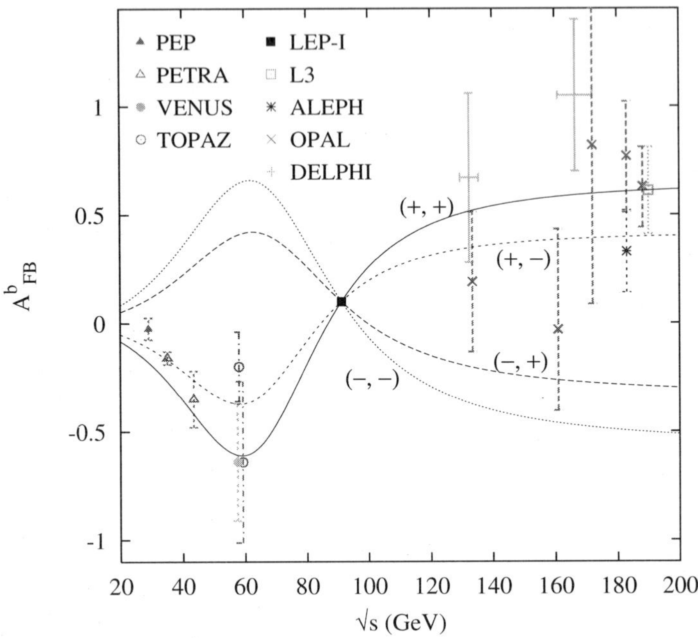

Fig. 4.3.   The forward-backward asymmetry of the bottom quark as a function of center-of-mass energy, showing data and the four possible sign combinations with magnitudes set by the SM (from Ref. [3]).

measure θ. If we define a forward particle to lie in the region $0 \leq \cos\theta \leq 1$ and a backward particle to be in the region $-1 \leq \cos\theta \leq 0$, we arrive at,

$$A^f_{FB} \equiv \frac{N_F - N_B}{N_F + N_B} = \frac{3}{4} \left(\frac{g^{e2}_L - g^{e2}_R}{g^{e2}_L + g^{e2}_R} \right) \left(\frac{g^{f2}_L - g^{f2}_R}{g^{f2}_L + g^{f2}_R} \right) = \frac{3}{4} A_e A_f \quad (4.29)$$

Since the cross section on the Z pole depends only on the square of couplings, we cannot actually get information about the absolute or relative signs of the couplings from polarized or forward-backward measurements. This is where the photon actually becomes a help instead of just a nuisance. From Eq. (4.4), we see that the interference terms contain single powers of Z couplings and single powers of electromagnetic couplings, and are sensitive to the relative signs. This is illustrated in Figure 4.3, which shows A^b_{FB} as a function of center of mass energy, for the four possible sign combinations with the magnitudes of g^b_L and g^b_R set as in the SM. The four curves meet at the Z mass, and diverge at higher and lower energies, driven by the different admixture of Z-γ interference.

4.1.6. *The Lone Resonance:* $\nu\bar{\nu} \to f\bar{f}$

To better understand the Z resonance without the complication of the photon exchange, let us consider a process which we cannot really do realistically in the lab. If we consider inelastic neutrino scattering $\nu\bar{\nu} \to f\bar{f}$, there is a single Feynman diagram, Fig 4.1 with only the Z boson exchanged. Since right-handed neutrinos (if they exist at all...) do not couple to the Z, the inclusive cross section simplifies to,

$$\sigma(\nu\bar{\nu} \to f\bar{f}) = \frac{g_L^{\nu 2}}{96\pi} \frac{s}{(s - M^2)^2 + M^2\Gamma^2}(g_L^{f2} + g_R^{f2}) . \qquad (4.30)$$

We've already discussed ad nauseum the Breit-Wigner behavior around $s \simeq M^2$. For $s \ll M^2$, we can Taylor-expand the propagator and find a cross section which grows with energy, but is suppressed by the large mass. For $s \gg M^2$, the cross section falls as $1/s$, consistent with the requirements of unitarity. To summarize,

$$\sigma(s) \sim \begin{cases} s/M^4 & (s \ll M^2) \\ s/[(s - M^2)^2 + M^2\Gamma^2] & (s \sim M^2) \\ 1/s & (s \gg M^2) \end{cases} \qquad (4.31)$$

When interference is not important, these three regimes summarize the most important behavior of an exchanged "resonance". When $s \ll M^2$, the resonance is effective integrated out, and the cross section represents the four-fermion interaction it leaves behind in the low energy theory.

4.2. Models with Resonances

A new resonance can be classified by its basic properties such as spin and transformation under the $SU(3)_c \times SU(2)_W \times U(1)_Y$ SM gauge symmetries. Below, I will run through some of our favorite models which lead to such objects, with some of their known LHC phenomenology interleaved as appropriate.

4.2.1. Z'

Experimentalists sometimes call any resonance (usually decaying into leptons) a Z', which is a useful classification for them because it organizes many models with similar signatures together. For theorists, the standard definition is a color singlet, electrically neutral, vector particle. As a vector particle, a Z' is a typical signal when the SM gauge groups are extended. That could occur in a simple way, such as promoting the gauge sector to $SU(3)_c \times SU(2)_W \times U(1)_Y \times U(1)'$, or we could enlarge either $SU(2)_L$ or $U(1)_Y$ (or both), or unify the entire SM gauge structure into a higher rank GUT such as $SO(10)$. In any of these options, extra massive gauge fields result when we break these larger symmetries down to the SM itself.

If we take a $U(1)'$ extension of the SM as a prototype, we have a well defined framework with well defined parameters. We can add terms to describe the propagation of the new vector (V_μ),

$$\mathcal{L} = -\frac{1}{4}(F'_{\mu\nu})^2 + \frac{\zeta}{4}F'_{\mu\nu}F_Y^{\mu\nu} + \frac{1}{2}M^2 V_\mu V^\mu \tag{4.32}$$

The first term is a usual kinetic term for a vector particle. The second term induces kinetic mixing between the Z' and hyper-charge, which causes the Z' to pick up some fraction of coupling to all SM fields proportional to their SM hypercharges.[4] The final term is actually a stand-in for an entire new Higgs sector which gives mass to the Z'. For colliders, this new Higgs sector usually boils down to just the mass of the Z', as we have written here. However, one may choose the mass of this new scalar to be light enough it can be produced at colliders, and couple it, say, to the usual SM Higgs, affecting Higgs physics.

A Standard Model matter field Ψ will couple to the Z' through its charge under $U(1)'$ in terms of an extended covariant derivative,

$$D_\mu \Psi \to D_\mu^{(SM)}\Psi - ig' z_\Psi V_\mu \Psi \tag{4.33}$$

where g' is the Z''s universal gauge coupling, which may be a free parameter or may be predicted if the Z' descends from a higher rank group in which the SM is also embedded. Every matter field may have its own charge z, which again may be predicted some models.

If we want the SM fermion masses to come from renormalizable inter-actions, we should choose the z's such that they permit us to write such terms down. For example, the top Yukawa coupling would require,

$$z_{t_R} - z_{Q_3} + z_H = 0 \; . \tag{4.34}$$

Note, however, that most of the SM fermion masses are counter-intuitively small, and by assigning charges which violate $U(1)'$, we can actually engineer their observed sizes to some degree - this is one particular UV realization of a Frogatt-Nielsen model of flavor.[5]

Many models[6] choose the Higgs charge z_H to be zero. This choice removes the danger that the ordinary Higgs VEV will induce Z-Z' mixing. Given the success of the electroweak fit, such mixing is generically limited at the 10^{-3} level,[1] though it is possible in a multi-Higgs model to play the VEVs against one another so that the mixing is small despite the charges being non-zero. That is how the canonical E_6 GUTs survive precision measurements.[7] Another set of bounds on the charges and mass come about from LEP-II, which studied the reaction $e^+e^- \to f\bar{f}$ above the Z-pole. We saw in the previous section that at energies far below the mass of the Z', its contribution to to the cross section goes like $\sim s/M^2$ (when the Z' is the only mediator of the reaction, the cross section goes like s/M^4, but in this case the interference with the SM photon and Z goes like the SM rate times $\sim s/M^2$. As a result, LEP-II is able to derive limits on M/g' for different patterns of couplings to fermions.[6,8]

A further guide to the charges of the fermions is provided by anomaly cancellation. In order for the gauge symmetry to survive at the quantum level, a model with an extra gauge symmetry should not have gauge anomalies. However, the fact that the symmetry is spontaneously broken raises the possibility that the anomalies may cancelled by additional fermions which are chiral under $U(1)'$ (and are usually vector-like under the SM gauge symmetries in order to preserve the success of anomaly cancellation in the SM). If such fermions have masses around the mass of the Z' itself, it is unlikely they will have much effect on the collider phenomenology of the Z'.

A final concern for the fermion charges comes from flavor-changing neutral currents (FCNCs). If the Z' charges are not universal over the three families, we will generically have tree level FCNCs, which in the least will require the Z' to be heavy to avoid low energy bounds.[9] Let's see how this works using the right-handed down-quarks as an example. If the charges are not Universal, then before EWSB we will have interactions with the Z' such as

$$g' \begin{bmatrix} \bar{d}_1 & \bar{d}_2 & \bar{d}_3 \end{bmatrix} \begin{bmatrix} z_{d_1} & 0 & 0 \\ 0 & z_{d_2} & 0 \\ 0 & 0 & z_{d_3} \end{bmatrix} \not{Z}' P_R \begin{bmatrix} d_1 \\ d_2 \\ d_3 \end{bmatrix} \equiv g' \bar{\mathcal{D}} \mathcal{Z}_d \not{Z}' P_R \mathcal{D} \qquad (4.35)$$

where $\mathcal{D}$ is a vector in flavor space, and $\mathcal{Z}_d$ the 3×3 diagonal matrix of the couplings to right-handed down quarks. After EWSB, there will be a change of basis to take us from the quarks from the weak to the mass eigenstates,

$$\mathcal{Q}_d \equiv \begin{bmatrix} d \\ s \\ b \end{bmatrix} = U_{dR} \begin{bmatrix} d_1 \\ d_2 \\ d_3 \end{bmatrix} = U_{dR}\mathcal{D} \tag{4.36}$$

where $\mathcal{Q}_d$ is the vector of the mass eigenstates. Applying this rotation to the Z' interactions in Eq. (4.36), we arrive at the quark mass eigenstate interactions,

$$g'\bar{\mathcal{Q}}_d \left(U_{dR}^\dagger \mathcal{Z}_d U_{dR} \right) \not{Z}' P_R \mathcal{Q}_d \tag{4.37}$$

where we see that the matrix $(U_{dR}^\dagger \mathcal{Z}_d U_{dR})$ need not be diagonal, even though $\mathcal{Z}_d$ was. The easiest way to insure that it there are no FCNCs is to assign a common charge for all three right-handed down-type quarks. In that case, $\mathcal{Z}_d$ is proportional to the unit matrix, and the product of unitary transformations $U_{dR}^\dagger U_{dR}$ result in the unit matrix. This is the same mechanism by which the SM Z was left without tree-level FCNC interactions.

Putting all of my personal prejudices together, we should have family universal couplings to fermions (to avoid FCNCs), no charge on the SM Higgs to avoid Z-Z' mixing, and I do not particularly care if the anomalies cancel with SM fermion content or not. When the dust has settled, I am left with parameters:

$$M_{Z'}, \Gamma_{Z'} \quad z_Q, z_u, z_d, z_L, z_e \tag{4.38}$$

technically there is also the gauge coupling g', but unless I am embedding into a GUT or other extended structure, I can just absorb it into the normalization of the the z's without any confusion. I left the width of the Z' as a free input to allow for decays into non-SM channels, including perhaps right-handed neutrinos. Obviously, if there are important observable decays into non-SM modes, I would have to add to this list accordingly.

4.2.1.1. *Z's at the LHC*

At the Tevatron or LHC, the obvious signal for a Z' would be a bump in an invariant mass distribution, much the way the Z appears as a bump against the continuum photon-mediated processes. A final state with a pair of charged leptons[10,11] is an obvious place to look, because they are easy to trigger on, have smaller backgrounds, and more precise reconstruction than jet final states. A Z' with mass below a few TeV and large branching ratio into leptons is a discovery that could be made with a relatively modest amount of LHC data.

We can transcribe much of our experience with the Z boson in the earlier chapter onto a Z'. At a hadron collider, we replace the e^+e^- initial state with a $q\bar{q}$ one. The rate at a hadron machine is estimated by taking this "partonic" reaction and convolving it with the parton distribution functions (PDFs). The rate of Z' production thus turns into a measurement of a sum of the quark charges (both right- and left-handed) with coefficients determined by the PDFs, and an over-all factor of the branching ratio into the final state of interest. Information about individual quark charges including right- versus left-handed couplings is a little more difficult, because the LHC, as a pp collider, makes it difficult to identify which direction along the beam is the direction of the quark, and which direction the anti-quark. However, using the difference of valence q versus $\bar{q}$ PDFs as well as clever observables can disentangle a lot of parameter space.[12]

4.2.2. *Topcolor*

A second class of theory which predicts an interesting resonance is the family of topcolor models.[13] Broadly, these models predict a composite Higgs boson which is a bound state of top quarks, and thus couples strongly to top as a residual of the binding force.[14] In practice, topcolor generally has difficulty getting the right amount of EWSB and the right top mass by itself, leading to extensions in which either there is more electroweak breaking from a technicolor sector[15] or there is an additional fermion participating in the strong dynamics whose mixing with top dials in the correct top mass.[16] The specific details of those models are not very important for our purposes here – all of them lead to resonances with similar phenomenology. So I will describe the simplest model[13] here.

The theory extends the color sector of the Standard Model to $SU(3)_1 \times SU(3)_2$, usually called "topcolor" and "protocolor", respectively. The top quark (both Q_3 and t_R) are triplets under $SU(3)_1$ and the light quarks are triplets under $SU(3)_2$. At the scale of a few TeV, this structure breaks down through some usually unspecified dynamics, resulting in an unbroken $SU(3)_c$ to play the role of ordinary color and a massive color octet of "topgluons" (g^1). The couplings are usually chosen such that the topgluons are mostly the original $SU(3)_1$ bosons, and the ordinary gluons are mostly the $SU(3)_2$ gluons. This choice results in the topgluons coupling strongly to top quarks and weakly to the light quarks. The unbroken $SU(3)_c$ symmetry

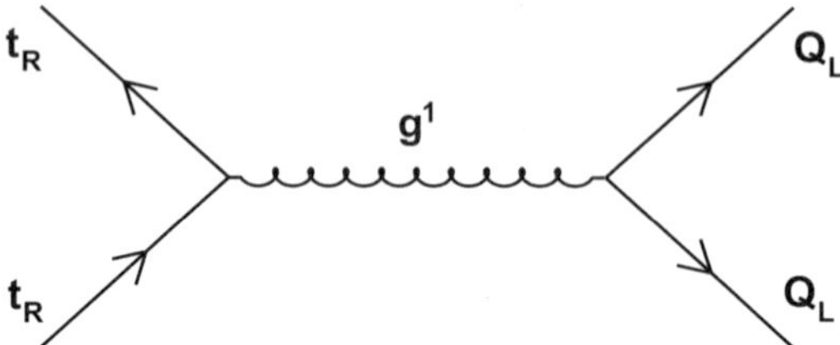

Fig. 4.4.   Feynman diagram showing how g^1 mediates a 4-top interaction at low energies.

guarantees that the massless bosons couple universally to all of the quarks. If you find this discussion confusing, you can go down to Section 4.3 and you will find it worked out in more detail.

At sub-TeV energies, we are in the regime where the topgluons can be integrated out, and look like a contact interaction (see Figure 4.4). Their important effect in this low energy effective theory is to produce a coupling between right-handed and left-handed top quarks,

$$-\frac{g^2}{M^2}\left[\bar{t}_R\gamma^\mu T^a t_R\right]\left[\bar{Q}_L\gamma_\mu T^a Q_L\right] = -\frac{g^2}{M^2}\left[\bar{t}_R Q_L\right]\left[\bar{Q}_L t_R\right] \qquad (4.39)$$

where g is the g^1 coupling to top quarks and M is its mass, Q_L is the third family quark doublet and the last equality is easily understood as a Fierz transformation. Four fermion interactions of this kind (but involving the light quarks) have been extensively studied as a low energy model of QCD and chiral symmetry breaking, the Nambu-Jona Lasinio (NJL) model.[17]

If g is large enough, this interaction results in scalar bound states. We can see how this works in more detail by taking our four-fermion interaction, and rewriting it in terms of an auxiliary field Φ,

$$\mathcal{L}(\Lambda) = \frac{1}{2}g_t\bar{Q}_L t_R\Phi + g_t^*\bar{t}_R Q_L\Phi^* - \Lambda^2|\Phi|^2 \qquad (4.40)$$

where Φ transforms as a SM Higgs, but has no kinetic terms. Integrating it out is thus trivial, and reproduces the original four fermion interaction provided $g_t^2/\Lambda^2 = g^2/M^2$. An obvious identification would be $g_t = g$ and $\Lambda = M$ (and at tree level, it makes no difference), but the fact that nothing guarantees that identification is an indication that our effective theory is sensitive to the cut-off[¶], and thus to the details of the UV dynamics responsible for binding Φ.

[¶]Another way to appreciate this fact is to notice that we are taking a non-renormalizable

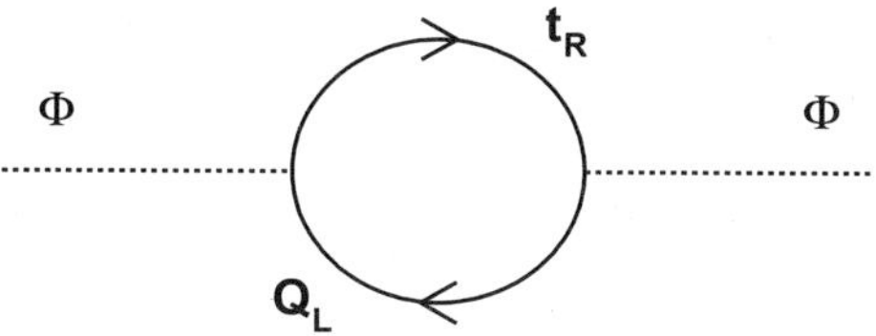

Fig. 4.5.   Feynman diagram showing how Φ develops kinetic terms at one loop.

Our effective theory lives at the scale Λ. Below that scale, kinetic terms (and a quartic interaction) for Φ will be induced at one loop, and the parameters Λ^2 and g_t will be renormalized (see Figure 4.5). Just as in the case of the self-energy of the Z, these quantities are divergent and need to be renormalized. However, we know the effective theory at the scale Λ, and we can use this to set the renormalization constants,

$$Z(\Lambda) = 0 \tag{4.41}$$

$$m^2(\Lambda) = \Lambda^2 \tag{4.42}$$

The vanishing of the wave function renormalization at the compositeness scale is a generic feature of a composite field, because it implies that this field has no non-zero matrix elements between the vacuum and a single particle state.[18] At scales below Λ,

$$\mathcal{L}(\mu) = Z(\mu)\,(D^\mu\Phi)^\dagger\,(D_\mu\Phi) - m^2(\mu)|\Phi|^2 - \lambda(\mu)|\Phi|^4 \tag{4.43}$$
$$+\frac{1}{2}g_t(\mu)\bar{Q}_L t_R \Phi + H.c.$$

where,

$$Z(\mu) = N_c\frac{g_t^2}{16\pi^2}\log\left(\frac{\Lambda}{\mu}\right) + ... \tag{4.44}$$

$$m^2(\mu) = \Lambda^2 - N_c\frac{g_t^2}{8\pi^2}\Lambda^2 + ... \tag{4.45}$$

with similar expressions for $g_t(\mu)$ and $\lambda(\mu)$. Note that if

$$g_t^2 > \frac{8\pi^2}{N_c} \equiv g_{crit}^2 \tag{4.46}$$

---

interaction, Eq. (4.39) and writing it in renormalizable language. This arbitrariness is a representation of the non-renormalizability of the original theory as expressed in renormalizable language.

$m^2(\mu)$ becomes negative, and Φ develops a VEV, breaking the electroweak symmetry.

To analyze the physics of Φ, we rescale the kinetic term for Φ to canonical form by taking $\Phi \to Z^{-1/2}\Phi$. In terms of the canonically normalized field, $m^2 \to Z^{-1}m^2$, $\lambda \to Z^{-2}\lambda$, and $g_t \to Z^{-1/2}g_t$. A useful approximation to determine the physics at the electroweak scale is to fix the parameters in the canonically normalized theory such that $m^2(\Lambda) \sim \infty$ and $g_t(\Lambda) \sim \lambda(\Lambda) \sim \infty$ (since all of these are finite in the unnormalized theory at Λ and going to the canonical normalization divides each by a positive power of $Z(\Lambda) \simeq 0$ at that scale), and then use the renormalization group to determine the parameters in the canonically normalized theory at scales $\mu \leq \Lambda$.

At energies much below Λ, our theory is nothing more than the Standard Model itself, but with the Higgs potential parameters m^2 and λ and the top Yukawa coupling g_t all predicted in terms of the original parameters Λ and g_t (provided $g_t > g_{crit}$ so that we have EWSB). The constraint required so that $m^2 < 0$ is the reason this theory has trouble fitting the right top mass – it turns out that for g_t large enough for $m^2 < 0$, we are in a regime where the top mass turns out to be too large.

4.2.2.1. *Topcolor at the LHC*

Since topcolor looks like the Standard Model at low energies, to distinguish it from the Standard Model, we need to turn to the high energy behavior of the theory. The generic feature of topcolor is the need to invoke the extended $SU(3) \times SU(3)$ symmetry in order to generate the four-top contact interaction at low energies, resulting in the composite Higgs. So a very generic feature at high energies is the existence of the massive color octet vector particle g^1 that couples strongly to top and weakly to light quarks.

At the LHC, g^1 can be produced by $q\bar{q}$ annihilation in the initial state. While the coupling of g^1 to light quarks is not huge, the fact that the light quark PDFs are sizeable usually renders this the dominant production mechanism. Once produced, the large coupling to top dictates that g^1 decays into a $t\bar{t}$ pair with a very high branching ratio. So the signature is a resonant structure in the invariant mass of $t\bar{t}$ pairs. While not as clean as a decay into leptons, decays into tops still have a lot of potential compared to backgrounds, because top decays produce jets enriched with bottom quarks which can be tagged and also a fair fraction of leptonic W decays. The primary challenge for high mass resonances is that they result in very boosted tops, whose decay products become highly collimated, which can be challenging to reconstruct properly as jets merge and leptons end up buried inside them[19] This is an interesting and active subject in collider phenomenology and analysis technique.[20]

4.2.2.2. *Topcolor versus Technicolor*

I often get asked what the difference between topcolor and technicolor is. They get confounded in people's minds because both use strong dynamics to trigger electroweak symmetry-breaking and further confusion arises because both are families of theories, as opposed to single models. But the systematic differences are that topcolor is a model of a composite Higgs. Technicolor is a model of no Higgs.

One way to keep the definitive difference straight is to think about how each model explains the high energy perturbative unitarity of $WW \to WW$ scattering. Topcolor has a Higgs which does the job much the way the Higgs does it in the Standard Model. Technicolor has no Higgs, and perturbative unitarity is maintained by the existence of a weak triplet of vector particles (usually called "techni-rhos" in analogy with the massive vectors of QCD). They look like a Z' and a pair of W's with significant coupling to the SM Ws and Z (and perhaps weak coupling to fermions). So technicolor is another model predicting new resonances! We will talk about particles much like the technirhos below when we discuss topflavor.

This brings up an important point: the SM Higgs is an example of a resonance! But you have many lectures devoted just to the Higgs itself, so that is all I will say about it here.

4.2.3. *Topflavor*

Topflavor[21] is a similar construction to topcolor in many ways. The difference is that instead of taking the SM $SU(3)$ interaction and identifying it with the diagonal subgroup of two $SU(3)$'s, we take the SM $SU(2)$ interaction, and promote it to $SU(2)_1 \times SU(2)_2$. The left-handed fermions of the third generation are charged under $SU(2)_1$ and the light fermions under $SU(2)_2$. This arrangement automatically takes care of anomaly cancellation, because the $SU(2)$ anomalies cancel within an entire generation of SM fermions.

When the symmetry breaks, $SU(2) \times SU(2) \to SU(2)$, a massive Z' and pair of W's results, with enhanced coupling to the third generation. The unbroken $SU(2)$ is identified with the usual weak interaction, and has

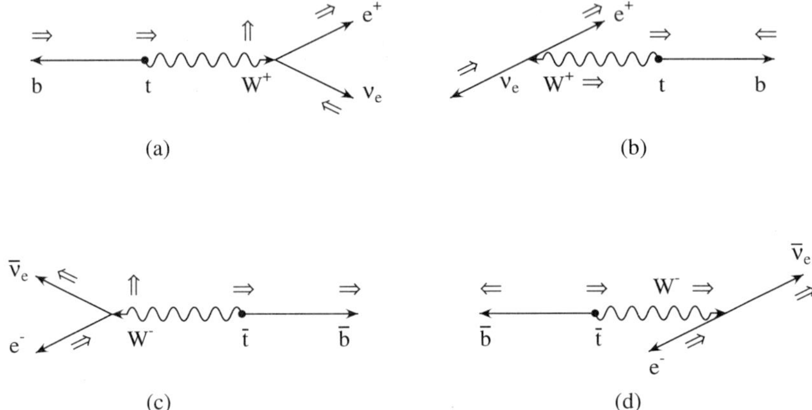

Fig. 4.6. Cartoon explanation for why the charged lepton in a top (anti-top) decay tends to go in the same (opposite) direction as the top spin. The arrows above each particle indicate their spin. Figures (a) and (c) cover the longitudinally polarized W decay cases, and figures (b) and (d) cover the transverse W cases. From Ref. [24]

approximately universal coupling to all of the SM fermions. In its many incarnations topflavor has been helpful as a model of top mass generation on top of strong dynamics models,[21] raised the light Higgs mass above the LEP-II bound in supersymmetric models[22] and its phase transition in the early Universe may drive baryogenesis.[23]

The Z' can be produced at the LHC by light quark annihilation. Once produced, it tends to decay into third family fermions: tops, bottoms, tau leptons, and their neutrinos. We've already covered top resonances. The resonance in $b\bar{b}$ is challenging to distinguish over backgrounds. The τ resonance is challenging to reconstruct because τ decays include missing energy and are usually into hadrons, but nonetheless it is an interesting and probably viable channel (with large statistics). The topflavor Z' will also decay into ordinary Ws. This decay is usually rare, but can be covered by high mass SM Higgs searches. Because of the chiral structure of topflavor, the decay into top quarks is mostly into left-handed tops. Top polarization can be reconstructed, for example by looking at the direction of the charged lepton from a semi-leptonic top decay (see Figure 4.6 for a cartoon explanation as to why the charged lepton tends to move in the direction the top spin was pointing).

The topflavor W' will decay largely into $t\bar{b}$ and $\tau\nu$. The decay into a top quark appears as a resonance against the continuum background of s-channel single top production.[24-26] Again, the coupling is left-handed and polarization observables can help distinguish the W' of topflavor from other theories.

4.2.4. $SU(2)_R$

$SU(2)_R$ models have the gauge structure $SU(2)_L \times SU(2)_R \times U(1)_X$ followed by a breakdown of $SU(2)_R \times U(1)_X \to U(1)_Y$ and were briefly covered in the lectures by David Kaplan. The fermion generations couple universally, with the left-handed doublets charged as usual under $SU(2)_L$ and the right-handed fermions assembled (together with right-handed neutrinos) into $SU(2)_R$ doublets. This gauge structure descends naturally from $SO(10)$ GUTs, and $SU(2)_R$ is often invoked in other contexts because it acts as a custodial symmetry which prevents large contributions to the Peskin-Takeuchi T parameter.[27]

Because the W' and Z' couple in a family-universal manner, searches typically look for electrons and muons from their decays. While decays into tops are not particularly enhanced, they are still useful to measure the fact that the couplings are right-handed.

4.2.5. *Little Higgs*

Little Higgs theories are an interesting solution to the little hierarchy problem. There are too many versions and too many details that go into constructing such a model, so I restrict myself to a few remarks here. There are nice lectures from a previous TASI by Martin Schmaltz,[30] and a good review article[31] which is more easily accessible online.

Little Higgs theories postulate that the Higgs is a pseudo-Nambu-Goldstone boson in order to protect its mass and solve the little hierarchy problem. From a nuts and bolts point of view, they invoke W's and Z's to cancel the one-loop quadratic divergence induced on the Higgs mass by the SM W and Z, and a heavy t' quark to cancel the divergence induced by the SM top. The W's and Z's have phenomena similar to the ones we have already described. The heavy quark will be dominantly produced singly through its mixing with the ordinary top[28] and decays into Wb, Zt, and ht, forming a (fermion) resonance in those channels.

Modern little Higgs theories often incorporate a symmetry (T-parity) to provide a dark matter candidate and lessen contraints from precision electroweak data[29] In this class of models, the signatures involve missing energy and were covered by Howie Baer.

4.2.6. *Extra Dimensions*

Bogdan Dobrescu provided nice lectures about extra dimensions, explaining how the Kaluza-Klein decomposition results in massive copies of any particles propagating in the extra dimension. You can also take a look at previous TASI lectures by Graham Kribs[32] or Csaba Csaki.[33] The heavy KK modes provide many potential new resonances: gravtons, gluons, weak bosons, Higgs, etc....

4.3. Effective Theory Descriptions

We can write the low energy physics of a wide class of models with extended gauge structures using effective field theory. A common theme among the models of the previous section was a structure like $G_1 \times G_2 \to G_{SM}$ where G_{SM} is any of the SM gauge groups, $SU(3)_c$ (as in topcolor), $SU(2)_W$ (as in topflavor, $SU(2)_R$), or $U(1)_Y$. This structure is also common in Little Higgs theories, and mocks up the lowest KK mode of an extra dimension through dimensional deconstruction.[34]

In fact, it is more general than even those examples would indicate. If we have vector particles transforming as adjoints under a SM group, the combined requirements of gauge invariance under the SM symmetries with the need for a hidden gauge symmetry for the heavy particles to insure consistency, leads us to $G_1 \times G_2 \to G_{SM}$ in every case. Indeed $SU(2) \times SU(2)$ is the effective theory description of the technirho[35] and also of the ordinary rho meson of QCD.[36]

So let's work out an example in detail. The specific example I will use is motivated by the topcolor model, with $SU(3)_1 \times SU(3)_2 \to SU(3)_c$, but one can pretty easily transcribe it into any $SU(N) \times SU(N) \to SU(N)$, and with minimal headaches into $U(1) \times U(1)$. A "moose" or "quiver" diagram[37] is shown in Figure 4.7, and specifies the gauge groups ($SU(3)_1 \times SU(3)_2$) as the two circles in the diagram. Each group has its own gauge coupling, g_1 and g_2. I have chosen ψ_1 to be a left-handed fundamental of

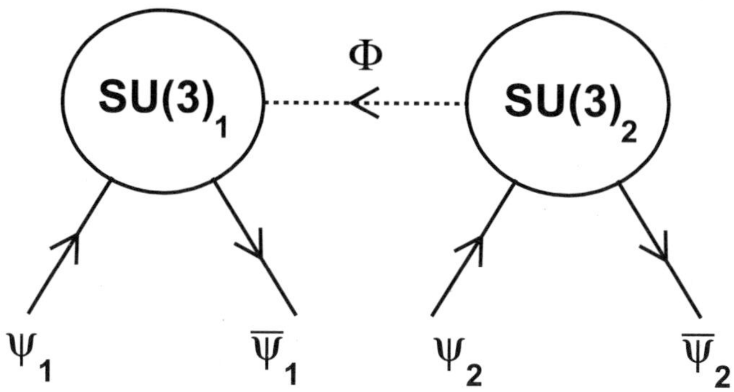

Fig. 4.7.   Moose diagram for the $SU(3) \times SU(3)$ model.

Table 4.1.   Representations for the matter fields.

Field	Spin	$SU(3)_1$	$SU(3)_2$
ψ_1	1/2	3	1
$\bar{\psi}_1$	1/2	$\bar{3}$	1
ψ_2	1/2	1	3
$\bar{\psi}_2$	1/2	1	$\bar{3}$
Φ	0	3	$\bar{3}$

$SU(3)_1$, indicated by its arrow going into that group. $\bar{\psi}_1$ is a right-handed fundamental (or if you like, a left-handed anti-fundamental). ψ_2 and $\bar{\psi}_2$ form another vector-like pair, fundamental under $SU(3)_2$. The arrows make it very easy to keep track of anomaly cancellation. The dashed line Φ going between the two groups is a scalar field which is a fundamental under $SU(3)_1$ and an anti-fundamental under $SU(3)_2$. The gauge assignments are written out in Table 4.1, but it is worthwhile to take the time to learn how to read the moose diagram. It may seem awkward at first, but once you get used to it, it becomes second-nature.

The Lagrangian follows from the gauge structure. The gauge assignments dictate the kinetic terms,

$$-\frac{1}{4}\left(F^1_{\mu\nu}\right)^2 - \frac{1}{4}\left(F^2_{\mu\nu}\right)^2 + i\bar\psi_1\left(\slashed{D} - m_1\right)\psi_1 + i\bar\psi_2\left(\slashed{D} - m_2\right)\psi_2 \quad (4.47)$$

$$+\mathrm{Tr}\left(D^\mu\Phi\right)^\dagger\left(D^\mu\Phi\right) - V(\Phi) + y\Phi\bar\psi_2\psi_1 + H.c.$$

where the F's are the usual $SU(3)$ field strengths, built from the gauge fields A^1_μ and A^2_μ, and we have used the fact that we chose the ψ's to be vectorlike to pass immediately to four component language. If we had chosen a chiral theory, we would have just used 2-component language or put in projectors to end up where we wanted. Since I chose a vectorlike theory, I went ahead and wrote down masses for the fermions. If they had had chiral electroweak quantum numbers, as the SM fermions, the $SU(2)_W \times U(1)_Y$ gauge symmetries would forbid such masses. The Tr on the kinetic term for Φ is a way of tracing over the (double) gauge indices when I represent the field as a 3×3 matrix with the rows representing the $SU(3)_1$ index and the columns representing the $SU(3)_2$ index. The covariant derivatives are given by,

$$D^\mu\psi_1 = \partial_\mu\psi_1 - ig_1 A^{1a}_\mu T^a \psi_1 \quad\quad\quad (4.48)$$

$$D^\mu\psi_2 = \partial_\mu\psi_1 - ig_2 A^{2a}_\mu T^a \psi_2 \quad\quad\quad (4.49)$$

$$D^\mu\Phi = \partial_\mu\Phi - ig_1 A^{1a}_\mu T^a \Phi + ig_2 A^{2a}_\mu \Phi T^a \quad\quad (4.50)$$

in terms of the two gauge couplings and the generators of $SU(3)$ in the fundamental representation, T^a. We have also written down a potential for Φ, and the Yukawa interaction between it and the fermions allowed by gauge invariance.

Our theory thus has parameters,

$$g_1, g_2 \quad m_1, m_2, y \quad\quad\quad\quad\quad (4.51)$$

plus the parameters needed to describe the potential of Φ. The potential parameters will be important for determining the VEV of Φ, and also the masses of the Higgs bosons associated with it. These new Higgs bosons are interesting physics, but usually hard to access at near-future colliders, so in many cases we can ignore them when we talk about phenomena at the LHC. Since our theory is a *generic* description of a massive color octet vector particle, these parameters tell us what we are aiming for when we discover an octet vector and want to measure its properties. Once we pin them down, we can either try to fit them into a bigger framework for the UV physics, or we may see that the framework does not describe all of the phenomena we see – in which case we need to extend the effective theory itself.

We assume that the potential for Φ induces a VEV that is "diagonal" in $SU(3)_1$-$SU(3)_2$ color space,

$$\langle \Phi_\alpha^i \rangle = u \, \delta_\alpha^i \tag{4.52}$$

where u is the magnitude of the VEV and i is the gauge index of $SU(3)_1$ and α the index of $SU(3)_2$. In practice, it is not hard to write down a potential which results in the vacuum that we want. Inserting this VEV in the kinetic term for Φ yields,

$$u^2 \mathrm{Tr} \left[\left(-g_1 A_\mu^{1a} T^a + g_2 A_\mu^{2a} T^a \right) \left(-g_1 A_{1a}^\mu T^a + g_2 A_{2a}^\mu T^a \right) \right] \tag{4.53}$$
$$= \frac{1}{2} u^2 \left[g_1^2 A^1 \cdot A^1 - 2 g_1 g_2 A^1 \cdot A^2 + g_2^2 A^2 \cdot A^2 \right]$$

where $\cdot$ denotes contraction of the Lorentz and gauge indices. In analogy with the electroweak theory, we rewrite the two gauge couplings,

$$g_1 \equiv \frac{g}{\sin \phi} \,, \qquad g_2 \equiv \frac{g}{\cos \phi} \,. \tag{4.54}$$

In terms of these new parameters, the mass terms for the gauge fields can be written,

$$\frac{1}{2} \frac{g^2}{\sin^2 \phi \cos^2 \phi} u^2 \, [A_1^{a\mu} \ A_2^{a\mu}] \begin{bmatrix} \cos^2 \phi & -\sin \phi \cos \phi \\ -\sin \phi \cos \phi & \sin^2 \phi \end{bmatrix} \begin{bmatrix} A_{1\mu}^a \\ A_{2\mu}^a \end{bmatrix} \tag{4.55}$$

The mass matrix is now easily diagonalized, and yields mass eigenstates,

$$\begin{aligned} g_\mu^a &= \ \ \sin \phi \, A_\mu^{1a} + \cos \phi \, A_\mu^{2a} \\ A_\mu^a &= -\cos \phi \, A_\mu^{1a} + \sin \phi \, A_\mu^{2a} \end{aligned} \tag{4.56}$$

where g^a is the massless mode, which we identify with the usual SM gluon, and A^a is the massive color octet with mass,

$$M^2 = \frac{g^2}{\sin^2 \phi \cos^2 \phi} u^2 \,. \tag{4.57}$$

The fermion couplings to g_μ^a are universal, as required by the unbroken residual gauge invariance,

$$g_1 A_\mu^{a1} \bar{\psi}_1 \gamma^\mu T^a \psi_1 \rightarrow g \left(-\cot \phi A_\mu^a + g_\mu^a \right) \bar{\psi}_1 \gamma^\mu T^a \psi_1 \tag{4.58}$$
$$g_2 A_\mu^{a2} \bar{\psi}_2 \gamma^\mu T^a \psi_2 \rightarrow g \left(\tan \phi A_\mu^a + g_\mu^a \right) \bar{\psi}_2 \gamma^\mu T^a \psi_2 \tag{4.59}$$

with g the QCD coupling. The fermions have different couplings to the massive octet, depending whether they were originally charged under $SU(3)_1$

208   *T.M.P. Tait*

or $SU(3)_2$. In topcolor, we could take ϕ to be a small angle, in which case ψ_1 could represent the top quark, and ψ_2 could represent the light quarks.

Through the Yukawa interactions, the symmetry-breaking can also induce mixing between ψ_1 and ψ_2. The mass matrix is,

$$\begin{bmatrix} \bar{\psi}_1 & \bar{\psi}_2 \end{bmatrix} \begin{bmatrix} m_1 & yu \\ y^*u & m_2 \end{bmatrix} \begin{bmatrix} \psi_1 \\ \psi_2 \end{bmatrix} \tag{4.60}$$

The mass matrix is diagonalized (in general) by a separate rotation of the left-handed and right-handed fields, which can be found by considering the matrices $M^\dagger M$ and $MM^\dagger$. The mass eigenstates will continue to have universal couplings to the zero mass vector, but will have a mixture of ψ_1 and ψ_2 couplings to the massive vector.

4.4. Closing Thoughts

Theories of physics beyond the Standard Model can show resonances in almost any pair of Standard Model particles we can imagine, with the possibilities far out-stripping our ability to cover all of them. The discovery of any resonance at the LHC will raise similar questions - what is its mass, how wide is its width, how is it produced, and what does it decay into? Having established the answers to those questions, we can then try to fit the resonance into a deeper picture of organizing principles and symmetries. And finding the answers themselves will be fun!

Many thanks to the lead organizers Tao and Robin (both examples that spooky action at a distance can sometimes be effective∥), and to the local organizers KT, Tom, and the whole Boulder team. They made this TASI another success. The students did their part and kept us lecturers on our toes, through discussions both during and after the lectures.

References

1. [ALEPH Collaboration and DELPHI Collaboration and L3 Collaboration and], Phys. Rept. **427**, 257 (2006) [arXiv:hep-ex/0509008].
2. For a nice review of the optical theorem see Chapter 7.3 of, M. E. Peskin and D. V. Schroeder, *Reading, USA: Addison-Wesley (1995) 842 p.*
3. D. Choudhury, T. M. P. Tait and C. E. M. Wagner, Phys. Rev. D **65**, 053002 (2002) [arXiv:hep-ph/0109097].
4. J. Kumar and J. D. Wells, Phys. Rev. D **74**, 115017 (2006) [arXiv:hep-ph/0606183].

∥Robin had a better excuse!

5. C. D. Froggatt and H. B. Nielsen, Nucl. Phys. B **147**, 277 (1979).

6. M. S. Carena, A. Daleo, B. A. Dobrescu and T. M. P. Tait, Phys. Rev. D **70**, 093009 (2004) [arXiv:hep-ph/0408098].

7. J. L. Hewett and T. G. Rizzo, Phys. Rept. **183**, 193 (1989).

8. [LEP Collaboration and ALEPH Collaboration and DELPHI Collaboration and], arXiv:hep-ex/0312023.

9. See, for example, M. Bona *et al.* [UTfit Collaboration], JHEP **0803**, 049 (2008) [arXiv:0707.0636 [hep-ph]].

10. E. Eichten, I. Hinchliffe, K. D. Lane and C. Quigg, Rev. Mod. Phys. **56**, 579 (1984) [Addendum-ibid. **58**, 1065 (1986)].

11. A. Leike, Phys. Rept. **317**, 143 (1999) [arXiv:hep-ph/9805494].

12. F. Petriello and S. Quackenbush, Phys. Rev. D **77**, 115004 (2008) [arXiv:0801.4389 [hep-ph]].

13. C. T. Hill, Phys. Lett. B **266**, 419 (1991).

14. W. A. Bardeen, C. T. Hill and M. Lindner, Phys. Rev. D **41**, 1647 (1990).

15. C. T. Hill, Phys. Lett. B **345**, 483 (1995) [arXiv:hep-ph/9411426]; K. D. Lane and E. Eichten, Phys. Lett. B **352**, 382 (1995) [arXiv:hep-ph/9503433].

16. B. A. Dobrescu and C. T. Hill, Phys. Rev. Lett. **81**, 2634 (1998) [arXiv:hep-ph/9712319]; R. S. Chivukula, B. A. Dobrescu, H. Georgi and C. T. Hill, Phys. Rev. D **59**, 075003 (1999) [arXiv:hep-ph/9809470]; H. J. He, C. T. Hill and T. M. P. Tait, Phys. Rev. D **65**, 055006 (2002) [arXiv:hep-ph/0108041].

17. Y. Nambu and G. Jona-Lasinio, Phys. Rev. **122**, 345 (1961).

18. To brush up on the field theory which inspires these arguments, see the fourth chapter of L. S. Brown, *Cambridge, UK: Univ. Pr. (1992) 542 p.*

19. K. Agashe, A. Belyaev, T. Krupovnickas, G. Perez and J. Virzi, Phys. Rev. D **77**, 015003 (2008) [arXiv:hep-ph/0612015]; B. Lillie, L. Randall and L. T. Wang, JHEP **0709**, 074 (2007) [arXiv:hep-ph/0701166]; U. Baur and L. H. Orr, Phys. Rev. D **77**, 114001 (2008) [arXiv:0803.1160 [hep-ph]].

20. J. Thaler and L. T. Wang, JHEP **0807**, 092 (2008) [arXiv:0806.0023 [hep-ph]].; D. E. Kaplan, K. Rehermann, M. D. Schwartz and B. Tweedie, Phys. Rev. Lett. **101**, 142001 (2008) [arXiv:0806.0848 [hep-ph]]; M. Cacciari, J. Rojo, G. P. Salam and G. Soyez, JHEP **0812**, 032 (2008) [arXiv:0810.1304 [hep-ph]]; Y. Bai and Z. Han, arXiv:0809.4487 [hep-ph]; L. G. Almeida, S. J. Lee, G. Perez, G. Sterman, I. Sung and J. Virzi, arXiv:0807.0234 [hep-ph].

21. X. Li and E. Ma, Phys. Rev. Lett. **47**, 1788 (1981); R. S. Chivukula, E. H. Simmons and J. Terning, Phys. Rev. D **53**, 5258 (1996) [arXiv:hep-ph/9506427]; D. J. Muller and S. Nandi, Phys. Lett. B **383**, 345 (1996) [arXiv:hep-ph/9602390]; E. Malkawi, T. M. P. Tait and C. P. Yuan, Phys. Lett. B **385**, 304 (1996) [arXiv:hep-ph/9603349].

22. P. Batra, A. Delgado, D. E. Kaplan and T. M. P. Tait, JHEP **0402**, 043 (2004) [arXiv:hep-ph/0309149].

23. J. Shu, T. M. P. Tait and C. E. M. Wagner, Phys. Rev. D **75**, 063510 (2007) [arXiv:hep-ph/0610375].

24. T. M. P. Tait and C. P. P. Yuan, Phys. Rev. D **63**, 014018 (2001) [arXiv:hep-ph/0007298].

25. E. H. Simmons, Phys. Rev. D **55**, 5494 (1997) [arXiv:hep-ph/9612402].

26. Z. Sullivan, Phys. Rev. D **66**, 075011 (2002) [arXiv:hep-ph/0207290].

27. M. E. Peskin and T. Takeuchi, Phys. Rev. D **46**, 381 (1992); M. E. Peskin and T. Takeuchi, Phys. Rev. Lett. **65**, 964 (1990).

28. T. Han, H. E. Logan, B. McElrath and L. T. Wang, Phys. Rev. D **67**, 095004 (2003) [arXiv:hep-ph/0301040].

29. H. C. Cheng and I. Low, JHEP **0309**, 051 (2003) [arXiv:hep-ph/0308199].

30. M. Schmaltz, *Prepared for Theoretical Advance Study Institute in Elementary Particle Physics (TASI 2004): Physics in D ¿= 4, Boulder, Colorado, 6 Jun - 2 Jul 2004*

31. M. Schmaltz and D. Tucker-Smith, Ann. Rev. Nucl. Part. Sci. **55**, 229 (2005) [arXiv:hep-ph/0502182].

32. G. D. Kribs, arXiv:hep-ph/0605325.

33. C. Csaki, arXiv:hep-ph/0404096; C. Csaki, J. Hubisz and P. Meade, arXiv:hep-ph/0510275.

34. N. Arkani-Hamed, A. G. Cohen and H. Georgi, Phys. Rev. Lett. **86**, 4757 (2001) [arXiv:hep-th/0104005]; C. T. Hill, S. Pokorski and J. Wang, Phys. Rev. D **64**, 105005 (2001) [arXiv:hep-th/0104035].

35. J. Hirn, A. Martin and V. Sanz, JHEP **0805**, 084 (2008) [arXiv:0712.3783 [hep-ph]].

36. M. Bando, T. Kugo, S. Uehara, K. Yamawaki and T. Yanagida, Phys. Rev. Lett. **54**, 1215 (1985).

37. H. Georgi, Nucl. Phys. B **266**, 274 (1986).

Chapter 5

Collider Signal II:
Missing E_T Signatures and Dark Matter Connection

Howard Baer

*Homer L. Dodge Department of Physics and Astronomy,
University of Oklahoma, Norman, OK 73019, USA*

These lectures give an overview of aspects of missing E_T signatures from new physics at the LHC, along with their important connection to dark matter physics. Mostly, I will concentrate on supersymmetric (SUSY) sources of $\not{E}_T$, but will also mention Little Higgs models with T-parity (LHT) and universal extra dimensions (UED) models with KK-parity. Lecture 1 covers SUSY basics, model building and spectra computation. Lecture 2 addresses sparticle production and decay mechanisms at hadron colliders and event generation. Lecture 3 covers SUSY signatures at LHC, along with LHT and UED signatures for comparison. In Lecture 4, I address the dark matter connection, and how direct and indirect dark matter searches, along with LHC collider searches, may allow us to both discover and characterize dark matter in the next several years. Finally, the interesting scenario of Yukawa-unified SUSY is examined; this case works best if the dark matter turns out to be a mixture of axion/axino states, rather than neutralinos.

5.1. Introduction: $\not{E}_T$ collider signatures and the dark matter connection

I have been assigned the topic of discussing missing transverse energy signatures at hadron colliders, especially the CERN LHC, which should begin producing data in 2009. Every collider event– be they lepton-lepton, lepton-hadron or hadron-hadron interactions– will contain some amount of missing energy ($\not{E}$) just due to the fact that energy measurements are not perfectly precise. Hadronic events will usually contain more $\not{E}$ than leptonic events since the energy resolution of hadronic calorimeters is not as precise as is electromagnetic resolution. In addition, collider detectors are imperfect devices, often containing un-instrumented regions, cracks, hot or dead calorimeter cells, and only partial coverage of the 4π steradians surrounding the interaction region: always, an allowance must be made at least to allow the beam pipe to enter the detector. Also, missing energy can arise from cosmic rays entering the detector, or beam-gas collisions, multiple scattering etc.

At e^+e^- colliders, $\not{E}$ can be directly measured since the incoming beam energies are well-known (aside from bremsstrahlung/beamstrahlung effects at very high energy e^+e^- colliders). At $p\bar{p}$ or pp colliders, the hard scattering events which are likely to produce new physics are initiated by quark-quark, quark-gluon, or gluon-gluon collisions, and the partonic constituents of the proton carry only a fraction of the beam energy. The proton constituents not participating in the hard scattering will carry an unknown fraction of beam energy into the un-instrumented or poorly instrumented forward region, so that for hadron colliders, only missing *transverse* energy ($\not{E}_T$) is meaningful. In these lectures, we will concentrate on pp colliders, in anticipation of the first forthcoming collisions at the CERN LHC.

Along with these imperfect detector effects, $\not{E}_T$ can arise in Standard Model (SM) production processes wherein neutrinos are produced, either directly or via particle decays. For instance, the key signature for W boson production at colliders was the presence of a hard electron or muon balanced by missing energy coming from escaping neutrinos produced in $W^- \to \ell^- \bar{\nu}_\ell$ (or charge conjugate reaction) decay (here, $\ell = e$, μ or τ). In addition, neutrinos will be produced by semi-leptonic decays of heavy flavors and τs, and indeed $\not{E}_T$ was an integral component of $t\bar{t}$ production events followed by $t \to bW$ decays at the Fermilab Tevatron, and led to discovery of the top quark.

In these lectures, we will focus mainly on *new physics* reactions which lead to events containing large amounts of $\not{E}_T$. The major motivation nowadays that $\not{E}_T$ signatures from new physics should appear at LHC comes from cosmology. A large array of data– coming from measurements of galactic rotation curves, galaxy cluster velocity profiles, gravitational lensing, hot gas in galaxy clusters, light element abundances in light of Big Bang nucleosynthesis, large scale structure simulations and measurements and especially measurements of anisotropies in the cosmic microwave background radiation (CMB)– all point to a consistent picture of a universe which is constructed of

- baryons: $\sim 4\%$
- dark matter: $\sim 21\%$

- dark energy: $\sim 75\%$
- νs, γs: a tiny fraction.

Thus, particles present in the SM of particle physics constitute only $\sim 4\%$ of the universe's energy budget, while the main portion is comprised of dark energy (DE) which causes an accelerating expansion of the universe and dark matter (DM). There is an ongoing program of experiments under way or under development which will probe the dark energy. The goal is to try to tell if DE comes from Einstein's cosmological constant (which theoretical prejudice says really ought to be there), or something quite different. Measurements focus on the parameters entering the dark energy equation of state.

For dark matter, there exist good reasons to believe that it will be connected to new physics arising at the weak scale: exactly the energy regime to be probed by the LHC. The density of dark matter is becoming known with increasing precision. The latest WMAP5 measurements[1] find

$$\Omega_{CDM} h^2 = 0.110 \pm 0.006 \qquad (5.1)$$

where Ω_{CDM} is the cold DM density divided by the critical closure density of the universe ρ_c, and h is the scaled Hubble constant: $h = 0.73^{+0.04}_{-0.03}$. While the cosmic DM density is well-known, the identity of the DM particle(s) is completely unknown. Nevertheless, we do know some properties of the dark matter: it must have mass, it must be electrically neutral (non-interacting with light) and likely color neutral, and it must have a non-relativistic velocity profile, *i.e.* be what is termed *cold* dark matter (CDM). The reason it must be cold is that it must seed structure formation in the universe, *i.e.* it must clump. That means its velocity must be able to drop below its escape velocity; otherwise, it would disperse, and not seed structure formation. We already know of one form of dark matter: cosmic neutrinos. However, the SM neutrinos are so light that they must constitute *hot* dark matter.

The theoretical literature contains many possible candidate DM particles. A plot of some of these is shown in Fig. 5.1 (adapted from L. Roszkowski) in the mass versus interaction cross section plane. Some of these candidates emerge from attempts to solve longstanding problems in particle physics. For instance, the axion arises from the Peccei-Quinn solution to the strong CP problem. The weakly interacting massive particles (WIMPs) frequently arise from models which attempt to explain the origin of electroweak symmetry breaking. WIMP dark matter candidates include the lightest neutralino of models with weak scale supersymmetry, while Kaluza-Klein photons arise in models with universal extra dimensions, and lightest *T*-odd particles arise in Little Higgs models with a conserved *T*-parity.

Some Dark Matter Candidate Particles

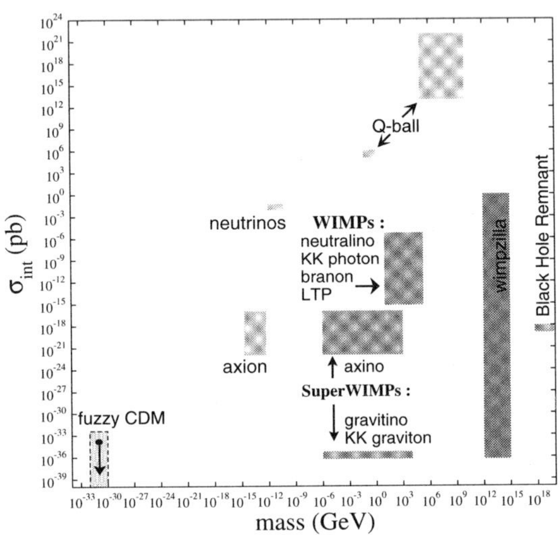

Fig. 5.1. *Dark matter candidates in the mass versus interaction strength plane, taken from Ref. [2]: http://www.science.doe.gov/hep/hepap_reports.shtm. See also, L. Roszkowski, Ref. [3].*

WIMP particles have an additional motivation, which has recently been coined the "WIMP miracle".[4] The idea here is to assume the existence of a dark matter particle which was once in thermal equilibrium at high temperatures in the early universe. The equilibrium abundance is easily calculated from thermodynamics and provides one boundary condition. As the universe expands and cools, the DM particle will drop out of thermal equilbrium (freeze-out); while DM will no longer be produced thermally, it can still annihilate away. The subsequent time evolution of the dark matter density is governed by the Boltzmann equation as formulated for the Friedman-Robertson-Walker universe:

$$dn/dt = -3Hn - \langle \sigma v_{rel} \rangle (n^2 - n_0^2) \qquad (5.2)$$

wherein the $-3Hn$ terms gives rise to dilution due to the Hubble expansion, and the $\langle \sigma v_{rel} \rangle$ term involves the dark matter annihilation cross section. The Boltzmann equation is solved in various textbooks,[5] and the solution is given approximately by

$$\Omega h^2 = \frac{s_0}{\rho_c/h^2} \left(\frac{45}{\pi g_*} \right)^{1/2} \frac{x_f}{M_{Pl}} \frac{1}{\langle \sigma v \rangle},$$ (5.3)

where s_0 is the current entropy density, g_* is the effective degrees of freedom and x_f is the freeze-out temperature $\sim m_{\widetilde{Z}_1}/20$. The quantities entering ths expression are unrelated to one another and several have large exponents. Filling in numerical values, they conspire to give

$$\Omega h^2 \sim \frac{0.1 \text{ pb}}{\langle \sigma v \rangle},$$ (5.4)

so that in order to obtain the measured dark matter abundance $\Omega h^2 \sim 0.11$, we expect $\langle \sigma v \rangle$ to be a *weak scale sized cross section*! Inputing a typical weak scale cross section,

$$\sigma = \frac{\alpha^2}{8\pi} \frac{1}{m^2}$$ (5.5)

and setting $\sigma \sim 1$ pb, we find $m \sim 100$ GeV! The lesson here is an important one: that the measured dark matter density is consistent with the annihilation cross section expected of a weakly interacting particle with *mass of order the weak scale*. This auspicious relation amongst dark matter density, annihilation cross section and particle mass is sometimes refered to as the *WIMP miracle*, and seemingly offers independent *astrophysical* evidence for new physics at or around the weak scale: in this case the new physics is associated with the cosmic density of dark matter. If such a WIMP particle exists, then it should be produced at large rates at the CERN LHC. Since it is dark matter, once produced in a collider experiment, it would escape normal detection. If produced in association with other SM particles (such as quarks and gluons), then it will carry away missing (transverse) energy, and will give rise to new physics signatures including jets plus $\not{E}_T$. It is this connection of weak scale physics and dark matter which makes the jets $+\not{E}_T$ signature so important and exciting at the LHC.

5.2. Lecture 1: SUSY basics: model building and spectra calculation

5.2.1. *The Wess-Zumino model*

We begin with a toy model due to Wess and Zumino, from 1974.[6] The model Lagrangian is given by

$$\mathcal{L} = \mathcal{L}_{\text{kin.}} + \mathcal{L}_{\text{mass}} \tag{5.6}$$

with

$$\mathcal{L}_{\text{kin.}} = \frac{1}{2}(\partial_\mu A)^2 + \frac{1}{2}(\partial_{mu} B)^2 + \frac{i}{2}\bar{\psi}\,\partial\!\!\!/\psi + \frac{1}{2}(F^2 + G^2)$$

$$\mathcal{L}_{\text{mass}} = -m[\frac{1}{2}\bar{\psi}\psi - GA - FB],$$

where A and B are real scalar fields, ψ is a 4-component Majorana spinor field and F and G are non-propagating *auxiliary* fields. The auxiliary fields may be eliminated from the Lagrangian via the Euler-Lagrange equations: $F = -mB$ and $G = -mA$.

But instead, impose the transformations

$$A \to A + \delta A \text{ with } \delta A = i\bar{\alpha}\gamma_5\psi,$$
$$\delta B = -\bar{\alpha}\psi,$$
$$\delta\psi = -F\alpha + iG\gamma_5\alpha + \partial\!\!\!/\gamma_5 A\alpha + i\,\partial\!\!\!/B\alpha,$$
$$\delta F = i\bar{\alpha}\,\partial\!\!\!/\psi,$$
$$\delta G = \bar{\alpha}\gamma_5\,\partial\!\!\!/\psi.$$

Note that the transformation mixes bosonic and fermionic fields. It is called a *supersymmetry* transformation. Using Majorana bilinear re-arrangements (such as $\bar{\psi}\chi = -\bar{\chi}\psi$), the product rule $\partial_\mu(f \cdot g) = \partial_\mu f \cdot g + f \cdot \partial_\mu g$ and some algebra, we find that $\mathcal{L} \to \mathcal{L} + \delta\mathcal{L}$ with

$$\delta\mathcal{L}_{\text{kin}} = \partial^\mu(-\frac{1}{2}\bar{\alpha}\gamma_\mu\,\partial\!\!\!/B\psi + \frac{i}{2}\bar{\alpha}\gamma_5\gamma_\mu\,\partial\!\!\!/A\psi + \frac{i}{2}F\bar{\alpha}\gamma_\mu\psi + \frac{1}{2}G\bar{\alpha}\gamma_5\gamma_\mu\psi),$$

$$\delta\mathcal{L}_{\text{mass}} = \partial^\mu(mA\bar{\alpha}\gamma_5\gamma_\mu\psi + imB\bar{\alpha}\gamma_\mu\psi).$$

Since the Lagrangian changes by a total derivative, the *action* $S = \int \mathcal{L}d^4x$ is invariant! (owing to Gauss' law in 4-d $\int_V d^4x\partial_\mu\Lambda^\mu = \int_{\partial V} d\sigma\Lambda^\mu n_\mu$). Under supersymmetry transformations, the Lagrangian always transforms into a total derivative, leaving the action invariant.

The Wess-Zumino (WZ) model as presented in 1974 exhibited a qualitatively new type of symmetry transformation with astonishing properties:

one being that the usual quantum field theory quadratic divergences associated with scalar fields all cancel! Theories with this sort of property might allow for a solution of the so-called "gauge hierarchy problem" of grand unified theories (GUTs).

5.2.2. *Superfield formalism*

While the WZ model was very compelling, it was constructed by hand, and offered no insight into how to construct general supersymmetric models. A short while later, Salam and Strathdee[7] developed the superfield formalism, which did provide rules for general SUSY model building. They first expanded the usual four-component spacetime x^μ to *superspace*: $x^\mu \to (x^\mu, \theta_a)$, where θ_a ($a = 1 - 4$) represents four *anti-commuting* (Grassmann vaued) dimensions arranged as a Majorana spinor. Then they introduced *general* superfields:

$$\hat{\Phi}(x,\theta) = \mathcal{S} - i\sqrt{2}\bar{\theta}\gamma_5\psi - \frac{i}{2}(\bar{\theta}\gamma_5\theta)\mathcal{M} + \frac{1}{2}(\bar{\theta}\theta)\mathcal{N} + \frac{1}{2}(\bar{\theta}\gamma_5\gamma_\mu\theta)V^\mu$$
$$+ i(\bar{\theta}\gamma_5\theta)[\bar{\theta}(\lambda + \frac{i}{\sqrt{2}}\,\partial\!\!\!/\psi)] - \frac{1}{4}(\bar{\theta}\gamma_5\theta)^2[\mathcal{D} - \frac{1}{2}\Box\mathcal{S}],$$

left chiral scalar superfields (LCSSF):

$$\hat{\mathcal{S}}_L(x,\theta) = \mathcal{S}(\hat{x}) + i\sqrt{2}\bar{\theta}\psi_L(\hat{x}) + i\bar{\theta}\theta_L\mathcal{F}(\hat{x}) \tag{5.7}$$

(where $\hat{x}_\mu = x_\mu + \frac{i}{2}\bar{\theta}\gamma_5\gamma_\mu\theta$) and right chiral scalar superfields (RCSSF):

$$\hat{\mathcal{S}}_R(x,\theta) = \mathcal{S}(\hat{x}^\dagger) - i\sqrt{2}\bar{\theta}\psi_R(\hat{x}^\dagger) - i\bar{\theta}\theta_R\mathcal{F}(\hat{x}^\dagger). \tag{5.8}$$

The multiplication table for superfields is as follows:

- $LCSSF \times LCSSF = LCSSF$
- $RCSSF \times RCSSF = RCSSF$
- $LCSSF \times RCSSF = $ a general superfield.

Moreover, under a supersymmetry transformation, the D-term of a general superfield and the F term of a LCSSF or a RCSSF always transforms into a total derivative.

This last observation gave the key insight allowing for construction of general supersymmetric models. The superpotential $\hat{f}$ is a function of LCSSFs only, and hence by the multiplication table above is itself a LCSSF. Its F term transforms into a total derivative under SUSY, and hence is a candidate Lagrangian. Also, the Kahler potential K is introduced: it is a function of products of RCSSFs times LCSSFs: it is therefore a general superfield and its D-term transforms into a total derivative under SUSY, and hence its D-term is a candidate Lagrangian. The choice $K = \sum_{fields} \hat{S}^\dagger \hat{S}$ is completely general for a renormalizable theory. Augmenting the above superfields with *gauge* superfields (ones containing spin-1 gauge bosons and spin-$\frac{1}{2}$ gauginos all transforming under the adjoint representation) allows one to construct locally gauge invariant, renormalizable field theories which are also invariant under SUSY transformations.

A Master Lagrangian can then be constructed. Its form is given in Ch. 5 of Ref. [8]. It allows for a recipe for SUSY model building:

- select the gauge group (simple or direct product),
- select the matter and Higgs representations,
- write down a superpotential (gauge invariant and renormalizable),
- plug all terms into the Master Lagrangian, calculate the physical states and Feynman rules, and you are good to go!

5.2.3. *Minimal Supersymmetric Standard Model (MSSM)*

The superfield formalism[8–10] allows for the construction of a supersymmetric version of the Standard Model, known as the Minimal Supersymmetric Standard Model, or MSSM. For each quark and lepton of the SM, the MSSM necessarily includes spin-0 superpartners $\tilde{q}_L$ and $\tilde{q}_R$ along with $\tilde{\ell}_L$ and $\tilde{\ell}_R$, whose gauge quantum numbers are fixed to be the known gauge quantum numbers of the corresponding SM fermions. Thus, for example, the right-handed up quark scalar (usually denoted by $\tilde{u}_R$) is a color-triplet, weak isosinglet with the same weak hypercharge 4/3 as the right-handed up-quark. The MSSM thus includes a plethora of new scalar states: $\tilde{e}_L$, $\tilde{e}_R$, $\tilde{\nu}_{eL}$, $\tilde{u}_L$, $\tilde{u}_R$, $\tilde{d}_L$, $\tilde{d}_R$ in the first generation, together with analogous states for the other two generations. Spin-zero *squark* partners of quarks with large Yukawa couplings undergo left-right mixing: thus, the $\tilde{t}_L$ and $\tilde{t}_R$ states mix to form mass eigenstates – $\tilde{t}_1$ and $\tilde{t}_2$ – ordered from lowest to highest mass.

The spin-0 Higgs bosons are embedded in Higgs superfields, so that the MSSM also includes spin-$\frac{1}{2}$ higgsinos. Unlike in the SM, the same Higgs doublet cannot give a mass to both up- and down- type fermions without catastrophically breaking the underlying supersymmetry. Thus the MSSM includes two Higgs doublets instead of one as in the SM. This gives rise to a richer spectrum of physical Higgs particles, including neutral light h and heavy H scalars, a pseudoscalar A and a pair of charged Higgs bosons $H^\pm$.

The gauge sector of the MSSM contains gauge bosons along with spin-half gauginos in the adjoint representation of the gauge group: thus, along with eight colored gluons, the MSSM contains eight colored spin-$\frac{1}{2}$ gluinos. Upon electroweak symmetry breaking, the four gauginos of $SU(2)_L \times U(1)_Y$ mix (just as the $SU(2)_L$ and $U(1)_Y$ gauge bosons mix) amongst themselves *and* the higgsinos, to form charginos – $\widetilde{W}_1^\pm$ and $\widetilde{W}_2^\pm$ – and neutralinos – $\widetilde{Z}_1$, $\widetilde{Z}_2$, $\widetilde{Z}_3$ and $\widetilde{Z}_4$. The $\widetilde{Z}_1$ state, the lightest neutralino, is often the lightest supersymmetric particle (LSP), and turns out to be an excellent WIMP candidate for CDM in the universe.

If nature is perfectly supersymmetric, then the *spin-0 superpartners would have exactly the same mass* as the corresponding fermions. Charged spin-0 partners of the electron with a mass of 0.51 MeV could not have evaded experimental detection. Their non-observation leads us to conclude that SUSY must be a *broken* symmetry. In the MSSM, SUSY is broken *explicitly* by including so-called soft SUSY breaking (SSB) terms in the Lagrangian. The SSB terms preserve the desirable features of SUSY, such as the stabilization of the scalar sector in the presence of radiative corrections, while lifting the superpartner masses in accord with what is necessary from experiment. It is important to note that the equality of dimensionless couplings between particles and their superpartners is still preserved (modulo small effects of radiative corrections): in particular, phenomenologically important gauge interactions of superpartners and the corresponding interactions of gauginos remain (largely) unaffected by the SSB terms.

5.2.4. *Supergravity*

The addition of the SSB Lagrangian terms may seem *ad-hoc* and ugly. It would be elegant if instead supersymmetry could be spontaneously broken. But it was recognized in the early to mid-1980's that models where global SUSY is spontaneously broken at the weak scale ran into serious difficulties. The situation is very different if we elevate SUSY from a global symmetry to a *local* one. In local SUSY, we are *forced to include the graviton/gravitino super-multiplet* into the theory, in much the same way that we have to include spin-1 gauge fields to maintain local gauge invariance of Yang-Mills theories. Theories with local SUSY are known as *supergravity* (SUGRA) theories because they are supersymmetric and necessarily include gravity. Moreover, the gravitational sector of the theory reduces to general relativity in the classical limit. Within the framework of SUGRA, it is possible to add an additional sector whose dynamics spontaneously breaks SUSY but which interacts with SM particles and their superpartners only via gravity (the so-called hidden sector). The spontaneous breakdown of supersymmetry results in a mass for the gravitino in the same way that in local gauge theories gauge bosons acquire mass by the Higgs mechanism. This is, therefore, referred to as the *super-Higgs mechanism*. The remarkable thing is that because of the gravitational couplng between the hidden and the MSSM sectors, the effects of spontaneous supersymmetry breaking in the hidden sector are conveyed to the MSSM sector, and (provided the SUSY-breaking scale in the hidden sector is appropriately chosen) weak scale SSB terms that lift the undesirable degeneracies between the masses of SM particles and their superpartners are automatically induced. Indeed, in the limit where $M_{\rm Pl} \to \infty$ (keeping the gravitino mass fixed), we recover a global SUSY theory along with the desired SSB terms! The gravitino typically has a weak scale mass and in many cases decouples from collider experiments because of its tiny gravitational couplings. For reasons that we cannot discuss here, these locally supersymmetric models are free[8–10] of many of the difficulties that plague globally supersymmetric models.

5.2.5. *SUGRA GUTs*

Motivated by the successful unification of gauge couplings at a scale $M_{\rm GUT} \sim 2 \times 10^{16}$ GeV in the MSSM, we are led to construct a grand unified theory (GUT) based on local supersymmetry. In this case, the theory renormalized at $Q = M_{\rm GUT}$ contains just one gaugino mass parameter $m_{1/2}$. Renormalization effects then split the physical gaugino masses in the same way the measured values of the gauge couplings arise from a single unified GUT scale gauge coupling. In general, supergravity models give rise to complicated mass matrices for the scalar superpartners of quarks and leptons, with concomitant flavor violation beyond acceptable levels. However, in models with *universal* soft SUSY breaking terms, a super-GIM mechanism suppresses flavor violating processes.[11] In what has come to be known as the minimal supergravity (mSUGRA) model, a universal scalar mass m_0 and also a universal SSB scalar coupling A_0 are assumed to exist at a high scale $Q \sim M_{\rm GUT} - M_{\rm Pl}$. The physical masses of squarks and sleptons are split after renormalization, and can be calculated using renormalization group techniques. Typically, in the mSUGRA model, we have $m_{\tilde{q}} \overset{>}{\sim} m_{\tilde{\ell}_L} \overset{>}{\sim} m_{\tilde{\ell}_R}$. Although the Higgs scalar mass parameters also start off at the common value m_0 at the high scale, the large value of the top quark Yukawa coupling drives the corresponding squared mass parameter to negative values and EWSB is *radiatively broken*. Within this framework, the masses and couplings required for phenomenology are fixed by just a handful of parameters which are usually taken to be,

$$m_0, \ m_{1/2}, \ A_0, \ \tan\beta, \ \text{and} \ sign(\mu). \tag{5.9}$$

Here, $\tan\beta$ is the ratio of the vacuum expectation values of the Higgs fields that give masses to up and down type fermions, and μ is the supersymmetric higgsino mass parameter whose magnitude is fixed to reproduce the measured value of M_Z. If all parameters are real, then potentially large CP-violating effects are suppressed as well. Computer codes such as Isajet, SuSpect, SoftSUSY and Spheno that calculate the full spectrum of sparticle and Higgs boson masses are publicly available.[12]

5.2.6. *Some realistic SUSY models*

The mSUGRA model (sometimes referred to as the constrained MSSM or CMSSM) serves as a paradigm for many SUSY phenomenological analyses. However, it is important to remember that it is based on many assumptions that can be tested in future collider experiments but which may prove to be incorrect. For instance, in many GUT theories, it is common to get *non-universal* SSB parameters. In addition, there are other messenger mechanisms besides gravity. In gauge-mediated SUSY breaking models (GMSB),[13] a special messenger sector is included, so gravitinos may be much lighter than all other sparticles, with implications for both collider physics and cosmology. In anomaly-mediated SUSY breaking (AMSB) models,[14] gravitational anomalies induce SSB terms, and the gravitino can be much heavier than the weak scale. There are yet other models[15] where SSB parameters get comparable contributions from gravity-mediated as well as from anomaly-mediated sources, and very recently, also from gauge-mediation.[16] The pattern of superpartner masses is sensitive to the mediation-mechanism, so that we can expect collider experiments to reveal which of the various mechanisms that have been proposed are actually realized in nature. We also mention that in both the GMSB and AMSB models, it is somewhat less natural (but still possible!) to obtain the required amount of SUSY dark matter in the Universe. Although these are all viable scenarios, they have not been as well scrutinized as the mSUGRA model.

5.3. Lecture 2: Sparticle production, decay and event generation

5.3.1. *The role of the CERN Large Hadron Collider*

The CERN Large Hadron Collider (LHC) turned on briefly in 2008 before an electrical mis-connection caused a helium leak, and a shut-down until the end of 2009. The LHC is a *pp* collider, expected to operate ultimately at center-of-mass energy $\sqrt{s} = 14$ TeV. Since protons are not fundamental particles, we may think instead of LHC as being a quark-gluon collider, where the quark and gluon constituents of the proton that undergo hard strong or electroweak scatterings only typically carry a small fraction of the proton energy.

For the hadronic reaction

$$A + B \rightarrow c + d + X, \tag{5.10}$$

where A and B are hadrons, c and d might be superpartners, and X is associated hadronic debris, we are really interested in the subproces reaction

$$a + b \rightarrow c + d, \tag{5.11}$$

where a is a partonic constituent of hadron A, and b is a partonic constituent of hadron B. The SM or MSSM or whatever effective theory we are working in provides us with the Feynman rules to calculate the subprocess cross-section $\hat{\sigma}(ab \rightarrow cd)$. To find the total hadron-hadron cross section, we must convolute the subprocess cross section with the *parton distribution functions* $f_{a/A}(x_a, Q^2)$ which gives the probability to find parton a in hadron A with momentum fraction x_a at energy-squared scale Q^2. Thus, we have

$$d\sigma(AB \rightarrow cdX) = \sum_{a,b} \int_0^1 dx_a \int_0^1 dx_b f_{a/A}(x_a, Q^2)\, f_{b/B}(x_b, Q^2) d\hat{\sigma}(ab \rightarrow cd). \tag{5.12}$$

where the sum extends over all initial partons a and b whose collisions produce the final state particles c and d. The parton-parton center-of-mass energy-squared $\hat{s} = x_a x_b s$ which enters the hard scatterng is only a small fraction of the proton-proton center-of-mass energy-squared s. Thus, to explore the TeV scale, a hadron-hadron collider of tens of TeV is indeed needed!

Around the 17 mile circumference LHC ring are situated four experiments: Atlas, CMS, LHC-B and Alice. Atlas and CMS are all-purpose detectors designed to detect almost all the energy emitted in hard scattering reactions. They will play a central role in the search for new physics at the LHC, and the search for dark matter production in the LHC environment.

5.3.2. *Cross section calculations for the LHC*

As noted above, if a perturbative model for new physics is available, then the relativistic-quantum mechanical amplitude $\mathcal{M}$ for the relevant scattering sub-process may be calculated, usually at a low order in perturbation theory. The probability for scattering is given by the square of the (complex) scattering amplitude $\mathcal{M}\mathcal{M}^\dagger$, and can often be expressed as a function of dot products of the external leg momenta entering the diagrams, after summing and averaging over spins and colors. Nowadays, several computer programs– such as CalcHEP, CompHEP and MadGraph– are available to automatically do tree level calculations of various $2 \to n$ processes.

To gain the parton level scattering cross section, one must multiply by suitable phase space factors and divide by the flux factor. The phase space integrals for simple sub-processes may be done analytically. More often, the final integrals are done using Monte Carlo (MC) techniques, which efficiently allow integration over multi-body phase space. The MC technique, wherein random integration variables are generated computationally, and summed over, actually allows for a *simulation* of the scattering process. This can allow one to impose cuts, or simulate detectors, to gain a more exact correspondence to the experimental environment.

5.3.3. *Sparticle production at the LHC*

Direct production of neutralino dark matter at the LHC ($pp \to \widetilde{Z}_1 \widetilde{Z}_1 X$, where X again stands for assorted hadronic debris) is of little interest since the high p_T final state particles all escape the detector, and there is little if anything to trigger an event record. Detectable events come from the production of the heavier superpartners, which in turn decay via a multi-step cascade which ends in the stable lightest SUSY particle (LSP).

In many SUSY models, the strongly interacting squarks and/or gluinos are among the heaviest states. Unless these are extremely heavy, they will have large production cross sections at the LHC. Strong interaction production mechanisms for their production include, 1. gluino pair production $\tilde{g}\tilde{g}$, 2. squark pair production $\tilde{q}\tilde{q}$ and 3. squark-gluino associated production $\tilde{q}\tilde{g}$. Note here that the reactions involving squarks include a huge number of subprocess reactions to cover the many flavors, types (left- and right-), and also the anti-squarks. The various possibilities each have different angular dependence in the production cross sections,[17] and the different flavors/types of squarks each have different decay modes.[18] These all have to be kept track of in order to obtain a reliable picture of the implications of SUSY in the LHC detector environment. Squarks and gluinos can also be produced in association with charginos and neutralinos.[19] Associated gluino production occurs via squark exchange in the t or u channels and is suppressed if squarks are very heavy.

If colored sparticles are very heavy, then electroweak production of charginos and neutralinos may be the dominant sparticle production mechanism at the LHC. The most important processes are pair production of charginos, $\widetilde{W}_i^{\pm}\widetilde{W}_j^{\mp}$ where $i,j = 1,2$, and chargino-neutralino production, $\widetilde{W}_i^{\pm}\widetilde{Z}_j$, with $i = 1,2$ and $j = 1-4$. In models with unified GUT scale gaugino masses and large $|\mu|$, $Z\widetilde{W}_1\widetilde{W}_1$ and $W\widetilde{Z}_2\widetilde{W}_1$ couplings are large so that $\widetilde{W}_1\widetilde{W}_1$ and $\widetilde{W}_1\widetilde{Z}_2$ production occurs at significant rates. The latter process can lead to the gold-plated trilepton signature at the LHC.[20] Neutralino pair production ($pp \to \widetilde{Z}_i\widetilde{Z}_j X$ where $i,j = 1 - 4$) is also possible. This reaction occurs at low rates at the LHC unless $|\mu| \simeq M_{1,2}$ (as in the case of mixed higgsino dark matter (MHDM)). Finally, we mention slepton pair production: $\tilde{\ell}^+\tilde{\ell}^-$, $\tilde{\nu}_\ell\tilde{\ell}$ and $\tilde{\nu}_\ell\bar{\tilde{\nu}}_\ell$, which can give detectable dilepton signals if $m_{\tilde{\ell}} \stackrel{<}{\sim} 300$ GeV.[21]

In Fig. 5.2 we show various sparticle production cross sections at the LHC as a function of $m_{\tilde{g}}$. Strong interaction production mechanisms dominate at low mass, while electroweak processes dominate at high mass. The associated production mechanisms are never dominant. The expected LHC integrated luminosity in the first year of running is expected to be around 0.1 fb^{-1}, while several tens of fb^{-1} of data is expected to be recorded in the first several years of operation. The ultimate goal is to accumulate around 500-1000 fb^{-1}, correponding to $10^5 - 10^6$ SUSY events for $m_{\tilde{g}} \sim 1$ TeV.

5.3.4. *Sparticle cascade decays*

In R-parity conserving SUSY models, sparticles decay to lighter sparticles until the decay terminates in the LSP.[18] Frequently, the direct decay to the LSP is either forbidden or occurs with only a small branching fraction. Since gravitational interactions are negligible, gluinos can only decay via $\tilde{g} \to q\tilde{q}$, where the q and $\tilde{q}$ can be of any flavor or type. If two body decay modes are closed, the squark will be virtual, and the gluino will decay via

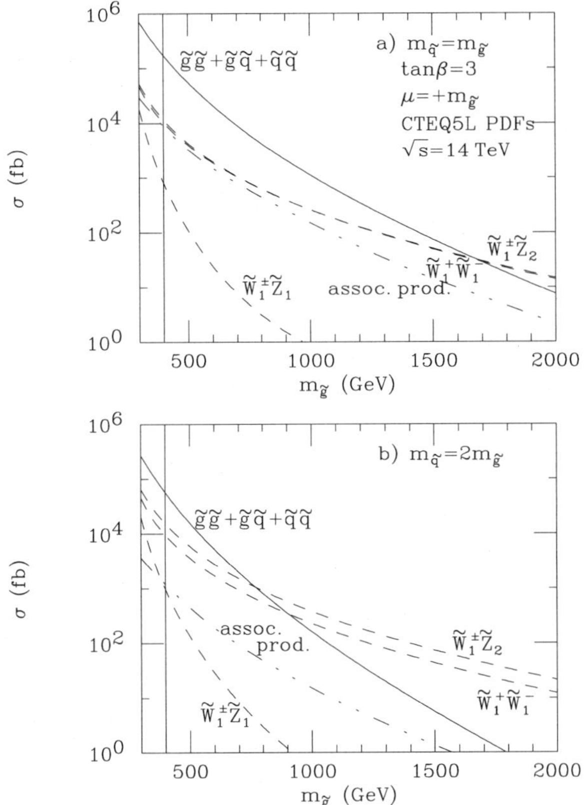

Fig. 5.2. *Cross sections for production of various sparticles at the LHC. Gaugino mass unification is assumed.*

three body modes $\tilde{g} \to q\bar{q}\widetilde{Z}_i$, $q\bar{q}'\widetilde{W}_j$. If squarks are degenerate, and Yukawa coupling effects negligible, three-body decays to the wino-like chargino and neutralino usually have larger branching fractions on account of the larger gauge coupling. If $|\mu| < M_2$, gluinos and squarks may thus decay most of the time to the heavier charginos and neutralinos, resulting in lengthy cascade decay chains at the LHC.

Squarks decay always to two-body modes: $\tilde{q} \to q\tilde{g}$ if it is kinematically allowed, or $\tilde{q}_L \to q'\widetilde{W}_i$, $q\widetilde{Z}_j$, while $\tilde{q}_R \to q\widetilde{Z}_j$ only, since right-squarks do not couple to charginos. Sleptons do not have strong interactions so cannot decay to gluinos. Their electroweak decays are similar to corresponding decays of squarks $\tilde{\ell}_L \to \ell'\widetilde{W}_i$, $\ell\widetilde{Z}_j$ while $\tilde{\ell}_R \to \ell\widetilde{Z}_j$ only.

Charginos may decay via two-body modes: $\widetilde{W}_i \to W\widetilde{Z}_j$, $\tilde{\ell}\nu_\ell$, $\ell\tilde{\nu}_\ell$, $Z\widetilde{W}_j$ or even to $\phi\widetilde{W}_j$ or $H^-\widetilde{Z}_j$, where $\phi = h, H, A$. If two-body modes are inaccessible, then three-body decays dominate: $\widetilde{W}_i \to \widetilde{Z}_j f\bar{f}'$, where f and f' are SM fermions which couple to the W. Frequently, the decay amplitude is dominated by the virtual W so that the three-body decays of $\widetilde{W}_1$ have the same branching fractions as those of the W. Neutralinos decay via $\widetilde{Z}_i \to W\widetilde{W}_j$, $H^+\widetilde{W}_j$, $Z\widetilde{Z}_j$, $\phi\widetilde{Z}_j$ or $f\tilde{f}$. If two body neutralino decays are closed, then $\widetilde{Z}_i \to \widetilde{Z}_j f\bar{f}$, where f are the SM fermions. In some models, the branching fraction for radiative decays $\widetilde{Z}_i \to \widetilde{Z}_j \gamma$ (that only occurs at the one-loop level) may be significant.[22] The cascade decay modes of neutralinos depend sensitively on model parameters.[23]

If $\tan\beta$ is large, then b and τ Yukawa coupling effects become important, enhancing three body decays of $\tilde{g}$, $\widetilde{W}_i$ and $\widetilde{Z}_j$ to third generation fermions.[24] For very large values of $\tan\beta$ these decays can even dominate, resulting in large rates for b-jet and τ-jet production in SUSY events.[25]

Finally, the various Higgs bosons can be produced both directly and via sparticle cascades at the LHC.[26] Indeed, it may be possible that h is first discovered in SUSY events because in a sample of events enriched for SUSY, it is possible to identify h via its dominant $h \to b\bar{b}$ decays rather than via its sub-dominant decay modes, as required for conventional searches.[26] The heavier Higgs bosons decay to a variety of SM modes, but also to SUSY particles if these latter decays are kinematically allowed, leading to novel signatures such as $H, A \to \widetilde{Z}_2\widetilde{Z}_2 \to 4\ell + \not{E}_T$.[27]

The cascade decays terminate in the LSP. In the case of a $\widetilde{Z}_1$ LSP, the $\widetilde{Z}_1$ is a DM candidate, and leaves its imprint via $\not{E}_T$. In the case of a weak scale $\tilde{G}$ or $\tilde{a}$ LSP, then $\widetilde{Z}_1$ will decay as $\widetilde{Z}_1 \to \gamma\tilde{G}$ or $\gamma\tilde{a}$. In these cases, the $\widetilde{Z}_1$ lifetime is long enough that it decays outside the detector, so one still expects large $\not{E}_T$ in the collider events. An exception arises for the case of super-light gravitinos (with masses in the eV to keV range) that are possible in GMSB models: see (5.15). Then, the decay may take place inside the detector, possibly with a large vertex separation. It is also possible that the NLSP is charged and quasi-stable, in which case collider events may include highly ionizing tracks instead of, or in addition to, $\not{E}_T$.

The decay branching fractions depend on the entire spectrum of SUSY particle masses and and their mixings. They are pre-programmed in several codes: Isajet,[28] SDECAY[29] and Spheno.[30]

5.3.5. *Event generation for LHC*

Once sparticle production cross sections and decay branching fractions have been computed, it is useful to embed these into event generator programs to simulate what SUSY collider events will look like at LHC. There are several steps involved:

- Calculate all sparticle pair production cross sections. Once all initial and final states are accounted for, this involves over a thousand individual subprocess reactions. In event generation, a particular reaction is selected on a probabilistic basis, with a weight proportional to its differential cross-section.
- Sparticle decays are selected probabilistically into all the allowed modes in proportion to the corresponding branching fractions.
- Initial and final state quark and gluon radiation are usually dealt with using the parton shower (PS) algorithm, which allows for probabilistic parton emission based on approximate collinear QCD emission matrix elements, but exact kinematics. The PS is also applied at each step of the cascade decays, which may lead to additional jet production in SUSY collider events.
- A hadronization algorithm provides a model for turning various quarks and gluons into mesons and baryons. Unstable hadrons must be further decayed.
- The beam remnants – proton constituents not taking part in the hard scattering – must be showered and hadronized, usually with an independent algorithm, so that energy deposition in the forward detector region may be reliably calculated.

At this stage, the output of an event generator program is a listing of particle types and their associated four-vectors. The resulting event can then be interfaced with detector simulation programs to model what the actual events containing DM will look like in the environment of a collider detector.

Several programs are available, including Isajet,[28] Pythia,[31] Herwig[32] and Sherpa.[33] Other programs such as Madevent,[34] CompHEP/CalcHEP[35] and Whizard[36] can generate various $2 \to n$ processes including SUSY particles. The output of these programs may then be used as input to Pythia or Herwig for showering and hadronization. Likewise, parton level Isajet SUSY production followed by cascade decays can be input to Pythia and Herwig via the Les Houches Event format.[37]

5.4. Lecture 3: SUSY, LHT and UED signatures at the LHC

5.4.1. *Signatures for SUSY particle production*

Unless colored sparticles are very heavy, the SUSY events at the LHC mainly result in gluino and squark production, followed by their possibly lengthy cascade decays. These events, therefore, typically contain very hard jets (from the primary decay of the squark and/or gluino) together with other jets and isolated electrons, muons and taus (identified as narrow one- and three-prong jets), and sometimes also photons, from the decays of secondary charginos and neutralinos, along with $\not{E}_T$ that arises from the escaping dark matter particles (as well as from neutrinos). In models with a superlight gravitino, there may also be additional isolated photons, leptons or jets from the decay of the NLSP. The relative rates for various n-jet + m-lepton + k-photon $+\not{E}_T$ event topologies is sensitive to the model as well as to the parameter values, and so provide a useful handle for phenomenological analyses.

Within the SM, the physics background to the classic $jets + \not{E}_T$ signal comes from neutrinos escaping the detector. Thus, the dominant SM backgrounds come from $W + jets$ and $Z + jets$ production, $t\bar{t}$ production, QCD multijet production (including $b\bar{b}$ and $c\bar{c}$ production), WW, WZ, ZZ production plus a variety of $2 \to n$ processes which are not usually included in event generators. These latter would include processes such as $t\bar{t}t\bar{t}$, $t\bar{t}b\bar{b}$, $t\bar{t}W$, WWW, WWZ production, *etc.* Decays of electroweak gauge bosons and the t-quark are the main source of isolated leptons in the SM. Various additional effects– uninstrumented regions, energy mis-measurement, cosmic rays, beam-gas events– can also lead to $\not{E}_T$ events.

In contrast to the SM, SUSY events naturally tend to have large jet multiplicities and frequently an observable rate for high multiplicity lepton events with large $\not{E}_T$. Thus, if one plots signal and background versus multiplicity of any of these quantities, as one steps out to large multiplicity, the expected SUSY events should increase in importance, and even dominate the high multiplicity channels in some cases. This is especially true of isolated multi-lepton signatures, and in fact it is convenient to classify SUSY signal according to lepton multiplicity:[38]

- zero lepton $+jets + \not{E}_T$ events,
- one lepton $+jets + \not{E}_T$ events,
- two opposite sign leptons $+jets + \not{E}_T$ events (OS),
    - same-flavor (OSSF),

  – different flavor (OSDF),

- two same sign leptons $+jets + \not{E}_T$ events (SS),
- three leptons $+jets + \not{E}_T$ events (3ℓ),
- four (or more) leptons $+jets + \not{E}_T$ events (4ℓ).

5.4.2. *LHC reach for SUSY*

Event generators, together with detector simulation programs can be used to project the SUSY discovery reach of the LHC. Given a specific model, one may first generate a grid of points that samples the parameter (sub)space where signal rates are expected to vary significantly. A large number of SUSY collider events can then be generated at every point on the grid along with the various SM backgrounds to the SUSY signal mentioned above. Next, these signal and background events are passed through a detector simulation program and a jet-finding algorithm is implemented to determine the number of jets per event above some $E_T(jet)$ threshold (usually taken to be $E_T(jet) > 50 - 100$ GeV for LHC). Finally, *analysis cuts* are imposed which are designed to reject mainly SM BG while retaining the signal. These cuts may include both topological and kinematic selection criteria. For observability with an assumed integrated luminosity, we require that the signal exceed the chance 5 standard deviation upward fluctuation of the background, together with a minimum value of ($\sim 25\%$) the signal to background ratio, to allow for the fact that the background is not perfectly known. For lower sparticle masses, softer kinematic cuts are used, but for high sparticle masses, the lower cross sections but higher energy release demand hard cuts to optimize signal over background.

In Fig. 5.3, we illustrate the SUSY reach of the LHC within the mSUGRA model assuming an integrated luminosity of 100 fb^{-1}. We show the result in the $m_0 - m_{1/2}$ plane, taking $A_0 = 0$, $\tan\beta = 10$ and $\mu > 0$. The signal is observable over background in the corresponding topology below the corresponding curve. We note the following.

(1) Unless sparticles are very heavy, there is an observable signal in several different event topologies. This will help add confidence that one is actually seeing new physics, and may help to sort out the production and decay mechanisms.
(2) The reach at low m_0 extends to $m_{1/2} \sim 1400$ GeV. This corresponds to a reach for $m_{\tilde{q}} \sim m_{\tilde{g}} \sim 3.1$ TeV.
(3) At large m_0, squarks and sleptons are in the $4 - 5$ TeV range, and are

too heavy to be produced at significant rates at LHC. Here, the reach comes mainly from just gluino pair production. In this range, the LHC reach is up to $m_{1/2} \sim 700$ GeV, corresponding to a reach in $m_{\tilde{g}}$ of about 1.8 TeV, and may be extended by $\sim$ 15-20% by b-jet tagging.[39]

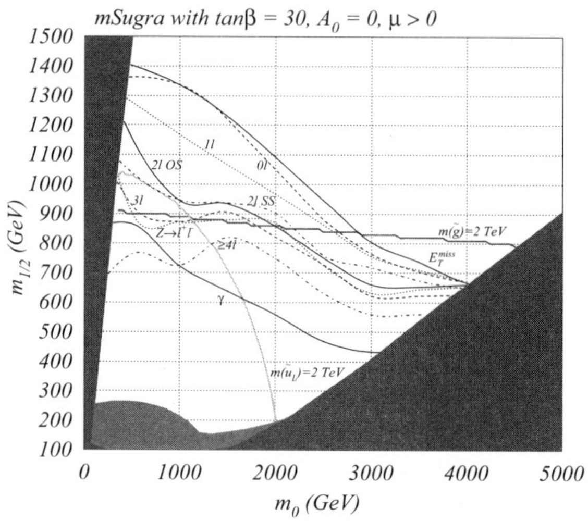

Fig. 5.3. *The 100 fb^{-1} fb reach of LHC for SUSY in the mSUGRA model. For each event topology, the signal is observable below the corresponding contour.*

In Fig. 5.4 we can see a comparison of the LHC reach (notice that it is insensitive to $\tan\beta$ and sign(μ)) with that of the Tevatron (for clean 3ℓ events with 10 fb^{-1}), and the proposed e^+e^- International Linear Collider (ILC), with $\sqrt{s} = 0.5$ or 1 TeV along with various dark matter direct detection (DD) and indirect detection (ID) search experiments. We remark that:

- While LHC can cover most of the relic density allowed region, the HB/FP region emerges far beyond the LHC reach.
- As will be discussed, the DD and ID experiments have the greatest sensitivity in the HB/FP region where the neutralino is MHDM. In this sense, DD and ID experiments *complement* LHC searches for SUSY.
- The ILC reach is everywhere lower than LHC, except in the HB/FP region. In this region, while gluinos and squarks can be extremely heavy, the μ parameter is small, leading to a relatively light spec-

trum of charginos and neutralinos. These are not detectable at the LHC because the visible decay products are too soft. However, since chargino pair production is detectable at ILC even if the energy release in chargino decays is small, the ILC reach extends beyond LHC in this region.[40]

Finally, we note here that while the results presented above are for the LHC reach in the mSUGRA model, the LHC reach (measured in terms of $m_{\tilde{g}}$ and $m_{\tilde{q}}$) tends to be relatively insensitive to the details of the model chosen, as long as gluino and squark production followed by cascade decays to the DM particle occur.

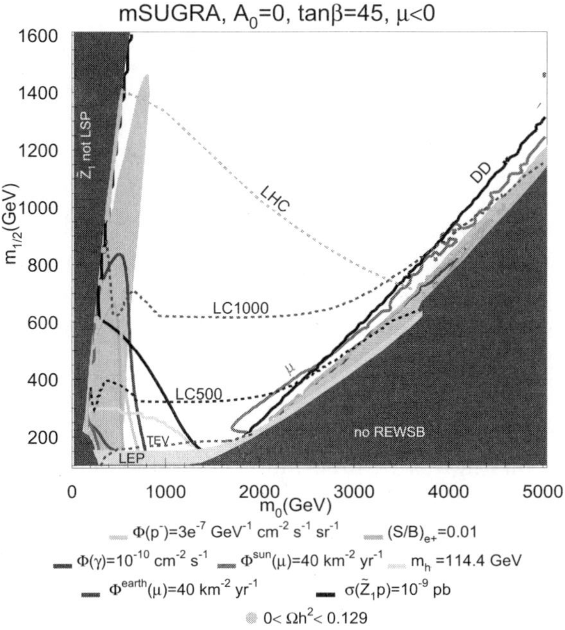

Fig. 5.4. *The projected reach of various colliders, direct and indirect dark matter search experiments in the mSUGRA model. For the indirect search results we have adopted the conservative default DarkSUSY isotropic DM halo density distribution. Plot is from Ref. [87].*

5.4.3. *Early discovery of SUSY at LHC without $\not{E}_T$*

The classic signature for SUSY at hadron colliders is the presence of jets, isolated leptons and especially large $\not{E}_T$. However, in the early period of LHC running, $\not{E}_T$ may be very difficult to measure reliably. In a real detector, there will be calorimetry calibration issues, dead or uninstrumented regions, "hot" cells, cosmic ray events, beam-gas collisions: all of these contribute to $\not{E}_T$ events, in additon to SM backgrounds and possibly new physics. As learned from Tevatron experiments, a reliable measurement of $\not{E}_T$ requires knowledge of the complete detector, and that may well take some time– possibly over a year– after start-up.

Can LHC search for SUSY even if $\not{E}_T$ searches are not viable? The answer seems to be yes, as long as lepton isolation cuts are possible. In Ref. [41], it was shown that requiring events with ≥ 4 jets plus a *high multiplicity* of isolated leptons: SS, OSSF, OSDF, 3ℓ, $\cdots$ would allow one to severely reduce SM background while maintaining large enough signal rates. It is claimed that an LHC reach of $m_{\tilde{g}} \sim 750$ GeV is possible with just 0.1 fb^{-1} of integrated luminosity, *without* using an $\not{E}_T$ cut, by requiring events with ≥ 3 isolated leptons (*es* or μs).

In addition, reliable electron identification may also be an issue in early LHC running. The problem here is differentiating the electron's EM shower from other energy depositions such as charged pions, or hard photons where in additon a charged track is nearby. In this case, a SUSY search may still be made by looking for multi-jet plus *multi-muon* production, again without any requirement on missing E_T.[42] Atlas and CMS are already reliably detecting cosmic muons. Also, muons enjoy an advantage over electrons in that they can be detected reliably to p_T values as low as 5 GeV.

The multiplicity of muons in ≥ 4 jet events is shown in Fig. 5.5. Here, we see that as one moves to high muon multiplicity, the SM background drops off sharply, and by $n_\mu \geq 3$, signal already far exceeds BG, at least for the SPS1a' signal point shown in the figure. In fact, by requiring *same-sign dimuon* plus multi-jet events, already signal in many cases will exceed background. The LHC should be able to discover gluinos with $m_{\tilde{g}} \sim 550$ (750) GeV in the SS dimuon plus ≥ 4 jets state with just 0.04 (0.1) fb^{-1} of integrated luminosity.

5.4.4. *Determination of sparticle properties*

Once a putative signal for new physics emerges at LHC, the next step is to establish its origin. This will entail detailed measurements of cross sections

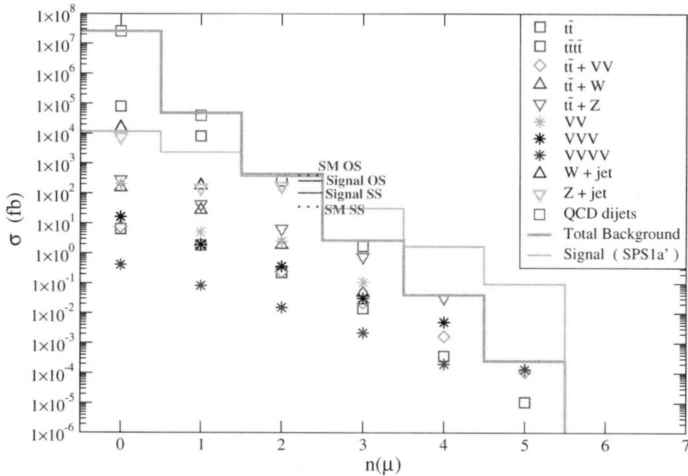

Fig. 5.5. *Multiplicity of isolated muons in ≥ 4 jets events, from signal point SPS1a' and various SM background processes.*

and distributions in various event topologies to gain insight into the identity of the new particles being produced, their masses, decay patterns, spins, couplings (gauge quantum numbers) and ultimately mixing angles. These measurements are not straightforward in the LHC environment because of numerous possible SUSY production reactions occurring simultaneously, a plethora of sparticle cascade decay possibilities, hadronic debris from initial state radiation and lack of invariant mass reconstruction due to the presence of $\not{E}_T$. All these lead to ambiguities and combinatoric problems in reconstructing exactly what sort of signal reactions are taking place. In contrast, at the ILC, the initial state is simple, the beam energy is tunable and beam polarization can be used to select out specific processes.

While it seems clear that the ILC is better suited for a systematic program of precision sparticle measurements, studies have shown (albeit in special cases) that interesting measurements are also possible at the LHC. We go into just a subset of all details here in order to give the reader an idea of some of the possibilities suggested in the literature.

One suggested starting point is the distribution of effective mass $M_{\text{eff}} = \not{E}_T + \sum E_T(jets)$ in the inclusive SUSY sample, which sets the approximate mass scale $M_{\text{SUSY}} \equiv \min(m_{\tilde{g}}, m_{\tilde{q}})$ for the strongly interacting sparticles are being produced,[43] and provides a measure of M_{SUSY} to 10-15%.

More detailed information on sparticle masses may be accessed by studying specific event topologies. For instance, the mass of dileptons from $\widetilde{Z}_2 \to \ell^+\ell^-\widetilde{Z}_1$ decays is bounded by $m_{\widetilde{Z}_2} - m_{\widetilde{Z}_1}$ (this bound is even more restrictive if $\widetilde{Z}_2$ decays via an on-shell slepton).[44] We therefore expect an OSSF invariant mass distribution to exhibit an edge at $m_{\widetilde{Z}_2} - m_{\widetilde{Z}_1}$ (or below) in any sample of SUSY events so long as the "spoiler" decay modes $\widetilde{Z}_2 \to \widetilde{Z}_1 Z$ or $\widetilde{Z}_1 h$ are closed. Contamination from chargino production can be statistically removed by subtracting out the distribution of OSDF dileptons. In MHDM models, there may be more than one visible mass edge because the $\widetilde{Z}_3$ may also be accessible in cascade decays.

In the happy circumstance where production of gluinos or a single type of squark is dominant, followed by a string of two-body decays, then further invariant mass edges are possible. One well-studied example comes from $\tilde{g} \to b\bar{\tilde{b}}_1 \to b\bar{b}\widetilde{Z}_2 \to b\bar{b}\ell\bar{\ell}\widetilde{Z}_1$; then one can try to combine a b-jet with the dilepton pair to reconstruct the squark-neutralino mass edge: $m(b\ell\bar{\ell}) < m_{\tilde{b}_1} - m_{\widetilde{Z}_1}$. Next, combining with another b-jet can yield a gluino-neutralino edge: $m(b\bar{b}\ell\bar{\ell}) < m_{\tilde{g}} - m_{\widetilde{Z}_1}$. The reconstruction of such a decay chain may be possible as shown in Ref. [43], where other sequences of two-body decays are also examined. In practice, such fortuitous circumstances may not exist, and there are many combinatoric issues to overcome as well. A different study[45] shows that end-point measurements at the LHC will make it possible to access the mass difference between the LSP and the stau in a mSUGRA scenario where the stau co-annihilation mechanism is operative.

These end-point measurements generally give mass differences, not masses. However, by an analysis of the decay chain $\tilde{q}_L \to q\widetilde{Z}_2 \to q\tilde{\ell}^{\pm}\ell^{\mp} \to q\ell^{\pm}\ell^{\mp}\widetilde{Z}_1$, it has been argued[46] that reconstruction of masses may be possible under fortuituous circumstances. More recently, it has been suggested that it may be possible to directly access the gluino and/or squark masses (not mass differences) via the introduction of the so-called m_{T2} variable. We will refer the reader to the literature for details.[47] It also may be possible to make a measurement of $m_{\tilde{g}}$ basd on total production cross section in cases where pure gluino pair production is dominant.[48]

Mass measurements allow us to check consistency of specific SUSY models with a handful of parameters, and together with other measurements can readily exclude such models. But these are not the only interesting measurements at the LHC. It has been shown that if the NLSP of GMSB models decays into a superlight gravitino, it may be possible to determine its lifetime, and hence the gravitino mass at the LHC.[49] This will then allow one to infer the underlying SUSY breaking scale, a scale at least as important as the weak scale! A recent study[50] suggests that this is possible even when the the decay length of the NLSP is too short to be measured. While linear collider experiments will ultimately allow the precision measurements that will directly determine the new physics to be softly broken supersymmetry,[51] it will be exciting to analyze real LHC data that will soon be available to unravel many of the specific details about how (or if) SUSY is actually implemented in nature.

5.4.5. *Measuring DM properties at LHC and ILC*

SUSY discovery will undoubtedly be followed by a program (as outlined in Sec. 5.4.4) to reconstruct sparticle properties. What will we be able to say about dark matter in light of these measurements? Such a study was made by Baltz *et al.*[52] where four mSUGRA case study points (one each in the bulk region, the HB/FP region, the stau coanihilation region and the A-funnel region) were examined for the precision with which measurements of sparticle properties that could be made at LHC, and also at a $\sqrt{s} = 0.5$ and 1 TeV e^+e^- collider. They then adopted a 24-parameter version of the MSSM and fit its parameters to these projected measurements. The model was then used to predict several quantities relevant to astrophysics and cosmology: the dark matter relic density $\Omega_{\widetilde{Z}_1} h^2$, the spin-independent neutralino-nucleon scattering cross section $\sigma_{SI}(\widetilde{Z}_1 p)$, and the neutralino annihilation cross section times relative velocity, in the limit that $v \to 0$: $\langle \sigma v \rangle|_{v \to 0}$. The last quantity is the crucial particle physics input for estimating signal strength from neutralino annihilation to anti-matter or gammas in the galactic halo. What this yields then is a *collider measurement* of these key dark matter quantities.

As an illustration, we show in Fig. 5.6 (taken from Ref. [52]) the precision with which the neutralino relic density is constrained by collider measurements for the LCC2 point which is in the HB/FP region of the mSUGRA model. Measurements at the LHC cannot fix the LSP composition, and so are unable to resolve the degeneracy between a wino-LSP solution (which gives a tiny relic density) and the true solution with MHDM. Determinations of chargino production cross sections at the ILC can easily resolve the difference. It is nonetheless striking that up to this degeneracy ambiguity, experiments at the LHC can pin down the relic density to within $\sim 50\%$ (a remarkable result, given that there are sensible models where the predicted relic density may differ by orders of magnitude!). This improves to 10-20% if we can combine LHC and ILC measurements.

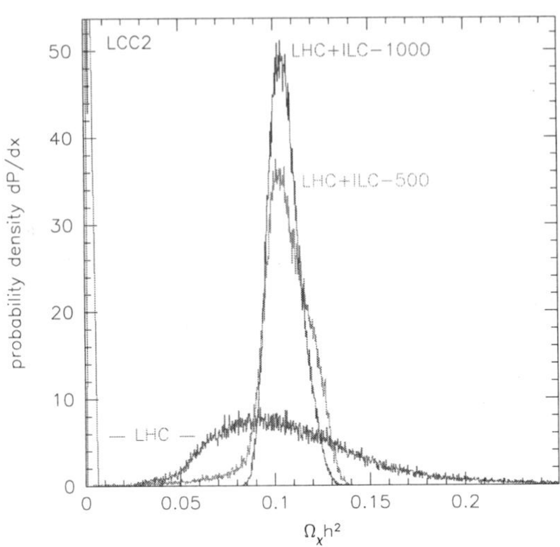

Fig. 5.6. *Determination of neutralino relic abundance via measurements at the LHC and ILC, taken from Ref. [52].*

This collider determination of the relic density is very important. If it agrees with the cosmological measurement it would establish that the DM is dominantly thermal neutralinos from the Big Bang. If the neutralino relic density from colliders falls significantly below (5.1), it would provide direct evidence for multi-component DM– perhaps neutralinos plus axions or other exotica. Alternatively, if the collider determination gives a much larger value of $\Omega_{\widetilde{Z}_1} h^2$, it could point to a long-lived but unstable neutralino and/or non-thermal DM.

The collider determination of model parameters would also pin down the neutralino-nucleon scattering cross section. Then if a WIMP signal is actually observed in DD experiments, one might be able to determine the local DM density of neutralinos and aspects of their velocity distribution based on the DD signal rate. This density should agree with that obtained from astrophysics if the DM in our Galaxy is comprised only of neutralinos.

Finally, a collider determination of $\langle\sigma v\rangle|_{v\to 0}$ would eliminate uncertainty on the particle physics side of projections for any ID signal from annihilation of neutralinos in the galactic halo. Thus, the observation of a gamma ray and/or anti-matter signal from neutralino halo annihilations would facilitate the determination of the galactic halo dark matter density distribution.

5.4.6. *Some non-SUSY WIMPs at the LHC*

5.4.6.1. *B_μ^1 state from universal extra dimensions*

Models with Universal Extra Dimensions, or UED, are interesting constructs which provide a foil for SUSY search analyses.[53] In the 5-D UED theory, one posits that the fields of the SM actually live in a 5-D brane world. The extra dimension is "universal" since *all* the SM particles propagate in the 5-D bulk. The single extra dimension is assumed to be compactified on a S_1/Z_2 orbifold (line segment). After compactification, the 4-D effective theory includes the usual SM particles, together with an infinite tower of Kaluza-Klein (KK) excitations. The masses of the excitations depend on the radius of the compactified dimension, and the first ($n = 1$) KK excitations can be taken to be of order the weak scale. In these theories, KK-parity $(-1)^n$ can be a conserved quantum number. If this so-called KK-parity is exact, then the lightest odd KK parity state will be stable and can be a DM candidate. At tree-level, all the KK excitations in a given level are essentially degenerate. Radiative corrections break the degeneracy, leaving colored excitations as the heaviest excited states and the $n = 1$ KK excitation of the SM $U(1)_Y$ gauge boson B_μ^1 as the lightest[54] KK odd state: in the UED case, therefore, the DM particle has spin-1. The splitting caused by the radiative corrections is also essential to assess how the KK excitations decay, and hence are crucial for collider phenomenology.[55]

The relic density of B_μ^1 particles has been computed, and found to be compatible with observation for certain mass ranges of B_μ^1.[56] Also, in UED, the colored excitations can be produced with large cross sections at the LHC, and decay via a cascade to the B_μ^1 final state. Thus, the collider signatures are somewhat reminiscent of SUSY, and it is interesting to ask whether it is possible to distinguish a *jets+leptons+$\not{E}_T$* signal in UED from that in SUSY. Several studies[57] answer affirmatively, and in fact provide strong motivation for the measurement of the spins of the produced new particles.[58] UED DM generally leads to a large rate in IceCube, and may also give an observable signal in anti-protons and possibly also in photons and positrons.[53,59] DD is also possible but the SI cross section is typically smaller than 10^{-9} pb.

5.4.6.2. *Little Higgs models*

Little Higgs models[60,61] provide an alternative method compared to SUSY to evade the quadratic sensitivity of the scalar Higgs sector to ultra-violet

(UV) physics. In this framework, the Higgs boson is a pseudo-Goldstone boson of a spontaneously broken global symmetry that is not completely broken by any one coupling, but is broken when *all* couplings are included. This then implies that there is quadratic sensitivity to UV physics, but only at the multi-loop level. Specific models where the quadratic sensitivity enters at the two-loop level should, therefore, be regarded as low energy effective theories valid up to a scale $\Lambda \sim 10$ TeV, at which a currently unknown, and perhaps strongly-coupled UV completion of the theory is assumed to exist. Models that realize this idea require new TeV-scale degrees of freedom that can be searched for at the LHC: new gauge bosons, a heavy top-like quark, and new spin-zero particles, all with couplings to the SM. These models, however, run into phenomenological difficulties with precision EW constraints, unless a discrete symmetry– dubbed T-parity[62]– is included. SM particles are then T-even, while the new particles are T-odd.

We will set aside the issue of whether T-parity conservation is violated by anomalies,[64] and assume that a conserved T-parity can be introduced.[65] In this case, the lightest T-odd particle A_H – the Little Higgs partner of the hypercharge gauge boson with a small admixture of the neutral W_{3H} boson – is stable and yields the observed amount of DM for a reasonable range of model parameters.[59] In this case, the DM particle has spin-1, though other cases with either a spin-$\frac{1}{2}$ or spin-0 heavy particle may also be possible. A_H can either annihilate with itself into vector boson pairs or $t\bar{t}$ pairs via s-channel Higgs exchange, or into top pairs via exchange of the heavy T-odd quark in the t-channel. Co-annihilation may also be possible if the heavy quark and A_H are sufficiently close in mass. Signals at the LHC[66,67] mainly come from pair production of heavy quarks $pp \to T\bar{T}X$ followed by $T \to tA_H$ decay, and from single production of the heavy quark in association with A_H. These lead to $t\bar{t} + \not{E}_T$ events at LHC.[68] The $\not{E}_T$ comes from the escaping A_H particle, which must be the endpoint of all T-odd particle decays.* If A_H is the dominant component of galactic DM, we will generally expect small DD and ID rates for much the same reasons that the signals from the bino LSP tend to be small:[59] see, however, Ref. [70] for a different model with large direct detection rate.

---

*We note here that it is also possible to construct so-called twin-Higgs models[69] where the Higgs sector is stabilized via new particles that couple to the SM Higgs doublet, but are singlets under the SM gauge group. In this case, there would be no obvious new physics signals at the LHC.

5.5. Lecture 4: $\not{E}_T$ and the dark matter connection

5.5.1. *Neutralino relic density*

Once a SUSY model is specified, then given a set of input parameters, it is possible to compute all superpartner masses and couplings necessary for phenomenology. We can then use these to calculate scattering cross sections and sparticle decay patterns to evaluate SUSY signals (and corresponding SM backgrounds) in collider experiments. We can also check whether the model is allowed or excluded by experimental constraints, either from direct SUSY searches, *e.g.* at LEP2 which requires that $m_{\widetilde{W}_1} > 103.5$ GeV, $m_{\tilde{e}} \gtrsim 100$ GeV, and $m_h > 114.4$ GeV (for a SM-like light SUSY Higgs boson h), or from indirect searches through loop effects from SUSY particles in low energy measurements such as $B(b \to s\gamma)$ or $(g - 2)_\mu$. We can also calculate the expected thermal LSP relic density. To begin our discussion, we will first assume that the lightest neutralino $\widetilde{Z}_1$ is the candidate DM particle.

As mentioned earlier, the relic density calculation involves solving the Boltzmann equation, where the neutralino density changes due to both the expansion of the Universe and because of neutralino annihilation into SM particles, determined by the thermally averaged $\widetilde{Z}_1\widetilde{Z}_1$ annihilation cross section. An added complication occurs if neutralino *co-annihilation* is possible. Co-annihilation occurs if there is another SUSY particle close in mass to the $\widetilde{Z}_1$, whose thermal relic density (usually suppressed by the Boltzmann factor $exp^{\frac{-\Delta M}{T}}$) is also significant. In the mSUGRA model, co-annihilation may occur from a stau, $\tilde{\tau}_1$, a stop $\tilde{t}_1$ or the lighter chargino $\widetilde{W}_1$. For instance, in some mSUGRA parameter-space regions the $\tilde{\tau}_1$ and $\widetilde{Z}_1$ are almost degenerate, so that they both have a significant density in the early universe, and reactions such as $\widetilde{Z}_1\tilde{\tau}_1 \to \tau\gamma$ occur. Since the electrically charged $\tilde{\tau}_1$ can also annihilate efficiently via electromagnetic interactions, this process also alters the equilibrium density of neutralinos. All in all, there are well over a thousand neutralino annihilation and co-annihilation reactions that need to be computed, involving of order 7000 Feynman diagrams. There exist several publicly available computer codes that compute the neutralino relic density: these include DarkSUSY,[71] MicroMegas[72] and IsaReD[73] (a part of the Isatools package of Isajet[28]).

As an example, we show in Fig. 5.7 the m_0 *vs.* $m_{1/2}$ plane from the mSUGRA model, where we take $A_0 = 0$, $\mu > 0$, $m_t = 171.4$ GeV and $\tan\beta = 10$. The red-shaded regions are not allowed because either the $\tilde{\tau}_1$

becomes the lightest SUSY particle, in contradiction to negative searches for long lived, charged relics (left edge), or EWSB is not correctly obtained (lower-right region). The blue-shaded region is excluded by LEP2 searches for chargino pair production ($m_{\widetilde{W}_1} < 103.5$ GeV). We show contours of squark (solid) and gluino (dashed) mass (which are nearly invariant under change of A_0 and $\tan\beta$). Below the magenta contour near $m_{1/2} \sim 200$ GeV, $m_h < 110$ GeV, which is roughly the LEP2 lower limit on m_h in the model. The thin green regions at the edge of the unshaded white region has $0.094 < \Omega_{\widetilde{Z}_1} h^2 < 0.129$ where the neutralino saturates the observed relic density. In the adjoining yellow regions, $\Omega_{\widetilde{Z}_1} h^2 < 0.094$, so these regions require multiple DM components. The white regions all have $\Omega_{\widetilde{Z}_1} h^2 > 0.129$ and so give too much thermal DM: they are excluded in the standard Big Bang cosmology.

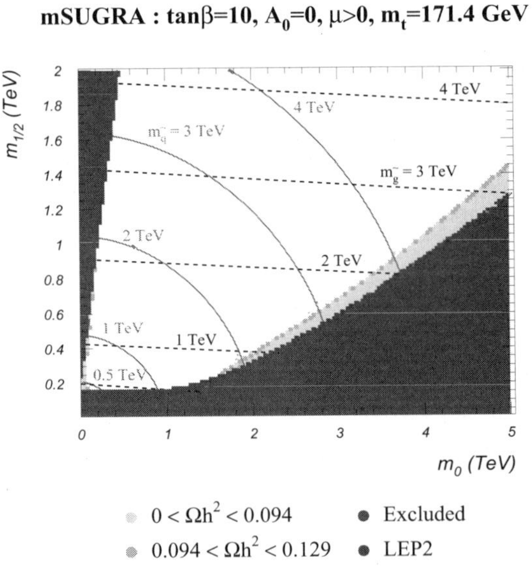

mSUGRA : tanβ=10, A$_0$=0, μ>0, m$_t$=171.4 GeV

Fig. 5.7. *DM-allowed regions in the $m_0 - m_{1/2}$ plane of the mSUGRA model for* $\tan\beta = 10$ *with* $A_0 = 0$ *and* $\mu > 0$.

The DM-allowed regions are classified as follows:

- At very low m_0 and low $m_{1/2}$ values is the so-called *bulk* annihilation region.[74] Here, sleptons are quite light, so $\widetilde{Z}_1\widetilde{Z}_1 \to \ell\bar{\ell}$ via t-channel slepton exchange. In years past (when $\Omega_{\mathrm{CDM}}h^2 \sim 0.3$ was quite consistent with data), this was regarded as the favored region. But today LEP2 sparticle search limits have increased the LEP2-forbidden region from below, while the stringent bound $\Omega_{CDM}h^2 \le 0.13$ has pushed the DM-allowed region down. Now hardly any bulk region survives in the mSUGRA model.

- At low m_0 and moderate $m_{1/2}$, there is a thin strip of (barely discernable) allowed region adjacent to the stau-LSP region where the neutralino and the lighter stau were in thermal equilibrium in the early universe. Here co-annihilation with the light stau serves to bring the neutralino relic density down to its observed value.[75]

- At large m_0, adjacent to the EWSB excluded region on the right, is the hyperbolic branch/focus point (HB/FP) region, where the superpotential μ parameter becomes small and the higgsino-content of $\widetilde{Z}_1$ increases significantly. Then $\widetilde{Z}_1$ can annihilate efficiently via gauge coupling to its higgsino component and becomes mixed higgsino-bino DM. If $m_{\widetilde{Z}_1} > M_W$, M_Z, then $\widetilde{Z}_1\widetilde{Z}_1 \to WW$, ZZ, Zh is enhanced, and one finds the correct measured relic density.[76]

We show the corresponding situation for $\tan\beta = 52$ in Fig. 5.8. While the stau co-annihilation and the HB/FP regions are clearly visible, we see that a large DM consistent region now appears.

- In this region, the value of m_A is small enough so that $\widetilde{Z}_1\widetilde{Z}_1$ can annihilate into $b\bar{b}$ pairs through s-channel A (and also H) resonance. This region has been dubbed the A-funnel.[77] It can be very broad at large $\tan\beta$ because the width Γ_A can be quite wide due to the very large b- and τ- Yukawa couplings. If $\tan\beta$ is increased further, then $\widetilde{Z}_1\widetilde{Z}_1$ annihilation through the (virtual) A^* is large all over parameter space, and most of the theoretically-allowed parameter space becomes DM-consistent. For even higher $\tan\beta$ values, the parameter space collapses due to a lack of appropriate EWSB.

It is also possible at low $m_{1/2}$ values that a light Higgs h resonance annihilation region can occur just above the LEP2 excluded region.[78] Finally, if A_0 is large and negative, then the $\tilde{t}_1$ can become light, and $m_{\tilde{t}_1} \sim m_{\widetilde{Z}_1}$, so that stop-neutralino co-annihilation[79] can occur.

Up to now, we have confined our discussion to the mSUGRA framework in which compatibility with (5.1) is obtained only over selected portions of

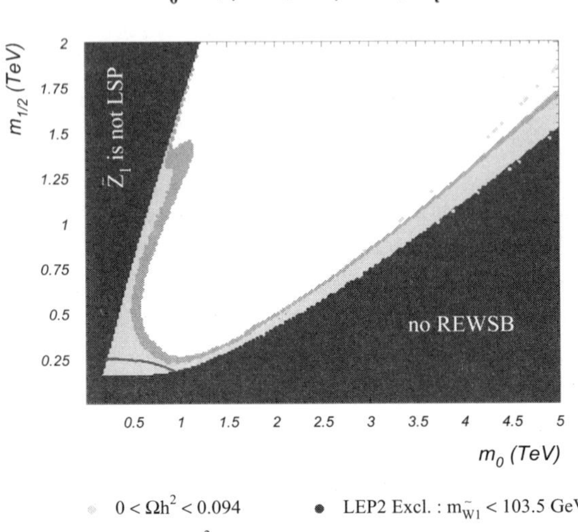

Fig. 5.8. *DM-allowed regions in the $m_0 - m_{1/2}$ plane of the mSUGRA model for* $\tan\beta = 52$ *with* $A_0 = 0$ *and* $\mu > 0$*. The various colors of shading is as in Fig. 5.7.*

the $m_0 - m_{1/2}$ plane. The reader may well wonder what happens if we relax the untested universality assumptions that underlie mSUGRA. Without going into details, we only mention here that in many simple one-parameter extensions of mSUGRA where the universality of mass parameters is relaxed in any one of the matter scalar, the Higgs scalar, or the gaugino sectors, *all points in the $m_0 - m_{1/2}$ plane become compatible with the relic density constraint* due to a variety of mechanisms: these are catalogued in Ref. [80]. Implications of the relic density measurement for collider searches must thus be drawn with care.

5.5.2. *Neutralino direct detection*

Fits to galactic rotation curves imply a local relic density of $\rho_{CDM} \sim 0.3$ GeV/cm^3. For a 100 GeV WIMP, this translates to about one WIMP per coffee mug volume at our location in the galaxy. The goal of DD experiments is to detect the very rare WIMP-nucleus collisions that should be occuring as the earth, together with the WIMP detector, moves through the DM halo.

DD experiments are usually located deep underground to shield the experimental apparatus from background due to cosmic rays and ambient radiation from the environment or from radioactivity induced by cosmic ray exposure. One technique is to use cryogenic crystals cooled to near absolute zero, and look for phonon and ionization signals from nuclei recoiling from a WIMP collision. In the case of the CDMS experiment[81] at the Soudan iron mine, target materials include germanium and silicon. Another technique uses noble gases cooled to a liquid state as the target. Here, the signal is scintillation light picked up by photomultiplier tubes and ionization. Target materials include xenon,[82] argon and perhaps neon. These noble liquid detectors can be scaled up to large volumes at relatively low cost. They have the advantage of fiducialization, wherein the outer layers of the detector act as an active veto against cosmic rays or neutrons coming from phototubes or detector walls: only single scatters from the inner fiducial volume qualify as signal events. A third technique, typified by the COUPP experiment,[83] involves use of superheated liquids such as CF^3I located in a transparent vessel. The nuclear recoil from a WIMP-nucleon collision then serves as a nucleation site, so that a bubble forms. The vessel is monitored visually by cameras. Background events are typically located close to the vessel wall, while neutron interactions are likely to cause several bubbles to form, instead of just one, as in a WIMP collision. This technique allows for the use of various target liquids, including those containing elements such as fluorine, which is sensitive to *spin-dependent* interactions.

The cross section for WIMP-nucleon collisions can be calculated, and in the low velocity limit separates into a coherent spin-independent component (from scattering mediated by scalar quarks and scalar Higgs bosons) which scales as nuclear mass squared, and a spin-dependent component from scattering mediated by the Z boson or by squarks, which depends on the WIMP and nuclear spins.[9] The scattering cross section per nucleon versus m_{WIMP} serves as a figure of merit and facilitates the comparison of the sensitivity of various experiments using different target materials.

In Fig. 5.9, we show the spin-independent $\widetilde{Z}_1 p$ cross section versus $m_{\widetilde{Z}_1}$ for a large number of SUSY models (including mSUGRA). Every color represents a different model. For each model, parameters are chosen so that current collider constraints on sparticle masses are satisfied, and further, that the lightest neutralino (assumed to be the LSP) saturates the observed relic abundance of CDM. Also shown is the sensitivity of current experiments together with projected sensitivity of proposed searches at superCDMS, Xenon-100, LUX, WARP and at a ton-sized noble liquid de-

Spin-independent Direct Detection

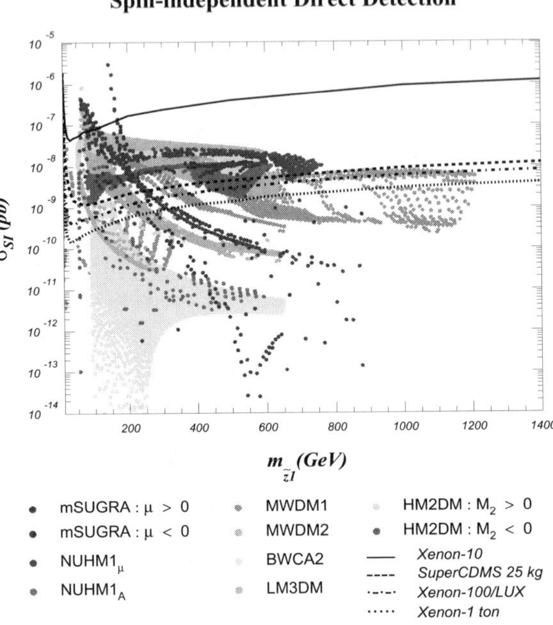

Fig. 5.9. *The spin-independent neutralino-proton scattering cross-section vs $m_{\widetilde{Z}_1}$ in a variety of SUSY models, compatible with collider constraints where thermally produced Big Bang neutralinos saturate the observed dark matter density.*

tector. The details of the various models are unimportant for our present purpose. The key thing to note is that while the various models have a branch where $\sigma_{\rm SI}(p\widetilde{Z}_1)$ falls off with $m_{\widetilde{Z}_1}$, there is another branch where this cross-section asymptotes to just under 10^{-8} pb.[80,84,85] Points in this branch (which includes the HB/FP region of mSUGRA), are consistent with (5.1) because $\widetilde{Z}_1$ has a significant higgsino component. Neutralinos with an enhanced higgsino content can annihilate efficiently in the early universe via gauge interactions. Moreover, since the spin-independent DD amplitude is mostly determined by the Higgs boson-higgsino-gaugino coupling, it is large in models with MHDM which has both gaugino and higgsino components. Thus the enhanced higgsino component of MHDM increases both the neutralino annihilation in the early universe as well as the spin-independent DD rate. The exciting thing is that the experiments currently being deployed– such as Xenon-100, LUX and WARP– will have the sensitivity to probe this class of models. To go further will require ton-size or greater target material.

We note here that if $m_{\mathrm{WIMP}} \lesssim 150$ GeV, then it may be possible to extract the WIMP mass by measuring the energy spectrum of the recoiling nuclear targets.[86] Typically, of order 100 or more events are needed for such a determination to 10-20%. For higher WIMP masses, the recoil energy spectrum varies little, and WIMP mass extraction is much more difficult. Since the energy transfer from the WIMP to a nucleus is maximized when the two have the same mass, DD experiments with several target nuclei ranging over a wide range of masses would facilitate the distinction between somewhat light and relatively heavy WIMPs, and so, potentially serve to establish the existence of multiple WIMP components in our halo.

5.5.3. *Indirect detection of neutralinos*

There are also a number of indirect WIMP search techniques that attempt to detect the decay products from WIMP annihilation at either the center of the sun, at the galactic center, or within the galactic halo.

5.5.3.1. *Neutrino telescopes*

Neutrino telescopes such as ANTARES or IceCube can search for high energy neutrinos produced from WIMP-WIMP annihilation into SM particles in the core of the sun (or possibly the earth). The technique involves detection of multi-tens of GeV muons produced by ν_μ interactions with polar ice (IceCube) or ocean water (ANTARES). The muons travel at speeds greater than the speed of light in the medium, thus leaving a tell-tale signal of Cerenkov light which is picked up by arrays of phototubes. The IceCube experiment, currently being deployed at the south pole, will monitor a cubic kilometer of ice in search of $\nu_\mu \to \mu$ conversions. It should be fully deployed by 2011. The experiment is mainly sensitive to muons with $E_\mu > 50$ GeV.

In the case of neutralinos of SUSY, mixed higgsino dark matter (MHDM) has a large (spin-dependent) cross-section to scatter from hydrogen nuclei via Z-exchange and so is readily captured. Thus, in the HB/FP region of mSUGRA, or in other SUSY models with MHDM, we expect observable levels of signal exceeding 40 events/km^2/yr with $E_\mu > 50$ GeV. For the mSUGRA model, the IceCube signal region is shown beneath the magenta contour labelled μ in Fig. 5.4.[87] These results were obtained using the Isajet-DarkSUSY interface.[71] Notice that DD signals are also observable in much the same region (below the contour labelled DD) where the neutralino is MHDM.

5.5.3.2. *Anti-matter from WIMP halo annihilations*

WIMP annihilation in the galactic halo offers a different possibility for indirect DM searches. Halo WIMPs annihilate equally to matter and anti-matter, so the rare presence of high energy anti-matter in cosmic ray events – positrons e^+, anti-protons $\bar{p}$, or even anti-deuterons $\bar{D}$ – offer possible signatures. Positrons produced in WIMP annihilations must originate relatively close by, or else they will find cosmic electrons to annihilate against, or lose energy via bremsstrahlung. Anti-protons and anti-deuterons could originate further from us because, being heavier, they are deflected less and so lose much less energy. The expected signal rate depends on the WIMP annihilation rate into anti-matter, the model for the propogation of the anti-matter from its point of origin to the earth, and finally on the assumed profile of the dark matter in the galactic halo. Several possible halo density profiles are shown in Fig. 5.10. We see that while the *local* WIMP density is inferred to a factor of $\sim$ 2-3 (we are at about 8 kpc from the Galactic center), the DM density at the galactic center is highly model-dependent close to the core. Since the ID signal should scale as the square of the WIMP density at the source, positron signals will be uncertain by a factor of a few with somewhat larger uncertainty for $\bar{p}$ and $\bar{D}$ signals that originate further away. Anti-particle propagation through the not so well known magnetic field leads to an additional uncertainty in the predictions. The recently launched Pamela space-based anti-matter telescope can look for e^+ or $\bar{p}$ events while the balloon-borne GAPS experiment will be designed to search for anti-deuterons. Anti-matter signals tend to be largest in the case of SUSY models with MHDM or when neutralinos annihilate through the A-resonance.[88]

5.5.3.3. *Gamma rays from WIMP halo annihilations*

As mentioned in the Introduction, high energy gamma rays from WIMP annihilation offer some advantages over the signal from charged antiparticles. Gamma rays would point to the source, and would degrade much less in energy during their journey to us. This offers the possibility of the line signal from $\widetilde{Z}_1 \widetilde{Z}_1 \to \gamma\gamma$ processes that occur via box and triangle diagrams. While this reaction is loop-suppressed, it yields monoenergetic photons with $E_\gamma \simeq m_{\text{WIMP}}$, and so can provide a measure of the WIMP mass. Another

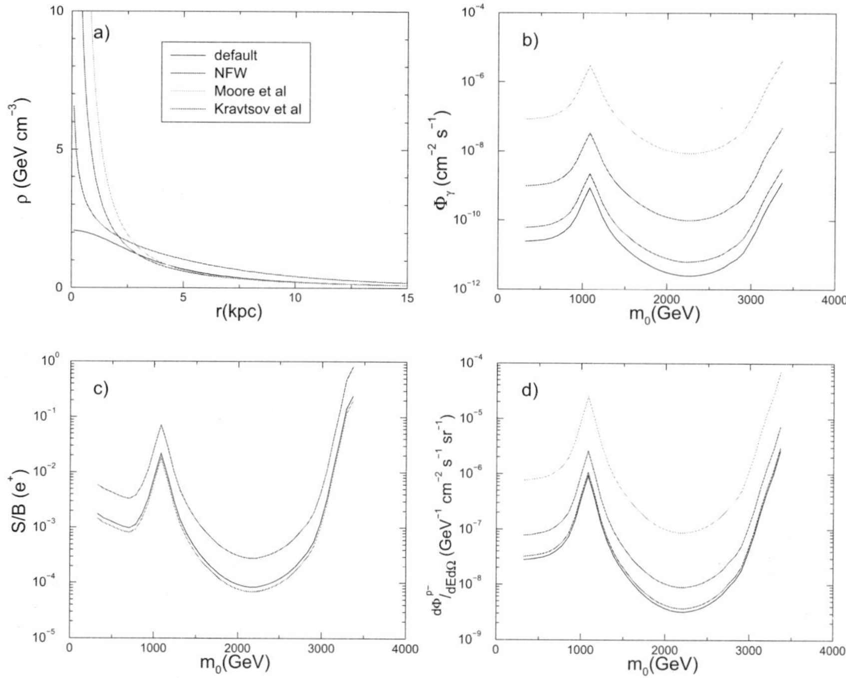

Fig. 5.10.   *Various predictions for the DM halo in the Milky Way as a function of distance from the galactic center. The earth is located at $r \sim 8$ kpc.*

possibility is to look for continuum gamma rays from WIMP annihilation to hadrons where, for instance, the gamma is the result of π^0 decays. Since the halo WIMPS are essentially at rest, we expect a diffuse spectrum of gamma rays, but with $E_\gamma < m_{\mathrm{WIMP}}$. Because gamma rays can traverse large distances, a good place to look at is the galactic center, where the WIMP density (see Fig. 5.10) is expected to be very high. Unfortunately, the density at the core is also very uncertain, making predictions for the gamma ray flux uncertain by as much as four orders of magnitude. Indeed, detection of WIMP halo signals may serve to provide information about the DM distribution in our galaxy.

Anomalies have been reported in the cosmic gamma ray spectrum. In one example, the Egret experiment[89] sees an excess of gamma rays with $E_\gamma > 1$ GeV. Explanations for the Egret GeV anomaly range from $\widetilde{Z}_1 \widetilde{Z}_1 \rightarrow b\bar{b} \rightarrow \gamma$ with $m_{\widetilde{Z}_1} \sim 60$ GeV,[90] to mis-calibration of the Egret calorimeter.[91] The GLAST gamma ray observatory is scheduled for lift-off in 2008 and should help resolve this issue, as will the upcoming LHC searches.[92]

5.5.4. *Gravitino dark matter*

In gravity-mediated SUSY breaking models, gravitinos typically have weak scale masses and, because they only have tiny gravitational couplings, are usually assumed to be irrelevant for particle physics phenomenology. Cosmological considerations, however, lead to the *gravitino problem*, wherein overproduction of gravitinos, followed by their late decays into SM particles, can disrupt the successful predictions of Big Bang nucleosynthesis. The gravitino problem can be overcome by choosing an appropriate range for $m_{\tilde{G}}$ and a low enough re-heat temperature for the universe after inflation[93] as illustrated in Fig. 5.11, or by hypothesizing that the $\tilde{G}$ is in fact the stable LSP, and thus constitutes the DM.[94]

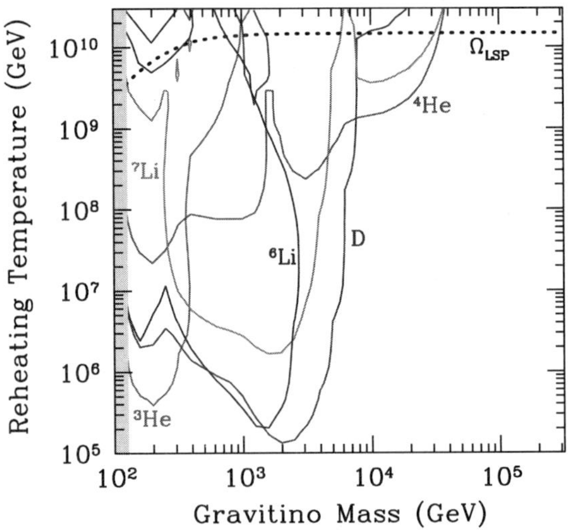

Fig. 5.11. *An illustration of constraints from Big Bang nucleosynthesis which require T_R to be below the various curves, for the HB/FP region of the mSUGRA model with $m_0 = 2397$ GeV, $m_{1/2} = 300$ GeV, $A_0 = 0$ and $\tan\beta = 30$, from Kohri et al.[93] to which we refer the reader for more details.*

Here, we consider the consequences of a gravitino LSP in SUGRA models. If gravitinos are produced in the pre-inflation epoch, then their number density will be diluted away during inflation. After the universe inflates, it enters a re-heating period wherein all particles can be thermally produced. However, the couplings of the gravitino are so weak that though gravitinos can be produced by the particles that do partake of thermal equilibrium, gravitinos themselves never attain thermal equilibrium: indeed their density is so low that gravitino annihilation processes can be neglected in the calculation of their relic density. The thermal production (TP) of gravitinos in the early universe has been calculated, and including EW contributions, is given by the approximate expression (valid for $m_{\tilde{G}} \ll M_i$[95]):

$$\Omega_{\tilde{G}}^{TP} h^2 \simeq 0.32 \left(\frac{10 \; GeV}{m_{\tilde{G}}} \right) \left(\frac{m_{1/2}}{1 \; \text{TeV}} \right)^2 \left(\frac{T_R}{10^8 \; \text{GeV}} \right) \tag{5.13}$$

where T_R is the re-heat temperature.

Gravitinos can also be produced by decay of the next-to-lightest SUSY particle, the NLSP. In the case of a long-lived neutralino NLSP, the neutralinos will be produced as usual with a thermal relic abundance in the early universe. Later, they will each decay as $\widetilde{Z}_1 \to \gamma \tilde{G}$, $Z\tilde{G}$ or $h\tilde{G}$. The total relic abundance is then

$$\Omega_{\tilde{G}} h^2 = \Omega_{\tilde{G}}^{TP} h^2 + \frac{m_{\tilde{G}}}{m_{\widetilde{Z}_1}} \Omega_{\widetilde{Z}_1} h^2. \tag{5.14}$$

The $\tilde{G}$ from NLSP decay may constitute warm/hot dark matter depending in the $\widetilde{Z}_1 - \tilde{G}$ mass gap, while the thermally produced $\tilde{G}$ will be CDM.[96]

The lifetime for neutralino decay to the photon and a gravitino is given by,[97]

$$\tau(\widetilde{Z}_1 \to \gamma \tilde{G}) \simeq \frac{48\pi M_P^2}{m_{\widetilde{Z}_1}^3} A^2 \frac{r^2}{(1-r^2)^3(1+3r^2)}$$

$$\sim 5.8 \times 10^8 \; \text{s} \left(\frac{100 \; \text{GeV}}{m_{\widetilde{Z}_1}} \right)^3 \frac{1}{A^2} \frac{r^2}{(1-r^2)^3(1+3r^2)} \; ,$$

where $A = (v_4^{(1)} \cos\theta_W + v_3^{(1)} \sin\theta_W)^{-1}$, with $v_{3,4}^{(1)}$ being the wino and bino components of the $\widetilde{Z}_1$,[8] M_P is the reduced Planck mass, and $r = m_{\tilde{G}}/m_{\widetilde{Z}_1}$. Similar formulae (with different mixing angle and r-dependence) hold for decays to the gravitino plus a Z or h boson. We see that – except when the gravitino is very much lighter than the neutralino as may be the case in GMSB models with a low SUSY breaking scale – the NLSP decays well after Big Bang nucleosynthesis. Such decays would inject high energy gammas and/or hadrons into the cosmic soup post-nucleosynthesis, which could break up the nuclei, thus conflicting with the successful BBN predictions of Big Bang cosmology. For this reason, gravitino LSP scenarios usually favor a stau NLSP, since the BBN constraints in this case are much weaker.

Finally, we remark here upon the interesting interplay of baryogenesis via leptogenesis with the nature of the LSP and NLSP. For successful thermal leptogenesis to take place, it is found that the reheat temperature of the universe must exceed $\sim 10^{10}$ GeV.[98] If this is so, then gravitinos would be produced thermally with a huge abundance, and then decay late, destroying BBN predictions. For this reason, some adherents of leptogenesis tend to favor scenarios with a gravitino LSP, but with a stau NLSP.[99]

5.5.5. *Axion/axino dark matter*

If we adopt the MSSM as the effective theory below M_{GUT}, and then seek to solve the strong CP problem via the Peccei-Quinn solution,[100] we must introduce not only an axion but also a spin-$\frac{1}{2}$ *axino* $\tilde{a}$ into the theory. The axino mass is found to be in the range of keV-GeV,[101] but its coupling is suppressed by the Peccei-Quinn breaking scale f_a, which is usually taken to be of order $10^9 - 10^{12}$ GeV: thus, the axino interacts more weakly than a WIMP, but not as weakly as a gravitino. Both the axion and axino can be compelling choices for DM in the universe.[102]

Like the gravitino, the axino will likely not be in thermal equilibrium in the early universe, but can still be produced thermally via particle scattering. The thermal production abundance is given by[102,103]

$$\Omega_{\tilde{a}}^{TP} h^2 \simeq 5.5 g_s^6 \log\left(\frac{1.108}{g_s}\right)\left(\frac{10^{11}\ \mathrm{GeV}}{f_a/N}\right)^2$$
$$\times \left(\frac{m_{\tilde{a}}}{100\ \mathrm{MeV}}\right)\left(\frac{T_R}{10^4\ \mathrm{GeV}}\right),$$

where f_a is the PQ scale, N is a model-dependent color anomaly factor that enters only as f_a/N, and g_s is the strong coupling at the reheating scale.

Also like the gravitino, the axino can be produced non-thermally by NLSP decays, where the NLSP abundance is given by the standard relic density calculation. Thus,

$$\Omega_{\tilde{a}} h^2 = \Omega_{\tilde{a}}^{TP} h^2 + \frac{m_{\tilde{a}}}{m_{NLSP}} \Omega_{\mathrm{NLSP}} h^2. \tag{5.15}$$

In this case, the thermally produced axinos will be CDM for $m_{\tilde{a}} \overset{>}{\sim} 0.1$ MeV,[102] while the axinos produced in NLSP decay will constitute hot/warm DM.[96] Since the PQ scale is considerably lower than the Planck scale, the lifetime for decays such as $\widetilde{Z}_1 \to \gamma\tilde{a}$ are of order $\sim 0.01 - 1$ sec– well before BBN. Thus, the axino DM scenario is much less constrained than gravitino DM.

Note also that if axinos are the CDM of the universe, then models with very large $\Omega_{\widetilde{Z}_1} h^2 \sim 100-1000$ can be readily accommodated, since there is a huge reduction in relic density upon $\widetilde{Z}_1$ decay to the axino. This possibility occurs in models with multi-TeV scalars (and hence a multi-TeV gravitino) and a bino-like $\widetilde{Z}_1$. In this case– with very large $m_{\widetilde{G}}$– there is no gravitino problem as long as the re-heat temperature $T_R \sim 10^6 - 10^8$ GeV. This range of T_R is also what is needed to obtain successful *non-thermal* leptogenesis (involving heavy neutrino N production via inflaton decay)[104] along with the correct abundance of axino dark matter.[108]

5.5.5.1. *Yukawa-unified SUSY with mixed axion/axino dark matter*

The gauge group $SO(10)$ is very highly motivated in that it unifies all matter particles of each generation into a spinorial **16** dimensional representation. Included in the **16** is a state containing a SM gauge singlet right-hand neutrino. SUSY $SO(10)$ theories may also allow for third generation Yukawa coupling unification. In Ref. [105,106], it was found via scans over SUSY parameter space that in fact it is possible to find models with Yukawa coupling unification, but only for certain choices of GUT scale SSB boundary conditions. Yukawa unified solutions can be found if

- $m_{16} \sim 3 - 15$ TeV,
- $m_{1/2}$ is very small,
- $\tan\beta \sim 50$,
- $A_0 \sim -2m_{16}$ with $m_{10} \sim 1.2m_{16}$,
- $m_{H_d} > m_{H_u}$ at M_{GUT}.

In this case, models with an inverted scalar mass hierarchy are found. The spectra is characterized by 1. first/second generation scalars in the 3-15 TeV range, 2. third generation scalars, m_A and μ in the few TeV range and 3. gluinos around 350-500 GeV, with charginos $\sim 100 - 160$ GeV and $\widetilde{Z}_1 \sim 50 - 90$ GeV.

A problem emerges in that the calculated neutralino relic density is about $10^2 - 10^5$ times its measured value. A solution has been invoked that the axino and axion are instead the DM particles. In this case, neutralinos would still be produced at large rates in the early universe, but each neutralino would decay after a fraction of a second (slightly before the onset of BBN) to axino $\tilde{a}$ plus photon. Then the relic axino density would be $(m_{\tilde{a}}/m_{\widetilde{Z}_1})\Omega_{\widetilde{Z}_1}h^2$, and the mass ratio out in front reduces the relic density by a factor of $10^{-2} - 10^{-5}$!. The neutralino is still long lived enough that it will give rise to missing energy at the LHC. In fact, since gluinos are so light, we would expect a gluino pair cross section of order 10^5 fb, along with decays $\tilde{g} \to b\bar{b}\widetilde{Z}_2$, $b\bar{b}\widetilde{Z}_1$ and $t\bar{b}\widetilde{W}_1 + c.c.$.[107] These new physics reactions should be easily seen via isolated multi-muon plus jet production in the early stages of LHC running.

The Yukawa-unified SUSY scenario is also very appealing cosmologically. We would expect the gravitino mass to be of order m_{16}: 3-15 TeV. If it is heavier than about 5 TeV, then the gravitino decays with lifetime around 1 sec: right at the onset of BBN! This eliminates the BBN gravitino problem as long as re-heat temperature $T_R \lesssim 10^9$ GeV. Several scenarios of mixed axion/axino dark matter in Yukawa unified SUSY have been examined in Refs. [108,109]. It is found that one may accommodate either dominant axion or axino DM, but m_{16} at the high end. Values of $m_{16} \gtrsim 10$ TeV are preferred, as are high values for the PQ breaking scale: $f_a \gtrsim 10^{12}$ GeV. The values of T_R allowed can accommodate several baryogenesis mechanisms, for instance, non-thermal leptogenesis.

5.6. Conclusion

The union of particle physics, astrophysics and cosmology has reached an unprecedented stage. Today we are certain that the bulk of the matter in the universe is non-luminous, not made of any of the known particles, but instead made of one or more *new physics* particles that do not appear in the SM. And though we know just how much of this unknown dark matter there is, we have no idea *what* it is. Today, many theoretical speculations seek to answer one of the most pressing particle physics puzzles, "What is the origin of EWSB and how can we embed this into a unified theory of particle interactions?" The answer may automatically also point to a resolution of this 75 year old puzzle as to what the dominant matter component of our universe might be. Particle physicists have made many provocative suggestions for the origin of DM, including supersymmetry and extra spatial dimensions, ideas that will completely change the scientific paradigm if they prove to be right.

The exciting thing is that many of these speculations will be *directly tested* by a variety of particle physics experiments along with astrophysical and cosmological searches. The Large Hadron Collider, scheduled to commence operation in 2009, will directly study particle interactions at a scale of 1 TeV where new matter states are anticated to exist for sound theoretical reasons. These new states may well be connected to the DM sector, and so in this way the LHC can make crucial contributions to not only particle physics, but also to cosmology. If indeed the LHC can make DM particles or their associated new physics states, then a large rate for signal events with jets, leptons and $\not{E}_T$ is expected.

Any discovery at LHC of new particles at the TeV scale will make a compelling case for the construction of a lepton collider to study the properties of these particles in detail and to elucidate the underlying physics. Complementary to the LHC, there are a variety of searches for signals from relic dark matter particles either locally or dispersed throughout the galactic halo. The truly unprecedented thing about this program is that if our ideas connecting DM and the question of EWSB are correct, measurements of the properties of new particles produced at the LHC (possibly complemented by measurements at an electron-positron linear collider) may allow us to independently infer just how much DM there is in the universe, and quantitatively predict what other searches for DM should find.

Acknowledgments

I take this opportunity to thank Tao Han for organizing an exceptional TASI workshop, and K. T. Mahantappa for his hospitality and patience in organizing many TASI workshops.

References

1. D. N. Spergel *et al.* (WMAP Collaboration), *Astrophys. J. Supp.*, **170** (2007) 377.
2. Report of the DMSAG panel on http://www.science.doe.gov/hep /hepap_reports.shtm. See also, L. Roszkowski,*Pramama*, **62** (2004) 389.
3. For reviews, see *e.g.* C. Jungman, M. Kamionkowski and K. Griest,*Phys. Rept.* **267** (1996) 195; A. Lahanas, N. Mavromatos and D. Nanopoulos, *Int. J. Mod. Phys.* **D 12** (2003) 1529; M. Drees, hep-ph/0410113; K. Olive, "Tasi Lectures on Astroparticle Physics", astro-ph/0503065; G. Bertone, D. Hooper and J. Silk, *Phys. Rept.* **405** (2005) 279.
4. J. Feng and J. Kumar, arXiv:0803.4196 [hep-ph].
5. R. Kolb and M. Turner, *The Early Universe*, (Addison-Wesley, 1990).
6. J. Wess and B. Zumino, *Nucl. Phys.* **B 70** (1974) 39.
7. A. Salam and J. Strathdee, *Nucl. Phys.* **B 76** (1974) 477.
8. H. Baer and X. Tata, *Weak Scale Supersymmetry: From Superfields to Scattering Events*, (Cambridge University Press, 2006).
9. M. Drees, R. Godbole and P. Roy, *Sparticles*, (World Scientific, 2004).
10. P. Binétruy, *Supersymmetry* (Oxford, 2006).
11. S. Dimopoulos and H. Georgi,*Nucl. Phys.* **B 193** (1981) 150.
12. B.C. Allanach, S. Kraml and W. Porod, *J. High Energy Phys.* **03** (2003) 016; G. Belanger, S. Kraml and A. Pukhov, *Phys. Rev.* **D 72** (2005) 015003;

S. Kraml and S. Sekmen in: *Physics at TeV Colliders 2007, BSM working group report*, in prep.; see http://cern.ch/kraml/comparison.

13. M. Dine, A. Nelson, Y. Nir and Y. Shirman, *Phys. Rev.* **D 53** (1996) 2658; for a review, see G. Giudice and R. Rattazzi, *Phys. Rept.* **322** (1999) 419.

14. L. Randall and R. Sundrum, *Nucl. Phys.* **B 557** (1999) 79; G. Giudice, M. Luty, H. Murayama and R. Rattazzi, *J. High Energy Phys.* **9812** (1998) 027.

15. S. Kachru, R. Kallosh, A. Linde and S. P. Trivedi, *Phys. Rev.* **D 68** (2003) 046005; K. Choi, A. Falkowski, H. P. Nilles, M. Olechowski and S. Pokorski, *J. High Energy Phys.* **0411** (2004) 076; K. Choi, A. Falkowski, H. P. Nilles and M. Olechowski, *Nucl. Phys.* **B 718** (2005) 113; K. Choi, K-S. Jeong and K. Okumura, *J. High Energy Phys.* **0509** (2005) 039; H. Baer, E. Park, X. Tata and T. Wang, *J. High Energy Phys.* **0706** (2007) 033, and references therein.

16. L. Everett, I.-W. Kim, P. Ouyang and K. Zurek, *Phys. Rev. Lett.* **101** (2008) 101803 and *J. High Energy Phys.* **0808** (2008) 102.

17. H. Baer and X. Tata, *Phys. Lett.* **B 160** (1985) 159.

18. H. Baer, J. Ellis, G. Gelmini, D. V. Nanopoulos and X. Tata, *Phys. Lett.* **B 161** (1985) 175; G. Gamberini, *Z. Physik* **C 30** (1986) 605; H. Baer, V. Barger, D. Karatas and X. Tata, *Phys. Rev.* **D 36** (1987) 96; H. Baer, X. Tata and J. Woodside, *Phys. Rev.* **D 45** (1992) 142.

19. H. Baer, D. Dzialo-Karatas and X. Tata, *Phys. Rev.* **D 42** (1990) 2259.

20. H. Baer, C. H. Chen, F. Paige and X. Tata, *Phys. Rev.* **D 50** (1994) 4508.

21. H. Baer, C. H. Chen, F. Paige and X. Tata, *Phys. Rev.* **D 49** (1994) 3283

22. H. E. Haber and D. Wyler, *Nucl. Phys.* **B 323** (1989) 267; S. Ambrosanio and B. Mele, *Phys. Rev.* **D 53** (1996) 2541 and *Phys. Rev.* **D 55** (1997) 1399 [Erratum-ibid. D**56**, 3157 (1997)]; H. Baer and T. Krupovnickas, *J. High Energy Phys.* **0209** (2002) 038.

23. H. Baer and X. Tata, *Phys. Rev.* **D 47** (1993) 2739.

24. H. Baer, C. Chen, M. Drees, F. Paige and X. Tata, *Phys. Rev. Lett.* **79** (1997) 986.

25. H. Baer, C. Chen, M. Drees, F. Paige and X. Tata, *Phys. Rev.* **D 59** (1999) 015010.

26. H. Baer, M. Bisset, X. Tata and J. Woodside, *Phys. Rev.* **D 46** (1992) 303.

27. H. Baer, M. Bisset, D. Dicus, C. Kao and X. Tata, *Phys. Rev.* **D 47** (1993) 1062; H. Baer, M. Bisset, C. Kao and X. Tata, *Phys. Rev.* **D 50** (1994) 316.

28. ISAJET, by H. Baer, F. Paige, S. Protopopescu and X. Tata, hep-ph/0312045; see also H. Baer, J. Ferrandis, S. Kraml and W. Porod, *Phys. Rev.* **D 73** (2006) 015010.

29. M. Muhlleitner, A. Djouadi and Y. Mambrini, *Comput. Phys. Commun.***168** (2005) 46.

30. W. Porod, *Comput. Phys. Commun.***153** (2003) 275.

31. T. Sjostrand, S. Mrenna and P. Skands, *J. High Energy Phys.* **0605** (2006) 026.

32. G. Corcella *et al.*, *J. High Energy Phys.* **0101** (2001) 010.

33. T. Geisberg *et al.*, arXiv:0811.4622 (2008).

34. F. Maltoni and T. Stelzer, *J. High Energy Phys.* **0302** (2003) 027; J. Alwall *et al.*, *J. High Energy Phys.* **0709** (2007) 028.

35. A. Pukhov *et al.*, hep-ph/9908288.

36. W. Kilian, T. Ohl and J.Reuter, arXiv:0708.4233.

37. J. Alwall *et al.*, *Comput. Phys. Commun.***176** (2007) 300.

38. H. Baer, X. Tata and J. Woodside, Ref. [18].

39. P. Mercadante, J. K. Mizukoshi and X. Tata, *Phys. Rev.* **D 72** (2005) 035009; S. P. Das *et al.*, *Eur. Phys. J.* **C 54** (2008) 645; R. Kadala, P. Mercadante, J. K. Mizukoshi and X. Tata, *Eur. Phys. J.* **C 56** (2008) 511.

40. H. Baer, A. Belyaev, T. Krupovnickas and X. Tata, *J. High Energy Phys.* **0402** (2004) 007; H. Baer, T. Krupovnickas and X. Tata, *J. High Energy Phys.* **0406** (2004) 061.

41. H. Baer, H. Prosper and H. Summy, *Phys. Rev.* **D 77** (2008) 055017.

42. H. Baer, A. Lessa and H. Summy, arXiv:0809.4719 (2008).

43. I. Hinchliffe *et al.*, *Phys. Rev.* **D 55** (1997) 5520 and *Phys. Rev.* **D 60** (1999) 095002.

44. H. Baer, K. Hagiwara and X. Tata, *Phys. Rev.* **D 35** (1987) 1598; H. Baer, D. Dzialo-Karatas and X. Tata, *Phys. Rev.* **D 42** (1990) 2259; H. Baer, C. Kao and X. Tata, *Phys. Rev.* **D 48** (1993) 5175; H. Baer, C. H. Chen, F. Paige and X. Tata, *Phys. Rev.* **D 50** (1994) 4508.

45. R. Arnowitt *et al.* *Phys. Lett.* **B 639** (2006) 46 and *Phys. Lett.* **B 649** (2007) 73.

46. H. Bachacou, I. Hinchliffe and F. Paige, *Phys. Rev.* **D 62** (2000) 015009; Atlas Collaboration, LHCC 99-14/15.

47. C. Lester and D Summers, *Phys. Lett.* **B 463** (1999) 99; A. Barr, C. Lester and P. Stephens, J. Phys. bf G29 (2003) 2343; C. Lester and A. Barr, *J. High Energy Phys.* **0712** (2007) 102; W. Cho, K. Choi, Y. Kim and C. Park, *J. High Energy Phys.* **0802** (2008) 035.

48. H. Baer, V. Barger, G. Shaughnessy and H. Summy, *Phys. Rev.* **D 75** (2007) 095010.

49. K. Kawagoe *et al.* *Phys. Rev.* **D 69** (2004) 035003; S. Ambrosanio *et al.* *J. High Energy Phys.* **0101** (2001) 014.

50. H. Hamaguchi, S. Shirai and T. Yanagida, arXiv:0712.2462.

51. J. Feng *et al.* *Phys. Rev.* **D 52** (1995) 1418; M. Nojiri, K. Fujii and T. Tsukamoto, *Phys. Rev.* **D 54** (1996) 6756.

52. E. Baltz, M. Battaglia, M. Peskin and T. Wizansky, *Phys. Rev.* **D 74** (2006) 103521. See also R. Arnowitt *et al.*, *Phys. Rev. Lett.* **100** (2008) 231802 for a similar study in the stau co-annihilation region.

53. For a review, see D. Hooper and S. Profumo, *Phys. Rept.* **453** (2007) 29.

54. H. C. Cheng, K. Matchev and M. Schmaltz, *Phys. Rev.* **D 66** (2002) 036005.

55. T. Rizzo, *Phys. Rev.* **D 64** (2001) 095010; C. Macescanu, C. McMullen and S. Nandi, *Phys. Rev.* **D 66** (2002) 015009.

56. G. Servant and T. Tait,*Nucl. Phys.* **B 650** (2003) 391; K. Kong and K. Matchev, *J. High Energy Phys.* **0601** (2006) 038.

57. H. C. Cheng, K. Matchev and M. Schmaltz, *Phys. Rev.* **D 66** (2002) 056006; A. Datta, K. Kong and K. Matchev, *Phys. Rev.* **D 72** (2005) 096006.

58. A. Alves, O. Eboli and T. Plehn, *Phys. Rev.* **D 74** (2006) 095010.

59. A. Birkedal, A Noble, M. Perelstein and A. Spray, *Phys. Rev.* **D 74** (2006) 035002; D. Hooper and G. Zaharijas, *Phys. Rev.* **D 75** (2007) 035010.

60. N. Arkani-Hamed, A. Cohen and H. Georgi,*Phys. Lett.* **B 513** (2001) 232; N. Arkani-Hamed, A. Cohen, E. Katz and A. Nelson, *J. High Energy Phys.* **07** (2002) 034.

61. For reviews, see M. Schmaltz, *Ann. Rev. Nucl. Part. Sci.***55** (2005) 229 and M. Perelstein, *Prog. Part. Nucl. Phys.***58** (2007) 247.

62. H. C. Cheng and I. Low, *J. High Energy Phys.* **0309** (2003) 051.

63. J. Hubisz and P. Meade, *Phys. Rev.* **D 71** (2005) 035016.

64. C. Hill and R. Hill *Phys. Rev.* **D 75** (2007) 115009.

65. H.-C. Cheng, arXiv:0710.3407 [hep-ph].

66. C.-S. Chen, K. Cheung and T. C. Yuan, *Phys. Lett.* **B 664** (2007) 158; T. Han, H. Logan and L.-T. Wang, *J. High Energy Phys.* **0601** (2006) 099.

67. H. C. Cheng, I. Low and L. T. Wang, *Phys. Rev.* **D 74** (2006) 055001; S. Matsumoto, M. Nojiri and D. Nomura, *Phys. Rev.* **D 75** (2007) 055006; A. Belyaev, C. Chen, K. Tobe and C. P. Yuan, *Phys. Rev.* **D 74** (2006) 115020; M. Carena, J. Hubisz, M. Perelstein and P. Verdier, *Phys. Rev.* **D 75** (2007) 091701.

68. T. Han, R. Mahbubani, D. Walker and L. T. Wang, axViv:0803.3820 (2008).

69. Z. Chacko, H-S. Goh and R. Harnik, *Phys. Rev. Lett.* **96** (2006) 231802; see E. Dolle and S. Su, *Phys. Rev.* **D 77** (2008) 075013 for an analysis of DM in a twin-Higgs scenario.

70. Y. Bai, *Phys. Lett.* **B 666** (2008) 332.

71. P. Gondolo, J. Edsjo, P. Ullio, L. Bergstrom, M. Schelke and E. A. Baltz, JCAP **0407** (2004) 008.

72. G. Belanger, F. Boudjema, A. Pukhov and A. Semenov, *Comput. Phys. Commun.***174** (2006) 577; *Comput. Phys. Commun.***176** (2007) 367.

73. IsaRED, by H. Baer, C. Balazs and A. Belyaev, *J. High Energy Phys.* **0203** (2002) 042.

74. H. Baer and M. Brhlik, *Phys. Rev.* **D 53** (1996) 597; V. Barger and C. Kao, *Phys. Rev.* **D 57** (1998) 3131.

75. J. Ellis, T. Falk and K. Olive, *Phys. Lett.* **B 444** (1998) 367; J. Ellis, T. Falk, K. Olive and M. Srednicki, *Astropart. Phys.* **13** (2000) 181; M.E. Gómez, G. Lazarides and C. Pallis, *Phys. Rev.* **D 61** (2000) 123512 and *Phys. Lett.* **B 487** (2000) 313; A. Lahanas, D. V. Nanopoulos and V. Spanos, *Phys. Rev.* **D 62** (2000) 023515; R. Arnowitt, B. Dutta and Y. Santoso, *Nucl. Phys.* **B 606** (2001) 59; see also Ref. [73].

76. K. L. Chan, U. Chattopadhyay and P. Nath, *Phys. Rev.* **D 58** (1998) 096004; J. Feng, K. Matchev and T. Moroi, *Phys. Rev. Lett.* **84** (2000) 2322 and *Phys. Rev.* **D 61** (2000) 075005; see also H. Baer, C. H. Chen, F. Paige and X. Tata, *Phys. Rev.* **D 52** (1995) 2746 and *Phys. Rev.* **D 53** (1996) 6241; H. Baer, C. H. Chen, M. Drees, F. Paige and X. Tata, *Phys. Rev.* **D 59** (1999) 055014; for a model-independent approach, see H. Baer, T. Krupovnickas, S. Profumo and P. Ullio, *J. High Energy Phys.* **0510** (2005) 020.

77. M. Drees and M. Nojiri, *Phys. Rev.* **D 47** (1993) 376; H. Baer and M. Brhlik, *Phys. Rev.* **D 57** (1998) 567; H. Baer, M. Brhlik, M. Diaz, J. Ferrandis, P. Mercadante, P. Quintana and X. Tata, *Phys. Rev.* **D 63** (2001) 015007; J. Ellis, T. Falk, G. Ganis, K. Olive and M. Srednicki, *Phys. Lett.* **B 510** (2001) 236; L. Roszkowski, R. Ruiz de Austri and T. Nihei, *J. High Energy Phys.* **0108** (2001) 024; A. Djouadi, M. Drees and J. L. Kneur, *J. High Energy Phys.* **0108** (2001) 055; A. Lahanas and V. Spanos, *Eur. Phys. J.* **C 23** (2002) 185.

78. R. Arnowitt and P. Nath, *Phys. Rev. Lett.* **70** (1993) 3696; H. Baer and M. Brhlik, Ref. [74]; A. Djouadi, M. Drees and J. Kneur, *Phys. Lett.* **B 624** (2005) 60.

79. C. Böhm, A. Djouadi and M. Drees, *Phys. Rev.* **D 30** (2000) 035012; J. R. Ellis, K. A. Olive and Y. Santoso, *Astropart. Phys.* **18** (2003) 395; J. Edsjö, *et al.*, JCAP **0304** (2003) 001.

80. H. Baer, A. Mustafayev, E. Park and X. Tata, *J. High Energy Phys.* **0805** (2008) 058.

81. Z. Ahmed *et al.*, arXiv:0802:3530 [astro-ph].

82. J. Angle *et al.*, *Phys. Rev. Lett.* **100** (2008) 021303.

83. E. Behnk *et al.*, arXiv:0804:2886 [astro-ph]

84. H. Baer, A. Mustafayev, E. Park and X. Tata, JCAP**0701**, 017 (2007).

85. D. Feldman, Z. Liu and P. Nath, *Phys. Lett.* **B 662** (2008) 190.

86. R. Schnee, (CDMS Collaboration); A. M. Green, JCAP **0708** (2007) 022; C-L. Shan and M. Drees and C. L. Shan, JCAP**0806** (2008) 012.

87. H. Baer, A. Belyaev, T. Krupovnickas and J. O'Farrill, JCAP **0408** (2004) 005.

88. H. Baer and J. O'Farrill, JCAP**0404**, 005 (2004); H. Baer, C. Balazs, A. Belyaev and J. O'Farrill, JCAP **0309**, (2003) 007.

89. P. Sreekumar *et al.* [EGRET Collaboration], Astrophys. J. **494**, 523 (1998) [arXiv:astro-ph/9709257].

90. W. de Boer, M. Herold, C. Sander, V. Zhukov, A. V. Gladyshev and D. I. Kazakov, arXiv:astro-ph/0408272.

91. F. W. Stecker, S. D. Hunter and D. A. Kniffen, *Astropart. Phys.* **29** (2008) 25.

92. H. Baer, A. Belyaev and H. Summy, *Phys. Rev.* **D 77** (2008) 095013.

93. S. Weinberg, *Phys. Rev. Lett.* **48** (1982) 1303; R. H. Cyburt, J. Ellis, B. D. Fields and K. A. Olive, *Phys. Rev.* **D 67** (2003) 103521; K. Jedamzik, *Phys. Rev.* **D 70** (2004) 063524; M. Kawasaki, K. Kohri and T. Moroi, *Phys. Lett.* **B 625** (2005) 7 and *Phys. Rev.* **D 71** (2005) 083502; K. Kohri, T. Moroi and A. Yotsuyanagi, *Phys. Rev.* **D 73** (2006) 123511; M. Kawasaki, K. Kohri, T. Moroi and A. Yotsuyanagi, arXiv:0804.3745 (2008).

94. H. Pagels and J. Primack, *Phys. Rev. Lett.* **48** (1982) 223; J. Feng, A. Rajaraman and F. Takayama, *Phys. Rev. Lett.* **91** (2003) 011302 and *Phys. Rev.* **D 68** (2003) 085018.

95. M. Bolz, A. Brandenburg and W. Buchmuller, *Nucl. Phys.* **B 606** (2001) 518; J. Pradler and F. Steffen, *Phys. Rev.* **D 75** (2007) 023509.

96. K. Jedamzik, M. LeMoine and G. Moultaka, JCAP**0607** (2006) 010.

97. J. Feng, S. Su and F. Takayama, *Phys. Rev.* **D 70** (2004) 075019.

98. W. Buchmuller, P. Di Bari and M. Plumacher, Annal. Phys. **315** (2005) 305.

99. W. Buchmuller, L. Covi, J. Kersten, K. Schmidt-Hoberg, JCAP**0611** (2006) 007; W. Buchmuller, L. Covi, K. Hamaguchi, A. Ibarra and T. Yanagida, *J. High Energy Phys.* **0703** (2007) 037.

100. R. Peccei and H. Quinn, *Phys. Rev. Lett.* **38** (1977) 1440 and *Phys. Rev.* **D 16** (1977) 1791; S. Weinberg, *Phys. Rev. Lett.* **40** (1978) 223; F. Wilczek, *Phys. Rev. Lett.* **40** (1978) 279.

101. J. E. Kim and H. P. Nilles, *Phys. Lett.* **B 138** (1984) 150.

102. L. Covi, J. E. Kim and L. Roszkowski, *Phys. Rev. Lett.* **82** (1999) 4180; L. Covi, H. B. Kim, J. E. Kim and L.Roszkowski, *J. High Energy Phys.* **0105** (2001) 033; L. Covi, L. Roszkowski and M. Small, *J. High Energy Phys.* **0207** (2002) 023.

103. A. Brandenburg and F. Steffen, JCAP**0408** (2004) 008.

104. G. Lazarides and Q. Shafi, *Phys. Lett.* **B 258** (1991) 305; K. Kumekawa, T. Moroi and T. Yanagida, *Prog. Theor. Phys.* **92** (1994) 437; T. Asaka, K. Hamaguchi, M. Kawasaki and T. Yanagida, *Phys. Lett.* **B 464** (1999) 12.

105. H. Baer and J. Ferrandis, *Phys. Rev. Lett.* **87** (2001) 211803; D. Auto, H. Baer, C. Balazs, A. Belyaev, J. Ferrandis and X. Tata, *J. High Energy Phys.* **0306** (2003) 023; D. Auto, H. Baer and A. Belyaev and T. Krupovnickas, *J. High Energy Phys.* **0410** (2004) 066; H. Baer, S. Kraml, S. Sekmen and H. Summy, JHEP**0803**, 056 (2008).

106. T. Blazek, R. Dermisek and S. Raby, *Phys. Rev. Lett.* **88** (2002) 111804 and *Phys. Rev.* **D 65** (2002) 115004.

107. H. Baer, S. Kraml, S. Sekmen and H. Summy, *J. High Energy Phys.* **0810** (2008) 079.

108. H. Baer and H. Summy, *Phys. Lett.* **B 666** (2008) 5;

109. H. Baer, M. Haider, S. Kraml, S.Sekmen and H. Summy, arXiv:0812.2693 (2008).

PART 2
LHC Experimentation

Chapter 6

A Short Guide to Accelerators, Detectors, Objects, and Searches

Peter Wittich

Department of Physics, Cornell University,
Ithaca, NY 14853, USA
`wittich@cornell.edu`

A very short guide to LHC-like detectors for students in theoretical particle physics interested in LHC phenomenology.

6.1. Introduction

This document is intended for the audience of the 2008 TASI summer school - physics graduate students in particle theory who are early in their career. The goal of the experimental lectures at the school were to give students interested in the Large Hadron Collider (LHC) at the European Center for Nuclear Research (CERN) a foundation for understanding the experimental results that will be published in the coming years. The first lectures covered how particles interact with matter and how the difference interactions are exploited to build a detector that allows you to determine the properties and identities of many of the final-state particles that result from proton-proton collisions. The next topic I covered was a description of the typical HEP detector, and finally some details of the LHC experiments, focusing on the general purpose detectors ATLAS and CMS. Ideally these lectures were intended to give theory students the tools to read an experimental paper or attend an experimental seminar and not only understand the presentation but also have some tools to judge for themselves the validity of the claims made.

This is not intended as a complete reference guide. It will not tell you enough to derive any of the claims made and is not intended for an experimentalist audience. There will be a lot of simplifications and many details will be ignored. However, I hope that reading this will help you commu-

261

nicate with your experimental colleagues and allow us to work together better.

6.2. Determining particle properties

The goal of a particle physics experiment is to measure some particle's properties. The particle may be previously unknown, in which case its existence may be the only property we care about, or we may be testing the standard model and therefore want to measure a mass, coupling or cross section in order to compare our measurement to a theoretical prediction.

In most cases, we are interested in understanding what happened in a high-momentum-squared interaction. However, most of the time, the quantities we extract directly from our detectors are only related to the quantities we want to measure. Our detector reports to us mainly energy deposits at a particular position and time. These deposits typically come from secondary particles, for instance, from the results of hadronization of the initial-state quarks produced in the hard scatter. The bulk of the work of the experimentalist is to use the information our detector gives us and use it to get back at the four-momenta of the initial hard scatter. Once that work is done, we can try to compare our results to theoretical expectations, and hopefully find an exciting discrepancy that points towards new physics.

In this paper I will not discuss all the aspects of reconstructing particle properties from the low-level information returned by the detector; however, I will try to give you some understanding of the physical processes involved and how they affect how well we can ultimately do the reconstruction. For instance, in measuring the energy of hadronic jets, a limitation comes from the fact that the energy deposits in the calorimeter from these jets contain both an electromagnetic and a hadronic component, and some calorimeters do not have a symmetric response to electromagnetic and hadronic show-ers.* As you cannot directly tell them apart, this can lead to a fundamental limitation on your jet energy resolution. In discussing these limitations, I want to give you an understanding of how well particle properties can be measured.

6.2.1. *Process of measurement*

The first step in any physical measurement is to reconstruct the four-vectors of the final state particles in the detector. In the language of hadron collider

---

*In the lingo, their e/h or e/π ratio differs from unity, or, they are *not compensating*.

physics, this means assigning a particle identification and reconstructing the transverse momentum p_T, the pseudo-rapidity η and azimuthal angle ϕ, and the transverse energy E_T. The final state particles that can be identified are

- electrons (e^+ and e^-),
- muons (μ^+ and μ^-),
- taus (τ^+ and τ^-), either via $\tau \to \ell \nu \bar{\nu}$ or via $\tau \to h, 3h$, where the h is a charged kaon or pion,
- hadronic jets as the manifestation of quarks and gluons. In this category we can also identify b or c quarks (so-called 'heavy flavor'), but we can't tell the difference between q and $\bar{q}$,
- neutral stable particles such as neutrinos or WIMPs, with some caveats ($\not{E}_T$).

This list is incomplete; we can, in fact, identify many more intermediate particles such as J/ψ's or W gauge bosons, for instance, by combining some from the list above with selections on things like invariant mass. Likewise, we can sometimes distinguish kaons from pions, which is often of interest. However, the *final state particles* we can identify are to a good approximation limited to those in the list above.

The next step is to select a set of collisions which have final state particles consistent with the types of process one is trying to study. For instance, consider the case where we want to study the top quark. The top quark is produced mostly in pairs, and decays almost 100% of the time via $t \to Wb$. The W, in turn, can either decay leptonically ($W \to \ell \nu$) or hadronically ($W \to q'\bar{q}$). Therefore, one might select events with a final-state muon, some large missing transverse momentum and a b quark jet. This defines your signal sample. Typically, there will be some other process that also satisfy the same criteria but are not the processes you are interested in; these are called *backgrounds* and must be either eliminated or modeled to high precision such that they not ruin your sensitivity to the process you are interested in.

Having thus assembled our data sample and made our best estimate at its composition, we now move to extract the relevant physics parameter. This might be as simple as counting the number of events in our sample to extract a cross section or might involve a complicated statistical analysis involving neural networks and shape fits to extract coupling constants or masses.

6.2.2. *Accelerators*

The tool of the trade for experimental particle physics is the accelerator. The two main types are circular and linear. The obvious advantage of a circular accelerator is that you get many chances to accelerate and to collide the counter-circulating particles. However, *synchrotron radiation* puts a lower limit on the radius of curvature for light particles. The power radiated is a function of γ^4, *i.e.*, the ratio of $(E/m)^4$. Therefore, the energy loss ratio for a 100 GeV electron to a proton is $(m_p/m_e)^4 \sim 10^{13}$! As the cost of constructing a large tunnel is one of the limiting factors for making an accelerator, circular accelerators for electrons are thought to be impractical much above the energies that LEP achieved in the early 21st century ($\sqrt{s} \approx 200$ GeV). For lepton colliders, the alternatives are to go to heavier leptons (e.g., muons) or to linear colliders. Each has its own set of challenges, the discussion of which is beyond the scope of this article.

In a modern particle physics accelerator, the particles are accelerated by a high-frequency standing wave which is excited in an accelerating cavity. The frequency of these waves is typically radio frequency, and hence the cavities are often called radio-frequency (RF) cavities. The oscillations of the standing waves is timed such that particles receive a kick in the right direction when they enter the cavity. To eliminate resistive losses in the walls, the cavities are often made of superconducting materials. Therefore, the cavities must be cooled to low temperatures (typically liquid helium temperatures, *i.e.*, around 4K.) As can be seen in Fig. 6.1, this requires the cavity to be immersed in a cryogenic bath. The entire assembly, containing, among other things, the cavity, cryogenic modules, and RF antenna, are often called a *cryomodule*. Typical materials used in modern SRF applications are niobium (chemical symbol Nb) or niobium-tin alloys. A remarkable aspect of these resonant cavities is their extremely high quality factor Q_0. As a reminder, the quality factor compares the natural frequency of a resonator to the rate at which the resonator dissipates energy. For SRF cavities, quality factors in the range of $Q_0 \geq 10^{10}$ and above are not unusual. In addition to the SRF accelerating cavities, circular accelerators require bending dipole magnets and focusing quadrupole magnets. There magnets, too, are made of superconducting materials to achieve the high fields required to bend ultra-high energy beams.

For an end-users of the collisions, one of the most important number that quantifies the performance of an accelerator is how much data it has delivered. We use the quantity called *luminosity* for this purpose. More

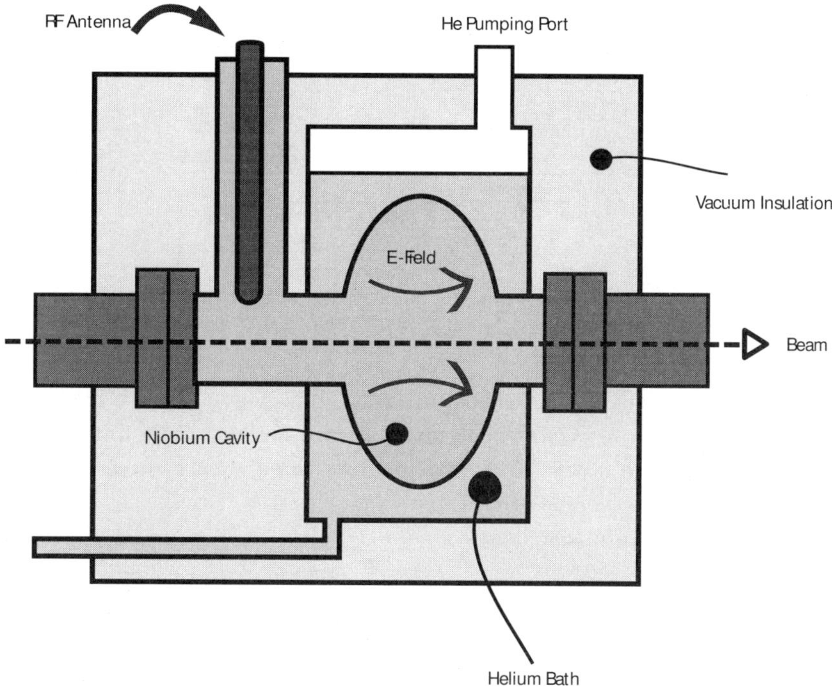

Fig. 6.1. Schematic diagram of a superconduction radiofrequency cavity.

specifically, we talk about *instantaneous* luminosity as the amount of data per unit time an accelerator is delivering, and *integrated* luminosity as the total size of a data sample.

Luminosity $\mathcal{L}$ is defined as follows for a collider.

$$\mathcal{L} = f\frac{n_1 n_2}{4\pi\sigma_x\sigma_y} = f\frac{n_1 n_2}{4\sqrt{\epsilon_x\beta_x^\star\epsilon_y\beta_y^\star}}, \tag{6.1}$$

where f is the revolution frequency, n_1, n_2 are the number of particles in each counter-rotating beam, and σ is a number that describes the size of the beam. In slightly different notation preferred by beams physicists (the second version above), the σ terms are replaced by terms ϵ and $\beta^\star$. The emittance ϵ is a function of the bunch preparation, *i.e.*, it depends on how the particles are accelerated and bunched when they are injected into the accelerator. $\beta^\star$, called 'beta star', depends on the arrangement of the beam optics, *i.e.*, the magnetic lenses that steer and accelerate the beams.

In Table 6.1 I list the relevant quantities for the LHC. The beam structure is such that beam consists of groups of particles called 'bunches.'

Table 6.1. LHC machine parameters

rotation frequency	$f = 11.25$ kHz
number of bunches	$n_{\text{bunch}} = 2808$
protons per bunch	$n_{\text{p}} = 1.7 \times 10^{11}$
normalized emittance	$\epsilon_n = 3.75\ \mu m$
beta star	$\beta^* = 0.55$ m
bunch length	7.5 cm
bunch spacing	25 ns (75 ns)

The number of particles n_1 and n_2 above are thus broken out into $n_1 = n_{\text{bunch}} \times n_{\text{p}}$. Multiplying these together leads to $\mathcal{L} \approx 10^{34}$ cm^{-2} s^{-1} as the design luminosity of the machine. The bunch length and spacing are not required for calculating the luminosity; however, they are interesting pieces of information. Two bunch spacings are listed; in the initial phase, bunches will collide every 75 ns; when more bunches are introduced, the bunch spacing will decrease to 25 ns.

The units of luminosity are

$$[\mathcal{L}] = 1/\text{cm}^2 \times 1/s, \tag{6.2}$$

i.e., the number of particles per unit area per unit time. We usually replace the cm^2 with a unit more appropriate for atomic cross sections: 1 b $=$ 10^{-24}cm^2, pronounced 'barn.'[†] The units of integrated luminosity, or of the total data sample, is just the integral of the above over time:

$$\left[\int \mathcal{L} dt\right] = 1/\text{cm}^2 = 1/\text{cross section}. \tag{6.3}$$

These somewhat unnatural-looking units allow for easy conversion between data set size to number of events for a given process. For example, in 100/pb of data, LHC will produce $N = \sigma \times \int \mathcal{L}\, dt = 800$ pb $\times$ 100/pb $= 80,000$ top quark pairs.

6.2.3. *Units of HEP measurements*

In HEP we use a set of units optimized for the scale of the problems we are studying. We measure energies in units of GeV, or 10^9 electron-volts. We measure momenta in GeV/c (though we often drop the c), distances in cm, and time in ns. We measure cross sections in picobarns. The space coordinates we use x, y and z, where z traditionally lies along the beam

---

[†]The name 'barn' is said to come from the expression, 'as big as a barn,' indicating that 10^{-24}cm^2 is a rather large cross section even in nuclear physics times. Today we talk about picobarns and smaller.[6]

axis. Alternatives are ϕ, the azimuthal angle, θ, the polar angle (though we typically transform it to η – see Sec. 6.2.4 below), and z.

6.2.4. *Pseudo-rapidity*

Rapidity y is defined as follows.

$$y \equiv \frac{1}{2} \ln \left(\frac{E + p_z}{E - p_z} \right)$$
$$= \tanh^{-1} \left(\frac{p_z}{E} \right),$$

where p_z is the momentum along the beam (z) axis. The number of particles

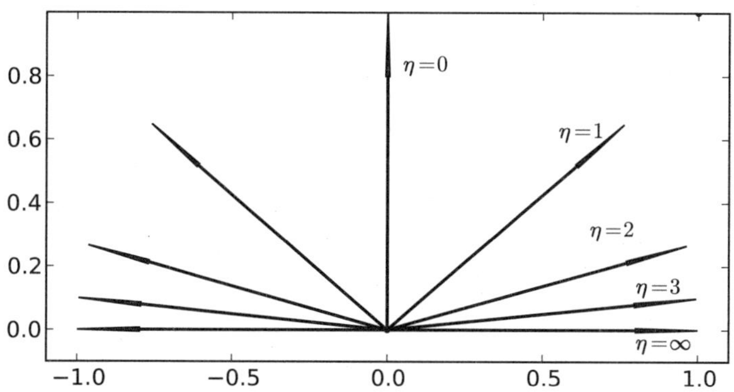

Fig. 6.2.   Representative values of η. In this figure, the beam runs along the x-axis and collisions would occur at $(0, 0)$.

per unit rapidity, dN/dy, is constant under boost, as are differences in rapidities. We define a related quantity: $\eta \equiv - \ln \tan \frac{\theta}{2}$. For $p \gg m, \eta \approx y$. η is more useful in an experimental situation because we can calculate it without knowing m or p, i.e., without knowing the mass or momentum of the particle. Since we most often do not know the mass of the particles before we need to know something about its coordinates, η is what we usually use.

Typical values for η, with their designations, are below.

- $|\eta| < 2$ ("central")
- $2 < |\eta| < 5$ ("forward")

Detector coverage for LHC experiments typically end around $\eta \sim 5$. Figure 6.2 demonstrates the irregular spacing of η in Cartesian space.

6.2.5. *The dominance of the transverse plane*

Most of the interesting, high momentum-squared physics physics occurs around $\eta = 0$. We therefore find it interesting to find objects that have a large fraction of their momentum in this area, *i.e.*, in the transverse plane. We define the projection of the momentum on the transverse plane $p_T = p \sin \theta$. Most detectors have solenoidal magnetic fields with field lines parallel to the beamline to allow measurement of this p_T. We also define an analogous quantity, the transverse energy, E_T, as the projection of the energy on the unit vector along the direction of motion of the particle, $E_T \equiv E \sin \theta$.

As an example, here is a mass quantity defined entirely with quantities defined in the transverse plane.

$$m_T \equiv \sqrt{E_T^2 - p_T^2} \tag{6.4}$$

The transverse mass is used, for instance, in measuring the W boson mass.

Another reason the transverse plane is critical is that it is the only one where we have a good understanding of the initial momentum distributions. Since we are colliding the constituent partons of the proton, not the protons themselves, we do not know what fraction of the momentum of the proton the colliding parton carries. However, we do know that to good approximation, the parton's momentum lies along the z axis, *i.e.*, along the beam. Therefore, we know that the total initial momentum of the proton-proton system is $\sum_j p_T^j = 0$, where j runs over all particles. Since we know that momentum is conserved, the same must be true in the final state, and any non-zero measured $\sum_j p_T^j$ of the final-state particles is indicative of an escaping neutral particle, such as a neutrino or, more interestingly, a dark matter candidate. This is the only way we can detect a neutrino, and p_T is the only thing we can measure of its kinematic properties. Due to reasons that will be explained in Sec. 6.4.4, we call the measured sum of final state momenta *missing E_T*, often denoted as $\not{E}_T$. With this definition we can rewrite the m_T definition from Eq. (6.4) above for the case of a decay $W \to \mu\nu$ as follows:

$$m_T \equiv \sqrt{E_T^2 - p_T^2} = \sqrt{2 p_T^\mu \not{E}_T (1 - \cos \Delta\phi)} \tag{6.5}$$

Here, $\not{E}_T$ is a measure of the transverse momentum of the escaping neutrino.

We see the usefulness of m_T in the case where we do not know the p_z of the neutrino.

6.2.6. ΔR

Another important quantity frequently used is ΔR.

$$\Delta R \equiv \sqrt{\Delta \eta^2 + \Delta \phi^2} \tag{6.6}$$

As can be seen, this is like a distance in $\eta - \phi$ space. We think of this like distance in Cartesian space. We use it to measure distance between objects. For instance, we use it to group objects for jet reconstruction; to measure the proximity of objects to electrons or muons to determine the isolation of a particle. Typical values for jet algorithms are $\Delta R < 0.5, 0.7$. Note also that it does not look like a circle on the detector, due to the irregular distances of η at increasing value.

6.2.7. *Parton Distribution Functions (PDFs)*

The parton model allows factorization of QCD into two parts, long and short distance. The short-distance part is calculable and describes the hard scatter. In this approximation, the interaction is between one constituent of each proton - the other constituents can be treated as spectators which do not interfere with the hard scatter.

The long-distance part of the parton model describes the dynamics of the hadrons in the quark.

In the LHC we collide beams of protons. Since these protons are not fundamental particles, we have to consider their constituents when we calculate the collision cross sections. In the collision, the hadrons are highly Lorentz-contracted and time-dilated and the interaction probes a frozen configuration of the quarks and gluons that make up the proton. The interaction consists of a hard scatter between a parton from each proton. The others form remnant hadronic particles,[‡] which can be expressed mathematically.

$$\sigma(pp \to CX) = \sum_{ij} \int f_i^p(x_i, Q^2) f_j^p(x_j, Q^2) \hat{\sigma}(ij \to C) dx_i dx_j \tag{6.7}$$

In Eq. (6.7), $f_i(x, Q^2)$ is what is usually called the *parton density function*. The cross section describes the probability for a proton-proton collision to result in a particle C and any other beam remnant particles X. x_i is the

---

[‡]This is the so-called *factorization argument*.

momentum fraction of the incoming proton that is probed by the transfer: $x = Q^2/(2p \cdot q)$. i and j refer to the constituents of each proton, and the sum runs over all constituents. The integral runs over all values of x. Note that f depends on both the momentum fraction x and on the momentum transfer Q^2. $\hat{\sigma}(ij \to C)$ describes the cross section of parton of the process parton i + parton j to C.

The parton density functions are determined with input from experiments such as neutrino deep inelastic scattering experiments, electron-proton collisions, and other hadron collider experiments.

6.2.8. *Hadronization*

The Feynman diagrams used to calculate the final state of interest at the LHC often have quarks as final-state particles, however, we know that quarks are never observed as free particles but only in color-neutral bound hadronic states called mesons or baryons. The process of transformation from a single free quark to a bound state of mesons or baryons is called hadronization. As hadronization is a strong process, it is described by quantum chromodynamics (QCD). However, hadronization occurs in the regime where QCD becomes strongly interacting and hence regular QCD perturbation theory approaches are not applicable. Instead, various models are used to give an approximate description of the process. These models offer a good description of existing experimental data but are understood not to offer a complete theoretical description.

In words, a description of these models is that when quark and anti-quark fly apart, they are connected by a color-filled flux tube. You can think of the size of this tube as approximately 1 GeV in appropriate units. The amount of energy stored in the tube is $\mathcal{O}(1 \text{ GeV/fb})$. As the quarks fly further apart, the potential energy in the string increases until the tube breaks, creating new quark-antiquark pair that 'dresses' the quark into a colorless bound state. This process continues repeatedly until the ultimate manifestation of quarks in the detector is a collimated spray of particles, most of which are neutral- and charged- pions and kaons. We call these sprays 'jets.' A considerable experimental challenge comes from trying to group particles together that belong to the same parton ("jet reconstruction"), and then to extract the initial parton energy from the measured jet energy ("jet energy scale determination".)

6.2.9. *Underlying event*

In addition to the hard interaction we are interested in, there can be additional soft interactions from other proton-proton pairs in the colliding bunches. This can include radiation from particles in the initial and final state. Additionally, you can see beam remnants, *i.e.*, beam particles that have undergone small-angle scattering and are no longer confined to the beam pipe but traversing the sensitive volume of the detector. Particles from both sources tend to be rather soft (transverse momenta below a few GeV).

We have now introduced the bulk of the information we need to proceed. We can now direct our attention to how we detect these particles. We will first discuss the dominant ways high-energy particles loose energy in interactions with matter and then describe how we can exploit these energy loss mechanisms to build powerful detectors that allow us to determine the energy and identity of many of the initial particles created in the hard scatter.

6.3. Passage of particles through matter

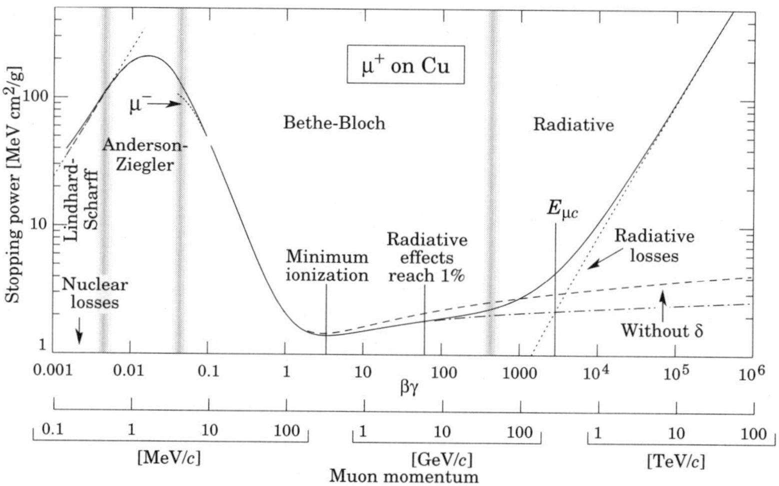

Fig. 6.3.   Energy loss for positive muons as a function of $\beta\gamma$.[1]

To understand how particle detection works, a basic understanding of how elementary particles interact with matter is required. There are many texts dedicated to this subject, for instance, see.[1,2] Here we give a very brief overview and invite you to read the references above for more information.

Figure 6.3 shows the average energy loss $-dE/dx$ for positive muons over nine orders of magnitude. The Figure is broken into four different ranges labeled by the dominant loss mechanism. From this plot, we can see that for most charged particles in the energy range of interest, the energy loss is mainly in the regime labeled 'Bethe-Bloch' in the Figure. For muons, for instance, energy loss via Bethe-Bloch ionization covers the energy range from roughly 10 MeV/c to 100 GeV/c.

$$-\frac{dE}{dx} = kz^2 \frac{Z}{A} \frac{1}{\beta^2} \left[\frac{1}{2} \ln \frac{2m_e c^2 \beta^2 \gamma^2 T_{\max}}{I^2} - \beta^2 - \frac{\delta}{2} \right] \tag{6.8}$$

In Eq. (6.8), z refers to the incident particle charge in units of electron charge, $T_{\max}$ is the maximum kinetic energy that can be transferred in a collision, δ is a density correction, I is the mean excitation energy, and k is a constant that describes the electron. The formula returns the mean energy loss; however, the distribution is highly asymmetrical so the most probably loss is actually.

Bottom line:

- most particles are MIP's in the detector when it comes to non-nuclear energy losses.

However, this does not tell the full story. There is a difference between the so-called electromagnetic particles (electrons and photons), muons, and hadrons due their quantum numbers and their masses. Electrons, photons and muons do not carry color and hence do not interact via the strong force. This is an important difference between them and the hadrons, which interact both via the electromagnetic force as well as via the strong force. Electrons and photons are different from muons due to their mass difference: with electrons and photons you cannot ignore radiative interactions in the energy regime we care about, while you can for all but the most energetic muons at the LHC. For electrons at all but the lowest energies, the dominant energy loss mechanism is *bremsstrahlung*, while for photons, it is *pair production*. The energy at which these processes become dominant is often called the *critical energy* E_c, though the exact definition of E_c is not uniformly defined. In the Figure, E_C is at between $\beta\gamma = 1,000 - 10,000$. In radiative energy loss processes, a shower develops. The secondary particles radiated off the primaries in turn interact and produce another set of secondary particles. This process continues until there is a cascade, or shower, of secondary particles, eventually absorbing all the energy of the original incident particles.

It is useful to define a length scale over which a beam of electrons or photons would have reduced its energy by $1/e$. This length scale is usually called the *radiation length* X_0. An approximate relation is given by

$$-\frac{dE}{dx} = \frac{E}{X_0},$$

(6.9)

and for this definition

$$X_0 = \frac{716.4 \text{ g cm}^{-2} A}{Z(Z+1)\ln(287/\sqrt{Z})}.$$

(6.10)

In the Equation, Z refers to the number of protons and A is the usual mass number. Typical values are given in Tab. 6.2.

Table 6.2. Some typical values of X_0 and E_c for electrons.

Material	X_0[cm]	E_c[MeV]
Pb	0.56	7.4
Fe	1.76	20.7

For LHC purposes, muons are stable particles. Their proper decay length $c\tau = 700$ m is much longer than the typical size of modern HEP detectors (even the big ones!) so most moderately energetic muons tend to decay outside the sensitive volume of the detector. Their dominant energy loss mechanism is through ionization. The critical energy E_c, above which radiation becomes dominant, scales with m^2, leading to a much higher energy for muons ($(m_\mu/m_e)^2 = 40,000$.) For example, the critical energy in iron is $E_c^{\text{Fe}} = 890$ GeV. As a result, muons are very penetrating particles and can go through large amounts of material. We will use this property to detect muons.

Colored particles also interact via the strong force. This provides a second energy loss mechanism for charged hadronic particles. In this process, particles interact with the nuclei of the detector material, and in these interactions, secondary strongly interactions are produced. This process repeats itself. As with the electromagnetic particles, we can define an interaction length here, usually called the *nuclear interaction length* λ_a such that a beam of hadrons would be attenuated as $N = N_0 \exp(-x/\lambda_a)$. The attenuation length can be related to the total inelastic cross section σ_{inel}

as follows:[2]

$$\lambda_a = \frac{A}{N_A \times \rho \times \sigma_{\text{inel}}}, \tag{6.11}$$

where σ_{inel} is a characteristic cross section for inelastic processes, A is the atomic number of the target nucleus, N_A is Avogadro's Number and ρ is defined such that $\lambda_a \times \rho$ is the usual surface mass density. These interactions result in hadronic showers, where secondary hadronic particles are created, but there is also an electromagnetic component, which can be created by processes such as $\pi_0 \to \gamma\gamma$. So the typical shower initiated by a hadronic particle has both an electromagnetic component as well as a hadronic particle. Typically, $X_0 < \lambda_a$, and therefore electromagnetic showers develop more quickly than hadronic showers. Table 6.3 shows some typical values of X_0 and λ_a and their relationship to one another.

Table 6.3. Typical values of X_0 and λ_a.

Material	$X_0 [\text{g/cm}^2]$	$\lambda_a [\text{g/cm}^2]$
H_2	63	52
Al	24	106
Fe	14	132
Pb	6	192

6.4. Building Detectors

Remembering that our goal is to reconstruct the momenta and particle types of the particles from the initial hard scatter, we can now understand why modern particle detectors are built as they are.

6.4.1. *Charged Particle Tracking System*

Most modern collider experiments are built with a tracking system. It usually consists of a detector that has relatively low mass and reports many positions along a particle's trajectory. The detector is usually immersed in a powerful solenoidal magnetic field which is coaxial with the incoming beam direction. The magnetic field leads to a bending of the particle in the plane transverse to the beamline. The amount of bending depends on the projection of the momentum on the transverse plane. There are many books dedicated to this topic, please see them for more details.[4,5]

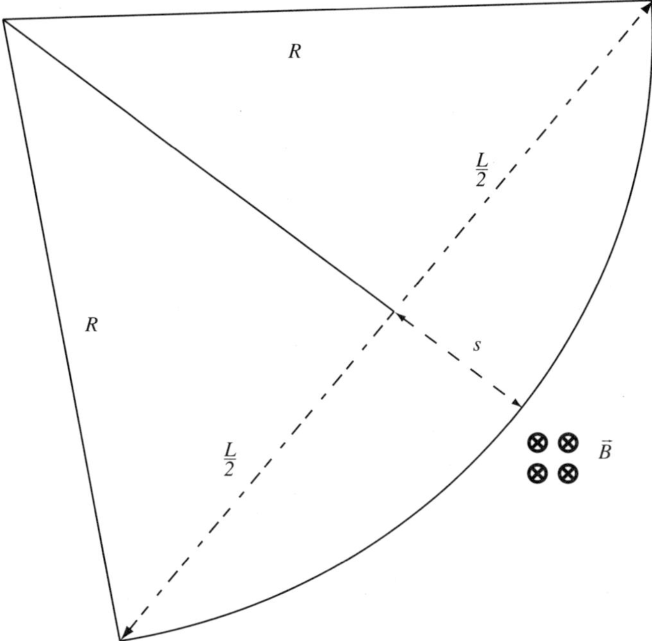

Fig. 6.4. The sagitta s is defined as the maximum excursion of a piece of the circle over the corresponding chord.

By measuring the trajectory at several points, one can reconstruct the radius of curvature of the track and extract the sagitta s as defined in Fig. 6.4. For small values of s, the sagitta is given by

$$s \simeq \frac{L^2}{8R} \propto \frac{BL^2}{p_T}$$

since the radius curvature is proportional to the momentum.[4] This shows that the size of the sagitta is proportional to the magnetic field and to the square of the size of the tracking system, and inversely proportional to the momentum of the particle. By propagation of errors we see that $\sigma_s/s = -\sigma_{p_T}/p_T$. In the limit where multiple scattering is unimportant (true at high momenta), the measurement error on individual track positions is constant and therefore

$$\frac{dp_T}{p_T} \propto p_T = c \times p_T. \tag{6.12}$$

The resolution therefore gets *worse* at higher momenta, which puts a limit on the momentum determination of very high energy electrons and muons.

The sign of the curvature allows you to determine the sign of the electric charge of the particle. Typically we assume that $q = \pm 1 \times e$, *i.e.*, we assume particles are singly charged.

The technology used in these chambers is either gaseous or solid-state. In a gaseous detector, a charged particle traverses a region of gas, ionizing it. The gaseous region contains wires at different voltages, called 'potential wires' and 'sense wires.' The field wires generate an electrostatic potential which guides the ionized particles towards the sense wires. The charge from the sense wires is collected, as well as the arrival time of the charge at the sense wires. By comparing the pattern of hit wires as well as the arrival times, one can group the hits from one particle ("pattern recognition") and reconstruct the particle trajectory ("track reconstruction") with high efficiency.

In solid-state detectors, the ionizing gas is replaced by a depleted pn junction acting as a diode. When a charged particle traverses the region, electron-hole pairs are created which do not recombine but instead migrate along field lines whence they are can be detected. Ignoring difference in technology, solid-state detectors come in two general flavors: strips and pixels. Strip detectors are extended objects in z, and usually only report a high-accuracy hit position in the r and ϕ coordinates. Pixel detectors are short in the z dimension and provide an accurate space point in all three dimensions.

The advantages of a gaseous detector are that they have a long radiation length, and therefore allow most particles to reach the calorimeter without radiating or showering. However, they cannot achieve comparable resolution to solid-state detectors. Additionally, at very high particle densities, the track density degrades the pattern recognition efficiency due to the relatively low segmentation of typical gaseous detectors compared to solid-state detectors.

The disadvantages of the solid state detectors compared to gaseous detectors include more material that particles must pass through en route to the calorimeter, thereby causing more radiation and scattering. Additionally, the complexity of large silicon trackers with tens of millions of channels will be explored for the first time at the LHC. Table 6.4 shows some achieved and projected tracking performance for hadron collider experiments. In the table, c refers to the constant in Eq. (6.12).

Table 6.4. Some resolutions for tracking systems from CDF[7] and DØ.[8]

Type	c
CDF open-cell drift chamber (COT)	0.15%
CDF combined Si and COT	0.07%
D0 Central Tracking	0.14%

6.4.2. *Calorimeters*

The goal of a calorimeter in a particle physics experiment is to measure the total energy of a particle. This is usually achieved by stopping the particle in the detector and measuring the resultant energy deposited by the shower created in the stopping process. All energy is collected above some cutoff energy. This type of calorimeter is called a 'total absorption calorimeter.' In addition to measuring the total energy, one can learn something about the particles which initiated the shower by looking at the lateral and longitudinal shower shape, *i.e.*, how wide the shower is at its maximum and how quickly it develops along the direction of travel of the particle.

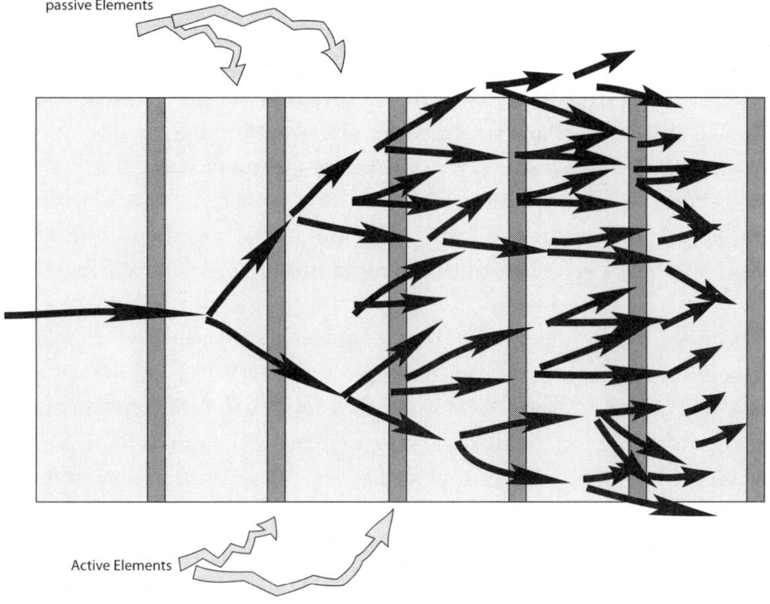

Fig. 6.5. Showering diagram

                                        P. Wittich

Fig. 6.6.   CLEO CsI crystal. Courtesy of the CLEO collaboration.

Most calorimeters can be classified as either sampling or homogeneous calorimeters. In a sampling calorimeter, the particles transit across two different materials, usually arrayed in a sandwich construction. The first layer is a passive radiator and consists of a material with high Z. The second layer contains the active detector and is some low-X_0 material which can be used to detect the shower particles created in the active scintillator. Examples of active detector include wire chambers or scintillators. Figure 6.5 shows a schematic view of such a calorimeter. The particle enters the calorimeter from the left. A shower develops in the passive layer, and the shower enters the active detector at several places where the energy of the shower is collected. By appropriate segmentation both along and perpendicular to the direction of the entering particle, one can determine how quickly the shower develops along the direction of motion as well as the total energy. Typical resolutions for sampling calorimeters are $\sigma_E/E \sim 10\%$.

A homogeneous calorimeter is one where the active material and radiator is combined into one. An example is the CLEO calorimeter, which consists of thallium-doped Cesium Iodide crystals. Charged particles entering the detector ionize atoms in the crystal, which emit light when they de-excite. The light travels down the crystal to four silicon diode photodetectors which are mounted on a lucite window on the rear face of the crystal.

Due to the difference in characteristic lengths of the showering processes described in Tab. 6.3, electromagnetic particles tend to shower earlier than hadronic particles. Therefore, a typical HEP calorimeter is segmented along the longitudinal shower direction into two parts. The *electromagnetic calorimeter* is optimized for electromagnetic particles and is closest to the interaction region. The *hadronic calorimeter* follows and typically is made of a material with a much shorter hadronic interaction length λ_a to fully capture the hadronic particles within its volume.

Calorimeters are segmented into cells in η and ϕ, which have dimensions of similar size to the size of the showers they are designed to measure. For electrons, this means that the $\eta - \phi$ segmentation is ideally of the scale of lateral shower size (Molière radius), which is typically a few centimeters in most detectors.

6.4.3. *Parametrization of Calorimeter resolution*

The resolution for electromagnetic calorimeters can typically be characterised as follows.

$$\frac{\sigma_E}{E} = \frac{a}{\sqrt{E}} \oplus b \oplus \frac{c}{E} \tag{6.13}$$

The fractional response σ_E/E is characterized by three terms. The first term comes from the statistical nature of the electromagnetic showering process. It is characterized by the a term in Eq. (6.13). The overall energy is related to the number of showering objects; therefore you can think of the first term as having a dependence like a counting experiment. Therefore, $E \sim n$ where n is the number of photons counted to establish the energy. The uncertainty on a counting experiment is given by $\sqrt{n}$ for a normal distribution; therefore the stochastic term has a dependence on $\sqrt{E}$. This term gets smaller at higher energies.

The next term, b, is a constant term. Its origin typically comes from calibration uncertainties, shower leakage at high energies, and other similar effects. This term typically becomes more important at higher energies, where the stochastic term gets small due to its $1/\sqrt{E}$ dependence. For hadronic calorimeters, this term also includes the difference between the electronic and hadronic response, *i.e.*, if the calorimeter is compensating or not. Note that this term *does not* get smaller at higher energies.

The final term, c, is related to electronic noise and is typically a constant ($\sigma_{\mathrm{noise}} = \mathrm{const}$).

The final resolution is obtained by adding the three terms in quadrature. It is important to note that the resolution typically gets *better* at higher energies for measurements done with a calorimeter. The showering process which is the basis for the calorimeter measurement is fundamentally a sta-

Table 6.5. Typical electromagnetic calorimeter resolutions.

Material		Depth $[X_0]$	a, b, c
$PbWO_4$ (CMS)	(homogeneous)	25	$3\%, 0.5\%, 0.2$
Scintillator/Pb (CDF)	(sampling)	17	14%
Liquid Ar/Pb (Atlas)	(sampling)	25	$10\%, 0.4\%, 0.3$

tistical process. The law of large numbers ensures that for higher energy processes, the energy that is measured approaches the true value.

Table 6.5 shows some typical resolutions from several current and future collider experiments.

6.4.4. *Missing E_T*

Neutrinos and neutrino-like particles are not directly measurable in a collider detector experiment. Neutrinos can be detected experimentally - many interesting physics results in the last ten years have come from experiments that directly measure the position and energy of neutrinos from accelerator and non-accelerator sources. However, these experiments rely on large fluxes and very large detectors (kilotons of target material, if not more) to allow them to get a sizable number of neutrino events. Even so, these detectors have a typical detection efficiency of $\mathcal{O}(10^{-18})$! This approach clearly would not suffice for a collider experiment - we need to be able to detect the presence of neutrinos on every event, since the most compelling evidence for 'physics beyond the standard model' suggests that dark matter will be part of what's out there and even large data samples will not have more than thousands of events. We can get evidence of the passage of stable weakly interacting particles by applying the laws of conservation of momentum. To a good approximation, we know that the momentum of the initial-state particles is only along the beam line. Therefore, we know that in the final state, the sum of the momenta in the plane *transverse* to the beam out to be zero. By measuring the total momentum of the visible particles, we can infer the sum of the transverse momenta of the invisible particles, such as neutrinos or supersymmetric neutralinos.

$$p_T^i = p_T^f \tag{6.14}$$

$$0 = \overset{\text{visible}}{\sum_i p_T^i} + \overset{\text{invisible}}{\sum_j p_T^j} \tag{6.15}$$

$$\overset{\text{invisible}}{\sum_j p_T^j} = - \overset{\text{visible}}{\sum_i p_T^i} \tag{6.16}$$

For reasons that will soon become clear we call this quantity the *missing energy* ($\not{E}_T$), though it might more aptly be called the missing momentum. The name comes from the fact that the simplest experimental definition of missing energy is the two-dimensional vector sum of the transverse energy deposits in the calorimeter.

$$\not{E}_T \equiv - \sum_i E_T^i \mathbf{n}_i \tag{6.17}$$

Since muons do not deposit more than a minimal amount of energy in the calorimeter, this quantity needs to be corrected for muon momenta

$$\not{E}_T \equiv - \sum_i E_T^i \mathbf{n}_i - \sum_{\mu's} p_T \tag{6.18}$$

where the second sum runs over muons in the event.

An alternate definition considers not all energy sums in the calorimeter but only those associated with identified particles (typically jets, electrons or photons.) This quantity is called *missing H_T* or $\not{H}_T$, though this definition is not always used uniformly. $\not{H}_T$ is less sensitive to events with large amounts of energy deposits not associated with the hard scatter, such as you might find in a high-luminosity scenario during later running at the LHC. The extra energy deposits come from additional interactions described in Sec. 6.2.9. Also, spurious high energy deposits can come from malfunctioning calorimeter cells. Ignoring these makes $\not{H}_T$ less susceptible to detector problems than $\not{E}_T$ and are the reason they are sometimes used in the trigger.

6.4.5. *Muon systems*

As was mentioned in Sec. 6.3, we can consider muons as stable particles that interact only minimally in our detectors. Since they are charged, they will leave hits in our tracking chambers. However, they do not deposit a significant amount of their energy in the calorimeters. Therefore, to a good approximation, any particle that exits the outer radius of the calorimeter is a muon. By surrounding this region with a tracking system, we can identify muons.

Typically, the tracking detectors that play a part in muon systems are immersed in a magnetic field to allow a momentum and charge determination for the muons. The field can either be the return field from the main solenoid or an auxiliary magnetic field created just for this purpose. The tracking systems are usually interspersed with additional shielding to stop any hadronic particles that leak out of the calorimeter.

6.4.6. *Trigger*

Finally, one needs a way of selecting events that harbor evidence of a hard scatter from the large background of soft interactions. The total cross section for the LHC's pp collisions is roughly 50 mb. Most of this physics has been studied previously and is relatively well understood. The expected cross section for high-energy processes from new physics interaction is typically as much as ten orders of magnitude smaller than this cross section, and even the standard candles such as $pp \to W$ production, with $\sigma(pp \to W) = 160$ nb, six orders of magnitude smaller than the total cross section. As can be seen, the different between the inelastic production cross section and the $t\bar{t}$ production cross section is about eight orders of magnitude at $\sqrt{s} = 14$ TeV. Rarer processes such as Higgs production for a heavy standard model Higgs Boson ($m_H = 500$ GeV) is smaller by another three orders of magnitude. Simultaneously, the expense and technical complexity of storing every one of the 40 MHz of collision at the LHC makes it impossible to record all events. To solve these problems, we use a trigger to rapidly decide which events to store. Figure 6.7 shows the cross section as a function of center-of-mass energy in pp collisions for several processes of interest.[§]

This extreme rejection ratio of 1:40,000 is typically achieved in a multi-step process. In the first stage, hardware processors look for evidence of a high-momentum-transfer interaction such as large transverse energy deposits, or the presence of high-momentum charged leptons. The hardware processors use low-resolution detector information that can be transmitted off the detector to the counting room very quickly - typically within a few

---

[§]In the Figure, Higgs(1-3) refer to standard model Higgs bosons, with masses of $m_H = 110, 300$ GeV, and 500 GeV, respectively. All cross sections are leading-order and calculated with the Pythia package.[3]

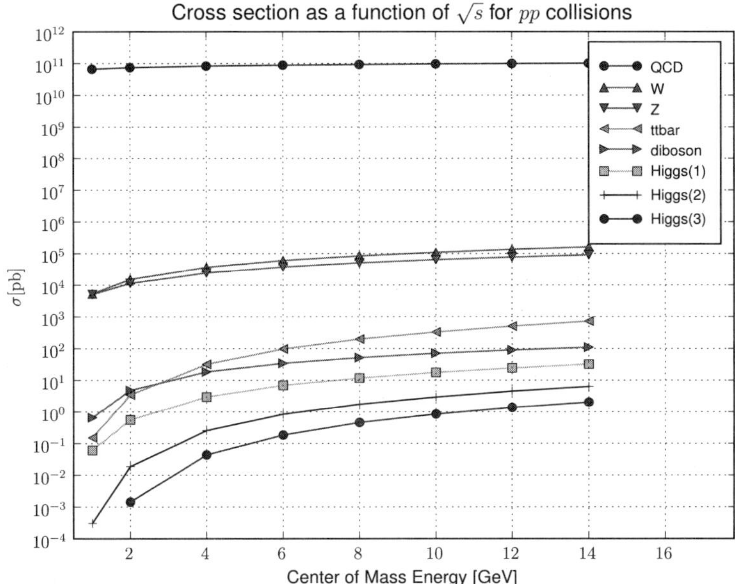

Fig. 6.7.   Cross sections for various processes in pp collisions for $\sqrt{s}$ between $1-14$ TeV.

microseconds. Later stages or processing are usually either a combination of hardware and software processors or pure software processors. These stages use higher-resolution information to make a more precise determination of energies of the particles identified at the first stage of rejection.

Table 6.6.   Overview of CMS and ATLAS trigger systems.

Level	CMS	ATLAS
L1	40 MHz $\to$ 100 kHz	40 MHz $\to$ 75 kHz
	dedicated hardware processors	
L2	None	75 kHz $\to$ 3 kHz
		software farm, c++ code, regional info
HLT	100 kHz $\to$ 300 Hz	3 kHz $\to$ 300Hz
	software farm running c++ code; global info	

Table 6.6 and Figure 6.8 show a brief overview of the CMS and AT-LAS trigger systems. Both systems must manage to reject all but a few

                                    P. Wittich

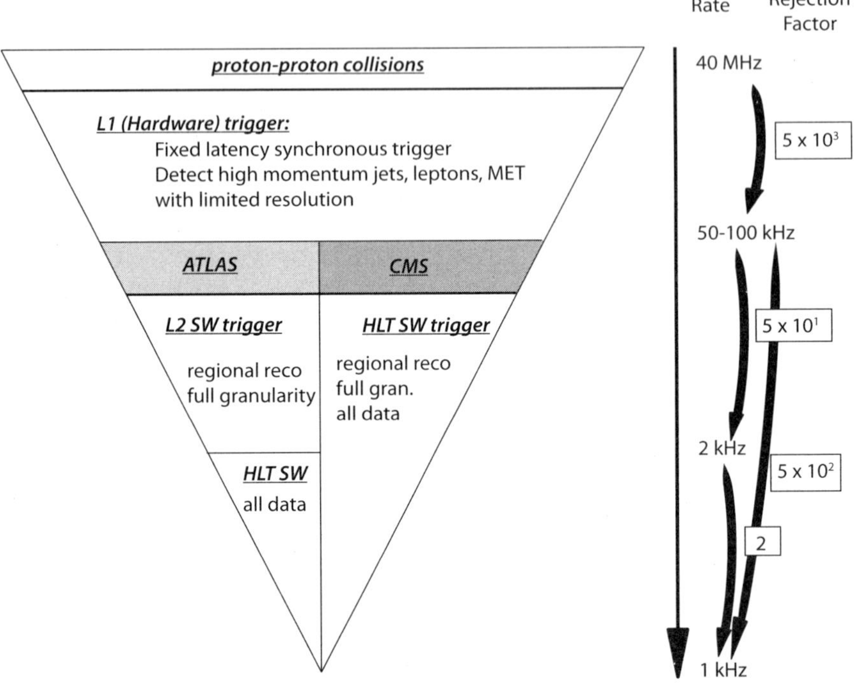

Fig. 6.8.   Conceptual view of trigger levels for CMS and ATLAS.

hundred Hz of the collisions within a very short time after the collision. In both cases, the trigger system looks for the evidence of a high-momentum collision which might indicate the presence of new physics, or of interesting high-p_T standard model physics. A significant difference between the two experiments is CMS's lack of a L2 filtering stage. In both experiments, the data pulled from the detector after L1, *i.e.*, at a rate of about 100 kHz; however, in ATLAS the regional parts of the information is selectively processed in a L2 stage, which consists of a processor farm. In both experiments, the entire event is then sent to a processor farm which looks at global event information to make the final rejection to achieve a rate to tape of a few hundred events per second.

It is important to note that the job of the trigger is not to determine the energies of particles or to measure them to high resolution; the trigger's job is to determine if the energy of the particle is *above* threshold. For instance, if a muon's true momentum is 400 GeV/c but the trigger reports it as 450 GeV, it does not matter as long as the reported value is properly above

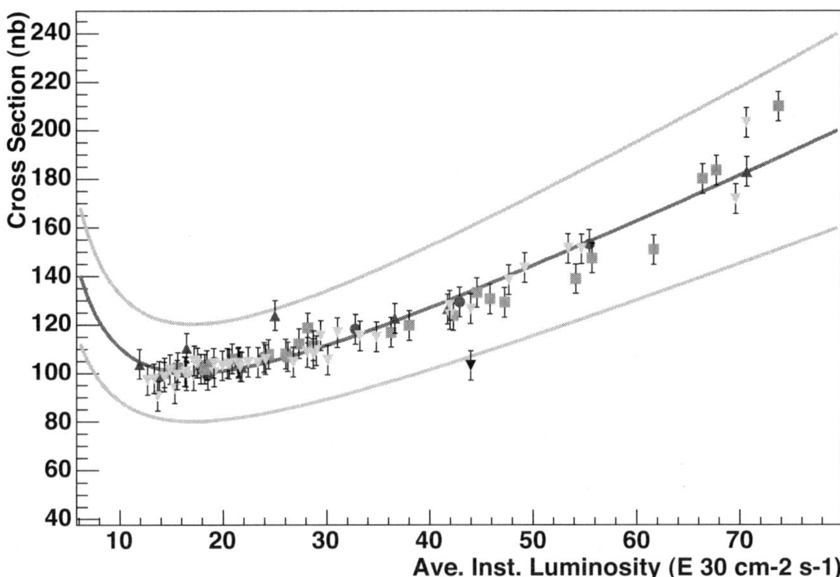

Fig. 6.9. An example of a trigger cross section's dependence on the instantaneous luminosity. The points refer to different running periods.

threshold. As such the job of the trigger software is significantly different from the job of the reconstruction software. The goal of the trigger is to get the right events on tape; the goal of the final reconstruction software is to determine the best estimate of the true kinematical properties of the reconstructed particles.

For a theorist, it is important to realize that the currency of a trigger system is bandwidth: how many events (of a given size) you can push through the system at any given point in time. A reasonable expectation is that the rate of events a trigger system collects would scale simply by the luminosity. Since the rate is $R = \sigma_{\text{trig}} \times \mathcal{L}$, it stands to reason that the trigger rate scales only with the instantaneous luminosity $\mathcal{L}$. However, the trigger cross section is *not* independent of the instantaneous luminosity: $d\sigma_{\text{trig}}/d\mathcal{L} \neq 0$. The trigger cross section exhibits so-called growth terms:

$$\sigma_{\text{trig}} = \sigma_{\text{phys}} + f(\mathcal{L}) + g(\mathcal{L}^2) + ...,$$

where σ_{phys} is the usual cross section and f and g are functions that depend linearly and quadratically on the instantaneous luminosity. Figure 6.9

shows an example of the trigger cross section versus instantaneous luminosity from the CDF muon trigger. We see a strong dependence. Most of this effect is due to accidental coincidences of various other particles that mimic muons. Other reasons can be insufficient resolution in the momentum distribution, for instance. The upshot of this effect is that trigger performance (and therefore an experiment's ability to collect data) can degrade more quickly with luminosity than one might naively expect.

Another trigger-related topic that experimentalists will frequently spend hours discussing in the context of trigger are prescales. A prescale allows the trigger to record only a subset of the events that would otherwise pass the trigger condition. This is typically employed for signals that would otherwise break the bandwidth limit of the system. For instance, at design luminosity, the 'calibration signal' $pp \to W \to e\nu$ has a sufficiently large cross section that recording every single W decay would absorb a significant fraction of the total bandwidth. The W has been studied extensively at the Tevatron and at LEP, and the measurements we want to make at the LHC are typically systematically limited. Therefore, only collecting a fraction of the $W \to e\nu$ decays would be a reasonable choice given the physics goals of the LHC experiments.

Generally, it is preferable to prescale a trigger rather than change other requirements such as momentum threshold. It is easier to analyze data if all of it is collected with the same trigger requirements.

The prescale for a trigger might not be constant: during a run, the instantaneous luminosity will decrease; however, the bandwidth of the trigger system is constant. Therefore, for a constant trigger configuration, the amount of bandwidth used will decrease as the instantaneous luminosity decreases. Since unused bandwidth is wasted bandwidth, one can adjust a prescale downwards during a run to keep the total bandwidth constant. Figure 6.10 shows an example of changing prescales during a physics run with the CDF experiment. As time passes (and the luminosity drops), the prescales are adjusted by an automatic process to keep the L1 accept rate, and therefore the bandwidth, mostly filled. The last dramatic jump in L1 rate at about 8:30 AM on the figure indicates the turn-on of a special kind of prescale used in CDF, the ultra-dynamic prescale, where b physics events are accepted based on microsecond-timescale downward fluctuations of the regular trigger rate.

In using a data sample collected with a prescaled trigger, we have to take into account the reduction of the effective integrated luminosity since we have not kept all events that would have passed our trigger. This is simple

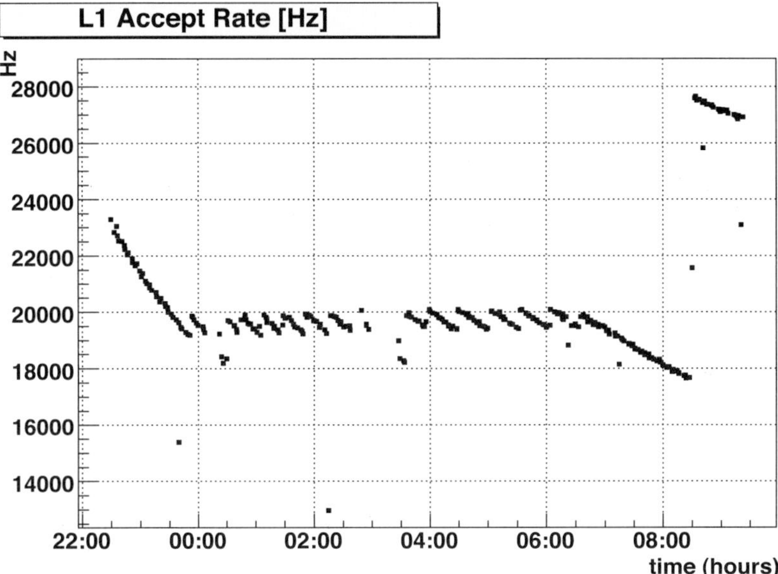

Fig. 6.10. Rate of L1-accepted triggers as a function of time in the CDF experiment showing the effect of L1 prescale changes.

if the prescale is constant throughout the data-taking period, and can get complicated in cases such as the ultra-dynamic prescale mentioned in the previous paragraph. However, if one does not need to know the integrated luminosity, for instance, in measuring a branching ratio, we would not have to worry about this complication.

6.4.7. *Central Data Processing*

Finally, once the data has been collected, it must be made available to a world-wide collaboration of scientists. The LHC experiments expect to store about 15 Petabytes (15×10^{15} bytes, or 15,000 Terabytes) of data each year. This data is then distributed in a tiered computing model, as shown in Fig. 6.11. There is one Tier-0 center at CERN. Below that is one Tier-1 in each region. Data samples will be stored at specific Tier-1 centers, which will also participate in the secondary processing of the data. The data will then be made available to regional Tier-2 centers, where the physicist can then do his analysis. From an end-user's point of view, the goal is to make this process 'transparent', *i.e.*, invisible to the end-user.

Here are the typical stages of data processing. In a central processing

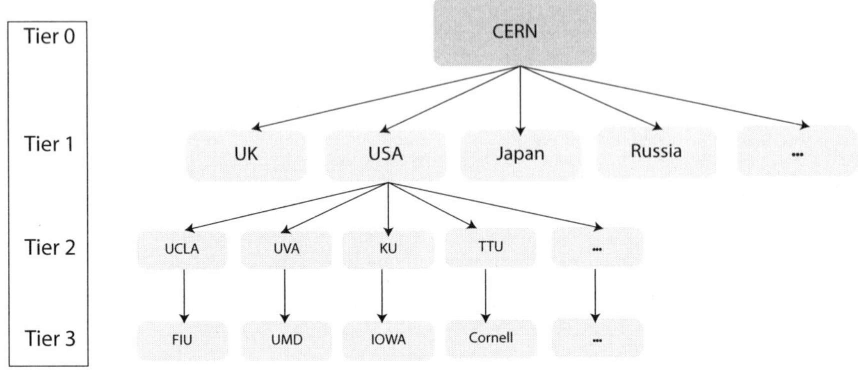

Fig. 6.11.   Tiered LHC computing model.

pass, we derive calibration and alignment constants which are fed into the first pass reconstruction. Here, we enumerate electron, muon, tau, jet, and heavy-flavor jet candidates, and global event quantities such as missing E_T. These quantities are not yet at the level of four-vectors; they are just 'candidates.' The data is then split into sets based on trigger criteria and distributed to the tiered data centers described above.

This process is repeated periodically, when better calibration and alignment procedures become available or when the reconstruction algorithms are improved.

6.5. Reconstructing Particles

Particle reconstruction is a very lengthy topic, and has a lot of subtleties that are not of interest to the average phenomenologist. I will describe in broad terms how electrons, muons, taus and jets are reconstructed, and point out some of the difficulties.

Electrons   As was discussed in Sec. 6.4.2, electrons will deposit most of their energy in the electromagnetic calorimeter. In electron reconstruction we look for a narrowly collimated cluster of energy which has a longitudinal and lateral shower profile consistent with that of an electron. We then match the cluster to a trajectory that has been measured in the tracking system. Typical identification variables for electrons include looking at the ratio of the energy in the electromagnetic and hadronic calorimeters (H/E out to be small for real electrons); the lateral shower profile has a

Table 6.7. Tau decay branching fractions.

Decay ($h^\pm$ only)	Branching Fraction
$\tau^\pm \to \ell^\pm \nu_\ell \nu_\tau$	$2 \times 17\%$
$\tau^\pm \to h^\pm \nu_\tau X^0$	49%
$\tau^\pm \to h^\pm h^\mp h^\pm \nu_\tau X^0$	15%

width consistent with the Molière radius of electrons in that material; we test that the momentum as measured in the tracker is consistent with the energy deposit for an effectively massless particle ($E/P \sim 1$). Electron identification tends to work best with electrons that are well-isolated from other particles in the event; therefore, we often apply isolation requirements to reject jets mimicking electrons. Typical cuts compare the ratio of the electron's energy to the energy found in a cone in ΔR around the energy in the calorimeter or the tracker and reject the electron candidate if this ratio is great than 5-10%.

Photons are similar, though in this case we veto photon candidates in the presence of a track pointing at the cluster in the electromagnetic calorimeter.

Since the calorimeter resolution increases with higher energy (see Sec. 6.4.3), the fractional energy resolution of electrons tends to improve as the electron energy improves.

Muons Sec. 6.4.5 describes the hardware systems used for muon detection. For reconstruction, we simply match tracks in the muon system with tracks in the tracking chambers. Additionally we look that the energy deposits in the calorimeters are consistent with a minimum ionizing particles - typically a fixed quantity of a few GeV for all but the highest-energy muons.

Since the main detectors used in determining the muon's momentum (and hence its energy) is the tracking system, it is the resolution of this system that dominates the energy resolution for muons. As discussed in Sec. 6.4.1, is resolution scales inversely with transverse momentum, so therefore muon resolution degrades at higher energies.

Taus Taus are the most complicated leptons. The tau distance a tau will travel before decaying is small (order $c\tau = 87.11$ μm.) Therefore, we can only find taus via their decay products. The decays can be classified as leptonic, single-prong hadronic, or triple-prong hadronic. Single-prong and

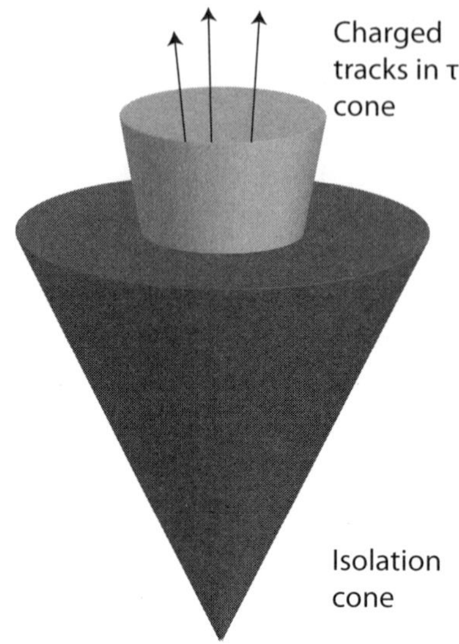

Fig. 6.12.　Isolation cones used in τ identification.

triple-prong corresponds to the number of charged tracks the tau decays to. The branching fractions for the relevant processes are shown in Tab. 6.7, where the second and third lines correspond to the single- and triple-prong events, respectively. The X^0 in the table corresponds to zero or more additional neutral particles, mostly K^0 or π^0.

Leptonic tau decays usually do not require any additional work to reconstruct. You simply look for electrons or muons. However, it is hard to determine if you have a primary e or μ or from a τ decay. One avenue that can be explored on a statistical basis is to look for missing energy colinear with the leptons; however, typically, this is not done and instead the leptons from tau decays are lumped in with primary leptons.

To get a significant fraction of the taus, however, you need to reconstruct the hadronic tau decays. As can be seen in the Table, there are two types of decays: ones with one charged particles, and one with three charged particles. These particles leave collimated deposits in the tracking chamber. In addition, there are often some number of neutral particles in a narrow cone around the charged particles. The signature for a tau decay is therefore a narrow energy deposit of neutral and charged electromagnetic and hadronic particles. This suggests that a way of reconstructing particles hadronic tau decays is to look for either one or three narrowly collimated tracks that point to a narrow energy deposit of energy (a narrow jet) in the calorimeter. Figure 6.12 shows an example of a three-pronged tau decay. The central tau cone shows three tracks in a narrow cone. The ratio of energy deposits in the central cone to the isolation cone ensures that the tau jet is narrow.

Taus identification are the most challenging of the charged lepton reconstruction algorithms. However, tau identification has become very mature at the Tevatron and as such we can expect this to be an effective tool at the LHC.

Jet reconstruction As discussed in Sec 6.2.8, jets are the experimental manifestation of quarks and gluons in our detector. Jet reconstruction is conceptually simple. We want to gang together all particles that come from the hadronization process and then extract the original parton properties. Experimentally, this means that we try to gang together energy deposits in the calorimeter. There are many different algorithms for this task, and many experimental challenges.

Here is a sample algorithm, courtesy of Rick Field.

(1) Order all charged particles according to their p_T.
(2) Start with the highest p_T particle above some threshold (a *seed*) and include in the "jet" all particles within the "radius" $\Delta R = 0.7$ (considering each particle in the order of decreasing p_T and recalculating the centroid of the jet after each new particle is added to the jet).
(3) Go to the next highest p_T particle (not already included in a jet) and include in the "jet" all particles (not already included in a jet) within the radius $\Delta R = 0.7$.
(4) Continue until all particles are in a "jet."

This very basic algorithm gives a flavor of how clustering is done, though there are many variations and many complications that make the actual algorithms a lot more complicated than the one outlined above. Two such complications are shown in Fig. 6.13. The top row shows *infra-red* safety. Imagine a situation where two jets are relatively close by, as is shown in left column. Now imagine that an additional soft particle is radiated in the area between the two jets, as shown in the second column. This addition can cause the two jets to overlap and now be merged into one jet. This sensitivity to soft radiation is not a desirable feature.

The second row show *colinear safety.* In this case, two colinear particles are both below the threshold used to seed the jet clustering, and therefore

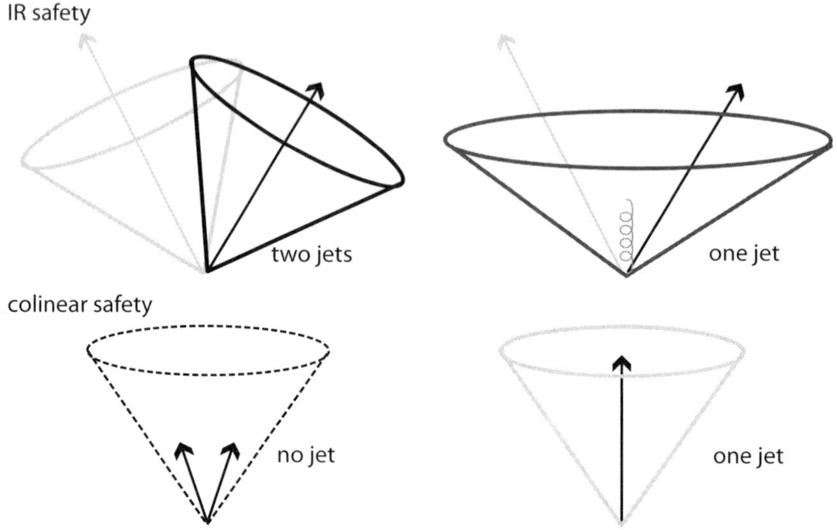

Fig. 6.13.   Issues with jet clustering algorithms: infra-red safety and colinear safety.

no jet will be created, as is indicated in the left column. In the left column, imagine that the two colinear particles land within the same calorimeter region and are recorded as a single energy deposit. In this case, the sum is above threshold and now a single jet is reconstructed. Again, the dependence on the details of the detector segmentation and the thresholds are not desirable features.

Details such as these make up the main difference between the many jet clustering algorithms, and are crucial to allow clean comparison between experimental measurements and theoretical calculations.

Heavy Flavor Jet Reconstruction   To determine if a jet is a light flavor (u, d, s) or heavy flavor (b, c) jet, we use the fact that B hadrons have a long lifetime. The decay length $c\tau = 450$ μm, combined with a large boost $\beta\gamma \gtrsim 5$, makes the distance traveled by a B hadron in the lab frame before decaying measurable with modern silicon detectors. In the decay, many charged tracks are produced, which can be reconstructed to find the secondary vertex where they originated from. Figure 6.14 shows a B hadron produced at the primary vertex. It travels a distance L_{xy} along the path shown by the dashed line to where it decays. By looking at the tracks' intersection point of the tracks from the decay, one can reconstruct

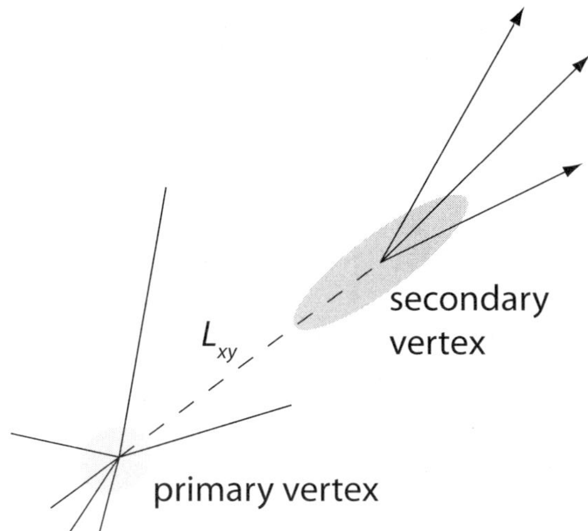

Fig. 6.14.   Reconstructing B hadron decays.

the secondary vertex. In the Figure, the ellipses show the uncertainty of the primary and secondary vertices, and we see that we can clearly distinguish tracks from the primary interaction from those from the B decay. By looking at L_{xy} and the ratio of L_{xy} to its uncertainty, we can find tracks associated with jets that correspond to the decay heavy flavor mesons.

There are other methods too; however, the secondary vertex determination is the most powerful single technique.

6.6. The Large Hadron Collider

The Large Hadron Collider (LHC) is a collider scheduled to start taking data this year (2009). Ultimately, it will collide protons at a center-of-mass energy of $\sqrt{s} = 14$ TeV. The LHC ring runs 100 m beneath the Swiss and French countryside, straddling the border between the two countries. The ring has a circumference of 27 km, and four major experiments populate its interaction regions. The LHC was initially proposed in 1988, and was originally slated to start taking data in 1997, at around the same time as the Superconduction Super Collider (SSC) was to turn on. A main part of the proposal was to re-use the LEP tunnel to reduce civil construction costs. Original plans included an electron-positron collider, with the electrons coming from the LEP ring. That plan was dropped; however, in addition to the proton running, the accelerator will also run in a heavy ion mode,

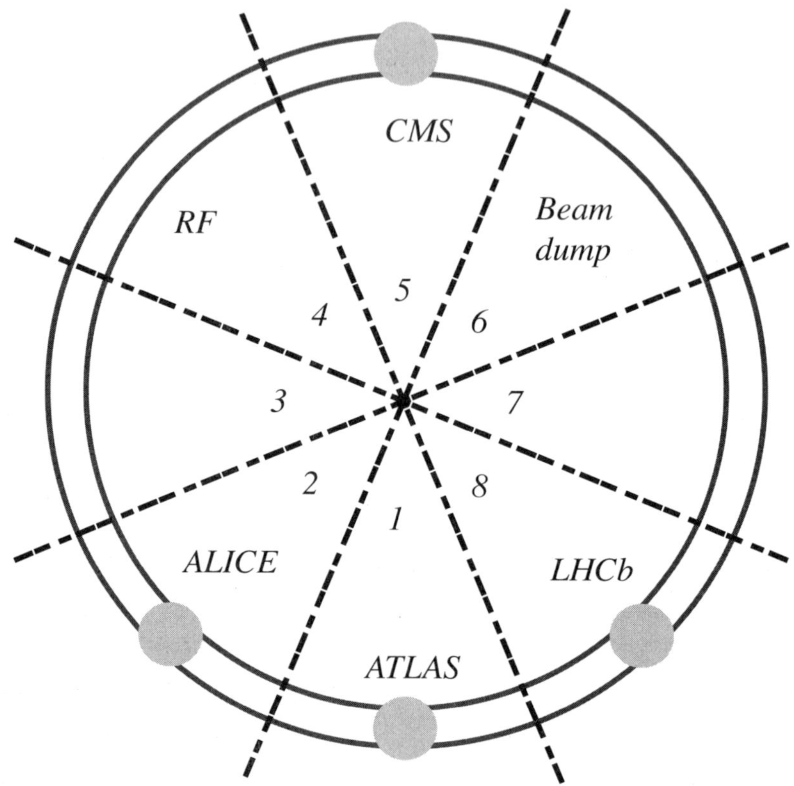

Fig. 6.15.   The LHC complex.

colliding gold atoms.

Two of the experiments, ATLAS (**A T**oroidal **LHC A**ppartu**S**) and CMS (Compact Muon Solenoid), are general purpose experiments designed to search for new physics beyond the standard model. LHCb is designed to study B mesons and the CKM matrix to new levels of precision. ALICE is a dedicated experiment to study heavy ion collisions. With apologies to my LHCb and ALICE colleagues, I will not discuss their experiments in this paper. A schematic view of the complex is shown in Fig. 6.15. In addition to the four experiments, we see the RF accelerating station in Sector 4 and the beam dump in Sector 6.

ATLAS is the largest of the four experiments. Measuring 45 m long and 25 m long, it weight an astounding 7 000 t. In most respects, ATLAS is a traditional particle physics detector, with an inner detector comprised

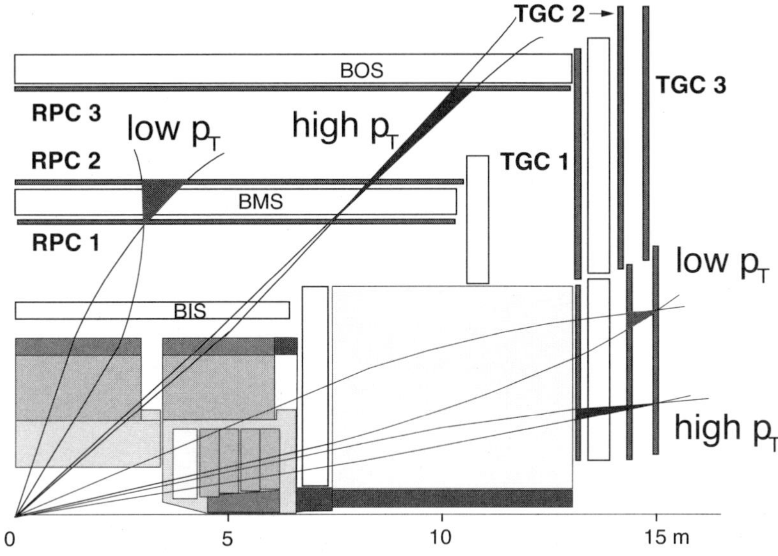

Fig. 6.16. ATLAS cross-sectional view showing effect of toroidal magnetic field.[9]

of a tracking system (silicon pixels, strips, and a straw tracker). The inner
detector is immersed in a 2 T solenoidal magnetic field. The solenoid is
encased in electromagnetic and hadronic calorimeters. The outermost part
of the detector consists of a series of muon detectors. As the name indicates,
ATLAS has a toroidal magnetic field in addition to the solenoidal field. This
air-core magnet allows a bending field for use in the muon system with a
minimum of material, thereby allowing momentum determination at large
radii with a minimum of multiple-scattering. Figure 6.16 is a cross-sectional
view of one quarter of the ATLAS detector, and shows the deflection path
of muons in the toroidal field, in both the central and forward regions of
the detector.

CMS is the second general-purpose detector. Though the acronym contains 'compact,' the experiment is by no means small. Figure 6.17 shows
a schematic view of the experiment. At 21 m long and 16 m tall, the experiment is large on all scales except maybe compared to ATLAS. Dense
might be a better descriptor, as, at 12 500 t, it weighs almost twice as
much as its bigger colleague. Like ATLAS, CMS is built like a traditional
HEP detector. The inner detector consists of two solid-state tracking devices, a pixel detector and a strip detector. Unlike ATLAS, there are no

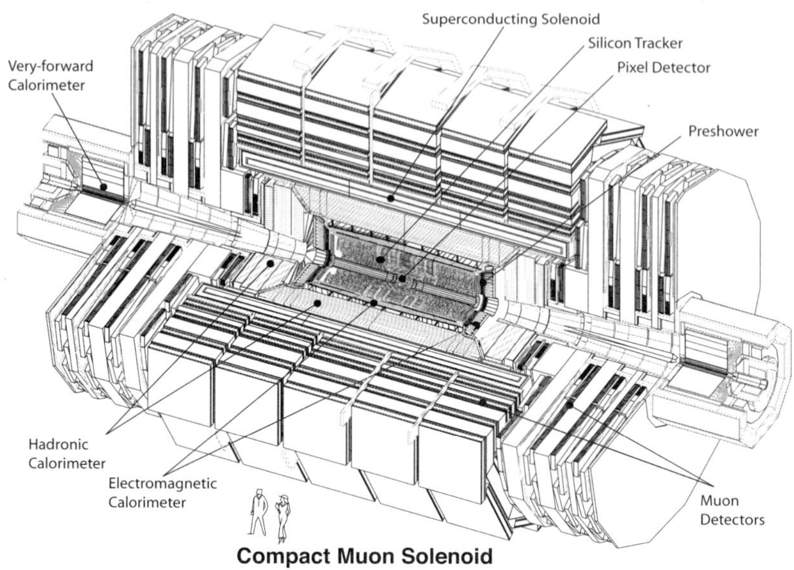

Fig. 6.17.   A schematic view of the CMS detector.[11]

gaseous tracking detectors in the central region in CMS - the strip detector covers an unprecedented 210 m^2 of area cover tracking needs even at large radii. Immediately following the tracking system are electromagnetic and hadronic calorimeters. CMS's inner detector is bathed in a 4 T magnetic field. The 2 T return field lines run through the muon system, which cause muons in CMS to undergo a pronounced 's' curve in their trajectory when crossing from the central field to the return field. This 's' is shown in the CMS logo.

Most of the material in this Section comes from the technical design reports (TDR's) from the experiments themselves.[9–12]

Tracking Resolution   As mentioned above, the ATLAS tracking system consists of a solid-state pixel detector, a solid-state strip detector and a gaseous straw chamber. The CMS tracking system is entirely a solid-state and covers 210 m^2. The absolute momentum scale will be determined by $pp \to Z \to \mu^+\mu^-$ at the Z resonance.

Electromagnetic calorimeter resolution   Both CMS and ATLAS have high-performance electromagnetic calorimeters (ECAL) for measuring the

Table 6.8. CMS and ATLAS tracking resolution, as estimated by the collaborations.

Type	c
CMS	0.01%
ATLAS	0.04%

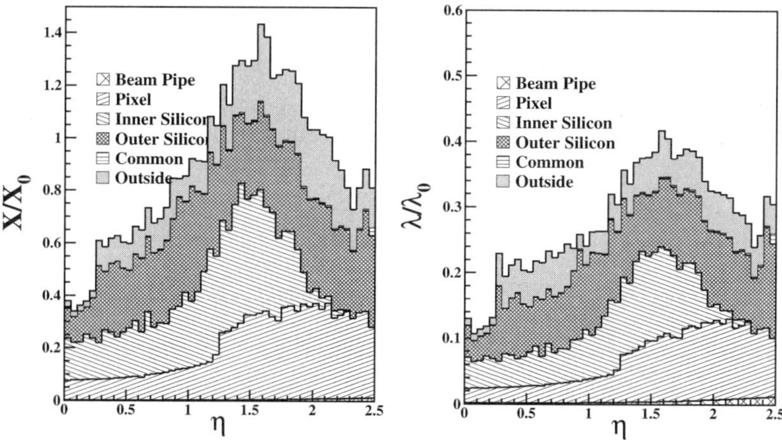

Fig. 6.18. Material in front of the CMS ECAL.[11]

energy of EM particles. ATLAS' calorimeter consists of a segmented liquid Argon (LAr) calorimeter. CMS' calorimeter is a homogeneous $PbWO_4$ system. One can parametrize the resolution as discussed in Sec. 6.4.3. As one might expect due to the advantages of a homogeneous calorimeter, the CMS resolution is better than ATLAS'.

$$\frac{\sigma}{E} = \frac{10.1\%}{\sqrt{E}} \oplus 0.20\% \quad \text{(ATLAS)}$$

$$\frac{\sigma}{E} = \frac{2.8\%}{\sqrt{E}} \oplus \frac{0.12}{E} \oplus 0.30\% \quad \text{(CMS)}$$

In both experiments, the ECAL resolution is limited by the amount of material in the silicon detectors and tracking systems. Figure 6.18 shows the amount of material in front of the CMS tracker in units of radiation lengths X_0 and λ_a. Photons and electrons passing through this material will radiate or pair-produce due to interactions with the material, leading to a degradation of the resolution unless the radiated particles can be successfully identified.

Determining the absolute electromagnetic energy scale   The most basic energy scale is determined from test beam data with beams of known energy. Cosmic rays can be used to inter-calibrate different parts of the detector. As minimum ionizing particles, they deposit a constant amount of energy in each calorimeter tower. In this method the relative response of different parts of the detector can be equalized.

Another *in-situ* method for determining the electromagnetic energy scale is Z decays. Di-electron events can be selected with small amount of bias, with an invariant mass m_{ee} in the area around the known m_Z mass. The measured energy can then be tuned to match the known Z mass. Similar calibrations can be made at low energies with J/ψ and Υ mesons, though these are more challenging with electrons and the CMS and ATLAS detectors have not been optimized for the measurement of these low energies.

Hadronic calorimeter resolution   Both CMS and ATLAS have similar composition in their hadronic calorimeter: steel and brass, followed by scintillators to collect the shower. The CMS hadronic calorimeter is thinner than the ATLAS calorimeter. The thickness of the CMS HCAL detector at $\eta = 0$ is $5.3 \times \lambda$, while ATLAS is $7.4 \times \lambda$. The CMS HCAL is contained inside the solenoid, which restricts the size (and hence the depth) of the detector. Some of the particles of high-energy showers in CMS leak out of the HCAL. In order to minimize this, a *tail-catcher* calorimeter is installed in the central region to measure any hadronic particles that leak past the solenoid. Figure 6.19 shows the simulated response of the CMS calorimeter to a 200 GeV pion, with and without the tail-catcher. The Figure shows a significant reduction of the tail of the resolution on the low end.

$$\frac{\sigma}{E} = \frac{52\%}{\sqrt{E}} \oplus \frac{1.59}{E} \oplus 3.02\% \quad (\text{ATLAS}[9])$$

$$\frac{\sigma}{E} = \frac{91\%}{\sqrt{E}} \oplus 3.8\% \quad (\text{CMS}[11])$$

These numbers are based on test-beam data and simulations.

Determining the hadronic energy scale   The absolute jet energy scale is determined by a variety of methods. Before installation, modules of the

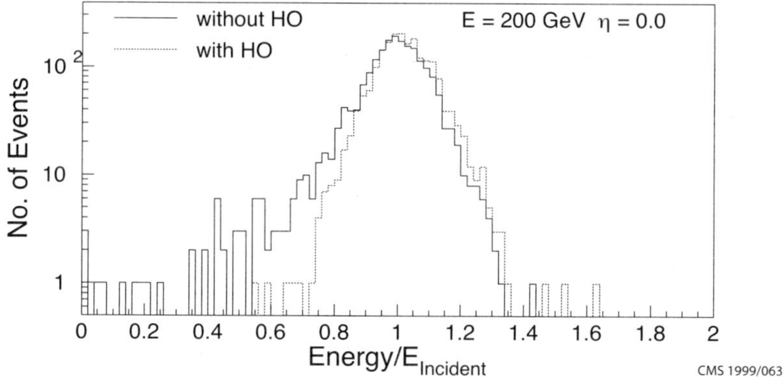

Fig. 6.19.   CMS HCAL response with and without the tail-catcher.[14]

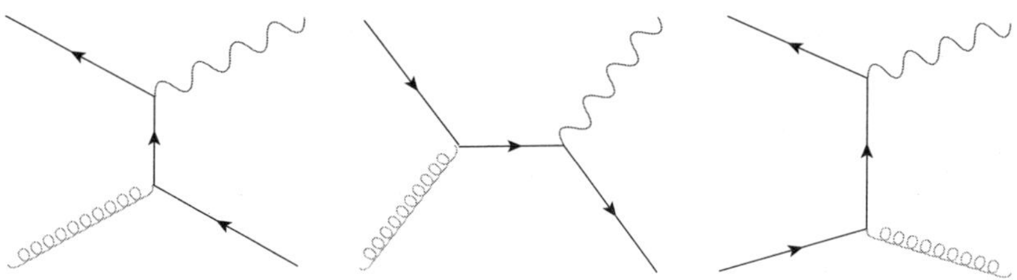

Fig. 6.20.   Processes used in γ-jet balancing.

calorimeter can be placed in a test beam of hadronic particles of known energies. This allows a good determination of the initial calibration; however, an in-situ calibration is preferable to track changes in the device.

The calibration task can be broken down into a relative and absolute correction. To get the right relative response, you can use $q\bar{q} \to g \to q\bar{q}$ events. Due to momentum conservation, we know that the energies of the two outgoing jets should balance one another. In this way, one can assure that the response of the detector is uniform across η and ϕ.

To determine the absolute scale, we can use the absolute electromagnetic energy scale. Consider the two interactions shown in Fig. 6.20. In each case, the final state shows a photon balanced against a quark or a gluon. Using momentum conservation, we can determine the energy scale of the hadronic particle given the calibrated electromagnetic scale determined as described above.

The corrections to the jet energy scale is done in stages. The steps can be broken down in different ways, but typically follow a series like the one outlined below.

(1) Offset: removal of pile-up and residual electronic noise.
(2) Relative (η): variations in jet response with η relative to control region.
(3) Absolute (p_T): correction to particle level versus jet p_T in control region.
(4) Flavor: correction to particle level for different types of jet (b, τ, etc.)
(5) Underlying Event: luminosity independent spectator energy in jet removed.
(6) Parton: correction to parton level.

The corrections are semi-independent - not all have to be applied at the same time.

6.7. Measurements at the Large Hadron Collider

Many papers have been written estimating the sensitivity of the ATLAS and CMS experiments to various new physics models for differing estimates of integrated luminosity. Rather than try to summarize them here, I will give a broad overview of what kinds of measurements one might expect from early LHC data. The approach is similar in nature to those by previous TASI speakers and some papers.[13] For estimates on the experiments' sensitivity to specific signatures, see the physics technical design reports ATLAS and CMS collaborations have published.[9–12]

Early measurements at the LHC experiments will be the 'easy' ones. They will use objects that are easier to reconstruct and will be *self-calibrating*. They will not require an understanding of the absolute energy scale, or the total integrated luminosity. They will be facilitated by the early availability of calibration samples, such as $pp \to Z \to \mu^- \mu^+$.

The easiest objects to understand are muons, followed by jets and electrons. Traditionally, taus identification has not been a high priority for hadron collider experiments, but it looks like this will not be the case for the LHC experiments and taus might be used relatively early in the run.

The early measurements will use single-object triggers. In the early data-taking phases (where we expect a lower instantaneous luminosity), the trigger bandwidth will likely not be filled and therefore more complicated triggers will not be necessary. Additionally, it will be easier to

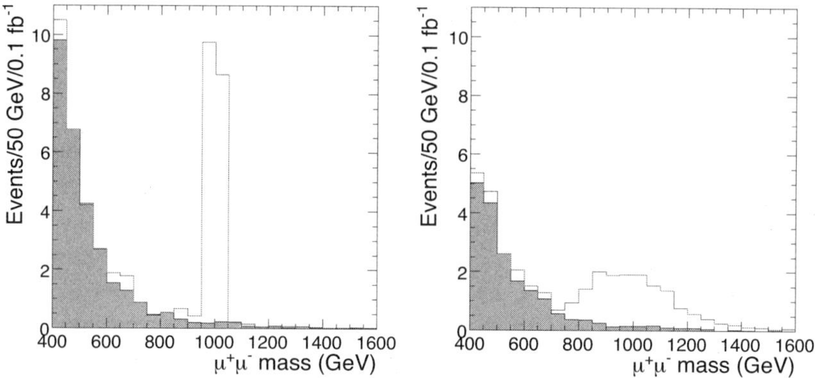

Fig. 6.21. Effect on misalignment on mass resolution for high-energy dimuon events.

understand the efficiency and turn-on curves for these triggers. Missing E_T ($\not{E}_T$) triggers are the sum of all the activity in the event, and as such are the ultimate 'multi-object' triggers. Additionally, any malfunction such as detector dead spots, beam halo, or reconstruction problems can create false $\not{E}_T$. Understanding the behavior of $\not{E}_T$ triggers is likely a more challenging task and therefore is unlikely to be one of the first measurements to emerge from the LHC.

6.7.1. *Early results*

I list here some of the results that are easiest to do and hence likely to be among the first reported.

Bump hunts A *bump hunt* is a search in the invariant mass spectrum of di-object pairs for a resonance. For example, a dimuon bump hunt would be sensitive to $pp \to X \to \mu^+\mu^-$. This Drell-Yan spectrum is typically steeply falling except for in the presence of a resonance. Figure 6.21 shows an example of such a search for the decay of a Z_η boson with a mass of 1 TeV at the CMS detector with 100 pb^{-1}.[12] The Figure at left shows the expected information given a perfect detector and indicates a clear excess above the shaded background. The Figure at right shows the same mass peak but applying early (lack of) understanding of the calibration and alignment. As can be seen, the peak is severely washed out at right, showing that even these early searches benefit from calibrating and aligning the detector.

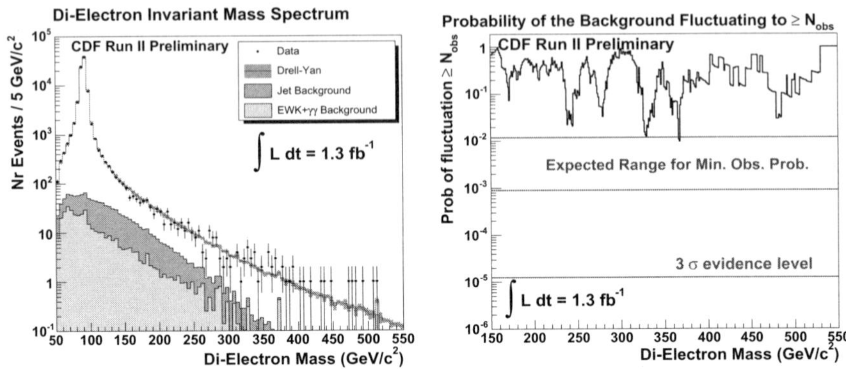

Fig. 6.22. At left: CDF $p\bar{p} \to Z \to e^+e^-$ bump hunt. At right: significance of local variations.[15]

However, if you look long enough, you will find a bump; it is important to remember the statistical significance of the bump in light of the many trials. Figure 6.22 shows such a search in the CDF experiment. The Figure at left shows the invariant mass spectrum. The Figure at right shows the significance of the excesses; a three-sigma excess would have a probability of 1 in 10^{-5}, given the number of trials, rather than 0.003% one might expect from naive application of the Gaussian distribution.

Counting Experiments Another early result one might expect is a counting experiment. For instance, the large cross sections of strongly produced new physics, combined with the large branching fraction to hadronic final states, suggests that the $\not{E}_{\mathrm{T}}$ + jets final state will be an early search to be conducted at the LHC. However, a challenge here is that these searches are often triggered with a $\not{E}_{\mathrm{T}}$ trigger. As mentioned previously, $\not{E}_{\mathrm{T}}$ is the ultimate multi-object trigger, and therefore understanding its performance will be challenging. These searches will use of data-driven background estimate techniques. These techniques allow you to use the data to extract a background estimate. A trivial example is the *bifurcated analysis* technique. To apply this technique, you choose two uncorrelated variables such as x and y (See Fig. 6.23). For instance, in estimating the amount of fake electrons in a $W \to e\nu$ sample, the two variables often used are $\not{E}_{\mathrm{T}}$ and the electron isolation. The C region is the signal region; the other regions are devoid of signal and dominated by background. For the W example, region C is large $\not{E}_{\mathrm{T}}$ and isolated electrons, while the background-dominated region has low $\not{E}_{\mathrm{T}}$ and unisolated electrons. The number of background

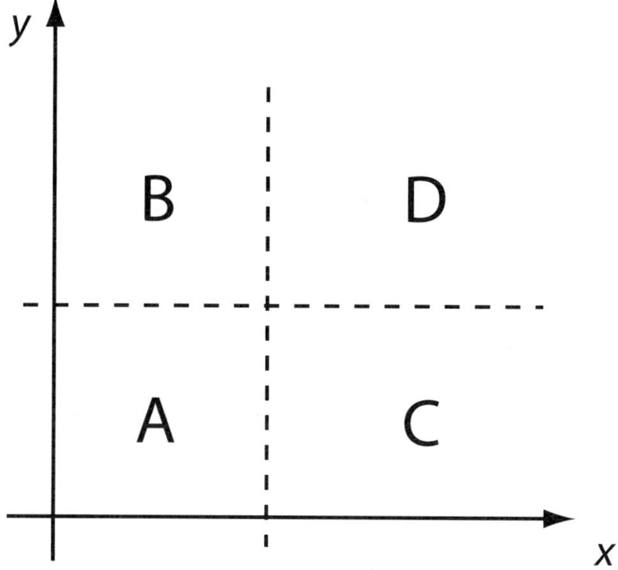

Fig. 6.23. Bifurcated analysis technique.

events in region C is then given by $N_C = N_D \times N_A/N_B$. This technique allows you to extrapolate into the signal region.

The advantage of these techniques is that it obviates the need to make sure the simulation agrees with the actual detector performance.

Role of Simulation Many differential background estimates are accomplished via Monte Carlo calculations. In this process, you first generate four-vector level events using generators such as Pythia.[3] This takes care of the hard scatter and the hadronization of the colored particles, as described above. Next, the particle trajectories are evolved to interact with the detector. The simulation here includes a full interaction of active and passive elements of the detector, including the hardware that records the detector response. Finally, the same reconstruction programs are used to produce electron, muon, jet etc candidates as are used on the data.

The generator-level first step is the "best possible" detector response; the output of the simulation is our best estimate of what the data coming from the detector looks like. The Monte Carlo programs must be checked and tuned to match the data; this process takes a long time and often needs statistically limited calibration data sets to complete.

6.8. Conclusion

This short report is intended to give a brief overview of LHC physics to the students of the TASI 2008 summer school. I encourage you to read the bibliography items to deepen your understanding of topics that have been raised here in this chapter.

References

1. C. Amsler *et al.* (Particle Data Group), Phys. Lett. **B667**, 1 (2008).
2. C. Grupen and B. A. Shwartz, *Particle Detectors.* (Cambridge University Press, Cambridge, 2008).
3. T. Sjöstrand, P. Edén, C. Friberg, L. Lönnblad, G. Miu, S. Mrenna and E. Norrbin, Computer Phys. Commun. **135** (2001) 238 (LU TP 00-30, arxiv:hep-ph/0010017).
4. W. Blum and L. Rolandi, *Paticle Detection with Drift Chambers.* (Springer Verlag, Berlin, 1994).
5. H. Spieler, *Semiconductor Detector Systems.* (Oxford University Press, New York, 2005).
6. Mike Perricone, *Symmetry* Magazine **03** 01 (2006).
7. CDF Collaboration, D. Acosta *et al.*, Phys. Rev. D **71**, 032001 (2005).
8. DØ Collaboration, V. M. Abazov *et al.*, Nucl. Instrum. Meth. A **565**:463-537 (2006).
9. ATLAS collaboration, ATLAS Detector and Physics Performance, Technical Design Report, Volume I. CERN/LHCC 99-14 (1999).
10. ATLAS collaboration, ATLAS Detector and Physics Performance, Technical Design Report, Volume II. CERN/LHCC 99-15 (1999).
11. CMS collaboration, CMS Physics TDR: Volume I, Detector Performance and Software. CERN/LHCC 2006-001.
12. CMS collaboration, CMS Physics TDR: Volume II, Physics Performance. CERN/LHCC 2006-021.
13. F. Gianotti and M.L. Mangano, LHC Physics: The First One–Two Year(s), CERN-PH-TH/2005-072 arXiv:hep-ph/0504221v1 (2005).
14. S. Banerjee and S. Banerjee, CMS internal Note 1999/063 (1999).
15. CDF Collaboration, T. Aaltonen *et al.*, Phys. Rev. Lett. **99**, 171802 (2007).

Chapter 7

Dealing with Data: Signals, Backgrounds, and Statistics

Luc Demortier

Laboratory of Experimental High Energy Physics, Rockefeller University,
New York, NY 10065, U.S.A.
E-mail: luc.demortier@rockefeller.edu
www.rockefeller.edu

We review the basic statistical tools used by experimental high energy physicists to analyze, interpret, and present data. After an introduction on the meaning of probability, we describe approaches to hypothesis testing, interval estimation, and search procedures.

Keywords: Bayes; Frequentism; Hypothesis testing; Interval estimation; Search procedures.

7.1. Introduction

The primary goal of these lectures is to review the basic statistical concepts needed to understand and interpret experimental results in the high energy physics literature. These results are typically formulated in terms of point estimates, intervals, p values, likelihood functions, Bayes factors, and/or posterior probabilities. Knowing the meaning of these quantities, their limitations, and the rigorous methods needed to extract them from data, will help in evaluating the reliability of published results. A secondary goal is to provide some tools to phenomenologists who would like to estimate the sensitivity of a particular experiment to a model of new physics.

These goals are facilitated by the availability of many web resources. For example, several experimental collaborations have formed internal statistics committees whose purpose is to make recommendations on proper statistical methods, to act as consultants on specific data analyses, and to help with the comparison and combination of experimental results from different experiments; some of these committees have public web pages with plenty

305

of useful information.[1-3] In addition, high energy physicists and astrophysicists regularly meet with professional statisticians to discuss problems and methods. These so-called PhyStat meetings have their own webpages and proceedings.[4-9] Finally, there is a repository of statistics software and other resources at `http://phystat.org`, and professional statistics literature is available online through `http://www.jstor.org`.

We begin our review with a discussion of the frequentist and Bayesian concepts of probability in section 7.2. This is followed by sections on hypothesis testing and interval estimation. Section 7.5 combines these two methodologies in the design of search procedures, which are at the heart of everyone's hopes for the success of the LHC program. Finally, section 7.6 contains some remarks about systematic uncertainties.

7.2. What Is Probability?

There is a long-standing philosophical dispute on the appropriate definition of probability, between two contenders known as frequentism and Bayesianism. This dispute has interesting implications both for the interpretation of scientific measurements and for the determination of quantum states.

7.2.1. *Frequentism*

Frequentists attempt to define probabilities as relative frequencies in sequences of trials. This corresponds to the common-sense intuition that if, for example, we toss a coin a large number of times, we can use the fraction of times it falls heads up as an estimate of the probability of "heads up", and this estimate becomes more accurate as the total number of tosses increases. To physicists this is a very attractive aspect of the frequentist definition: probabilities are postulated to be real, objective quantities that exist "outside us" and can be measured just as the length of a table or the weight of a book. Unfortunately it is very difficult to formulate a rigorous, non-circular definition of probability in terms of sequences of trials.[10] One possibility is to define probability as the limiting relative frequency in an infinite sequence of trials, or as the limiting relative frequency which would be obtained if the sequence of trials were extended to infinity. However, unlike a table or a book, infinite sequences are unobservable to finite beings like us. Furthermore, they may not even be empirically relevant. If at some point very far into an infinite sequence, the probability of interest suddenly changes by a discrete amount, this will affect the "infinite-sequence" value of the probability, but why should we care if we do not get to live until that point? Thus from a practical point of view it would seem more sensible to define the probability of an event as the relative frequency of that event in a *sufficiently long* sequence of trials. This is clearly a much weaker definition though. Indeed, given a finite number of trials, *every* sequence has a non-zero probability of occurring, and therefore also every probability value allowed by the discreteness of the measurement. The only way to resolve the difficulties in the frequentist definition of probability is to assume that the trials in the defining sequence are independent and equally probable. Hence the circularity: we need the concept of equal probability in order to be able to define probability.

Setting aside these foundational problems, the frequentist definition of probability seriously constrains the type of inferences that can be made. Indeed, according to frequentism, a random variable is a physical quantity that fluctuates from one observation to the next. Hence it is not possible to assign a meaningful probability value to a statement such as "the true mass M_H of the Higgs boson is between 150 and 160 GeV/c^2", since M_H is a fixed constant of nature. Frequentism therefore needs an additional, separate concept to describe the reliability of inferences: this is the concept of confidence. As applied to interval estimates of M_H, confidence represents the probability that the measurement *procedure* will yield an interval that contains the true value of M_H if the experiment is repeated a large number of times; it does *not* represent the probability that the numerical interval actually obtained from the data at hand contains that true value. Thus, even though confidence is defined in terms of probability, it should not be confused with the latter since it is applied to statements to which a (non-trivial) frequentist probability value cannot be assigned.

The objective of frequentist statistics is then to transform measurable probabilities of observations into confidence statements about physics parameters, models, and hypotheses. This transformation is not unique however. In the great variety of measurement situations, frequentism offers many "ad hoc" rules and procedures. In contrast with Bayesianism, to be described next, there is no unique frequentist principle that guides the process of drawing inferences.

7.2.2. *Bayesianism*

Bayesianism makes a strict distinction between propositions and probabilities.[11] Propositions include statements such as "the Higgs mass is between

150 and 160 GeV/c^2", and "it will rain tomorrow". These are either true or false. On the other hand, Bayesian probabilities are degrees of belief about the truth of some proposition. They are themselves not propositions and are therefore neither true nor false. In contrast with frequentist probability, which claims to be a measurable physical reality, Bayesian probability is a logical construct.

It can be shown that *coherent* degrees of belief satisfy the usual rules of probability theory. The Bayesian paradigm is therefore entirely based on the latter, viewed as a form of extended logic:[12] a process of reasoning by which one extracts uncertain conclusions from limited information. This process is guided by Bayes' theorem, which prescribes how degrees of belief about a parameter $\theta \in \Theta$ are to be updated when new data x become available:

$$\pi(\theta \,|\, x) \;=\; \frac{p(x \,|\, \theta)\,\pi(\theta)}{m_{prior}(x)}. \tag{7.1}$$

On the left-hand side, the quantity $\pi(\theta \,|\, x)$ represents the posterior probability density of θ, after having observed data value x. It is expressed as a function of the prior probability density $\pi(\theta)$ and the likelihood function $p(x \,|\, \theta)$, which is the probability density of the data x for a given value of θ, viewed as a function of θ; to emphasize this view, the likelihood is sometimes written as $\mathcal{L}(\theta \,|\, x)$. Finally, the denominator $m_{prior}(x)$ is the marginal distribution of x, also called prior-predictive distribution, or evidence, depending on the context:

$$m_{prior}(x) \;\equiv\; \int_{\Theta} p(x \,|\, \theta)\,\pi(\theta)\,d\theta. \tag{7.2}$$

All the basic tools of Bayesian statistics are direct applications of probability theory. A typical example is marginalization. Suppose we have a model for some data that depends on two parameters, θ and λ, but that we are only interested in θ. The posterior density of θ can then be obtained from the joint posterior of θ and λ by integration:

$$\pi(\theta \,|\, x) \;=\; \int_{\Lambda} \pi(\theta, \lambda \,|\, x)\,d\lambda. \tag{7.3}$$

Another useful example involves prediction. Suppose we observe data x and wish to predict the distribution of future data y. This can be done via the posterior-predictive distribution:

$$m_{post}(y \,|\, x) \;=\; \int_{\Theta} p(y \,|\, \theta)\,\pi(\theta \,|\, x)\,d\theta. \tag{7.4}$$

We emphasize that the output of a Bayesian analysis is always the *full* posterior distribution (7.1). The latter can be summarized in various ways, by providing point estimates, interval estimates, hypothesis probabilities, predictions for new data, etc., but the summary should not be substituted for the "whole story".

7.2.2.1. *Bayesian Priors: Evidence-Based Constructions*

The elicitation of prior probabilities on an unknown parameter or incompletely specified model is often difficult work, especially if the parameter or model is multidimensional and prior correlations are present. In particle physics we can usually construct so-called evidence-based priors for parameters such as the position of a detector element, an energy scale, a tracking efficiency, or a background level. Such priors are derived from subsidiary data measurements, Monte Carlo studies, and theoretical beliefs.

If for example the position of a detector is measured to be $x_0 \pm \Delta x$, and Δx is accurately known, it will be sensible to make the corresponding prior a Gaussian distribution with mean x_0 and standard deviation Δx. On the other hand, for an energy scale, which is usually a positive quantity, it will be more natural to use a gamma distribution, and for an efficiency bounded between 0 and 1 a beta distribution should be appropriate. In each of these cases, other functional forms should be tried to assess the robustness of the final analysis result to changes in prior shape. Note that evidence-based priors are always proper, that is, they integrate to unity.

7.2.2.2. *Bayesian Priors: Formal Constructions*

In physics data analysis we often need to extract information about a parameter θ about which very little is known a priori, or perhaps we would like to pretend that very little is known for reasons of objectivity. How do we apply Bayes' theorem in this case? How do we construct the prior $\pi(\theta)$?

Historically, this problem is the main reason for the development of alternative statistical paradigms: frequentism, likelihoodism, fiducial probability, and others. Even Bayesianism has come up with its own solution, known as objective Bayes. In general, results from these different methods tend to agree on large data samples, but not necessarily on small samples (discovery situations). For this reason, statistics committees in various experiments recommend data analysts to cross-check their results using alternative methods.

At its most optimistic, objective Bayesianism tries to find a completely coherent, objective Bayesian methodology for "letting the data speak for themselves". A much more modest goal is to provide a collection of useful methods to learn from the data as part of a robustness study. There are in fact several approaches to objective Bayesianism, all of which attempt to construct prior distributions that are minimally informative in some sense. Some approaches make use of concepts from information theory, others exploit the group invariance properties of some problems, and still others try to produce posterior distributions for which Bayesian credibilities can be matched with frequentist confidence statements. Bayesian analyses in high energy physics tend to err on the side of simplicity by using flat priors for parameters about which nothing is known a priori. The naive justification for flat priors is that they give the same weight to all parameter values and therefore represent ignorance. However, flat priors are not invariant under parameter transformations and they sometimes lead to improper posterior distributions and other kinds of problems.

Objective priors are also known as neutral, formal, or conventional priors. Although they are often improper when the parameter space is unbounded, they must lead to proper posteriors in order to make sense. A very important example of objective Bayesian prior is due to Harold Jeffreys. Suppose the data X have a distribution $p(x\,|\,\theta)$ that depends on a single continuous parameter θ; Jeffreys' prior is then:

$$\pi_J(\theta) \equiv \left\{ -\mathbb{E}\left[\frac{\partial^2}{\partial\theta^2} \ln p(x\,|\,\theta) \right] \right\}^{1/2}, \tag{7.5}$$

where the expectation is with respect to the data distribution $p(x\,|\,\theta)$. This prior illustrates how formal priors depend on the model assumed for the data; however, they do not depend on the data themselves. When θ is multidimensional, Jeffreys' prior tends to misbehave and must be replaced by the more general reference analysis prescription.[13]

7.2.3. *Quantum Probabilities*

An argument that is sometimes made is that frequentism must be the correct approach to data analysis because quantum mechanical probabilities are frequentist.[14] This argument is specious however, because the process by which we learn from our observations is logically distinct from the process that generates these observations. Furthermore, advances in quantum information science have shown that it is possible to interpret quantum mechanical probabilities as states of knowledge, i.e. as Bayesian.[15]

Part of the motivation for research into whether quantum probabilities are frequentist or Bayesian comes from EPR-style arguments. Suppose two systems A and B are prepared in some entangled quantum state and then spatially separated. By measuring one of two observables on A alone, one can immediately write down a new state for B. If one accepts that the "real, objective state of affairs" at B cannot depend on measurements made at A, then the simplest interpretation of the new state for B is that it is a *state of knowledge*.

It is possible to develop this idea of quantum states as states of knowledge in a fully consistent way. There are many aspects to this:[15]

- Subjective probability assignments must follow the standard quantum rule for probabilities (Gleason's theorem).
- The connection between quantum probability and long-term frequency still holds, but is a non-trivial consequence of Gleason's theorem and the concept of maximal information in quantum theory.
- Even quantum certainty (probability-1 predictions for pure states) is always some agent's certainty. Any agent-independent certainty about a measurement outcome would correspond to a pre-existing system property and would be in conflict with locality.[11]

Aside from providing yet another interpretation of quantum mechanics, do Bayesian quantum probabilities have any practical consequence? This is very much an open question. It may be for example, that vacuum fluctuations represent a Bayesian uncertainty rather than a real, physical phenomenon. If so, we do not need to worry about their contribution to the cosmological constant. Arguments for the physical reality of vacuum fluctuations are usually based on the experimental observations of spontaneous emission, the Lamb shift, and the Casimir effect. However E.T. Jaynes showed that spontaneous emission and the Lamb shift can both be derived without the need for vacuum fluctuations,[16] and R. L. Jaffe proved this for the Casimir effect.[17]

7.2.4. *Data Analysis: Frequentist or Bayesian?*

With some reasonable care, frequentist and Bayesian inferences generally agree in large samples. Disagreements tend to appear in small samples, where prior assumptions play a more important role both for frequentists and Bayesians. For a small number of problems, the Bayesian and frequentist answers agree exactly, even in small samples.

An often fruitful approach is to start with a Bayesian method, and then verify if the solution has any attractive frequentist properties. For example, if a Bayesian interval is calculated, does the interval contain the true value of the parameter of interest sufficiently often when the measurement is repeated? This approach has been formally studied by professional statisticians and is quite valuable.

On the other hand, if one starts with a purely frequentist method, it is also important to check its Bayesian properties for a reasonable choice of prior.

In experimental HEP we often use a hybrid method: a frequentist method to handle the randomness of the primary observation, combined with Bayesian techniques to handle uncertainties in auxiliary parameters. This is not easy to justify from a foundational point of view, but if the auxiliary parameter uncertainties are small, the overall measurement result may exhibit acceptable frequentist coverage.

7.3. Testing a Hypothesis

Hypothesis testing in high energy physics comes up in two very different contexts. The first one is when we wish to decide between two hypotheses, in such a way that if we repeat the same testing procedure many times, the rate of wrong decisions will be fully controlled in the long run. For example, when selecting good electron candidates for a measurement of the mass of the W boson, we need to minimize background contamination and signal inefficiency. The second context is when we wish to characterize the evidence provided by the data against a given hypothesis. In searching for new phenomena for example, we need to establish that an observed enhancement of a given background spectrum is evidence against the background-only hypothesis, and we need to quantify that evidence.

Traditionally, the first problem is solved by Neyman-Pearson theory and the second one by the use of p values, likelihood ratios, or Bayes factors.

7.3.1. *The Neyman-Pearson Theory of Testing*

Suppose we wish to decide which of two hypotheses, H_0 (the "null") or H_1 (the "alternative"), is more likely to be true given some observation X. The frequentist strategy is to minimize the probability of making the wrong decision over the long run. However, that probability depends on which hypothesis is actually true. There are therefore two types of error that can be committed:

- Type-I error: Rejecting H_0 when H_0 is true;
- Type-II error: Accepting H_0 when H_1 is true.

To fix ideas, suppose that the hypotheses have the form:

$$H_0 : X \sim f_0(x) \quad \text{versus} \quad H_1 : X \sim f_1(x), \tag{7.6}$$

by which one means that the observation X has probability density $f_0(x)$ under H_0 and $f_1(x)$ under H_1. For the test to be meaningful, f_0 and f_1 must be distinguishable given the measurement resolution. In other words, there must be a region C in sample space (the space of all possible data X) where the observation is much more likely to fall if H_1 is true than if H_0 is true. This region is called the critical region of the test and is used as follows: if the observation X falls inside C, we decide to reject H_0, otherwise we decide to accept it. The Type-I error probability α and the Type-II error probability β are then given by:

$$\alpha = \int_C f_0(x)\, dx \quad \text{and} \quad \beta = 1 - \int_C f_1(x)\, dx. \tag{7.7}$$

The probability of correctly accepting the alternative hypothesis equals $1 - \beta$ and is known as the power of the test.

In general the critical region C is constructed so as to achieve a suitably small Type-I error rate α, but there are many possible critical regions that will yield the same α. The idea of the Neyman-Pearson theory is to choose the C that minimizes β for a given α. In the above example, the distributions f_0 and f_1 are fully specified before the test (this is known as "simple versus simple testing"). In this case it can be shown that, in order to minimize β for a given α, C must be of the form:

$$C = \{x : f_0(x)/f_1(x) < c_\alpha\}, \tag{7.8}$$

where c_α is a constant depending on α. This result is known as the Neyman-Pearson lemma, and the quantity $y \equiv f_0(x)/f_1(x)$ is known as the likelihood ratio statistic.

Unfortunately, f_0 and/or f_1 are often composite, meaning that they depend on an unknown, possibly multidimensional parameter $\theta \in \Theta$. This happens when the measurement is affected by systematic uncertainties (in which case θ or one of its components could be an imperfectly known detector energy scale or tracking efficiency) or when the alternative hypothesis does not fully specify the value of a parameter of interest (as when θ or one of its components represents the production cross section for a new physics process and one is testing whether that cross section is exactly zero or strictly positive). The likelihood ratio is then defined as:

$$\lambda \equiv \frac{\sup\limits_{\theta \in \Theta_0} f_0(x_{obs} \mid \theta)}{\sup\limits_{\theta \in \Theta} f_1(x_{obs} \mid \theta)}, \tag{7.9}$$

where $\Theta_0 \subset \Theta$ is the subspace of θ values allowed by the null hypothesis. Although the Neyman-Pearson lemma does not generalize to this composite situation, the likelihood ratio remains an extremely useful test statistic. This is partly due to Wilks' theorem, which states that for large samples the distribution of $-2\ln\lambda$ under H_0 is that of a chisquared variate with number of degrees of freedom equal to the difference between the dimensionality of Θ and that of Θ_0. Under some rather general conditions, this theorem can be used to construct approximate critical regions for finite samples (however, see section 7.3.4).

As already stated, the Neyman-Pearson theory of testing is most useful in data quality control applications, when a given test has to be repeated on a large sample of identical items. In HEP we use this technique to select events of a given type. For example, if we want to select a sample of events to measure the mass of the top quark, we define H_0 to be the hypothesis that a given event contains a top quark, and try to minimize the background contamination β for a given signal efficiency $1 - \alpha$.

On the other hand, this approach to testing is not very satisfactory when dealing with one-time testing situations, for example when testing a hypothesis about a new phenomenon such as the Higgs boson or SUSY. This is because the result of a Neyman-Pearson test is either "accept H_0" or "reject H_0", without consideration for the strength of evidence contained in the data. In fact, the level of confidence in the decision resulting from the test is already known *before* the test: it is either $1 - \alpha$ or $1 - \beta$. One would like a way to quantify evidence from observed data, *after* the test. The frequentist solution to this problem uses p values exclusively, whereas the Bayesian one works with p values, Bayes factors and posterior hypothesis probabilities.

7.3.2. *The p Value Method for Quantifying Evidence*

Suppose we collect some data $\mathbf{X}$ and wish to characterize the evidence contained in $\mathbf{X}$ against a hypothesis H_0 about the distribution $f(\mathbf{x} \mid \theta)$ of the population from which $\mathbf{X}$ was drawn. A general approach is to construct a test statistic $T(\mathbf{X})$ such that large observed values of T are evidence against H_0 in the direction of some alternative of interest H_1. Often a good choice for T is $1/\lambda$, where λ is the likelihood ratio statistic defined in Eq. (7.9). In general, different testing problems require different test statistics, and the observed values of these test statistics cannot be directly compared across problems. We therefore need a method for *calibrating* the evidence provided by T. One way to do this is to calculate the probability for observing $T = t_{\mathrm{obs}}$ *or a larger value* under H_0; this tail probability is known as the p value of the test:

$$p = \mathbb{P}(T \geq t_{\mathrm{obs}} \mid H_0). \qquad (7.10)$$

Thus, small p values are evidence against H_0. Typically one will reject H_0 if $p \leq \alpha$, where α is some predefined, small error rate. This α has essentially the same interpretation as in the Neyman-Pearson theory of testing, but the emphasis here is radically different: with p values we wish to characterize *post-data* evidence, a concept which plays no role whatsoever in the Neyman-Pearson theory. Indeed, the only output of the latter is a report of acceptance or rejection of H_0, together with *pre-data* expectations of long-run error rates.

Clearly, the usefulness of p values for *calibrating* evidence against a null hypothesis H_0 depends on their null distribution being known to the experimenter and being the same in all problems considered. In principle, the very definition (7.10) of a p value guarantees that its distribution under H_0 is uniform. In practice however, this guarantee is rarely fulfilled exactly, either because the test statistic is discrete or because of the presence of nuisance parameters. The following terminology then characterizes the true null distribution of p values:

$$
\begin{aligned}
p \text{ exact} \quad &\Leftrightarrow \mathbb{P}(p \leq \alpha \mid H_0) = \alpha, \\
p \text{ conservative} &\Leftrightarrow \mathbb{P}(p \leq \alpha \mid H_0) < \alpha, \\
p \text{ liberal} \quad &\Leftrightarrow \mathbb{P}(p \leq \alpha \mid H_0) > \alpha.
\end{aligned}
$$

Compared to an exact p value, a conservative p value tends to understate the evidence against H_0, whereas a liberal p value tends to overstate it.

In spite of the apparent simplicity of the motivation and definition of p values, their correct interpretation in terms of evidence is notoriously subtle. In fact, p values themselves are controversial. Here is a partial list of caveats:

(1) P values are neither frequentist error rates nor confidence levels.
(2) P values are not hypothesis probabilities.

(3) Equal p values do not necessarily represent equal amounts of evidence (for example, sample size also plays a role).

Because of these and other caveats, it is better to treat p values as nothing more than useful "exploratory tools," or "measures of surprise." In any search for new physics, a small p value should only be seen as a first step in the interpretation of the data, to be followed by a serious investigation of an alternative hypothesis. Only by showing that the latter provides a better explanation of the observations than the null hypothesis, can one make a convincing case for discovery.

7.3.2.1. *The 5σ Discovery Threshold*

A small p value has little intuitive appeal, so it is conventional to map it into the number N_σ of standard deviations a normal variate is from zero when the probability outside $\pm N_\sigma$ equals $k \cdot p$, where $k = 1$ or 2:

$$p = \frac{2}{k} \int_{N_\sigma}^{+\infty} \frac{e^{-x^2/2}}{\sqrt{2\pi}} \, dx = \frac{1}{k} \left[1 - \mathrm{erf}(N_\sigma/\sqrt{2}) \right]. \tag{7.11}$$

Experiments at the LHC set $k = 2$. This choice is not universal however.

The threshold for discovery is typically set at $N_\sigma = 5$. This convention can be traced back to a 1968 paper by A. Rosenfeld,[18] where the author argued that, given the number of histograms examined by high energy physicists every year, one should expect several 4σ claims per year. He therefore recommended that experimental groups publish any tantalizing effect that passes the 3σ threshold, as a recompense for the time and funds invested in their experiments, but that they take additional data in the amount needed to confirm a real effect at the 5σ level. As for theorists, they should always wait for 5σ (or nearly 5σ) effects.

Rosenfeld's argument was based on what is known as the look-elsewhere effect, and according to which the probability of a significant background fluctuation scales with the number of places one looks in. This is a 40-year old calculation however, and it is legitimate to ask whether the discovery threshold should be adjusted for the increase in the number and scope of searches for new physics that have been performed every year since then. A purely empirical answer is that at the present time there is still no evidence that the rate of false 5σ claims is running out of control. Sure, there is the occasional false alarm,[19] but this is balanced by the increased sophistication of experimental methods, in particular a better understanding of particle interactions inside detectors, the investment of large amounts of computer power in the modeling of background processes and systematic uncertainties, and the use of "safer" statistical techniques such as blind analysis.[20] In any case, professional statisticians are usually surprised by the stringency of our discovery threshold, and few of them would trust our ability to model the tails of distributions beyond 5σ. Thus, raising the current discovery threshold could not be justified without first demonstrating our understanding of such extreme tails.

7.3.3. *The Problem of Nuisance Parameters in the Calculation of p Values*

Often the distribution of the test statistic, and therefore the p value (7.10), depends on parameters that model various uninteresting background processes and instrumental features such as calorimeter energy scales and tracking efficiencies. The values of these parameters usually have uncertainties on them, known as systematic uncertainties, and since this complicates the evaluation of p values the corresponding parameters are referred to as "nuisance parameters". There is obviously considerable interest in methods for calculating p values that eliminate the dependence on nuisance parameters while taking into account the corresponding systematic uncertainties. In fact there are many such methods, but before we discuss them, it is useful to list some desiderata that we might wish them to satisfy:

(1) *Uniformity:* the method should preserve the uniformity of the null distribution of p values. If exact uniformity is not achievable in finite samples, then asymptotic uniformity should be aimed for.
(2) *Monotonicity:* for a fixed value of the observation, systematic uncertainties should decrease the significance of null rejections.
(3) *Generality:* the method should not depend on the testing problem having a special structure, but should be applicable to as wide a range of problems as possible.
(4) *Power:* all other things being equal, more power is better.

Keeping these criteria in mind, in the following subsections we discuss four classes of methods for eliminating nuisance parameters: structural, supremum, bootstrap, and predictive. Only the first three of these methods are compatible with frequentism; the last one requires a Bayesian concept of probability.

7.3.3.1. *Structural Methods*

We label "structural" any purely frequentist method that requires the testing problem to have a special structure in order to eliminate nuisance parameters. A classical example is the pivotal method introduced by W. S. Gossett. Assume we have $n \geq 2$ observations X_i from a Gaussian distribution with mean μ and standard deviation σ, both unknown, and suppose we wish to test $H_0 : \mu = \mu_0$ versus $H_1 : \mu \neq \mu_0$, for a given value μ_0. The obvious test statistic here is the average $\bar{X}$ of all observations, but it can't be used because its distribution depends on the unknown parameter σ. However, Gosset discovered that the quantity

$$T \equiv \frac{\bar{X} - \mu_0}{S/\sqrt{n}}, \quad \text{where} \quad S \equiv \sqrt{\frac{1}{n-1} \sum_{i=1}^{n} (X_i - \bar{X})^2}, \qquad (7.12)$$

is a pivot, i.e. a function of both data and parameters whose distribution under H_0 is itself independent of unknown parameters:

$$T \sim \frac{\Gamma(n/2)}{\sqrt{(n-1)\,\pi}\,\Gamma((n-1)/2)} \left(1 + \frac{t^2}{n-1}\right)^{-n/2}. \qquad (7.13)$$

Thus, if we evaluate T for our observed data, we can use the above distribution to calculate a p value and perform the desired test.

Another interesting example is the conditioning method: suppose that we have some data X and that there exists a statistic $C = C(X)$ such that the distribution of X given C is independent of the nuisance parameter(s). Then we can use that conditional distribution to calculate p values. A simple illustration of this idea involves observing a number of events N from a Poisson distribution with mean $\mu + \nu$, where μ represents a signal rate of interest, whereas ν is a nuisance parameter representing the rate of a background process. Without further knowledge about ν it is not possible to extract information from N about μ and hence to test the null hypothesis that $\mu = 0$. Suppose however that we perform a subsidiary experiment in which we observe M events from a Poisson distribution with mean $\tau\nu$, where τ is a known calibration constant. We have then:

$$N \sim \text{Poisson}(\mu + \nu) \quad \text{and} \quad M \sim \text{Poisson}(\tau\nu). \qquad (7.14)$$

It turns out that this problem has the required structure for applying the conditioning method, if we use as conditioning statistic $C \equiv N+M$. Indeed, the probability of observing $N = n$ given $C = n + m$ is binomial under H_0:

$$\mathbb{P}(N = n \,|\, C = n + m) = \frac{\mathbb{P}(N = n \,\&\, C = n + m)}{\mathbb{P}(C = n + m)}$$

$$= \frac{\mathbb{P}(N = n \,\&\, M = m)}{\mathbb{P}(C = n + m)} = \frac{[\nu^n e^{-\nu}/n!]\,[(\tau\nu)^m e^{-\tau\nu}/m!]}{(\nu + \tau\nu)^{n+m} e^{-\nu-\tau\nu}/(n+m)!}$$

$$= \binom{n+m}{n}\left(\frac{1}{1+\tau}\right)^n \left(1 - \frac{1}{1+\tau}\right)^m. \qquad (7.15)$$

The dependence on ν has disappeared in the final expression for this probability, allowing one to compute a conditional p value:

$$p_{cond} = \sum_{i=n}^{n+m} \binom{n+m}{i}\left(\frac{1}{1+\tau}\right)^i \left(1 - \frac{1}{1+\tau}\right)^{n+m-i}. \qquad (7.16)$$

Since m/τ is the maximum likelihood estimate of ν from the subsidiary measurement, this p value is based on defining as more extreme those observations that have a larger N value and simultaneously a lower background estimate than the actual experiment. This method is sometimes used to evaluate the significance of a bump on top of a background spectrum, where "sidebands" provide a subsidiary measurement of the background level in the signal window. Fluctuations in both the signal window and the sidebands are Poisson.

In Fig. 7.1 we study the uniformity of the conditional p value (7.16) under H_0, for several values of τ and the true background magnitude ν_{true}. In all cases the p value turns out to be conservative, and the conservativeness increases as τ decreases, i.e. as the uncertainty on the background estimate increases. Note that if the problem only involved continuous statistics instead of the discrete N and M, the conditional p value would be exact.

7.3.3.2. *Supremum Methods*

Structural methods have limited applicability due to their requirement that the testing problem have some kind of structure. A much more general technique consists in maximizing the p value with respect to the nuisance parameter(s):

$$p_{\mathrm{sup}} = \sup_{\nu} p(\nu). \qquad (7.17)$$

This is a form of worst-case analysis: one reports the largest p value, or the smallest significance, over the whole parameter space. By construction p_{sup} is guaranteed to be conservative, but may yield the trivial result $p_{\mathrm{sup}} = 1$

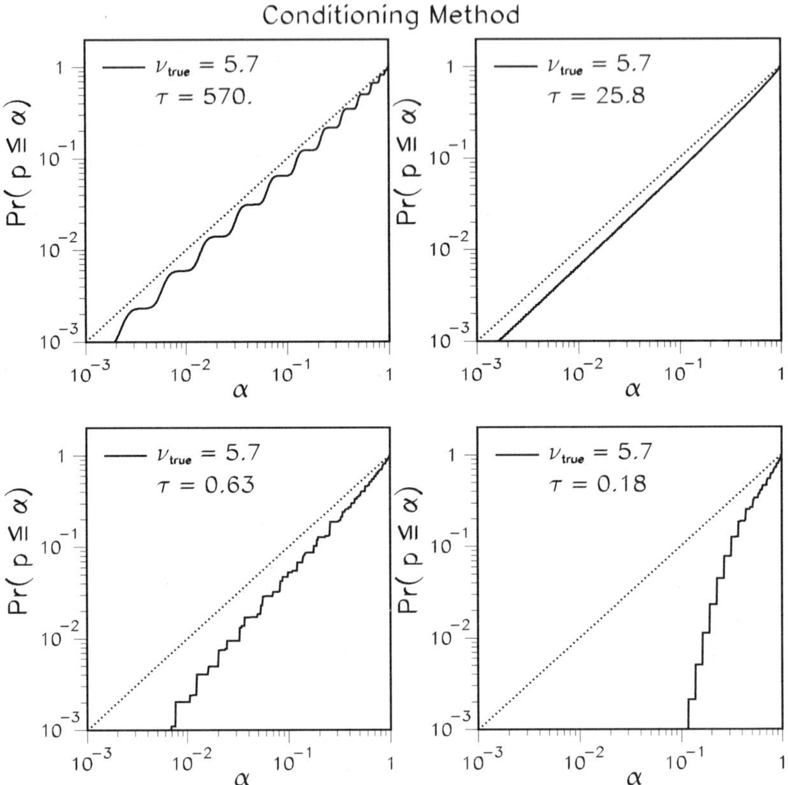

Fig. 7.1.   Solid lines: cumulative probability distribution of conditional p values under the null hypothesis, $\mathbb{P}(p \leq \alpha \mid H_0)$ as a function of α. The dotted lines indicate a uniform distribution, $\mathbb{P}(p \leq \alpha \mid H_0) = \alpha$. Note the log-log scale.

if one is not careful in the choice of test statistic. In general the likelihood ratio is a good choice. For an example, we consider again the Poisson problem from the previous section, but this time with a Gaussian distribution with mean ν and standard deviation $\Delta\nu$ for the subsidiary measurement:

$$N \sim \text{Poisson}(\mu + \nu) \quad \text{and} \quad X \sim \text{Gauss}(\nu, \Delta\nu). \tag{7.18}$$

The joint likelihood is:

$$\mathcal{L}(\nu, \mu \mid n, x) = \frac{(\nu + \mu)^n \, e^{-\nu-\mu}}{n!} \, \frac{e^{-\frac{1}{2}\left(\frac{x-\nu}{\Delta\nu}\right)^2}}{\sqrt{2\pi}\,\Delta\nu}, \tag{7.19}$$

and the likelihood ratio statistic is (compare Eq. (7.9)):

$$\lambda = \frac{\sup\limits_{\nu \geq 0,\, \mu=0} \mathcal{L}(\nu, \mu \mid n, x)}{\sup\limits_{\nu \geq 0,\, \mu \geq 0} \mathcal{L}(\nu, \mu \mid n, x)}. \tag{7.20}$$

Small λ is evidence against H_0. It can be shown that for large values of ν, the quantity $-2 \ln \lambda$ has the following distribution under H_0:

$$\begin{aligned}
\mathbb{P}(-2\ln\lambda = 0) &= \frac{1}{2}, \\
\mathbb{P}(-2\ln\lambda > x) &= \frac{1}{2} \int_x^\infty \frac{e^{-t/2}}{\sqrt{2\pi x}}\, dx = \frac{1}{2}\left[1 - \mathrm{erf}\left(\sqrt{\frac{x}{2}}\right)\right].
\end{aligned} \tag{7.21}$$

For small ν however, the distribution of $-2\ln\lambda$ depends on ν and is a good candidate for the supremum method. Here the supremum p value can be rewritten as:

$$p_{\mathrm{sup}} = \sup_{\nu \geq 0} \mathbb{P}(\lambda \leq \lambda_0 \mid \mu = 0) \tag{7.22}$$

A great simplification occurs when $-2\ln\lambda$ is stochastically increasing * with ν, because then $p_{\mathrm{sup}} = p_\infty \equiv \lim_{\nu \to \infty} p(\nu)$ and we can still use (7.21). Unfortunately this is not generally true, and is often difficult to check. When $p_{\mathrm{sup}} \neq p_\infty$, p_∞ will tend to be liberal. Figure 7.2 shows the cumulative distribution of p_∞ under H_0, for problem (7.18) and several values of ν_{true} and $\Delta\nu$. It is seen that the p_∞ approximation to p_{sup} is generally conservative, except at low $\Delta\nu$, where some minor, localized liberalism can be detected.

The supremum method has two important drawbacks. Computationally, it is often difficult to locate the global maximum of the relevant tail probability over the entire range of the nuisance parameter ν. Secondly, the very data one is analyzing often contain information about the true value of ν, so that it makes little sense to maximize over *all* values of ν. A simple way around these drawbacks is to maximize over a $1 - \gamma$ confidence set C_γ for ν (see section 7.4.1), and then to correct the p value for the fact that γ is not zero:

$$p_\gamma = \sup_{\nu \in C_\gamma} p(\nu) + \gamma. \tag{7.23}$$

---

*A statistic X with cumulative distribution $F(x \mid \theta)$ is stochastically increasing with the parameter θ if $\theta_1 > \theta_2$ implies $F(x \mid \theta_1) \leq F(x \mid \theta_2)$ for all x and $F(x \mid \theta_1) < F(x \mid \theta_2)$ for some x. In other words, X tends to be larger for larger values of θ.

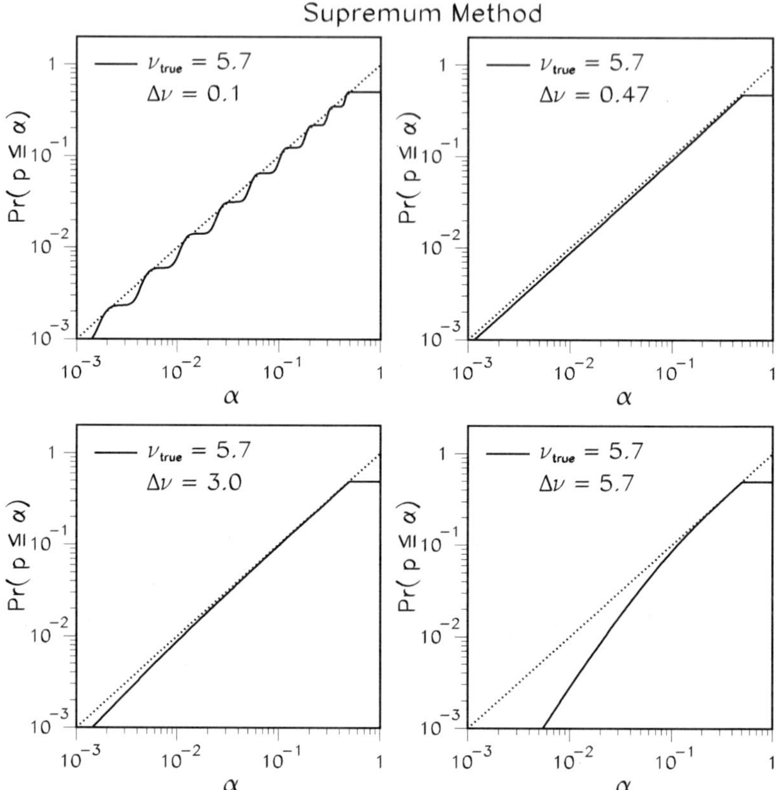

Fig. 7.2.　Cumulative probability distribution, under the null hypothesis, of the asymptotic approximation to the supremum p value, for a Poisson event count with Gaussian measurement of the mean.

This time the supremum is restricted to all values of ν that lie in the confidence set C_γ. It can be shown that p_γ, like p_{sup}, is conservative:

$$\mathbb{P}(p_\gamma \leq \alpha) \leq \alpha \quad \text{for all } \alpha \in [0, 1]. \tag{7.24}$$

Although there is a lot of flexibility in the choice of γ and C_γ, both should be chosen *before* looking at the data.

7.3.3.3. *Bootstrap Methods*

The first bootstrap method we consider is the plug-in. It gets rid of unknown parameters by estimating them, using for example a maximum-likelihood estimate, and then substituting the estimate in the calculation of the p value. For example (7.18) with likelihood function (7.19), the maximum-likelihood estimate of ν under H_0 is obtained by setting $\mu = 0$ and solving $\partial \ln \mathcal{L}/\partial \nu = 0$ for ν. This yields:

$$\hat{\nu}(x,n) = \frac{x - \Delta\nu^2}{2} + \sqrt{\left(\frac{x - \Delta\nu^2}{2}\right)^2 + n\,\Delta\nu^2}. \qquad (7.25)$$

The plug-in p value is then:

$$p_{plug}(x,n) \equiv \sum_{k=n}^{+\infty} \frac{\hat{\nu}(x,n)^k \, e^{-\hat{\nu}(x,n)}}{k!}. \qquad (7.26)$$

In principle two criticisms can be leveled at the plug-in method. Firstly, it makes double use of the data, once to estimate the nuisance parameters under H_0, and then again to calculate a p value. Secondly, it does not take into account the uncertainty on the parameter estimates. The net effect is that plug-in p values tend to be too conservative. The adjusted plug-in method attempts to overcome this.

If we knew the exact cumulative distribution function F_{plug} of plug-in p values under H_0, then the quantity $F_{plug}(p_{plug})$ would be an exact p value since its distribution is uniform by construction. In general however, F_{plug} depends on one or more unknown parameters and can therefore not be used in this way. The next best thing we can try is to substitute estimates for the unknown parameters in F_{plug}. Accordingly, one defines the adjusted plug-in p value by:

$$p_{plug,adj} \equiv F_{plug}(p_{plug} \,|\, \hat{\theta}), \qquad (7.27)$$

where $\hat{\theta}$ is an estimate for the unknown parameters collectively labeled by θ. This adjustment algorithm is known as a double parametric bootstrap and can also be implemented in Monte Carlo form.

Some cumulative distributions of the plug-in and adjusted plug-in p values are plotted in Fig. 7.3 for example (7.18). The adjusted plug-in p value provides a strikingly effective correction for the overconservativeness of the plug-in p value.

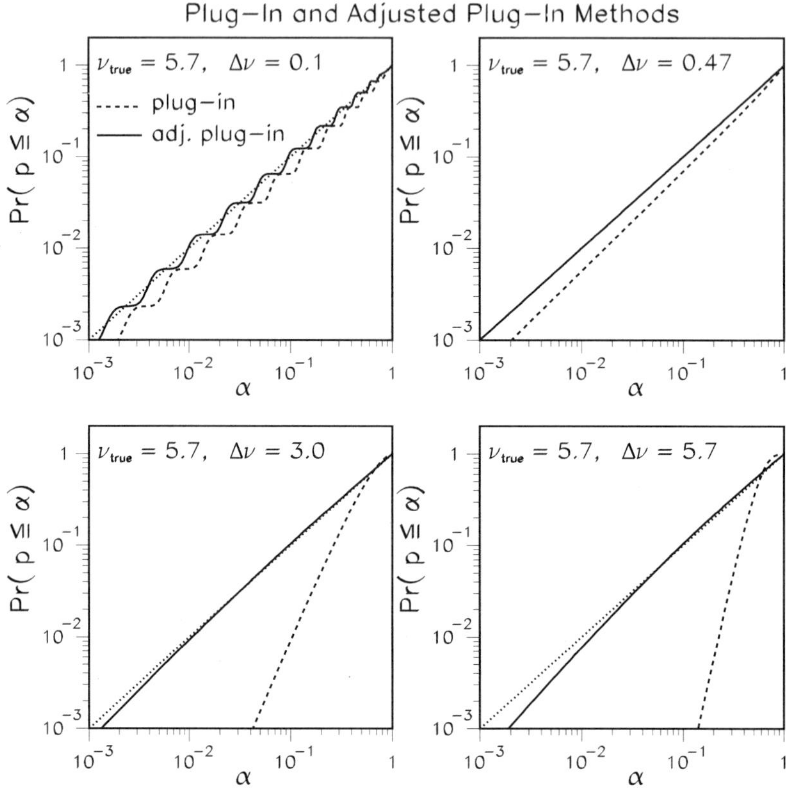

Fig. 7.3. Cumulative probability distribution of the plug-in (dashed lines) and adjusted plug-in (solid lines) p values under the null hypothesis for a Poisson event count with Gaussian measurement of the mean.

7.3.3.4. *Predictive Methods*

So far we have assumed that information about the nuisance parameter comes from a subsidiary measurement. This allows one to treat the problem of eliminating nuisance parameters in a purely frequentist way. The structural, supremum, and confidence interval methods are guaranteed to be conservative. The asymptotic approximation to the supremum method and the bootstrap methods do not provide this guarantee but are still frequentist. We now turn to the situation where information about the nuisance parameter comes in the form of a Bayesian prior. We discuss two approaches, known as prior-predictive and posterior-predictive.

The prior-predictive distribution of a test statistic T is the predicted distribution of T before the measurement:

$$m_{prior}(t) = \int p(t \mid \theta) \, \pi(\theta) \, d\theta, \tag{7.28}$$

where $\pi(\theta)$ is the prior probability density of θ. After having observed $T = t_0$ we can quantify how surprising this observation is by referring t_0 to m_{prior}, e.g. by calculating the prior-predictive p value:

$$
\begin{aligned}
p_{prior} = \mathbb{P}_{m_{prior}}(T \geq t_0 \mid H_0) &= \int_{t_0}^{\infty} m_{prior}(t) \, dt \\
&= \int \pi(\theta) \left[\int_{t_0}^{\infty} p(t \mid \theta) \, dt \right] d\theta,
\end{aligned} \tag{7.29}
$$

where the last equality follows from interchanging two integral signs. This last expression for p_{prior} shows that the prior-predictive p value can be interpreted as the average of the usual p value over the prior for the unknown parameter.

The posterior-predictive distribution of a test statistic T is the predicted distribution of T *after* measuring $T = t_0$:

$$m_{post}(t \mid t_0) = \int p(t \mid \theta) \, \pi(\theta \mid t_0) \, d\theta. \tag{7.30}$$

The posterior-predictive p value estimates the probability that a *future* observation will be at least as extreme as the current observation if the null hypothesis is true:

$$
\begin{aligned}
p_{post} = \mathbb{P}_{m_{post}}(T \geq t_0 \mid H_0) &= \int_{t_0}^{\infty} m_{post}(t \mid t_0) \, dt \\
&= \int \pi(\theta \mid t_0) \left[\int_{t_0}^{\infty} p(t \mid \theta) \, dt \right] d\theta.
\end{aligned} \tag{7.31}
$$

As the last expression on the right shows, the posterior-predictive p value can also be written as an average, this time over the posterior for the unknown parameter. Note the double use of the observation t_0 in p_{post}: first to compute the posterior for θ, and then again in the tail probability calculation. We encountered the same feature in the definition of the plug-in p value, and the same effect will be observed here, namely that the posterior-predictive p value is overly conservative.

What about the uniformity of p_{prior} and p_{post}? How well calibrated are these predictive p values? The answer depends on the distribution of the test statistic T under the null hypothesis. One can argue that this should be the prior-predictive distribution (7.28), since this distribution is fully specified and is available before observing the data. It is clear that, by construction, p_{prior} will be uniform with respect to the prior-predictive distribution. On the other hand, because of its double-use of the data, p_{post} will be conservative.

Frequentists will argue that the prior-predictive distribution is not frequentist and therefore does not provide a valid reference ensemble to check the uniformity of p_{prior} and p_{post}. If the testing problem of interest is purely frequentist, a different approach is in fact possible. Consider for example the Poisson+Gauss problem of Eq. (7.18). One way to apply a predictive method to this problem is to construct a posterior for the subsidiary Gaussian measurement of ν, and then use this posterior as a prior for ν when calculating a predictive p value for the Poisson event count N. We still need a prior for the subsidiary measurement however, and in the absence of further information about ν, it is appropriate to use an objective rule such as Jeffreys'. For a Gaussian likelihood with unknown mean, the Jeffreys' prior (7.5) is a constant. Thus the subsidiary posterior is:

$$\pi_{\text{sub.}}(\nu \,|\, x) = \frac{e^{-\frac{1}{2}\left(\frac{\nu-x}{\Delta\nu}\right)^2}}{\sqrt{2\pi}\,\Delta\nu\,\frac{1}{2}\left[1 + \operatorname{erf}\left(\frac{x}{\sqrt{2}\,\Delta\nu}\right)\right]}, \tag{7.32}$$

where the normalization comes from the requirement that ν, being a Poisson mean, is a positive parameter. We can use this posterior as a prior to construct p_{prior} and p_{post}. Furthermore, for every value of ν we now have a frequentist reference ensemble to check the uniformity of these p values, namely the set of all (X, N) pairs where X is a Gaussian variate with mean ν and standard deviation $\Delta\nu$, and N is an independent Poisson variate with mean ν. Contrast this with the reference ensemble represented by the prior-predictive distribution, which is defined for every value of x rather than every value of ν, and is the set of (ν, N) pairs where ν is a Gaussian variate with mean x and standard deviation $\Delta\nu$, and N is a *dependent* Poisson variate whose mean is the ν value in the same pair. Because of the random nature of the parameter ν in this ensemble, it is clearly Bayesian. Figure 7.4 shows the cumulative distributions of p_{prior} and p_{post} with respect to the frequentist ensemble, for several values of $\Delta\nu$. Both p values appear to be (mostly) conservative, and p_{post} much more so than p_{prior}, especially at large $\Delta\nu$.

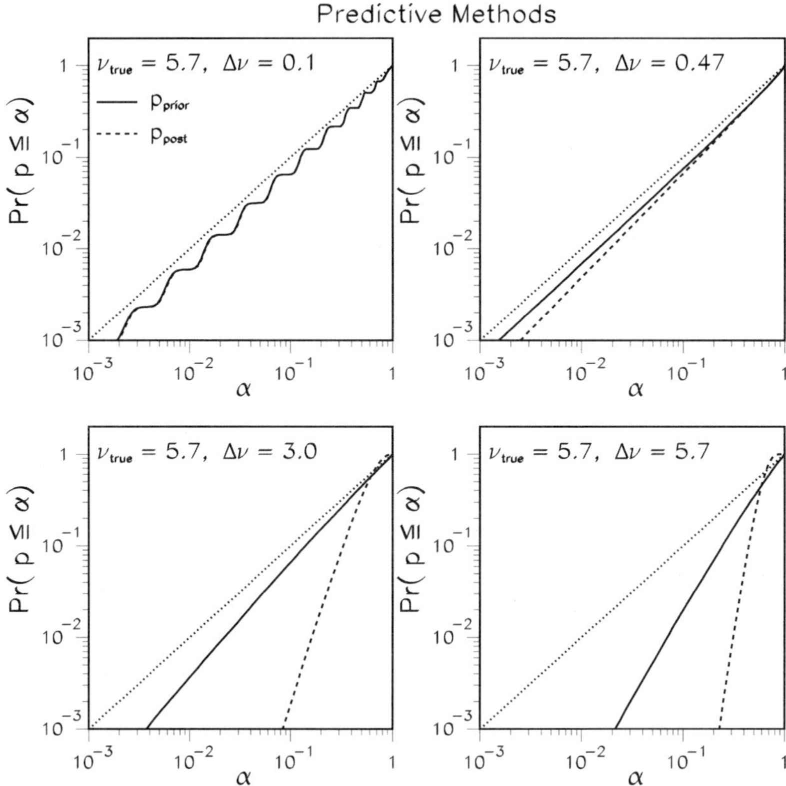

Fig. 7.4.　Cumulative distributions of the prior-predictive (solid lines) and posterior-predictive (dashed lines) p values for a Poisson event count with a Gaussian uncertainty on the mean. The dotted lines correspond to exact p values.

We end this discussion of predictive p values with some general comments:

- Prior-predictive p values cannot be defined for improper priors; in this case, posterior-predictive p values often provide a solution.
- Posterior-predictive p values can be calculated for discrepancy variables (i.e. functions of data *and* parameters) in addition to test statistics.
- Rather than simply reporting a predictive p value, it may be more

Table 7.1. P values for a Poisson observation of $n_0 = 3893$ events over an estimated background of $x_0 = 3234 \pm \Delta\nu$ events, where $\Delta\nu = 10$ or 100. For the confidence interval p value a 6σ upper limit was constructed for the nuisance parameter ($\gamma = 9.87 \times 10^{-10}$).

Method	$\Delta\nu = 10$		$\Delta\nu = 100$	
	P value	N_σ	P value	N_σ
Supremum	1.16×10^{-28}	11.05	9.81×10^{-9}	5.62
Confidence Interval	9.87×10^{-10}	6.00	1.23×10^{-8}	5.58
Plug-In	8.92×10^{-28}	10.86	1.86×10^{-3}	2.90
Adjusted Plug-In	1.13×10^{-28}	11.05	9.90×10^{-9}	5.61
Prior-Predictive	1.23×10^{-28}	11.04	9.85×10^{-9}	5.61
Posterior-Predictive	5.27×10^{-27}	10.70	1.35×10^{-2}	2.21

informative to plot the observed value of the test statistic against the appropriate predictive distribution.

- There are other types of predictive p values, which avoid some of the problems of the prior- and posterior-predictive p values.[21]

7.3.3.5. *Summary of p Value Methods*

To guide our summary of the various nuisance parameter elimination methods just described, we return to the desiderata listed at the beginning of section 7.3.3.

Figures 7.1 to 7.4 indicate quite a variation in uniformity, or rather lack thereof, of p value distributions under the null hypothesis. For the examples studied, the adjusted plug-in and supremum methods perform quite well, but this behavior depends strongly on the choice of test statistic. The likelihood ratio is generally a good choice. Our examples also show that uniformity tends to be violated on the conservative side, but this is only guaranteed for fully frequentist methods such as conditioning, supremum, and confidence interval. For other methods uniformity will have to be checked explicitly for the problem at hand. This is of course important if one wants to avoid overestimating the significance of a result.

An interesting point to note is that some p values tend to converge in the asymptotic limit. This is numerically illustrated for example (7.18) in Table 7.1, which shows that the supremum, adjusted plug-in, and prior-predictive p values give almost identical results on a data sample of thousands of events. Whenever possible, it is always instructive to compare the results of different methods.

Figure 7.5 compares the power functions of the supremum, adjusted plug-in, and prior-predictive p values for problem (7.18). There is not much difference between the curves, except perhaps at high $\Delta\nu$, where the prior-predictive p value seems somewhat less powerful. Note that as the signal strength goes to zero, the power function converges to α if the p value is exact.

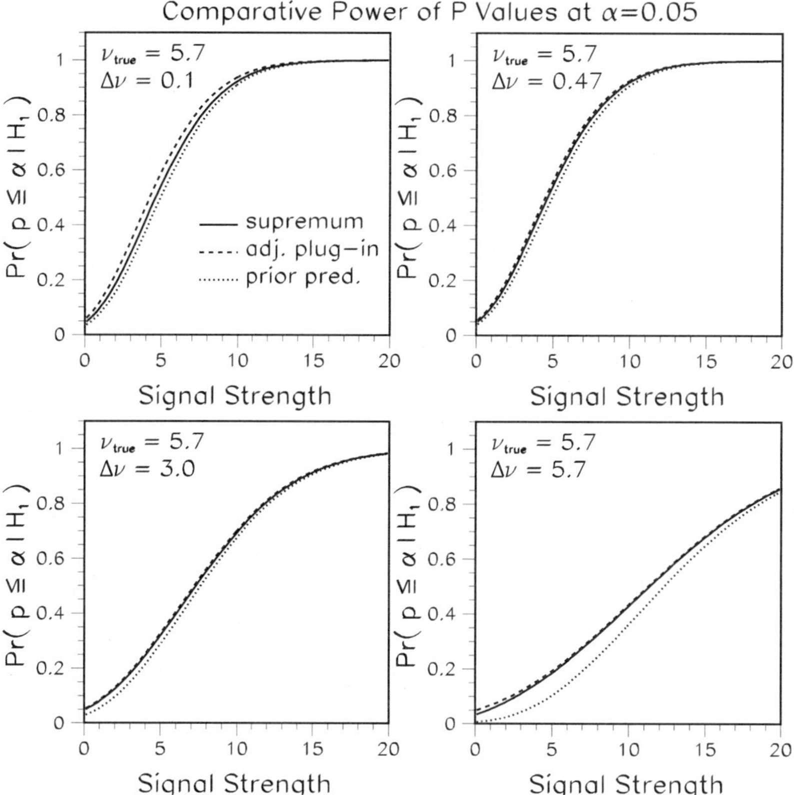

Fig. 7.5. Power functions of the supremum (solid), adjusted plug-in (dashed), and prior-predictive (dotted) p values for testing for the presence of a Poisson signal on top of a Poisson background whose mean ν_{true} has a Gaussian uncertainty $\Delta\nu$. The power is calculated for a test level of $\alpha = 0.05$ and is plotted as a function of true signal strength.

Finally, we comment on the monotonicity property: for the examples and methods studied here, it is true that the p value increases with the magnitude of the systematic uncertainty. In other words, significance claims are degraded by the presence of systematics. However, in practical problems not covered by this review, monotonicity will have to be checked explicitly.

7.3.4. *Caveats about the Likelihood Ratio Statistic*

As mentioned previously, the likelihood ratio λ defined in (7.9) is often a good choice of test statistic, because it is intuitively sensible, and is even optimal in the special case of simple vs. simple testing. Although this optimality does not transfer to the testing of composite hypotheses,[22] λ remains popular in that case due to Wilks' theorem, which gives the asymptotic distribution of $-2\ln\lambda$ under the null hypothesis as that of a chisquared (see section 7.3.1). Unfortunately, the conditions for this theorem to be applicable do not always hold in high energy physics problems. What follows are some examples where these regularity conditions are violated.

- One of the regularity conditions is that the tested hypotheses must be nested, i.e. H_0 must be obtainable by imposing parameter restrictions on the model that describes H_1. A counter-example is a test that compares two new-physics models that belong to separate families of distributions.
- Another regularity condition is that H_0 should not be on the boundary of the model that describes H_1. A typical violation of this condition is when θ is a positive signal magnitude and one is testing $H_0 : \theta = 0$ versus $H_1 : \theta > 0$.
- A third condition is that there must not be any nuisance parameters that are defined under H_1 but not under H_0. Suppose for example that we are searching for a signal peak on top of a smooth background. The location, width, and amplitude of the peak are unknown. In this case the location and width of the peak are undefined under H_0, i.e. when the amplitude is zero. Hence $-2\ln\lambda$ will not have a chisquared distribution under H_0.

There does exist some analytical work on the distribution of the likelihood ratio when the above regularity conditions are violated; however, these results are not always easy to apply and still require some numerical calculations. Physicists aware of the limitations of Wilks' theorem usually prefer to estimate the distribution of $-2\ln\lambda$ with the help of a Monte Carlo calculation. The advantage of this approach is that it allows one to incorporate all the relevant details of the experimental data analysis; the disadvantage is that it sometimes requires enormous amounts of CPU time.

7.3.5. *Expected Significances*

Probably the most useful way to describe the sensitivity of a model of new physics, given specific instrumental conditions, is to calculate the integrated luminosity for which there is a 50% probability of claiming discovery at the 5σ level. The calculation can be done as follows:

(1) Compute (or simulate) the distribution of p values under the new physics model and assuming a fixed integrated luminosity.
(2) Find the median of the p value distribution from (1).
(3) Repeat steps (1) and (2) for several values of the integrated luminosity and interpolate to find the integrated luminosity at which the median p value is 2.7×10^{-7} (5σ).

To determine the most sensitive method, or the most sensitive test statistic for discovering new physics, another useful measure is the expected significance level (ESL), defined as the observed p value averaged over the new physics hypothesis. If the test statistic X has density $f_i(x)$ under H_i, and if $p = 1 - F_0(X) \equiv 1 - \int_{-\infty}^{X} f_0(t)\, dt$, then:

$$\mathrm{ESL} \equiv \mathbb{E}(p \,|\, H_1) = \int [1 - F_0(x)]\, f_1(x)\, dx = \int F_1(x)\, f_0(x)\, dx. \quad (7.33)$$

The integral on the right is easy to estimate by Monte Carlo, since it represents the probability that $X \geq Y$, where X and Y are independent random variables distributed according to F_0 and F_1, respectively.

7.3.6. *Combining Significances*

When searching for new physics in several different channels, or via different experiments, it is sometimes desired to summarize the search by calculating a combined significance. This is a difficult problem. The best approach is to combine the likelihood functions for all the channels and derive a p value from the combined likelihood ratio statistic. However, it may not always be possible or practical to do such a calculation. In this case, if the individual p values are independent, another possibility is to combine the p values directly.[23] Unfortunately there is no unique way of doing this. The general idea is to choose a rule $S(p_1, p_2, p_3, \ldots)$ for combining individual p values $p_1,\, p_2,\, p_3, \ldots$, and then to construct a combined p value by calculating the tail probability corresponding to the observed value of S. Some plausible combination rules are:

(1) The product of p_1, p_2, p_3,... (Fisher's rule);
(2) The smallest of p_1, p_2, p_3,... (Tippett's rule);
(3) The average of p_1, p_2, p_3,...;
(4) The largest of p_1, p_2, p_3,....

This list is by no means exhaustive. To narrow down the options, there are some properties of the combined p value that one might consider desirable. For example:

(1) If there is strong evidence against the null hypothesis in at least one channel, then the combined p value should reflect that, by being small.
(2) If none of the individual p values shows any evidence against the null hypothesis, then the combined p value should not provide such evidence.
(3) Combining p values should be associative: the combinations $((p_1, p_2), p_3)$, $((p_1, p_3), p_2)$, $(p_1, (p_2, p_3))$, (p_1, p_2, p_3), should all give the same result.

Now, it turns out that property 1 eliminates rules 3 and 4; property 2 is satisfied by all four rules, and property 3, called evidential consistency, is satisfied by none. This leaves Tippett's and Fisher's rules as reasonable candidates. Actually, it appears that Fisher's rule has somewhat more uniform sensitivity to alternative hypotheses of interest in most problems. So Fisher's rule is quite popular.

Here is a simple mathematical trick to combine n p-values by Fisher's rule: take twice the negative logarithm of their product and treat it as a chisquared variate for $2n$ degrees of freedom (this is valid because the cumulative distribution of a chisquared variate for 2 d.o.f. is $1 - e^{-x/2}$, and chisquared variates are additive). The general result is that

$$p_{\mathrm{comb}} \equiv \Pi \sum_{j=0}^{n-1} \frac{(-\ln \Pi)^j}{j!}, \qquad \text{where } \Pi \equiv \prod_{j=1}^{n} p_j, \tag{7.34}$$

will have a uniform distribution under H_0 if the individual p_i are uniform. One situation in which the p_i will not be uniform is if they are derived from discrete test statistics. In this case the formula will give a combined p value that is larger than the correct one, and therefore conservative.

The literature on combining p values is extensive; see Ref. 24 for an annotated bibliography.

7.3.7. *Bayesian Hypothesis Testing*

The Bayesian approach to hypothesis testing is to calculate posterior probabilities for all hypotheses in play. When testing H_0 versus H_1, Bayes' theorem yields:

$$\pi(H_0 \,|\, x) \;=\; \frac{p(x \,|\, H_0)\,\pi_0}{p(x \,|\, H_0)\,\pi_0 \;+\; p(x \,|\, H_1)\,\pi_1}, \tag{7.35}$$

$$\pi(H_1 \,|\, x) \;=\; 1 \;-\; \pi(H_0 \,|\, x), \tag{7.36}$$

where π_i is the prior probability of H_i, $i = 0, 1$. If $\pi(H_0 \,|\, x) < \pi(H_1 \,|\, x)$, one rejects H_0 and the posterior probability of error is $\pi(H_0 \,|\, x)$. Otherwise H_0 is accepted and the posterior error probability is $\pi(H_1 \,|\, x)$.

In contrast with frequentist Type-I and Type-II errors, which are known *before* looking at the data, Bayesian error probabilities are fully conditioned on the observations. They do depend on the prior hypothesis probabilities however, and it is often interesting to look at the evidence against H_0 provided by the data alone. This can be done by computing the ratio of posterior odds to prior odds and is known as the Bayes factor:

$$B_{01}(x) \;=\; \frac{\pi(H_0 \,|\, x)/\pi(H_1 \,|\, x)}{\pi_0/\pi_1} \tag{7.37}$$

In the absence of unknown parameters, $B_{01}(x)$ is a likelihood ratio.

Often the distributions of X under H_0 and H_1 will depend on unknown parameters θ, so that posterior hypothesis probabilities and Bayes factors will involve marginalization integrals over θ:

$$\pi(H_0 \,|\, x) \;=\; \frac{\displaystyle\int p(x \,|\, \theta, H_0)\,\pi(\theta \,|\, H_0)\,\pi_0\,d\theta}{\displaystyle\int \left[p(x \,|\, \theta, H_0)\,\pi(\theta \,|\, H_0)\,\pi_0 \;+\; p(x \,|\, \theta, H_1)\,\pi(\theta \,|\, H_1)\,\pi_1 \right] d\theta} \tag{7.38}$$

$$\text{and:} \quad B_{01}(x) \;=\; \frac{\displaystyle\int p(x \,|\, \theta, H_0)\,\pi(\theta \,|\, H_0)\,d\theta}{\displaystyle\int p(x \,|\, \theta, H_1)\,\pi(\theta \,|\, H_1)\,d\theta} \tag{7.39}$$

Suppose now that we are testing $H_0 : \theta = \theta_0$ versus $H_1 : \theta > \theta_0$. Then:

$$B_{01}(x) \;=\; \frac{p(x \,|\, \theta_0)}{\displaystyle\int p(x \,|\, \theta, H_1)\,\pi(\theta \,|\, H_1)\,d\theta} \;\geq\; \frac{p(x \,|\, \theta_0)}{p(x \,|\, \hat{\theta}_1)} \;=\; \lambda, \tag{7.40}$$

Table 7.2.   Verbal description of standards of evidence provided by Bayes factors.

$2 \ln B_{10}$	B_{10}	Evidence against H_0
0 to 2	1 to 3	Not worth more than a bare mention
2 to 6	3 to 20	Positive
6 to 10	20 to 150	Strong
> 10	> 150	Very strong

where $\hat{\theta}_1$ maximizes $p(x \mid \theta, H_1)$. Thus, the ratio between the Bayes factor and the corresponding likelihood ratio is larger than unity. It is sometimes called the Ockham's razor penalty factor: it penalizes the evidence against H_0 for the introduction of an additional degree of freedom under H_1, namely θ.[25]

The smaller B_{01}, or equivalently, the larger $B_{10} \equiv 1/B_{01}$, the stronger the evidence against H_0. A rough descriptive statement of standards of evidence provided by Bayes factors against a hypothesis is given in Table 7.2.[26] There is at present not much experience with Bayes factors in high energy physics.

For a hypothesis of the form $H_0 : \theta = \theta_0$ versus $H_1 : \theta \neq \theta_0$, a Bayesian test can be based directly on the posterior distribution of θ. First calculate an interval for θ, containing an integrated posterior probability β. Then, if θ_0 is outside that interval, reject H_0 at the $\alpha = 1 - \beta$ credibility level. An exact significance level can be obtained by finding the smallest α for which H_0 is rejected. There is a lot of freedom in the choice of posterior interval. A natural possibility is to construct a highest posterior density (HPD) interval. If the lack of parametrization invariance of HPD intervals is a problem, there are other choices (see section 7.4.4).

If the null hypothesis is $H_0 : \theta \leq \theta_0$, a valid approach is to calculate a lower limit θ_L on θ and exclude H_0 if $\theta_0 < \theta_L$. In this case the exact significance level is the posterior probability of $\theta \leq \theta_0$.

7.4. Interval Estimation

Suppose that we make an observation $X = x_{obs}$ from a distribution $f(x \mid \mu)$, where μ is a parameter of interest, and that we wish to make a statement about the location of the true value of μ, based on our observation x_{obs}. One possibility is to calculate a point estimate $\hat{\mu}$ of μ, for example via the maximum-likelihood method:

$$\hat{\mu} \;=\; \arg\max_{\mu} f(x_{obs}\,|\,\mu). \tag{7.41}$$

Although such a point estimate has its uses, it comes with no measure of how confident we can be that the true value of μ equals $\hat{\mu}$.

Bayesianism and Frequentism both address this problem by constructing an interval of μ values believed to contain the true value with some confidence. However, the interval construction method and the meaning of the associated confidence level are very different in the two paradigms.

On the one hand, frequentists construct an interval $[\mu_1, \mu_2]$ whose boundaries μ_1 and μ_2 are random variables that depend on X in such a way that if the measurement is repeated many times, a fraction γ of the produced intervals will cover the true μ; the fraction γ is called the confidence level or coverage of the interval construction.

On the other hand, Bayesians construct the posterior probability density of μ and choose two values μ_1 and μ_2 such that the integrated posterior probability between them equals a desired level γ, called credibility or Bayesian confidence level of the interval.

7.4.1. *Frequentist Intervals: the Neyman Construction*

The Neyman construction is the most general method available for constructing interval estimates that have a guaranteed frequentist interpretation. The principal steps of the construction are illustrated in Fig. 7.6 for the simplest case of a one-dimensional continuous observation X whose probability distribution depends on an unknown one-dimensional continuous parameter μ. The procedure can be described as follows:

Step 1: Make a graph of the parameter μ versus the data X, and plot the density distribution of X for several values of μ (plot a);

Step 2: For each value of μ, select an interval of X values that has a fixed integrated probability, for example 68% (plot b);

Step 3: Connect the interval boundaries across μ values (plot c);

Step 4: Drop the "scaffolding", keeping only the two lines drawn at step 3; these form a *confidence belt* that can be used to construct an interval $[\mu_1, \mu_2]$ for the true value of μ every time you make an observation x_{obs} of X (plot d).

To see why this procedure works, refer to Fig. 7.7. Suppose that $\mu^{\star}$ is the true value of μ. Then $\mathbb{P}(x_1 \leq X \leq x_2\,|\,\mu^{\star}) = 68\%$ by construction.

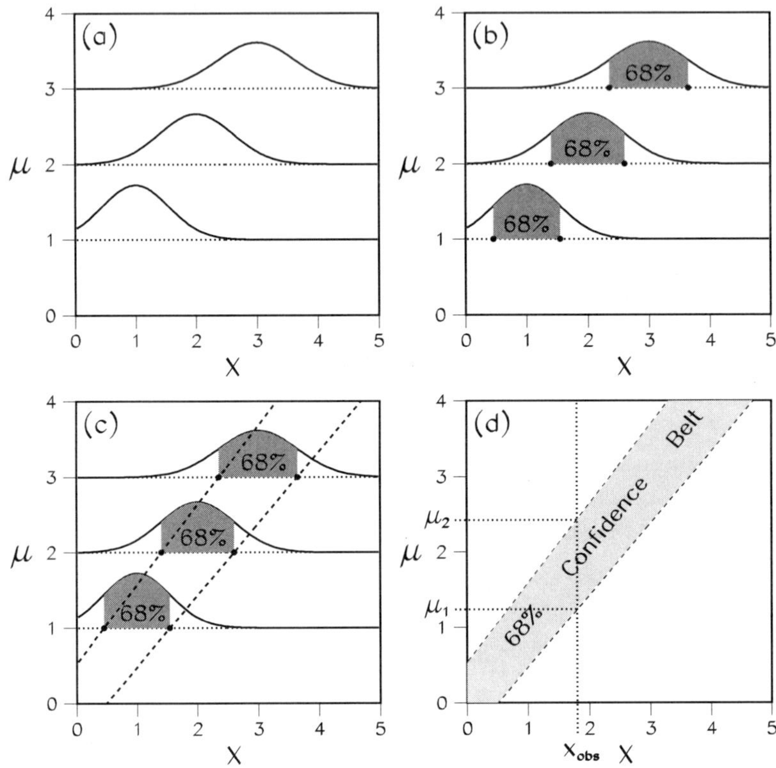

Fig. 7.6. Four steps in the Neyman construction of confidence intervals (see text).

Furthermore, for every $X \in [x_1, x_2]$, the reported μ interval will contain $\mu^\star$ and for every $X \notin [x_1, x_2]$, the reported μ interval will *not* contain $\mu^\star$. Therefore, the probability of covering $\mu^\star$ is exactly 68%, and this holds regardless of the value of $\mu^\star$. For problems with discrete statistics (such as Poisson event counts), the construction yields intervals that are conservative, i.e. which cover above the nominal level for some parameter values.

There are four basic ingredients in the Neyman construction: an estimator $\hat{\mu}$ of the parameter of interest μ, an ordering rule, a reference ensemble, and a confidence level. We now take a look at each of these individually.

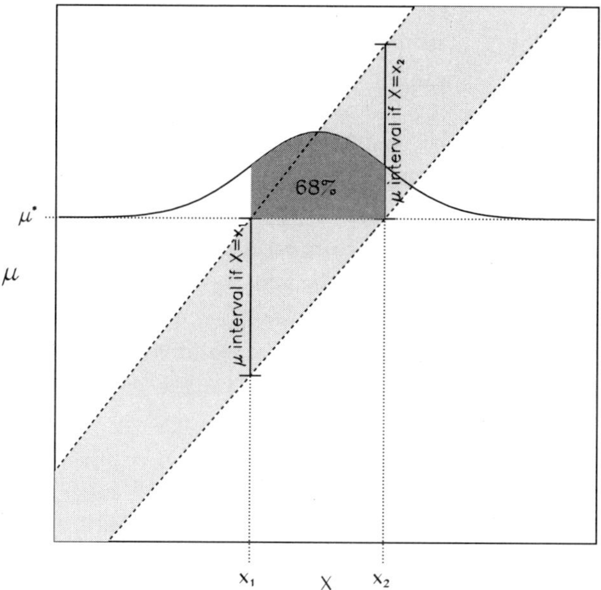

Fig. 7.7. Why the Neyman construction works (see text).

7.4.1.1. *Ingredient 1: the Estimator*

The estimator is the quantity plotted along the abscissae in the Neyman construction plot. Suppose for example that we collect n independent measurements x_i of the mean μ of a Gaussian distribution with known standard deviation. Then clearly we should use the average $\bar{x}$ of the x_i as an estimate of μ, since $\bar{x}$ is a sufficient statistic [†] for μ. On the other hand, if μ is constrained to be positive, then it would make sense to use either $\hat{\mu} = \bar{x}$ or $\hat{\mu} = \max\{0, \bar{x}\}$. These two estimators lead to intervals with very different properties. We will come back to this example in section 7.4.5.

[†]A statistic $T(X)$ is sufficient for μ if the conditional distribution of the sample X given the value of $T(X)$ does not depend on μ. In a sense, $T(X)$ captures all the information about μ contained in the sample.

7.4.1.2. *Ingredient 2: the Ordering Rule*

The ordering rule is the rule we use to decide which X values to include in the interval at step 2 of the construction. The only constraint on that interval is that it must contain 68% of the X distribution (or whatever confidence level is desired for the overall construction). For example, we could start with the X value that has the largest probability density and then keep adding values with lower and lower probability density until we cover 68% of the distribution. Another possibility is to start with $X = -\infty$ and add increasing values of X, again until we reach 68%. Of course, in order to obtain a smooth confidence belt at the end, we should choose the ordering rule consistently from one μ value to the next. In this sense it is better to formulate the ordering rule in terms of μ rather than X. This emphasizes the inferential meaning of the resulting intervals: an ordering rule is a rule that orders parameter values according to their perceived compatibility with the observed data. Here are some examples, all assuming that we have observed data x and are interested in a 68% confidence interval $[\mu_1, \mu_2]$ for a parameter μ whose maximum likelihood estimate is $\hat{\mu}(x)$:

- Central ordering
  $[\mu_1, \mu_2]$ is the set of μ values for which the observed data falls between the 16th and 84th percentiles of its distribution.
- Probability density ordering
  $[\mu_1, \mu_2]$ is the set of μ values for which the observed data falls within the 68% most probable region of its distribution.
- Likelihood ratio ordering
  $[\mu_1, \mu_2]$ is the set of μ values for which the observed data falls within a 68% probability region R, such that any point x inside R has a larger likelihood ratio $\mathcal{L}(\mu \,|\, x)/\mathcal{L}(\hat{\mu}(x) \,|\, x)$ than any point outside R.
- Upper limit ordering
  $]-\infty, \mu_2]$ is the set of μ values for which the observed data is at least as large as the 32nd percentile of its distribution.
- Minimal expected length
  This rule minimizes the average interval length $(\mu_2(X) - \mu_1(X))$ over the sample space.

7.4.1.3. *Ingredient 3: the Reference Ensemble*

This refers to the replications of a measurement that are used to calculate coverage. In order to specify these replications, one must decide which random and non-random aspects of the measurement are relevant to the inference of interest. When measuring the mass of a short-lived particle for example, it may be that its decay mode affects the measurement resolution. Should we then refer our measurement to an ensemble that includes all possible decay modes, or only the decay mode actually observed?

For simplicity assume that the estimator X of the mass μ is normal with mean μ and standard deviation σ, and that there is a $p = 50\%$ probability that the particle will decay hadronically, in which case $\sigma \equiv \sigma_h = 10$; otherwise the particle decays leptonically and $\sigma \equiv \sigma_\ell = 1$. As interval ordering rule we'll use minimal expected length. Since the decay mode is observable, one can proceed in two ways:

- Unconditional minimization;
  The reference ensemble includes all decay modes. We report $x \pm \delta_h$ if the decay is hadronic and $x \pm \delta_\ell$ if it is leptonic, where δ_h and δ_ℓ are constants that minimize the expected interval length, $2[p\delta_h + (1 - p)\delta_\ell]$, subject to the constraint of 68% coverage over the whole reference ensemble. Substituting the given numbers, this yields $\delta_h = 5.06$, $\delta_\ell = 2.20$, and an expected length of 7.26.
- Conditional minimization;
  The reference ensemble includes only the observed decay mode. We report $x \pm \sigma_h$ in the hadronic case and $x \pm \sigma_\ell$ in the leptonic one; the expected interval length is $2[p\sigma_h + (1 - p)\sigma_\ell] = 11.0$.

The expected interval length is quite a bit larger for the conditional method than for the unconditional one. If one were to repeat the measurement a large number of times, one would find that in the conditional analysis the coverage of the interval is 68% both within the subensemble of hadronic decays and within the subensemble of leptonic decays. On the other hand, in the unconditional analysis the coverage is 39% for hadronic decays and 97% for leptonic decays, correctly averaging to 68% over all decays combined. Qualitatively, by shifting some coverage probability from the hadronic decays to the higher precision leptonic ones, the unconditional construction is able to reduce the average interval length.

The above problem is an adaptation to high-energy physics of a famous example in the statistics literature,[27,28] used to discuss the merits of conditioning versus power (or interval length).

7.4.1.4. *Ingredient 4: the Confidence Level*

The confidence level labels a family of intervals; some conventional values
are 68%, 90%, and 95%. It is very important to remember that a confidence
level does *not* characterize single intervals; it only characterizes families of
intervals. The following example illustrates this.

Suppose we are interested in the mean μ of a Gaussian population with
unit variance. We have two observations, x and y, so that the maximum
likelihood estimate of μ is $\hat{\mu} = (x + y)/2$. Consider the following two
intervals for μ:

$$I_1 : \hat{\mu} \pm 1/\sqrt{2} \quad \text{and} \quad I_2 : \hat{\mu} \pm \sqrt{\max\{\, 0,\ 4.60 - (x - y)^2/4\}}$$

Both I_1 and I_2 are centered on the maximum likelihood estimate of μ.
Interval I_1 uses likelihood ratio ordering, is never empty, and has 68%
coverage. Interval I_2 uses probability density ordering, is empty whenever
$|x - y| \geq 4.29$, and has 99% coverage. Suppose next that we observe
$x = 10.00$ and $y = 14.05$. It is easy to verify that the corresponding I_1 and
I_2 intervals are numerically identical and equal to 12.03 ± 0.71. Thus, the
same numerical interval can have two very different coverages (confidence
levels), depending on which ensemble it is considered to belong to.

7.4.2. *Handling of Nuisance Parameters in the Neyman Construction*

In principle the Neyman construction can be performed when there is more
than one parameter; it simply becomes a multidimensional construction,
and the confidence belt becomes a "hyperbelt". If some parameters are
nuisances, they can be eliminated by projecting the final confidence region
onto the parameter(s) of interest at the end of the construction. This is a
difficult problem: the ordering rule has to be designed so as to minimize
the amount of overcoverage introduced by projecting.

There are simpler solutions. A popular one is to eliminate the nuisance
parameters ν from the data probability density function (pdf) first, by
integrating them over proper prior distributions:

$$f(x \,|\, \mu, \nu) \;\rightarrow\; \tilde{f}(x \,|\, \mu) \;\equiv\; \int f(x \,|\, \mu, \nu)\, \pi(\nu)\, d\nu \qquad (7.42)$$

This is a Bayesian step: the data pdf it yields depends only on the parame-
ter(s) of interest and can then be used in a standard Neyman construction.

Another possibility is to eliminate the nuisance parameters by profiling
the pdf. This is particularly useful if one has an independent measurement

y of ν, with pdf $g(y \,|\, \nu)$:

$$f(x \,|\, \mu, \nu) \;\rightarrow\; \check{f}(x \,|\, \mu) \;\propto\; \max_{\nu}\Big\{ f(x \,|\, \mu, \nu)\, g(y \,|\, \nu) \Big\} \tag{7.43}$$

The profiled pdf is then used in a Neyman construction.

Note that the coverage of the simpler solutions is not guaranteed! However, if necessary it is sometimes possible to "recalibrate" these methods in such a way that coverage *is* achieved. Recall the Neyman construction of a γ-level confidence interval:

$$C_\gamma(x_{obs}) \;=\; \Big\{ \mu : \; x_{obs} \in S_\gamma(\mu) \Big\}, \tag{7.44}$$

where $S_\gamma(\mu)$ is a subset of sample space that satisfies:

$$\mathbb{P}_{\mu,\nu}\Big[X \in S_\gamma(\mu) \Big] \;\geq\; \gamma \quad \text{for all } \mu \text{ and } \nu, \tag{7.45}$$

or equivalently:

$$\min_{\nu} \mathbb{P}_{\mu,\nu}\Big[X \in S_\gamma(\mu) \Big] \;\geq\; \gamma \quad \text{for all } \mu. \tag{7.46}$$

In the recalibrated profile likelihood method, one sets:

$$S_\gamma(\mu) \;=\; \Big\{ x : \; \lambda_\mu(x) \equiv \frac{\mathcal{L}(\mu, \hat{\nu}_\mu(x) \,|\, x)}{\mathcal{L}(\hat{\mu}(x), \hat{\nu}(x) \,|\, x)} \;\geq\; c_\gamma(\mu) \Big\}, \tag{7.47}$$

where $\hat{\nu}_\mu(x)$ maximizes $\mathcal{L}(\mu, \nu \,|\, x)$ for given μ and x, and $(\hat{\mu}(x), \hat{\nu}(x))$ maximizes $\mathcal{L}(\mu, \nu \,|\, x)$ for given x. For each μ one adjusts $c_\gamma(\mu)$ to satisfy (7.46).

7.4.3. *Other Frequentist Interval Construction Methods*

In practice, a popular method for constructing intervals is via *test inversion*. Suppose we are interested in some parameter $\theta \in \Theta$, and that for each allowed value θ_0 of θ we can construct an exact p value to test $H_0 : \theta = \theta_0$. We then have a family $\{p_\theta\}$ of p values indexed by the θ value of the corresponding test, and we can use this family to construct one- and two-sided γ confidence-level intervals for θ:

$$C_{1\gamma} \;=\; \Big\{ \theta : p_\theta \geq 1 - \gamma \Big\} \quad \text{and} \quad C_{2\gamma} \;=\; \Big\{ \theta : \frac{1 - \gamma}{2} \leq p_\theta \leq \frac{1 + \gamma}{2} \Big\}. \tag{7.48}$$

To describe the one-sided construction for example, one would say that a γ confidence limit for θ is obtained by collecting all the θ values that are not rejected at the $1 - \gamma$ significance level by the p value test. Indeed:

$$\mathbb{P}[\theta_{\text{true}} \in C_{1\gamma}] \;=\; \mathbb{P}[p_{\theta_{\text{true}}} \geq 1 - \gamma] \;=\; 1 - \mathbb{P}[p_{\theta_{\text{true}}} < 1 - \gamma]$$

$$= \; 1 - (1 - \gamma) \;=\; \gamma, \tag{7.49}$$

where the first equality follows from the definition of $C_{1\gamma}$ and the third one from the uniformity of p values under the tested hypothesis.

In general one can expect the properties of a family of p values to be reflected in the properties of the resulting family of intervals. A conservative p value will lead to conservative intervals, and a powerful p value will result in short intervals.

Another popular interval construction method is based on the likelihood function.[29] In one dimension, an approximate 68% confidence interval can be obtained by collecting all the parameter values for which the log-likelihood is within half a unit from its maximum. The validity of this approximation tends to increase with sample size.

Finally, another method explored by statisticians is based on objective Bayesian ideas. Objective priors can be designed in such a way that the resulting posterior intervals have a frequentist coverage that matches their Bayesian credibility to some order in $1/\sqrt{n}$, n being the sample size. When there are no nuisance parameters and the parameter of interest is one-dimensional, the matching prior to $\mathcal{O}(1/n)$ for one-sided intervals is Jeffreys' prior (7.5). Results are harder to come by in higher dimensions, but it is believed that reference analysis offers the best hope.[13] A major advantage of this approach is that it automatically yields intervals with Bayesian credibility, meaning intervals that are relevant for the actually observed data.

7.4.4. *Bayesian Interval Constructions*

As emphasized in section 7.2.2, the output of a Bayesian analysis is *always* the complete posterior distribution for the parameter(s) of interest. However, it is often useful to summarize the posterior by quoting a region with a given probability content. Such a region can be an interval or a union of intervals. Several schemes are available:

- Highest probability density regions;
  Any parameter value inside such a region has a higher posterior probability density than any parameter value outside the region, guaranteeing that the region will have the smallest possible length (or volume). Unfortunately this construction is not invariant under reparametrizations, and there are examples where this lack of invariance results in zero coverage for a subset of parameter values (of course this would only be of concern to a frequentist or an objective Bayesian).
- Central intervals;

These are intervals that are symmetric around the median of the posterior distribution. For example, a 68% central interval extends from the 16th to the 84th percentiles. Central intervals are parametrization invariant, but they can only be defined for one-dimensional parameters. Furthermore, if a parameter is constrained to be non-negative, a central interval will usually not include the value zero; this may be problematic if zero is a value of special physical significance.

- Upper and lower limits;
  For one-dimensional posterior distributions, these one-sided intervals can be defined using percentiles.

- Likelihood regions;
  These are standard likelihood regions where the likelihood ratio between the region boundary and the likelihood maximum is adjusted to obtain the desired posterior credibility. Such regions are metric independent and robust with respect to the choice of prior. In one-dimensional problems with physical boundaries and unimodal likelihoods, this construction yields intervals that smoothly transition from one-sided to two-sided.

- Intrinsic credible regions;
  These are regions of parameter values with minimum reference posterior expected loss[30] (a concept from Bayesian reference analysis).

High energy physicists using Bayesian procedures are generally advised to check the sensitivity of their result to the choice of prior, and its behavior under repeated sampling (coverage).

7.4.5. *Examples of Interval Constructions*

The effect of a physical boundary on frequentist and Bayesian interval constructions is illustrated in Figures 7.8 and 7.9 for the measurement of the mean μ of a Gaussian with unit standard deviation. The mean μ is assumed to be positive. All intervals are based on a single observation x. In general intervals have many properties that are worth studying: here we only examine the Bayesian credibility of frequentist constructions and the frequentist coverage of Bayesian constructions.

Figure 7.8 shows only frequentist constructions; Feldman-Cousins intervals[31] use x as estimator of μ and are based on a likelihood ratio ordering rule, whereas Mandelkern-Schultz intervals[32] use $\max\{0, x\}$ as estimator of μ and are based on a central ordering rule. The central, Feldman-Cousins, and upper limit confidence sets have very low credibility when the observation X is a large negative number. Mandelkern-Schultz intervals avoid this problem by reporting the same result for any negative X as for zero X, resulting in excess credibility at negative X.

Figure 7.9 shows central, highest posterior density, intrinsic, and upper limit Bayesian constructions, using Jeffreys' rule as prior for μ. They generally have good frequentist coverage, except near $\mu = 0$, where the curves for central and intrinsic intervals dip to zero.

Note how frequentist coverage and Bayesian credibility always agree with each other when one is far enough from the physical boundary.

7.5. Search Procedures

Search procedures combine techniques from hypothesis testing and interval construction. The basic idea is to test a hypothesis about a new physics model, and then characterize the result of the test by computing point and interval estimates. We discuss both the frequentist and Bayesian approaches to this problem.

7.5.1. *Frequentist Search Procedures*

The standard frequentist procedure to search for new physics processes is as follows:

(1) Calculate a p value to test the null hypothesis that the data were generated by standard model processes alone.
(2) If $p \leq \alpha_1$ claim discovery and calculate a two-sided, α_2 confidence level interval on the production cross section of the new process.
(3) If $p > \alpha_1$ calculate an α_3 confidence level upper limit on the production cross section of the new process.

Typical confidence levels are $\alpha_1 = 2.9 \times 10^{-7}$, $\alpha_2 = 0.68$, and $\alpha_3 = 0.95$.

There are a couple of issues regarding this procedure. The first one is coverage: since the procedure involves one p value and two confidence intervals, an immediate question concerns the proper frequentist reference ensemble for each of these objects. The second issue arises when one fails to claim a discovery and calculates an upper limit. The stated purpose of this limit is to exclude cross sections that the experiment is sensitive to and did not detect. How then does one avoid excluding cross sections that the experiment is *not* sensitive to? We take a closer look at these two issues in the following subsections.

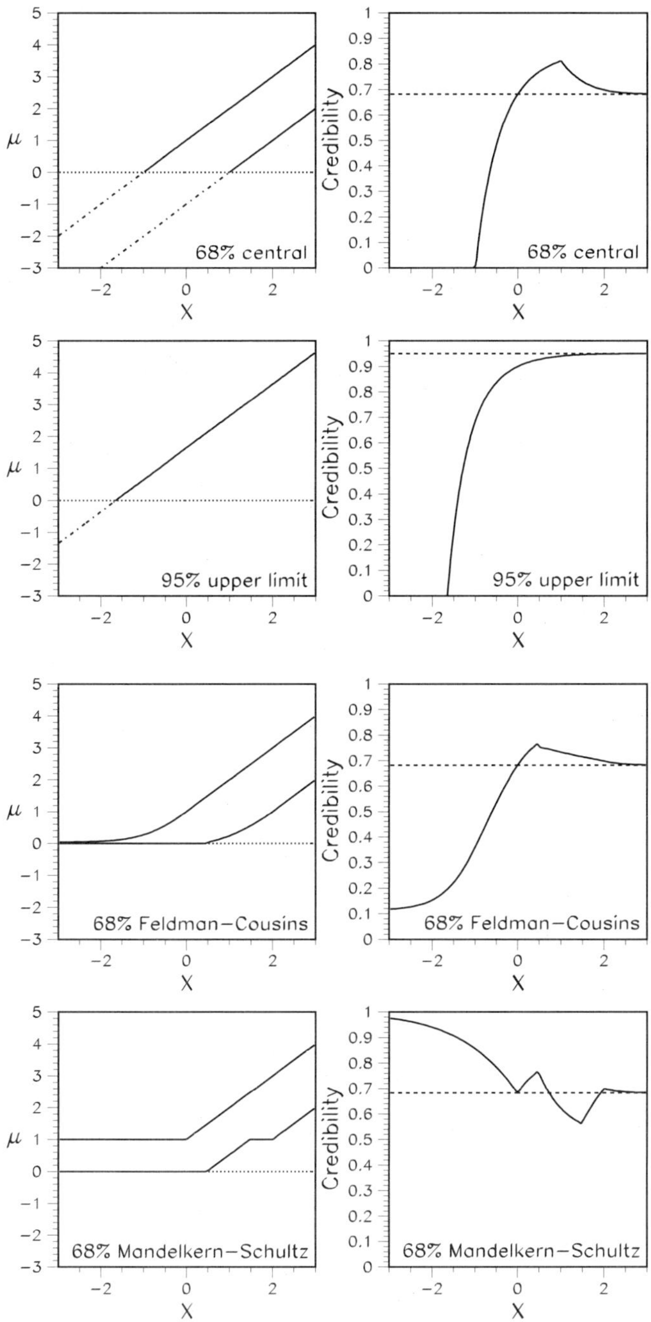

Fig. 7.8. Frequentist interval constructions. Left: graphs of μ versus X. Right: Bayesian credibility levels based on Jeffreys' prior; dashed lines indicate the frequentist coverage.

  L. Demortier

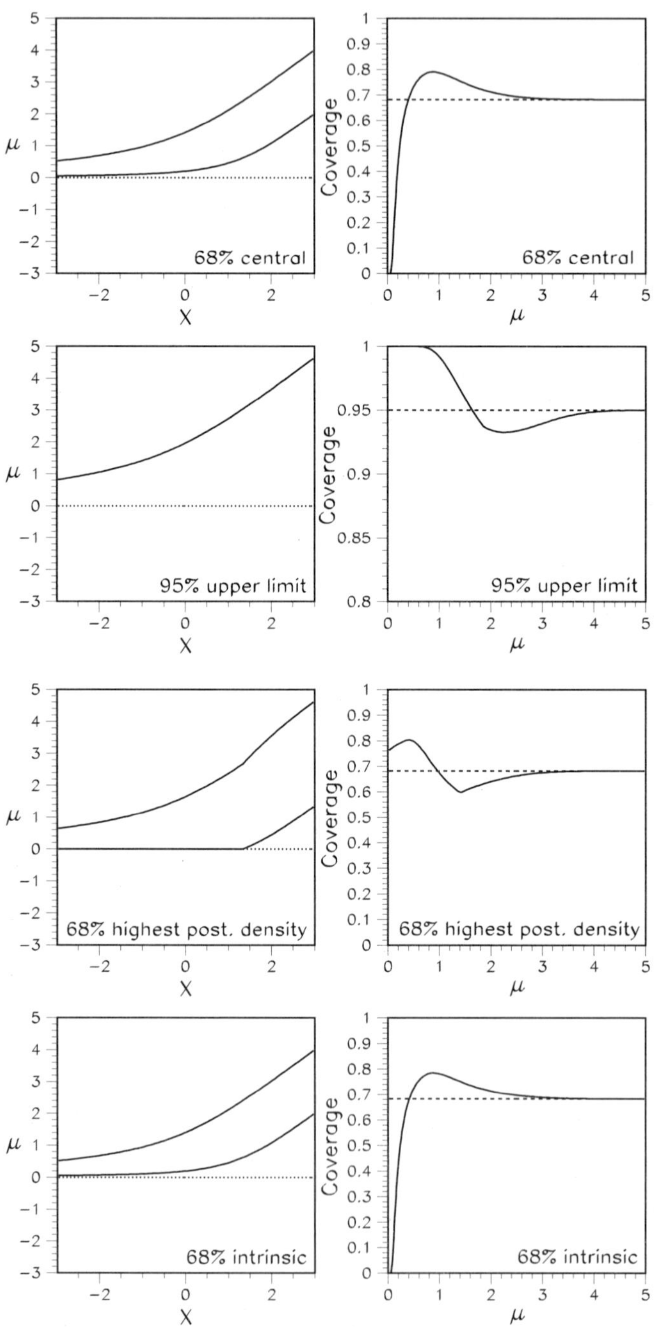

Fig. 7.9. Bayesian interval constructions. Left: graphs of μ versus X. Right: frequentist coverage levels; dashed lines indicate the Bayesian credibility.

7.5.1.1. *The Coverage Issue*

In a 1998 paper on frequentist interval constructions,[31] Feldman and Cousins characterize as *flip-flopping* the procedure by which some experimenters decide whether to report an upper limit or a two-sided interval on a physics parameter that is constrained to be non-negative (such as a mass or a mean event rate). In flip-flopping, this decision is based on first inspecting the data; an upper limit is then reported if the data is less than 3σ away from the physical boundary, and a two-sided interval otherwise. Because the initial data inspection is not taken into account in constructing the intervals, the flip-flopping procedure undercovers and is therefore invalid from a frequentist point of view.

In the frequentist search procedure just described, the decision to calculate a two-sided interval or an upper limit is based on the significance of the observed data with respect to the null hypothesis, a clear case of flip-flopping. One way to solve this problem would be to construct a Feldman-Cousins interval, since the latter transitions automatically from a one-sided to a two-sided interval as the data exhibits increasing evidence against the null hypothesis (see Fig. 7.8, top right). Unfortunately the Feldman-Cousins construction requires $\alpha_1 = \alpha_2 = \alpha_3$; this is unsatisfactory because it leads either to intervals that are too wide or test levels that are too low.

Another possibility is to construct *conditional* frequentist intervals: If $p \leq \alpha_1$, calculate a two-sided α_2 confidence level interval conditional on the observation that $p \leq \alpha_1$; otherwise, calculate an α_3 confidence level upper limit conditional on the observation that $p > \alpha_1$. What this means practically, in terms of the Neyman construction of each interval, is that the estimator X along the horizontal axis must be constrained to live within the region of sample space selected by the test, i.e. $p \leq \alpha_1$ or $p > \alpha_1$. The distribution of X must be appropriately truncated and renormalized in each case. An example of such a construction is shown in Fig. 7.10, for a simple search that involves testing whether the mean μ of a Gaussian distribution is zero (the null hypothesis) or greater than zero (the alternative). The data consists of a single sample X from that Gaussian, and can be negative or positive. The Gaussian width is assumed known. The plot shows that the conditional upper limit diverges as the discovery threshold is approached from the left, indicating that, so close to discovery, it becomes impossible to exclude *any* non-zero value of μ. On the other hand, as the threshold is approached from the right, the conditional two-sided interval turns into

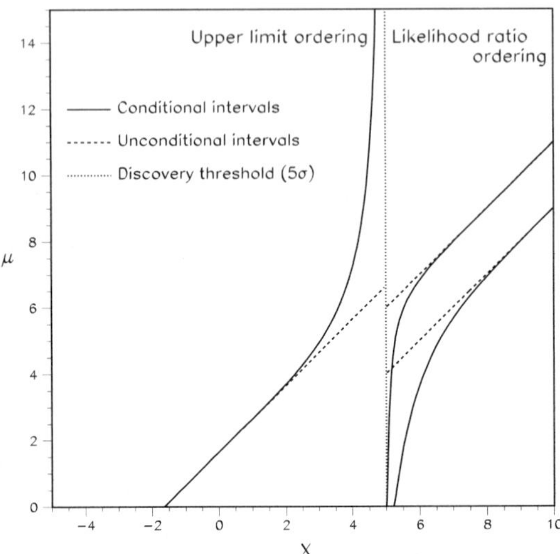

Fig. 7.10. Neyman construction of conditional intervals (solid lines) for the positive mean μ of a Gaussian, after having tested at the 5σ level whether $\mu = 0$. On the left of the discovery threshold, a 95% confidence level upper limit is shown, and on the right a 68% confidence level interval. Dashed lines indicate the corresponding unconditional intervals.

an upper limit, indicating that, so close to failing to make a discovery, it is possible that the true value of μ is zero and that the observed effect is just a background fluctuation. Note that a likelihood-ratio ordering rule was used here, in order to avoid creating a region of X values for which the reported μ interval is empty. For a central ordering rule for example, a small such region appears just above the discovery threshold.

In general physicists using the frequentist search procedure do not bother with the conditional construction. It is a more complicated calculation, and in any case its results coincide with those of the unconditional construction if one is far enough from the rejection threshold. Presumably one could argue that eventually, as more data are collected, one *will* be far enough.

So far our discussion of frequentist search procedures is based on a strict error-rate interpretation of measurement results. An alternative approach, not widely known in HEP, is to adopt an evidential interpretation.[33] This approach is centered around the p value, and the reported intervals serve to quantify the actual severity with which the hypothesis test has probed deviations from the null hypothesis. Suppose for example that we are testing $H_0 : \mu = \mu_0$ versus $H_1 : \mu > \mu_0$. If the p value against H_0 is not small, this is regarded as evidence that the true value of μ must be less than $\mu_0 + \delta$ for some δ. Thus one may examine the probability $\beta(\delta)$ of observing a worse fit of the data if the true value of μ is $\mu_0 + \delta$. If that probability is near one, the data are good evidence that $\mu < \mu_0 + \delta$. What physicists do in practice is to solve $\beta(\delta) = \alpha_3$ for δ, and report that all values of μ above $\mu_0 + \delta$ are excluded at the $1 - \alpha_3$ confidence level. A similar reasoning can be followed to justify the reporting of a two-sided interval for μ when the p value against H_0 is small.

7.5.1.2. *The Sensitivity Issue*

Suppose the result of a test of H_0 is that it cannot be rejected: we find $p_0 > \alpha_1$, where the subscript 0 on the p value emphasizes that it is calculated *under the null hypothesis*. A natural question is then: what values of the new physics cross section μ can we actually exclude? This is answered by calculating an α_3 C.L. upper limit on that cross section, and the easiest way to do this is by inverting a p value test: exclude all μ values for which $p_1(\mu) \leq 1 - \alpha_3$, where $p_1(\mu)$ is the p value under the alternative hypothesis that μ is the true value.

If our measurement has little or no sensitivity for a particular value of μ, this means that the distribution of the test statistic is (almost) the same under H_0 and H_1. In this case $p_0 \sim 1 - p_1$, and under H_0 we have:

$$\mathbb{P}_0(p_1 \leq 1 - \alpha_3) \sim \mathbb{P}_0(1 - p_0 \leq 1 - \alpha_3) = \mathbb{P}_0(p_0 \geq \alpha_3)$$
$$= 1 - \mathbb{P}_0(p_0 < \alpha_3) = 1 - \alpha_3. \quad (7.50)$$

For example, if we calculate a 95% C.L. upper limit, there will be a $\sim 5\%$ probability that we will be able to exclude μ values for which we have no sensitivity. Some experimentalists consider that 5% is too much; to avoid this problem they only exclude μ values for which

$$\frac{p_1(\mu)}{1 - p_0} \leq 1 - \alpha_3. \quad (7.51)$$

For historical reasons, the ratio of p values on the left-hand side is known as CL_s. The resulting upper limit procedure *over*covers.

It is often useful to examine plots of p_1 versus p_0 for a given experimental resolution.[34] If $F_i(x)$ is the cumulative distribution function of the test statistic X under H_i, then we have $p_1 = F_1(x)$ and $p_0 = 1 - F_0(x)$ (assuming that large values of X are evidence against H_0). Hence, $p_1 = F_1[F_0^{-1}(1 - p_0)]$. This is illustrated in Fig. 7.11 for the simple case where $F_i(x)$ is Gaussian with mean μ_i and known width σ. The horizontal dashed line

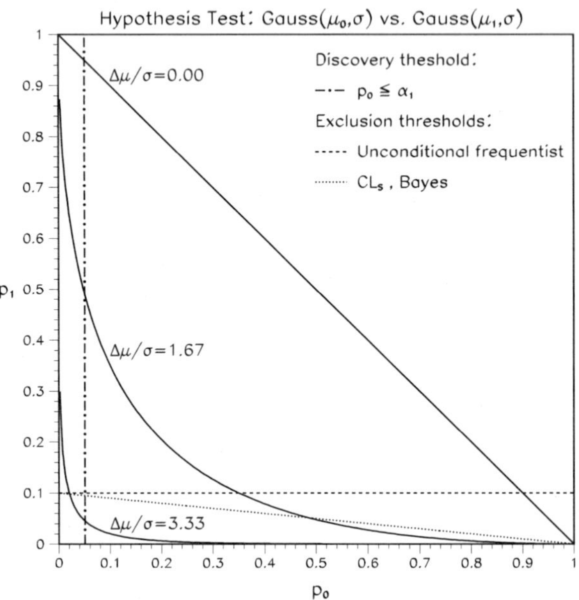

Fig. 7.11. Plot of p_1 versus p_0 in a test of $H_0 : \mu = \mu_0$ versus $H_1 : \mu = \mu_1$, where μ is the mean of a Gaussian of known width σ. The experimental resolution is $\Delta\mu/\sigma$, with $\Delta\mu = |\mu_1 - \mu_0|$.

in the plot is the standard frequentist exclusion threshold: any μ value for which $p_1(\mu)$ is below that line will be excluded at the α_3 confidence level. In the lower right-hand corner of the plot, one sees that even for experiments with no resolution ($\Delta\mu/\sigma = 0$) p_1 can dip below the horizontal line, leading to the rejection of some values of μ. This is avoided by the CLs procedure (7.51), represented by the slanted line of dots. Interestingly, Bayesian upper limits coincide with CLs limits for this problem. As the measurement resolution $\Delta\mu/\sigma$ increases, the corresponding p_1 versus p_0 contour approaches the lower left-hand corner of the plot, with the result that the probability of rejecting a false H_0 increases, and conversely, the probability of excluding a given μ value if H_0 is true also increases.

This last observation provides an interesting way to quantify *a priori* the sensitivity of a search procedure when the new physics model depends on a parameter μ, namely by reporting the set S of μ values for which

$$1 - \beta(\alpha_1, \mu) \geq \alpha_3, \tag{7.52}$$

where $\beta(\alpha_1, \mu)$ is the frequentist Type-II error rate corresponding to a discovery threshold α_1 and a value μ for the parameter under the alternative hypothesis. The set S has a couple of valuable interpretations:[35]

(1) If the true value of μ belongs to S, the probability of making a discovery is at least α_3, by definition of β.
(2) If the test does not result in discovery, it will be possible to exclude *at least* the entire sensitivity set with confidence α_3. Indeed, if we fail to reject H_0 at the α_1 level, then we can reject any μ in H_1 at the $\beta(\alpha_1, \mu)$ level, so that $p_1(\mu) \leq \beta(\alpha_1, \mu)$; furthermore, if $\mu \in S$, then $\beta(\alpha_1, \mu) \leq 1 - \alpha_3$ and therefore $p_1(\mu) \leq 1 - \alpha_3$, meaning that μ is excluded with confidence α_3.

In general the sensitivity set depends on the event selection and the choice of test statistic. Maximizing the size of the sensitivity set provides a criterion for optimizing the event selection and choice of test statistic. The appeal of this criterion is that it optimizes the result regardless of the outcome of the test.

7.5.2. *Bayesian Search Procedures*

The starting point of a Bayesian search is the calculation of a Bayes factor. For a test of the form $H_0 : \theta = \theta_0$ versus $H_1 : \theta > \theta_0$, this can be written as:

$$B_{01}(x) = \frac{p(x \mid \theta_0)}{\int p(x \mid \theta, H_1) \, \pi(\theta \mid H_1) \, d\theta}, \tag{7.53}$$

and points to an immediate problem: what is an appropriate prior $\pi(\theta \mid H_1)$ for θ under the alternative hypothesis? Ideally one would be able to elicit some kind of proper "consensus" prior representing scientific knowledge prior to the experiment. If this is not possible, one might want to use an "off the rack" objective prior, but such priors are typically *improper*, and therefore only defined up to a multiplicative constant, rendering the Bayes factor totally useless.

A possible objective solution is to use the so-called *intrinsic* or *expected posterior* prior construction:[25]

- Let $\pi^O(\theta)$ be a good estimation objective prior (for example a reference prior), and $\pi^O(\theta \,|\, x)$ the corresponding posterior.
- Then the intrinsic prior is

$$\pi^I(\theta) \equiv \int \pi^O(\theta \,|\, y)\, p(y \,|\, \theta_0)\, dy, \qquad (7.54)$$

where $p(y \,|\, \theta_0)$ is the pdf of the data under H_0. The dimension of y (the sample size) should be the smallest one for which the posterior $\pi^O(\theta \,|\, y)$ is well defined.

The idea is that if we were given separate data y, we would compute the posterior $\pi^O(\theta \,|\, y)$ and use it as a proper prior for the test. Since we are *not* given such data, we simply compute an average prior over all possible data.

In addition to the Bayes factor we need prior probabilities for the hypotheses themselves. An "objective" choice is the impartial $\pi(H_0) = \pi(H_1) = 1/2$. The posterior probability of H_0 is then

$$\pi(H_0 \,|\, x) \;=\; \frac{B_{01}}{1 + B_{01}}, \qquad (7.55)$$

and the complete outcome of the search is this probability $\pi(H_0 \,|\, x)$, plus the posterior distribution of θ under the alternative hypothesis, $\pi(\theta \,|\, x, H_1)$. Often it will be useful to summarize the posterior distribution of θ under H_1 by calculating an upper limit or a two-sided interval.

7.6. Systematic Uncertainties

Although we have mentioned systematic uncertainties in our treatment of nuisance parameters in section 7.3.3, they deserve some additional remarks in a separate section. To begin, systematic uncertainties should be distinguished from statistical uncertainties, which are due to random fluctuations resulting from the finite size of the data sample. Systematic uncertainties are associated with the measuring apparatus, assumptions made by the experimenter, and the model used to draw inferences. Whereas statistical uncertainties from different samples are independent, this is not usually the case with systematics, which tend to be correlated across samples.

One can distinguish three types of systematic uncertainties:[36]

(1) Systematics that can be constrained by ancillary measurements and can therefore be treated as statistical uncertainties. As example, consider the measurement of the mass of the top quark in a $t\bar{t}$ channel where at

least one top quark decays hadronically, i.e. $t \to Wb \to j_1 j_2 b$, where j_1 and j_2 are light-quark jets; since these come from the decay of the W, the known W mass can be used to constrain the jet energy scale.

(2) Systematics that cannot be constrained by existing data and are due to poorly understood features of the model used to draw inferences. Here, examples include background composition and shape, gluon radiation, higher-order corrections, and fragmentation parameters.

(3) Sources of uncertainty not easily modeled in a standard probabilistic setup, such as unknown experimenter bias.

In general a measurement result f is affected by several systematic uncertainties simultaneously. Assuming that these are all Type-2 systematics and that we adopt a Bayesian framework, we can find a prior $\pi(\mu, \nu, \ldots)$ for the corresponding nuisance parameters. The variance of f due to these uncertainties is then:

$$V[f] = \int \left[f(\mu, \nu, \ldots) - f(\mu_0, \nu_0, \ldots) \right]^2 \pi(\mu, \nu, \ldots) \, d\mu \, d\nu \ldots, \qquad (7.56)$$

where μ_0, ν_0, $\ldots$, are the nominal values of the nuisance parameters. In HEP however, we usually quantify the effect of these systematics on f by summing independent variations in quadrature:

$$\begin{aligned}
S^2 = {} & \left[f(\mu_0 + \sigma_\mu, \nu_0, \ldots) - f(\mu_0, \nu_0, \ldots) \right]^2 \\
& + \left[f(\mu_0, \nu_0 + \sigma_\nu, \ldots) - f(\mu_0, \nu_0, \ldots) \right]^2 + \ldots \qquad (7.57)
\end{aligned}$$

This procedure is called OFAT, for "One Factor At a Time", and only takes into account linear terms in the dependence of f on the nuisance parameters. This may be a mistake, as there often are quadratic (μ^2, ν^2, $\ldots$), mixed ($\mu\nu$), and even higher order terms that should be included in the calculation of the variance of f.

Techniques exist to estimate these higher-order effects by order of importance — this is called DOE, for "Design Of Experiments". The idea is to vary several systematics simultaneously instead of just one by one. DOE techniques are not much used in current experimental high energy physics. However, it is believed that these are valuable ideas that should be kept in mind as the complexity of data analyses continues to increase.[37,38]

References

1. The BABAR statistics committee web page: `http://www.slac.stanford.edu/BFROOT/www/Statistics`

2. The CDF statistics committee web page: `http://www-cdf.fnal.gov/physics/statistics/statistics_home.html`

3. The CMS statistics committee web page: `https://twiki.cern.ch/twiki/bin/view/CMS/StatisticsCommittee`

4. Workshop on *Confidence Limits*, (CERN, Geneva, Switzerland, 2000); `http://doc.cern.ch/cernrep/2000/2000-005/2000-005.html`.

5. Workshop on *Confidence Limits*, (Fermilab, Batavia, Illinois, 2000); `http://conferences.fnal.gov/cl2k/`.

6. Conference on *Advanced Statistical Techniques in Particle Physics*, (University of Durham, UK, 2002); `http://www.ippp.dur.ac.uk/old/Workshops/02/statistics/`.

7. Conference on *Statistical Problems in Particle Physics, Astrophysics and Cosmology (PHYSTAT2003)*, (Stanford Linear Accelerator Center, Stanford, California, 2003); `http://www-conf.slac.stanford.edu/phystat2003/`.

8. Conference on *Statistical Problems in Particle Physics, Astrophysics and Cosmology (PHYSTAT05)*, (Oxford, UK, 2005); `http://www.physics.ox.ac.uk/phystat05/index.htm`.

9. PHYSTAT LHC Workshop on *Statistical Issues for LHC Physics*, (CERN, Geneva, Switzerland, 2007); `http://phystat-lhc.web.cern.ch/phystat-lhc/`.

10. D. M. Appleby, "Probabilities are single-case, or nothing," arXiv:quant-ph/0408058v1 (8 Aug 2004).

11. C. M. Caves, C. A. Fuchs, and R. Schack, "Subjective probability and quantum certainty," arXiv:quant-ph/0608190v2 (26 Jan 2007).

12. E. T. Jaynes, "Probability theory: the logic of science," (ed. by G. L. Bretthorst), Cambridge University Press, 2003 (727pp).

13. J. M. Bernardo and A. F. M. Smith, "Bayesian theory," John Wiley & Sons, 1994 (586pp).

14. F. James, "Introduction and statement of the problem," in *Proceedings of the 1st workshop on confidence limits, 17–18 January 2000, CERN, Geneva, Switzerland*, L. Lyons, Y. Perrin, F. James, eds., CERN Yellow Report cernrep/2000-005.

15. C. M. Caves, C. A. Fuchs, and R. Schack, "Quantum probabilities as Bayesian probabilities," Phys. Rev. A **65**, 022305 (2002).

16. E. T. Jaynes, "Probability in quantum theory," `http://bayes.wustl.edu/etj/articles/prob.in.qm.pdf` (1990).

17. R. L. Jaffe, "The Casimir effect and the quantum vacuum," Phys. Rev. D **72**, 021301 (2005).

18. A. H. Rosenfeld, "Are there any far-out mesons or baryons?," in *Meson Spectroscopy. A collection of articles*, C. Baltay and A. H. Rosenfeld, eds., W.A. Benjamin, Inc., New York, Amsterdam, 1968, pg. 455.

19. S. Stepanyan *et al.* (CLAS Collaboration), "Observation of an exotic $S = +1$

baryon in exclusive photoproduction from the deuteron," Phys. Rev. Lett. **91**, 252001 (2003); B. McKinnon *et al.* (CLAS Collaboration), "Search for the Θ^+ pentaquark in the reaction $\gamma d \to p K^- K^+ n$," Phys. Rev. Lett. **96**, 212001 (2006).

20. A. Roodman, "Blind analysis in particle physics," in *Proceedings of the Phy-Stat2003 Conference*, SLAC, Stanford, California, September 8–11, 2003, pg. 166; also at arXiv:physics/0312102v1 (17 Dec 2003).

21. M. J. Bayarri and J. O. Berger, "P-values for composite null models [with discussion]," J. Amer. Statist. Assoc. **95**, 1127 (2000).

22. E. L. Lehmann, "On likelihood ratio tests," arXiv:math/0610835v1 [math.ST] (27 Oct 2006).

23. C. Goutis, G. Casella, and M. T. Wells, "Assessing evidence in multiple hypotheses," J. Amer. Statist. Assoc. **91**, 1268 (1996).

24. R. D. Cousins, "Annotated bibliography of some papers on combining significances or p-values," arXiv:0705.2209v1 [physics.data-an] (15 May 2007).

25. J. Berger, "A comparison of testing methodologies," CERN Yellow Report CERN-2008-001, pg. 8; see `http://phystat-lhc.web.cern.ch/phystat-lhc/proceedings.html`

26. R. E. Kass and A. E. Raftery, "Bayes factors," J. Amer. Statist. Assoc. **90**, 773 (1995).

27. D. R. Cox, "Some problems connected with statistical inference," Ann. Math. Statist. **29**, 357 (1958).

28. J. V. Bondar, Discussion of "Conditionally acceptable frequentist solutions," by G. Casella, in "Statistical decision theory and related topics IV," Vol. 1, S. S. Gupta and J. O. Berger, eds., Springer-Verlag 1988, pg. 91.

29. F. James, "Statistical methods in experimental physics," 2nd ed., World Scientific Publishing Co., 2006 (345pp).

30. J. Bernardo, "Intrinsic credible regions: an objective Bayesian approach to interval estimation," Test **14**, 317 (2005); see also `http://www.uv.es/~bernardo/2005Test.pdf`.

31. G. J. Feldman and R. D. Cousins, "Unified approach to the classical statistical analysis of small signals," Phys. Rev. D **57**, 3873 (1998).

32. M. Mandelkern and J. Schultz, "The statistical analysis of Gaussian and Poisson signals near physical boundaries," J. Math. Phys. **41**, 5701 (2000).

33. D. G. Mayo and D. R. Cox, "Frequentist statistics as a theory of inductive inference," IMS Lecture Notes — Monograph Series: 2nd Lehmann Symposium — Optimality, Vol. 49, pg. 77-97 (2006); arXiv:math/0610846v1 [math.ST] (27 Oct 2006); see also D. G. Mayo's comment on J. O. Berger, "Could Fisher, Jeffreys and Neyman have agreed on testing?," Statist. Science **18**, 1 (2003).

34. The suggestion to study plots of p_1 versus p_0 was made by Louis Lyons.

35. G. Punzi, "Sensitivity of searches for new signals and its optimization," in *Proceedings of the PHYSTAT2003 Conference*, SLAC, Stanford, California, September 8–11, 2003, pg. 79; also at arXiv:physics/0308063v2 (4 Dec 2003).

36. P. Sinervo, "Definition and treatment of systematic uncertainties in high energy physics and astrophysics," in *Proceedings of the PHYSTAT2003 Con-*

ference, SLAC, Stanford, California, September 8–11, 2003, pg. 122.
37. J. T. Linnemann, "A pitfall in evaluating systematic errors," CERN Yellow Report CERN-2008-001, pg. 94.
38. N. Reid, "Some aspects of design of experiments," CERN Yellow Report CERN-2008-001, pg. 99.

PART 3
Advanced Theoretical Topics

Chapter 8

Introduction to Supersymmetry and Supersymmetry Breaking

Yuri Shirman

Department of Physics and Astronomy
University of California
Irvine, CA 92697-4575

These lectures, presented at TASI 08 school, provide an introduction to supersymmetry and supersymmetry breaking. We present basic formalism of supersymmetry, supersymmetric non-renormalization theorems, and summarize non-perturbative dynamics of supersymmetric QCD. We then turn to discussion of tree level, non-perturbative, and metastable supersymmetry breaking. We introduce Minimal Supersymmetric Standard Model and discuss soft parameters in the Lagrangian. Finally we discuss several mechanisms for communicating the supersymmetry breaking between the hidden and visible sectors.

8.1. Introduction

8.1.1. *Motivation*

In this series of lectures we will consider introductory topics in the study of supersymmetry (SUSY) and supersymmetry breaking. There are many motivations which make SUSY a worthwhile subject of research. Since the topic of this TASI school is LHC, we will concentrate only on the motivation most closely related to physics at the TeV scale that will be probed by LHC experiments — the gauge hierarchy problem.

The Standard Model of particle physics is a consistent quantum field theory that may be valid up to energies as high as M_{Pl}. On the other hand, it is also characterized by some intrinsic energy scales such as Λ_{QCD} and the scale of electroweak symmetry breaking (EWSB), $m_Z \sim 100\,\mathrm{TeV}$, both of which are much smaller than M_{Pl}. There is nothing disturbing about the smallness of the ratio $\Lambda_{\mathrm{QCD}}/M_{\mathrm{Pl}}$. Indeed, Λ_{QCD} is generated by dimensional transmutation from the dimensionless parameter g^2, QCD coupling constant. The value of $\Lambda_{\rm QCD}$ is exponentially sensitive to the value of the gauge coupling at the cutoff scale M, $\Lambda_{\rm QCD} \sim M \exp(-8\pi^2/b_o g^2(M))$, where b_o is a one loop β-function coefficient. A small Λ_{QCD} can be obtained with $\mathcal{O}(1)$ coupling at the cutoff scale, even when the cutoff is taken to be $M_{\rm Pl}$. On the other hand, Z mass is determined by the Higgs vacuum expectation value (vev). This poses a conceptual problem since quantum corrections generically make mass parameters in the scalar field Lagrangian as large as the cutoff scale of the theory. Let us illustrate this with a simple example. Consider a Yukawa model with a massless scalar field:

$$\mathcal{L} = \frac{1}{2}(\partial_\mu \varphi)^2 + i\bar{\psi}\gamma^\mu \partial_\mu \psi - \frac{\lambda}{4!}\varphi^4 - y\varphi\bar{\psi}\psi \,. \tag{8.1}$$

We can easily calculate renormalization of the scalar mass squared. At one loop order there are two contributions arising from the scalar and fermion loops shown in Fig. 8.1. Both contributions are individually quadratically divergent:

$$\begin{aligned}
-i\delta m^2\big|_{\text{scalar loop}} &\sim -i\frac{\lambda}{32\pi^2}M^2 \,, \\
-i\delta m^2\big|_{\text{fermion loop}} &\sim i\frac{4y^2}{16\pi^2}M^2 \,.
\end{aligned} \tag{8.2}$$

Generically (for $\lambda \neq y^2/8$) this leads to a quadratically divergent contribution to scalar mass squared, just one loop below the cutoff of the theory

$$m^2 \sim \frac{\lambda/2 - 4y^2}{16\pi^2}M^2 \,. \tag{8.3}$$

If our goal is to construct a low energy theory with a light scalar particle, $m \ll M$, we need to make sure this correction cancels the bare mass in the classical Lagrangian to a very high precision. This is equivalent to fine-tuning the Lagrangian parameters m_0, λ, and y. Even if we do so at one loop, two loop corrections will be quadratically divergent again. In general, adjusting coupling constants so that the leading contribution appears only at n-loop order simply suppresses the mass squared correction by a factor of the order $(1/4\pi)^n$ relative to the UV scale M (assuming order one coupling constants). The need for such a cancellation implies that the low energy physics is sensitive to arbitrarily high energy scales. The presence of additional heavy particles with masses of order M can modify λ and y and affect cancellations of quadratic divergencies that low energy theorist worked so hard to arrange. As a result the mass of our light scalar φ will sensitively depend on the physics at arbitrarily high energy scales. In the Standard Model a similar problem, usually referred to as a gauge

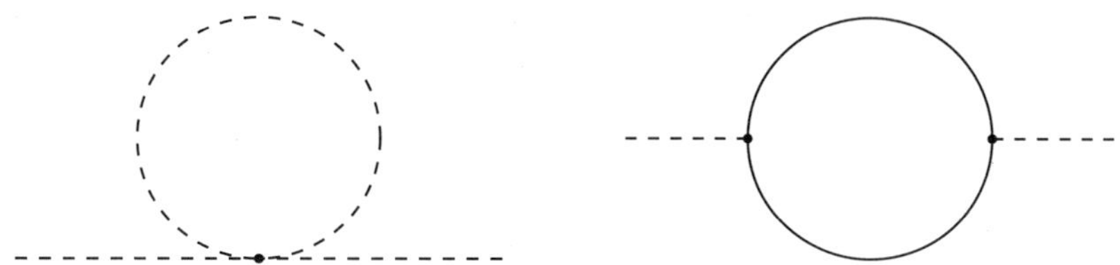

Fig. 8.1. One loop diagrams contributing to scalar mass squared in Yukawa theory.

hierarchy problem, requires an explanation of a hierarchy of some 17 orders of magnitude between the scale of electroweak symmetry breaking and the Planck scale.

It is useful to compare mass renormalization of a scalar field to that of a fermion. Theories of massless fermions possess a chiral symmetry which forbids mass terms. Thus mass terms can not be generated radiatively unless the symmetry is broken – for example by tree level masses. In such a case radiative corrections to fermion masses are proportional to their tree level values and can be at most logarithmically divergent. While the one loop contribution to the fermion mass is enhanced by large logs, it can remain small even when the cutoff scale is as large as $M_{\rm Pl}$. This mechanism does not explain the origin of a hierarchy between fermion masses and UV scales in the theory, but once introduced into the theory the hierarchy is not destabilized by radiative effects. Thus mass hierarchies in the fermion sector are at least *technically natural*. In fact, in some cases fermion masses arise dynamically. For example, proton and neutron masses are largely determined by strong QCD dynamics and naturally are of the order Λ_{QCD} which, in turn, can easily be small compared to the Planck scale. In this case, not only is the fermion mass stable against small changes in parameters of the theory, it is also naturally small — QCD dynamics explains the *origin* of the hierarchy between baryon masses and Planck scale.

We would like to find similar explanations for the origin and stability of the EWSB scale. In particular, we would like to find an extension of the Standard Model with new physics at the TeV. The presence of new fields and interactions would provide a cutoff for the Standard Model calculation of quantum corrections to the Higgs mass. If the scale of new physics is stable against radiative corrections, the technical naturalness problem would be resolved. We would further like to find a theory which also explains the origin of the hierarchy between TeV and Planck scales. Whether this situation is realized in nature or the specific value of electroweak scale is just an "accident" will eventually be answered by experimental data.

Over the years significant effort was invested into the investigation of possible solutions of the naturalness problem. Supersymmetry, the symmetry relating particles with different spins, emerged as one of the leading candidates for such a solution. If the Standard Model were supersymmetric, one could easily explain the stability of gauge hierarchy. Indeed, in a supersymmetric model the Higgs boson would be related by symmetry to a spin 1/2 particle (particles related by SUSY are called superpartners, and the superpartner of the Higgs boson is referred to as higgsino), and this symmetry would guarantee that the masses of bosonic and fermionic partners are equal. On the other hand, corrections to the fermion mass are at most logarithmically divergent — thus in a supersymmetric theory scalar particles may be naturally light. Of course, we have not yet observed even a single elementary scalar particle, let alone a particle with the same mass as the mass of any known fermion. Therefore at low energies supersymmetry may only be an approximate symmetry. The low energy effective Lagrangian must contain SUSY breaking terms. On the other hand, adding an arbitrary SUSY breaking term to the Lagrangian would be dangerous – one must take care not to reintroduce quadratic divergencies. Therefore, while supersymmetry can not be a symmetry of the ground state, it must remain a symmetry of the Lagrangian; it must be broken spontaneously rather than explicitly.

As we will see later, supersymmetry is extremely difficult to break. It can be broken either at tree level or, in some theories, non-perturbatively. In the latter case the scale of dynamical effects leading to supersymmetry breaking can naturally be low providing an explanation not only for the stability but also for the origin of gauge hierarchy.

This series of lectures is intended as a first introduction to supersymmetry and supersymmetry breaking. We begin by briefly reviewing Weyl fermions and introducing SUSY algebra. In section 8.2 we construct supersymmetric Lagrangians step by step starting with Wess-Zumino model and progressing to non-abelian SUSY gauge theories. In section 8.3 we discuss non-renormalization theorems which provide powerful tools in theoretical studies of SUSY. We also review non-perturbative dynamics of non-abelian SUSY gauge theories. In section 8.5 we discuss spontaneous and dynamical SUSY breaking. We then introduce Minimal Supersymmetric Standard Model in section 8.6. Finally, in section 8.7 we discuss interactions between MSSM sector and so called hidden sector where SUSY must be broken.

8.1.2. *Weyl fermions*

Let us briefly review the description of theories with fermions in a language of two component Weyl spinors. Consider the theory of a free Dirac fermion

$$\mathcal{L} = i\overline{\Psi}\partial_\mu\gamma^\mu\psi - m\overline{\Psi}\Psi.$$ (8.4)

It is convenient to work in a chiral basis where γ-matrices take the form

$$\gamma^\mu = \begin{pmatrix} 0 & \sigma^\mu \\ \bar{\sigma}^\mu & 0 \end{pmatrix}, \quad \gamma^5 = \begin{pmatrix} -1_2 & 0 \\ 0 & -1_2 \end{pmatrix},$$

$$\sigma^\mu = (1_2, \sigma^i), \quad \bar{\sigma}^\mu = (1_2, -\sigma^i),$$ (8.5)

and σ^i are Pauli matrices

$$\sigma^1 = \begin{pmatrix} 0 & 1 \\ 1 & 0 \end{pmatrix}, \quad \sigma^2 = \begin{pmatrix} 0 & -i \\ i & 0 \end{pmatrix}, \quad \sigma^3 = \begin{pmatrix} 1 & 0 \\ 0 & -1 \end{pmatrix}.$$ (8.6)

We can be explicit about left and right handed components of Ψ by writing

$$\Psi = \begin{pmatrix} \eta^\alpha \\ \bar{\chi}^*_{\dot\alpha} \end{pmatrix}.$$ (8.7)

Here η^α and $\bar{\chi}^*_{\dot\alpha}$ are left and right handed two-component fermions and $\alpha, \dot\alpha = 1, 2$. If Ψ has well defined transformations under some local or global symmetry, charges of η and $\bar{\chi}^*$ under this symmetry are identical. On the other hand, η and $\bar{\chi}$ are both left-handed but transform in conjugate representations under all the symmetries[a].

It is convenient to lower and raise spinor indices with ϵ-tensors $\epsilon^{\alpha\beta}$ and $\epsilon_{\dot\alpha\dot\beta}$. Lorentz scalars can be written as

$$\eta^\alpha\eta_\alpha = \epsilon^{\alpha\beta}\eta_\alpha\eta_\beta, \qquad \eta^*_{\dot\alpha}\eta^{*\dot\alpha} = \epsilon_{\dot\alpha\dot\beta}\eta^{*\dot\alpha}\eta^{*\dot\beta},$$
$$\eta\bar{\chi} = \bar{\chi}\eta = \epsilon^{\alpha\beta}\eta_\alpha\bar{\chi}_\beta, \quad \eta^*\bar{\chi}^* = \bar{\chi}^*\eta^* = \epsilon_{\dot\alpha\dot\beta}\eta^{\dot\alpha}\bar{\chi}^{\dot\beta}.$$ (8.8)

Left-handed and right-handed spinors can be combined into Lorentz vectors as

$$\eta^*_{\dot\alpha}\bar{\sigma}^{\mu\dot\alpha\alpha}\eta_\alpha = -\eta^\alpha\sigma^\mu_{\alpha\dot\alpha}\eta^{*\dot\alpha}.$$ (8.9)

We can now write the Lagrangian in terms of left-handed Weyl fermions η and $\bar{\chi}$

$$\mathcal{L} = i\eta^*\partial_\mu\bar{\sigma}^\mu\eta + i\bar{\chi}^*\partial_\mu\bar{\sigma}^\mu\bar{\chi} - m\bar{\chi}\eta - m\bar{\chi}^*\eta^*,$$ (8.10)

---

[a]In these lectures bar above the quantum field denotes a conjugate representation under (relevant) symmetry rather than complex conjugation. For example, χ and $\bar{\chi}$ represent two different fields with opposite charges, while χ^* and $\bar{\chi}^*$ are antiparticles of χ and $\bar{\chi}$ respectively.

where we have integrated by parts to obtain identical kinetic terms for η and $\bar{\chi}$. Clearly we can write down theories with both gauge and Yukawa interactions in terms of Weyl fermions. In fact Weyl fermions represent the most natural language for the description of chiral theories.

8.1.3. *A first look at supersymmetry*

Our goal is to construct a quantum field theory with the symmetry relating fermions and bosons. This symmetry is referred to as *supersymmetry*. The generator of the symmetry must relate two types of particles:

$$Q\,|\text{fermion}\rangle = |\text{boson}\rangle\,, \quad Q\,|\text{boson}\rangle = |\text{fermion}\rangle\,. \tag{8.11}$$

It follows that Q must be a spinor. Furthermore, in 4-dimensional spacetime the minimal spinor is a Weyl spinor and therefore the minimal supersymmetry has 4 supercharges.

In fact, under quite general assumptions there exist a unique non-trivial extension of the Poincare symmetry[3] which includes spinorial generators. This extension forms graded-Lie algebra[4] defined by the usual commutation relations of the Poincare symmetry together with the new anti-commutation relations:

$$\begin{aligned}
&\{Q_\alpha, \bar{Q}_{\dot{\beta}}\} = 2\sigma^\mu_{\alpha\dot{\beta}} P_\mu\,, \\
&\{Q_\alpha, Q_\beta\} = \{\bar{Q}_{\dot{\beta}}, \bar{Q}_{\dot{\beta}}\} = 0\,, \\
&[Q_\alpha, P_\mu] = [\bar{Q}_{\dot{\beta}}, P_\mu] = 0\,,
\end{aligned} \tag{8.12}$$

where P is the translation generator.

Even before we begin our study of supersymmetric Lagrangians the algebra (8.12) leads to important consequences. Taking the trace on both sides of the first equation in (8.12) we find

$$Q_1\bar{Q}_{\dot{1}} + \bar{Q}_{\dot{1}}Q_1 + Q_2\bar{Q}_{\dot{2}} + \bar{Q}_{\dot{2}}Q_2 = 4P^0\,. \tag{8.13}$$

Any state in the theory that is invariant under a symmetry is annihilated by symmetry generators. In particular, if the ground state $|0\rangle$ is supersymmetric, it is annihilated by SUSY generators. In general for an energy eigenstate we have

$$\langle E|\,Q_1\bar{Q}_{\dot{1}} + \bar{Q}_{\dot{1}}Q_1 + Q_2\bar{Q}_{\dot{2}} + \bar{Q}_{\dot{2}}Q_2\,|E\rangle = \langle E|\,4P^0\,|E\rangle = 4E\,. \tag{8.14}$$

We, therefore, conclude that the energy of a supersymmetric ground state must be zero. On the other hand, if supersymmetry is spontaneously broken, the vacuum energy is positive definite.

Our last statement is strictly correct only in the limit $M_{Pl} \to \infty$. In a theory with dynamical gravity supersymmetry must be a local symmetry. In this case energy of supersymmetric ground states (*i.e.* cosmological constant) is non-positive but not necessarily zero. This means that while ground states with a positive energy can not be supersymmetric, non-supersymmetric ground states do not necessarily have positive energy.

8.2. Constructing supersymmetric Lagrangians

8.2.1. *Wess-Zumino Model*

Our first goal is to construct the simplest possible supersymmetric field theory. Such a theory must contain at least one Weyl fermion, η, a minimal spinor in 4 dimensions. As its superpartner we need either a complex field, φ, or a vector field, A_μ. Let us use the complex scalar for our first example of supersymmetry. An on-shell Weyl fermion contains two degrees of freedom. Therefore, its spin 0 superpartner must have two degrees of freedom, *i.e.* it must be a complex scalar. However, an off-shell Weyl fermion has 4 degrees of freedom while a complex scalar still has two degrees of freedom. If we want to maintain explicit supersymmetry while performing calculations involving off-shell particles, number of bosonic and fermionic degrees of freedom must match both on-shell and off-shell. Thus we need to add an additional, *auxiliary*, complex bosonic field, F, which has 2 degrees of freedom off-shell and no on-shell degrees of freedom. This can be achieved by introducing a field with purely algebraic equations of motion, that is a field without kinetic terms. Let us try the simplest non-interacting Lagrangian with all the required degrees of freedom (this theory[5] is known as free Wess-Zumino model):

$$\mathcal{L} = \int d^4x |\partial_\mu \varphi|^2 + i\eta^* \partial_\mu \bar{\sigma}^\mu \eta + |F|^2 \,. \tag{8.15}$$

Note that the F equation of motion is indeed algebraic and sets $F = 0$ in the ground state of the theory. We now specify how supersymmetry transformations act on the fields:

$$\delta\varphi = \epsilon^\alpha \eta_\alpha \,, \qquad \delta\varphi^* = \epsilon^*_{\dot\alpha} \eta^{*\dot\alpha} \,,$$
$$\delta\eta_\alpha = -i\sigma^\mu_{\alpha\dot\alpha} \epsilon^{*\dot\alpha} \partial_\mu \varphi + \epsilon_\alpha F \,, \qquad \delta\eta^*_{\dot\alpha} = i\epsilon^\alpha \sigma^\mu_{\alpha\dot\alpha} \partial_\mu \varphi^* + \epsilon^*_{\dot\alpha} F^* \,, \tag{8.16}$$
$$\delta F = -i\epsilon^*_{\dot\alpha} \bar{\sigma}^{\mu\dot\alpha\alpha} \partial_\mu \eta \,, \qquad \delta F^* = i\partial_\mu \eta^* \bar{\sigma}^{\mu\dot\alpha\alpha} \epsilon_\alpha \,.$$

One can easily check that under (8.16) the Lagrangian shifts by a total derivative, equations of motion are unaffected, and the transformation

(8.16) is a symmetry of the theory.

A simple generalization of (8.15) leads to our first example of supersymmetry breaking. Since F transforms into a total derivative, the Lagrangian remains supersymmetric after the addition of the term

$$\Delta \mathcal{L} = \mu^2 F + h.c..\qquad (8.17)$$

On the other hand, solving the F-term equation of motion we find $F = -\mu^2$. Substituting this result into (8.16) we find that the ground state is not invariant and SUSY is spontaneously broken. We thus establish that F-term is an order parameter for SUSY breaking. This is also consistent with our earlier argument that the vacuum energy is such an order parameter because in our model $E = |F|^2$.

Finding SUSY breaking in a theory described by (8.15) and (8.17) may seem puzzling at first. Indeed, in a limit $M_{\rm Pl} \to \infty$ we are still dealing with a non-interacting theory which only differs from our original example by a non-observable shift in zero-point energy. Inclusion of gravity, however, would resolve the puzzle by turning both examples into interacting theories with SUSY broken spontaneously in the second model. In a locally supersymmetric theory, F-terms (and D-terms which will be introduced later) still play the role of order parameters for SUSY breaking. On the other hand, the vacuum energy of a non-supersymmetric ground state is arbitrary while vacuum energy of a supersymmetric ground state is non-positive.

Having constructed our first (if trivial) example of a supersymmetric field theory, we would like to proceed to study interacting models. Given supersymmetry transformations (8.16) this is relatively straightforward. Let us begin by adding an arbitrary interaction to the Lagrangian, for example, $-\varphi\eta\eta + h.c.$ term. Its variation under SUSY transformations is given by:

$$\delta(-\varphi\eta\eta) = 2i\varphi\eta\partial_\mu\varphi\sigma^\mu\epsilon - 2\varphi\eta\epsilon F\,,\qquad (8.18)$$

where we suppressed spinor indices. To cancel this variation, additional Lagrangian terms are required. A reasonable guess is an interaction term $F\varphi^2 + h.c.$:

$$\delta(F\varphi^2) = -i\epsilon^*\bar{\sigma}^\mu\partial_\mu\eta\varphi^2 + 2F\varphi\eta\epsilon\,.\qquad (8.19)$$

It is easy to see that the sum of the two variations is a total derivative and the theory is supersymmetric. We can now write down the simplest interacting SUSY theory, an interacting Wess-Zumino model

$$\mathcal{L} = |\partial_\mu\varphi|^2 + i\eta^\dagger\partial_\mu\bar{\sigma}^\mu\eta + |F|^2 + (\lambda F\varphi^2 - \lambda\varphi\eta\eta + h.c.)\,,\qquad (8.20)$$

where λ is a coupling constant.

8.2.2. *Superfield formalism*

It is possible to extend the procedure discussed in the previous section to more complicated theories but it becomes increasingly complicated, moreover some interactions can not appear in a supersymmetric Lagrangian. Therefore it is useful to introduce a new formalism which will allow us to treat all superpartners as a single field (or *superfield*). Scalars and fermions related by supersymmetry should simply correspond to different components of a single superfield very much like spin up and spin down states are different components of a single fermion. To arrive at the desired superfield formalism it is convenient to introduce the notion of the superspace by extending 4 commuting spacetime coordinates $\{x_\mu\}$ to 4 commuting and 4 anti-commuting coordinates $\{x_\mu, \theta^\alpha, \bar{\theta}_{\dot{\alpha}}\}$, where $\bar{\theta}_{\dot{\alpha}} = (\theta^\alpha)^*$. The new coordinates satisfy anti-commutation relations

$$\{\theta_\alpha, \bar{\theta}_{\dot{\beta}}\} = \{\theta_\alpha, \theta_\beta\} = \{\bar{\theta}_{\dot{\alpha}}, \bar{\theta}_{\dot{\beta}}\} = 0. \tag{8.21}$$

We can also define integrals over the superspace

$$\int d\theta = \int d\bar{\theta} = \int d\theta \bar{\theta} = \int d\bar{\theta}\theta = 0,$$

$$\int d\theta^\alpha \theta_\beta = \delta^\alpha_\beta, \qquad \int d\bar{\theta}_{\dot{\alpha}} \bar{\theta}^{\dot{\beta}} = \delta^{\dot{\beta}}_{\dot{\alpha}},$$

$$\int d^2\theta \theta^2 = \int d^2\bar{\theta}\bar{\theta}^2,$$

$$\int d^4\theta \theta^2 \bar{\theta}^2 = 1, \tag{8.22}$$

where

$$d^2\theta \equiv -\frac{1}{4}\epsilon_{\alpha\beta} d\theta^\alpha d\theta^\beta,$$

$$d^2\bar{\theta} \equiv -\frac{1}{4}\epsilon^{\dot{\alpha}\dot{\beta}} d\theta_{\dot{\alpha}} d\theta_{\dot{\beta}},$$

$$d^4\theta \equiv d^2\bar{\theta} d^2\theta. \tag{8.23}$$

Functions of superspace coordinates are quite simple — the expansion in power series terminates at order $\theta^2\bar{\theta}^2$. Furthermore, integration and differentiation over superspace coordinates lead to the same results.

We can now express any supermultiplet as a single *superfield* which depends on superspace coordinates. Expanding in the Taylor series we have for the most general scalar superfield (*i.e.* superfield whose lowest component is a scalar field)

$$\Phi(x_\mu, \theta, \bar{\theta}) = \varphi(x_\mu) + \theta\eta(x_\mu) + \bar{\theta}\chi^\dagger(x_\mu) + \bar{\theta}\bar{\sigma}^\mu\theta V_\mu(x_\mu)$$
$$+ \theta^2 F(x_\mu) + \bar{\theta}^2 \bar{F}(x_\mu) + \ldots + \theta^2\bar{\theta}^2 D(x_\mu). \tag{8.24}$$

While Φ depends on a finite number of component fields, it has many more components than is necessary to reproduce the simplest free supersymmetric theory described in section 8.2. It turns out that the most general superfield (8.24) gives a *reducible* representation of supersymmetry. To describe the Wess-Zumino model (8.20), we will construct an irreducible representation of SUSY by imposing additional conditions on Φ. To that end we will consider *chiral* and *antichiral* superfields Φ and $\Phi^\dagger$ respectively satisfying conditions

$$\bar{D}_{\dot{\alpha}}\Phi = 0, \qquad D_\alpha\Phi^\dagger = 0, \tag{8.25}$$

where

$$D_\alpha = \frac{\partial}{\partial\theta^\alpha} - i\sigma^\mu_{\alpha\dot{\alpha}}\bar{\theta}^{\dot{\alpha}}\frac{\partial}{\partial_\mu}, \qquad \bar{D}_{\dot{\alpha}} = -\frac{\partial}{\partial\bar{\theta}^{\dot{\alpha}}} + i\theta^\alpha\sigma^\mu_{\alpha\dot{\alpha}}\frac{\partial}{\partial_\mu}. \tag{8.26}$$

It is convenient to introduce new variables, $y^\mu = x^\mu + i\bar{\theta}\bar{\sigma}^\mu\theta$ and $y^{\mu\dagger} = x^\mu - i\bar{\theta}\bar{\sigma}^\mu\theta$. Note that $\bar{D}_{\dot{\alpha}}y^\mu = D_\alpha y^{\dagger\mu} = 0$. Therefore, a superfield defined by

$$\Phi(y^\mu) = \varphi(y^\mu) + \sqrt{2}\theta\eta(y^\mu) + \theta^2 F(y^\mu) \tag{8.27}$$

is chiral, $\bar{D}_{\dot{\alpha}}\Phi(y^\mu) = 0$, while its hermitian conjugate is antichiral. Expanding $\Phi(y^\mu)$ in powers of superspace coordinates, we find

$$\Phi = \varphi(x) - i\theta\sigma^\mu\bar{\theta}\partial_\mu\varphi(x) - \frac{1}{4}\theta^2\bar{\theta}^2\partial^2\varphi(x)$$
$$+ \sqrt{2}\theta\eta + \frac{i}{\sqrt{2}}\theta^2\partial_\mu\eta\sigma^\mu\bar{\theta} + \sqrt{2}\theta^2 F(x). \tag{8.28}$$

As we can see, chiral superfield only depends on three component fields and is a good candidate to describe our supersymmetric theory. To write supersymmetric Lagrangians using chiral superfields recall that F transforms into a total derivative and therefore all the F-terms in the Lagrangian are invariant under SUSY transformations. The supersymmetric Lagrangian term (8.17) can be written simply as

$$\Delta\mathcal{L} = \int d^2\theta\mu^2\Phi + \text{h.c.} = \mu^2 F(x) + \mu^{\dagger 2} F^\dagger(x). \tag{8.29}$$

It is easy to check that any analytic function of a chiral superfield, $W(\Phi)$, is also a chiral superfield and its θ^2 component of transforms into a total derivative. $W(\Phi)$, referred to as a superpotential, together with its hermitian conjugate $W(\Phi^\dagger)$ gives rise to supersymmetric interactions. After trivial generalization to a theory with several chiral superfields we can write:

$$\mathcal{L}_W = \int d^2\theta\, W(\Phi_i) + \text{h.c.}. \tag{8.30}$$

Superpotential allows us to introduce a broad set of supersymmetric interactions. However, $W|_{\theta^2}$ does not contain spacetime derivatives, and therefore, does not lead to kinetic terms. It turns out that a $\theta^2\bar{\theta}^2$ component of a real function of chiral superfields, a Kähler potential, is also invariant under the supersymmetry transformations. In fact the simplest Kähler potential gives rise to canonical kinetic terms:

$$\begin{aligned}
K &= \sum_i \Phi_i^\dagger \Phi_i\,, \\
\mathcal{L}_K &= \int d^4\theta\, K = \sum_i \left(|\partial_\mu \varphi_i|^2 + i\eta_i^* \partial_\mu \bar{\sigma}^\mu \eta_i + |F_i|^2 \right).
\end{aligned} \tag{8.31}$$

A general Kähler potential leads to more complicated terms in the action

$$\mathcal{L} = \int d^4\theta\, K \supset g^{ij} \left(\partial_\mu \varphi_i^* \partial^\mu \varphi_j + i\eta_i^* \bar{\sigma}^\mu \partial_\mu \eta_j + F_i^* F_j \right), \tag{8.32}$$

where $g^{ij} = \partial^2 K/(\partial \Phi_i^\dagger \partial \Phi_j)|_{\Phi=\varphi}$ is a Kähler metric which implicitly depends both on the fields and parameters of the theory. The Kähler metric determines the normalization of the kinetic terms and at a quantum level it contains information about wave-function renormalization.

We are now ready to write down a general form of the Lagrangian in an interacting theory of chiral superfields

$$\begin{aligned}
\mathcal{L} &= \int d^4\theta\, K(\Phi_i) + \int d^2\theta W(\Phi_i) + \int d^2\bar{\theta} W(\Phi_i^\dagger) \\
&= g^{ij} \left(\partial \varphi_i^* \partial \varphi_j + i\eta_i^* \partial_\mu \bar{\sigma}^\mu \eta_j + F_i^* F_j \right) \\
&\quad - \left(\frac{1}{2} \frac{\partial^2 W}{\partial \Phi_i \partial \Phi_j} \eta_i \eta_j - \frac{\partial W}{\partial \Phi_i} F_i + \text{h.c.} \right) + \ldots,
\end{aligned} \tag{8.33}$$

where dots represent possible higher order terms. By solving F-term equation of motion we arrive at the scalar potential of the theory

$$V = \frac{\partial \overline{W}}{\partial \Phi_i^\dagger} g_{ij} \frac{\partial W}{\partial \Phi_j}, \tag{8.34}$$

where $g_{ij} = (g^{ij})^{-1}$. We will generally assume that the Kähler potential is non-singular and therefore extrema of the superpotential correspond to supersymmetric ground states of the theory. However, even in the supersymmetric vacuum information about the spectrum of the theory requires knowledge of the Kähler potential.

As an explicit example of the superfield formalism let us write down Lagrangian of interacting Wess-Zumino model.[5] We will assume canonical Kähler potential (8.31) and the superpotential

$$W = \frac{m}{2}\Phi^2 + \frac{\lambda}{3}\Phi^3 \, . \tag{8.35}$$

Lagrangian in terms of component fields takes the form

$$\mathcal{L} = |\partial_\mu\varphi|^2 + i\eta^\dagger\partial_\mu\bar{\sigma}^\mu\eta + |F|^2 + \left(mF\varphi + \lambda F\varphi^2 - \frac{m}{2}\eta\eta - \lambda\varphi\eta\eta + \text{h.c.} \right) \, . \tag{8.36}$$

We now briefly discuss generalization to local supersymmetry or supergravity (SUGRA). In SUGRA the scalar potential becomes (neglecting D-terms that will appear in gauge theories):

$$V = \exp\left(\frac{K}{M_{\text{Pl}}} \right) \left((g^{ij}(D_iW)(D_jW)^* - \frac{3|W|^2}{M_{\text{Pl}}^2} \right) \, , \tag{8.37}$$

where D_i is a covariant supergravity derivative

$$D_iW = \partial_iW + K_iW/M_{\text{Pl}} \, . \tag{8.38}$$

The F-type order parameters for SUSY breaking now involve covariant derivatives $F_i = D_iW$. In supersymmetric vacua $D_iW = 0$ but as advertized earlier cosmological constant may be either zero or negative depending on the vev of W. In phenomenological applications, one is interested in vacua with zero cosmological constant and broken supersymmetry — this can always be achieved by shifting the superpotential by a constant, $W(\Phi_i) \rightarrow W(\Phi_i) + W_0$ and adjusting W_0 to cancel D- and F-term cotributions to vacuum energy.

Another important consequence of promoting supersymmetry to a local symmetry is the requirement that there exists a spin-3/2 superpartner of the graviton, gravitino. We will not write down the full gravitino Lagrangian carefully but will note one important term

$$\mathcal{L}_{\text{gravitino}} \supset e\exp\left(\frac{K}{2M_{\text{Pl}}^2} \right) \left(\frac{\overline{W}}{M_{\text{Pl}}^2}\psi_\mu\sigma^{\mu\nu}\psi_\nu + \frac{W}{M_{\text{Pl}}^2}\psi_\mu^\dagger\bar{\sigma}^{\mu\nu}\psi_\nu^\dagger \right) \, , \tag{8.39}$$

where e is a vierbein. We see that the gravitino is massive whenever the superpotential has non-vanishing vev. In particular, it is massive in supersymmetric vacua with negative cosmological constant — but this is actually required by SUSY in anti-de Sitter spacetime. Supersymmetric vacua in Minkowski spacetime imply vanishing $\langle 0|W|0 \rangle$ and a massless gravitino. Once supersymmetry is broken and the vev of the superpotential is tuned to obtain the flat background, gravitino mass is determined by SUSY breaking parameters. In particular, when SUSY is broken by an F-term vev the gravitino mass becomes

$$m_{3/2}^2 = e^{K/M_{\mathrm{Pl}}^2} \frac{F_i^* g^{ij} F_j}{3 M_{\mathrm{Pl}}^2} \,. \tag{8.40}$$

8.2.3. *Vector Superfield*

Our next task is to construct a Lagrangian of a supersymmetric gauge theory. A gauge field A_μ has 2 on-shell degrees of freedom and 3 off-shell degrees of freedom. We already know that a Weyl fermion has 2 on-shell degrees of freedom, and thus is a good candidate for being the superpartner of the gauge boson, gaugino. Off-shell, however, degrees of freedom do not match. Just as in the case of a chiral superfield, we need to introduce an auxiliary scalar field with one off-shell degree of freedom. This field, denoted by D, must be a real scalar without a kinetic term.

SUSY transformations for the components of the vector multiplet are given by

$$\delta A_\mu^a = -\frac{1}{\sqrt{2}} \left(\epsilon^\dagger \bar{\sigma}_\mu \lambda^a + \lambda^{\dagger a} \bar{\sigma}^\mu \epsilon \right) \,,$$

$$\delta \lambda_\alpha^a = -\frac{1}{2\sqrt{2}} \left(\sigma^\mu \bar{\sigma}^\nu \epsilon \right)_\alpha F_{\mu\nu}^a + \frac{1}{\sqrt{2}} \epsilon_\alpha D^a \,,$$

$$\delta \lambda_\alpha^{\dagger a} = -\frac{1}{2\sqrt{2}} \left(\epsilon^\dagger \bar{\sigma}^\mu \sigma^\nu \right)_{\dot\alpha} F_{\mu\nu}^a + \frac{1}{\sqrt{2}} F_{\mu\nu}^a + \frac{1}{\sqrt{2}} \epsilon_{\dot\alpha}^\dagger D^a \,, \tag{8.41}$$

$$\delta D^a = -\frac{i}{\sqrt{2}} \left(\epsilon^\dagger \bar{\sigma}^\mu D_\mu \lambda^a - D_\mu \lambda^{\dagger a} \bar{\sigma}^\mu \epsilon \right) \,.$$

To give a superfield description of a vector multiplet containing A_μ, λ, and D consider a real superfield

$$V = V^\dagger \,. \tag{8.42}$$

In components this superfield can be written as

$$V = \frac{1}{2}C + i\theta\chi + \frac{i}{2}\theta^2(M + iN) + \theta\sigma^\mu\bar\theta A_\mu$$
$$+ i\theta^2\bar\theta(\lambda^\dagger - \frac{i}{2}\bar\sigma^\mu\partial_\mu\chi) + \frac{1}{4}\theta^2\bar\theta^2(D(x) + \frac{1}{2}\Box C(x)) + \text{h.c.} .$$

(8.43)

The vector superfield contains more degrees of freedom than we hoped for. However, these additional degrees of freedom are auxiliary fields required by gauge invariance in a theory of the massless vector field that we wish to formulate. To see this, introduce a chiral superfield Λ:

$$\Lambda = \frac{\alpha(y_\mu) + i\beta(y_\mu)}{2} + \theta\frac{\xi(y_\mu)}{\sqrt{2}} + \frac{1}{2}\theta^2 f(y_\mu) .$$

(8.44)

If we shift the vector superfield according to

$$V \to V + i(\Lambda - \Lambda^\dagger) .$$

(8.45)

we find that its vector component is gauge transformed:

$$A_\mu \to A_\mu + \partial_\mu\alpha$$

(8.46)

and the shift by Λ represents a SUSY generalization of a regular gauge transformation. Other components of a vector superfield transform according to

$$C \to C - \beta ,$$
$$\chi \to \chi + \xi ,$$
$$M + iN \to M + iN + 2f ,$$
$$\lambda \to \lambda ,$$
$$D \to D .$$

(8.47)

We would like to maintain gauge invariance as an explicit symmetry of the Lagrangian, however we can use the remaining components of Λ to set all the auxiliary fields other than D in a vector multiplet to zero. This is equivalent to a gauge choice and is referred to as Wess-Zumino gauge. The Wess-Zumino gauge is very convenient in practice despite the fact that in this gauge SUSY is not fully manifest.

8.2.4. *Supersymmetric U(1) gauge theory*

To write down kinetic terms for the vector superfields we define a chiral spinor superfield

$$\mathcal{W}_\alpha = -\frac{1}{4}\bar D^2 D_\alpha V .$$

(8.48)

As usual, analytic functions of chiral superfields are themselves chiral superfields. Therefore, θ^2 component of $\mathcal{W}^\alpha \mathcal{W}_\alpha$ transforms into a total derivative. In additon $\mathcal{W}^\alpha \mathcal{W}_\alpha$ contains gauge kinetic terms allowing us to write supersymmetric Lagrangian for $U(1)$ theory as

$$\mathcal{L} = \left(\int d^2\theta \frac{1}{4g^2} \mathcal{W}^\alpha \mathcal{W}_\alpha + \int d^2\bar{\theta} \frac{1}{4g^2} \mathcal{W}^\dagger_{\dot\alpha} \mathcal{W}^{\dagger\dot\alpha} \right)$$
$$= \frac{1}{4g^2} F_{\mu\nu} F^{\mu\nu} + \frac{i}{g^2} \lambda^\dagger \partial_\mu \bar{\sigma}^\mu \lambda + \frac{1}{2g^2} D^2 \, . \tag{8.49}$$

Notice that we have chosen a non-canonical normalization for kinetic terms — as we will see later, this normalization is very convenient in SUSY gauge theories.

A chiral superfield of charge q transforms under gauge transformations according to

$$\Phi \to e^{-iq\Lambda} \Phi \, . \tag{8.50}$$

A Kähler potential of the form

$$K = K\left(\Phi^\dagger, e^{qV} \Phi \right) \tag{8.51}$$

is gauge invariant. In particular, canonical Kähler potential

$$K = \Phi^\dagger e^{qV} \Phi \tag{8.52}$$

contains regular gauge interactions as well as new interactions of matter fields and gauginos

$$-\sqrt{2} \left(\phi^* \eta \lambda + \lambda^\dagger \eta^\dagger \phi \right) \, . \tag{8.53}$$

Combining this with kinetic terms for vector superfield and introducing a general gauge invariant superpotential we can write the Lagrangian for a theory with several charged chiral multiplets

$$\mathcal{L} = \left(\int d^2\theta \frac{1}{4g^2} \mathcal{W}^\alpha \mathcal{W}_\alpha + \text{h.c.} \right) + \int d^4\theta \sum_i \Phi_i^\dagger e^{q_i V} \Phi_i^\dagger$$
$$+ \left(\int d^2\theta W(\Phi_i) + \int d^2\bar{\theta} \overline{W}(\Phi_i^\dagger) \right) \, . \tag{8.54}$$

Integrating over the superspace coordinates leads to the component Lagrangian. It is especially useful to look at D-terms:

$$\mathcal{L}_D = \frac{1}{2g^2} D^2 + D \left(\sum_i q_i |\varphi_i|^2 \right)^2 \, . \tag{8.55}$$

Integrating out D-term we find a new contribution to the scalar potential

$$V_D = \frac{g^2}{2}\left(\sum_i q_i |\varphi_i|^2\right)^2 . \tag{8.56}$$

It is easy to see that anomaly free $U(1)$ gauge theories necessarily have directions in the field space along which V_D vanishes. These directions are referred to as *D*-flat. As we will see shortly, some chiral non-abelian gauge theories do not have *D*-flat directions. Moreover, some of the D-flat directions may be lifted by the F-term potential V_F. However, it is quite generic in SUSY gauge theories that there exist directions in the field space along which both V_D and V_F vanish. Submanifold of the field space satisfying both *D*- and *F*-flatness conditions represents (classical) vacua in the theory and is often referred to as a (classical) *moduli space*. Field fluctuations along the moduli space are massless and are called moduli. In section 8.4 we will see that classical moduli space may be modified or completely lifted by non-perturbative dynamics, however, even then it plays a useful role in the analysis of dynamical properties of the theory.

Before moving on to a discussion of non-abelian gauge theories, we should consider one more supersymmetric and gauge invariant term, the Fayet-Illiopoulus D-term, that can arise only in an abelian case. According to (8.45) vector superfield V is not invariant under SUSY transformations nor are its components invariant under $U(1)$ gauge transformation. However, both D and λ are neutral under the gauge symmetry. Furthermore, the θ^4 component of the vector superfield is a scalar and shifts only by a total derivative. Therefore the following SUSY and gauge invariant term can be added to the Lagrangian

$$\int d^4\theta \xi^2 V . \tag{8.57}$$

Upon integrating out an auxiliary D-term, the D-term potential in the theory becomes

$$V_D = \frac{g^2}{2}\left(\sum_i q_i |\varphi_i|^2 + \xi^2\right)^2 . \tag{8.58}$$

8.2.5. *Non-abelian gauge theory*

It is easy to extend our discussion to the case of non-abelian gauge theories. Gauge transformation take the form

$$V \to e^{-i\Lambda} V e^{i\Lambda} , \tag{8.59}$$

where V transforms in an adjoint representation of the gauge group and $\Lambda = \Lambda^a T^a$.

The Lagrangian for a supersymmetric Yang-Mills theory with matter fields is a simple generalization of (8.54). In particular, the gauge kinetic terms become

$$\mathcal{L}_{\text{SYM}} = \int d^2\theta \frac{1}{2g^2} \text{Tr}\mathcal{W}^\alpha \mathcal{W}_\alpha + \text{h.c.} \, . \tag{8.60}$$

To discuss physics of the theory with matter fields, it is convenient to consider a specific example of an $SU(N)$ gauge theory with F pairs of chiral superfields in fundamental and anti-fundamental representations, Q and $\bar{Q}$ respectively. We will refer to this matter content as F flavors. Quantum numbers of the matter fields under local and non-anomalous global symmetries are given by

$$\begin{array}{c|c|cccc}
 & SU(N) & SU(F)_L & SU(F)_R & U(1)_B & U(1)_R \\
\hline
Q & N & F & 1 & 1 & \frac{N_f - N_c}{N_f} \\
\bar{Q} & \bar{N} & 1 & \bar{F} & -1 & \frac{N_f - N_c}{N_f}
\end{array} \tag{8.61}$$

The D-term potential has the form

$$V_D = \frac{g^2}{2} \left(q^\dagger T^a q - \bar{q} T^a \bar{q}^\dagger \right)^2 \tag{8.62}$$

and has many classical flat directions. It is possible to show that it vanishes when squark vevs satisfy the following condition[6,7]

$$q^{\dagger i f} q_{jf} - \bar{q}^{if} \bar{q}^\dagger_{jf} = \alpha \delta^i_j \, , \tag{8.63}$$

where i and f are color and flavor indices respectively and α is an arbitrary constant. In a theory with $F < N$ one can use gauge and global symmetry transformations to write the vevs in the form

$$Q = \bar{Q} = \begin{pmatrix} v_1 & & & 0 \\ & v_2 & & \\ & & \ddots & \\ & & & v_F \\ & & & \\ 0 & \cdots & & 0 \end{pmatrix} \, . \tag{8.64}$$

When $F \geq N$ flat directions can be parameterized by

$$
Q = \begin{pmatrix} v_1 & & & & 0 \\ & v_2 & & & \vdots \\ & & \ddots & & \\ 0 & \cdots & & v_F & 0 \end{pmatrix}, \quad
\bar{Q} = \begin{pmatrix} \bar{v}_1 & & & & 0 \\ & \bar{v}_2 & & & \vdots \\ & & \ddots & & \\ 0 & \cdots & & \bar{v}_F & 0 \end{pmatrix},
\tag{8.65}
$$

where $|v_i|^2 - |\bar{v}_i|^2 = \alpha$.

An equivalent description of the moduli space can be obtained if we note that all states related to (8.64) by global symmetry transformations may be parameterized in terms of gauge invariant composites

$$
M_{ff'} = Q_f \bar{Q}_{f'} .
\tag{8.66}
$$

We will refer to $M_{ff'}$ as mesons. Additional flat directions arising when $F \geq N$ can be described in terms of baryons and anti-baryons

$$
\begin{aligned}
B_{f_{N+1}\ldots f_F} &= \epsilon_{f_1\ldots f_F} \epsilon^{a_1\ldots a_n} Q_{a_1}^{f_1} \ldots Q_{a_N}^{f_N}, \\
\bar{B}_{f_{N+1}\ldots f_F} &= \epsilon_{f_1\ldots f_F} \epsilon^{a_1\ldots a_n} \bar{Q}_{a_1}^{f_1} \ldots \bar{Q}_{a_N}^{f_N} .
\end{aligned}
\tag{8.67}
$$

Not all of these gauge invariant composites are independent. When $F = N$ we have a classical relation

$$
\det M = B\bar{B} .
\tag{8.68}
$$

Similarly, for $F > N$

$$
\begin{aligned}
\det M \, M_{ff'}^{-1} &= B_f \bar{B}'_f , \\
B^f M_{ff'} &= M_{ff'} \bar{B}^{f'} = 0 .
\end{aligned}
\tag{8.69}
$$

8.3. Non-renormalization theorems

Clearly the presence of an additional symmetry must simplify calculation of quantum corrections to the Lagrangian. At the very least SUSY requires that counterterms for interactions related by symmetry are identical. For example, in the Wess-Zumino model (8.36) counterterms for both $F\varphi^2$ and $\varphi\eta\eta$ must be the same. Similarly, in SUSY gauge theories renormalization of the gauge coupling, gaugino-scalar-fermion coupling, and quartic scalar coupling in the D-term potential must be related. However, it turns out that SUSY imposes much more powerful constraints on supersymmetric Lagrangians. Namely, only Kähler potential terms are renormalized to all orders in perturbation theory. On the other hand, superpotential terms are not renormalized while gauge coupling and the Fayet-Illiopolous D-term are not renormalized beyond one loop order. One can prove these statements, known as non-renormalization theorems, order by order in perturbation theory.[8] Instead we will consider a much slicker derivation due to Seiberg[9] which uses the symmetries and holomorphy of SUSY Lagrangians. This derivation will also allow us to see how non-perturbative dynamics may affect low energy physics of SUSY gauge theories.

8.3.1. *R-symmetry*

An important role in our discussion will be played by an R-symmetry — a symmetry that rotates superspace coordinates θ by a phase, $\theta \to e^{i\alpha}\theta$. Defining the R-charge of θ to be $R_\theta = 1$ and using the equation (8.22) we find $R_{d\theta} = -1$. Lagrangian terms arising from the Kähler potential are always invariant under the R-symmetry since both $d^4\theta$ and the Kähler potential are real. On the other hand, the invariance of the full Lagrangian is neither guaranteed nor required. R-symmetry is only a symmetry of the Lagrangian if the superpotential transforms with the charge 2, $W \to e^{2i\alpha}W$. Imposing these transformation properties on the superpotential determines R-symmetry charges of the superfields. It is possible that a consistent assignment of R-charges does not exist and then R-symmetry is explicitly broken by some terms in the Lagrangian.

For example, consider an interacting massless Wess-Zumino model. If we assign an R-charge of 2/3 to Φ, the superpotential has R-charge 2 and the Lagrangian is invariant under R-symmetry. Alternatively, we can consider free theory with a non-vanishing mass. In this case an R-charge of Φ under R-symmetry must be 1. On the other hand, in a massive interactive Wess-Zumino model there is no charge assignment which leaves the Lagrangian invariant under R-symmetry.

R-symmetry is quite unusual. Unlike all other global symmetries, it acts on superspace coordinates θ and $\bar{\theta}$. As a result different components of the superfields transform differently under R-symmetry. Consider, for example, a chiral superfield with R-charge R. Its lowest component has the same R-charge as the superfield itself, its θ component has R-charge $R-1$ and its θ^2 component has R-charge $R-2$.

R-charges of matter fields may depend on the model under consideration. However, R-charges of the fields in a vector multiplet are uniquely fixed. Indeed $\mathcal{W}^\alpha \mathcal{W}_\alpha$ has R-charge 2. Since gaugino λ is the lowest component of $\mathcal{W}^\alpha$ its R-charge is 1 while the D-term and the gauge field A_μ are neutral (as should have been expected for real fields).

8.3.2. *Superpotential terms*

To prove non-renormalization theorems we will use all the symmetries of
the SUSY field theories, including an R-symmetry. Moreover, we will be
able to use R-symmetry even in models where it is explicitly broken by the
superpotential interactions. To this end we will promote the parameters
of the Lagrangian to background superfields.[9] Consider, for example, the
Wess-Zumino model (8.35). We will interpret this model as an effective
low energy description of a more fundamental theory in which parameters
m and λ arise as vacuum expectation values of heavy superfields. This
interpretation enhances apparent symmetries of the theory. The model
now has a $U(1) \times U(1)_R$ global symmetry which is spontaneously broken by
expectation values of the spurions m and λ. The charges of the dynamical
superfield Φ and spurions under the symmetries of the theory are given by

	$U(1)_R$	$U(1)$
Φ	1	1
m	0	-2
λ	-1	-3

$$(8.70)$$

In this approach superpotential must be described by a holomorphic
function of both the dynamical and background superfields. On the other
hand, the Kähler potential is still a general real function of superfields and
spurions:

$$
\begin{aligned}
K &= K(\Phi^\dagger, \Phi, m^\dagger, m, \lambda^\dagger, \lambda)\,, \\
W &= W(\Phi, m, \lambda)\,.
\end{aligned}
\tag{8.71}
$$

The requirement that the renormalized superpotential is holomorphic
and has correct transformation properties under global symmetries restricts
its form to be

$$
W = \frac{m}{2}\Phi^2 f\left(\frac{\lambda\Phi}{m}\right)\,.
\tag{8.72}
$$

In the weak coupling limit the effective superpotential should approach
the classical one and therefore there should exist a Taylor series expansion
of f in $\lambda\Phi/m$:

$$
W = \frac{m}{2}\Phi^2\left(1 + \frac{2}{3}\frac{\lambda\Phi}{m} + \mathcal{O}\left(\frac{\lambda^2\Phi^2}{m^2}\right)\right) = \frac{m}{2}\Phi^2 + \frac{\lambda}{3}\Phi^3 + \mathcal{O}\left(\lambda^2\right)\,.
\tag{8.73}
$$

Thus

$$
f\left(\frac{\lambda\Phi}{m}\right) = 1 + \frac{2}{3}\frac{\lambda\Phi}{m} + \mathcal{O}\left(\lambda^2\right)\,.
\tag{8.74}
$$

Furthermore, the $m \to 0$ limit must be regular, and therefore W should not contain negative powers of m. Thus (8.35) is exact.[9] No higher dimension terms are generated. This, in particular, means that there are no counterterms leading to renormalization of m or λ. To make the latter conclusion obvious, let's assume, for example, that a counterterm δ_m is generated at some order in the perturbation theory. Then $\delta_m \sim \lambda^n$ — but as we just argued no such powers of λ can appear in the superpotential.

The Kähler potential K, on the other hand, is real and can be a function of $|m|^2$ as well as of $|\lambda|^2$ both of which are invariant under all the symmetries. After renormalization Kähler potential becomes

$$K = \Phi^\dagger \Phi \to Z \left(\frac{\mu^\dagger \mu}{m^\dagger m}, \lambda^\dagger \lambda \right) \Phi^\dagger \Phi \,. \tag{8.75}$$

The renormalization of the Kähler potential implies that the physical coupling constants, m and λ, are renormalized in contrast to holomorphic parameters in the superpotential. Nevertheless an RG evolution of coupling constants is completely determined by wave function renormalization.

8.3.3. *Gauge coupling renormalization*

To discuss the renormalization of the gauge coupling constant in SUSY gauge theories we will once again promote the gauge coupling function to a background superfield whose lowest component is

$$\tau = \frac{8\pi^2}{g^2} + i\theta_{\rm YM} \,. \tag{8.76}$$

The gauge field Lagrangian then becomes

$$\frac{1}{4g^2} \int d^2\theta \mathcal{W}^\alpha \mathcal{W}_\alpha + \text{h.c.} \to \frac{1}{32\pi^2} \int d^2\theta \, \tau \mathcal{W}^\alpha \mathcal{W}_\alpha + \text{h.c.} \supset \frac{1}{4g^2} F_{\mu\nu}^2 + \frac{\theta_{YM}}{32\pi^2} F\tilde{F} \,. \tag{8.77}$$

$F\tilde{F}$ is a total derivative and does not affect local equations of motion. An abelian theory is invariant under shifts $\theta_{\rm YM} \to \theta_{\rm YM} + \text{const}$. In a non-abelian theory the action must be a periodic function of $\theta_{\rm YM}$ with a period 2π. To see that recall that in a non-abelian theory there exist topologically non-trivial gauge configurations whose contribution to the action is

$$\frac{\theta_{YM}}{32\pi^2} \int d^4x F\tilde{F} = n\theta_{YM} \,, \tag{8.78}$$

where n is a winding number of the field configuration. To calculate correlation functions one needs to sum over all n and periodicity in $\theta_{\rm YM}$ follows immediately.

These arguments imply that renormalization can at most shift the coefficient of the gauge kinetic term by a constant which corresponds to one-loop renormalization. No higher order corrections are allowed. The one loop coefficient of the β-function is given by

$$b_0 = 3C(G) - \sum_r C_2(r)\,. \tag{8.79}$$

As an example, consider an $SU(N)$ SUSY gauge theory with F flavors. In this theory renormalization group evolution of a *holomorphic* gauge coupling is given by

$$b_0 = 3N - F\,,$$
$$\frac{8\pi^2}{g^2(\mu)} = \frac{8\pi^2}{g^2(M)} + b_0 \ln \frac{\mu}{M}\,. \tag{8.80}$$

However, we have seen in the Wess-Zumino model that even as holomorphic parameters are not renormalized, physical coupling constants are renormalized to all orders in the perturbation theory due to the wave function renormalization. Similarly, physical gauge couplings in supersymmetric gauge theories are renormalized to all orders in the perturbation theory. However, any renormalization beyond one loop is due to wavefunction renormalization and an exact β-function can be written in terms of anomalous dimensions of the matter fields:[10,11]

$$\beta(\alpha) \equiv \frac{d\alpha(\mu)}{d \ln \mu} = -\frac{\alpha^2}{2\pi} \frac{3C(G) - \sum_r C_2(r)(1 - \gamma_r)}{1 - \frac{\alpha}{2\pi}C(G)}\,, \tag{8.81}$$

where

$$\gamma_r = \frac{\partial \ln Z_r(\mu)}{\partial \ln \mu}\,. \tag{8.82}$$

8.3.4. *D-term renormalization*

We conclude the discussion of non-renormalization theorems by considering Fayet-Illiopoulos D-terms (8.57). It was shown in [12] that D-term is renormalized at most at one loop. Once again spurion formalism is the most straightforward way to derive this result[13] (see also [11]). Recall that while D-term of the $U(1)$ vector superfield is invariant both under gauge and supersymmetry transformations, the full superfield V is not. If the D-term coefficient ξ depends on coupling constants in the theory, it becomes superspace valued once we promote couplings to spurions. Then performing superspace integral in (8.57) results in gauge non-invariant terms in the

Lagrangian. Therefore, ξ must be a pure number independent of all the coupling constants in the theory. There are, however, one loop quadratically divergent diagrams generating a tadpole for D. These arise from the D-term coupling to matter superfields in (8.58) and individually are quadratically divergent. The result, however, is only non-vanishing if the sum of $U(1)$ charges in a theory is non-vanishing, *i.e.* in theories where low energy Lagrangian suffers from gravitational anomalies.

8.4. Non-perturbative dynamics in SUSY QCD

8.4.1. *Affleck-Dine-Seiberg superpotential*

Our discussion of non-renormalization theorems made use of a requirement that the description of the theory is non-singular in weak coupling and massless limits. This immediately lead to a constraint that the superpotential does not contain negative powers of light superfields. While Lagrangian terms with negative powers of fields are never generated in the perturbation theory, it is known that such terms may arise due to non-perturbative dynamics. Therefore, non-renormalization theorems may be violated by non-perturbative effects.

As an example consider $SU(N)$ SUSY gauge theory with F flavors of matter fields in a fundamental representation introduced in section 8.2.5. When $F < N$ symmetries of (8.61) allow the appearance of non-perturbative superpotential[6,7,14]

$$W = C_{N,F} \left(\frac{\Lambda^{3N-F}}{\det Q\bar{Q}} \right)^{\frac{1}{N-F}} . \tag{8.83}$$

The theory is strongly coupled near the origin of the moduli space and in general the coefficient $C_{N,F}$ is not calculable. However, if $C_{N,F}$ is non-zero, the superpotential (8.83) forces Q and $\bar{Q}$ to run away and $SU(N)$ is broken to an $SU(N-F)$ subgroup. There are no light charged matter fields left in the low energy physics (all the components of Q and $\bar{Q}$ charged under unbroken group are eaten by the super-Higgs mechanism). The low energy effective field theory is described by a pure super Yang-Mills theory. It is expected that non-perturbative dynamics in pure SYM leads to a gaugino condensate $\langle 0|\lambda\lambda|0\rangle = \Lambda_L^3$. To verify that (8.83) is consistent with this expectation let us consider the evolution of holomorphic gauge coupling constants in the low and high energy theory and require that they match at the scale of gauge boson masses. Denoting this scale by $v \sim \sqrt{Q\bar{Q}}$ we can write

$$\frac{8\pi^2}{g_L^2(\mu)} = \frac{1}{g_H^2(\Lambda_{\mathrm{UV}})} + (3N - F)\ln\left(\frac{v^2}{\Lambda_{\mathrm{UV}}}\right) - 3(N - F)\ln\left(\frac{\mu^2}{v^2}\right). \quad (8.84)$$

This allows us to match the dynamical scales of two theories

$$\begin{aligned}
\Lambda_L^{3(N-F)} &= \mu^{3(N-F)}\exp\left(-\frac{8\pi^2}{g_L^2(\mu)}\right) \\
&= \frac{\Lambda_{\mathrm{UV}}^{3N-F}\exp\left(-\frac{8\pi^2}{g_H^2(\Lambda_{\mathrm{UV}})}\right)}{v^{2F}} = \frac{\Lambda_H^{3N-F}}{v^{2F}}.
\end{aligned} \quad (8.85)$$

We see that superpotential (8.83) can be expressed in terms of the parameters of low energy physics and has the form expected of gaugino condensate

$$W = C_{N-F,0}\Lambda_L^3. \quad (8.86)$$

Moreover we conclude that $C_{N,F} = C_{N-F,0} = C_{N-F}$.

This argument provides non-trivial evidence that the superpotential (8.83) is non-vanishing, however, we still have not calculated C_{N-F}. Fortunately, there is one case where C_{N-F} is calculable. In a model with $F = N - 1$ the gauge group completely broken at a generic point on the classical moduli space. This allows one to perform an explicit instanton calculation[6,7,15] and find $C_{N,N-1} = C_1 = 1$.

We can now derive C_{N-F} for other values of F. To do so, let us add mass term for one quark flavor

$$W = C_{N-F}\left(\frac{\Lambda^{3N-F}}{\det Q\bar{Q}}\right)^{\frac{1}{N-F}} + mQ_F\bar{Q}_F. \quad (8.87)$$

When $m \gg \Lambda$, heavy superfields decouple and low energy physics is described by a theory with $F - 1$ flavors. Solving Q_F and $\bar{Q}_F$ equations of motion we obtain an effective superpotential

$$W = C_{N-(F-1)}\left(\frac{m\Lambda^{3N-F}}{\det' Q\bar{Q}}\right)^{\frac{1}{N-(F-1)}}, \quad (8.88)$$

where prime implies that the determinant is taken only over $F - 1$ light flavors and $C_{N-(F-1)}$ is determined is determined by C_{N-F}, N, and F.

Similarly to (8.85) we can match the dynamical scales of two effective descriptions

$$\Lambda_L^{3N-(F-1)} = m\,\Lambda^{3N-F}. \quad (8.89)$$

We thus establish that (8.88) is indeed equivalent to (8.83). Finally, the knowledge of C_1 allows us to find $C_{N-F} = N - F$.

To conclude our discussion of theories with $F < N$ we note that when all flavors are massive, the supersymmetric ground state is found at finite vevs and is given by

$$\langle 0|Q_i\bar{Q}_j|0\rangle = \left(\det m\,\Lambda^{3N-F}\right)^{1/N} m_{ij}^{-1}\,. \tag{8.90}$$

8.4.2. *Quantum modified moduli space*

When the number of colors equals the number of flavors, there is no obvious superpotential that can arise dynamically. However, one can still perform an instanton calculation and find that the classical constraint (8.68) is modified:[16]

$$\det M - B\bar{B} = \Lambda^{2N}\,. \tag{8.91}$$

To interpret this result let us recall that in the classical theory the Kähler potential becomes singular at the origin of the field space, indicating the appearance of additional massless degrees of freedom — gauge bosons of the restored $SU(N)$ symmetry. Non-perturbative effects lead to a spectacular result — the origin of the field space does not belong to the moduli space and the Kähler potential is non-singular everywhere. The gauge symmetry is not restored anywhere on the moduli space and the fluctuations of M, B, and $\bar{B}$ satisfying the constraint (8.91) are the only massless particles in the quantum theory.

It is convenient to parameterize the dynamics that led to (8.91) by introducing an auxiliary Lagrange multiplier superfield A and the superpotential

$$W = A(\det M - B\bar{B} - \Lambda^{2N})\,. \tag{8.92}$$

There are several checks that can be performed to verify that this superpotential leads to the correct description of physics. For example, adding a mass term for a single flavor results in a low-energy theory with $N-1$ flavors. Using (8.92) and integrating out heavy flavor we indeed obtain the ADS superpotential (8.83) appropriate for this theory.

8.4.3. *s-confinement*

Adding one more flavor to the theory leads to a model which exhibits confinement without chiral symmetry breaking.[16] In this model the classical constraint is not modified quantum mechanically. The low energy effective field theory can be described by the superpotential

$$W = \frac{1}{\Lambda^{2N+1}}\left(BM\bar{B} - \det M\right)\,. \tag{8.93}$$

All the degrees of freedom in M, B, and $\bar{B}$ are physical. At the origin of the field space the full global symmetry is restored and all components of M, B, and $\bar{B}$ become massless. This is in contrast to the interpretation of the singularity in the classical description where the origin of the field space corresponds to an appearance of massless gluons.

The validity of this description is supported by the fact that 't Hooft anomaly conditions match between the UV and IR descriptions for the full global symmetry. One can also verify the validity of the description by perturbing the theory with a mass term — such a perturbation leads to a theory with the quantum modified moduli space we described earlier.

8.4.4. *Dualities in SUSY QCD*

The story becomes even more interesting as we further increase the number of flavors, $F > N + 1$. In this regime the infrared physics of an (electric) gauge theory has a dual description in terms of a magnetic theory with the same global symmetries but different gauge symmetry. It is convenient to start our discussion with an example of large N and F with $F < 3N$. We will choose $\epsilon = 1 - F/(3N)$ to be positive and small. In this case, the one loop β-function coefficient is small and the interplay between one and two loop running leads to a weakly coupled infrared fixed point.[18] Seiberg[16,17] argued that the physics at the ifrared fixed point has a dual (magnetic) description in terms of $SU(\tilde{N})$ gauge group with $\tilde{N} = F - N$ colors. The dual theory has the same global symmetries and contains dual quarks, q and $\bar{q}$, as well as elementary mesons $\widetilde{M}$ related to the gauge invariant composites of the electric description by

$$\widetilde{M} \sim \frac{1}{\mu} M = \frac{1}{\mu}(Q\bar{Q})\,, \tag{8.94}$$

where μ is the matching scale of the two descriptions. In addition, there exist a correspondence between the baryon operators in two theories

$$\begin{aligned}
B^{i_i \cdots i_N} &\sim \epsilon^{i_1 \cdots i_N j_1 \cdots j_{\tilde{N}}} b_{j_1 \cdots j_{\tilde{N}}}\,, \\
\bar{B}^{i_i \cdots i_N} &\sim \epsilon^{i_1 \cdots i_N j_1 \cdots j_{\tilde{N}}} \bar{b}_{j_1 \cdots j_{\tilde{N}}}\,.
\end{aligned} \tag{8.95}$$

If the superpotential term

$$W = \tilde{M} q\bar{q} \sim \frac{1}{\mu} M q\bar{q} \tag{8.96}$$

is added to the Lagrangian, the magnetic theory flows to the (strongly interacting) infrared fixed point that is identical to the fixed point of the electric theory. The dynamical scales of two descriptions are related by

$$\Lambda^{3N-F}\tilde{\Lambda}^{3\tilde{N}-F} = (-1)^{F-N}\mu^F\,. \tag{8.97}$$

It is instructive to consider various deformations of the two theories. We will perturb the electic theory by the mass term for one flavor of quarks. The number of quark flavors in low energy physics is reduced by one and the theory flows to a new (slightly more strongly coupled) infrared fixed point. In the magnetic description the perturbation corresponds to adding a tadpole for the meson field

$$W = mQ\bar{Q} = mM \sim m\mu\widetilde{M}\,. \tag{8.98}$$

This forces one flavor of dual quarks to acquire a vev, breaking the magnetic gauge group to $SU(\tilde{N}-1)$ and reducing the number of flavors by one. The magnetic theory flows to a new (less strongly coupled) infrared fixed point. Infrared physics of the two descriptions remains equivalent. As we continue this procedure the ifrared fixed point in the electric theory moves to strong coupling while the infrared fixed point in the magnetic theory moves to weak coupling. When the number of flavors in the electric theory becomes $F \leq 3N/2$, the conformal fixed point disappears. The magnetic description now has $F \geq 3\tilde{N}$ and asymptotic freedom is lost. Nevertheless the infrared duality still holds as long as $N + 2 \leq F \leq 3N/2$.

8.5. Supersymmetry breaking

We now turn our attention to theories with spontaneous supersymmetry breaking. We will be especially interested in models where supersymmetry is broken dynamically, *i.e.* theories where at the classical level potential posseses supersymmetric ground states, yet dynamical quantum effects modify the potential and the full quantum theory either does not have any supersymmetric ground states or at least has long lived local minima with spontaneously broken supersymmetry. Due to the non-renormalization theorems discussed earlier, the existence of a SUSY vacuum at the classical level implies its existence to all orders in the perturbation theory and dynamical supersymmetry breaking (DSB) is always a non-perturbative effect.

8.5.1. *O'Raifeartaigh model*

The simplest example[19] of spontaneous supersymmetry breaking in an interacting theory is a model with three chiral superfields and the superpotential

$$W = X\left(\frac{\lambda}{2}\Phi^2 - \mu^2\right) + m\Phi Y\,. \tag{8.99}$$

Note that the model possesses an R-symmetry under which fields carry the following charges

$$R_X = 2\,, \quad R_Y = 2\,, \quad R_\Phi = 0\,. \tag{8.100}$$

The F-term equations are

$$\begin{aligned}
\frac{\partial W}{\partial X} &= \left(\frac{\lambda}{2}\Phi^2 - \mu^2\right) = 0\,,\\[4pt]
\frac{\partial W}{\partial Y} &= m\Phi = 0\,,\\[4pt]
\frac{\partial W}{\partial \Phi} &= \lambda \Phi X + mY = 0\,.
\end{aligned} \tag{8.101}$$

The first two of these equations are incompatible and the scalar potential of the theory has no supersymmetric ground state:

$$V = \left|\frac{\partial W}{\partial X}\right|^2 + \left|\frac{\partial W}{\partial Y}\right|^2 + \left|\frac{\partial W}{\partial \Phi}\right|^2 > 0\,. \tag{8.102}$$

On the other hand, the last equation in (8.101) always has a solution, $X = -mY/(\lambda\Phi)$, leaving the vev of X arbitrary. It is a flat direction of the tree level potential, however, it is quite different from the moduli space of supersymmetric vacua. As we shall see shortly, this flat direction is lifted in perturbation theory due to SUSY breaking. We will therefore refer to X as a pseudo-modulus.

Let us first analyze the properties of the ground states of the theory. To simplify the analysis, let us assume that m is large so that the extrema of the classical potential are found at $\Phi = Y = 0$ with an arbitrary X. This implies that $F_\Phi = F_Y = 0$ and $F_X = \mu^2$. The spectrum of states in the theory depends on the vev of X:

$$\begin{aligned}
\text{scalars:}\quad & 0, 0, \frac{1}{2}\left(2m^2 + \lambda^2|X|^2 - \lambda\mu^2 \pm D(1)\right),\\[4pt]
& \frac{1}{2}\left(2m^2 + \lambda^2|X|^2 - \lambda\mu^2 \pm D(-1)\right),\\[4pt]
\text{fermions:}\quad & 0, \frac{1}{2}\left(2m^2 + \lambda^2|X|^2 \pm D(0)\right),
\end{aligned} \tag{8.103}$$

where $D(s) = \sqrt{4\lambda^2 m^2|X|^2 + (\lambda^2|X|^2 + s\lambda\mu^2)^2}$. It is easy to see that

$$\mathrm{Tr}[M^2_{\text{scalars}}] = \mathrm{Tr}[M^2_{\text{fermions}}]\,. \tag{8.104}$$

which can be expressed in terms of a *supertrace*

$$\mathrm{STr}M^2 = \mathrm{Tr}(-1)^F M^2 = 0. \tag{8.105}$$

In fact the supertrace condition is a general tree-level property of theories with spontaneous supersymmetry breaking.

Our next step is to calculate leading perturbative contributions to the X mass by calculating one loop corrections to the vacuum energy. A quartically divergent contribution vanishes since the theory has an equal number of bosonic and fermionic states. The quadratically divergent contribution vanishes due to the vanishing supertrace (8.105). We then have

$$V_{\mathrm{eff}} = \frac{1}{64\pi^2}\mathrm{STr}\mathcal{M}^4 \log\frac{\mathcal{M}^2}{\Lambda^2}. \tag{8.106}$$

It is easy to verify that while $\mathrm{STr}M^4$ is non-vanishing it is finite and X-independent. Therefore, the logarithmically divergent contribution to the potential also vanishes. We are left with a finite one loop correction to the pseudo-modulus potential and see that it acquires a positive mass squared

$$V(X) = \frac{\lambda^2}{48\pi^2}\frac{\mu^4}{m^2}|X|^2 + \mathcal{O}\left(|X|^4\right). \tag{8.107}$$

The vacuum is found at the origin of the field space and as a result the R-symmetry remains unbroken. As was recently shown in [20] an unbroken R-symmetry is a general property of O'Raifeartaigh models as long as all the fields in the theory have R-charges 0 and 2.

We now consider generalizations of the O'Raifeartaigh model that will be of interest later in these lectures. As a first step, let us consider a theory with a global $SU(F)$ symmetry and chiral superfields transforming according to

$$
\begin{array}{|c|ccc|}
\hline
 & SU(F) & U(1)_B & U(1)_R \\
\hline
B & F & 1 & 0 \\
\bar{B} & \bar{F} & -1 & 0 \\
M & \mathrm{Adj}+1 & 0 & 2 \\
\hline
\end{array}
\tag{8.108}
$$

For reasons that will become clear later we will refer to these fields as baryon, anti-baryon and meson. We can write the following superpotential consistent with the global symmetry

$$W = \lambda \bar{B}_i M_{ij} B_j + f^2 \mathrm{Tr}M. \tag{8.109}$$

The F-term equations for the meson fields have the form

$$\begin{cases} \lambda \bar{B}_i B_j + f^2 \delta_{ij} = 0 & i = j \\ \lambda \bar{B}_i B_j = 0 & i \neq j. \end{cases} \tag{8.110}$$

It is easy to verify directly that these equations do not have a solution and therefore SUSY must be broken. One often refers to this as a supersymmetry breaking by rank condition. Indeed, by performing a global symmetry transformation we can guarantee that only one component of B, say B_1, has a vev. We see that the matrix $\bar{B}_i B_j$ has rank one while rank F would be needed to cancel all the F-terms arising from the linear term in the superpotential.

A further generalization involves a model with an $SU(\tilde{N}) \times SU(F)$ global symmetry[b] and the following matter content

	$SU(\tilde{N})$	$SU(F)$	$U(1)_R$
q	N	F	0
$\bar{q}$	$\bar{N}$	$\bar{F}$	0
M	1	$\mathrm{Adj} + 1$	2

$$(8.111)$$

The most general renormalizable superpotential consistent with the symmetries is given by

$$W = \lambda \bar{q}_a^i M_i^j q_j^a + f^2 \mathrm{Tr} M , \qquad (8.112)$$

where a and i are $SU(\tilde{N})$ and $SU(F)$ indices respectively. We can use the global symmetry to rotate vevs of q so that $\langle 0|q_a^i|0\rangle = v_i \delta_a^i$ for $i \leq \tilde{m}in(F, \tilde{N})$ and $\langle 0|q_a^i|0\rangle = 0$ otherwise. This implies that the rank of the $\bar{q}q$ matrix can not exceed $min(F, \tilde{N})$ while there still are F non-trivial contributions to the F-terms arising from the linear term in the superpotential. We see that for $\tilde{N} < F$ supersymmetry must be broken.

The analysis of the spectrum in models with rank condition supersymmetry breaking is more involved but is similar to that in a simple O'Raifeartaigh model. In addition to several pseudo-moduli analogous to the field X of a simple O'Raifeartaigh model there exist true Goldstone bosons arising from the spontaneous breakdown of global symmetries. Nevertheless the same conclusion holds — all pseudo-moduli obtain positive mass squareds and there exist a stable non-supersymmetric ground state in the theory.[21] In particular the mass of $\mathrm{Tr}\,M$ is

$$m_{\mathrm{Tr}\,M}^2 = \frac{\log 4 - 1}{8\pi^2} N |\lambda^2 f^2| , \qquad (8.113)$$

in the ground state $\mathrm{Tr}\,M = 0$, and the R-symmetry remains unbroken.

8.5.2. *Dynamical supersymmetry breaking*

O'Raifeartaigh models of SUSY breaking can be used to construct phenomenological solutions of the technical hierarchy problem. However, they can not explain the origin of the hierarchy. This is because the vacuum energy in such models is an input parameter in the supersymmetric Lagrangian. On the other hand, if the supersymmetry breaking scale were determined by an energy scale associated with non-perturbative dynamics, it could be naturally small. Thus models with dynamical supersymmetry breaking are of great interest.

There are several guidelines in the search for dynamical SUSY breaking. The most important criterion is the value of the Witten index[22] given by the difference between the number of bosonic and fermionic states in a theory:

$$\mathrm{Tr}(-1)^F \equiv n_B^0 - n_F^0 \,. \qquad (8.114)$$

In fact, supersymmetry guarantees that the numbers of fermionic and bosonic states with non-zero energy are equal. Therefore, the value of the Witten index is determined purely by zero energy states in the theory. Moreover, the Witten index is a topological invariant of the theory. Once it is calculated for some choice of the parameters (for example in a weakly coupled regime) the result is valid quite generally. For example, varying parameters of the theory may lift some of the ground states — but only an equal number of fermionic and bosonic states. This has important consequences for the analysis of supersymmetry breaking. If the Witten index is non-zero, then there exists at least one zero-energy state and supersymmetry is unbroken. On the other hand, if the Witten index vanishes, there may either be no zero-energy states or their number is even. In the former case, supersymmetry must be broken. Witten calculated the value of the index in several theories and found that it is non-zero in a pure super Yang-Mills. Therefore, pure SYM theory does not break SUSY dynamically. Furthermore, in non-chiral theories one can take all masses to be large so that low energy physics is described by super Yang-Mills. This leads us to a conclusion that non-chiral theories in general do not break SUSY. The Witten index, however, may change if the asymptotic behavior of the potential changes as some of the parameters are taken to zero or infinity. While the examples are rare, it is indeed possible for SUSY to be broken in a vector-like theory.[23]

The next step in the model-building process involves the study of classical moduli space. As we know it can only be lifted by non-perturbative dynamics. Generically, non-perturbative effects lifting the moduli space lead to runaway behavior of the scalar potential (we will discuss important counterexamples later). Thus the most promising candidates for dynamical supersymmetry breaking are represented by models without classical flat directions: if the moduli space at infinity is lifted by a tree level term in the scalar potential, the interplay between the tree level and non-perturbative effects may lead to SUSY breaking.

We now turn to the analysis of global symmetries. It was argued in [7,26] that a theory without classical flat directions and with spontaneously broken global symmetry must break supersymmetry. To see that, recall that spontaneously broken global symmetry leads to the appearance of a Goldstone boson. Unbroken supersymmetry requires that a second real scalar field living in the same supermultiplet as the Goldstone has no potential. Changes in the vev of this scalar describe the motion along the flat direction — contradicting our initial assumption.

In model building, two additional conditions are often imposed — calculability and genericity. In this context calculability means that by choice of parameters the SUSY breaking scale can be made arbitrarily small compared to the scale of strong gauge dynamics; thusthe gauge dynamics can be integrated out and the low energy physics may be described by a Wess-Zumino model. Genericity means that once all the non-perturbative effects are taken into account the superpotential is the most generic holomorphic function of the superfields consistent with symmetries of microscopic theory. In this class of models, spontaneously broken R-symmetry is a sufficient condition of dynamical supersymmetry breaking.[24]

We will now consider several explicit examples of dynamical supersymmetry breaking. We begin by introducing the 3–2 model,[7] probably the simplest calculable model of dynamical supersymmetry breaking. We then give an example of strongly interacting $SU(5)$ theory where supersymmetry breaking can be established by several arguments[26,27] but details of the low energy physics are not calculable. Finally we discuss an Intriligator-Thomas-Izawa-Yanagida (ITIY) model[23] which breaks supersymmetry despite violating several of our guidelines.

$3 - 2$ *Model of dynamical supersymmetry breaking*

Consider a theory[7] with an $SU(3) \times SU(2)$ gauge group and matter fields transforming under gauge and global symmetries according to[c]

$$
\begin{array}{c|cccc}
 & SU(3) & SU(2) & U(1)_Y & U(1)_R \\
\hline
Q & 3 & 2 & 1/3 & -1 \\
\bar{u} & \bar{3} & 2 & -4/3 & -8 \\
\bar{d} & \bar{3} & 1 & 2/3 & 4 \\
L & 1 & 2 & -1 & -3
\end{array}
\tag{8.115}
$$

We will also add tree level superpotential

$$
W_{\text{tree}} = \lambda Q \bar{d} L \,.
\tag{8.116}
$$

In the limit $\Lambda_3 \gg \Lambda_2$, the non-perturbative effects are captured by including the Affleck-Dine-Seiberg superpotential generated by $SU(3)$ dynamics

$$
W = \frac{\Lambda_3^7}{\det\left(Q\bar{Q}\right)} \,.
\tag{8.117}
$$

It is easy to check that the tree level superpotential lifts all classical D-flat directions while presence of the non-perturbative term guarantees that the ground state is found away from the origin of the field space. A simple scaling argument shows that at the minimum

$$
Q \sim \bar{u} \sim \bar{d} \sim L \sim \frac{\Lambda}{\lambda^{1/7}}, \qquad E \sim \lambda^{10/7}\Lambda^4 \,.
\tag{8.118}
$$

This is sufficient to conclude that the R-symmetry and, therefore, supersymmetry is broken. For small Yukawa coupling λ the vacuum is found at large field vevs and the theory is weakly coupled and completely calculable. An explicit minimization of the potential can be performed to confirm the existence of SUSY breaking ground state and calculate the spectrum of light degrees of freedom.

In vicinity of the ground state the Kähler potential is nearly canonical in terms of elementary fields while light degrees of freedom are given by projections of these fields onto D-flat directions of the theory. Therefore, it is often convenient to work in terms of gauge invariant composites

$$
X_1 = Q\bar{d}L, \quad X_2 = Q\bar{u}L, \quad Y = \det(\bar{Q}Q) \,.
\tag{8.119}
$$

In terms of these variables the superpotential becomes

$$
W = \frac{\Lambda_3^7}{Y} + \lambda X_1 \,.
\tag{8.120}
$$

---

[c]It is interesting to note that $U(1)_Y$ could be gauged provided a new field, $\bar{e}$, with charges $(1, 1, 2)$ is added to the theory. This addition would not affect our discussion of supersymmetry breaking, yet, curiously enough, turn our simplest example of DSB into a one generation version of supersymmetric Standard Model.

The Kähler potential is a bit more complicated:[7,25]

$$K = 24\frac{A + Bx}{x^2},\tag{8.121}$$

where

$$A = \frac{1}{2}(X_1^\dagger X + X_2^\dagger X_2),$$

$$B = \frac{1}{3}\sqrt{Y^\dagger Y},\tag{8.122}$$

$$x \equiv 4\sqrt{B}\cos\left(\frac{1}{3}\arccos\frac{A}{B^{3/2}}\right).$$

We see that in low energy effective field theory the supersymmetry breaking is described in terms of a simple, albeit somewhat unusual, O'Raifeartaigh model with negative powers of fields in the superpotential.

DSB in strongly interacting models

It is possible to find strongly interacting gauge theories which satisfy our general guidelines for supersymmetry breaking. As an example, consider an $SU(5)$ theory with one superfield in 10 and one in $\bar{5}$ representation of the gauge group.[26,27] The theory also possesses global $U(1) \times U(1)_R$ symmetry under which fields carry charges $(1, 1)$ and $(-3, -9)$ respectively. No classical superpotential can be written down but the D-term potential does not have flat directions. Both of these conclusions follow from the fact that one can not form gauge invariant operators out of single 10 and $\bar{5}$. If a supersymmetric ground state exist, it must be located near the origin of the field space where both global symmetries are unbroken.

On the other hand, it is expected that the theory confines and at low energies the physics is described by gauge invariant degrees of freedom. The consistency of the theory requires that these light composites reproduce triangle anomalies of microscopic physics. In [26] it was shown that the anomaly matching conditions require a rather large set of massless fermions: at least five if charges are required to be less than 50. This makes it quite implausible that the full global symmetry remains unbroken. But if the global symmetry is broken, so is supersymmetry.

An independent argument for supersymmetry breaking may be obtained by deforming the $SU(5)$ model.[28] Specifically, one can add an extra generation of fields in 5 and $\bar{5}$ representations. In the perturbed theory with the most general renormalizable superpotential (incuding a small mass m for additional fields), one can show that supersymmetry is broken. One can then take the limit $m \to \infty$ and arrive at the original $SU(5)$ model. Assuming the absence of a phase transition as a function of mass one concludes that SUSY is broken.

DSB on quantum modified moduli space

In our last example of dynamical supersymmetry breaking we will discuss ITIY model[23] which illustrates the possibility of DSB in non-chiral theories as well as in theories with classical flat directions. Consider an $SU(2)$ gauge theory with 4 gauge doublet and 6 gauge singlet superfields, Q_i and S_{ij} respectively. We will assume that singlets S transform in an antisymmetric representation of $SU(4)_F$ flavor symmetry. We will write down a classical superpotential in terms of elementary degrees of freedom

$$W = \lambda S_{ij} Q_i Q_j \, . \tag{8.123}$$

This superpotential lifts all D-flat directions of the $SU(2)$ gauge theory, however flat directions associated with singlets S remain. Thus the model does not satisfy our guidelines for dynamical supersymmetry breaking: it is non-chiral and it has classical flat directions.

At the non-perturbative level the model possesses a quantum modified moduli space. In an $SU(2)$ gauge theory quantum modified constraint (8.68) can be implemented with the following superpotential

$$W = A(\text{Pf}M - \Lambda^4) \, , \tag{8.124}$$

where M_{ij} represent the 6 mesons that can be formed out of four gauge doublets, and Λ is a dynamical scale of microscopic theory. Writing down the classical superpotential (8.123) in terms of mesons M we can see that the low energy physics is described by an O'Raifeartaigh model of supersymmetry breaking:

$$W = A(\text{Pf}M - \Lambda^4) + \lambda S_{ij} M_{ij} \, . \tag{8.125}$$

As usual, there is a flat direction and we need to verify that there is no runaway behaviour as $S \to \infty$. To do so we consider non-perturbative dynamics at a generic point on the moduli space. Let us assume that singlets S obtain large vevs giving mass to all quark superfields. We can integrate out heavy superfields and describe the low energy physics in terms of pure super Yang Mills theory with the dynamical scale

$$\Lambda_L^6 = \lambda \text{Pf} S \, \Lambda^4 \, , \tag{8.126}$$

where Λ_L is a dynamical scale of low energy SYM theory. Gaugino condensation in low energy effective theory generates the superpotential

$$W = \Lambda_L^2 = \lambda \, (\mathrm{Pf} S)^{1/2} \, \Lambda^2 \, . \tag{8.127}$$

It is easy to verify that the scalar potential is independent of the modulus field $\tilde{S} = (\mathrm{Pf} S)^{1/2}$ and is non-vanishing, $V = \lambda^2 \Lambda^4$.

Furthermore, at large S corrections to the scalar potential are perturbative[29] and the pseudo-flat direction is lifted

$$V = \frac{\lambda^2}{Z_S} \Lambda^4 = \lambda^2 \Lambda^4 \left(1 + \frac{2}{16\pi^2} \ln \frac{|\tilde{S}|^2}{M_{\mathrm{UV}}} + \mathcal{O}\left(\lambda^4\right) \right) \, . \tag{8.128}$$

The analysis of the Coleman-Weinberg potential near the origin of the moduli space is more complicated since a priori the strong coupling dynamics may be important. This analysis was performed in [30] and it was found that uncalculable corrections due to strong dynamics are negligible and the $\tilde{S}$ potential is

$$V = \frac{5\lambda^4 \Lambda^2}{16\pi^2} (2\ln 2 - 1)|\tilde{S}|^2 + \mathcal{O}\left(S^4\right) \, . \tag{8.129}$$

We conclude that the potential of the ITIY model is calculable both for small and large $\tilde{S}$ and a ground state is found at $S = 0$. On the other hand the approximations made in the above calculations break down when $\lambda S \sim \Lambda$ leaving the possibility that another minimum of the potential exists with $S \sim \Lambda/\lambda$.

This model can be generalized to other examples with quantum modified moduli space, most straightforwardly to $SU(N)$ theories with $F = N$ flavors and $SP(2N)$ theories with $N + 1$ flavors.

8.5.3. *Metastable SUSY breaking*

Our discussion of dynamical supersymmetry breaking makes it clear that this is not a generic effect in SUSY gauge theories. Moreover, once the DSB sector is coupled to SUSY extensions of the Standard Model, one typically finds that supersymmetric vacua reappear elsewhere on the field space while SUSY breaking minima survive only as metastable, if long-lived, vacua.

If metastability is inevitable, then it is reasonable to accept it from the start. Indeed it was shown recently by Intriligator, Seiberg, and Shih that metastable minima of the potential with broken SUSY are quite generic[21] and often arise in very simple models. Probably the simplest example is SUSY QCD with N colors and $F = N + 1$ massive flavors, a theory we already discussed in section 8.4. In the presence of the mass term the full superpotential of the model, including effects of the non-perturbative dynamics is

$$W = \frac{1}{\Lambda^{2N-1}}\left(BM\bar{B} - \det M\right) + m\mathrm{Tr}M\,, \qquad (8.130)$$

where m is the quark mass term, Λ is the strong coupling scale of the theory while M, B, and $\bar{B}$ are mesons (8.66) and baryons (8.67). Near the origin of the field space baryons and mesons are weakly coupled degrees of freedom with canonical kinetic terms. This means that we can read off dimensions of the operators directly from the superpotential (8.130). We see that $BM\bar{B}$ term in the superpotential is a marginal operator in the infrared while the quark mass term is a relevant operator, $m\Lambda\mathrm{Tr}\,M$, in terms of canonically normalized meson field M. On the other hand, $\det M$ remains irrelevant (except in the case of $N=2$) and can be neglected in the discussion of dynamics near the origin of the moduli space. It is now easy to notice that the low energy dynamics of this theory is well described by an O'Raifeartaigh model (8.108) with an identification $f^2 = m\Lambda$. Near the origin of field space and as long as mass term m is sufficiently small effects of strong gauge dynamics are negligible compared to terms in Coleman-Weinber potential and the analysis of section 8.5.1 remains valid. On the other hand, in the UV (or at large field vevs) the theory deconfines and presence of the $\det M$ in the superpotential leads to restoration of supersymmetry at

$$\langle 0|M|0\rangle = \left(m\Lambda^{2N-1}\right)^{1/N}\mathbb{1}_F\,. \qquad (8.131)$$

The effects of the strong gauge dynamics become important when $m \sim \Lambda$ and it is reasonable to restrict ourselves to small masses $m \ll \Lambda$. In this case $\langle 0|M|0\rangle \ll \Lambda$ and the perturbative calculations are reliable not only near the origin of the fields space but also in the vicinity of supersymmetric ground states.

Following [21] we can generalize this example by gauging the global $SU(\widetilde{N})$ symmetry of the model defined by (8.111) and (8.112). We will also choose $F \geq 3\widetilde{N}$ and identify this model with the magnetic description of an asymptotically free $SU(N)$ theory. The fields of our model, M, q, and $\bar{q}$, are then mesons, as well as magnetic quarks and antiquarks. The first term in (8.112) is generated by the strong dynamics in the electric theory while the second term corresponds to the quark mass in the electric description, $f^2\mathrm{Tr}\,M \sim m\mathrm{Tr}\,Q\bar{Q}$. To verify that the analysis of section 8.5.1 remains valid in the presence of newly introduced gauge dynamics, we note that for our choice of F and $\widetilde{N}$ the magnetic description is IR free, gauge dynamics of the $SU(\widetilde{N})$ theory is weakly coupled near the origin of the field space and we are still justified in performing perturbative calculation. We must include contributions of gauge supermultiplet in our calculation of Coleman-Weinberg potential, however, to leading order in SUSY breaking parameter it has a supersymmetric spectrum and our earlier results are not modified.

On the other hand, from the analysis of electric description we know that supersymmetric vacua exist in this theory. They can also be found in magnetic description by carefully examining the effects of gauge dynamics at large field vevs. Indeed, at large M magnetic quarks become massive and can be integrated out, leading to gaugino condensation in magnetic theory and the effective superpotential, $W = \Lambda_L^3$, where Λ_L is a strong coupling scale of a low energy $SU(\widetilde{N})$ SYM theory. Using the fact that holomorphic gauge coupling evolves only at one loop, this superpotential can be written as

$$
\begin{aligned}
W = \Lambda_L^3 &= \left(\mu^{3\widetilde{N}} e^{-8\pi^2/g_L^2(\mu)} \right)^{1/\widetilde{N}} \\
&= \left(\Lambda_{UV}^{3\widetilde{N}-F} \det M e^{-8\pi^2/\tilde{g}^2(\Lambda_{UV})} \right)^{1/\widetilde{N}} = \frac{\det M}{\widetilde{\Lambda}^{F-3\widetilde{N}}} ,
\end{aligned}
\tag{8.132}
$$

where $\widetilde{\Lambda}$ is the scale of the Landau pole in the magnetic theory. Together with the superpotential (8.112) this dynamical term leads to restoration of supersymmetry. Just as in our previous example, for sufficiently small mass terms in electric theory, the potential is fully calculable both near supersymmetric and supersymmetry breaking minima.

It is important for phenomenological applications that the metastable non-supersymmetric vacua are sufficiently long-lived. The semi-classical decay probability of a false vacuum is given[31] by $\exp(-S)$, where S is a bounce action. For the models discussed in this section the bounce action was estimated in [21]:

$$
S \sim \left(\frac{\Lambda}{m} \right)^{2(F-N)/(F-N)} \gg 1 .
\tag{8.133}
$$

This suggests that generically it is possible to achieve sufficiently long lifetime of the vacuum by appropriate choice of the parameters. One still needs to explain why the non-supersymmetric ground state is chosen in the early Universe. It was argued in [32] that a non-supersymmetric vacuum is generically preferred over a supersymmetric one in the early Universe due to thermal effects. This conclusion holds as long as a metastable vacuum is closer to the origin of the field space than a supersymmetric one.

8.5.4. *Fayet-Illiopolous model*

For completeness, we will briefly introduce an example of spontaneous SUSY breaking with a non-vanishing D-term.[33] Consider a $U(1)$ theory with the Lagrangian

$$
\begin{aligned}
\mathcal{L} = \int d^4\theta \Phi_+^\dagger e^V \Phi_+ + \Phi_-^\dagger e^{-V} \Phi + (\int d^2\theta \frac{1}{4g^2} \mathcal{W}^\alpha \mathcal{W}_\alpha + \text{h.c.}) \\
+ \int d^4\theta \xi^2 V + (\int d^2\theta \, m\Phi_+\Phi_- + \text{h.c.}).
\end{aligned}
\tag{8.134}
$$

In this model the D-term equation of motion

$$
D = g(|\Phi_+|^2 - |\Phi_-|^2 + \xi^2) = 0
\tag{8.135}
$$

requires non-zero vev for Φ_-. On the other hand thr F-term conditions

$$
F_{\Phi_+} = m\Phi_- = 0, \quad F_{\Phi_-} = m\Phi_+ = 0
\tag{8.136}
$$

require that both fields vanish. Thus the potential at the minimum is non-vanishing and SUSY is broken. It is straightforward to verify that the tree level spectrum of the theory satisfies the supertrace condition, $\text{STr}\mathcal{M}^2 = 0$.

8.5.5. *Goldstino*

According to the Goldstone theorem spontaneously broken symmetries must always lead to an appearance of massless particles with derivative interactions. In theories of spantaneously broken supersymmetry the broken generator is a spinor and the corresponding massless particle is a fermion, goldstino. Indeed, presence of massless goldstino is required by the goldstino theorem which can be derived quite easily for a general supersymmetric model.[34,35]

The goldstino is always a fermion in the supermultiplet that leads to SUSY breaking, in other words it is a fermion in a supermultiplet with a non-vanishing F- or D-term. If several supermultiplets acquire F- and D-term vevs in the ground state of the theory, goldstino is a linear combination of fermions living in these supermultiplets. To illustrate the appearance of a goldstino, consider a theory with gauginos λ^a, and matter fermions χ_i. In all examples of SUSY breaking we have considered so far, fermion mass matrix is not affected by supersymmetry breaking to leading order in SUSY breaking parameters F and D. It has the form

$$\begin{pmatrix} 0 & \sqrt{2}g_a\varphi_i^*\lambda^a T_{ij}^a\chi_j \\ g_a\varphi_i^*\lambda^a T_{ij}^a\chi_j & W_{ij} \end{pmatrix} , \tag{8.137}$$

and off-diagonal termes appear whenever gauge symmetry is broken by vev of φ. When SUSY is broken this mass matrix has at least one zero eigenvalue whose eigenvector is given by

$$\begin{pmatrix} D^a/\sqrt{2} \\ F_i \end{pmatrix} . \tag{8.138}$$

This eigenvector is non-trivial whenever at least one F- or D-term is non-vanishing and corresponds to the goldstino which can be written as

$$\tilde{G} = \frac{1}{F_{\tilde{G}}}\left(\frac{D^a}{\sqrt{2}}\lambda^a + F_i\chi_i \right) , \tag{8.139}$$

where

$$F_{\tilde{G}} = \sqrt{\sum_a \frac{(D^a)^2}{2} + \sum_i |F_i|^2} . \tag{8.140}$$

The goldstino effective Lagrangian can be written as

$$\mathcal{L}_{\text{goldstino}} = i\tilde{G}^\dagger \partial_\mu \bar{\sigma}^\mu \tilde{G} + \frac{1}{F_{\tilde{G}}}\tilde{G}^\alpha \partial_\mu j_\alpha^\mu , \tag{8.141}$$

where the supercurrent j_α^μ is

$$j_\alpha^\mu = (\sigma^\nu\bar{\sigma}^\mu\eta_i)_\alpha D_\nu\varphi^{*i} - \frac{1}{2\sqrt{2}}(\sigma^\nu\bar{\sigma}^\mu\sigma^\rho\lambda^{\dagger a})_\alpha F_{\nu\rho}^a . \tag{8.142}$$

Finally, we would like to mention the role of a goldstino in locally supersymmetric theories. As we know such theories require the existence of a gravitino, a spin 3/2 superpartner of the graviton. Once supersymmetry is broken, the gravitino becomes massive by eating the goldstino, in a complete analogy to Higgs mechanism where the gauge boson becomes massive by eating a Goldstone boson.

8.6. Minimal Supersymmetric Standard Model

8.6.1. *Matter content and interactions*

We will now study a Minimal Supersymmetric Standard Model (MSSM). Supersymmetry requires that all the Standard Model particles are accompanied by superpartners. In the gauge sector we must include gauginos in the adjoint representation for each of the Standard Model gauge groups. The matter content is given by chiral superfields with the following charge assignments

$$\begin{array}{c|ccc}
 & SU(3) & SU(2) & U(1)_Y \\
\hline
Q_f & 3 & 2 & 1/3 \\
\bar{u}_f & \bar{3} & 1 & -4/3 \\
\bar{d}_f & \bar{3} & 1 & 2/3 \\
L_f & 1 & 2 & -1 \\
\bar{E}_f & 1 & 1 & 2 \\
H_u & 1 & 2 & 1 \\
H_d & 1 & 2 & -1
\end{array} \tag{8.143}$$

where $f = 1..3$ is a generation index. With a little abuse of notation we will use the same symbol both for the superfields and Standard Model particles (i.e. fermions and Higgses) while denoting the superpartners (sfermions and Higgs superpartners, higgsinos)· with a tilde. An important feature of matter content (8.143) is the presence of two Higgs multiplets. There are two reasons for this. First of all, higgsinos carry $SU(2) \times U(1)_Y$ quantum numbers and contribute to anomalies. Since the SM is anomaly free in the absence of higgsinos, two higgsinos with opposite charges represents a minimal extension without new contributions to anomalies. Furthermore, two Higgs supermultiplets are required to reproduce all the SM Yukawa couplings.

Indeed, in the Standard Model, due to the fact that $\bar{2}$ and 2 representations of the $SU(2)$ gauge group are equivalent, a single Higgs boson is sufficient to write down both the up and down type Yukawa matrices. In MSSM, however, Yukawas must arise from holomorphic terms in the superpotential and complex conjugate of 2 representation can not appear in the superpotential. Thus full set of Yukawa couplings is only possible in the presence of two Higgs doublets and is contained in the following superpotential

$$W_{\text{Yukawa}} = \lambda^u_{ff'} H_u Q_f \bar{u}'_f + \lambda^d_{ff'} H_d Q_f \bar{d}'_f + \lambda^L_{ff'} H_u L_f \bar{e}'_f \,. \tag{8.144}$$

For example, the top Yukawa coupling, $\lambda_t H_u Q_3 \bar{t}$, is contained in the first term while the bottom Yukawa, $\lambda_b H_d Q_3 \bar{b}$, is contained in the second.

In addition to the superpotential of eqn. (8.144), two more types of terms are allowed by the symmetries. First, we can write a supersymmetric Higgs mass term

$$W_H = \mu H_u H_d \,. \tag{8.145}$$

As we will see shortly this term is important for generating electroweak symmetry breaking but at the same time it leads to a well-known μ-problem.

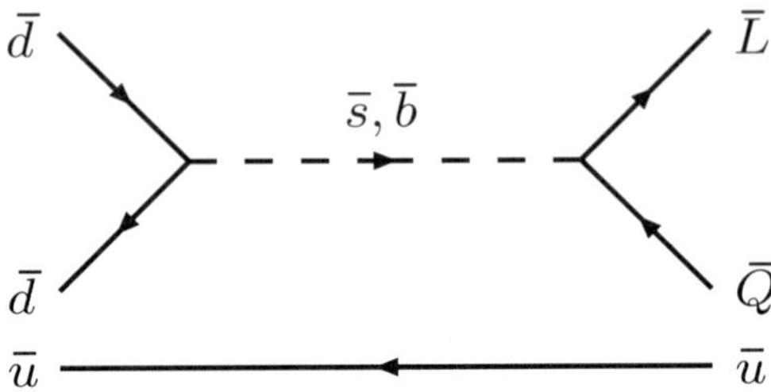

Fig. 8.2.   Proton decay diagram arising from R-symmetry violating interactions.

Another set of allowed superpotential interactions:

$$W_{\not R} = \alpha^{ijl} Q_i L_j \bar{d}_k + \beta^{ijk} L_i L_j \bar{E}_k + \gamma^i L_i H_u + \delta^{ijk} \bar{d}_i \bar{d}_j \bar{u}_k \,, \qquad (8.146)$$

where β and δ are antisymmetric under the interchange $(i \leftrightarrow j)$ is much more dangerous. Terms in $W_{\not R}$ are renormalizable and violate both lepton and baryon numbers. They immediately lead to proton decay through the diagram depicted in Figure 8.2. A rough estimate of the proton lifetime gives

$$\Gamma_p \sim \frac{|\alpha\delta|^2}{8\pi^2} \frac{m_p^5}{m_{\tilde q}^4} \,,$$

$$\tau_p = \frac{1}{\Gamma_p} \sim \frac{1}{|\alpha\delta|^2} \left(\frac{m_{\tilde q}}{1\,\text{TeV}}\right)^4 2 \times 10^{-11} s \,, \qquad (8.147)$$

where $m_{\tilde q}$ is a squark mass. Comparing this result with experimental limits on proton lifetime we see that either coupling constants in $W_{\not R}$ must be extremely suppressed, $|\alpha\delta| < 10^{-25}$, or the SUSY breaking scale (parameterized here by $m_{\tilde q}$) is very large, $m_{\tilde q}^2 > 10^{31}\,\text{TeV}^2$.

One usually approaches this problem is by introducing a new discrete symmetry which forbids dangerous couplings. Such a symmetry can be thought of as a discrete subgroup of an R-symmetry, called R-parity. Under R-parity all the Standard Model particles (including both Higgs boson doublets) are even while all the superpartners are odd. Interestingly, R-parity can be introduced without reference to R-symmetry by defining charges of particles under R parity as a combination of their fermion number and $B - L$ charge

$$R = (-1)^{(B-L)+F} \,. \qquad (8.148)$$

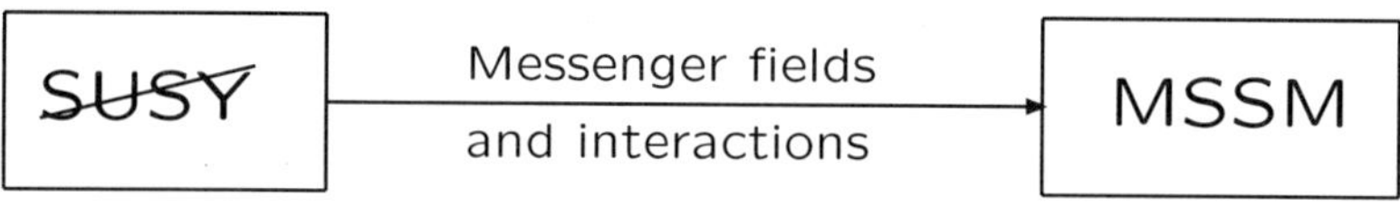

Fig. 8.3. Typical structure of supersymmetric extensions of the Standard Model.

In addition to suppressing proton decay, R-parity leads to several important consequences:

- At colliders superpartners are produced in pairs;
- The lightest superpartner is stable and if it happens to be neutral, provides an excellent dark matter candidate;
- Each sparticle other than LSP will decay to an odd number of LSP's (plus ordinary particles).

We would like to comment that one can consider R-parity violating extensions of the SM as long as dangerous couplings are fine-tuned to be small. While this class of models may lead to interesting experimental signatures its is beyond the scope of these lectures.

8.6.2. *Soft SUSY breaking*

We now introduce SUSY breaking into the MSSM. The supersymmetry breaking must be *soft*, that is it should not reintroduce quadratic divergencies. This could be achieved if SUSY breaking is a spontaneous symmetry breaking. One could attempt to construct extensions of the MSSM whith spontaneously broken supersymmetry. As we know tree level spectrum of such models satisfies the supertrace condition, $\mathrm{STr}\,\mathcal{M} = 0$. While the supertrace condition is modified by quantum effects, within MSSM alone such modifications are small since the Standard Model is a weakly interacting theory at EWSB scale. Since the Standard Model fermions are generally light, the supertrace condition requires the existence of new light bosons which have not been observed experimentally.[36] We conclude that SUSY must be broken in a different, *hidden*, sector of the theory.

We will consider a scenario of SUSY breaking depicted in Fig. 8.3. We will imagine that supersymmetry is broken in a hidden sector by one of the mechanisms described in section 8.5. We will then have to introduce interactions between the hidden and visible sectors that will communicate SUSY breaking to the MSSM fields and generate superpartner masses. While there are several different mechanisms that could mediate SUSY breaking to the Standard Model sector, a general form of soft-breaking terms can be obtained from the following argument. Imagine integrating out hidden sector physics completely and obtaining an effective Lagrangian in MSSM sector. In such a Lagrangian coupling constants become functions of integrated out hidden sector superfields. We can now use a familiar trick of promoting Lagrangian parameters to background superfields. To include the effects of spontaneous SUSY breaking in the hidden sector, we will simply allow these background superfields to have non-vanishing F-terms[d]. As an example consider (8.145). Promoting μ to a superfield $\mu \to \mu + B\theta^2$ leads to the soft SUSY breaking terms in the Lagrangian

$$\mathcal{L}_B = B h_u h_d + \text{h.c.} \, . \tag{8.149}$$

This term represents the new SUSY breaking masses in the Higgs potential and together with the μ-term plays an important role in electroweak symmetry breaking. More generally, we can promote all the Yukawa and gauge couplings to functions of background superfields

$$\lambda^u_{ff'} \to \lambda^u_{ff'}(\Sigma/M), \quad \lambda^d_{ff'} \to \lambda^d_{ff'}(\Sigma/M), \quad \lambda^L_{ff'} \to \lambda^L_{ff'}(\Sigma/M),$$
$$\frac{1}{g_a^2} \to \frac{1}{g_a^2 \, (\Sigma/M)}, \tag{8.150}$$

where M is a characteristic scale of interactions between the Standard Model and SUSY breaking sector and Σ represents hidden sector superfields. In the simplest case, Σ is a single superfield and the flavor structure of soft parameters is completely encoded in Yukawa matrices $\lambda_{ff'}$. More generally, the Yukawa matrices and gauge coupling functions could depend non-trivially on several hidden sector superfields. For example one can have $\lambda^{u,d}_{ff'} = \Sigma^{u,d}_{ff'}/M$ and $1/g_a^2 = \Sigma^a/M$. In this case, suppressing flavor indices and writing $\langle 0|\Sigma|0\rangle = \sigma + F_\Sigma \theta^2$ we can write Yukawa couplings as

$$\lambda^u = \frac{\sigma^u}{M}, \quad \lambda^d = \frac{\sigma^d}{M}, \quad \lambda^L = \frac{\sigma^L}{M}, \quad \frac{1}{g_a^2} = \frac{\sigma^a}{M}, \tag{8.151}$$

as well as trilinear scalar interactions, so called A-terms,

$$\mathcal{L}_A = A^u h_u \tilde{Q}\tilde{\bar{u}} + A^d h_d \tilde{Q}\tilde{\bar{d}} + A^L h_u \tilde{L}\tilde{\bar{e}} + \text{h.c.} \, , \tag{8.152}$$

where

$$A^u = \frac{F^u_\Sigma}{M}, \quad A^d = \frac{F^d_\Sigma}{M}, \quad A^L = \frac{F^L_\Sigma}{M} \, . \tag{8.153}$$

---

[d]Hidden sector D-terms may also be considered.

Similarly, gauge couplings and gaugino masses may be written in terms of the spurion vevs

$$\frac{1}{g_a^2} = \frac{\sigma^a}{M}, \quad M^a = \frac{F_\Sigma^a}{M}. \tag{8.154}$$

We still need to generate scalar masses. Let us look at the Kähler potential. Generically non-renormalizable interactions between the hidden and visible sectors will appear in the Kähler potential of the effective theory even if they are absent in the microscopic description. As an example Kähler potential of squark superfields may take the form

$$K = \left(\delta_{ff'} + c_{ff'}\frac{\Sigma^\dagger \Sigma}{M^2}\right) Q_f^\dagger Q_f'. \tag{8.155}$$

This results in soft squark masses

$$\tilde{m}_{ff'}^2 = \frac{c_{ff'}|F_\Sigma|^2}{M^2}. \tag{8.156}$$

Similar soft masses are generated for other sfermions as well as scalar components in the Higgs multiplets.

8.6.3. *Higgs Sector*

We now turn to the question of electroweak symmetry breaking in MSSM with softly broken SUSY. The Higgs potential is given by

$$V = V_D + V_F + V_{SUSY}, \tag{8.157}$$

where V_D is the D-term potential

$$V(H_u, H_d) = \frac{g^2 + g'^2}{8} \left(|H_u^0|^2 + |H_u^+|^2 - |H_d^0|^2 - |H_d|^2\right)^2$$
$$+ \frac{g^2}{2} \left|H_u^+ H_d^{0*} + H_u^0 H_d^{-*}\right|^2, \tag{8.158}$$

V_F is the F-term potential

$$V_F = \mu^2 \left(|H_u^0|^2 + |H_u^+|^2 + |H_d^0|^2 + |H_d^-|^2\right), \tag{8.159}$$

and $V_{\rm SUSY}$ represents SUSY breaking terms in the potential

$$V_{\rm SUSY} = \tilde{m}_u^2 \left(|H_u^0|^2 + |H_u^+|^2\right) + \tilde{m}_d^2 \left(H_d^0|^2 + |H_d^-|^2\right)$$
$$+ B \left(H_u^+ H_d^- - H_u^0 H_d^0\right) + B \left(H_u^{+*} H_d^{-*} - H_u^{0*} H_d^{0*}\right), \tag{8.160}$$

where $\tilde{m}_u^2$ and $\tilde{m}_d^2$ are soft Higgs masses. Electroweak symmetry breaking requires that the Higgs potential is bounded from below and has a minimum at non-vanishing vevs. The Higgs mass matrix will satisfy these conditions if

$$|B|^2 > (\tilde{m}_u^2 + |\mu|^2)(\tilde{m}_d^2 + |\mu|^2)\,,$$
$$2\mu^2 + \tilde{m}_u^2 + \tilde{m}_d^2 > 2|B|\,. \tag{8.161}$$

Typically these conditions are not satisfied at the SUSY breaking scale. However, RG evolution modifies relations between superpartner masses and may lead to radiative electroweak symmetry breaking. The dominant effect arises from the Higgs interactions with the third generation

$$\frac{d}{dt}\begin{pmatrix}\tilde{m}_u^2 \\ \tilde{m}_t^2 \\ \tilde{m}_{Q_3}^2\end{pmatrix} = -\lambda_t^2\begin{pmatrix}3 & 3 & 3 \\ 2 & 2 & 2 \\ 1 & 1 & 1\end{pmatrix} - A_t^2\begin{pmatrix}3 \\ 2 \\ 1\end{pmatrix}\,. \tag{8.162}$$

We can see that H_u receives the largest negative contribution and once its mass is driven negative electroweak symmetry is broken.

By requiring that the parameters in the Higgs potential lead to experimentally observed Z and W mass we obtain relations between soft parameters which must be satisfied at the weak scale:

$$\mu^2 = \frac{\tilde{m}_u^2 - \tilde{m}_d^2 \tan^\beta}{\tan^2\beta - 1} - \frac{1}{2}M_Z^2\,,$$
$$B = \frac{(\tilde{m}_u^2 + \tilde{m}_d^2 + 2\mu^2)\sin 2\beta}{2}\,, \tag{8.163}$$

where $\tan\beta = v_u/v_d$.

From this expression we see that naturalness requires that both μ and B are electroweak scale parameters. However, the μ-term is a supersymmetric term in the Lagrangian and could take any value between EWSB and Planck scales. This leads to the so-called μ-problem. Only models where the μ-term arises as a result of SUSY breaking are expected to avoid fine-tuning. Even then, the absence of fine-tuning is not guaranteed. To illustrate this issue, let's replace the μ-term with a vacuum expectation value of a new gauge singlet field X coupled to the Higgses

$$W_X = \lambda_X X H_u H_d\,. \tag{8.164}$$

It is possible to construct models where $\langle 0|X|0\rangle$ is only generated as a result of SUSY breaking. Even in these models the coupling constant λ often needs to be small to guarantee the correct magnitude of the μ-term. Additionally, $\langle 0|F_X|0\rangle$ is often generated and leads to the B-term. Typically one finds

$F_X \sim X^2$ and the ratio between μ^2 and B terms is given by

$$\frac{B}{\mu^2} \sim \frac{\lambda_X F_X}{\lambda_X^2 X^2} \sim \frac{F_X}{\lambda_X X^2} \sim \frac{1}{\lambda_X}. \tag{8.165}$$

In other words, if a μ-term of the correct magnitude results in a phenomenologically unacceptable B term. This result holds quite generically in theories were small parameters are used to generate the soft terms from the fundamental scale of SUSY breaking.

8.6.4. *Flavor problem*

The most general set of R-parity invariant soft terms leads to a model with 105 parameters in addition to those in the Standard Model itself. One would like to find an organizing principle which reduces number of parameters and makes the model predictive. Furthermore, generic points on the MSSM parameter space are ruled out by existing experiments. The most stringent limits arise from constraints on flavor violating processes.

For example, consider $K - \bar{K}$ mixing. In the Standard Model, GIM mechanism ensures that the leading contribution to this process, arising through a diagram in Figure 8.4, starts at order $\mathcal{O}\left(m_{\text{quark}}^2\right)$:

$$\mathcal{M}_{K\bar{K}}^{SM} \approx \alpha_2^2 \frac{m_c^2}{M_W^2} \sin^2 \theta_c \cos^2 \theta_c, \tag{8.166}$$

where θ_c is the Cabbibo angle.

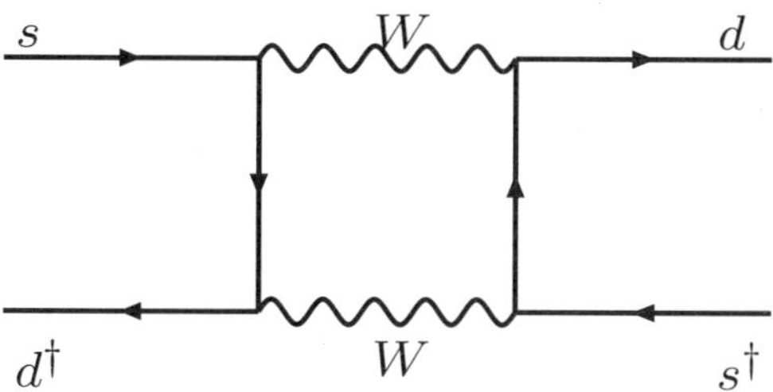

Fig. 8.4.   Standard Model contribution to KK mixing.

In the MSSM additional contributions arise due to processes in Figure 8.5:

$$\mathcal{M}_{K\bar{K}}^{MSSM} \approx 4\alpha_3^2 \left(\frac{\Delta \tilde{m}_Q^2}{M_{SUSY}^2}\right) \frac{1}{M_{SUSY}^2}, \tag{8.167}$$

where M_{SUSY} is a typical scale of the soft MSSM parameters. As we can see this contribution is formally enhanced compared to the Standard Model amplitude by a factor of the order $(\alpha_3/\alpha_2)^2$. On the other hand, any new physics contribution can't be large since the Standard Model result is consistent with experimental observations. This implies the following relation

$$\left(\frac{\Delta\tilde{m}_Q^2}{M_{SUSY}^2}\right) < 4 \times 10^{-3} \frac{M_{SUSY}}{550\mathrm{TeV}} . \tag{8.168}$$

We conclude that the squark mass matrix must be diagonal in flavor space in the same basis as the quark mass matrix or SUSY breaking scale is much larger than electroweak scale.

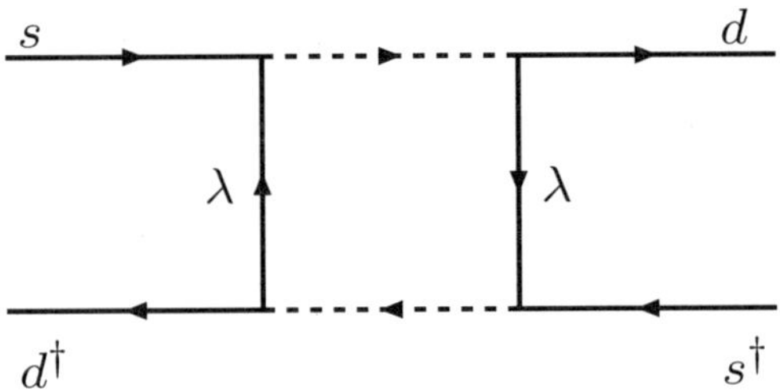

Fig. 8.5.   MSSM contributions to $K - \bar{K}$ mixing.

There also exist strong constraints on flavor violation in the slepton sector. In addition to the Standard Model muon decay $\mu \to e\nu\nu^*$ the generic choice of MSSM parameters introduces a new decay channel, $\mu \to e\gamma$, which proceeds through the diagram in Figure 8.6. It is not difficult to estimate the branching ratio

$$\frac{\Gamma_{\mu\to e\gamma}}{\Gamma_{\mu\to e\nu\nu^*}} \approx 10 \times 10^{-4} \left(\frac{500\mathrm{TeV}}{M_{SUSY}}\right)^4 \times \left(\frac{\Delta\tilde{m}_L^2}{M_{SUSY}^2}\right) . \tag{8.169}$$

Experimentally this ratio is less than 10^{-11}. Once again either $\Delta\tilde{m}_L^2$ is nearly diagonal in the same basis as the charged lepton mass matrix or the SUSY breaking scale is extremely high.

As we just indicated the flavor problem could be resolved if the fermion and sfermion mass matrices are diagonal in the same basis, in other words if a super-GIM mechanism is operational in MSSM. This can be achieved in models with flavor symmetries, see for example.[37] Another resolution would require that mechanism mediating SUSY breaking between hidden

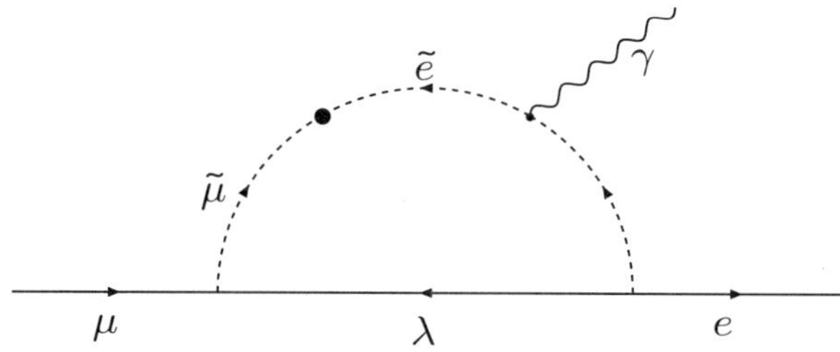

Fig. 8.6.   SUSY contribution to muon decay.

and visible sectors is flavor blind. We will discuss some realizations of this idea in the next section.

8.7. Mediation of SUSY breaking

We finally come to the discussion of mechanisms that can communicate SUSY breaking between hidden and visible sectors. We will only discuss the three most popular mechanisms out of several interesting possibilities: supergravity, gauge, and anomaly mediation.

8.7.1. *SUGRA mediation*

The most minimal approach to mediating supersymmetry breaking between hidden and visible sectors is through supergravity interactions. Generically one should expect that the most general interactions consistent with the symmetries of both hidden and visible sector will be generated in an effective theory with Planck suppressed couplings. Thus formulas of section 8.6.2 will apply with a messenger scale $M = M_{\mathrm{Pl}}$. Using M_{Pl} in formulas (8.153), (8.154), and (8.156) and requiring that soft SUSY breaking parameters in the Standard Model sector are of the order TeV, implies that the fundamental SUSY breaking scale is of order 10^{11} TeV. This also leads to a gravitino mass of the order TeV.

Is the spectrum of gravity mediation consistent with FCNC constraints? At first it appears natural to assume that all soft scalar masses are universal since gravity couples universally to all fields. Similarly, one could expect universality for A-terms as well as gaugino masses. However, this assumption is not fully justified. Indeed, one should expect that the microscopic description of fundamental theory may contain new particles with order 1 couplings both to the hidden and visible sectors and masses of order M_{Pl}.

These particles do not necessarily belong to the gravity multiplet and as such do not have to couple universally to all the Standard Model fields. Integrating out these particles leads to low energy effective description with order one flavor violations in sfermion mass matrices.

Thus one needs to impose additional conditions to guarantee the compatibility of theoretical predictions with the existing experimental observations. In the gravity mediation approach one simply assumes universality at the matching scale. Namely, one assumes universal gaugino and sfermion masses while A-terms are taken to be proportional to Yukawa matrices. The soft terms in the Lagrangian are determined by 4 parameters

$$m_{1/2} = f\frac{F}{M_{\rm Pl}}, \quad m_0^2 = k\frac{|F|^2}{M_{\rm Pl}^2}, \quad A_0 = \alpha\frac{F}{M_{\rm Pl}}, \quad B = \beta\frac{F}{M_{\rm Pl}}. \qquad (8.170)$$

At the matching scale sfermion masses as well as soft Higgs masses are given by

$$\tilde{m}^2_{Qff'} = \tilde{m}^2_{\bar{u}ff'} = \tilde{m}^2_{\bar{d}ff'} = \tilde{m}^2_{Lff'} = \tilde{m}^2_{\bar{e}ff'} = \tilde{m}^2_u = \tilde{m}^2_d = m_0^2\delta_{ff'}. \qquad (8.171)$$

Tirlinear couplings are

$$A^u_{ff'} = \lambda^u_{ff'}A_0, \quad A^d = \lambda^d_{ff'}A_0, \quad A^L = \lambda^L_{ff'}A_0. \qquad (8.172)$$

Finally gaugino masses are unified at the matching scale

$$M_3 = M_2 = M_1 = m_{1/2}. \qquad (8.173)$$

One usually assumes that the hidden sector does not have light fields and decouples at the matching scale. As a result low energy values of soft masses are determined by the equations (8.171), (8.172), (8.173) and renormalization group evolution between the matching and electroweak scales. In particular, renormalization group evolution drives H_u mass squared negative according to (8.162). However, it is possible that hidden sector is both strongly interacting and contains particles much lighter that matching scale. In this case, effect of the hidden sector RG evolution can not be neglected.[38] This situation is not unique to supergravity and effects of hidden sector renormalization on soft parameters may be significant in other mediation mechanisms.

To conclude this section we briefly mention the status of μ problem in gravity mediation. It turns out that gravity mediation allows for a rather elegant solution of the μ problem.[39] First, it is quite easy to forbid appearance of a large μ-term by imposing some symmetry on the Lagrangian. This can be an R-symmetry, PQ-symmetry or a discrete symmetry. One can then introduce Planck suppressed interactions between hidden and visible sectors which generate μ and B terms of comparable size once supersymmetry is broken. For simplicity, let us assume that SUSY is broken by an F-term of a gauge singlet hidden sector field X. The most general Kähler potential allowed by symmetries is then

$$\mathcal{L}_{B\mu} = \int d^4\theta \left(a\frac{X^\dagger}{M_{\rm Pl}} H_u H_d + b\frac{X^\dagger X}{M_{\rm Pl}^2} H_u H_d + \text{h.c.} \right) . \tag{8.174}$$

Note that while these terms must respect a global symmetry imposed to forbid a large μ, the symmetry is broken by the F-term of X. It is easy to see that (8.174) generates μ and B given by

$$\mu \sim a\frac{F^\dagger}{M_{\rm Pl}}, \quad B \sim b\frac{|F|^2}{M_{\rm Pl}^2} . \tag{8.175}$$

So far our discussion is very similar to the argument at the end of section 8.6.3. However, in the gravity mediation coupling constants a and b are both naturally of the order one. Combining this with an observation that (8.170) and (8.175) depend on the same dimensionful parameters, we conclude that the μ-problem is solved in this model.

8.7.2. *Gauge Mediation*

Minimal gauge mediation

To avoid the possibility that the Planck scale physics leads to observable flavor violation one could postulate that SUSY is broken the low energies and SUGRA contributions to the soft terms are negligible. To communicate SUSY breaking to the Standard Model fields, one then needs to introduce non-gravitational interactions between the hidden and visible sectors. If the two sectors interact only through the Standard Model gauge interactions, FCNC problem does not arise. This mechanism[41-44] is known as gauge mediated supersymmetry breaking (GMSB). It is instructive to start with a bottom-up approach to gauge mediation. We need to introduce new multiplets charged under all the Standard Model gauge groups. To avoid existing experimental constraints these messenger fields must be sufficiently heavy (which means that they must come in vector-like representations). To communicate SUSY breaking the spectrum of messenger multiplets should be non-supersymmetric. This is achieved by assuming that messengers couple to the SUSY breaking sector directly, either through Yukawa couplings or hidden sector gauge interactions. We will also choose messengers in complete representations of the $SU(5)$ gauge group. While this choice has the benefit of maintaining successful gauge coupling unification, it is not strictly required and the $SU(5)$ language that we will use in the rest of the discussion is largely a convenient book-keeping device. The simplest messenger content will contain N flavors of messengers Q and $\bar{Q}$ in 5 and $\bar{5}$ representations of $SU(5)$. The simplest way to parameterize messenger interactions with the SUSY breaking sector is by introducing a coupling to the SUSY-breaking spurion $X = M + \theta^2 F$:

$$W_{mess} = XQ\bar{Q}. \tag{8.176}$$

This form of the messenger spectrum is not the most general one, and the reader should consult the literature for examples of many interesting non-minimal models. To construct a complete GMSB model with dynamical supersymmetry breaking in the hidden sector one usually promotes the spurion X to a dynamical gauge singlet superfield. One then introduces interactions between X and the fields in the DSB sector that generate X and F_X vevs.

The assumption that the messengers only interact with the Standard Model fields through gauge interactions implies that holomorphic soft terms, *i.e.* A-terms and B-term are parametrically small in GMSB models. On the other hand, the Standard Model gauge interactions generate superpartner masses through processes shown in Figure 8.7. The resulting masses are given by:[50]

$$M_a = \frac{\alpha_a}{4\pi} N \frac{F}{M} g(x) \,,$$

$$\tilde{m}^2 = 2 \left| \frac{F}{M} \right|^2 \sum_a \left(\frac{\alpha_a}{4\pi} \right)^2 C_a N f(x) \,, \tag{8.177}$$

where $a = 1, 2, 3$ for $SU(3)$, $SU(2)$, and $U(1)_Y$ respectively, C_a is a quadratic Casimir of a relevant scalar, $x = F/M^2$, and

$$g(x) = \frac{1}{x^2} \left((1+x) \log(1+x) + (1-x) \log(1-x) \right) \,,$$

$$f(x) = \frac{1+x}{x^2} \left(\log(1+x) - 2\mathrm{Li}_2 \left(\frac{x}{1+x} \right) + \frac{1}{2}\mathrm{Li}_2 \left(\frac{2x}{1+x} \right) \right) \tag{8.178}$$
$$+ (x \to -x) \,.$$

It is often sufficient and convenient to work in the limit of small SUSY

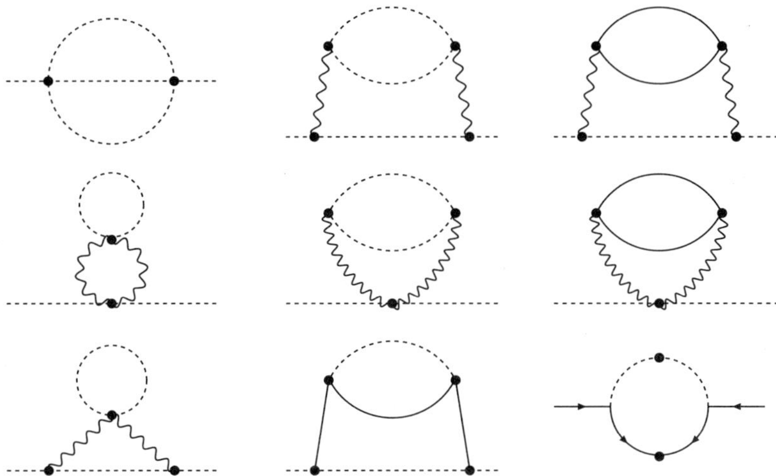

Fig. 8.7.   Diagrams giving rise to superpartner masses in gauge mediation.

breaking splitting within a messenger multiplet, $F \ll M^2$

$$\tilde{m}^2 = \sum_a \left(\frac{\alpha_a}{4\pi}\right)^a C_a N \left|\frac{F}{M}\right|^2 ,$$
$$M_a = \frac{\alpha_a}{4\pi} N \frac{F}{M} . \tag{8.179}$$

From (8.177) we see that sfermion and gaugino masses are generated at the same order in gauge couplings. Furthermore, requiring that the superpartner masses are at the electroweak scale we obtain a relation between the parameters of the messenger sector

$$F/M \sim 100\,\mathrm{TeV} . \tag{8.180}$$

On the other hand, gauge mediation allows a large range for the fundamental SUSY breaking scale, $100\,\mathrm{TeV} < F_{\mathrm{DSB}} < 10^{10}\,\mathrm{TeV}$. The lower bound arises from the requirement that messenger mass squareds are positive and therefore $F > M^2$. Combining this with (8.180) we conclude that the lower bound on both the messenger mass and the splitting within the messenger multiplet is of the order $100\,\mathrm{TeV}$. The requirement that the SUGRA contributions to soft masses are small compared to GMSB masses imposes an upper bound on the fundamental scale of SUSY breaking in the hidden sector, $F_{\mathrm{DSB}} < 10^{10}\,\mathrm{TeV}$. Note that even if the messenger masses are near the lower bound, fundamental scale of SUSY breaking may be significantly higher: if messengers couple to the DSB sector weakly it is quite possible that $F \ll F_{\rm DSB}$. The discussion of SUSY breaking scale allows us to determine expression for gravitino masses in GMSB models

$$m_{3/2} = \left(\frac{\sqrt{F_{\rm DSB}}}{100\,{\rm TeV}} \right)^2 2.4 {\rm eV} \,. \tag{8.181}$$

We chose messenger fields in the complete representations of the $SU(5)$ group to preserve one of the attractive features of the MSSM, gauge coupling unification. Insisting that the unification is perturbativity imposes an additional requirement — if the messengers are light (with masses of $\mathcal{O}\,(100{\rm TeV})$) the number of messengers is restricted to be no more than five to avoid the Landau pole below the GUT scale.

Direct gauge mediation

Generating the necessary spectrum for messenger fields is non-trivial. As we mentioned earlier, this can be achieved by promoting the spurion X to a dynamical field and introducing interactions of X with the DSB sector. Models of this type are often very complicated. Another interesting approach involves attempts to construct models of direct gauge mediation where messengers themselves play an essential role in SUSY-breaking dynamics. Realistic models of this type can be constructed if the DSB sector possesses a large global symmetry. Then one gauges an $SU(3) \times SU(2) \times U(1)$ subgroup of the flavor symmetry and identifies it with the MSSM. Several viable examples of direct gauge mediation exist in the literature.[45–48]

Let us illustrate direct gauge mediation with an explicit example.[48] This model takes an advantage of the recent discovery of metastable SUSY breaking. As a DSB sector we will choose a massive SUSY QCD with $\tilde{N}$ colors and F flavors in the magnetic description. The global $SU(F)$ symmetry of the theory is broken in a non-supersymmetric vacuum to an $SU(F-\tilde{N})$ subgroup. We will embed the Standard Model gauge group into the unbroken global symmetry of the DSB sector. One can easily see that the contribution of the DSB sector to the Standard Model β-functions is F. To avoid Landau poles as long as possible we choose minumal values for $F = 6$ and $\tilde{N} = 1$. With this choice, the infrared physics of the DSB sector is described by an s-confined QCD rather than magnetic gauge theory. The electric gauge group, $SU(N)_{DSB}$, has $N = F - \tilde{N} = 5$ colors. Let us write down the matter content of the model in the magnetic description:

$$
\begin{array}{|c|cccc c ccc|}
\hline
 & \phi & \bar\phi & \psi & \bar\psi & M & N & \bar N & X \\
\hline
SU(5)_{\mathrm{SM}} & \square & \bar\square & 1 & 1 & \mathbf{Adj}+\mathbf{1} & \square & \bar\square & 1 \\
\hline
\end{array}
\tag{8.182}
$$

where $SU(5)_{\mathrm{SM}}$ is an unbroken subgroup of the global $SU(6)$ symmetry. We will identify its $SU(3)\times SU(2)\times U(1)$ subgroup with the Standard Model. We choose the superpotential

$$
W = \bar\phi M\phi + \bar\psi X\psi + \bar\phi N\psi + \bar\psi \bar N\phi - f^2\mathrm{Tr}(M+X)\,.
\tag{8.183}
$$

Clearly this model has the same matter content and superpotential as the one given in (8.108) with the identifications:

$$
\begin{aligned}
B &\to (\phi,\psi)\,, \\
\bar B &\to (\bar\phi,\bar\psi)\,, \\
M &\to \begin{pmatrix} X & N \\ \bar N & M \end{pmatrix}\,.
\end{aligned}
\tag{8.184}
$$

We conclude that supersymmetry is broken. DSB sector fields, ϕ, $\bar\phi$, N, $\bar N$ and M, are responsible for SUSY breaking but, once $SU(3)\times SU(2)\times U(1)$ subgroup of the global symmetry is weakly gauged, they also serve as messengers. However, the model is not fully realistic at this stage. This is due to the fact that the Coleman-Weinberg potential in this model leads to the ground state at $M=0$ and an unbroken accidental R-symmetry. On the other hand, R-symmetry breaking is required to generate gaugino masses (it is also required to generate masses for fermions in M multiplet).

We need to modify the model so that M acquires vev in the ground state and breaks the R-symmetry. This can be achieved by introducing new fields S, $\bar S$, Z, and $\bar Z$ with interactions[e]

$$
W = (d\,\mathrm{Tr}M + m)\bar S S + m'(\bar Z S + S\bar Z)\,.
\tag{8.185}
$$

Since new Coleman-Weinberg contributions to the potential of $\mathrm{Tr}M$ favor the minimum at $\mathrm{Tr}M = -m/d$ it is possible to choose the parameters of the Lagrangian so that the minimum is shifted away from the origin. Another elegant way of breaking R-symmetry through gauge interactions was proposed in [46]. To make the SUSY breaking in this model fully dynamical, additional dynamics must be introduced to generate all mass terms in the superpotential.[48,49]

Finally, we should discuss gauge coupling evolution in this model. Formally gauge couplings unify since the messengers come in complete GUT

---

[e]This superpotential breaks the R-symmetry explicitly.

representations. However, the effective number of messengers below the confinement scale of the DSB sector is 7. Above the confinement scale the effective number of messengers is 5. It is therefore clear that the QCD coupling will hit the Landau pole at a scale of the order of $10^{11} - 10^{12}$TeV.

One of the goals in GMSB model building is finding theories with a very low SUSY breaking scale. In the model we just described this happens both by design and out of necessity — unless SUSY breaking scale is low, gaugino masses are too small. The model predicts new light particles which could potentially be observable at future colliders. On the other hand, the model is quite complicated and, most importantly, it can not be valid up to the GUT scale. These problems are general and often arise in other direct gauge mediation models based on metastable SUSY breaking that have been constructed recently.[46,47]

General gauge mediation

While the microscopic physics describing various GMSB models may be quite different, their low energy phenomenology is usually very similar and can be described in terms of two parameters, an effective SUSY breaking scale, $\Lambda = F/M$ and an effective number of messengers N. As we saw, models of direct low energy gauge mediation introduce new interesting features — additional light particles which, if we are lucky, may be accessible at colliders. More recently, it was realized[40] that GMSB phenomenology may be a lot richer. We will only briefly discuss these results here. Following [40] we will define general gauge mediation (GGMSB) as a class of models where SUSY breaking sector decouples from MSSM in the limit of vanishing MSSM gauge couplings, $\alpha_a \to 0$. This definition includes models of minimal and direct gauge mediation discussed earlier but it also includes strongly interacting theories. In such models, the perturbative calculations of superpartner masses are not reliable since the messengers themselves are strongly coupled. The authors of [40] analyzed GMSB contribution in terms of the correlation functions of gauge supercurrents and reached several important conclusions

- The description of the most general gauge mediation model requires three complex parameters describing gaugino masses and three real parameters describing contributions to sfermion masses from each of the Standard Model gauge groups[f].

---

[f]In addition there is a possibility for D-term contribution to sfermion masses proportional to their hypercharge quantum numbers. However, such a contribution is dangerous since it generically leads to tachyonic slepton masses. It can be forbidden, for example, by

- Gaugino masses formally arise at tree level while sfermion masses squared arise at one loop. As a result it may be possible to construct feasible models with a fundamental scale of supersymmetry breaking as low as 10 TeV. Unfortunately, such models are necessarily strongly coupled and calculation of the GGMSB parameters from the microscopic theory is not currently viable. On the other hand, weakly coupled theories will generally have an additional suppression resulting in the usual scaling of superpartner masses with the Standard Model gauge couplings but the SUSY breaking scale must be at least 100 TeV.

- As a consequence of the new scaling of superpartner masses with the Standard Model gauge couplings, there may exist a hierarchy between gaugino and sfermion masses. Therefore, the definition of general gauge mediation encompasses gaugino mediated supersymmetry breaking.[52] Moreover, in existing gaugino mediation models, supersymmetry is broken at a relatively high scale, and the renromalization group evolution leads to comparable sfermion and gaugino masses at the EWSB scale. On the other hand, since general GMSB models may break supersymmetry at very low energies, they could lead to a "true" gaugino mediated spectrum.

μ problem in gauge mediation

To illustrate the nature of the μ-problem in GMSB we can review the argument at the end of section 8.6.3 and identify the superfield X in (8.164) with the spurion that generates messenger mass. This means that in typical GMSB models (that is in the models with $F/X \sim 100\,\text{TeV}$) the coupling constant λ may be at most of the order $1/(16\pi^2)$. As we have seen such a small λ implies unacceptably large B-term. There are several viable examples where μ and B terms of the right size are generated without significant fine-tuning.[44,53] Several new ideas have been proposed recently.[54] However, one can not say that a fully satisfactory solution for the μ-problem in gauge mediation exists.

8.7.3. *Anomaly Mediation*

While the assumption of low scale supersymmetry breaking is attractive, it is not the only mechanism which may suppress flavor changing neutral currents. Even with the gravitino mass of the order TeV or larger, FCNCs

---

invoking messenger parity.[51]

may be suppressed if the Lagrangian has a sequestered form

$$K = -3M_{\mathrm{Pl}}^2 \ln \left(1 - \frac{f_{\mathrm{vis}}}{3M_{\mathrm{Pl}}^2} - \frac{f_{\mathrm{hid}}}{3M_{\mathrm{Pl}}^2} \right),$$

$$W = W_{\mathrm{vis}} + W_{\mathrm{hid}} + W_0,$$

(8.186)

where f_{vis} and f_{hid} are real functions of hidden and visible sector superfields respectively. Indeed, this form of the Lagrangian leads to vanishing of the soft terms at tree level. However, as we will see shortly, gaugino masses and A-terms are generated at one loop while the scalar mass squared are generated at two loops. This approach[55] to communication of SUSY breaking is referred to as the anomaly mediated supersymmetry breaking (AMSB).

The sequestered form of the Lagrangian may be achieved in one of two ways. The first approach[55] is based on the following assumptions:

- The fundamental theory lives in a 5-dimensional spacetime with one direction compactified on S^1/Z_2.
- The hidden and visible sector fields are localized on different boundaries of the extra dimension.
- There are no light bulk fields except for the fields in the supergravity multiplet.

With these assumptions, the locality of the low energy effective field theory guarantees the sequestered form of the effective Lagrangian. While originally the 5D construction was suggested in the context of the flat 5D backgrouns, it may also be implemented within the Randal-Sundrum scenario. The AdS/CFT correspondence then suggests that there should exist a 4-dimensional realization of the theory. Such a realization was found in [56]. The hidden sector is assumed to be nearly conformal. One can then treat interactions between the hidden and visible sectors in (8.155) as small perturbations of strong conformal dynamics in the hidden sector. As the hidden sector approaches the infrared fixed point, coupling constants, $c_{ff'}$, become negligibly small as a consequence of RG flow.

Given the sequestered form of the Lagrangian, one can integrate out dynamics of the hidden sector and parameterize supersymmetry breaking by an F-term of an auxiliary superfield in the supergravity multiplet, referred to as a compensator superfield[g]

$$\Phi = 1 + F_\Phi \theta^2.$$

(8.187)

---

[g]It is conventional to work in units of M_{Pl} and the F_Φ has an unusual dimension one, in fact $F_\Phi = m_{3/2}$.

The compensator superfield couples to the MSSM fields according to

$$\mathcal{L} = \int d^4\theta \Phi^\dagger \Phi K(Q^\dagger, e^V Q) + \left(\int d^2\theta \Phi^3 W(Q) + \text{h.c.} \right). \tag{8.188}$$

We can see the effects of sequestering here — the visible sector Lagrangian appears completely supersymmetric if we perform holomorphic field redefinition to write it in terms of a rescaled superfield $\tilde{Q} = \Phi Q$. This result is a consequence of scale invariance of MSSM Lagrangian at the tree level. However, at the quantum level scale invariance is lost. To maintain formal scale invariance we need to rescale not only light fields but also the cutoff scale of the theory, $\Lambda_{UV} \to \Phi \Lambda_{UV}$. This last rescaling results in the appearance of soft terms in the visible sector.

The most straightforward way to derive soft masses relies on a method of analytic continuation to the superspace.[57] Let us begin with gaugino masses. Writing down gauge kinetic terms with rescaled cutoffs and expanding the gauge coupling function in powers of θ^2 we obtain

$$\int d^2\theta \frac{1}{4g_i^2(\mu/\Lambda_{\text{UV}}\Phi)} \mathcal{W}^\alpha \mathcal{W}_\alpha = \int d^2\theta \left(\frac{1}{4g^2(\mu/\Lambda_{\text{UV}})} - \frac{b_i}{32\pi^2} \ln \Phi \right) \mathcal{W}^\alpha \mathcal{W}_\alpha, \tag{8.189}$$

where b_i is a one loop β-function coefficient. Expanding the log in the last term and performing superspace integral we obtain gaugino mass[55,58]

$$m_{\lambda_i}(\mu) = \frac{b_i}{2\pi} \alpha_i(\mu) F_\Phi. \tag{8.190}$$

Soft scalar masses are obtained by starting with an expression for the renormalized Kähler potential[55]

$$\int d^4\theta Z \left(\frac{\mu}{\Lambda_{\text{UV}}(\Phi^\dagger \Phi)^{1/2}} \right) Q^\dagger Q. \tag{8.191}$$

Expanding in powers of θ^2 leads to

$$\begin{aligned}
\tilde{m}_f^2(\mu) &= -\frac{1}{4} \frac{\partial \gamma_f(\mu)}{\partial \ln \mu} |F_\Phi|^2 \\
&= \frac{1}{4} \left(\frac{b_i}{2\pi} \alpha_i^2 \frac{\partial \gamma_f}{\partial \alpha_i} + \frac{b_\lambda}{2\pi} \frac{\lambda^2}{(4\pi)^2} \frac{\partial \gamma_f}{\partial \alpha_\lambda} \right) |F_\Phi|^2
\end{aligned} \tag{8.192}$$

where

$$\gamma_f(\mu) = \frac{\partial \ln Z(\mu)}{\partial \ln \mu} \tag{8.193}$$

is the anomalous dimension of the sfermion, λ is the Yukawa coupling, b_λ and b_i are one loop coefficients of gauge and Yukawa couplings respectively, and one needs to sum over all gauge and Yukawa couplings of the field in question. One can similarly obtain trilinear scalar soft terms[55]

$$
\begin{aligned}
A^u_{ff'} &= \frac{1}{2}\left(\gamma_u(\mu) + \gamma_f(\mu) + \gamma_{f'}(\mu)\right)\lambda^u_{ff'}F_\Phi\,, \\
A^d_{ff'} &= \frac{1}{2}\left(\gamma_d(\mu) + \gamma_f(\mu) + \gamma_{f'}(\mu)\right)\lambda^d_{ff'}F_\Phi\,,
\end{aligned}
\tag{8.194}
$$

where $\gamma_u(\mu)$ and $\gamma_d(\mu)$ are anomalous dimensions of H_u and H_d respectively.

As we can easily see, AMSB is an extremely predictive theory — the soft parameters are given in terms of F_Φ and the Standard Model gauge and Yukawa couplings at the TeV scale. The minimal model is insensitive to UV physics. Let us illustrate this by adding[h] to MSSM a set of heavy fields, a vector-like fourth generation with mass m_H:

$$
\mathcal{L}_H = \int d^4\theta\,\Phi^\dagger\Phi K(Q_H^\dagger, e^V Q_H) + \left(\int d^2\theta\,\Phi^3 m_H Q_H \bar{Q}_H + \text{h.c.}\right)\,. \tag{8.195}
$$

We have included coupling to the compensator field Φ. Naively, it appears that new fields might affect AMSB predictions since they modify β-functions at high scales $\mu \gg m_H$. However, the tree level Lagrangian of heavy superfields depends on Φ even after holomorphic rescaling — the spectrum of the heavy supermultiplet is not supersymmetric. In fact, identifying $m_H\Phi$ with the spurion X of GMSB models, we see that heavy superfields play a role of messengers. The soft masses in the infrared are given by the sum of the high energy AMSB contribution and gauge mediated contribution of the new fields. It is easy to check that to the leading order in F_Φ the soft parameters in the IR, $\mu \ll m_H$, are completely determined by β-functions and coupling constants of the low energy theory.

Unfortunately, the minimal AMSB model can be immediately ruled out. Slepton masses squared given by (8.192) are negative! It turns out to be extremely difficult to modify AMSB models to fix the slepton mass problem — the difficulty is due to celebrated UV-insensitivity of anomaly mediation. A number of solutions[59] to this problem were proposed over the years. However, while many of these solutions are viable, none of them seem sufficiently compelling as they are typically quite complicated and almost necessarily sacrifice the UV-insensitivity of the anomaly mediation.

---

[h]To slightly simplify the argument we will assume that new fields do not couple to MSSM in the superpotential.

Acknowledgments

I would like to thank Tao Han and K. T. Mahantappa, as well as students and lecturers at TASI 2008 for providing great and stimulating atmosphere at. This work was supported in part by NSF grant No. PHY-0653656.

References

1. J. Wess and J. Bagger, *Princeton, USA: Univ. Pr. (1992) 259 p*
2. H. K. Dreiner, H. E. Haber and S. P. Martin, arXiv:0812.1594 [hep-ph].
3. S. R. Coleman and J. Mandula, Phys. Rev. **159** (1967) 1251. R. Haag, J. T. Lopuszanski and M. Sohnius, Nucl. Phys. B **88**, 257 (1975).
4. Yu. A. Golfand and E. P. Likhtman, JETP Lett. **13**, 323 (1971) [Pisma Zh. Eksp. Teor. Fiz. **13**, 452 (1971)].
5. J. Wess and B. Zumino, Nucl. Phys. B **78**, 1 (1974).
6. I. Affleck, M. Dine and N. Seiberg, Nucl. Phys. B **241**, 493 (1984).
7. I. Affleck, M. Dine and N. Seiberg, Nucl. Phys. B **256**, 557 (1985).
8. M. T. Grisaru, W. Siegel and M. Rocek, Nucl. Phys. B **159**, 429 (1979).
9. N. Seiberg, Phys. Lett. B **318**, 469 (1993) [arXiv:hep-ph/9309335].
10. V. A. Novikov, M. A. Shifman, A. I. Vainshtein and V. I. Zakharov, Nucl. Phys. B **229**, 381 (1983); V. A. Novikov, M. A. Shifman, A. I. Vainshtein and V. I. Zakharov, Nucl. Phys. B **260**, 157 (1985) [Yad. Fiz. **42**, 1499 (1985)]; V. A. Novikov, M. A. Shifman, A. I. Vainshtein and V. I. Zakharov, Phys. Lett. B **166**, 329 (1986) [Sov. J. Nucl. Phys. **43**, 294.1986 YAFIA,43,459 (1986 YAFIA,43,459-464.1986)]; M. A. Shifman and A. I. Vainshtein, Nucl. Phys. B **359**, 571 (1991).
11. M. A. Shifman and A. I. Vainshtein, Nucl. Phys. B **277**, 456 (1986) [Sov. Phys. JETP **64**, 428 (1986 ZETFA,91,723-744.1986)].
12. W. Fischler, H. P. Nilles, J. Polchinski, S. Raby and L. Susskind, Phys. Rev. Lett. **47**, 757 (1981).
13. M. Dine, arXiv:hep-ph/9612389; S. Weinberg, Phys. Rev. Lett. **80**, 3702 (1998) [arXiv:hep-th/9803099].
14. A. C. Davis, M. Dine and N. Seiberg, Phys. Lett. B **125**, 487 (1983).
15. S. F. Cordes, Nucl. Phys. B **273**, 629 (1986); D. Finnell and P. Pouliot, Nucl. Phys. B **453**, 225 (1995) [arXiv:hep-th/9503115].
16. N. Seiberg, Phys. Rev. D **49**, 6857 (1994) [arXiv:hep-th/9402044].
17. N. Seiberg, Nucl. Phys. B **435**, 129 (1995) [arXiv:hep-th/9411149].
18. T. Banks and A. Zaks, Nucl. Phys. B **196**, 189 (1982).
19. L. O'Raifeartaigh, Nucl. Phys. B **96**, 331 (1975).
20. D. Shih, JHEP **0802** (2008) 091 [arXiv:hep-th/0703196].
21. K. A. Intriligator, N. Seiberg and D. Shih, JHEP **0604**, 021 (2006) [arXiv:hep-th/0602239].
22. E. Witten, Nucl. Phys. B **202**, 253 (1982).
23. K. A. Intriligator and S. D. Thomas, Nucl. Phys. B **473**, 121 (1996)

[arXiv:hep-th/9603158]; K. I. Izawa and T. Yanagida, Prog. Theor. Phys. **95**, 829 (1996) [arXiv:hep-th/9602180].

24. A. E. Nelson and N. Seiberg, Nucl. Phys. B **416**, 46 (1994) [arXiv:hep-ph/9309299].

25. J. Bagger, E. Poppitz and L. Randall, Nucl. Phys. B **455**, 59 (1995) [arXiv:hep-ph/9505244].

26. I. Affleck, M. Dine and N. Seiberg, Phys. Lett. B **137**, 187 (1984).

27. Y. Meurice and G. Veneziano, Phys. Lett. B **141**, 69 (1984).

28. H. Murayama, Phys. Lett. B **355**, 187 (1995) [arXiv:hep-th/9505082].

29. Y. Shirman, Phys. Lett. B **389**, 287 (1996) [arXiv:hep-th/9608147]; N. Arkani-Hamed and H. Murayama, Phys. Rev. D **57**, 6638 (1998) [arXiv:hep-th/9705189].

30. Z. Chacko, M. A. Luty and E. Ponton, JHEP **9812**, 016 (1998) [arXiv:hep-th/9810253].

31. S. R. Coleman, Phys. Rev. D **15**, 2929 (1977) [Erratum-ibid. D **16**, 1248 (1977)].

32. S. A. Abel, C. S. Chu, J. Jaeckel and V. V. Khoze, JHEP **0701**, 089 (2007) [arXiv:hep-th/0610334]; N. J. Craig, P. J. Fox and J. G. Wacker, Phys. Rev. D **75**, 085006 (2007) [arXiv:hep-th/0611006]; W. Fischler, V. Kaplunovsky, C. Krishnan, L. Mannelli and M. A. C. Torres, JHEP **0703**, 107 (2007) [arXiv:hep-th/0611018].

33. P. Fayet and J. Iliopoulos, Phys. Lett. B **51** (1974) 461.

34. A. Salam and J. A. Strathdee, Phys. Lett. B **49**, 465 (1974).

35. E. Witten, Nucl. Phys. B **188**, 513 (1981).

36. S. Dimopoulos and H. Georgi, Nucl. Phys. B **193**, 150 (1981).

37. Y. Nir and N. Seiberg, Phys. Lett. B **309**, 337 (1993) [arXiv:hep-ph/9304307]; M. Leurer, Y. Nir and N. Seiberg, Nucl. Phys. B **420**, 468 (1994) [arXiv:hep-ph/9310320].

38. A. G. Cohen, T. S. Roy and M. Schmaltz, JHEP **0702**, 027 (2007) [arXiv:hep-ph/0612100].

39. G. F. Giudice and A. Masiero, Phys. Lett. B **206**, 480 (1988).

40. P. Meade, N. Seiberg and D. Shih, arXiv:0801.3278 [hep-ph].

41. M. Dine, W. Fischler and M. Srednicki, Nucl. Phys. B **189**, 575 (1981); S. Dimopoulos and S. Raby, Nucl. Phys. B **192**, 353 (1981); M. Dine and W. Fischler, Phys. Lett. B **110**, 227 (1982); C. R. Nappi and B. A. Ovrut, Phys. Lett. B **113**, 175 (1982); L. Alvarez-Gaume, M. Claudson and M. B. Wise, Nucl. Phys. B **207**, 96 (1982); S. Dimopoulos and S. Raby, Nucl. Phys. B **219**, 479 (1983).

42. M. Dine and A. E. Nelson, Phys. Rev. D **48**, 1277 (1993) [arXiv:hep-ph/9303230].

43. M. Dine, A. E. Nelson and Y. Shirman, Phys. Rev. D **51**, 1362 (1995) [arXiv:hep-ph/9408384].

44. M. Dine, A. E. Nelson, Y. Nir and Y. Shirman, Phys. Rev. D **53**, 2658 (1996) [arXiv:hep-ph/9507378].

45. E. Poppitz and S. P. Trivedi, Phys. Rev. D **55**, 5508 (1997) [arXiv:hep-ph/9609529]; N. Arkani-Hamed, J. March-Russell and H. Murayama, Nucl.

Phys. B **509**, 3 (1998) [arXiv:hep-ph/9701286]; S. Dimopoulos, G. R. Dvali, R. Rattazzi and G. F. Giudice, Nucl. Phys. B **510**, 12 (1998) [arXiv:hep-ph/9705307]; H. Murayama, Phys. Rev. Lett. **79**, 18 (1997) [arXiv:hep-ph/9705271]; M. A. Luty, Phys. Lett. B **414**, 71 (1997) [arXiv:hep-ph/9706554]; K. I. Izawa, Y. Nomura, K. Tobe and T. Yanagida, Phys. Rev. D **56**, 2886 (1997) [arXiv:hep-ph/9705228]; Y. Shirman, Phys. Lett. B **417**, 281 (1998) [arXiv:hep-ph/9709383].

46. M. Dine and J. Mason, Phys. Rev. D **77**, 016005 (2008) [arXiv:hep-ph/0611312];

47. R. Kitano, H. Ooguri and Y. Ookouchi, Phys. Rev. D **75**, 045022 (2007) [arXiv:hep-ph/0612139]; H. Murayama and Y. Nomura, Phys. Rev. Lett. **98**, 151803 (2007) [arXiv:hep-ph/0612186]; O. Aharony and N. Seiberg, JHEP **0702**, 054 (2007) [arXiv:hep-ph/0612308]; N. Haba and N. Maru, Phys. Rev. D **76**, 115019 (2007) [arXiv:0709.2945 [hep-ph]].

48. C. Csaki, Y. Shirman and J. Terning, JHEP **0705**, 099 (2007) [arXiv:hep-ph/0612241].

49. M. Dine, J. L. Feng and E. Silverstein, Phys. Rev. D **74**, 095012 (2006) [arXiv:hep-th/0608159].

50. S. P. Martin, Phys. Rev. D **55**, 3177 (1997) [arXiv:hep-ph/9608224].

51. S. Dimopoulos and G. F. Giudice, Phys. Lett. B **393**, 72 (1997) [arXiv:hep-ph/9609344].

52. D. E. Kaplan, G. D. Kribs and M. Schmaltz, Phys. Rev. D **62**, 035010 (2000) [arXiv:hep-ph/9911293]; Z. Chacko, M. A. Luty, A. E. Nelson and E. Ponton, JHEP **0001**, 003 (2000) [arXiv:hep-ph/9911323]; M. Schmaltz and W. Skiba, Phys. Rev. D **62**, 095004 (2000) [arXiv:hep-ph/0004210]; M. Schmaltz and W. Skiba, Phys. Rev. D **62**, 095005 (2000) [arXiv:hep-ph/0001172].

53. G. R. Dvali, G. F. Giudice and A. Pomarol, Nucl. Phys. B **478**, 31 (1996) [arXiv:hep-ph/9603238]; M. Dine, Y. Nir and Y. Shirman, Phys. Rev. D **55**, 1501 (1997) [arXiv:hep-ph/9607397]; T. Yanagida, Phys. Lett. B **400**, 109 (1997) [arXiv:hep-ph/9701394]; S. Dimopoulos, G. R. Dvali and R. Rattazzi, Phys. Lett. B **413**, 336 (1997) [arXiv:hep-ph/9707537].

54. A. Delgado, G. F. Giudice and P. Slavich, Phys. Lett. B **653**, 424 (2007) [arXiv:0706.3873 [hep-ph]]; T. S. Roy and M. Schmaltz, Phys. Rev. D **77**, 095008 (2008) [arXiv:0708.3593 [hep-ph]]; H. Murayama, Y. Nomura and D. Poland, Phys. Rev. D **77**, 015005 (2008) [arXiv:0709.0775 [hep-ph]]; G. F. Giudice, H. D. Kim and R. Rattazzi, Phys. Lett. B **660**, 545 (2008) [arXiv:0711.4448 [hep-ph]]; C. Csaki, A. Falkowski, Y. Nomura and T. Volansky, Phys. Rev. Lett. **102**, 111801 (2009) [arXiv:0809.4492 [hep-ph]]; Z. Komargodski and N. Seiberg, JHEP **0903**, 072 (2009) [arXiv:0812.3900 [hep-ph]].

55. L. Randall and R. Sundrum, Nucl. Phys. B **557**, 79 (1999) [arXiv:hep-th/9810155].

56. M. A. Luty and R. Sundrum, Phys. Rev. D **65**, 066004 (2002) [arXiv:hep-th/0105137]; M. Luty and R. Sundrum, Phys. Rev. D **67**, 045007 (2003) [arXiv:hep-th/0111231].

57. N. Arkani-Hamed, G. F. Giudice, M. A. Luty and R. Rattazzi, Phys. Rev. D

58 (1998) 115005 [arXiv:hep-ph/9803290].

58. G. F. Giudice, M. A. Luty, H. Murayama and R. Rattazzi, JHEP **9812**, 027 (1998) [arXiv:hep-ph/9810442].

59. A. Pomarol and R. Rattazzi, JHEP **9905**, 013 (1999) [arXiv:hep-ph/9903448]; Z. Chacko, M. A. Luty, I. Maksymyk and E. Ponton, JHEP **0004**, 001 (2000) [arXiv:hep-ph/9905390]; E. Katz, Y. Shadmi and Y. Shirman, JHEP **9908**, 015 (1999) [arXiv:hep-ph/9906296]; I. Jack and D. R. T. Jones, Phys. Lett. B **482**, 167 (2000) [arXiv:hep-ph/0003081]; M. S. Carena, K. Huitu and T. Kobayashi, Nucl. Phys. B **592**, 164 (2001) [arXiv:hep-ph/0003187]; B. C. Allanach and A. Dedes, JHEP **0006**, 017 (2000) [arXiv:hep-ph/0003222]; Z. Chacko, M. A. Luty, E. Ponton, Y. Shadmi and Y. Shirman, Phys. Rev. D **64**, 055009 (2001) [arXiv:hep-ph/0006047]; D. E. Kaplan and G. D. Kribs, JHEP **0009**, 048 (2000) [arXiv:hep-ph/0009195]; A. E. Nelson and N. T. Weiner, arXiv:hep-ph/0210288; R. Sundrum, Phys. Rev. D **71**, 085003 (2005) [arXiv:hep-th/0406012].

Chapter 9

Strings for Particle Physicists

Gary Shiu

Department of Physics, University of Wisconsin, Madison, WI 53706
shiu@physics.wisc.edu

This is a set of four lectures on string phenomenology delivered at the TASI 2008 Summer School. These lectures are devoted to introducing some basic notions of string theory, with an emphasis on their applications to particle physics model building. I will discuss different ways of building four-dimensional string vacua with $\mathcal{N} \leq 1$ supersymmetry, and various attempts to construct realistic models. Focussing on D-brane models, I will describe how several beyond the Standard Model ideas such as supersymmetry, extra dimensions (large or warped), and technicolor-like theories can arise in string theory.

9.1. Introduction

The theme of this year's TASI is "The Dawn of the LHC Era", so perhaps the first question that comes to your mind is, "What does string theory have to do with the LHC?" The purpose of these lectures is to provide at least a partial answer to this question.

String theory is by far our best developed quantum theory of gravity. As such, it provides a consistent framework to address questions about early universe cosmology and black hole physics. As other lecturers in this school have discussed, these issues are of relevance to the LHC, at least indirectly. But what about more traditional particle physics questions, such as those we hope to unveil from the LHC? Does string theory have anything to say about physics beyond the Standard Model (BSM)?

Many of us have our own favorite scenario (or scenarios) of BSM physics. Only experiment can tell, and it may well be something that nobody has thought about. However, it is fair to say that the main contenders of BSM physics are:

- Supersymmetry
- Extra Dimensions
- Strong dynamics (technicolor)

Let's see how string theory score on this front. Supersymmetry was discovered to some extent first in the context of string theory, as a way to introduce fermions and to remove the unwanted tachyons (see, e.g., recent textbooks [1–5]). The idea of extra dimensions dated back to Kaluza and Klein so string theory cannot claim credit for their discovery. However, theories with extra dimensions are not renormalizable. To make sense of these theories, we need to complete them in the UV. String theory provides such a UV completion. In fact, string theory *requires* extra dimensions. Moreover, as we will see in these lectures, there are stringy constraints on extra dimensional physics that are not apparent from a low-energy bottom-up approach. So, one might hope that string theory can shed light on what kinds of extra dimensional scenarios are more likely to be realized in nature. Finally, technicolor,[6,7] just like extra dimensions, was introduced without any input from string theory. Nevertheless, one of the earlier proponents of this idea is a string theorist! More importantly, the advent of the AdS/CFT correspondence[8–11] provides a dual gravity description of such theories, and have offered insights into technicolor model building. In contrast to what we might have thought, string theory has a lot to say about BSM physics. More generally, the driving force behind the studies of physics beyond the Standard Model is arguably the hierarchy problem which is intrinsically about the existence of a high cutoff scale. String theory is one of the best motivated theories at work at such high energies. At any rate, string theory has shown to be a rather rich scenario generator. While we are still in the dark waiting for the dawn, string theory ideas such as branes and different extra dimensional scenarios may shed light on what to anticipate at the LHC.

Therefore, my lectures will be centering around string constructions with a view towards these BSM ideas. I will first introduce the five different superstring theories which are formulated in 10 dimensions and discuss their low energy effective theories upon compactication. As you will see, D-brane models allow for the possibility of a much lower fundamental scale and so are of interests to the LHC. I then discuss different ways in which chirality can arise in D-brane models, and apply these results to building "semi-realistic" models. Finally, I will discuss how warped models can be constructed from string theory. These models are string theory realizations of the Randall-Sundrum scenario.[12,13] They can also be thought of as the gravity duals of technicolor theories.

My lectures are drawn heavily from my review article [14] and several excellent lectures notes on the subject.[15,16] I refer the readers to the review and these lecture notes for details that I have omitted and for references therein. Due to severe page limitations, the list of references in this set of lecture notes will be kept to a minimal. I apologize to those whose important work has not been properly referred to as a result. Fortunately, the referencing in the aforementioned articles is rather exhaustive and so the readers should be able to start from the references given there and follow the trails. Finally, these lectures are to some extent complementary to that of Bogdan Dobrescu in this school who focused on the phenomenological aspects of extra dimensions.

9.2. String Theory Scenarios

9.2.1. *The Pre D-brane Era*

First, let us go back in time to the mid 1980s (when many of the students in the school were still babies!) and assess the state of string phenomenology. It was known that there are five consistent string theories, all formulated in ten dimensions: Type I theory with gauge symmetry $SO(32)$ (open and closed strings), Type IIA and IIB closed strings, and two closed heterotic strings with gauge symmetries $E_8 \times E_8$ and $SO(32)$ respectively. The massless spectrum of these five string theories are shown in the table. The yet unknown extension of string theory, M-theory, adds a further theory to these five, which has the massless spectrum of 11-dimensional supergravity g_{MN}, C_{MNP} (plus the fermionic partners). One can appreciate the difficulty in constructing a realistic string model by comparing the spectrum in this table to that of the Standard Model.

A few observations are in order:

- Among the five string theories, three of them (Type I and the two heterotic strings) already contain gauge bosons in ten dimensions. They seem more promising as a starting point to construct the Standard Model and chiral fermions upon dimensional reduction.
- For this reason, Type II theories look much less interesting. There was even a no-go theorem which forbids them to produce the Standard Model at low energies.[17] We will revisit this no-go theorem again after we discuss D-branes.

Table 9.1. The massless bosonic spectrum for the five consistent string theories in 10-dimensions. In Type I and II theories, the spectrum is splitted into an NS-NS sector corresponding to states built out of two worldsheet bosonic states in the NS-R formulation of the theories and a R-R sector corresponding to those constructed from two worldsheet fermionic states. The number of supersymmetries is related to the number of supercharges. 16 supercharges correspond to $\mathcal{N} = 1$ supersymmetry in 10 dimensions, leading to $\mathcal{N} = 4$ supersymmetry in 4D upon dimensional reduction on a torus, whereas 32 supercharges correspond to $\mathcal{N} = 2$ and $\mathcal{N} = 8$ in 10D and 4D respectively.

Theory	Strings	Supercharges	Bosonic Spectrum
Heterotic $E_8 \times E_8$	Closed Oriented	16	$g_{\mu\nu}$, $B_{\mu\nu}$, ϕ $A_\mu^{i\bar{j}}$ in adjoint rep.
Heterotic $SO(32)$	Closed Oriented	16	$g_{\mu\nu}$, $B_{\mu\nu}$, ϕ $A_\mu^{i\bar{j}}$ in adjoint rep.
Type I $SO(32)$	Open Closed Unoriented	16	NS-NS: $g_{\mu\nu}$, ϕ R-R: C_2 Open: $A_\mu^{i\bar{j}}$ in adjoint rep.
Type IIA	Closed Oriented	32	NS-NS: $g_{\mu\nu}$, $B_{\mu\nu}$, ϕ R-R: C_1, C_3
Type IIB	Closed Oriented	32	NS-NS: $g_{\mu\nu}$, $B_{\mu\nu}$, ϕ R-R: C_0, C_2, C_4

- The heterotic $E_8 \times E_8$ attracted much of the attention because it seems the most promising for phenomenology: upon compactification to 4D, it can give rise to chiral $\mathcal{N} = 1$ supersymmetric models with familiar gauge and matter content. The observable sector comes from the first E_8 which contains the Standard Model gauge symmetry

$$E_8 \supset SU(3) \times SU(2) \times U(1) \tag{9.1}$$

and several families of mater fields. The second E_8 gives rise to a hidden sector, which fits perfectly with attempts of supersymmetric model building prior to string theory. A hidden sector was often proposed to break supersymmetry at an intermediate scale $\sim 10^{12}$ GeV and gravity plays the role of messenger of supersymmetry breaking to the observable sector, which feels the breaking of supersymmetry just above the electroweak scale $\sim 10^3$ GeV.

- Type I and the $SO(32)$ heterotic string can in principle also lead to realistic gauge and matter content at low energies. However, the connection to GUT model building and the hidden sector paradigm is less direct.

Therefore, in the mid 1980s, the standard paradigm of string phenomenology was the $E_8 \times E_8$ heterotic string. A great deal of effort was dedicated to constructing and studying different compacitifications that lead to realistic models. The properties of the compact 6D manifold determine the low-energy physics. Hence, Calabi-Yau manifolds (or its singular limits known as orbifolds) are often chosen because they preserve $\mathcal{N} = 1$ supersymmetry in 4D, as well as permitting the existence of chiral fermions. In the simplest case (known as the standard embedding), the observable E_8 symmetry group is broken to E_6 which was a natural grand unified group, with the hidden E_8 unbroken. Also, the number of chiral families is given by the topological Euler number of the manifold. This deep connection between geometry with physics results in a great flourish of activities in string model building.

9.2.1.1. *Energy Scales in Heterotic String Theories*

Instead of discussing details of these heterotic string constructions, let us take a first look at a more basic property, namely, the fundamental energy scale. In heterotic string theory, both gravity and gauge fields come from the same source which are closed strings. So, both kinds of fields propagate in the full 10 spacetime dimensions. In 10D, the low energy effective action takes the form:

$$S_{10d} = M_s^8 \int d^{10}x \sqrt{-g} e^{-2\phi} \left(\mathcal{R} + \frac{1}{4} M_s^{-2} F_{MN}^2 + \dots \right) \qquad (9.2)$$

where $M_s = 1/\sqrt{\alpha'}$ is the string scale and ϕ is the dilaton field. We have suppressed the fermionic part of the effective action. Upon compactification to 4D, each of the two terms above will receive a volume factor coming from the integration of the 6 extra dimensions (assuming a factorized geometry $\mathcal{M}_{10} = \mathcal{M}_4 \times \mathcal{M}_6$). Comparing with the 4D effective action for gravity and Yang-Mills gauge fields:

$$S_{4d}^{eff} = \int d^4x \sqrt{-g} \left(M_P^2 \mathcal{R}_4 + \frac{1}{4 g_{YM}^2} F_{\mu\nu}^2 + \dots \right) \qquad (9.3)$$

we obtain the expression for the gravitational and gauge couplings:

$$M_P^2 = \frac{1}{g_s^2} M_s^8 V_6, \qquad g_{YM}^{-2} = \frac{1}{g_s^2} M_s^6 V_6 \qquad (9.4)$$

where V_6 is the overall volume of the extra dimensions and $g_s = \langle e^\phi \rangle$. (Precise numerical factors can be found in Polchinski's textbook [2]). Taking the ratio of these expressions, the volume and the dilaton factors cancel and we obtain:

$$M_P^2 \simeq g_{YM}^{-2} M_s^2 \qquad (9.5)$$

Therefore for $\alpha_{YM} \equiv g_{YM}^2/(4\pi)$ not too different from $1/25$ (as expected from low energies), we have the fundamental scale M_s to be not far from the Planck scale $M_P \sim 10^{19}$ GeV, i.e.,

$$M_s \simeq g_{YM} M_P \sim (10^{17} - 10^{18}) GeV \qquad (9.6)$$

Note that:

- M_s is very high, which makes the heterotic scenario very hard to test directly.
- In addition, the compactification scale $M_c \equiv V_6^{-1/6}$ cannot be too small, because M_c is the scale of KK modes of the Standard Model gauge bosons and matter.
- To trust the low energy supergravity approximation, $M_c < M_s$, this gives $g_{YM} < g_s$ and so we are in the perturbative limit.
- Since the relevant energy scales (M_c and M_s) are much bigger than M_{EW}, we need to protect Standard Model physics from large radiative corrections. Usually, supersymmetry is invoked to do the job.
- Although our present discussion assumes V_6 and g_s are parameters dialable at will, they are actually vevs of scalar fields known as moduli. These moduli parametrizes the size and shape of the extra dimensions and are stabilized by a potential. Moduli stabilization is a subject on its own and we will discuss some recent ideas in the last lecture.

9.2.2. *The Post D-brane Era*

String phenomenology has taken a drastic turn in the mid 1990s. The discovery of string dualities suggested that the five consistent string theories together with 11D SUGRA are different manifestations of the same underlying M-theory. It has also become clear that higher dimensional surfaces known as D-branes play a key role in string theory and its applications to phenomenology, giving support to the "brane world" idea.

The techniques of constructing realistic models have been very much increased. We can now start from any of the 6 descriptions of M-theory and obtain models with features of the Standard Model.

(1) *11D SUGRA:* Just as the requirement of $\mathcal{N} = 1$ supersymmetry in 4D singles out Calabi-Yau compactifications of string theory, G_2 manifolds provide a geometric construction if we start from 11D. These G_2 manifolds are much less understood but it has been shown that it is not possible to obtain chiral fermions except for singular points. General and elegant results exist for these constructions but no explicit compact models have been constructed in this way so far (except in the weak coupling limit when such compactifications reduce to known string theory constructions).

A simpler way to obtain $\mathcal{N} = 1$ models from 11D SUGRA is in fact closely related to the heterotic string theory we just discussed. In the strong coupling limit, an extra dimension opens up. In the Horava-Witten construction, the 11-th dimension is an interval (an S_1/Z_2 orbifold). One can think of the two E_8 as living at the endpoints of the interval which are 10D surfaces. Further compactification on Calabi-Yau manifolds give rise to interesting chiral models. Some explicit models have been constructed although their phenomenological properties are difficult to extract given the mathematical complexity and the fact that we do not actually know what is the full completion of 11D SURGRA. Still interesting results continued to be obtained in this direction. Incidentally, an additional dimension accessible only to gravity and not to gauge and matter fields helps solve the discrepancy between the unification scale and the Planck scale.

(2) *Type I, IIA, IIB Strings:* The main new ingredient in these models is the existence of D-branes, which are surfaces on which open strings can end. D-branes can carry gauge and matter fields within their world volume. They come in various dimensions: a D_p brane has 1 time and p space dimensions. The subject of these lectures is to introduce model building techniques with D-branes. Details will be discussed further as we go along, but as a preview, let us mention that to get chiral models, the branes have to placed at singular points of a manifold, or to intersect at non-trivial angles and chiral matter lives only at the intersection, or to turn on world-volume flux.

Further generalizations of these D-brane constructions include the so called F-theory (proposed by Vafa). In Type IIB string theory, the two scalar fields of the 10D theory, the dilaton and axion are combined into one complex field $S = a + i\phi$, which realizes the S-duality symmetry

$SL(2, \mathbf{Z})$:

$$S \to \frac{aS + b}{cS + d} \qquad a, b, c, d \in \mathbf{Z} \text{ with } ad - bc = 1 \qquad (9.7)$$

which is similar to how the modular parameter of a torus transforms. Therefore, one can "geometrize" this varying axio-dilaton background as compactification on a 4 complex dimensional Calabi-Yau manifold which is locally a product of this torus with a six-dimensional (non Calabi-Yau) base B_3 under which the Type II theory is compactified. The four-fold is then said to be an elliptically fibered Calabi-Yau[a]. These compactifications naturally incorporate D7-branes, which are given by points in the base[b] where the elliptic fibration degenerates.

Let us take stock of these new insights we gain in the post D-brane era. The common feature in these new constructions is the fact that our world can be a brane – either a D-brane or the end-of-the-world brane in the Horava-Witten construction, or possibiliy a surface at the singularity of a G_2 or elliptically fibered Calabi-Yau four-fold in M and F-theory respectively. The brane world scenario has been a subject of intense investigations during the past 10 years and new mechanisms have been proposed to solve longstanding problems with the Standard Model, such as the hierarchy problem, gauge coupling unification, neutrino masses, the strong CP problem etc, without necessarily referring to string theory. Some of these topics will be discussed in Bodgan Dobrescu's lectures. Here we will concentrate only on string theoretical realizations.

One of the interesting properties of this scenario is that it allows for a fundamental scale of nature to be much lower than the Planck scale and therefore closer to the energies accessible to experiments. We wil discuss explicit realizations of this scenario in the next lecture.

9.2.2.1. *D-brane Scenarios*

Before we discuss the construction of D-brane models, let us revisit the relation between the gravitational and gauge couplings with the fundamental scale.

If the Standard Model is localized on the worldvolume of D_p branes,

---

[a]In particular, B_3 itself can be thought of as a local product of K_3 and P_1.
[b]More precisely, points in P_1.

the low energy effective action in 4D takes the form:

$$S_{4d} = \int d^4x \sqrt{-g} \left(M_s^8 V_6 e^{-2\phi} \mathcal{R}_4 + \frac{1}{4} M_s^{p-3} V_{p-3} e^{-\phi} F_{\mu\nu}^2 + \dots \right) \quad (9.8)$$

There are two crucial differences in comparison to the heterotic case. First, the power of the dilaton is different for the gravity and the gauge kinetic terms. Second, the total volume V_6 enters on the gravity part but only the volume of the $p - 3$ cycle (i.e., $p - 3$ dimensional subspace) of the internal manifold V_{p-3} that the p-branes wrap around appears in the gauge part of the action. In particular for a $D3$-brane, there is no volume factor contribution to the gauge coupling. The gravitational and gauge couplings are then related to the fundamental string scale as follows:

$$M_P^2 = \frac{M_s^8}{(2\pi)^7} \frac{V_6}{g_s^2} , \qquad \alpha_{YM}^{-1} = 4\pi \frac{M_s^{p-3} V_{p-3}}{g_s} \quad (9.9)$$

where we have restored factors of 2 and π. We can easily see that if the Standard Model fits inside a D3-brane, for instance, we may have M_s substantially smaller than M_P as long as the volume of the extra dimensions are large enough, without affecting the gauge couplings. More generally, the above relations illustrate the "brane world" effect where the dimensions transverse to the branes (which are not necessarily D3-branes) if large can lower the fundamental string scale. In the heterotic case, setting the volume very large would make the gauge couplings extremely small, which is unrealistic.

The different power of the dilaton in the gravity and gauge kinetic terms also give rise to added flexibilities. Even if the Standard Model is within a set of D9-branes which fill all space, for sufficiently weak string coupling, the string scale can be significantly lower than the Planck scale because of the difference in dilaton factor. However, this dilaton factor can only help us in getting a few orders of magnitude suppression (say $M_s = M_{GUT}$ rather than $M_s = M_{EW}$) since otherwise the gauge couplings will be too weak.

The possibility of having large dimensions add a "geometrical" view of the hierarchy problem, though we still need to explain why the size of the dimensions are stabilized to a large value. Several scenarios have been proposed depending on the value of the fundamental scale. The main scenarios at present are:

(1) $M_s \sim M_P$. This is analogous to the heterotic scenario and similar
    comments apply.

(2) $M_s \sim M_{GUT} \sim 10^{16}$ GeV. This corresponds to a compactification scale $r \equiv V_6^{1/6} \sim 10^{-30}$cm or equivalently $M_c \sim 10^{14}$ GeV. This is analogous to the Horava-Witten construction and allow the possibility of the unification of gauge *and* gravitational couplings (if the Standard Model is realized on the same set of branes).

(3) $M_s \sim M_I \sim 10^{10-12}$ GeV $\sim \sqrt{M_{EW} M_P}$. If the Standard Model is realized on a set of D3-branes, this corresponds to a compactification scale $r \sim 10^{-23}$ cm. This proposal was based on the special role played by the intermediate scale M_I in different issues beyond the Standard Model. Examples include the scale of SUSY breaking in graviy mediated SUSY breaking scenario and the scale of the axion field introduced to solve the strong CP problem. This then allows one to identify the string scale with the SUSY breaking scale and opens up the possibility for non-SUSY string models to be relevant at low energies, solving the hierarchy problem. A simple example is to have a set of $\overline{D3}$ branes breaking the supersymmetry preserved by the background geometry and the SUSY breaking effects are transmitted to the observable sector by Planck suppressed interactions.

(4) $M_s \sim M_{EW} \sim$ TeV. This is the string theory realization[18] of the large extra dimensions scenario a la ADD.[19–21] We will discuss concrete models in the next lecture. If only two of the extra dimensions are large, the corresponding compactification scale is about a mm, which is the extreme case of the "brane world" scenario. However, the hierarchy problem is not totaly solved. We still need to explain why the compactification size is so large.

Thus, D-branes offer many more scenarios with $M_{EW} < M_s < M_P$ in contrast to the heterotic string where the fundamental string scale is fixed by the low energy couplings. Which of these scenarios is actually realized depends on how the moduli (size and shape of the extra dimensions) are stabilized.

Without going into details of the construction, a few observations can already be made:

- Gauge unification is not necessarily realized in D-brane models. The tree-level gauge coupling depends on the volume of the cycles the branes wrap around and can be different for different gauge factors of the Standard Model.

- We may or may not consider SUSY in the D-brane constructions because the fundamental string scale can be much lower than the Planck scale. However, stable D-brane configurations realizing non-supersymmetric models are harder to construct since they usually come with (i) closed string tachyons, and (ii) runaway potentials for moduli (dilaton, volume, etc). So, for the purpose of these lectures, we wil focus on $\mathcal{N} = 1$ supersymmetric examples. In particuar, one of our goals is to construct $\mathcal{N} = 1$ supersymmetric extensions of the Standard Model.

- D-brane models can be easily combined with other ingredients such as sources of moduli stabilization and supersymetry breaking such as background fluxes. These background fluxes are generalizations of the electromagnetic fluxes in string theory, and will be the subject of the last lecture.

- Although we will focus on D-brane models, much of our discussions can be generalized to other brane constructions in M/F theory models.

- D-brane models have been used to motivate many new BSM ideas. Just like the heterotic case, they are also useful in realizing more traditional scenarios such as hidden sector and gravity mediated SUSY breaking, but now in a more geometrical and stringy way.

9.3. D-branes and Chirality

We now turn to the main point of these lectures, which is the construction of realistic D-brane models. In addition to the $SU(3) \times SU(2) \times U(1)$ gauge structure, an important property of the Standard Model is that it is chiral. Before we can appreciate issue involved in introducing chirality to D-brane models, let us recall a few facts about D-branes:

9.3.1. *D-brane Primer*

D-branes are interesting objects for constructing particle physics models because their worldvolumes support non-Abelian gauge fields. What is interesting is that not only are they solutions to the SUGRA equations of motion, they admit a full string description as boundaries on which open strings can end. For the purpose of this lecture, there are several properties of D-branes we need to recall:

- D-branes are dynamical objects – the open strings ending on them describe the collective coordinates of the D-branes. The worldvolume of a D_p brane contains a $U(1)$ gauge field, $9 - p$ scalars and the corresponding fermion partners. The scalars correspond to the Goldstone modes of the part of the Poincare symmetry broken by the presence of the brane. The fermions are Goldstinos for supersymmetry. The D-brane breaks half of the supersymmetries. Therefore in flat space, after toroidal compactification, one D-brane carries the spectrum of an $\mathcal{N} = 4$ supersymmetric vector superfield.

- Furthermore, D-branes are BPS objects for which a no force condition applies. One can understand this condition as follows. Both D-branes have the same positve tension and therefore are naturally attracted to each other by gravitational interactions. The exchange of the dilaton field has the same effect of an attraction. However, D-branes are also charged under the antisymmetric Ramond-Ramond fields for which the interaction is repulsive, given both branes have the same charge. It turns out the combined effect of these three interactions cancels exactly if both of them are D-branes. This can be seen explicitly by computing a one-loop open string diagram corresponding to a cylinder (see, e.g., [2]). An anti D-brane carries the opposite RR charge as a D-brane and so there is a net attractive force between them.

- This gives rise to an interesting phenomenon which is essentially (but not quite) the inverse Higgs effects. Open strings with both endpoints on one brane give rise to the $U(1)$ gauge field for that brane. In the presence of a second brane, besides having the second $U(1)$ there are now pairs of strings with endponts on each of the two branes. These correspond to massive states with mass proportional to the separation of the branes. We may identify one string with a particle like W^+ (of the Standard Model) and the string with opposite orientation with W^-. The important point is that when both branes overlap these particles become massless and enhance the $U(1) \times U(1)$ symmetry to the full $U(2)$ symmetry. This is then the way to obtain non-Abelian supersymmetric theories on the branes. The bi-fundamental matter fields are also enhanced to adjoints.

- There are additional properties of D-branes which are useful for model building and we will discuss them along the way. (Other properties are given in the appendix for completeness).

9.3.2. *D-branes and Chirality*

However, the D-brane configurations we have considered so far are too simple to incorporate chirality, a key property of the Standard Model. This is because of the large amount of supersymmetry preserved by the D-brane configurations considered. In flat space, a D-brane preserve $\mathcal{N} = 4$ supersymmetry since D3-branes are 1/2 BPS. However, changing the compactification space from a torus to an orbifold or a Calabi-Yau does not improve the situation since this is a local issue at the location of the brane.

To make this point more precise, consider a D3-brane sitting at a point P in the extra dimensional space $\mathcal{M}_6$, with background fluxes. The background fluxes can in principle break the supersymmetry to $\mathcal{N} = 1$ or $\mathcal{N} = 0$ so chirality is possible. However, one can continuously deform $\mathcal{M}_6$ and dilute the fluxes to reach again flat space. Along this deformation, the gauge group does not change, so gauge protected quantities, like the number of chiral families should not change. Therefore, the spectrum in the original configuration is non-chiral. (A more stringy argument goes as follows. At P, we see flat space and constant fluxes. Around P, we start seeing deviations from flat space and non-constant fluxes. From the D3-brane point of view, this comes from closed strings running in loops, etc. So, we have the initial D3-brane theory together with perturbative corrections. If the initial theory is non-chiral, so will be the final theory.)

In general, an open string with both ends on a stack of N D-branes tansforms in the adjoint representation of $U(N)$, hence the theory is non-chiral. Let us consider instead a stack of N_{D3} D3-branes and N_{D7} D7-branes. If we place them on top of each other, we have

$$U(N_{D3}) \times U(N_{D7}) \tag{9.10}$$

gauge groups and matter in the representations:

$$m(N_{D3}, \overline{N}_{D7}) + m'(\overline{N}_{D3}, N_{D7}) \tag{9.11}$$

where m, m' are multiplicities. But since we can separate them by a distance ℓ the strings between D3 and D7 will have a minimal mass of ℓ/α'. Hence $m = m'$ and the theory is non-chiral.

How could one obtain chirality? Four dimensional chirality is a violation of four-dimensional parity. In string theory, the chirality in 4D is correlated with the chirality in the 6 extra dimensions. Hence to achieve 4D chirality, the D-brane configuration must violate 6D parity. The D-brane configurations we consided so far are too simple to introduce a preferred orientation in 6D.

This observation also suggests how one can construct D-brane configurations which admit 4D chiral fermions. The requirement is that the configuration introduces a preferred orientation in the 6 transverse dimensions. There are several ways to achieve this. These seemingly different strategies are in fact related.

(1) **D-branes at singularities:** We can consider placing D-branes in spaces that are not smooth. Chirality can arise if the D-branes are sitting at a singularity. A simple example is to consider a stack of D3-branes sitting at an orbifold singularity. An orbifold is a discrete identification of space and this defines a preferred orientation.

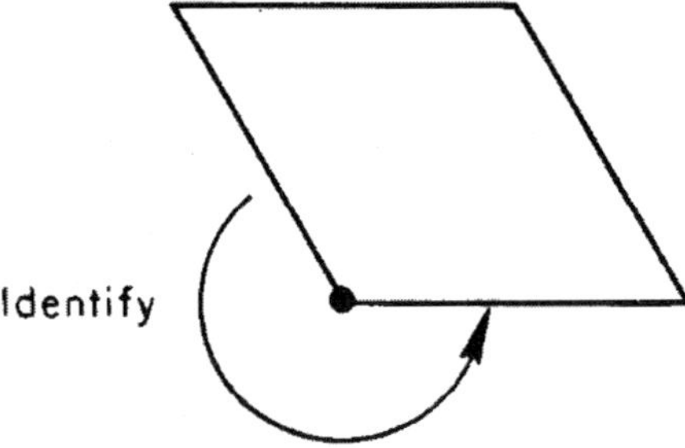

Fig. 9.1. An orbifold identifies points in space under a discrete symmetry.

(2) **Intersecting branes:** We can also consider pairs of D-branes that cannot be separated from one another. Intersecting D-branes lead to chiral fermions in the sector of open strings stretched between different kinds of D-branes. (Again, the angle between one stack of branes with respect to another defines a preferred orientation).

(3) **Magnetized D-branes:** Finally, chirality also arises when we turn on a non-trivial field strength background for the worldvolume $U(1)$ gauge fields. The magnetic fields introduce a preferred orientation in the internal dimensions through the wedge product $F \wedge F \wedge F$ as the volume form.

We will discuss these three methods in this particular order. The main point we will try to make is that these constructions are "modular" in the sense that we can locally obtain the Standard Model witout having to know all the details of the compactification. This is of great importance

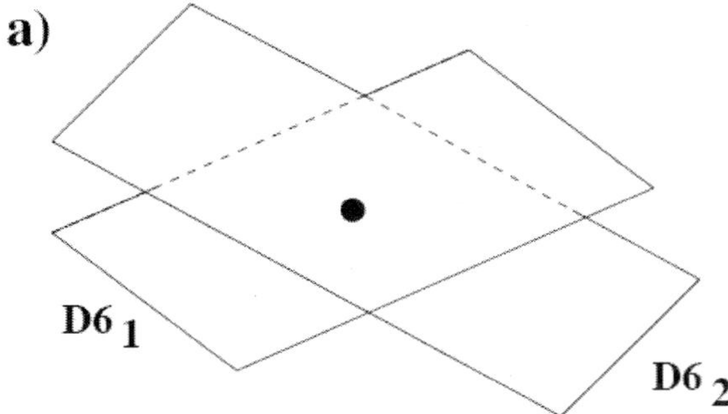

Fig. 9.2.   Intersecting branes.

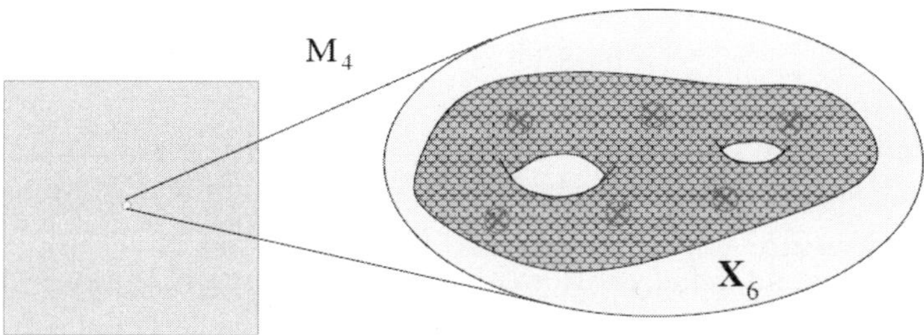

Fig. 9.3.   Magnetized D-branes.

because we can follow a bottom-up approach instead of looking at random compactifications that could give rise to the Standard Model.

We will present this bottom-up approach for building realistic models. Most of the important details of the model, such as the gauge group, chiraliy, number of chiral families, etc, will depend only on the structure of the singularities that the branes sit at or the way the branes intersect. This can happen in all sorts of spaces and therefore we can keep the main properties whether we are talking about a complicated Calabi-Yau space or a simple toroidal orbifold compactification. This makes the models constructed more robust. This is the main practical advantage of D-brane model building over the heterotic string.

Before we go on, let me point out that these methods of obtaining chirality are in fact related by dualities. The simplest way to see this is to note that turning on a gauge bundle on the world volume of D-branes induces lower-dimensional D-brane charges. For example, one can think of a D_p brane with n units of magnetic flux on a torus:

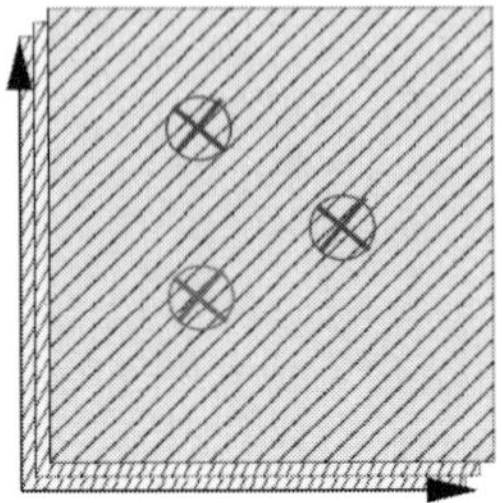

as a bound state of a D_p brane and n D_{p-2} brane. Now there is a remarkable symmetry of string theory known as T-duality:

$$R \to \alpha'/R \tag{9.12}$$

which maps a universe of enormous universe to a universe which is incredibly small in size. Here $\alpha' = 1/\ell_s$ where ℓ_s is the string length. This symmetry has the effect of exchanging the momentum and winding states. The mass spectrum of string theory is therefore the same under this large $\leftrightarrow$ small radius interchange. However, the dimension of D-branes changes: A D_p brane with its world volume extended along the T-dual direction will become a D_{p-1} brane whereas a D_p brane with its world volume transverse to the T-dual direction will become a D_{p+1} brane. Now let's T-dualize along one direction:

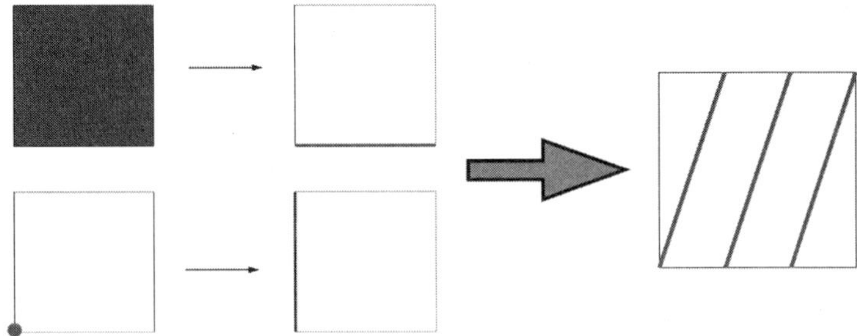

Fig. 9.4.   Intersecting branes and magnetized D-branes are related by T-duality.

The D_p and the D_{p-2} branes both turn into a D_{p-1} brane but they orient along different directions. Thus branes with magnetic fluxes become branes at angles. Likewise, we will see later that the D3-branes at singularities which give rise to chiral fermions, known as fractional branes, can be thought of higher dimensional branes wrapping around a collapsed cycle with some gauge bundle on them.

So, why do we discuss these approaches separately if the setups are dual to one another? It turns out that in some situations, one side of the duality is simpler than the other. Furthermore, we will in the last lecture consider D-brane models with fluxes. Under duality, the background flux turns into a non-trivial metric background and the two descriptions are no longer equivalent at least in this simple form. So it is useful to have an intuition about each of these approaches independently.

9.4. D-branes at Singularities

First, let us discuss how one can construct chiral models from D-branes at singularities. We will begin with a quick review of orbifolds.

- A manifold $\mathcal{M}_6$ is by definition locally like $\mathbb{R}^6$, but string theory is defined even on manifolds that are not smooth. The extra dimensions can contain singularities, like orbifold or conifold singularities.
- Orbifolds are spaces that locally look like $\mathbb{R}^6$ or $\mathbb{R}^6/\Gamma$ where Γ is a discrete subgroup of $SO(6)$, the rotation group of the 6 extra dimensional space.
- Although the gravity background is singular, strings are well-behaved at orbifold singularities. This has been shown in the classic papers [22,23] for closed strings and [24] for open strings.
- What is an orbifold? Consider a simple case $T^2/\mathbb{Z}_2$. The $\mathbb{Z}_2$ orbifold symmetry acts on the two-dimensional torus as follows:

$$\theta : (x_1, x_2) \to (-x_1, -x_2) \tag{9.13}$$

There are 4 fixed points: $(0,0), (1/2,0), (0,1/2), (1/2,1/2)$ (if we normalize the radii of T^2 to 1).

Locally, the singularity looks like a cone and globally the $T^2/\mathbb{Z}_2$ orbifold is a tetrahedral (or ravioli) as show in Figure 9.6.

- Let's look at a slightly more non-trivial example: $T^2/\mathbb{Z}_3$. The discrete $\mathbb{Z}_3$ orbifold symmetry acts on the complex coordinates $z = x_1 + ix_2$ of the torus as follows:

$$\theta : z \to \alpha z \qquad \text{where} \qquad \alpha = e^{2\pi i/3} \tag{9.14}$$

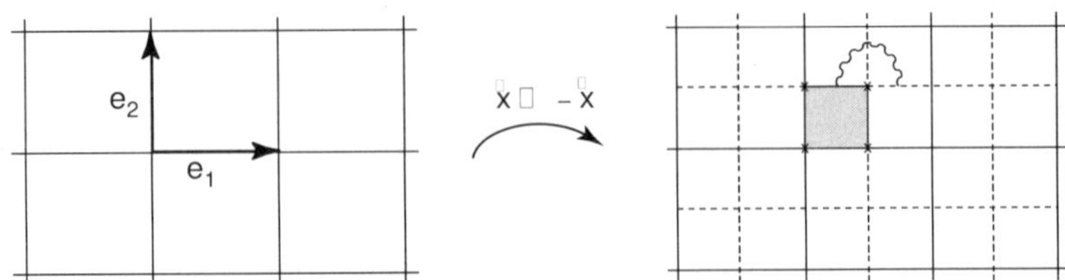

Fig. 9.5.   A $T^2/\mathbb{Z}_2$ orbifold.

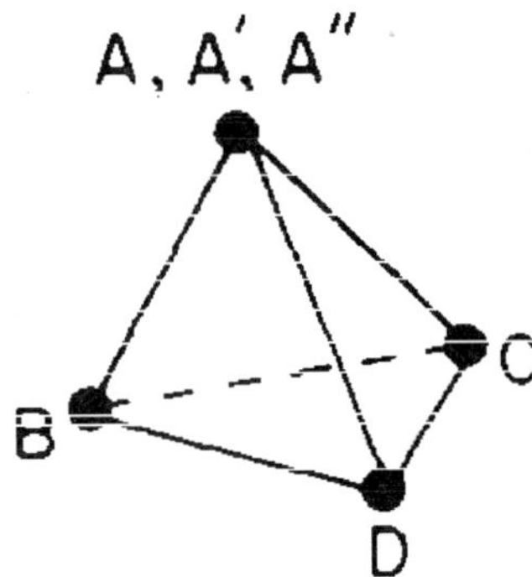

Fig. 9.6.   A $T^2/\mathbb{Z}_2$ orbifold is a tetrahedral.

There are three fixed points as shown in Figure 9.7.

Note that vectors undergo a non-trivial rotation when transported along a closed curve around the singularity (local holonomy).

- In general, we can consider a local singularity of the form $\mathbb{C}^3/\Gamma$ where $\Gamma = \mathbb{Z}_N$. The discrete symmetry $\mathbb{Z}_N$ acts on the complex coordinates of $\mathbb{C}^3$ as follows:

$$\theta : (z_1, z_2, z_3) \to (\alpha^{\ell_1} z_1, \alpha^{\ell_2} z_2, \alpha^{\ell_3} z_3) \tag{9.15}$$

where $\alpha = e^{2\pi i/N}$, $\ell_i \in \mathbb{Z}$ such that $\theta^N = 1$.

- One can check that if θ is a matrix of determinant 1:

$$\theta \in SU(3) \Leftrightarrow \ell_1 \pm \ell_2 \pm \ell_3 = 0 \quad (\text{mod } N) \tag{9.16}$$

for some choices of sign, then

$$\Gamma \subset SU(3) \subset SO(6) , \qquad \text{SUSY is preserved} \tag{9.17}$$

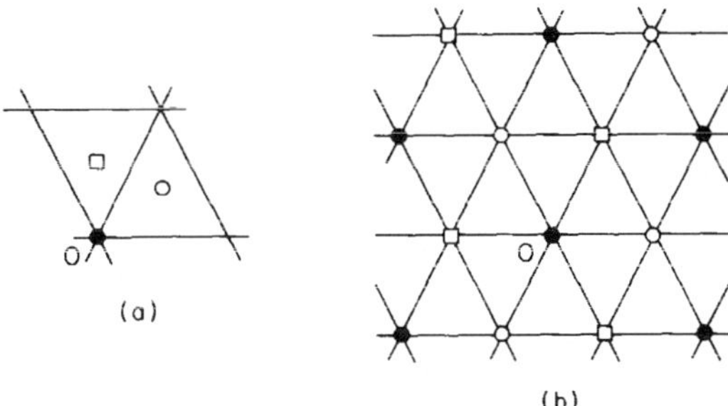

Fig. 9.7. A Z_3 orbifold.

If this condition is not satisfied, then we find closed tachyons in the closed string spectrum. For concreteness, we will take the choice of signs, $\ell_1 + \ell_2 + \ell_3 = 0 \pmod{N}$.

- Furthermore, only if we require:

$$\sum_i \ell_i = \text{even} \tag{9.18}$$

do we have fermions in the spectrum. In more technical terms, the orbifold is called a spin manifold.

- The orbifold action has the following effects on closed strings. It projects out states that are not invariant under the twist, reducing the number of states in the spectrum. It also increases the number of states in another way since it now includes the so called "twisted sector". An open string around a fixed point with its two endpoints lying at points which are identified under the orbifold is not included in the spectrum of states in the unorbifolded space but it is a valid closed string in the orbifold. In other words, there are two sectors:

  - Untwisted sector: strings closed on $\mathbb{C}^3$
  - Twisted sector: strings not closed on $\mathbb{C}^3$ but closed on $\mathbb{C}^3/\Gamma$ (confined to the singularity)

- Closed strings are not charged under the D-brane gauge group and so they will not give us the Standard Model particles. We will not analyze their spectrum. However, the closed string spectrum determines the types and number of moduli fields we have. They will be important later on when we discuss moduli stabilization.

- The open string spectrum is our main concern for particle physics model building. We know how to quantize open strings exactly to all orders

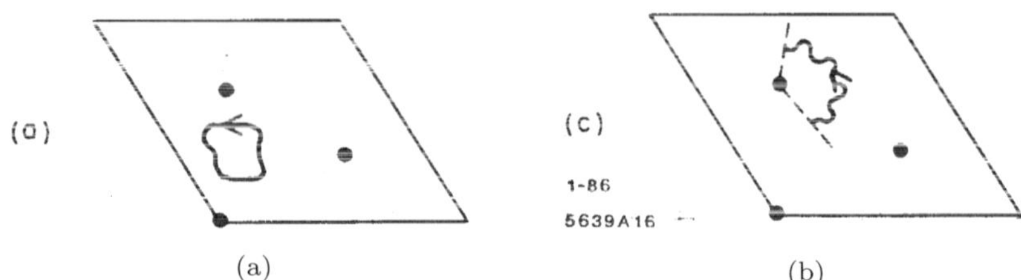

Fig. 9.8.   (a) An untwisted sector state (b) A twisted sector state. Figure taken from Ref. [25].

in α' in these backgrounds. We will however discuss only the massless spectrum.

- If we are away from the singularity, things work as in flat space. The branes are arranged in a $\mathbb{Z}_N$ invariant fashion. They are identified and the open string spectrum is that of $U(N)$ $\mathcal{N} = 4$ SYM (plus α' corrections).
- If the branes sit on the singularity, we have an "orbifolded" gauge theory. The spectrum is given by the open strings that are well defined at the singularity. This means open strings invariant under the action of θ.
- The orbifold twist θ has two actions

    - $\theta \in SO(6)$: R-symmetry group of $D = 4$, $\mathcal{N} = 4$ SYM.
    - $\theta \in U(N)$: "permutes" the D-brane positions.

The action of θ on the $U(N)$ gauge degrees of freedom is then given by an $N \times N$ matrix Γ_θ. We will elaborate on this shortly.

9.4.1. *Explicit Examples*

We now see explicitly how the spectrum of D-branes at a $\mathbb{Z}_N$ singularity, like the fixed points of orbifolds, become chiral. As discussed before, the spectrum is determined by the local properties so for simplicity, let us consider a D3-brane in flat 10D space with the six extra dimensions modded out by a $\mathbb{Z}_N$ twist θ. (There are too many N in this business! Here, we refer to N for the order of the twist, n for the number of overlapping D-branes, and $\mathcal{N}$ for the number of supersymmetries).

If we have a stack of n D-branes, the original gauge group is $U(n)$. The gauge degrees of freedom are represented by the Chan-Paton matrices λ^i_j, $i, j = 1, \ldots, n$, associated to the endpoints of the open strings and which belongs to the adjoint of $U(n)$. The action of the orbifold twist on the gauge degrees of freedom are given by:

$$\lambda \to \Gamma_\theta \lambda \Gamma_\theta^{-1} \tag{9.19}$$

where Γ_θ is of order N and can be diagonalized to take the simple form:

$$\Gamma_\theta = \begin{pmatrix} I_{n_0} & & & & \\ & \alpha I_{n_1} & & & \\ & & \alpha^2 I_{n_2} & & \\ & & & \cdot & \\ & & & & \cdot \\ & & & & & \alpha^{N-1} I_{n_{N-1}} \end{pmatrix} \tag{9.20}$$

Here I_{n_k} is the identity matrix in n_k dimensions and the integers n_k satisfy the constraint $\sum_k n_k = n$. (*i.e., n D-branes are split into groups of n_k's*).

Let us now see how the original $\mathcal{N} = 4$ vector multiplet transforms under the action of the twist defined by θ and Γ_θ. We can write the $\mathcal{N} = 4$ multiplet in terms of the $\mathcal{N} = 1$ multiplets:

- Vector Multiplet: $V \equiv (A_\mu, \lambda)$
- Chiral Multiplets: $\Phi_a \equiv (\phi_a, \psi_a)$, $a = 1, 2, 3$ (a labels the 3 complex extra dimensions)

Therefore Φ_a which has an a index feels both the action of θ and Γ_θ where V feels only the action through Γ_θ.

Remember that we have to keep only the states that are invariant under the twist. This means that $V = \Gamma_\theta V \Gamma_\theta^{-1}$. This breaks the gauge group to:

$$U(n) \to U(n_0) \times U(n_1) \times \cdots \times U(n_{N-1}) \tag{9.21}$$

with the number of factors equal to the order of the twist N. This means that if we want three gauge factors, we should have a $\mathbb{Z}_3$ twist and so on.

The surviving chiral superfields satisfy $\Phi_a = \alpha^{\ell_a} \Gamma_\theta \Phi_a \Gamma_\theta^{-1}$. The first factor being the action of θ. Therefore remembering that λ carries adjoint indices (which are composed of fundamentals and anti-fundamentals) we can easily see that the remaining matter fields transform as:

$$\sum_{a=1}^{3} \sum_{i=0}^{N-1} (\mathbf{n}_i, \overline{\mathbf{n}}_{i+\ell_a}) \tag{9.22}$$

Here the sum over i is understood to be mod N, and n_i means the fundamental of $U(n_i)$.

This is a typical spectrum in this class of models. The matter fields tend to come in bi-fundamentals of the product of gauge groups. These can be arranged into "quiver" diagrams (see figure). These diagrams are made out of one node per group factor, i.e., the i-th node corresponding to the gauge group $U(n_i)$. There are also arrows joining the nodes. An arrow going from the i-th to the j-node correspond to a chiral field the representation $(n_i, \overline{n}_j)$ (note the orientation). A closed triangle of arrow would indicate the existence of a gauge invariant cubic superpotetial for those fields.

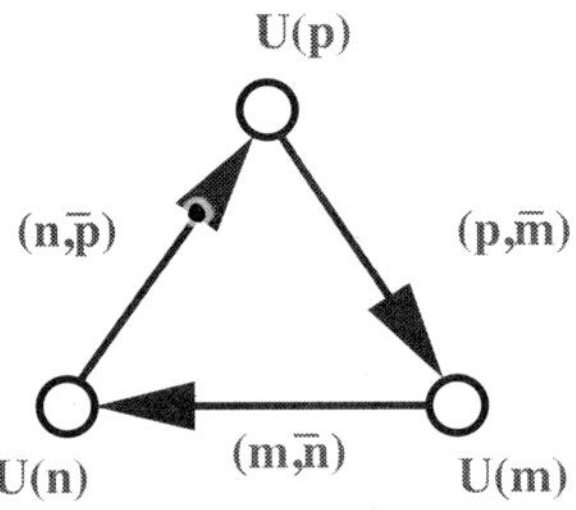

Fig. 9.9. The Z_3 quiver.

A D3-brane in a quiver node is called a "fractional" D3-brane, and it cannot be taken away from the singularity. This is why two fractional D3-branes in different nodes can host a chiral fermion. Besides putting D-branes at a $\mathbb{C}^2/\mathbb{Z}_N$ singularity, one can also consider other types of singularities such as the conifold.

From the generic chiral spectrum in eqn. (9.22), we can extract a very simple but powerful conclusion: Only for $\mathbb{Z}_3$ will we get the chiral matter spectrum in three identical copies or families. The reason is that only for that case we have $\ell_1 = \ell_2 = \ell_3 \mod N$, since $\ell_1 = \ell_2 = 1$ and $\ell_3 = -2 = 1 \mod 3$. Other twists given by $(1/N, 1/N, -2/N)$ will give rise to two families. Therefore three is not only the maximum number of families for this class of models but is obtained only for one twist, the $\mathbb{Z}_3$ twist. This is a rather remarkable result.

If we want to have the Standard Model we can consider $n_0 = 3$, $n_1 = 2$, $n_2 = 1$ to get the gauge group $U(3) \times U(2) \times U(1)$. The spectrum will then be:

$$3 \times \left[(\mathbf{3}, \mathbf{2}) + (\mathbf{1}, \mathbf{2}) + (\overline{\mathbf{3}}, \mathbf{1}) \right] \tag{9.23}$$

where we have suppressed the $U(1)$ quantum numbers. This gives the 3 families of left-handed quarks, right-handed up quarks, and leptons, just as in the Standard Model. However, we can easily see that we are missing at least the right-handed down quarks. Actually, the spectrum as it is, is anomalous. What happens from the string theory point of view is that there are uncancelled tadpoles for twisted sector fields. We will take care of this shortly and construct models that are fully consistent, but until then we can explore some general properties of the model as it stands now.

First, there are actually three $U(1)$'s. Only one combination of them is anomaly free and it is defined in general (for any N) by:

$$Q_Y = - \left(\frac{1}{3}Q_3 + \frac{1}{2}Q_2 + \sum_{s=1}^{N-2} Q_i^{(s)} \right) \tag{9.24}$$

In a general orbifold all other $N-1$ additional $U(1)$ factors are anomalous and therefore massive due to a version of the Green-Schwarz mechanism. The Green-Schwarz mechanism is the cancellation of one-loop diagrams by tree-level diagrams due to the exchange of p-forms. We will discuss the Green-Schwarz mechanism in more detail later. For now, we can quickly check that this $U(1)$ does correspond to the hypercharge, as expected since hypercharge is essentially the only non-anomalous $U(1)$ with the spectrum of the Standard Model. For instance, fields transforming in the $(\mathbf{3}, \mathbf{2})$ representation have Q_Y charge $-\frac{1}{3} + \frac{1}{2} = \frac{1}{6}$ as correspond to left-handed quarks. Fields transforming in the $(\mathbf{3}, \mathbf{1})$ (which necessarily have charge -1 under one of the $Q_1^{(s)}$ generators) have a Q_Y charge $-\frac{1}{3} + 1 = -\frac{2}{3}$, as corresponds to right-handed U quarks, etc.

It is worth noticing that the normalization of this hypercharge $U(1)$ depends on the order of the twist N. In fact, by normalizing $U(n)$ generators such that $\mathrm{Tr}_a^2 = \frac{1}{2}$, the normalization of the Y generator is fixed to be

$$k_1 = 5/3 + 2(N-2) \tag{9.25}$$

This amounts to a dependence on N in the Weinberg angle, namely

$$\sin^2 \theta_W = \frac{g_1^2}{g_1^2 + g_2^2} = \frac{1}{k_1 + 1} = \frac{3}{6N - 4} \tag{9.26}$$

Thus the weak angle decreases as N increases. Notice that the $SU(5)$ result $3/8$ is only obtained for a $\mathbb{Z}_2$ singularity. However in that case the D3-brane spectrum is necessarily vector-like and hence one cannot reproduce the Standard Model spectrum. For the interesting case for us, $N = 3$, we find $\sin^2 \theta_W = 3/14$.

Now, back to the uncancelled tadpoles. To cancel these tadpoles at the singularity it is necessary to have not only the D3-branes but also D7-branes.

There are three types of D7-branes that can be introduced depending on which 2 of the 3 complex dimensions they contain. The consistency condition (tadpole cancellation) can be written as

$$\mathrm{Tr}\Gamma_{\theta,7_3} - \mathrm{Tr}\Gamma_{\theta,7_1} - \mathrm{Tr}\Gamma_{\theta,7_2} + 3\mathrm{Tr}\Gamma_{\theta,3} = 0 \qquad (9.27)$$

This condition can be obtained by analyzing one-loop open string diagrams with boundaries on various combinations of D3 and D7-branes. More intuitively, we can understand this condition as equivalent to non-Abelian anomaly cancellation in the effective field theory (since $3 - 7$ strings introduce chiral matter fields charged under the D3-brane gauge groups). Notice that without the D7-branes we could not have the Standard Model on the D3-brane. Nevertheless, the choice $n_0 = n_1 = n_2 = 3$ gives an anomaly free spectrum even without introducing D7-branes, and so the trinification model with gauge group $U(3)^3$ can be realized.

The D7-branes will have extra gauge groups and matter fields living on the D7-brane which can be obtained in a similar way, with a matrix $\Gamma_{\theta,7_i}$ (with $i = 1, 2, 3$ labeling different D7-branes) acting on the gauge degrees of freedom of the D7-branes. There are also massless matter fields living at the intersection of the D7 and D3-branes, corresponding to open strings with one endpoint on the D3-branes and the other on the D7-branes. This will complete the spectrum of the Standard Model and render the model anomaly free at the singularity. The D7-brane gauge couplings depend on the volumes of the wrapped cycles. If the volumes are large, the D7 gauge groups act essentially as global symmetries. We can picture the Standard Model realized on the D3-D7 system as follows:

As an illustration, a particular example of configuration of D3 and D7 branes and the resulting spectrum is given in the following table:

A similar model can be constructed choosing $n_0 = 3$, $n_1 = 2$, and $n_2 = 2$ with

$$\Gamma_\theta = \mathrm{diag}\left(I_3, \alpha I_2, \alpha^2 I_2\right) \qquad (9.28)$$

giving rise to a left-right symmetric model with gauge group:

$$U(3) \times U(2)_L \times U(2)_R \qquad (9.29)$$

and three families of chiral matter:

$$3 \times \left[(\mathbf{3}, \mathbf{2}, \mathbf{1}) + (\overline{\mathbf{3}}, \mathbf{1}, \mathbf{2}) + (\mathbf{1}, \mathbf{2}, \mathbf{2}) + (\mathbf{1}, \mathbf{2}, \mathbf{1}) + (\mathbf{1}, \mathbf{1}, \mathbf{2})\right] \qquad (9.30)$$

where we suppressed the $U(1)$ charges above. Again, $B - L$ is the only anomaly free $U(1)$, the other $U(1)$'s acquire a mass by the Green-Schwarz mechanism.

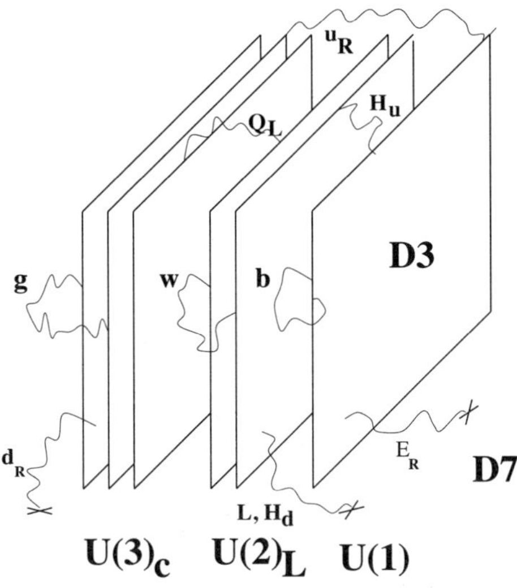

Fig. 9.10.   A bottom-up construction of the MSSM.

It is important to emphasize that this is not the full story for these models. Remember that we are building them step by step from a bottom-up approach. There are further issues involved in constructing compact models. So far we have concentrated only on a singularity in flat space modded out by the action of $\mathbb{Z}_N$. If we compactify the extra dimensions, the total RR charge of the D7-branes has to cancel, since there is no place for the RR flux to escape in the compact space (think of an analogous problem of an electric charge in electromagnetism). This will force us to add objects with negative RR charges like anti D7-branes or orientifold planes. For stability reasons, the anti D7-branes have to be separated from the D7-branes or else they will annihilate. They can be placed at different orbifold fixed points, for instance. Orientifolds are stringy objects that we will introduce in the next lecture.

An anti D7-brane breaks supersymmetry since it preserves the half of the supersymmetry that the D-brane breaks. Therefore if both are present, the full supersymmetry is broken. If the anti-branes are trapped at different fixed points, then only bulk fields can mediate the breaking of supersymmetry to the observable brane. This is a realization of the gravity mediated SUSY breaking scenario. In order to obtain a realistic spectrum of super-smmetric particles, the scale of SUSY breaking (string scale in this model) is typically the intermediate scale $M_I \sim 10^{11}$ GeV. The LR model men-

Table 9.2. Spectrum of $SU(3) \times SU(2) \times U(1)$ model. We present the quantum numbers under the $U(1)^9$ groups. The first three $U(1)$'s come from the D3-brane sector. The next two come from the D7$_r$-brane sectors, written as a single column with the understanding that e.g. fields in the **37**$_r$ sector are charged under the $U(1)$ in the **7**$_r$**7**$_r$ sector.

Matter fields	Q_3	Q_2	Q_1	$Q_{u_1^r}$	$Q_{u_2^r}$	Y
33 sector						
$3(3,2)$	1	-1	0	0	0	1/6
$3(\bar{3},1)$	-1	0	1	0	0	-2/3
$3(1,2)$	0	1	-1	0	0	1/2
37$_r$ sector						
$(3,1)$	1	0	0	-1	0	-1/3
$(\bar{3},1;2')$	-1	0	0	0	1	1/3
$(1,2;2')$	0	1	0	0	-1	-1/2
$(1,1;1')$	0	0	-1	1	0	1
7$_r$**7**$_r$ sector						
$3(1;2)'$	0	0	0	1	-1	0

tioned earlier is particularly interesting in this regard since the unification scale is also close to M_I. Therefore, one might realize gauge unification at the string scale and low energy supersymmetry breaking solving the hierarchy problem. (The intermediate scale M_I is also motivated from axion physics).

Finally, let us emphasize again the flexibility of this bottom-up approach in model building. Our discussions can be generalized to F-theory since the D3-brane gauge and matter content depends only on the local geometry. However, F-theory allows for more general D7-brane configurations, e.g., it allows D7-branes carrying both "electric" and "magnetic" charges to coexist in the model (more precisely, they are the so called (p,q) 7-branes). Furthermore, singularities beyond the simplest $\mathbb{Z}_N$ singularities discussed here have been considered in Ref. [26]. These include non-Abelian twists, orientifold singularities, and conifold singularities. For the case of non-Abelian singularities, an interesting model was proposed in [27] (see also [28]). There, a singularity of the type $\mathbb{C}^3/G$ with $G = \Delta_{27}$ was considered. The group $G = \Delta_{27}$ is one of the non-Abelian discrete subgroups of $SU(3)$ and thus preserves SUSY on the D3-brane, like in the $\mathbb{Z}_N$ cases. The Δ_{27}

group is actually one of the Δ_{3n^2} series whose action on $\mathbb{C}^3$ is given by:

$$e_1 : (z_1, z_2, z_3) \to (\omega_n z_1, \omega_n^{-1} z_2, z_3) \tag{9.31}$$
$$e_2 : (z_1, z_2, z_3) \to (z_1, \omega_n z_2, \omega_n^{-1} z_3)$$
$$e_3 : (z_1, z_2, z_3) \to (z_3, z_1, z_2)$$

One of the interesting properties of this model is that there is no need to introduce D7-branes to cancel the local tadpoles. The gauge group on the D3-branes is $U(3)^2 \times U(1)^9$ which can further be broken to the Standard Model with three families (due partly to the $\mathbb{Z}_3$ subgroup of Δ_{27}).

9.5. Intersecting Branes

Another mechanism to obtain $D = 4$ chiral fermions is to consider intersection of branes. Before we discuss all the details and subtleties, let us begin with the overall picture. We know by now that an open string carries indices in the adjoint representation of $U(n)$. The adjoint can be seen as the product of fundamental and anti-fundamental representation, therefore one end of the open string transform as the fundamental and the other as the anti-fundamental. When the two endpoints of the open string lie on the same stack of branes, we have particles like the gauge bosons in the adjoint. But when they lie on different stacks of branes it gives rise to bi-fundamentals. This is what happens at the intersections of two branes. The states corresponding to open strings ending on each of the two branes correspond to bi-fundamentals that can naturally lead to a chiral spectrum.

A way to obtain the Standard Model group and spectrum is to intersect several stacks of branes. One stack of three correspond to the strong interactions, it can intersect with a stack of two D-branes corresponding to $SU(2)_L$. At the intersection, we have then the quark doublets. At a different intersection point the stack of two D-branes will intersect with one brane carrying $U(1)$ and the leptons will be at the intersection and so on. As it turns out, we need to introduce minimally four stacks of branes to fully account for the quantum number of all the Standard Model particles. Pictorially, the building block looks something like this:

The four stacks of branes are named the baryonic branes, the left brane, the right brane, and the leptonic brane for obvious reasons.

If this is all it takes to construct the Standard Model from intersecting branes, this lecture will be very short. As you will see, there are further string theory constraints both in constructing a "local model" and in em-

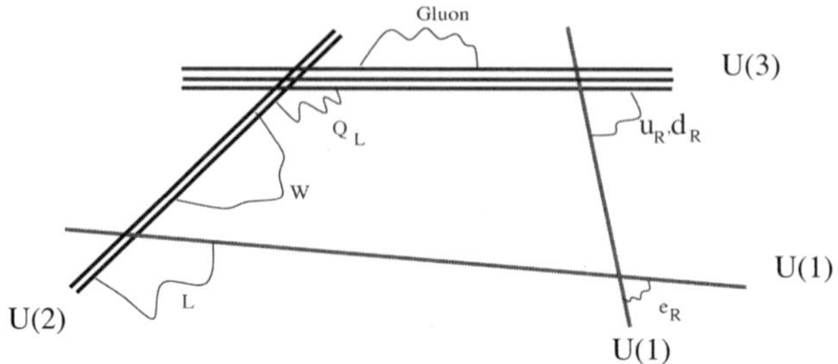

Fig. 9.11. A building block of the Standard Model.

bedding this setup in a compact setting. The purpose of this lecture is to discuss these subtleties.

9.5.1. *Local Geometry and Spectrum*

The basic configuration of intersecting D-brane models leading to 4D chiral fermions involve two stacks of D6-branes, each spanning our 4D space and three additional real dimensions.

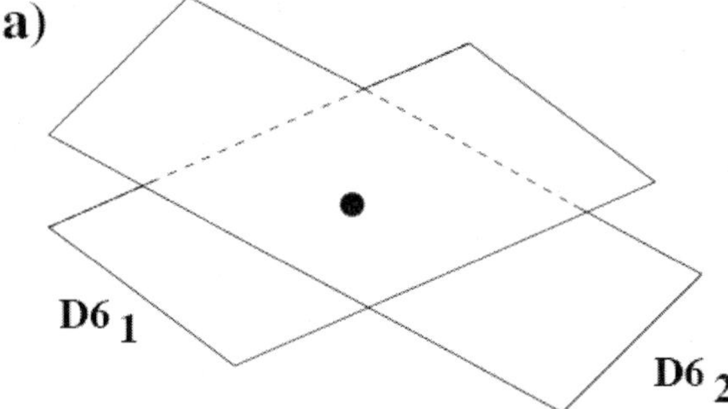

The local geometry is fully specified by the angles of rotations between the branes, which can be depicted as follows:

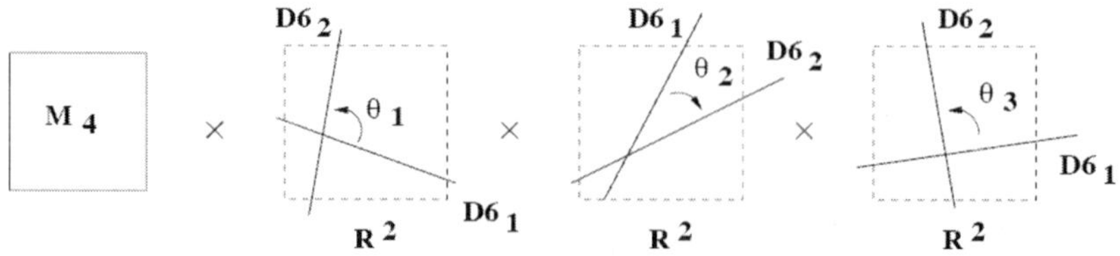

Fig. 9.12. The local geometry of intersecting branes.

As we discussed, chiral fermions in bi-fundamental representations are localized at the intersection of the brane world volumes, which is our usual 4D space. The appearance of chirality can be understood from the fact that the geometry of the two D-branes introduces a preferred orientation in the 6D space. We can see this by considering the relative rotation of the second D6-brane with respect to the first. This also explains why we consider configuration of D6-branes (and not other types of branes, like D4 and D5, etc). D6-branes are the only type of branes that intersect at a point and so the chiral fermions are confined in our four-dimensional space. They don't intersect at a line or a surface for instance, so one can define an orientation in the full 6D space.

The open string spectrum can also be obtained easily. In fact, one can quantize open strings in this intersecting brane background and obtain the full string spectrum and not only the massless states (see appendix) but we will skip over these details. As far as massless states go, the open strings ending on the same stack of D-branes provide the $U(N)$ gauge bosons, three real adjoint scalars and their superpartners propagating over the 7D world volume of the D6-branes. The open strings stretching between different kinds of branes lead to a 4D chiral fermion transforming in the bi-fundamental representation and localized at the intersection. The chirality is encoded in the orientation defined by the intersection.

This last point requires some elaboration. Notice that:

- Two intersecting D6-branes define 3 angles:

$$\vec{\theta}^{ab} = (\theta_1^{ab}, \theta_2^{ab}, \theta_3^{ab}) \tag{9.32}$$

This is because each 3-plane has an orientation, which differentiates between a D6-brane and an $\overline{D6}$-brane. Under a π rotation of any of these angles, a D6-brane becomes an $\overline{D6}$-brane:

$$\underset{D6}{(\theta_1^a, \theta_2^a, \theta_3^a)} \rightarrow \underset{\overline{D6}}{(\theta_1 + \pi, \theta_2^a, \theta_3^a)} \rightarrow \underset{D6}{(\theta_1^a + \pi, \theta_2^a + \pi, \theta_3)} \tag{9.33}$$

- We can always choose $-\pi \leq \theta_i^{ab} \leq \pi$. Then if $\theta_i^{ab} \neq 0, \pm\pi$,

$$\vec{\theta}^{ab} = -\vec{\theta}^{ba} \tag{9.34}$$

and so

$$\epsilon^{ab} \equiv \text{sign}(\theta_1^{ab}\theta_2^{ab}\theta_3^{ab}) = -\epsilon^{ba} \tag{9.35}$$

is a well defined quantity.

- If some $\theta_i^{ab} = 0$ or π, then ϵ^{ab} is not well defined, but the system is non-chiral since one can separate the branes, as shown in Figure 9.13. If they are separated by a length ℓ, the minimal mass is ℓ/α'.

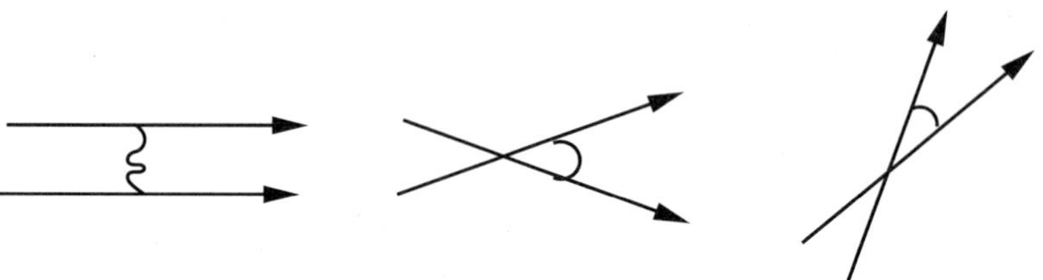

Fig. 9.13.   A non-chiral intersection of D-branes.

- If all $\theta_i^{ab} \neq 0, \pi$, then the intersection cannot be removed by deforming the D6-branes. Therefore, there could be a chiral fermion at the intersection.
- It was shown by [29] that there is indeed an $D = 4$ chiral fermion, of chirality ϵ^{ab}, at the intersection, as well as some light scalars (also in bi-fundamental representation) whose masses depend on the θ_i^{ab}'s. In string units, their masses are given by

$$\frac{1}{2\pi}(-\theta_1 + \theta_2 + \theta_3) \qquad \frac{1}{2\pi}(\theta_1 - \theta_2 + \theta_3)$$
$$\frac{1}{2\pi}(\theta_1 + \theta_2 - \theta_3) \qquad 1 - \frac{1}{2\pi}(-\theta_1 - \theta_2 - \theta_3) \qquad (9.36)$$

where for simplicity of notation, we drop the ab superscript. These light scalars can be massless, massive or tachyonic depending on the angles between the branes. This point will become clear after we analyze the SUSY preserved by the branes (see Section 9.5.3).

- Notice that:

  - An open string from a to b has quantum number $(N_a, \overline{N}_b)$ and chirality ϵ^{ab}.
  - An open string from b to a has quantum number $(\overline{N}_a, N_b)$ and chirality $-\epsilon^{ab}$.

They are anti-particles of one another and together gives the two fermionic degrees of freedom corresponding to one chiral Weyl fermion from a 4D spacetime point of view.

9.5.2. *Compactification*

Although intersecting D6-branes provide a mechanism to obtain 4D chiral fermions, the gauge bosons can propagate in the entire world volume of the D6-branes and so the gauge interactions remain 7D. Likewise, the gravitational interactions remain 10D before compactification. So, let us introduce the intersecting D-branes in a compact setting.

The general kind of configurations we will consider is string theory on a spacetime of the form $M_4 \times X_6$ where X_6 is compact.

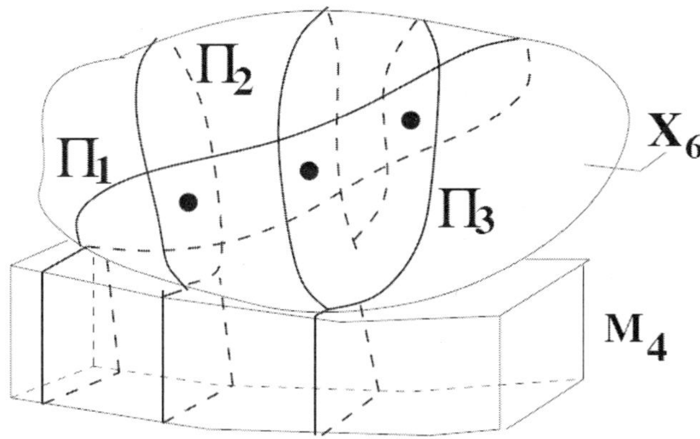

Fig. 9.14.   Compactification and intersecting brane models.

The D6-branes are space-filling and wrap 3-cycles of the compact space. The new feature is that two 3-cycles in the compact space intersect several times, leading to replicated families of chiral fermions.

- Consider N_a D6-branes on $\mathcal{M}_4 \times \Pi_3^a$ and N_b D6-branes on $\mathcal{M}_4 \times \Pi_3^b$, we have

  - $U(N_a) \times U(N_b)$ gauge group
  - One chiral fermion in $(N_a, \overline{N}_b)$ representation at each intersection

  since we have locally the setup of flat space.

- Now, different intersections may have different ϵ^{ab}'s, so different chiralities. The gauge protected quantity is the net number of chiral fermions (say left-handed):

$$(\#\text{Intersections with } \epsilon^{ab} > 0) - (\#\text{Intersections with } \epsilon^{ab} < 0) \quad (9.37)$$

- The above is a topological quantity, known as the intersection number of two 3-cycles:

$$I_{ab} = [\Pi_a] \cdot [\Pi_b] \tag{9.38}$$

- I_{ab} is topological because it does not depend on the specific embedding of Π_a, only on the topology (more precisely, the homology class $[\Pi_a]$). It does not change as we deform the background geometry or the D-branes, as illustrated in Figure 9.15.

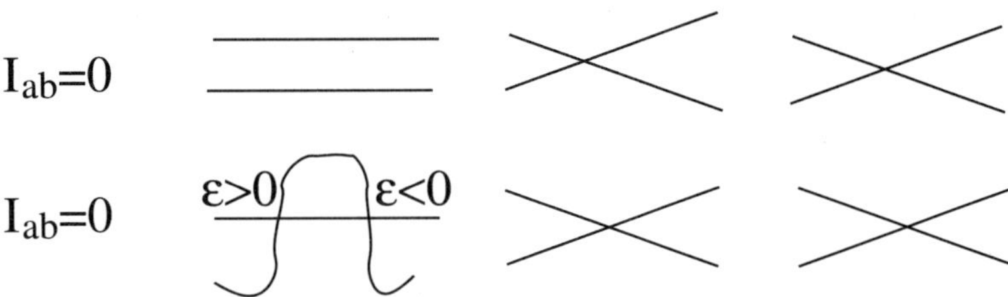

Fig. 9.15.   The intersection number is a topological quantity.

- The spectrum is thus:
  - Gauge Group: $\Pi_a U(N_a)$
  - Left chiral Fermions: $\sum_{a,b} I_{ab}(N_a, \overline{N}_b)$

  where in our convention, $I_{ab} < 0$ means right-handed chiral fermions. The possibility of $I_{ab} \neq 0, 1$ gives rise to an interesting mechanism of family replication.

- How do we compute the intersection number? Given $[\Pi_a]$, consider its Poincare dual defined as follows:

$$\int_{\Pi_a} \omega = \int_{X_6} \omega \wedge \alpha_a \, , \qquad \forall \omega \tag{9.39}$$

The intersection number is given by

$$I_{ab} = \int_{X_6} \alpha_a \wedge \alpha_b \tag{9.40}$$

- As an illustration, consider T^2 whose volume form is $dvol_{T^2} = dx \wedge dy$, then

$$[\Pi_a] = n_a[a] + m_a[b] \rightarrow \alpha_a = n_a dy - m_a dx \tag{9.41}$$

The intersection number is

$$I_{ab} = \int_{T^2} (n_a dy - m_a dx) \wedge (n_b dy - m_b dx) = (n_a m_b - n_b m_a) \tag{9.42}$$

Thus, the cycles shown in Figure 9.16 have intersection number $[(2,1)] \cdot [(0,1)] = 2$.

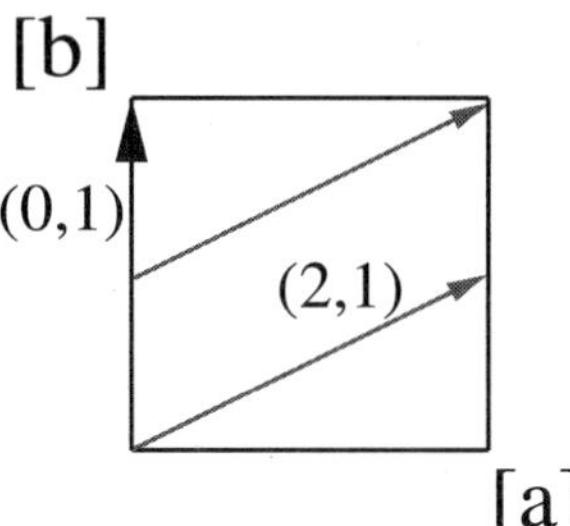

Fig. 9.16.   Examples of 1-cycles wrapping T^2.

- Now, consider a 6D torus which is factorizable, $X_6 = T^2 \times T^2 \times T^2$. We make a further simplification by assuming that the three-cycle are also factorizable, i.e., they can be expressed as products of 1-cycles on each T^2:

$$[\Pi_a] = [(n_a^1, m_a^1)] \otimes [(n_a^2, m_a^2)] \otimes [(n_a^3, m_a^3)] \tag{9.43}$$

This is not the most general form of 3-cycles.[c] In this case, the intersection number can be generalized from that for T^2 to:

$$I_{ab} = [\Pi_a] \otimes [\Pi_b] = \prod_{i=1}^{3} \left(n_a^i m_b^i - m_a^i n_b^i\right) \tag{9.44}$$

The intersection number I_{ab} is the intersection number in homology, and can be easly shown using the intersection of the basic homology cycles

$$[a_i] \circ [b_j] = \delta_{ij} \qquad [a_i] \circ [a_i] = [b_i] \circ [b_j] = 0 \tag{9.45}$$

and linearity and antisymmetry of the intersection pairing.

- Let's consider a few toroidal examples to check this formula and to illustrate the point that chirality arises when the branes cannot be separated from one another. First consider the following system:
  If we T-dualize on along all the y directions, we obtain:

$$D6_a \to D3 \tag{9.46}$$

$$D6_b \to D7 \tag{9.47}$$

---

[c]In fact, by brane recombination, one can start with two factorizable ones and construct a non-factorizable 3-cycle.

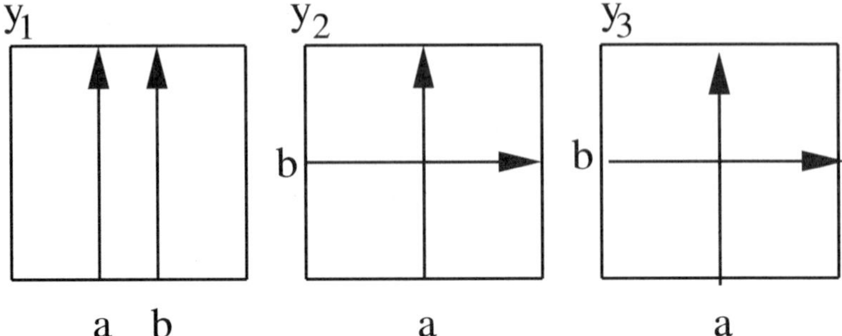

Fig. 9.17.   A non-chiral intersection of branes.

which is the non-chiral system previously considered. This is is consistent with $\theta_i^{ab} = 0$.

- If we consider instead the system:

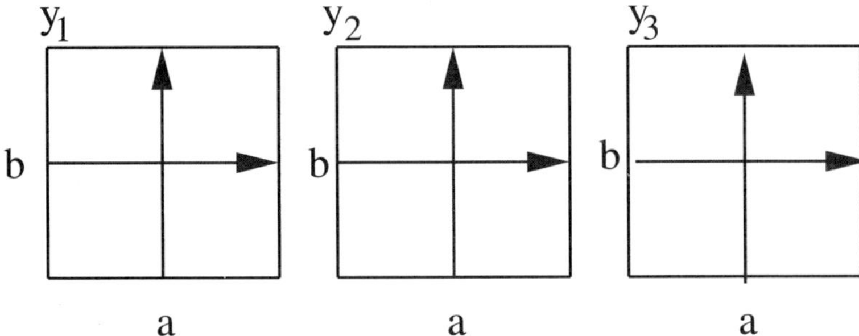

Fig. 9.18.   A chiral intersection of branes.

T-dualizing along all y directions give:

$$D6_a \to D3 \tag{9.48}$$
$$D6_b \to D9$$

We see that this is a chiral system (as the intersection number I_{ab} also shows) since we cannot separate a $D3$-brane from a $D9$ (same for $D5$ and $D7$).

- What about the following system?

  T-dializing along the y directions does not turn $D6_b$ into a pure $D9$, but rather a *bound state* of D-branes with different dimensions due to the world volume fluxes (see previously discussed T^2 example in Section 9.3). The intersection number has the interpretation of the index of a

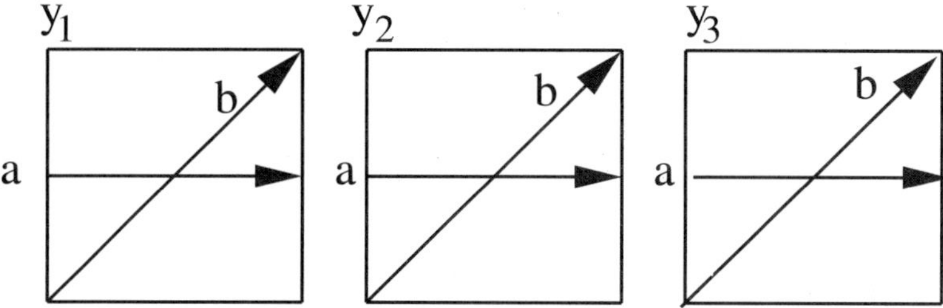

Fig. 9.19.   Another example of chiral intersection.

Dirac operator:

$$\text{index}_Q D = \int_{X_6} \text{ch}(F_a) \wedge \text{ch}(-F_b) \wedge \hat{A}(R) \qquad (9.49)$$

where $Q = (N_a, \overline{N}_b)$ and ch (F_a) is the Chern character, and $\hat{A}(R)$ is the A-roof genus. We will not need details of this expression. It is given here for completeness.

- Summarizing, compactification has the effect of family replication. The number of chiral families is topological:

  - **Type IIA:** Take two D6-branes, add and subtract intersection points. Number of chiral families is $I_{ab} = \int_{X_6} \alpha_a \wedge \alpha_b$.
  - **Type IIB:** A D_{2p+1}-brane and a $D_{2p'+1}$-brane intersect on submanifold $S_{2m} \subset X_6$. Number of chiral families is $\text{index}_Q D = \int_{S_{2m}} \text{ch}(F_a) \wedge \text{ch}(-F_b) \wedge \hat{A}(R)$.

9.5.3. *Supersymmetry for intersecting branes*

Intersecting D6-branes provide a particular setup for the brane world scenario:

- Gravity propagates in 10D
- Gauge bosons propagate in 7D
- Chiral matter propagates in 4D

So, in principle, we can consider both high and low string scale scenarios. In the former case, SUSY at the string scale helps to protect the Higgs mass from large radiative corrections. In the latter, one can consider the possibility of breaking SUSY at the string scale.

Nevertheless, we will focus on $\mathcal{N} = 1$ models whose effective theory is better understood. However, most of our results will be applicable to the $\mathcal{N} = 0$ case. SUSY models have the advantage that they are free of tachyons and the brane configurations are stable.

So, let us analyze the conditions for the D6-brane configuration to preserved SUSY. D-branes preserves $1/2$ of the supersymmetries but two D-branes can a priori preserve a different SUSY.

Let R be the $SO(6)$ rotation that takes the first D6-branes into the second. The condition that some supersymmetry is preserved by the combined system is that there exists a 6D spinor which is invariant under R. Such a spinor exists if and only if R belongs to an $SU(3)$ subgroup of $SO(6)$. The reason is that the spinor of $SO(6)$ which transform as a **4** decomposes under $SU(3)$ as

$$\mathbf{4} = \mathbf{3} \oplus \mathbf{1} \tag{9.50}$$

and the singlet is invariant under $SU(3)$ transformation. This condition can be more explicitly stated (locally at the intersection) as

$$\theta_1 + \theta_2 + \theta_3 = 0 \quad (\text{mod } 2\pi) \tag{9.51}$$

Indeed, one can check that the open string spectrum computed before is Bose-Fermi degenerate in such cases. In the generic case, there is no supersymmetry invariant under the two stacks of branes, and the open string sector at the intersection is non-supersymmetric. The configuration is at least $\mathcal{N} = 1$ supersymmetric if the sum of the angles is zero. $\mathcal{N} = 2$ supersymmetry arises if, in addition, one of the angles also vanishes while $\mathcal{N} = 4$ arises only for parallel stacks, i.e., $\theta_i = 0$.

More generally, for a Calabi-Yau manifold, the three-cycles that the D6-branes can wrap while preserving SUSY are the so-called special Lagrangian (sLag) cycles, which are defined as follows.

On a Calabi-Yau manifold there exist a covariantly constant holomorphic three-form, Ω_3, and a Kähler 2-form J. Locally, the holomorphic 3-form Ω_3 and the Kähler form J can be defined by

$$\Omega_3 = dz_1 \wedge dz_2 \wedge dz_3, \quad J = i \sum_{i=1}^{3} dz_i \wedge d\bar{z}_i. \tag{9.52}$$

A three-cycle π_a is called Lagrangian if the restriction of the Kähler form on the cycle vanishes

$$J\big|_{\pi_a} = 0. \tag{9.53}$$

If the three-cycle in addition is volume minimizing, which can be expressed as the property that the imaginary part of the three-form Ω_3 vanishes when restricted to the cycle,

$$\Im(e^{i\varphi_a}\,\Omega_3)|_{\pi_a} = 0, \tag{9.54}$$

then the three-cycle is called a sLag cycle. The parameter φ_a is a constant depending only on the homology class of π_a and determines which $\mathcal{N} = 1$ supersymmetry is preserved by the brane. Thus, different branes with different values for φ_a preserve different $\mathcal{N} = 1$ supersymmetries. One can show that (9.54) implies that the volume of the three-cycle is given by

$$\mathrm{Vol}(\pi_a) = \left| \int_{\pi_a} \Re(e^{i\varphi_a}\,\Omega_3). \right| \tag{9.55}$$

A shift of $\varphi_a \to \varphi_a + \pi$ corresponds to exchanging a D-brane by its anti-D-brane, where the D-brane really satisfies (9.55) without taking the absolute value. Therefore a supersymmetric cycle π_a is calibrated with respect to $\Re(e^{i\varphi_a}\Omega_3)$. To obtain a globally $\mathcal{N} = 1$ supersymmetric intersecting D-brane model all D6-branes have to wrap sLag three-cycles which are calibrated with respect to the same three-form.

Exercise: Show that for a torus, the condition that π_a is sLag reduces to the condition Eq. (9.51) derived before.

As mentioned earlier, the light scalars at the intersection can be massless, massive or tachyonic.

- The massless case corresponds to a situation with some unbroken supersymmetry. The massless scalar is a modulus whose vacuum expectation value parametrizes the possibility of recombining the two intersecting D-branes into a single smooth one.

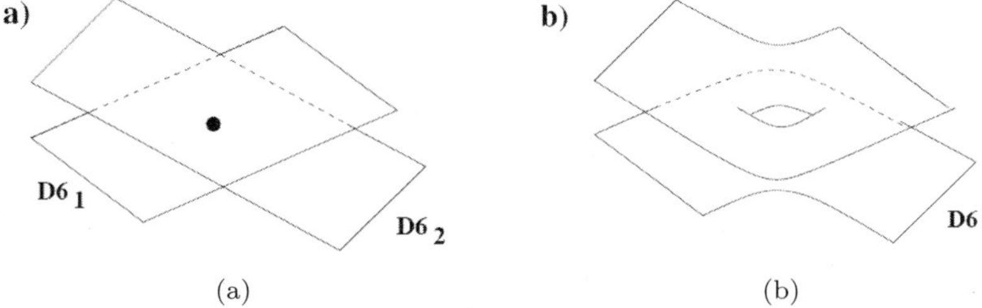

Fig. 9.20.   D-brane recombination.

- The configuration with tachyonic scalars corresponds to situations where this recombination is triggered dynamically, as a result of tachyon condensation at the intersection.

- The configuration where all light scalars have positive squared mass corresponds to a non-supersymmetric configuration, which is nevertheless dynamically stable against recombination. Namely, the recombined 3-cycle has a volume larger than the sum of the volumes of the intersecting 3-cycles.

9.5.4. *RR tadpole Cancellation*

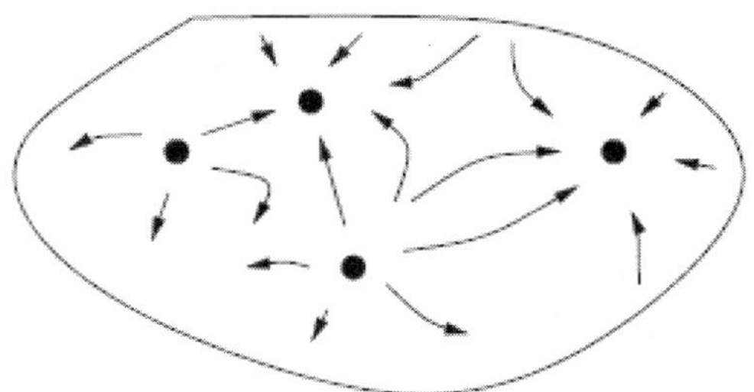

Fig. 9.21.   RR charges and Gauss's Law.

However, compactification also leads to new subtleties. This is because D-branes act as a source for RR fields via the disk coupling:

$$\int_{W^{p+1}} C_{p+1} \tag{9.56}$$

In a compact space, the total RR charge must vanish as required by Gauss's law.

The RR charges are characterized by the 3-cycles on which the branes are wrapped. Hence, the sum of the homological cycles wrapped by all the branes must be topologically trivial.

$$[\pi_{total}] = \sum_a N_a [\pi_a] = 0 \tag{9.57}$$

Equivalently, one can understand this constraint as the consistency requirement of the equations of motion for the RR fields. Keeping only terms in the spacetime action which depend on the RR 7-form C_7, we have:

$$S_{C7} = \int_{M_4 \times X_6} H_8 \wedge *H_8 + \sum_a N_a \int_{M_4 \times \pi_a} C_7$$

$$= -\int_{M_4 \times X_6} C_7 \times dH_2 + \sum_a N_a \int_{M_4 \times X_6} C_7 \wedge \delta(\pi_a) \tag{9.58}$$

where H_8 is the 8-form field strength, H_2 its Hodge dual, and $\delta(\pi_a)$ is a bump 3-form localized on the 3-cycle. The equation of motion reads:

$$dH_2 = \sum_a N_a \delta(\pi_a) \qquad (9.59)$$

Integrating this equation gives the tadpole constraint Eq. (9.57) on the homology classes.

9.5.5. *Anomaly Cancellation*

Cancellation of RR tadpoles in the underlying string theory is important because it implies the cancellation of chiral anomalies in the four dimensional effective field theory. These anomalies include the cubic non-Abelian anomalies, and mixed $U(1)$ non-Abelian anomalies, and mixed gravitational anomalies. Let us discuss them one by one.

(1) **Cubic non-Abelian anomalies**

The $SU(N_a)^3$ anomaly is proportional to the number of fundamental minus anti-fundamental representations of $SU(N)$. Since the matter fields transform as bifundamentals, the cubic anomaly it is proportional to the sum of the intersection number times the number of branes:

$$A_a = \sum_b I_{ab} N_b \qquad (9.60)$$

It is easy to check that this vanishes due to RR tadpole cancellation. By taking the intersection of the tadpole condition with any 3-cycle, we find

$$0 = [\pi_a] \circ \sum_b N_b [\pi_b] = \sum_b N_b I_{ab} \qquad (9.61)$$

as claimed. Note that the tadpole condition is slightly stronger than the absence of cubic anomalies because it must hold even for $N = 1, 2$ where no cubic anomaly exists. As we will see, this observation will turn out important in phenomenological model building.

(2) **Mixed anomalies**

What about mixed $U(1)_a - SU(N_b)^2$ anomalies? The usual field theory triangle diagram gives a non-zero contribution, even after using the RR tadpole conditions:

$$A_{ab} \simeq N_a I_{ab} \qquad (9.62)$$

However, in string theory there is an extra diagram known as the Green-Schwarz diagram, where the $U(1)$ gauge boson mixes with a 2-form which subsequently couples to two gauge bosons of $SU(N)$:

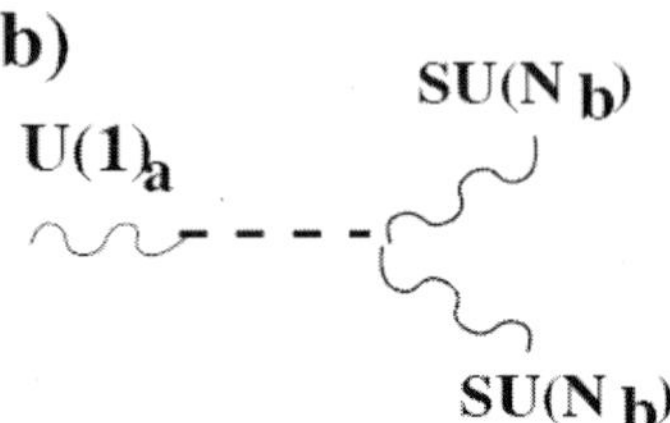

Fig. 9.22.   The Green-Schwarz term.

Exercise: Show that the following couplings arising in the KK reduction of the D6-brane world volume action

$$N_a \int_{D6_a} C_5 \wedge tr F_a, \qquad \int_{D6_b} C_3 \wedge tr F_b^2 \qquad (9.63)$$

give rise to the Green-Schwarz term which leads to a cancellation of the residual field theory triangle anomalies.

Solutions: Introducing a basis of 3-cycles $[\Lambda_k]$ and its dual $[\Lambda_{\tilde{\ell}}]$, we can define the KK reduced 4d fields

$$(B_2)_k = \int_{[\Lambda_k]} C_5, \qquad \phi_{\tilde{\ell}} = \int_{[\Lambda_{\tilde{\ell}}]} C_3 = -\delta_{k\tilde{\ell}} *_{4d} (B_2)_k \qquad (9.64)$$

The KK reduced 4d couplings read

$$N_a q_{ak} \int_{4d} (B_2)_k tr F_a, \qquad q_{b\tilde{\ell}} \int_{4d} \phi_{\tilde{\ell}} tr F_b^2 \qquad (9.65)$$

with $q_{ak} = [\pi_a] \circ [\Lambda_k]$, and similarly for $q_{b\tilde{\ell}}$. The total amplitude is proportional to

$$A_{ab}^{GS} = -N_a \sum_k q_{ak} q_{b\tilde{\ell}} \delta_{k\tilde{\ell}} = \cdots - N_a I_{ab} \qquad (9.66)$$

leading to cancellation between both kinds of contributions.

(3) **Mixed gravitational anomalies:** It is left as an exercise to show that for toroidal models, the mixed gravitational anomalies cancel automatically, without the Green-Schwarz contribution. **(Exercise)** This is no longer true for orbifolds/orientifolds.

An important observation is that any $U(1)$ gauge boson with $B \wedge F$ coupling gets massive, with mass roughly of the order of the string scale.

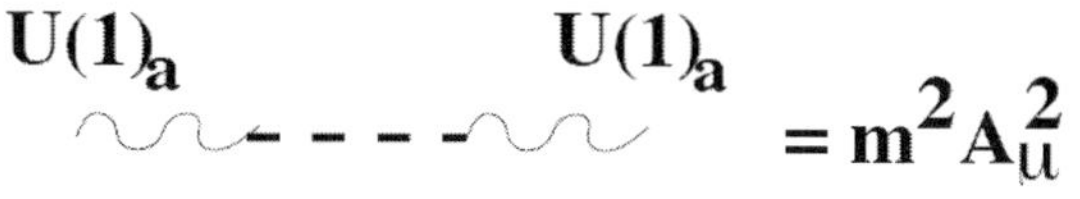

$$\text{U(1)}_a \qquad\qquad \text{U(1)}_a \qquad = \text{m}^2 \text{A}_\mu^2$$

Fig. 9.23. $B \wedge F$ coupling and massive $U(1)$.

There can be several of them, and they are not necessarily anomalous. Such $U(1)$'s disappear as gauge symmetries from the low energy effective theory, but remains as global symmetries, unbroken in perturbation theory. In constructing D-brane Standard Model, we need to make sure that the candidate for the hypercharge is not one of these massive $U(1)$.

9.5.6. *Phenomenological features*

Before we construct more realistic examples, let us briefly mention some phenomenological features we can extract from the toroidal examples studied so far. These features are natural in the general setup of intersecting brane models and are not restricted only to toroidal models.

- **Proton Stability:** First of all, in these models, the proton is perturbatively stable. This is because the $U(1)$ within the $U(3)$ plays the role of baryon number, and is preserved as a global symmetry, unbroken in perturbation theory though it can be broken by non-perturbative instanton effects. In some cases, the instanton effects breaking the Standard Model baryon number is calculable, e.g., as a Euclidean D2-brane wrapped on 3-cycles.
- **Gauge Unification:** These models do not have a natural gauge coupling unification, even at the string scale. Each gauge factor has a gauge coupling controlled by the volume of the wrapped 3-cycle. Gauge couplings are related to geometric volumes, hence the moduli controlling the sizes of these volumes are constrained by experiments.
- **Electroweak Symmetry Breaking:** These exists a geometric interpretation for the spontaneous electroweak symmetry breaking. In explicit models, the Higgs particle arises from the light scalar at the intersections, whose vev parametrizes the possibility of recombining two intersecting branes into a single smooth one. In this process, the

gauge symmetry is reduced, corresponding to a Higgs mechanism in the effective field theory.

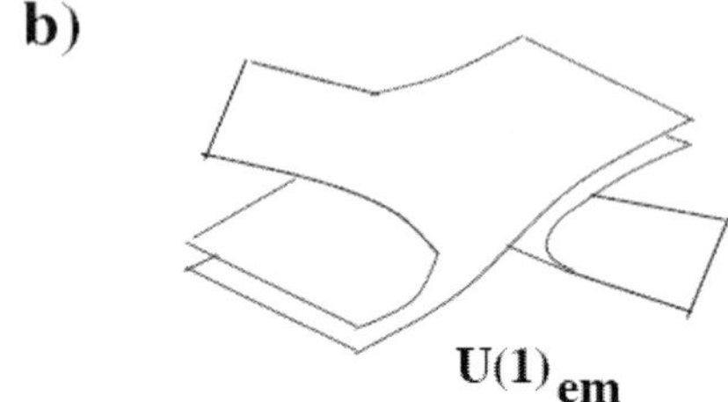

Fig. 9.24.   Electroweak symmetry breaking as brane recombination.

- **Yukawa Couplings:** There is a natural exponential hierarchy in the Yukawa couplings. Yukawa couplings among the scalar Higgs and chiral fermions arise at tree level in the string coupling from open string worldsheet instantons, namely from string worldsheet spanning the triangle with vertices at the intersections and sides on the D-branes.

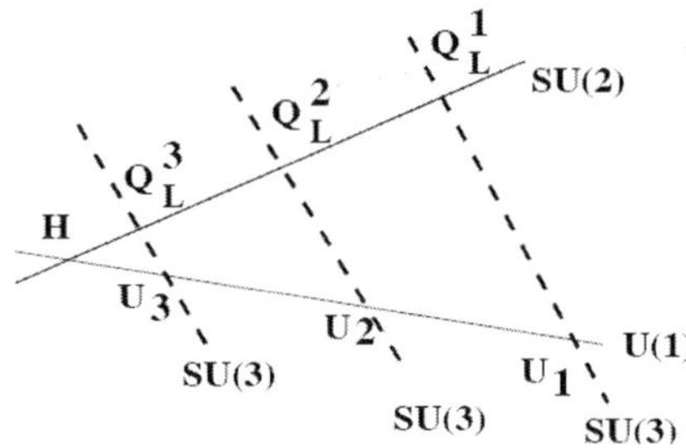

Fig. 9.25.   Yukawa couplings arise from worldsheet instantons.

Their values are exponentially suppressed by the area of the string worldsheet in string units. Different families are located at different intersections, leading to an exponential hierarchy in the Yukawa couplings between different families.

9.5.7. *Orientifold Models*

It turns out what we have introduced so far are not yet sufficient to construct a realistic model. We need an extra element, known as the orientifold planes.

Orientifold is a play of the word orbifold. Unlike an orbifold which acts only on the target space, an orientifold acts as on the worldsheet as well.

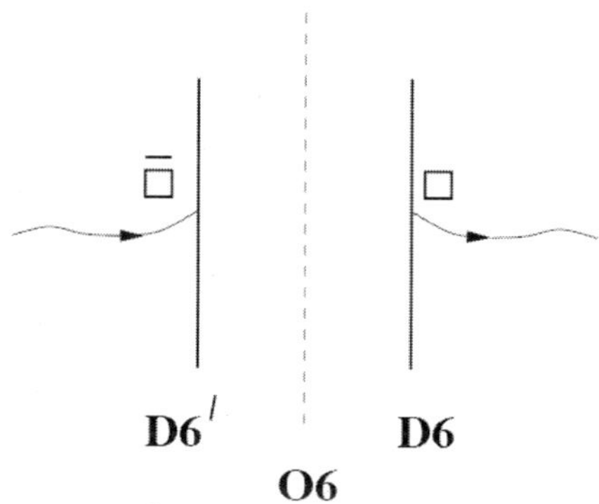

Fig. 9.26. An orientifold 6-plane.

An orientifold projection flips the orientation of the worldsheet and reflects the coordinates. The set of points fixed under the orientifold projection is called an orientifold plane.

So, why orientifolds? We have already seen that the total RR charge in a compact space must vanish. So, we need objects with negative charge. In order to preserve SUSY, these objects must also carry negative tension. We will see shortly that an orientifold plane has precisely these properties. Another reason to consider orientifolds is that as it turns out, intersecting D-brane models without orientifold planes are bounded to contain chiral exotics.

9.5.8. *Properties of O6-planes*

To see why this is the case, let us first review the properties of orientifold planes. To cancel the RR charges of the D6-branes, we need objects of the same dimensionality and hence O6-planes. Again, to start, consider the simplest case which is Type IIA string theory on 10d flat space and mod it out by the so called orientifold action $\Omega R(-1)^{F_L}$ where Ω is the worldsheet parity symmetry: $(\sigma, \tau) \to (-\sigma, \tau)$ which flips the orientation of the fundamental strings; R is a $\mathbb{Z}_2$ geometric action, acting locally as $(x^5, x^7, x^9) \to (-x^5, -x^7, -x^9)$; finally $(-1)^{F_L}$ is the left-moving worldsheet fermion number, introduced for technical reasons. One can argue for this

$(-1)^{F_L}$ factor from T-duality (**Exercise**).

The subspace in spacetime, fixed under the geometric part R of the above action, is known as an orientifold 6-plane. It is a 7d plane defined by $x^5 = x^7 = x^9 = 0$ and spanned by the remaining 7 coordinates. Physically, it corresponds to *a region of spacetime where the orientation of a string can flip*. The description of string theory in the presence of orientifold planes is modified by the inclusion of unoriented worldsheets, for instance with the topology of the Klein bottle.

Orientifold planes have some features similar to D-branes of the same dimension. For instance, an O_p plane carries tension and are charged under the RR $(p+1)$ form C_{p+1}.

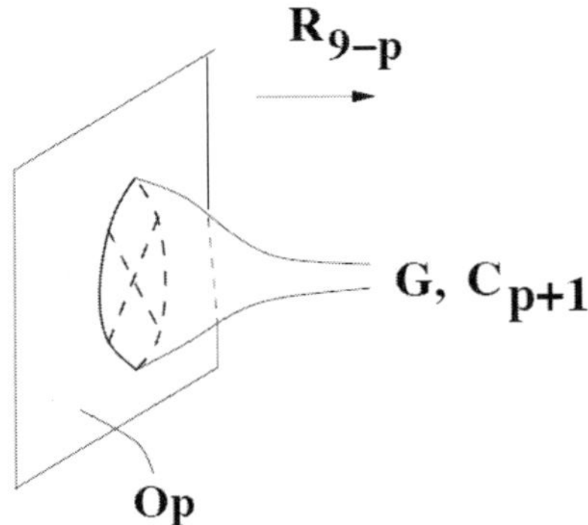

Fig. 9.27.   An orientifold plane carries tension and RR-charge.

In fact, one can compute its tension and charge by comparing the one-loop open string diagrams with and without the orientifold plane.

One finds that the charge of an O6-plane is -4 if the charge of a D6-brane is normalized to 1. (Actually, there are O6-planes which carry positive RR charges but to avoid confusion, let's ignore this possibility in what follows). Furthermore, O6-planes preserve the same supersymmetry as a D6-brane. This implies that there is a relation between the tension and the charge of an O-plane.

Hence, in a compact space, the cancellation of RR charges implies:

$$\sum_a N_a[\pi_a] + \sum_a N_a[\pi_{a'}] - 4 \times [\pi_{O6}] = 0 \tag{9.67}$$

where $[\pi_{a'}]$ is the homology class of the 3-cycle wrapped by the image D6-branes. *There are additional discrete constraints arising from cancellation*

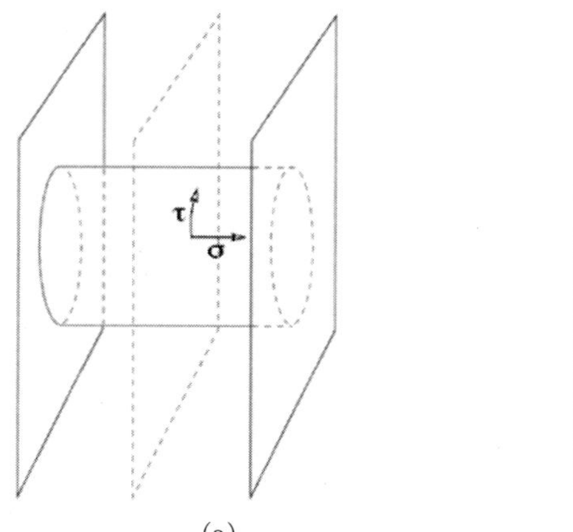
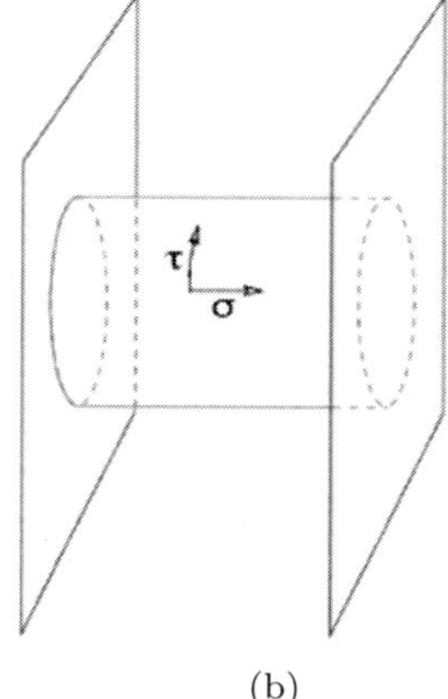

(a)   (b)

Fig. 9.28. (a) One-loop open string amplitude in the presence of O-planes, (b) without O-planes.

of $\mathbf{Z}_2$ valued K-theory charges. We will skip the discussion of such constraints for now and come back to them later when we construct concrete models.

There are however important differences between O-planes and D-branes. The most important one being that an orientifold plane has no world volume degrees of freedom since open strings are not stuck on them. So, unlike D-branes, they are not dynamical objects [d].

Although we have been discussing flat space and toroidal models so far, more generally, an O6-plane can be defined by the quotient $\Omega\overline{\sigma}(-1)^{F_L}$ where $\overline{\sigma}$ is an isometric, antiholomorphic involution of X_6, and acts on the Kahler class J and the holomorphic 3-form Ω_3 as:

$$\overline{\sigma}J = -J \qquad \overline{\sigma}\Omega_3 = e^{2\pi i\varphi}\overline{\Omega}_3 \tag{9.68}$$

with $\varphi \in \mathbb{R}$. For $\varphi = 0$ in local coordinates, this can be thought of as complex conjugation. The orientifold 6-plane is localized at the fixed point of $\overline{\sigma}$ which topologically is a three-cycle.

---

[d]In F-theory, which can be thought of as Type IIB orientifold compactification with varying dilaton, an orientifold planes can be interpreted as a bound state of mutually non-local 7-branes and so O-planes are dynamical in such a limit.

9.5.9. *Open String Spectrum with O6-planes and D6-branes*

To describe orientifold models, it is convenient to go to the covering space, and include the images of D-branes under the orientifold action. The spectrum of open strings in the orientifold construction can be obtained by simply computing the spectrum in the covering space, and then imposing the identification implied by the orientifold action.

First, consider configurations with parallel D6-branes and O6-planes.

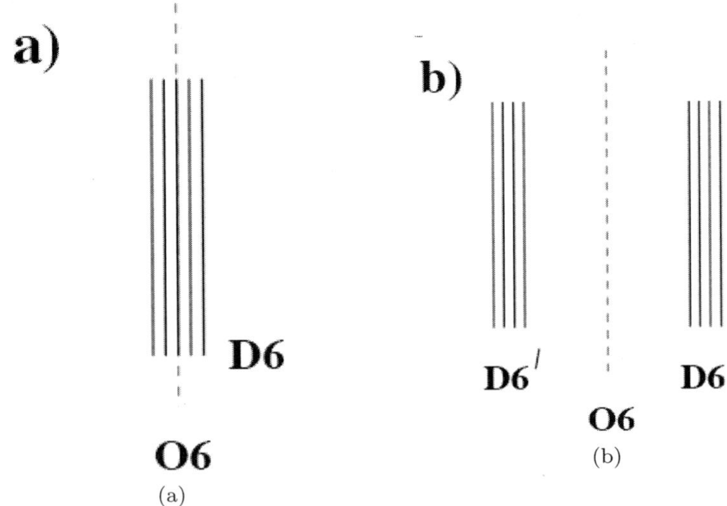

Fig. 9.29.   Parallel D6-branes and O6-planes: (a) on top of each other, (b) separated.

- If the D6-branes are on top of the O-plane, the orientifold action identifies the endpoints of the string:

$$|ab> \leftrightarrow \pm|ba> \tag{9.69}$$

  This identification breaks the $U(N)$ gauge group to SO or Sp depending on some subtle choices of sign.
- If the D6-branes are parallel but separated from the O-plane, the orientifold action maps a stack of D6-branes to its image. The two $U(N)$ factors are identified, only a linear combination survives. This agrees with the intuition that massless modes on D-branes are not sensitive to distant objects, hence the N D6-branes in the quotient do not notice, at the level of the massless spectrum, the distant O6-planes.

An important observation is that due to the orientation reversal, an open string starting on a stack of D6-branes is mapped to an open string ending on the image stack. This implies that a fundamental representation is mapped to an anti-fundamental representation and vice versa. This fact will be important later on.

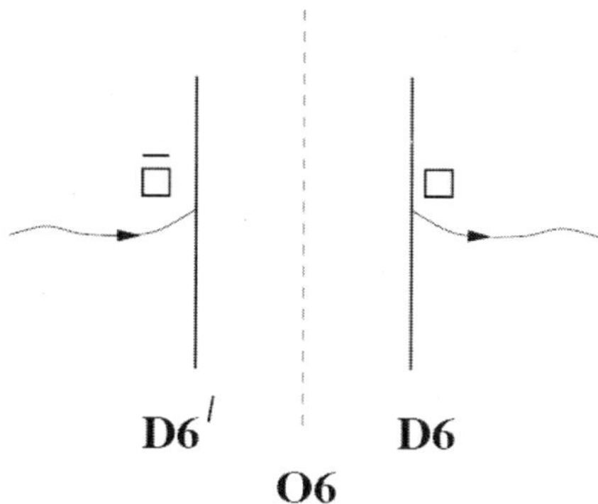

Fig. 9.30.   An O-plane reverses the orientation of an open string.

Now, consider situations with intersecting branes (and their images) in the presence of O-planes. There are several cases to consider (illustrated in Figures 9.31 and 9.32):

- If two stacks of D6-branes, labeled a and b, are intersecting away from the O-plane, the orientifold action simply maps these two stacks of branes to their images. After the identification,we are left with a $U(N) \times U(M)$ gauge group and a chiral fermion in the bi-fundamental representation.
- Consider now the intersection of D-branes in the a-stack with the orientifold image of the b-stack, the only difference is that the chiral fermion transform in the fundamental representations of both gauge groups.
- Finally, we can consider a stack of D-branes intersecting with its own image either on top of the O-plane or away from it. The open strings ending on the D6-branes and their images transform in the symmetric or antisymmetric representations of $U(N)$. If the intersection is right on top of the orientifold plane, only the antisymmetric representation survives. If the intersection is away from the O-plane, the orientifold action simply maps these two sets of symmetric plus anti-symmetric representations to each other.

    G. Shiu

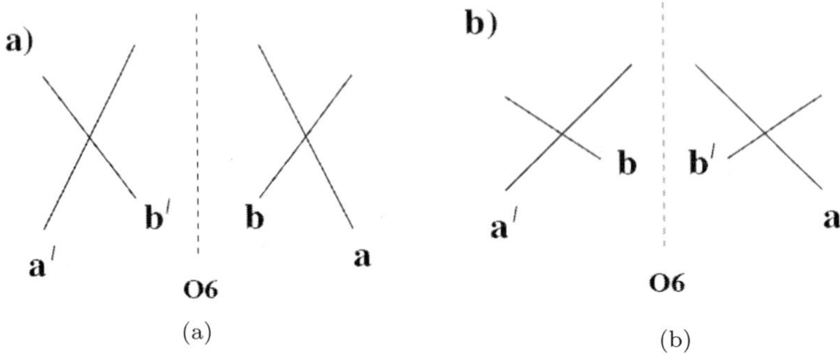

Fig. 9.31.   Intersection of different D6-planes in the presence of an O6-plane.

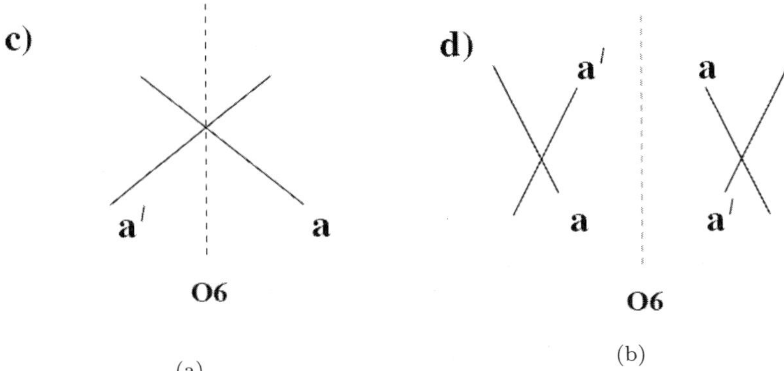

Fig. 9.32.   Intersection of a D6-plane and its image in the presence of an O6-plane.

To summarize, this is the chiral spectrum of an orientifold model.

Representation	Multiplicity
$\boxed{}\!\boxed{}_a$ (antisymmetric)	$\frac{1}{2}\left(\pi'_a \circ \pi_a + \pi_{O6} \circ \pi_a\right)$
$\boxed{}_a$ (symmetric)	$\frac{1}{2}\left(\pi'_a \circ \pi_a - \pi_{O6} \circ \pi_a\right)$
$(\overline{\square}_a, \square_b)$	$\pi_a \circ \pi_b$
$(\square_a, \square_b)$	$\pi'_a \circ \pi_b$

We will see later that the same chiral spectrum applies to more general
backgrounds such as orbifolds and even Calabi-Yau manifolds, *provided that*

we interpret these homology cycles correctly.

There is one more subtlety, involving orientfold planes, when we compactified the theory. The $\mathbb{Z}_2$ involution $\bar{\sigma}$ may only be a discrete symmetry for certain choices of the moduli. For example, $\bar{\sigma}=$ complex conjugate is a Z_2 involution on a torus only for two choices of complex structure moduli: a rectangular torus or a specifically "tilted" torus:

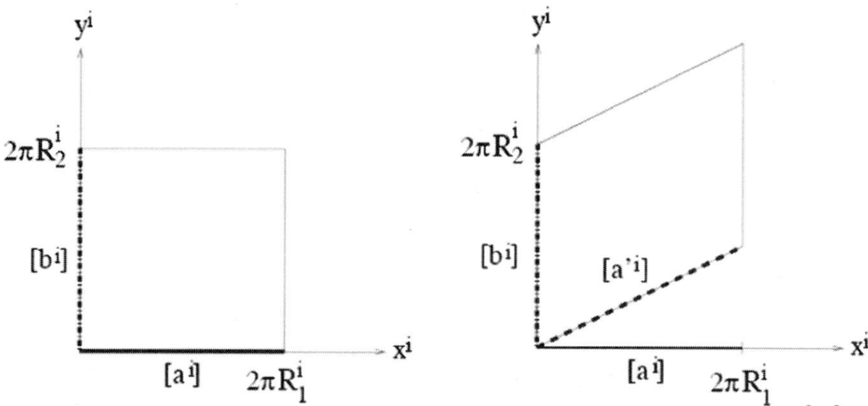

The basic homology cycles of the tilted torus can be expressed in terms of the untilted ones:

$$[a'] = [a] + \frac{1}{2}[b] \qquad [b'] = [b] \qquad (9.70)$$

The orientifold image of a D-brane is again given by the reflection along the O-plane.

9.5.10. *Getting just the Standard Model*

In addition to their importance in canceling the D-brane charges, there is in fact a general argument which shows that in the absence of O-planes, D-brane models necessarily contain $SU(2)$ chiral exotics. To see this, first notice that without the O-planes, the electroweak $SU(2)$ must belong to a $U(2)$ factor of the gauge group. An important point emphasized earlier is that the tadpole condition implies that the number of fundamentals and anti-fundamentals must be equal, even for $U(2)$ (where the 2 and the $\bar{2}$ are distinguished by their $U(1)$ charge). Now, since there are 3 families and the left-handed quarks transform as $(3,\bar{2})$, they contribute altogether 9 anti-fundamentals of $SU(2)$. The complete spectrum must necessarily contain 9 fundamentals, three of which may be interpreted as left-handed leptons;

the remaining six doublets are however exotic chiral fermions, beyond the spectrum of the SM.

There are two ways to avoid these exotics, both involving orientifolds.

9.5.10.1. *The U(2) case*

One possibility is to exploit the fact there are two kinds of bi-fundamental fields in an orientifold model, namely, $(N_a, \overline{N}_b)$ and (N_a, N_b). Consider realizing the three families of left-handed quarks as a combination of $(3, \overline{2}) + 2(3, 2)$. The number of $SU(2)$ doublets needed to cancel the tadpoles is 3 which is precisely number of left-handed leptons.

Exercise: Consider four stacks of branes denoted a, b, c, d (and their images), giving rise to a gauge group $U(3)_a \times U(2)_b \times U(1)_c \times U(1)_d$. If the intersection numbers between the corresponding 3-cycles are given by:

$$I_{ab} = 1 \qquad I_{ab'} = 2 \qquad I_{ac} = -3 \qquad I_{ac'} = -3$$
$$I_{bd} = 0 \qquad I_{bd'} = -3 \qquad I_{cd} = -3 \qquad I_{cd'} = 3 \qquad (9.71)$$

Show that the chiral spectrum has the non-Abelian quantum numbers of the Standard Model (plus three right-handed neutrinos). In order to reproduce exactly the Standard Model one also needs to require that the linear combination of $U(1)$'s

$$Q_Y = \frac{1}{6}Q_a - \frac{1}{2}Q_c + \frac{1}{2}Q_d \qquad (9.72)$$

to be massless.

However, several comments are in order. It is important to emphasize that at this level, we have not constructed any explicit model. In particular, we have only presented a set of intersection numbers. We haven't shown that there exist cycles on which the D-branes wrapped which lead to the given intersection numbers. Moreover, even if we manage to do so, the intersection numbers only define a local model. We still need to introduce additional hidden sector branes to cancel the tadpoles in constructing compact models. On the other hand, the intersection numbers are not restricted to toroidal models. In fact, it was shown that such topological data can indeed arise in more general approach such as Calabi-Yau orientifolds and Gepner constructions.

9.5.10.2. *The USp(2) Case*

Another way to avoid the $SU(2)$ exotics is to make use of the fact that $D6$-branes in the presence of orientifold planes can give rise to symplectic groups. For symplectic groups, all representations are real, and RR tadpole conditions do not impose any constraint on the number of doublets. Since $USp(2) \equiv SU(2)$, it is possible to realize the electroweak $SU(2)$ as a symplectic group, and thus circumvent the constraints on the number of doublets.

Again, as an exercise, you can show that the following set of branes with gauge groups $U(3)_a \times USp(2)_b \times U(1)_c \times U(1)_d$ and intersection numbers

$$I_{ab} = 3 \qquad I_{ab'} = 3 \qquad I_{ac} = -3 \qquad I_{ac'} = -3$$
$$I_{db} = 3 \qquad I_{db'} = 3 \qquad I_{dc} = -3 \qquad I_{dc'} = 3 \qquad I_{bc} = -1 \qquad I_{bc'} = 1 \tag{9.73}$$

give rise to the chiral spectrum of the Standard Model. The $U(1)$ that needs to be massless in order to reproduce the Standard Model hypercharge is

$$Q_Y = \frac{1}{6}Q_a - \frac{1}{2}Q_c - \frac{1}{2}Q_d \tag{9.74}$$

In fact, explicit models with D6-branes on 3-cycles and these intersection numbers have been constructed. We will discuss some examples later.

9.5.11. *Supersymmetric models*

Let us try to embed these D-brane configurations into a supersymmetric setup. We are interested in chiral models, which can arise in theories with $N = 1$ or less supersymmetry. The simplest way of reducing the supersymmetry to $N = 1$ is via orbifolding. For example, the $\mathbb{Z}_2 \times \mathbb{Z}_2$ orbifold whose generators acts on the three complex compactified coordinates as follows:

$$\theta : (z_1, z_2, z_3) \rightarrow (-z_1, -z_2, z_3)$$
$$\omega : (z_1, z_2, z_3) \rightarrow (z_1, -z_2, -z_3) \tag{9.75}$$

preserves $N = 1$ supersymmetry. Recall the orientifold projection introduced earlier:

$$\Omega R(-1)^{F_L} \tag{9.76}$$

Orbifolding introduces 3 new types of orientifold planes: $\Omega R\theta(-1)^{F_L}$, $\Omega R\omega(-1)^{F_L}$, and $\Omega R\theta\omega(-1)^{F_L}$. Hence, there are 4 kinds of O-planes, associated to the worldsheet parity couples with four different types of spatial reflection. The O6-planes are invariant under the $\mathbb{Z}_2$ involution, so we can depict them as follows:

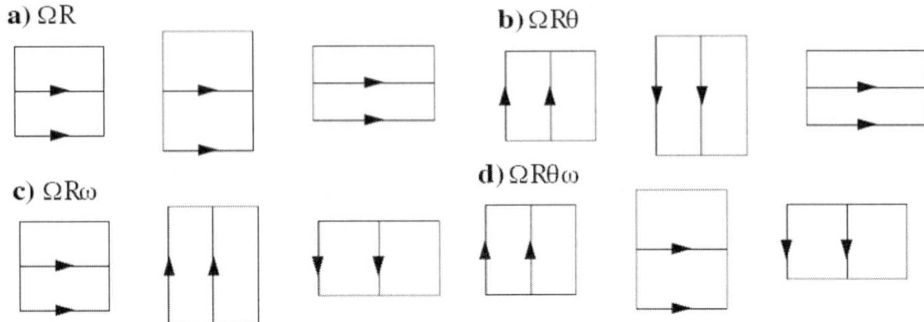

The chiral spectrum can be computed from the general results obtained earlier, except to keep in mind that the homology cycles are defined on the orbifold space instead of the torus.

To provide an illustrative example, consider the following model:

N_α	(n_α^1, m_α^1)	(n_α^2, m_α^2)	(n_α^3, m_α^3)
$N_a = 6$	$(1,0)$	$(3,1)$	$(3,-1)$
$N_b = 2$	$(0,1)$	$(1,0)$	$(0,-1)$
$N_c = 2$	$(0,1)$	$(0,-1)$	$(1,0)$
$N_d = 2$	$(1,0)$	$(3,1)$	$(3,-1)$
$N_{h1} = 2$	$(-2,1)$	$(-3,1)$	$(-4,1)$
$N_{h2} = 2$	$(-2,1)$	$(-4,1)$	$(-3,1)$
40	$(1,0)$	$(1,0)$	$(1,0)$

Up to some redefinition, the subset of branes in blue gives rise to the intersection numbers in the $USp(2)$ case introduced earlier.

The D-brane configurations preserve SUSY if the 3-cycles are sLag. For toroidal orbifolds (see homework), the condition can be simply stated as:

$$\theta_1 + \theta_2 + \theta_3 = 0 \tag{9.77}$$

or equivalently, in terms of the complex structure moduli $\chi_i = (R_2/R_1)_i$:

$$\sum_i \arctan(\chi_i \frac{m_i}{n_i}) = 0 \tag{9.78}$$

which admits solutions [e] for χ_i.

The spectrum is fairly complicated, but under D-brane recombination, the chiral spectrum is greatly simplified (see [30,31]):

---

[e]This formula is actually only valid for the case $n_a^i \geq 0$. See below for some other important cases.

Sector	Matter	$SU(4) \times SU(2)_L \times SU(2)_R$	Q_a	Q_h	Q'
(ab)	F_L	$3(4, 2, 1)$	1	0	$1/3$
(ac)	F_R	$3(\bar{4}, 1, 2)$	-1	0	$-1/3$
(bc)	H	$(1, 2, 2)$	0	0	0
(bh)		$2(1, 2, 1)$	0	-1	2
(ch)		$2(1, 1, 2)$	0	$+1$	-2

Fig. 9.33.   The full spectrum of the model in [30,31] after D-brane recombination.

Exercise: Show that unlike the toroidal model, the mixed gravitational anomalies in the $\mathbb{Z}_2 \times \mathbb{Z}_2$ orientifold are not canceled, even after the cancellation of RR charges. They are canceled by the Green-Schwarz mechanism.

In the strong coupling limit, these $\mathcal{N} = 1$ supersymmetric intersecting D-brane models are lifted to M theory on singular G_2 manifolds. This is an interesting subject which I have no time to discuss. For a brief discussion, see [14].

I hope these examples suffice to illustrate the concepts we have learned. Much of our discussions so far have centered on the open string sector since we are interested in getting the gauge and chiral spectrum of the Standard Model from string theory. The closed string sector is however also relevant for low energy dynamics, and leads to an additional set of important questions. This is the topic to which we now turn.

9.6. Warped Compactifications

The main such question that we have not yet addressed is the existence of moduli, massless fields with flat potential and hence undetermined vevs. Stabilization of these moduli (namely, providing them with a potential which fixes their vevs and give them masses) is an important question, of immediate relevance to phenomenology and cosmology. A proposal to attack this problem is to consider compactification with fluxes.[32–36] We will begin with a general discussion of flux compactifications and then discuss issues that arise when combining this mechanism of moduli stabilization with D-brane model building. For simplicity and concreteness, we consider Type IIB compactifications with 3-form fluxes. Similar analysis should (and could in principle) be carried out for other setups.

9.6.1. *Flux Compactification*

Flux compactification has developed into a very broad subject. Here, we will focus on aspects which are relevant for model building. General discussions can be found in some recent reviews [14,37,38]. The basic idea is that in string theory, there are many p-form gauge fields C_p in its massless spectrum. In particular, Type IIB contains NS-NS 0 and 2-forms, and RR 2 and 4 forms and their magnetic dual $8 - p$ forms. Analogous to electromagnetism, one can consider solutions with a non-vanishing flux $F_{p+1} = dC_p \neq 0$. Turning on fluxes has the following interesting consequences:

- The kinetic term

$$S_{kin} = \int_{\mathcal{M}} F \wedge *F \tag{9.79}$$

  induces a scalar potential in the 4D effective action, which in general depends on the moduli controlling the size of the cycles Σ the flux is running through, i.e.,

$$\int_{\Sigma} F \neq 0 \tag{9.80}$$

  This leads to *moduli stabilization*. This is analogous to the statement in electromagnetism that the electromagnetic energy $U = \frac{1}{8\pi} \int \left(E^2 + B^2 \right)$ which depends on the volume of space.
- The background flux provides a source term to the Einstein equation. Hence its backreaction warps the geometry of space and the metric is no longer Ricci flat.
- Fluxes carry D3-charge through Chern-Simons terms in the 10D action. They can affect the number of D-branes in compact examples.

In more details, consider Type IIB compactification with non-trivial NSNS and RR 3-form fluxes H_3 and F_3. These 3-form fluxes must satisfy the Bianchi identity

$$dF_3 = 0 \qquad dH_3 = 0 \tag{9.81}$$

and they should be properly quantized, namely, for any 3-cycle $\Sigma \subset X_6$

$$\frac{1}{(2\pi)^2\alpha'} \int_{\Sigma} F_3 \in \mathbb{Z} \qquad \frac{1}{(2\pi)^2\alpha'} \int_{\Sigma} H_3 \in \mathbb{Z} \tag{9.82}$$

An immediate consequence is that these background fluxes induce an effective D3 charge through the Chern-Simons terms in the Type IIB effective action:

$$\int_{M_4 \times X_6} H_3 \wedge F_3 \wedge C_4 \tag{9.83}$$

where C_4 is the IIB self-dual 4-form gauge potential. This coupling implies that upon compactification, the flux background contributes to a tadpole for C_4, and hence the 3-form fluxes carry D3-brane charges. With proper normalization factor restored, the flux-induced D3 charge is:

$$N_{flux} = \frac{1}{(4\pi^2\alpha')^2} \int_{X_6} H_3 \wedge F_3 = \frac{i}{(4\pi^2\alpha')^2} \int_{X_6} \frac{G_3 \wedge \overline{G}_3}{2\mathrm{Im}\tau} \tag{9.84}$$

where $\tau = a + i/g_2$ is the IIB complex dilaton, and we complexify the 3-form fluxes

$$G_3 = F_3 - \tau H_3 \tag{9.85}$$

for convenience. The kinetic term for G_3 is

$$V = \frac{1}{4\kappa_{10}^2 \mathrm{Im}\tau} \int_{X_6} d^6 y\, G_3 \wedge *_6 \overline{G}_3 \tag{9.86}$$

which induces a scalar potential. It can be written as:

$$V = \frac{1}{2\kappa_{10}^2 \mathrm{Im}\tau} \int_{X_6} d^6 y\, G_3^- \wedge *_6 \overline{G}_3^- - \frac{i}{4\kappa_{10}^2 \mathrm{Im}\tau} \int_{X_6} d^6 y\, G_3 \wedge \overline{G}_3 \tag{9.87}$$

Here $G_3^{\pm}$ is the imaginary self-dual/anti-self-dual (ISD/IASD) part of G_3, i.e.,

$$*_6 G_3^{\pm} = \pm i G_3^{\pm} \tag{9.88}$$

The second term in the potential is a topological term proportional to N_{flux}. They will be canceled by other sources of RR charges in a compact model. The first term is positive definite F-term potential which precisely vanishes if the flux is imaginary self dual. Thus, the equation of motion imposes self-duality of the 3-form flux.

It has been shown that this F-term potential V_F can be derived from the Gukov-Vafa-Witten superpotential:

$$W = \int_{X_6} G_3 \wedge \Omega \tag{9.89}$$

which depends only on the complex structure (shape) moduli and the dilaton and vanishes if these moduli are chosen such that G_3 is ISD. Thus all such moduli are generically stabilized. We can decompose the 3-form flux G_3 in terms of the Hodge cohomology according to the complex structure of the Calabi-Yau. The ISD condition implies that G_3 consists of only $(2,1)$ and $(0,3)$ forms. The $(2,1)$ component of the flux preserves $\mathcal{N} = 1$ SUSY whereas the $(0,3)$ component breaks SUSY while preserving the no-scale structure. We can see this from the F-term equation for the Kahler modulus ρ

$$0 = D_\rho W \sim W = \int G_3 \wedge \Omega \Rightarrow G = G^{(2,1)} \tag{9.90}$$

Indeed, $G^{(0,3)}$ induces a gravitino mass[f] of the order:

$$m_{3/2}^2 \sim \frac{|G_3 \wedge \Omega|^2}{Im\tau Vol(M_6)^2} \tag{9.91}$$

Although our discussion here is in the context of Type IIB string theory, there should be an alternative description in Type IIA string theory where intersecting D6-branes can be introduced. However, under duality, the three-form fluxes we consider here become metric fluxes on the Type IIA side and the underlying geometry become complicated (more precisely, non-Kahler). The types of 3-cycles that the D6-branes can wrap around in such geometries are not well understood. Therefore, we will restrict ourselves to D-brane model building in Type IIB theory, i.e., branes at singularities and magnetized D-branes. We will construct an example of each kind.

9.7. Warped Throats

Besides stabilizing moduli, another appealing feature of flux compactification is that the background fluxes backreact on the metric leading to a non-trivial warp factor. If the fluxes are localized in the compact space, strongly warped regions or warped throats can be generated.

Warped extra dimensions have been studied extensively in the past few years as a candidate for physics beyond the Standard Model. A prototypical example is the Randall-Sundrum scenario where the spacetime geometry is a slice of AdS_5 with two boundaries known as the Planck brane and the TeV brane, which can be interpreted as the UV and IR cutoff respectively. Localization of the SM fields (in particular the Higgs) at the strongly warped end (i.e., the IR brane) lead to an exponential suppression of the 4d scale, thus providing an interesting approach to the hierarchy problem.

---

[f]The factors in the denominator come from $e^{\mathcal{K}}$ where $\mathcal{K}$ is the Kahler potential.

Other than the generating the hierarchy between the Planck scale and the electroweak scale, there are several phenomenological appeals of warped throats. Warped throats are heavily used in the construction of string inflationary models. They are also useful in generating small numbers for supersymmetry breaking, its mediation, and in sequestering. From a string theory point of view, warped throats are interesting because the relevant interactions are peaked in the highly warped region, and so warping enables us to study physics in a local region of the compactification. The advantage is that a lot is known about local properties of string theory backgrounds such as their metrics and types of cycles that the branes can wrap around. So one can effectively use these results in the warped region to do concrete computations.

There are several motivations to embed warped phenomenology within string theory. First of all, the hierarchy set by the warp factor on the IR brane has so far been imposed by hand. We need a mechanism to stabilize this hierarchy. Secondly, as recent works have shown, details of the warped geometry could have significant effects on precision cosmology[39] and collider data.[40] Therefore, it is important to understand what kind of warped geometries can arise in string theory. Finally, inspired by the AdS/CFT correspondence in string theory, results in RS phenomenology have been interpreted in purely 4d terms by replacing the warped throat by a strongly interacting 4d conformal field theory. However, a lack of microscopic understanding of holography in the effective field theory approach prevents this picture to go beyond a qualitative rephrasing. Hence it is worthwhile to study the microscopic constructions of warped throats and their holographic description. To illustrate the point, let us consider an explicit realization of the Randall-Sundrum scenario in string theory, namely, the warped deformed conifold.

Calabi-Yau manifolds are generically non-singular, but at special values of the parameters, they can develop singularities. The most generic singular space is a conifold. Locally, it can be described as a submanifold of $\mathbb{C}^4$ defined by

$$w_1^2 + w_2^2 + w_3^2 + w_4^2 = 0 \qquad (9.92)$$

This submanifold is singular at $(w_1, w_2, w_3, w_4) = 0$ because the normal space is ill defined.[g] We can picture the conifold singularity as a cone whose base has the topology of $S^3 \times S^2$. The fact that it is a cone is obvious because

---

[g]In general, a variety defined by $f(w_i) = 0$ in $\mathbb{C}^4$ is singular at $f = df = 0$ since the normal space (associated with df) is ill defined.

if w_i satisfy the above equation, so does λw_i for any complex constant λ. To see why it has the topology of $S^3 \times S^2$, let $w_i = x_i + iy_i$, and introducing a new coordinate ρ, the defining equation for a conifold can be recast as 3 real equations:

$$\vec{x} \cdot \vec{x} - \frac{1}{2}\rho^2 = 0 \qquad \vec{y} \cdot \vec{y} - \frac{1}{2}\rho^2 = 0 \qquad \vec{x} \cdot \vec{y} = 0 \tag{9.93}$$

The first equation describes and S^3 with radius $\rho/\sqrt{2}$. Then the last 2 equations can be interpreted as describing an S^2 fibered over S^3. At the singular point, both the S^3 and the S^2 shrink to zero size.

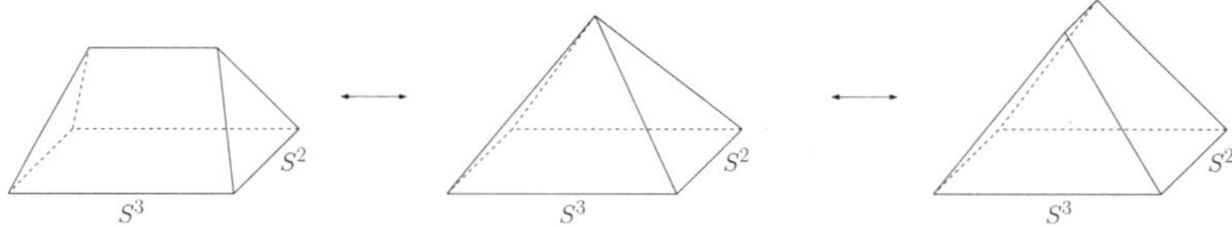

Fig. 9.34.   Conifold Transition.

The conifold can be smoothed into a nonsingular CY manifold in two ways. In the small resolution of the conifold, the S^2 is expanded to a finitie size. In the deformed conifold, the S^3 is expanded to finite size. Since we will be turning on 3-form fluxes in the background, it is the deformed conifold that will be relevant to us. The deformation corresponds to replacing 0 by z which can be assumed to be real and non-negative after a rescaling of ccordinates:

$$w_1^2 + w_2^2 + w_3^2 + w_4^2 = z \tag{9.94}$$

To see that the S^3 has a finite size after the deformation, decompose w_i into real and imaginary parts as before yields

$$z = \vec{x} \cdot \vec{x} - \vec{y} \cdot \vec{y} \tag{9.95}$$

Using the definition

$$\rho^2 = \vec{x} \cdot \vec{x} + \vec{y} \cdot \vec{y} \tag{9.96}$$

we see that the size of the S^3 is finite:

$$z \leq \rho^2 < \infty \tag{9.97}$$

The singularity at the origin is avoided for $z > 0$. As ρ^2 gets close to z the S^2 disappears leaving just an S^3 with finite size.

We now introduce fluxes to this geometry in order to stabilize the (complex structure) modulus controlling the size of S^3. Dirac quantization implies that these fluxes integrated over all 3-cycles of the CY be integers. In the vicinity of the conifold, there are 2 relevant cycles: one of which (the A-cycle) is the S^3 on which all the w_i are real. In general compact examples, there also exists a dual B-cycle which intersects A exactly once (and can be constructed by taking $w_{1,2,3}$ to be imaginary and w_4 real and positive). The warped deformed conifold (sometimes known as the Klebanov-Strassler solution) corresponds to putting M units of RR 3-form flux F_3 on the A-cycle. The field equation requires that NSNS 3-form flux H_3 be supported on the dual cycle to F_3, so let there be $-K$ units on the B-cycle:

$$\frac{1}{2\pi\alpha'} \int_A F_3 = 2\pi M \qquad \frac{1}{2\pi\alpha'} \int_B H_3 = -2\pi K \tag{9.98}$$

The D3-charge is therefore $N = MK$. We can now evaluate the flux induced superpotential

$$W = \int_{\mathcal{M}} G_3 \wedge \Omega = \int_{\mathcal{M}} (F_3 - \tau H_3) \wedge \Omega = \int_A F_3 \int_B \Omega + \tau \int_B H_3 \int_A \Omega \tag{9.99}$$

where the sign flip in the last term is due to the ordering of A and B. The integrals appearing here are the *periods* defining the complex structure of the conifold. In particular, the complex coordinate for the collapsing A cycle is defined by

$$z = \int_A \Omega \tag{9.100}$$

The integral over the B-cycle can be determined from the monodromy around $z = 0$ where the A-cycle shrinks. The result is:

$$\int_B \Omega = \mathcal{G}(z) = \frac{z}{2\pi i} \ln z + \text{holomorphic} \tag{9.101}$$

Putting things together, the superpotential is then

$$W = (2\pi)^2 \alpha' \left(M\mathcal{G}(z) - K\tau z \right) \tag{9.102}$$

Consider the F-term equation

$$0 = D_z W \propto M \partial_z \mathcal{G} - K\tau + \partial_z \mathcal{K}(M\mathcal{G} - K\tau z) \tag{9.103}$$

where $\mathcal{K}$ is the Kahler potential. In order to obtain a large hierarchy, we will take K/g_s to be large: this result in z being exponentially small. In this regime,

$$D_z W \propto \frac{M}{2\pi i} \ln z - i\frac{K}{g_s} + \mathcal{O}(1) \qquad (9.104)$$

It follows that for $K/Mg_s >> 1$, z is indeed exponentially small,

$$z \sim \exp(-2\pi K/Mg_s) \qquad (9.105)$$

Thus, we obtain a hierarchy of scales if, for example, $M = 1$ and K/g_s is of order 5.

To determine the actual warp factor requires solving the supergravity equation of motion, but one can estimate it as follows. The warp metric for a D3-brane is:

$$ds^2 = \left(\frac{r}{R}\right)^2 ds_4^2 + \left(\frac{R}{r}\right)^2 \left(dr^2 + r^2 d\Omega_5^2\right) \qquad \text{where} \quad R^4 = 4\pi g_s N(\alpha')^2 \qquad (9.106)$$

where r is the distance from the D3-branes located at $r = 0$. The deformation parameter sets a minimum for r and hence the warp factor $e^{A_{min}}$:

$$e^{A_{min}} \simeq r_{min} \simeq z^{1/3} \simeq \exp(-2\pi K/3Mg_s) \qquad (9.107)$$

To summarize, the background fluxes generate and stabilize an exponent hierarchy. The role of the IR and UV branes in RS are played by the S^3 of the deformed conifold and the bulk geometry.

The holographic dual of the warped deformed conifold is an $\mathcal{N} = 1$ SUSY gauge theory with an $SU(N) \times SU(N+M)$ gauge group, chiral multiplets A_i, B_i, $i = 1, 2$ in the $(N, \overline{N+M})$ and $(\overline{N}, N+M)$ representations and a superpotential $W = \epsilon^{ij}\epsilon^{kl}tr A_i B_k A_j B_l$. Along the RG flow to the infrared, the theory undergoes a cascade of Seiberg dualities in which the effective N decreases in step of M. Eventually at an infrared scale after $K = N/M$ steps, the theory confines and the running stops. The size of the S^3 has the interpretation as the infrared confinement scale. The warp factor is associated to the ratio between the UV and IR scales generated by the RG flow.

Thus, AdS-like warped throats with exponentially small but finite warp factor at the tip provide a stringy realization of the RS scenario. In order to make this more precise, it is desirable to construct warped throats with rich enough geometry at their tip to allow for chiral configurations of D-branes.

9.7.1. *Warped Throats and the Standard Model*

Now, one might wonder if the singularity which support the chiral spectrum of the Standard Model discussed earlier in Section 9.4 can arise at the tip of some warped throats. Unfortunately, the KS throat is too simple to generate chiral physics in the infrared. On the gravity side, the geometry is smooth after the deformation, while on the field theory side, the light degrees of freedom are simply the glueballs of the confining theory. Hence the KS throat does not lead to the SM degrees of freedom by putting D3-branes at the tip.

The simplest possibility is to consider throats which contain a singularity at the tip, e.g., a $\mathbb{Z}_3$ orbifold singularity. The first thing one might try is to construct the quotient of the deformed conifold by a $\mathbb{Z}_3$ action with isolated fixed points. Unfortunately, the deformed conifold does not admit such symmetries. For instance, we can change variables and describe the conifold as

$$xy - uv = z \tag{9.108}$$

There is a $\mathbb{Z}_3$ symmetry: $(x, u) \to e^{2\pi i/3}(x, u)$ and $(y, v) \to e^{-2\pi i/3}(y, v)$, which unfortunately is freely-acting[h]. Other possible $\mathbb{Z}_3$, such as $x \to e^{2\pi i/3}x$, $y \to e^{-2\pi i/3}y$ while leaving u, v invariant, have a whole complex curve of fixed points [i] so locally the singularity is $\mathbb{C} \times \mathbb{C}^2/\mathbb{Z}_3$. This singularity leads to $\mathcal{N} = 2$ world volume theories which are non-chiral and thus not interesting.

Progress in understanding warped throats for other geometries, generalizing the conifold, as well as their interpretation in terms of duality cascades and infrared confinement provides useful techniques to implement these ideas. Skipping the details, it suffices for our purposes to consider a particular example. Consider the so called suspended pinch point (SPP) singularity, which can be described as a hypersurface in $\mathbb{C}^4$ given by

$$xy - zw^2 = 0 \tag{9.109}$$

This geometry admits a complex deformation to the smooth geometry

$$xy - zw^2 = \epsilon w \tag{9.110}$$

which contains a finite size[j] S^3.

---

[h]The fixed point $x = y = z = w = 0$ does not solve the deformed conifold equation.
[i]defined by $uv = z$
[j]To see this, it is most convenient to change coordinates: $\rho = x/w$. The deformed SPP can now be written as $\rho y - zw = \epsilon$.

Moreover, the deformed geometry is invariant under a $\mathbb{Z}_3$ acting as

$$x \to \alpha x \qquad y \to \alpha y \qquad z \to \alpha z \qquad w \to \alpha^2 w \qquad (9.111)$$

with $\alpha = e^{2\pi i/3}$. Notice that the $\mathbb{Z}_3$ action leaves invariant the holomorphic 3-form $\Omega = \frac{dx\,dy\,dz}{zw}$ of the SPP, guaranteeing that the quotient is a new CY singularity. Furthermore, the $\mathbb{Z}_3$ action has the origin as the unique fixed point, and hence there is a left-over $\mathbb{Z}_3$ singularity after the deformation.

On general grounds, it is expected that turning on M units of RR flux on the finite size 3-cycle (as well as a suitable NSNS flux on its dual (non-compact) 3-cycle) leads to a warped throat. At its bottom the throat is cutoff by the finite size 3-cycle, leaving behind a $\mathbb{C}^3/\mathbb{Z}_3$ singularity. A chiral gauge theory is obtained by introducing a small set of D3-branes at the singularity. The latter are probes, and do not modify the structure of the throat significantly.

In fact, one can embed the local D-brane model we just described within this throat. The D7-branes can wrap the 4-cycle defined by $w = 0$ which pass through the D3-branes located at $x = y = z = w = 0$.

To summarize, we have succeeded in finding a throat with a semi-realistic D-brane sector at its tip. The warped geometry also admits a tractable holographic dual. It would be interesting to construct an explicit metric[k] for these types of warped throats since they allow us to study properties of the strongly coupled field theory dual. For example, the KK spectrum in such warped throat background tells us about the glueball masses of the dual field theory.

9.7.2. *Flux Vacua with Magnetized D-branes*

Finally, we construct another class of chiral D-brane models in warped compactifications Instead of branes at singularities, we consider magnetized D-branes. Given the materials on intersecting branes discussed earlier in Section 9.5, our strategy should be clear by now. Since we have developed already some useful results regarding intersecting branes in orientifold backgrounds, we will try to adopt these results here by T-dualizing the setup. The configuration of intersecting branes become sets of magnetized D-branes in the IIB theory. We can then combine these magnetized D-branes with 3-form fluxes directly within Type IIB string theory and look for vacuum solutions.

---

[k]The SPP singularity arises as a particular case of a cone over the $L^{a,b,c}$ families of Einstein-Sasaki metrics constructed in recent years. The warped deformed versions of such metrics have not been worked out yet.

To be concrete, consider the $\mathbb{Z}_2 \times \mathbb{Z}_2$ orientifold that we discussed before. Under T-duality, the 4 types of O6-planes turn into O3-planes and 3 sets of O7-planes. The O3-planes are points in the Calabi-Yau whereas the O7-planes spans different compact dimensions. Let us focus on the O3-planes. The RR tadpole cancellation, in the covering space, reads:

$$N_{D3} + N_{O3} = 0 \qquad \Leftrightarrow \qquad N_{D3} = 32 \qquad (9.112)$$

In the presence of fluxes, this tadpole condition is modified:

$$N_{D3} + N_{flux} = 32 \qquad (9.113)$$

where the charges appear above are defined in the covering space T^6.

What can N_{flux} be for the orbifold we are considering? It depends on the 3-cycles on which the fluxes are quantized. In addition to the 3-cycles that are inherited from the torus, there are also collapsed 3-cycles at the orbifold singularities. Since these collapsed cycles are not in the large volume regime, we cannot trust the supergravity analysis for fluxes supported on these cycles. Therefore we consider only untwisted fluxes, namely flux choices which exist in the parent toroidal theory and are invariant under the $\mathbb{Z}_2 \times \mathbb{Z}_2$ symmetry. This requires that the 3-forms have one leg along each of the T^2. Furthermore, we will center on ISD fluxes, which have only $(2,1)$ and $(0,3)$ components. Hence, in complex notation, G_3 is a linear combination of the forms:

$$d\bar{z}_1 d\bar{z}_2 d\bar{z}_3 \,, \qquad dz_1 dz_2 d\bar{z}_3 \,, \qquad dz_1 d\bar{z}_2 dz_3 \,, \qquad d\bar{z}_1 dz_2 dz_3 \qquad (9.114)$$

Of course, we have yet to impose the quantitization conditions of F_3 and H_3 in order for G_3 to be consistent. This fixes the dilaton and the complex structure moduli. In fact, the condition that flux quantization imposes on moduli is simply a rephrasing of the statement that a specific choice of the moduli minimizes the flux potential.

Now, in order for F_3 and H_3 to be properly quantized in the orbifold, they have to be quantized in multiples of N_{min} in the covering space where N_{min} is an integer depending on the order of the orbifold group. In the present case, N_{min} is 8 of which 4 comes from the order of the $\mathbb{Z}_2 \times \mathbb{Z}_2$ orbifold and a factor of 2 is due to the orientifold projection[1].

Back to the RR tadpoles. Due to the quantization we just discussed, N_{flux} is a multiple of 64. This seems to suggest that, no matter what kind

---

[1]To be precise, $N_{min} = 4$ or 8 depending on whether the cycle Σ in the covering space passes through an odd or even number of $O3^{(+,+)}$ planes. There are no such exotic $O3^{(+,+)}$ planes for the choice of discrete torsion (i.e., $(h^{1,1}, h^{2,1}) = (3, 51)$) we considered, and so $N_{min} = 8$.

of magnetized D-branes we use to embed the Standard Model, it is not possible to build consistent vacua without introducing some supersymmetry breaking anti D-branes.

However, this is not the case. It was realized that certain types of magnetized D9-branes may carry either anti D3-brane or anti D7-brane charge, while still preserving the $\mathcal{N} = 1$ supersymmetry of the orientifold background. This interesting fact was most apparent when we T-dualize the setup to Type IIA with intersecting branes, and in fact already implicit in one of the examples we discussed before. Consider a D6-brane wrapping the following 3-cycle:

$$[\pi] = (-2[a_1] + [b_1]) \circ (-3[a_2] + [b_2]) \circ (-4[a_3] + [b_3]) \tag{9.115}$$

It carries a D6 charge of the *same* sign as that of an orientifold plane along the $[a_1] \circ [a_2] \circ [a_3]$ direction but as usual the opposite sign from that of other orientifold planes. Yet, the 3-cycle is sLag if the complex structure moduli $\chi_i = (R_2/R_1)_i$ of the torus satisfy[m]:

$$\arctan(\chi_1/2) + \arctan(\chi_2/3) + \arctan(\chi_3/4) = \pi \tag{9.116}$$

T-dualizing this exotic D6-brane to the IIB picture, we have D9-branes with magnetic fluxes given by

$$\frac{m_a^i}{2\pi} \int_{T^2} F_a^i = n_a^i \tag{9.117}$$

The orientifold action ΩR [n] which defines the O3-planes map $m_a^i \to -m_a^i$ and hence a D9-brane to an anti-D9 brane. The brane-antibrane system usually break supersymmetry but with suitable choices of gauge bundles on their world volumes, they can preserve the same supersymmetry as the orientifold background and carry the $\overline{D3}$ brane charge that we need. The condition for the $D9 - \overline{D9}$ system to preserve SUSY is:

$$\arctan(\mathcal{A}_1/2) + \arctan(\mathcal{A}_2/3) + \arctan(\mathcal{A}_3/4) = \pi \tag{9.118}$$

where $\mathcal{A}_i$ is the area of $(T^2)_i$ in string units, which is nothing but the T-dual version of Eq. (9.118). Note that a $D9 - \overline{D9}$ pair breaks SUSY in the non-compact limit as expected, but can nevertheless preserves SUSY when the Kahler moduli take on some specific finite values.

---

[m]The RHS is π because the angles θ_i are bigger than π for $n_a^i \le 0$.
[n]where $R : z_i \to -z_i$

Other than the D3-branes, there are also RR tadpoles for the D7-branes:

$$N_{D3} + N_{flux} = Q_{O3} \Leftrightarrow 2\sum_a N_a n_a^1 n_a^2 n_a^3 + N_{flux} = 32$$

$$N_{D7_1} = Q_{O7_1} \Leftrightarrow 2\sum_a N_a m_a^1 m_a^2 n_a^3 = -32$$

$$N_{D7_2} = Q_{O7_2} \Leftrightarrow 2\sum_a N_a m_a^1 n_a^2 m_a^3 = -32$$

$$N_{D7_3} = Q_{O7_3} \Leftrightarrow 2\sum_a N_a n_a^1 m_a^2 m_a^3 = -32 \tag{9.119}$$

There is, however, a subtlety regarding the cancellation of RR tadpoles. In the presence of orientifold planes, D-branes may carry discrete $\mathbb{Z}_2$ charges, known as K-theory charges, which are invisible from one-loop divergences of the open string diagrams and from the cubic anomalies in the low energy spectrum so the above tadpole constraints do not guarantee the absence of these discrete charges. Just like the ordinary RR charges, however, these $\mathbb{Z}_2$ charges need to cancel globally in a consistent model. The uncanceled $\mathbb{Z}_2$ charges could give rise to the so called $SU(2)$ Witten anomalies – although $SU(2)$ is free of cubic anomalies, it suffers from a global gauge anomaly if there is an odd number of fermions charged in the fundamental representation. We can detect these global $SU(2)$ anomalies by introducing probe D-branes on top of the orientifold planes since this gives rise to a $USp(2)$ and hence $SU(2)$ gauge group on the probes. By demanding that the number of fundamental representations charged under each probe $SU(2)$ group to be even, we found some additional $\mathbb{Z}_2$ constraints:

$$\sum_a N_a m_a^1 m_a^2 m_a^3 \in 4\mathbb{Z}$$

$$\sum_a N_a n_a^1 n_a^2 m_a^3 \in 4\mathbb{Z}$$

$$\sum_a N_a n_a^1 m_a^2 n_a^3 \in 4\mathbb{Z}$$

$$\sum_a N_a m_a^1 n_a^2 n_a^3 \in 4\mathbb{Z} \tag{9.120}$$

These constraints tell us that the number of $D9 - \overline{D9}$ and $D5_i - \overline{D5}_i$ pairs must be even.

Collecting our results, an example satisfying these constraints is given as follows:

N_α	(n_α^1, m_α^1)	(n_α^2, m_α^2)	(n_α^3, m_α^3)
$N_a = 6$	$(1, 0)$	$(g, 1)$	$(g, -1)$
$N_b = 2$	$(0, 1)$	$(1, 0)$	$(0, -1)$
$N_c = 2$	$(0, 1)$	$(0, -1)$	$(1, 0)$
$N_d = 2$	$(1, 0)$	$(g, 1)$	$(g, -1)$
$N_{h1} = 2$	$(-2, 1)$	$(-3, 1)$	$(-4, 1)$
$N_{h2} = 2$	$(-2, 1)$	$(-4, 1)$	$(-3, 1)$
$8N_f$	$(1, 0)$	$(1, 0)$	$(1, 0)$

The brane configuration looks familiar. In fact, the part in blue is one of the MSSM modules introduced earlier. h_1 and h_2 label the exotic D9-branes which carry negative D3-charges. In order for the D-brane configuration to preserve SUSY, the Kahler moduli need to satisfy:

$$\mathcal{A}_2 = \mathcal{A}_3$$

$$\arctan(\mathcal{A}_1/2) + \arctan(\mathcal{A}_2/3) + \arctan(\mathcal{A}_3/4) = \pi \tag{9.121}$$

which have non-trivial solutions. It is worth noting that, although the open string sector in the presence of fluxes is not exactly solvable, the massless chiral sector is topological and hence protected so we can compute the chiral spectrum by the same index method before. On the other hand, non-chiral sectors can in general acquire mass terms induced by the fluxes and may disappear from the massless spectrum.

Given the brane configuration, the RR tadpole conditions become:

$$g^2 + N_f + 4n = 14 \tag{9.122}$$

where $N_{flux} = 64n$ with $n \in \mathbb{Z}$. There are several solutions, among them we find:

(1) $n = 0$, $g = 3$, $N_f = 5$
(2) $n = 1$, $g = 3$, $N_f = 1$
(3) $n = 2$, $g = 2$, $N_f = 2$
(4) $n = 3$, $g = 1$, $N_f = 1$

Let us look at these solutions in more detail. The first solution correspond to the 3-family model without the 3-form flux studied before. The last of these solutions is also interesting for another reason. The quantum of flux $N_{flux} = 3 \cdot 64$ can be achieved by considering the 3-form flux:

$$G_3 = \frac{8}{\sqrt{3}} e^{-\frac{\pi i}{6}} \left(d\bar{z}_1 dz_2 dz_3 + dz_1 d\bar{z}_2 dz_3 + dz_1 dz_2 d\bar{z}_3 \right) \tag{9.123}$$

which is well quantized at the particular value $\tau_1 = \tau_2 = \tau_3 = \tau = e^{\frac{2\pi i}{3}}$ for the untwisted complex structure moduli and the dilaton. These are indeed the values where those fields get fixed after the scalar potential generated by G_3 is minimized. Notice that the flux is a combination of $(2, 1)$ forms, and hence the closed string background as a whole preserves $\mathcal{N} = 1$ supersymmetry. Thus we find that it is actually possible to find chiral $\mathcal{N} = 1$ string theory vacua involving 3-form fluxes and magnetized D-branes. Since this model has only one Standard Model family, it shold be regarded as a toy model illustrating that it possible to get chirality and supersymmetry in explicit flux compactifications. Model building in general Calabi-Yau manifolds allows for more freedom to obtain realistic gauge sectors.

Exercise: Show that the above 3-form flux in Eq. (9.123) stabilizes the complex structure moduli and the dilaton to $\tau_1 = \tau_2 = \tau_3 = \tau = e^{\frac{2\pi i}{3}}$

If we relax the requirement that the fluxes preserve supersymmetry, more possibilties open up. Indeed the second solution gives rise to a 3-family MSSM-like spectrum. This is achieved by a 3-form flux with $N_{flux} = 64$:

$$G_3 = 2(d\bar{z}_1 dz_2 dz_3 + dz_1 d\bar{z}_2 dz_3 + dz_1 dz_2 d\bar{z}_3 + d\bar{z}_1 d\bar{z}_2 d\bar{z}_3) \qquad (9.124)$$

which contains a $(0, 3)$ component and hence breaks supersymmetry (although with vanishing cosmological constant at leading order due to the no-scale structure). A particular value of the moduli where such flux is well quantized is given by $\tau_1 = \tau_2 = \tau_3 = \tau = i$. Notice that this gives us $g_s = 1$ and the string perturbation theory may seem no longer realiable. It turns out that the scalar potential derived from the above flux has several flat directions. In particular, it vanishes when one imposes the complex structure moduli and the dilaton to be purely imaginary

$$\tau_i = it_i \ , t_i \in \mathbb{R}$$
$$\tau = i/g_s \qquad (9.125)$$

and to satisfy the constraint

$$g_s t_1 t_2 t_3 = 1 \qquad (9.126)$$

Exercise: Prove the above constraint by minimizing the scalar potential induced by G_3 in Eq. (9.124) explicitly.

So, in principle, one can find solutions at weak coupling. Of course, α' corrections may lift the flat directions left by Eq. (9.126), dynamically fixing g_s. Such an analysis is beyond our current scope.

There are several further interesting features about these models such as the Higgs sector, anomalous $U(1)$'s and perturbative global symmetries, the effects of background fluxes on the open string sector including flux induced SUSY breaking, D-brane moduli stabilization, Freed-Witten anomalies, and etc. I don't have time to discuss these interesting topics, but I refer the readers to [30,31] for a more detailed discussion.

Finally, let me emphasize that the purpose of presenting these explicit models is not that the models are fully realistic but that we want to illustrate the many subtle issues involved when one tries to construct the Standard Model from string theory. The consistency constraints on string theory models are highly intertwined. Contrary to what the landscape might naively suggest, not everything goes. Hopefully, these explicit constructions could point to more general lessons about physics beyond the Standard Model.

9.8. Final Thoughts

I hope these lectures have given you some flavor of string phenomenology, with an emphasis on D-brane models of particle physics. As you have seen, many interesting issues arise when one tries to embed the Standard Model within string theory. Hopefully, the general lessons we have learned will take us one step closer to the truth. At any rate, string theory has been a fountain of new ideas of physics beyond the Standard Model.

Acknowledgments

I would like to thank Tao Han and Robin Erbacher for organizing TASI 2008, and K.T. Mahanthappa for his hospitality. I also thank Paul McGuirk for proof-reading the lectures during their preparation and for his comments. This work has been supported in part by NSF CAREER Award No. PHY-0348093, DOE grant DE-FG-02-95ER40896, a Research Innovation Award and a Cottrell Scholar Award from Research Corporation, a Vilas Associate Award from the University of Wisconsin, and a John Simon Guggenheim Memorial Foundation Fellowship. I would also like to acknowledge support from the Ambrose Monell Foundation during my stay at the Institute for Advanced Study while parts of these lecture notes were written.

References

1. M. B. Green, J. H. Schwarz and E. Witten, "SUPERSTRING THEORY," Vol. 1 & 2, *Cambridge, Uk: Univ. Pr. (1987) (Cambridge Monographs On Mathematical Physics)*

2. J. Polchinski, "String theory," Vol. 1 & 2, *Cambridge, UK: Univ. Pr. (1998)*

3. C. V. Johnson, "D-Branes," *Cambridge, USA: Univ. Pr. (2003) 548 p*

4. K. Becker, M. Becker and J. H. Schwarz, "String theory and M-theory: A modern introduction," *Cambridge, UK: Cambridge Univ. Pr. (2007) 739 p*

5. E. Kiritsis, "String theory in a nutshell," *Princeton, USA: Univ. Pr. (2007) 588 p*

6. S. Weinberg, "Implications Of Dynamical Symmetry Breaking," Phys. Rev. D **13**, 974 (1976).

7. L. Susskind, "Dynamics Of Spontaneous Symmetry Breaking In The Weinberg-Salam Theory," Phys. Rev. D **20**, 2619 (1979).

8. J. M. Maldacena, "The large N limit of superconformal field theories and supergravity," Adv. Theor. Math. Phys. **2**, 231 (1998) [Int. J. Theor. Phys. **38**, 1113 (1999)] [arXiv:hep-th/9711200].

9. S. S. Gubser, I. R. Klebanov and A. M. Polyakov, "Gauge theory correlators from non-critical string theory," Phys. Lett. B **428**, 105 (1998) [arXiv:hep-th/9802109].

10. E. Witten, "Anti-de Sitter space and holography," Adv. Theor. Math. Phys. **2**, 253 (1998) [arXiv:hep-th/9802150].

11. O. Aharony, S. S. Gubser, J. M. Maldacena, H. Ooguri and Y. Oz, "Large N field theories, string theory and gravity," Phys. Rept. **323**, 183 (2000) [arXiv:hep-th/9905111].

12. L. Randall and R. Sundrum, "A large mass hierarchy from a small extra dimension," Phys. Rev. Lett. **83**, 3370 (1999) [arXiv:hep-ph/9905221].

13. L. Randall and R. Sundrum, "An alternative to compactification," Phys. Rev. Lett. **83**, 4690 (1999) [arXiv:hep-th/9906064].

14. R. Blumenhagen, M. Cvetic, P. Langacker and G. Shiu, "Toward realistic intersecting D-brane models," Ann. Rev. Nucl. Part. Sci. **55**, 71 (2005) [arXiv:hep-th/0502005].

15. A. Uranga, "String Compactifications and Model Building," TASI 2005 lectures.

16. F. Quevedo, "Phenomenology Aspects of D-branes,", ICTP Spring School on Superstrings and Related Matters, Trieste, March 2002.

17. L. J. Dixon, V. Kaplunovsky and C. Vafa, "On Four-Dimensional Gauge Theories From Type Ii Superstrings," Nucl. Phys. B **294**, 43 (1987).

18. G. Shiu and S. H. H. Tye, "TeV scale superstring and extra dimensions," Phys. Rev. D **58**, 106007 (1998) [arXiv:hep-th/9805157].

19. N. Arkani-Hamed, S. Dimopoulos and G. R. Dvali, "The hierarchy problem and new dimensions at a millimeter," Phys. Lett. B **429**, 263 (1998) [arXiv:hep-ph/9803315].

20. I. Antoniadis, N. Arkani-Hamed, S. Dimopoulos and G. R. Dvali, "New dimensions at a millimeter to a Fermi and superstrings at a TeV," Phys. Lett.

B **436**, 257 (1998) [arXiv:hep-ph/9804398].

21. N. Arkani-Hamed, S. Dimopoulos and G. R. Dvali, "Phenomenology, astrophysics and cosmology of theories with sub-millimeter dimensions and TeV scale quantum gravity," Phys. Rev. D **59**, 086004 (1999) [arXiv:hep-ph/9807344].

22. L. J. Dixon, J. A. Harvey, C. Vafa and E. Witten, "Strings On Orbifolds," Nucl. Phys. B **261**, 678 (1985).

23. L. J. Dixon, J. A. Harvey, C. Vafa and E. Witten, "Strings On Orbifolds. 2," Nucl. Phys. B **274**, 285 (1986).

24. M. R. Douglas and G. W. Moore, "D-branes, Quivers, and ALE Instantons," arXiv:hep-th/9603167.

25. M. E. Peskin, "INTRODUCTION TO STRING AND SUPERSTRING THEORY. 2," TASI 1986 lectures.

26. G. Aldazabal, L. E. Ibanez, F. Quevedo and A. M. Uranga, "D-branes at singularities: A bottom-up approach to the string embedding of the standard model," JHEP **0008**, 002 (2000) [arXiv:hep-th/0005067].

27. D. Berenstein, V. Jejjala and R. G. Leigh, "The standard model on a D-brane," Phys. Rev. Lett. **88**, 071602 (2002) [arXiv:hep-ph/0105042].

28. M. Grana, "MSSM parameters from supergravity backgrounds," Phys. Rev. D **67**, 066006 (2003) [arXiv:hep-th/0209200].

29. M. Berkooz, M. R. Douglas and R. G. Leigh, "Branes intersecting at angles," Nucl. Phys. B **480**, 265 (1996) [arXiv:hep-th/9606139].

30. F. Marchesano and G. Shiu, "MSSM vacua from flux compactifications," Phys. Rev. D **71**, 011701 (2005) [arXiv:hep-th/0408059].

31. F. Marchesano and G. Shiu, "Building MSSM flux vacua," JHEP **0411**, 041 (2004) [arXiv:hep-th/0409132].

32. K. Dasgupta, G. Rajesh and S. Sethi, "M theory, orientifolds and G-flux," JHEP **9908**, 023 (1999) [arXiv:hep-th/9908088].

33. B. R. Greene, K. Schalm and G. Shiu, "Warped compactifications in M and F theory," Nucl. Phys. B **584**, 480 (2000) [arXiv:hep-th/0004103].

34. K. Becker and M. Becker, "Compactifying M-theory to four dimensions," JHEP **0011**, 029 (2000) [arXiv:hep-th/0010282]; "M-Theory on Eight-Manifolds," Nucl. Phys. B **477**, 155 (1996) [arXiv:hep-th/9605053].

35. S. B. Giddings, S. Kachru and J. Polchinski, "Hierarchies from fluxes in string compactifications," Phys. Rev. D **66**, 106006 (2002) [arXiv:hep-th/0105097].

36. S. Kachru, R. Kallosh, A. Linde and S. P. Trivedi, "De Sitter vacua in string theory," Phys. Rev. D **68**, 046005 (2003) [arXiv:hep-th/0301240].

37. M. R. Douglas and S. Kachru, "Flux compactification," Rev. Mod. Phys. **79**, 733 (2007) [arXiv:hep-th/0610102].

38. R. Blumenhagen, B. Kors, D. Lust and S. Stieberger, "Four-dimensional String Compactifications with D-Branes, Orientifolds and Fluxes," Phys. Rept. **445**, 1 (2007) [arXiv:hep-th/0610327].

39. G. Shiu and B. Underwood, "Observing the Geometry of Warped Compactification via Cosmic Inflation," Phys. Rev. Lett. **98**, 051301 (2007) [arXiv:hep-th/0610151].

40. G. Shiu, B. Underwood, K. M. Zurek and D. G. E. Walker, "Probing the

Geometry of Warped String Compactifications at the LHC," Phys. Rev. Lett. **100**, 031601 (2008) [arXiv:0705.4097 [hep-ph]].

Chapter 10

Particle Physics in Extra Dimensions

Bogdan A. Dobrescu

Theoretical Physics Department, Fermilab, Batavia, IL 60510, USA

Any extra-dimensional field theory is equivalent to a 4-dimensional one that includes a series of heavy particles. The spectrum and interactions of these 'KK particles' depend on the boundary conditions and metric. If all fields propagate in a compact dimension with flat metric (referred to as a universal extra dimension), then the KK masses may be below the TeV scale, the lightest KK particle is a dark matter candidate, and the collider signals include cascade decays involving leptons, jets and missing energy, as well as narrow resonances. If the metric is warped, then the hierarchy between the electroweak and Plank scales is natural, while the collider signals involve heavy resonances. Field theories in extra dimensions are strongly coupled in the ultraviolet, so that their study could shed light on nonperturbative phenomena, with possible applications to dynamical electroweak symmetry breaking and compositeness.

10.1. Introduction

It is an obvious fact: we live in a space with three dimensions. However, not everything that is obvious is true, as the development of quantum mechanics has compellingly illustrated. The possibility of extra spatial dimensions has been the subject of metaphysical speculations dating back at least to the 19th century.[1] More serious, scientific inquiries into this issue have to address the question of how are the extra dimensions hidden from us. The first convincing proposal was suggested by Oskar Klein:[2] if an extra dimension is compactified on a circle, then its presence would not be observed unless the experiments have a resolution higher than the radius of the circle.

Nowadays we know that the circle compactification would not allow any particle discovered so far to propagate along the extra dimension. The reason is that any gauge field, such as the photon, would include a spin-0 partner of equal mass and couplings as the spin-1 particle. Furthermore, any fermion that propagates along the extra dimension would be a vectorlike fermion: its left- and right-handed components would have the same gauge charges, which is not true for any of the elementary fermions discovered so far. However, if the compactification is on an interval, then the unwanted vectorlike partners of the observed fermions and the spin-0 partners of the gauge fields may be eliminated by the boundary conditions at the end of the interval.

Any particle propagating through extra dimensions, whether compactified on a circle or an interval, would appear in experiments as a tower of massive particles in 3 spatial dimensions. These massive particles are usually called Kaluza-Klein (KK) modes. Their presence can be easily understood based on the usual particle-in-a-box problems in quantum mechanics: given that space along the extra dimensions is compact, the energy states are quantized. The kinetic energy due to motion along the extra dimensions manifests itself as mass in the usual 3 spatial dimensions. For an interval of length L, the mass of the lightest KK modes is $(\pi/L)\hbar/c$ (the natural unit system, $\hbar = c = 1$, is used in what follows). This mass is the compactification scale, and its inverse $R \equiv L/\pi$ is the 'radius' of the extra dimension.

The highest partonic energies attained so far in collider experiments (at the Tevatron) are typically below 1 TeV. Thus, it is expected that extra dimensions of size 10^{-19} m are consistent with all experimental observations. In practice the situation is more complicated, especially because of the possibility that not all known particles propagate along the extra dimensions. A scenario that received extensive attention at the end of the 20th century was based on the idea that only the graviton propagates in some flat extra dimensions.[3] In that case, measurements of gravity are sensitive to the presence of the extra dimensions, and the current upper limit on their radius is around 4×10^{-5} m.[4] The collider implications of that scenario involve emission of KK gravitons which appear as missing transverse momentum, and virtual exchange of KK gravitons which induce nonresonant modifications of various cross sections (the relevant Feynman rules are derived in [5]). In another scenario, all bosons propagate in extra dimensions while the fermions are localized at the end points of an interval.[6,7] As a result, the KK modes of the standard model gauge bosons mediate 4-fermion interactions at tree-level, and the constraints from a global fit to collider data set a very stringent limit, of $R \lesssim 0.3 \times 10^{-19}$ m.[8]

Universal extra dimensions (UED) are arguably the simplest kind of extra dimensions: all particles propagate along some flat compact extra dimensions. The remarkable feature of UED is that a remnant of translational invariance along the extra dimensions is preserved such that a single KK mode cannot couple at tree-level to zero modes.[9] As a result, the limits are relaxed by more than an order of magnitude compared to the extra dimensions accessible only to bosons: $R \lesssim 5 \times 10^{-19}$ m. Furthermore, UED lead to dramatically different phenomenological implications.

The search in collider experiments for KK modes having a spectrum and interactions consistent with a certain compactification is the best way of checking the existence of extra dimensions. Within the next few years, the ATLAS and CMS experiments at the Large Hadron Collider (LHC) are expected to discover KK modes associated with UED if the size of the extra dimensions is not far below 10^{-19} m.

If the metric along the extra dimensions is warped, then the observed weakness of gravity arises naturally due to the small wave function overlap of the graviton and standard model fields.[10] The graviton KK modes arising from a warped extra dimension could in principle have masses as low as the TeV scale.

Section 10.2 presents the derivation of the generic properties of KK modes in the case of a flat extra dimension compactified on the interval. Then the signatures of UED at the Tevatron and the LHC are reviewed in Sec. 10.3. Theories with a warped extra dimension are discussed in Sec. 10.4. Some comments about theories in more extra dimensions are included in Sec. 10.5.

10.2. Field theory on a flat compact dimension

Before discussing the phenomenology of extra dimensions, it is necessary to study the general features of quantum field theory in a flat extra dimension. The cases of spin 0, 1/2 and 1 are analyzed in turn.

10.2.1. *Scalar field on the interval*

Let us consider a five-dimensional (5D) spacetime: four spacetime dimensions of coordinates x^μ, $\mu = 0, 1, 2, 3$, form the usual Minkowski spacetime, and one transverse spatial dimension of coordinate x^4 is flat and compact, with $0 \leq x^4 \leq L$. Thus the extra dimension is an interval (see Fig. 10.1), and the boundary conditions at its end points determine the spectrum of KK modes.

Free scalar fields, $\Phi(x^\mu, x^4)$, are described by the following action:

$$S_\Phi = \int d^4x \int_0^L dx^4 \left(\partial_\alpha \Phi^\dagger \partial^\alpha \Phi - M_0^2 \Phi^\dagger \Phi\right) . \qquad (10.2.1)$$

The parameter M_0 is the 5D mass of Φ. We use letters from the beginning of the Greek alphabet to label the 5D coordinates $\alpha, \beta, \ldots = 0, 1, 2, 3, 4$, and letters from the middle of the Greek alphabet to label the Minkowski coordinates $\mu, \nu, \ldots = 0, 1, 2, 3$. Given that the action is dimensionless, and that the coordinates have mass dimension -1, the 5D bosons have mass dimension $+3/2$.

Under a variation of the field, $\delta\Phi(x^\mu, x^4)$, the variation of the action is given by

$$\delta S_\Phi = \delta S_\Phi^v + \delta S_\Phi^s , \qquad (10.2.2)$$

where the first term is a 'volume' integral,

$$\delta S_\Phi^v = -\int d^4x \int_0^L dx^4 \left(\partial^\alpha \partial_\alpha \Phi^\dagger + M_0^2 \Phi^\dagger\right) \delta\Phi , \qquad (10.2.3)$$

and the second term is a 'surface' integral,

$$\delta S_\Phi^s = \int d^4x \left(\partial_4 \Phi^\dagger \delta\Phi\big|_{x^4=L} - \partial_4 \Phi^\dagger \delta\Phi\big|_{x^4=0}\right) . \qquad (10.2.4)$$

Here we have assumed as usual that the field vanishes at $x^\mu \to \pm\infty$. Given that the action has to be stationary with respect to *any* variation of the field, the volume and surface terms must vanish independently. Requiring $\delta S_\Phi^v = 0$ implies that Φ is a solution to the 5D Klein-Gordon equation,

$$\left(\partial^\mu \partial_\mu - \partial_4^2 + M_0^2\right) \Phi = 0 , \qquad (10.2.5)$$

while $\delta S_\Phi^s = 0$ forces the boundary conditions that can be imposed on Φ to obey

$$(\partial_4 \Phi^\dagger)\delta\Phi\big|_{x^4=L} = (\partial_4 \Phi^\dagger)\delta\Phi\big|_{x^4=0} . \qquad (10.2.6)$$

Fig. 10.1.   The extra dimension of coordinate x^4 extends from $x^4 = 0$ to $x^4 = L$, and is transverse to the usual three spatial dimensions.

Given that the values of $\delta\Phi(x^\mu, x^4)$ at $x^4 = 0$ and $x^4 = L$ are in general not correlated (unless the two points are identified, which would not allow chiral fermions in the 4D theory), and Eq. (10.2.6) must be valid for any $\delta\Phi$, both the left- and right-handed sides of Eq. (10.2.6) must vanish. Therefore,

$$\partial_4\Phi\big|_{x^4=0} = 0 \quad \text{or} \quad \Phi(x^\mu, 0) = 0 \tag{10.2.7}$$

and

$$\partial_4\Phi\big|_{x^4=L} = 0 \quad \text{or} \quad \Phi(x^\mu, L) = 0 \ . \tag{10.2.8}$$

We now solve the 5D Klein-Gordon equation,

$$\left(\partial^\mu\partial_\mu - \partial_4^2 + M_0^2\right)\Phi = 0 \ , \tag{10.2.9}$$

subject to the boundary conditions (10.2.7) and (10.2.8). Since the boundary conditions are independent of x^μ, then Φ can be decomposed in Fourier modes as follows:

$$\Phi(x^\mu, x^4) = \sum_j \Phi^{(j)}(x^\mu) f^j(x^4) \ . \tag{10.2.10}$$

The 4D scalar fields $\Phi^{(j)}$ ('KK modes' or excitations), satisfy

$$\left(\partial^\mu\partial_\mu + M_0^2 + M_j^2\right)\Phi^{(j)}(x^\mu) = 0 \ , \tag{10.2.11}$$

where M_j^2 is a positive eigenvalue. The f^j functions are solutions to the one-dimensional equation,

$$\left(\partial_4^2 + M_j^2\right)f^j(x^4) = 0 \ . \tag{10.2.12}$$

A general solution to the above equation is

$$f^j(x^4) = C_+ e^{ijx^4/R} + C_- e^{-ijx^4/R} \ , \tag{10.2.13}$$

where $C_\pm$ are complex coefficients, and j is a real number such that

$$M_j = \frac{j}{R} \ , \tag{10.2.14}$$

and we defined the 'compactification radius'

$$R \equiv \frac{L}{\pi} \ . \tag{10.2.15}$$

The boundary conditions (10.2.7) and (10.2.8) impose a relation between the two coefficients, $C_- = \pm C_+$, and also restrict the values of j: $e^{4ij\pi} = 1$. Furthermore, the normalization condition,

$$\int_0^L dx^4 \left[f^j(x^4)\right]^* f^{j'}(x^4) = \delta_{jj'} \ , \tag{10.2.16}$$

determines the last coefficient up to a phase factor which we choose to be one. Explicitly, the solutions to Eq. (10.2.12) can be written as

$$f_0^j(x^4) = \frac{1}{\sqrt{L(1 + \delta_{j,0})}} \cos\left(\frac{jx^4}{R}\right) \tag{10.2.17}$$

for $\partial_4\Phi|_{x^4=0} = \partial_4\Phi|_{x^4=L} = 0$ (Neumann boundary conditions),

$$f_1^j(x^4) = \frac{1}{\sqrt{L}} \sin\left(\frac{jx^4}{R}\right) \tag{10.2.18}$$

for $\Phi(x^\mu, 0) = \Phi(x^\mu, L) = 0$ (Dirichlet boundary conditions),

$$f_2^j(x^4) = \frac{1}{\sqrt{L}} \sin\left(\frac{(j - 1/2)x^4}{R}\right) \tag{10.2.19}$$

for $\Phi(x^\mu, 0) = \partial_4\Phi|_{x^4=L} = 0$ (mixed boundary conditions),

$$f_3^j(x^4) = \frac{1}{\sqrt{L}} \cos\left(\frac{(j - 1/2)x^4}{R}\right) \tag{10.2.20}$$

for $\partial_4\Phi|_{x^4=0} = \Phi(x^\mu, L) = 0$ (mixed boundary conditions),

with j an integer called 'KK number'.

The functions f_n^j form a complete orthonormal set on the interval if

$$\sum_j \left[f_n^j(x^4)\right]^* f_n^j(x'^4) = \delta(x'^4 - x^4) \ . \tag{10.2.21}$$

The allowed values for j must be chosen such that the above completeness condition is satisfied. It is straightforward to check that $j \geq 0$ for $n = 0$, and $j \geq 1$ for $n = 1, 2, 3$. For $n = 0$ there is a state ($j = 0$) of zero momentum ('zero mode') along the compact dimension.

Exercise 10.2.1: Integrate the action (10.2.1) over x^4 and show that the 4D particles $\Phi^{(j)}(x^\mu)$ have masses

$$M^{(j)} = \sqrt{M_0^2 + \frac{j^2}{R^2}} \ , \tag{10.2.22}$$

if the boundary conditions are of the type $n = 0$ or 1, and

$$M^{(j)} = \sqrt{M_0^2 + \frac{(j - 1/2)^2}{R^2}} \ , \tag{10.2.23}$$

if $n = 2$ or 3.

In what follows we will concentrate on the KK functions f_0 and f_1, which are usually referred to as even and odd, respectively; the corresponding boundary conditions represent the so called S^1/Z_2 orbifold.

10.2.2. *Fermions on the interval: chiral boundary conditions*

We now turn to free spin-1/2 fields in five dimensions. The Clifford algebra is generated by five anti-commuting matrices: Γ^α, $\alpha = 0, 1, 2, 3, 4$. The minimal dimensionality of these matrices is 4×4. The Γ^α matrices can be used to construct a spinor representation of the $SO(1,4)$ Lorentz group, with the generators explicitly given by

$$\frac{\Sigma^{\alpha\beta}}{2} = \frac{i}{4}[\Gamma^\alpha, \Gamma^\beta] \ . \tag{10.2.24}$$

The spin-1/2 fermions in five dimensions have four components. In terms of the usual γ^μ matrices used in 4D field theory, one may take $\Gamma^\mu = \gamma^\mu$ and $\Gamma^4 = i\gamma_5$. The fact that $\Gamma^4 \propto \Gamma^0\Gamma^1\Gamma^2\Gamma^3$ implies that the $SO(1,4)$ Lorentz group has a single spin-1/2 representation, and thus the 5D fermions are vectorlike.

Upon compactification of x^4, the $SO(1,3)$ Lorentz symmetry generated by $\Sigma^{\mu\nu}/2$, $\mu, \nu = 0, 1, 2, 3$, remains unbroken. There are two chiralities under $SO(1,3)$, labeled as usual by L and R. These are projected by

$$P_{L,R} = \frac{1}{2}\left(1 \pm i\Gamma^4\right) \ . \tag{10.2.25}$$

A 5D fermion, Ψ, decomposes into two fermions of definite chirality under $SO(1,3)$:

$$\Psi(x^\mu, x^4) = \Psi_L(x^\mu, x^4) + \Psi_R(x^\mu, x^4) \ , \tag{10.2.26}$$

where

$$\Psi_{L,R} \equiv P_{L,R}\Psi \ . \tag{10.2.27}$$

As in Sec. 2, we consider the compactification on an interval: $0 \leq x^4 \leq L$. The action for a free 5D field of spin $1/2$ and mass zero is

$$S_\Psi = \int d^4x \int_0^L dx^4 \frac{i}{2}\left[\overline{\Psi}\Gamma^\alpha\partial_\alpha\Psi - \left(\partial_\alpha\overline{\Psi}\right)\Gamma^\alpha\Psi\right] \ . \tag{10.2.28}$$

Note that the 5D fermions have mass dimension $+2$. Under an arbitrary variation of the field, $\delta\Psi(x^\mu, x^4)$, the action has to be stationary both inside the interval and on its boundary:

$$\delta S_\Psi^v = -\int d^4x \int_0^L dx^4 i \left(\partial_\alpha \overline{\Psi}\right) \Gamma^\alpha \delta\Psi = 0 \ ,$$

$$\delta S_\Psi^s = \frac{i}{2} \int d^4x \int_0^L dx^4 \left(\overline{\Psi}\Gamma^4\delta\Psi\big|_{x^4=L} - \overline{\Psi}\Gamma^4\delta\Psi\big|_{x^4=0}\right)$$

$$= 0 \ . \tag{10.2.29}$$

The first equation implies that Ψ is a solution to the 5D Weyl equation, which can be decomposed into two equations:

$$\Gamma^\mu\partial_\mu\Psi_L = -\Gamma^4\partial_4\Psi_R \ ,$$

$$\Gamma^\mu\partial_\mu\Psi_R = -\Gamma^4\partial_4\Psi_L \ . \tag{10.2.30}$$

The second equation (10.2.29) restricts the values of Ψ on the boundary.

In the case of a fermion whose zero-mode is left-handed, the boundary conditions are as follows:

$$\partial_4\Psi_L(x^\mu,0) = \partial_4\Psi_L(x^\mu,L) = 0 \ ,$$

$$\Psi_R(x^\mu,0) = \Psi_R(x^\mu,L) = 0 \ . \tag{10.2.31}$$

The ensuing KK decomposition is given by

$$\Psi = \frac{1}{\sqrt{L}}\left\{ \Psi_L^{(0)}(x^\mu) + \sqrt{2}\sum_{j\geq 1}\left[\Psi_L^{(j)}(x^\mu)\cos\left(\frac{jx^4}{R}\right) \right.\right.$$

$$\left.\left. + \Psi_R^{(j)}(x^\mu)\sin\left(\frac{jx^4}{R}\right)\right]\right\} \ . \tag{10.2.32}$$

All fermion KK modes for $j \geq 1$ pair up to form vectorlike fermions of Dirac masses $M^{(j)}$ as given in Eq. (10.2.22). In the case of a fermion whose zero-mode is right-handed, the above equations apply with left- and right-handed labels interchanged. The conclusion is that the boundary conditions for the left- and right-handed fermions are forced by the stationary of the action to eliminate the zero mode for one of the chiralities. This is true for the interval compactification discussed here, while the compactification on a circle would preserve both the left- and right-handed zero modes.

Exercise 10.2.2: *Write down an Yukawa interaction between a bulk fermion and a bulk scalar, and show that the 4D Yukawa coupling λ (obtained after integration over x^4) is given in terms of the 5D Yukawa coupling*

λ_5 *by*

$$\lambda = \frac{\hat{\lambda}}{\sqrt{L}} \ . \tag{10.2.33}$$

10.2.3. *Gauge fields on the interval*

A 5D gauge boson has five components: $A_\mu(x^\nu, x^4)$, $\mu, \nu = 0, 1, 2, 3$, and $A_4(x^\nu, x^4)$ which corresponds to the polarization along the extra dimension. From the point of view of the 4D theory, A_4 is a tower of spinless KK modes.

Exercise 10.2.3: Write down the action for a 5D gauge boson, and then imposing the stationarity of the action under an arbitrary field variation, derive the field equations.

The boundary conditions consistent with gauge invariance are given by

$$\partial_4 A_\mu(x^\nu, 0) = \partial_4 A_\mu(x^\nu, L) = 0 \ ,$$

$$A_4(x, 0) = A_4(x, L) = 0 \ . \tag{10.2.34}$$

Solving the field equations with these boundary conditions yields the following KK expansions:

$$A_\mu = \frac{1}{\sqrt{L}} \left[A_\mu^{(0)}(x^\nu) + \sqrt{2} \sum_{j \geq 1} A_\mu^{(j)}(x^\nu) \cos\left(\frac{jx^4}{R}\right) \right] \ ,$$

$$A_4 = \sqrt{\frac{2}{L}} \sum_{j \geq 1} A_4^{(j)}(x^\nu) \sin\left(\frac{jx^4}{R}\right) \ . \tag{10.2.35}$$

The zero-mode $A_\mu^{(0)}(x^\nu)$ is the massless gauge boson associated with the 4D gauge transformations. Note that A_4 does not have a zero-mode. In the unitary gauge, the $A_4^{(j)}(x^\nu)$ KK modes are the longitudinal components of the heavy spin-1 KK modes $A_\mu^{(j)}(x^\nu)$.

Assuming that the gluon is a 5D field and that the quarks are localized at $x^4 = 0$, the terms of the 4D Lagrangian describing the interactions

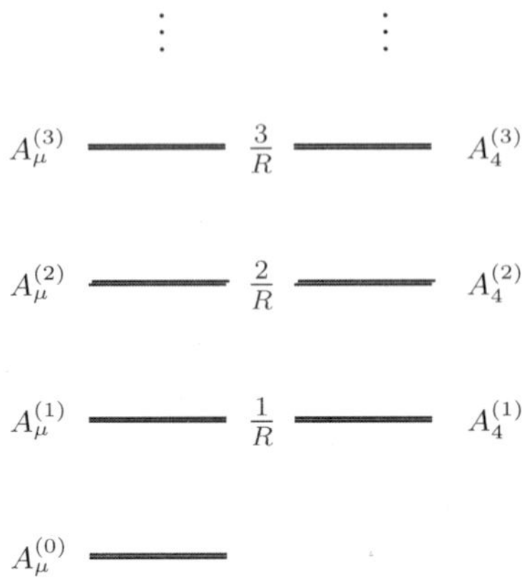

$$
A_\mu^{(3)} \quad\underline{\qquad}\quad \frac{3}{R} \quad\underline{\qquad}\quad A_4^{(3)}
$$

$$
A_\mu^{(2)} \quad\underline{\qquad}\quad \frac{2}{R} \quad\underline{\qquad}\quad A_4^{(2)}
$$

$$
A_\mu^{(1)} \quad\underline{\qquad}\quad \frac{1}{R} \quad\underline{\qquad}\quad A_4^{(1)}
$$

$$
A_\mu^{(0)} \quad\underline{\qquad}
$$

Fig. 10.2.   The spectrum of KK modes for the gauge boson associated with an unbroken gauge symmetry.

between gluon KK modes and quarks are given by

$$
\mathcal{L}_{4D} = \int_0^L dx^4 \, \hat{g} G_\mu^a(x^\nu, x^4) \left[\delta\left(x^4\right) \bar{q}(x^\nu) \gamma^\mu T^a q(x^\nu) \right]
$$

$$
= g_s \left(G_\mu^{(0)a} + \sqrt{2} \sum_{j \geq 1} G_\mu^{(j)a}(x) \right) \bar{q} \gamma^\mu T^a q
\tag{10.2.36}
$$

From the above equation follows that the 4D gauge coupling g_s is given in terms of the 5D gauge coupling $\hat{g}$ by $g_s = \hat{g}/\sqrt{L}$, and that the coupling of any gluon KK mode to quarks is larger than the QCD coupling by a factor of $\sqrt{2}$. This factor is a consequence of the normalization of the KK functions, which in turn follows from the canonical normalization of the kinetic terms.

If the gauge symmetry is broken by the vacuum expectation value, v, of a bulk scalar field Φ with Neumann boundary conditions, then the KK modes of the CP-odd components of Φ mix with the corresponding $A_4^{(j)}$. At each KK level one linear combination becomes the longitudinal degree of freedom of $A_\mu^{(j)}$ and the orthogonal one appears as a spin-0 particle. The masses of both $A_\mu^{(j)}$ and of the spin-0 particles are given by

$$
M_A^{(j)} = \sqrt{g^2 v^2 + \frac{j^2}{R^2}} \ .
\tag{10.2.37}
$$

where g is the 4D gauge coupling.

10.3. One universal extra dimension

Being equipped with the basics of field theory in five dimensions, we can now discuss the case where all standard model particles propagate along one flat extra dimension compactified on an interval, *i.e.*, one universal extra dimension.

Ignoring electroweak-symmetry breaking effects, the tree-level spectrum consists of equally spaced KK levels (of mass j/R), and on each level the KK modes for all standard model particles are degenerate. Each standard model chiral fermion has a tower of vectorlike modes. The KK spectrum of the $(t, b)_L$ doublet and t_R is illustrated in Fig. 10.3.

To understand the effects of electroweak symmetry breaking on the KK fermion spectrum, let us analyze the case of the top quark (the same applies to the other standard model fermions, except that electroweak symmetry breaking effects are suppressed by their small Yukawa couplings). Let us denote the bulk fermion whose zero mode is t_L (which is part of a weak doublet $\mathcal{Q}_3$) by $\mathcal{Q}_t(x^\mu, x^4)$, and the bulk fermion whose zero mode is t_R (weak singlet) by $\mathcal{U}_3(x^\mu, x^4)$. The terms of the 4D Lagrangian responsible for top KK masses come from the kinetic terms along the extra dimension and from the Yukawa couplings to the Higgs doublet:

$$\mathcal{L}_{4D} = \int_0^L dx^4 \left[\overline{\mathcal{Q}}_t i\gamma^4(-\partial_4)\mathcal{Q}_t + \overline{\mathcal{U}}_3 i\gamma^4(-\partial_4)\mathcal{U}_3 + \left(-\hat{\lambda}_t H \overline{\mathcal{Q}}\,\mathcal{U}_3 + \text{H.c.} \right) \right]$$

$$(10.3.1)$$

The bulk Higgs doublet, H, which is an even field, has a negative-squared 5D mass; this leads to a vacuum expectation value $v \simeq 174$ GeV only for

$$(\mathcal{Q}_{tL}^{(3)}, \mathcal{Q}_{bL}^{(3)}) \quad \text{---} \quad \tfrac{3}{R} \quad \text{---} \quad (\mathcal{Q}_{tR}^{(3)}, \mathcal{Q}_{bR}^{(3)}) \qquad \mathcal{U}_{3L}^{(3)} \quad \text{---} \quad \tfrac{3}{R} \quad \text{---} \quad \mathcal{U}_{3R}^{(3)}$$

$$(\mathcal{Q}_{tL}^{(2)}, \mathcal{Q}_{bL}^{(2)}) \quad \text{---} \quad \tfrac{2}{R} \quad \text{---} \quad (\mathcal{Q}_{tR}^{(2)}, \mathcal{Q}_{bR}^{(2)}) \qquad \mathcal{U}_{3L}^{(2)} \quad \text{---} \quad \tfrac{2}{R} \quad \text{---} \quad \mathcal{U}_{3R}^{(2)}$$

$$(\mathcal{Q}_{tL}^{(1)}, \mathcal{Q}_{bL}^{(1)}) \quad \text{---} \quad \tfrac{1}{R} \quad \text{---} \quad (\mathcal{Q}_{tR}^{(1)}, \mathcal{Q}_{bR}^{(1)}) \qquad \mathcal{U}_{3L}^{(1)} \quad \text{---} \quad \tfrac{1}{R} \quad \text{---} \quad \mathcal{U}_{3R}^{(1)}$$

$$(t_L, b_L) \quad \text{---} \qquad\qquad\qquad\qquad \text{---} \quad t_R$$

Fig. 10.3. The tree-level spectrum of top-quark KK modes, ignoring electroweak symmetry breaking.

the zero mode $H^{(0)}$. Inserting now the KK decompositions for $\mathcal{Q}_t$ and $\mathcal{U}_3$, which are analogous to Eq. (10.2.32), we obtain the following terms responsible for level-j top masses:

$$\frac{-2}{L}\int_0^L dx^4 \left\{ \overline{\mathcal{Q}}_{tR}^{(j)}\mathcal{Q}_{tL}^{(j)} \sin\frac{jx^4}{R}\, \partial_4 \cos\frac{jx^4}{R} + \overline{\mathcal{U}}_{3R}^{(1)}\mathcal{U}_{3L}^{(j)} \cos\frac{jx^4}{R}\, \partial_4 \sin\frac{jx^4}{R} \right.$$

$$\left. +\lambda_t v \left[\overline{\mathcal{Q}}_{tR}^{(j)}\mathcal{U}_{3L}^{(j)} \sin^2\left(\frac{jx^4}{R}\right) + \overline{\mathcal{U}}_{3R}^{(j)}\mathcal{Q}_{tL}^{(j)} \cos^2\left(\frac{jx^4}{R}\right) \right] \right\} + \text{H.c.}$$

$$(10.3.2)$$

where $\lambda_t = \hat{\lambda}_t/\sqrt{L} \approx 1$ is the standard model top Yukawa coupling. From the above equation it is clear that electroweak symmetry breaking leads to mixing between the towers of vectorlike fermions that have t_L and t_R as zero modes. After integration over x^4, we get the mass terms for the top quark KK modes at level j:

$$\left(\overline{\mathcal{Q}}_{tR}^{(j)}, \overline{\mathcal{U}}_{3R}^{(j)} \right) \begin{pmatrix} -\frac{j}{R} & \lambda_t v \\ \lambda_t v & \frac{j}{R} \end{pmatrix} \begin{pmatrix} \mathcal{Q}_{tL}^{(j)} \\ \mathcal{U}_{3L}^{(j)} \end{pmatrix}.$$

$$(10.3.3)$$

Due to the minus sign in the 11 element, the two eigenvalues are equal at tree level:

$$M_{t_1}^{(j)} = M_{t_2}^{(j)} = \sqrt{\frac{j^2}{R^2} + m_t^2}\,.$$

$$(10.3.4)$$

The mass degeneracy among the level-j modes of various standard model particles is further lifted by loop corrections.[12] An important property of any type of interaction in a 5D theory is that it grows with the energy, such that it becomes nonperturbative at a certain energy scale Λ. In the equivalent 4D description that includes a tower of KK modes, the number of KK modes grows with the energy scale such that the loop corrections are divergent; they need to be cutoff at a scale Λ, which means that the tower of modes is truncated. In practice, the QCD interactions become nonperturbative in the UV at scales roughly two orders of magnitude above the compactification scale. Hence, one can study perturbatively the effective theory below the scale Λ, but higher-dimensional operators suppressed by that scale are likely to be generated by physics at scales above Λ. Working within this effective theory, the loop corrections to KK masses are logarithmically dependent on Λ. For fermions, the leading 1-loop corrections may be obtained by substituting

$$\frac{j}{R} \rightarrow \frac{j}{R} \left[\sum_{i=1}^{3} 9 C_i \alpha_i - (3 - 2C_2) \frac{\lambda_f^2}{4\pi} \right] \frac{1}{4\pi} \ln(\Lambda R/j) \qquad (10.3.5)$$

in the tree-level mass formulas. Here α_i for $i = 1, 2, 3$ are the $U(1)_Y$, $SU(2)_W$ and $SU(3)_c$ coupling constants, $C_1 = y_f^2$ for fermions of hypercharge y_f (using the normalization where y_f is the electric charge for weak singlets), $C_2 = 3/4$ for $SU(2)_W$ doublets and 0 for singlets, $C_3 = 4/3$ for quarks and 0 for leptons, and λ_f is the fermion Yukawa coupling. Note that the $SU(2)_W$ loop corrections to the top KK masses split the diagonal elements in Eq. (10.3.3), and therefore $t_1^{(j)}$ and $t_2^{(j)}$ (which are the physical Dirac fermions representing the top modes at each level) end up with different masses.

For gauge bosons, the leading loop corrections are given by the following substitution in the tree-level masses:

$$\frac{j^2}{R^2} \rightarrow \frac{j^2}{R^2} C_i' \frac{\alpha_i}{4\pi} \ln(\Lambda R/j) , \qquad (10.3.6)$$

where C_i' equals to 23 for the gluon, to 15 for the $SU(2)_W$ bosons, and to $-1/3$ for the hypercharge boson B_μ. The different loop contributions to the W_μ^3 and B_μ KK masses have a dramatic effect on their mixing due to electroweak symmetry breaking: the mixing vanishes in the limit where $1/R \gg v$. For this reason, the KK modes of the photon are labelled in the literature by either $\gamma^{(j)}$ or $B^{(j)}$, and the Z modes are labelled by either $Z^{(j)}$ or $W^{3(j)}$.

Exercise 10.3.1: Compute the masses of $\gamma^{(j)}$ and $Z^{(j)}$ as a function of $1/R$, v, and Λ.

Exercise 10.3.2: Write down a Higgs mass term localized at both $x^4 = 0$ and $x^4 = L$, and determine its effects on the KK Higgs masses.

It is often assumed that higher-dimensional operators (in particular, kinetic terms localized on the boundary) and localized Higgs mass terms may be ignored at the scale Λ.[12] The lightest KK particle is then the first KK mode of the photon, and the heaviest particles at each level are the KK modes of the gluon and quarks. The mass spectrum of all level-1 particles is shown in Fig. 10.4. If the unspecified UV completion of this low-energy effective theory gives rise to operators localized at the ends of the interval, then the KK spectrum may change; this possibility is not considered in what follows.

Momentum conservation along the extra dimension is broken by the boundary conditions, but a remnant of it is left intact. This is reflected

B. Dobrescu

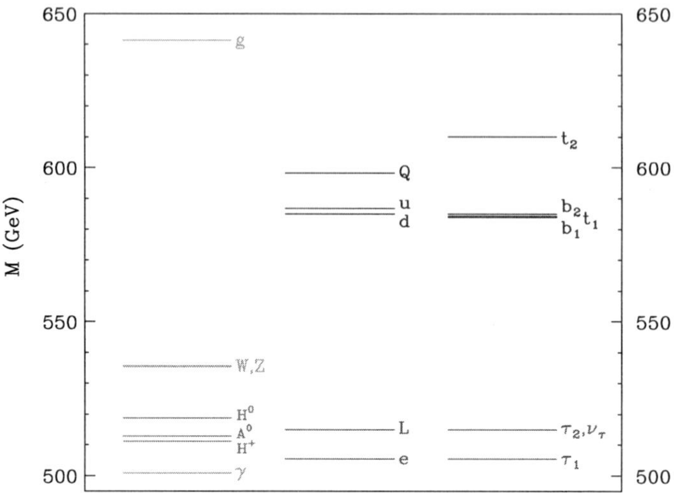

Fig. 10.4.   Mass spectrum of level-1 KK modes for a compactification scale of $1/R = 500$ GeV and $\Lambda R = 20$, from Ref. [12].

in a selection rule for the KK-numbers of the particles participating in any interaction. A vertex with particles of KK numbers $j_1, \ldots, j_p$ exists at tree level only if $j_1 \pm \ldots \pm j_p = 0$ for a certain choice of the $\pm$ signs. This selection rule has important phenomenological implications. First, it is not possible to produce only one KK 1-mode at colliders. Second, tree-level exchange of KK modes does not contribute to currently measurable quantities. Therefore, the corrections to electroweak observables are loop suppressed, and the limit on $1/R$ from electroweak measurements is rather weak, of the order of the electroweak scale.[9]

To derive the interactions among KK modes, one should start with the 5D Lagrangian, replace the 5D fields by their KK decomposition, integrate over x^4, and replace the 5D coupling parameters by their 4D counterparts which are identified by inspecting the interactions among 0-modes. The interactions of the KK modes with the massless gluon are fixed by the $SU(3)_c$ gauge invariance. For example, the coupling of a gluon to a pair of KK quarks is the same as for any quark pair, and the quartic coupling of two gluon 0-modes and two level-j KK gluons is the same as the quartic gluon coupling of usual QCD. However, the interactions of gluon KK modes with quarks are peculiar. As opposed to the 4D QCD interactions, which are vectorlike, the interactions of higher gluon modes distinguish between left- and right-handed quarks. This follows from the fact that the couplings

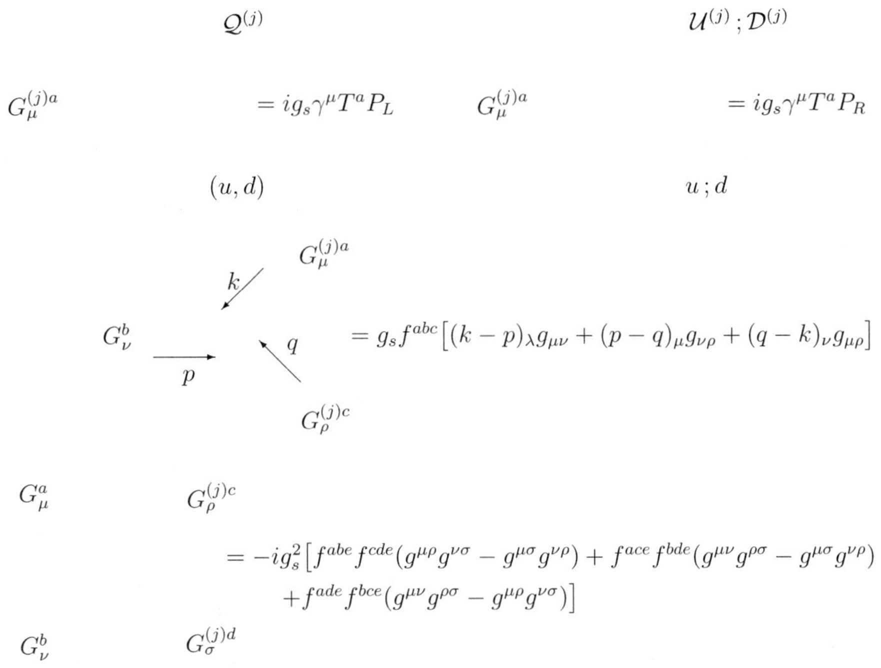

$$\mathcal{Q}^{(j)} \qquad\qquad \mathcal{U}^{(j)} \, ; \mathcal{D}^{(j)}$$

$$G_\mu^{(j)a} \qquad = ig_s\gamma^\mu T^a P_L \qquad G_\mu^{(j)a} \qquad = ig_s\gamma^\mu T^a P_R$$

$$(u,d) \qquad\qquad u \, ; d$$

$$
\begin{aligned}
&G_\mu^{(j)a} \\
k &\qquad\qquad\qquad = g_s f^{abc}\big[(k-p)_\lambda g_{\mu\nu} + (p-q)_\mu g_{\nu\rho} + (q-k)_\nu g_{\mu\rho}\big] \\
G_\nu^b \quad p \quad & q \\
&G_\rho^{(j)c}
\end{aligned}
$$

$$
\begin{aligned}
G_\mu^a \qquad G_\rho^{(j)c} \\
= -ig_s^2\big[f^{abe}f^{cde}(g^{\mu\rho}g^{\nu\sigma} - g^{\mu\sigma}g^{\nu\rho}) + f^{ace}f^{bde}(g^{\mu\nu}g^{\rho\sigma} - g^{\mu\sigma}g^{\nu\rho}) \\
+ f^{ade}f^{bce}(g^{\mu\nu}g^{\rho\sigma} - g^{\mu\rho}g^{\nu\sigma})\big] \\
G_\nu^b \qquad G_\sigma^{(j)d}
\end{aligned}
$$

Fig. 10.5. Feynman rules for vertices involving gluon KK modes ($G_\mu^{(j)a}$) and standard model particles. The couplings to quarks are chiral. G_μ^a is the massless gluon, $\mathcal{Q}^{(j)}$ are the $SU(2)_W$-doublet KK quarks, and $\mathcal{U}^{(j)}, \mathcal{D}^{(j)}$ are the $SU(2)_W$-singlet KK quarks.

of a spin-1 particle to fermions do not change the chirality, and thus a right-handed standard model quark couples only to the right-handed components of its KK modes (which are $SU(2)_W$ singlets, $\mathcal{U}^{(j)}$ or $\mathcal{D}^{(j)}$), while a left-handed standard model quark couples only to the left-handed components of its KK modes ($\mathcal{Q}^{(j)}$). These interactions are contained in the following terms of the 4D Lagrangian:

$$g_s G_\mu^{(j)a} \left(\overline{\mathcal{Q}}_L^{(j)} \gamma^\mu T^a (u_L, d_L)^\top + \overline{\mathcal{U}}_R^{(j)} \gamma^\mu T^a u_R + \overline{\mathcal{D}}_R^{(j)} \gamma^\mu T^a d_R + \text{H.c.} \right) , \tag{10.3.7}$$

where g_s is the QCD gauge coupling. The Feynman rules describing the interactions of gluon KK modes to 0-modes are shown in Fig. 10.5.

The 1-modes may be produced in pairs at colliders. At the Tevatron and the LHC, pair production of the colored KK modes has large cross sections[13,14] as long as $1/R$ is not too large. The colored KK modes suffer cascade decays[15] like the ones shown in Fig. 10.6. Note that at each

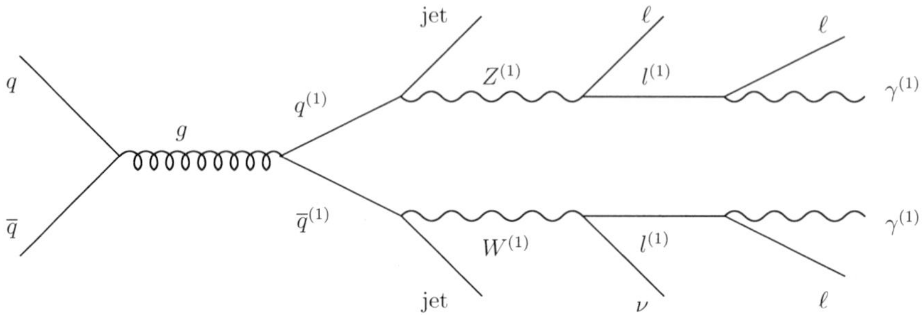

Fig. 10.6. $3\ell + \not{E}_T$ signal from UED. $\gamma^{(1)}$ is the dark matter candidate which escapes the detector.

vertex the KK-number is conserved, and the $\gamma^{(1)}$ escapes the detector. The signal is $\ell^+\ell^-\ell^\pm + 2j + \not{E}_T$. However, the approximate degeneracy of the KK modes implies that the jets may be relatively soft, and it could be challenging to distinguish them from the background. The leptons are also soft (with energies of a few percent of the compactification scale, as can be seen from the mass differences displayed in Fig. 10.4), but usually pass some reasonably chosen cuts. Using the Run I data from the Tevatron, the CDF Collaboration[16] searched for the $3\ell + \not{E}_T$ signal and has set a limit of $1/R > 280$ GeV at the 95% CL. The much larger Run II data set can be used to set a substantially improved limit, or alternatively, has a chance of leading to a discovery.

If a signal is seen at the Tevatron or LHC, then it is important to differentiate the UED models from alternative explanations, such as super-partner cascade decays.[15] Measuring the spins at the LHC would provide an important discriminant, but such measurements are challenging.[14,18] A more promising way is to look for second level KK modes. These can be pair produced as the first level modes. However, unlike the first level modes, the second level modes may decay into standard model particles. Such decays occur at one loop, via diagrams having the structure shown in Fig. 10.7.

This diagram demonstrates that in the presence of loop corrections, the selection rule for KK numbers of the particles interacting at a vertex becomes

$$j_1 \pm \ldots \pm j_p = 0 \bmod 2 \ . \tag{10.3.8}$$

This implies the existence of an exact Z_2 symmetry: the KK parity $(-1)^j$ is conserved. Its geometrical interpretation is invariance under reflections

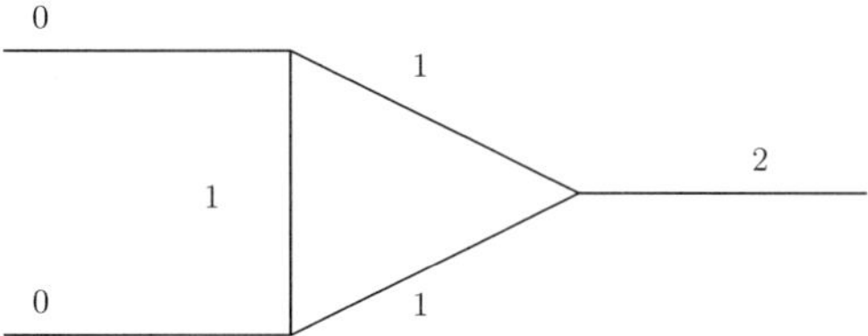

Fig. 10.7.   One-loop induced coupling of a 2-mode to two zero-modes.

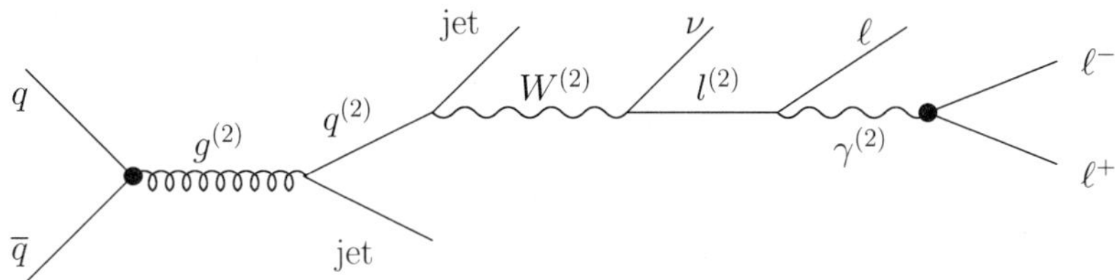

Fig. 10.8. s-channel production of the level-2 gluon followed by cascade decay, and $\gamma^{(2)}$ decays to e^+e^- and $\mu^+\mu^-$. The $\bullet$ represents an 1-loop effective vertex, as in Fig. 10.7.

with respect to the middle of the $[0, L]$ interval. Given that the lightest particle with j odd is stable, the $\gamma^{(1)}$ is a promising dark matter candidate. For $1/R$ in the 0.5 to 1.5 TeV range the $\gamma^{(1)}$ relic density fits nicely the dark matter density.[17] This whole range of compactification scales will be probed at the LHC.[15]

Another consequence of the loop-induced coupling of a 2-mode to two zero-modes is that the 2-mode can be singly produced in the s-channel. The typical signal will be the cascade decay shown in Fig. 10.8, followed by $\gamma^{(2)}$ decay into hard leptons. The cross section in this channel at the LHC is substantial for compactification scales up to 2 TeV (see Fig. 10.9).

10.4. A warped extra dimension

We have discussed so far the case of an extra dimension with a factorizable metric, namely the length of the line element along the 4D Minkowski space and the one along the extra dimension are independent. In particular we have assumed that the metric is flat, so that the line element is given by

$$ds^2 = \eta_{\mu\nu}\, x^\mu x^\nu - (x^4)^2 \, , \tag{10.4.1}$$

where $\eta_{\mu\nu} = \mathrm{diag}(1, -1, -1, -1)$ is the 4D Minkowski metric, with $\mu, \nu = 0, 1, 2, 3$. Gravity may be incorporated in this framework by writing the

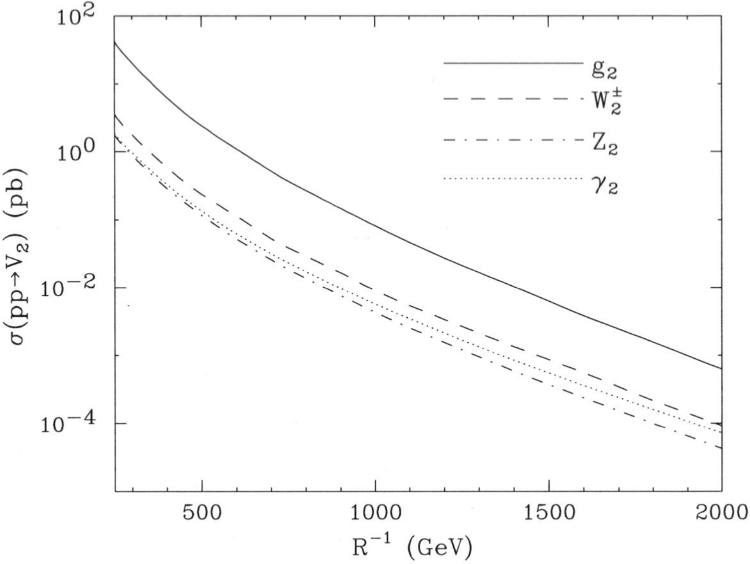

Fig. 10.9. Production cross section for level-2 gauge bosons at the 14 TeV LHC, from Ref. [18].

field equations for a spin-2 field.[20] These 5D Einstein equations may be solved by separating variables, as we have done for spin-0, 1/2 and 1 in Sec. 10.2. The result is that besides the graviton 0-mode, which is responsible for long-range gravitational interactions, there is a tower of KK modes of spin-2, with masses given by j/R and couplings given by $\sqrt{2}/\overline{M}_{\rm Pl}$, where the reduced Planck mass $\overline{M}_{\rm Pl}$ is related to the Newton constant G_N: $\overline{M}_{\rm Pl} = (8\pi G_N)^{-1/2} \approx 2.4 \times 10^{18}$ GeV. The only additional physical degree of freedom that originates from the 5D graviton is the 0-mode of its 44 component (called the radion field).

More complicated metrics may be considered, but they have to satisfy a nontrivial constraint: the 5D Einstein equations need to have a solution given certain boundary conditions. In this section we discuss a famous non-factorizable metric which provides a natural hierarchy between the Planck and electroweak scales.

10.4.1. *RS1 model*

An Anti-de-Sitter space along the 5th dimension (AdS_5) is described by the line element

$$ds^2 = e^{-2ky}\eta_{\mu\nu}x^{\mu}x^{\nu} - y^2 \ . \tag{10.4.2}$$

Here k is the AdS_5 curvature, and has dimensions of mass. The unit of length along the 4D Minkowski space depends on the position along $y \equiv x^4$. Randall and Sundrum[10] have shown that a slice of AdS_5 (space exists only for $0 \le y \le L$) is a solution to the 5D Einstein equations provided the vacuum energy $V(y)$ satisfies

$$V(0) = -V(L) = \frac{\Lambda}{k} \ , \tag{10.4.3}$$

where Λ is the 5D cosmological constant. They proposed the following set-up ("RS1"): gravity propagates in the 5D bulk, while all standard model fields are localized at $y = L$. The points $y = 0$ and $y = L$ are usually referred to as the Planck brane and the standard model brane, respectively. It is assumed that the curvature k is a couple of orders of magnitude below the fundamental 5D scale M_5, where 5D gravity becomes strongly coupled, so that the effective theory has a range of validity. The reduced Planck mass can be derived in terms of M_5 and $k \gg 1/L$:

$$\overline{M}_{\rm Pl} \approx \sqrt{\frac{M_5^3}{k}} \ , \tag{10.4.4}$$

and the 5D cosmological constant is given by $\Lambda = -24M_5^3 k^2$.

As a consequence of the metric, any mass parameter m_0 that appears in some terms of the 5D action which are localized at $y = L$ corresponds to a mass m in the effective 4D theory obtained after integrating over y, with

$$m = m_0\, e^{-kL} \ . \tag{10.4.5}$$

Thus, if the VEV that breaks the $SU(2)_W \times U(1)_Y$ symmetry in the 5D theory is $v_0 \sim O(k)$, then the corresponding VEV in the 4D effective theory is $v = v_0 e^{-kL}$. Taking $M_5 \approx 10k$ one finds $v \simeq 174$ GeV for $kL \approx 34$. This is a remarkable result: the huge ratio $\overline{M}_{\rm Pl}/v$ arises from a theory in which there are no large hierarchies between the input mass parameters (M_5, k, $1/L$, v_0).

Exercise 10.4.1: Change the Minkowski coordinates in Eq. (10.4.1) so that in terms of the new coordinates all distances are given as measured by an observer localized at $y = L$. Show that all input mass parameters are of

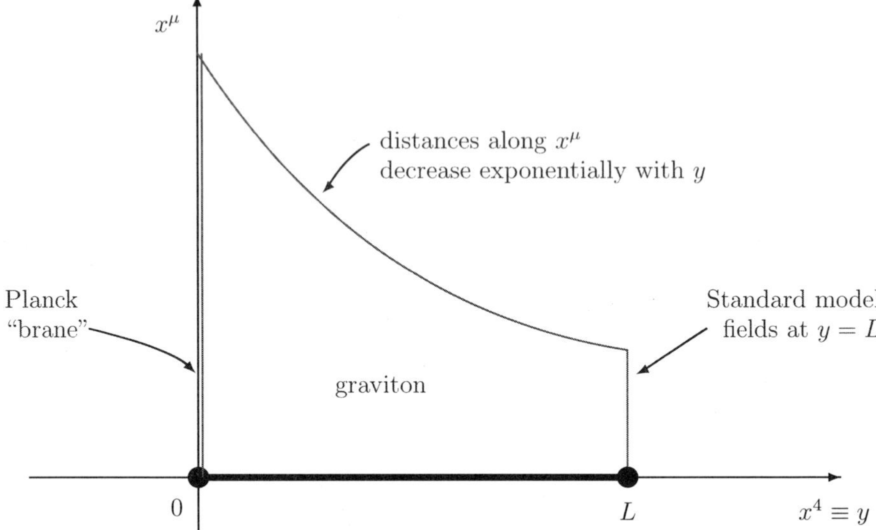

Fig. 10.10.   RS1 model: only the graviton propagates in the warped bulk.

order the TeV scale in this case, and that the observed weakness of gravity is due to the small wavefunction of the 0-mode graviton on the standard model brane.

The gravity action[19] in the warped bulk is given by

$$S_{\text{gravity}} = 2M_5^3 \int d^4x \int_0^L dy \, \sqrt{\det G} \left[G^{\alpha\beta} \left(\partial_\beta \Gamma^\delta_{\;\alpha\delta} - \partial_\delta \Gamma^\delta_{\;\alpha\beta} \right. \right.$$

$$\left. \left. + \; \Gamma^\eta_{\;\alpha\delta} \Gamma^\delta_{\;\eta\beta} - \Gamma^\eta_{\;\alpha\beta} \Gamma^\delta_{\;\eta\delta} \right) + 12k^2 \right] , \qquad (10.4.6)$$

where $G^{\alpha\beta}$ is the 5D metric, $\det G$ is the determinant of the metric, and $\Gamma^\delta_{\;\alpha\beta}$ is the 5D connection:

$$\Gamma^\delta_{\;\alpha\beta} = \frac{1}{2} G^{\delta\eta} \left(\partial_\alpha G_{\beta\eta} + \partial_\beta G_{\eta\alpha} - \partial_\eta G_{\alpha\beta} \right) . \qquad (10.4.7)$$

As in the previous sections, letters from the beginning of the Greek alphabet label the 5D coordinates ($\alpha, \beta, \delta, \eta, \ldots = 0, 1, 2, 3, 4$), and letters from the middle of the Greek alphabet label the Minkowski coordinates ($\mu, \nu, \rho, \sigma, \ldots = 0, 1, 2, 3$).

Small gravitational fluctuations are described by expanding the 5D metric about the warped background:

$$G_{\alpha\beta} = \begin{pmatrix} e^{-2ky}\,\eta_{\mu\nu} & 0 \\ 0 & -1 \end{pmatrix} + \frac{2}{M_5^{3/2}} \begin{pmatrix} e^{-2ky}\,h_{\mu\nu}(x^\rho, y) & h_{\mu 4}(x^\rho, y) \\ h_{4\nu}(x^\rho, y) & h_{44}(x^\rho, y) \end{pmatrix}.$$

$$(10.4.8)$$

Here $h_{\mu\nu}$ is the 5D graviton polarized along the Minkowski coordinates, while $h_{\mu 4}$ and h_{44} are the polarizations along the extra dimension of the 5D graviton; $h_{\mu 4}$ form a 5D spin-1 field called graviphoton, and $h_{\mu 4}$ is a 5D spin-0 field called the graviscalar. The 5D graviton $h_{\alpha\beta}$ is a symmetric tensor field, and therefore it has 15 components. However, not all of its components are physical. Imposing a gauge fixing condition that $h_{\alpha\beta}$ is traceless and transverse,

$$h_\alpha^\alpha = 0 \ , \quad \partial^\alpha h_{\alpha\beta} = 0 \ , \tag{10.4.9}$$

eliminates 10 of the components. The remaining five components are physical degrees of freedom, divided as follows: two in $h_{\mu\nu}$ (a traceless and transverse 4×4 symmetric tensor), two in $h_{\mu 4}$ (a massless gauge field with 4-components), and one in h_{44}.

Exercise 10.4.2: Keeping only the quadratic terms in $h^{\alpha\beta}$, derive the Lagrangian that describes the propagation of the graviton in the warped background.

The graviton must have a massless 0-mode, $h_{\mu\nu}^{(0)}$, in order to generate the observed long-range gravitational effects. Hence, the graviton is an even field, *i.e.*, its derivative with respect to y must vanish at $y = 0$ and $y = L$. The KK decomposition for the tensor components of the 5D graviton field is

$$h_{\mu\nu}(x^\rho, y) = \frac{1}{\sqrt{L}} \left[h_{\mu\nu}^{(0)}(x^\rho) + \sum_{j \geq 1} h_{\mu\nu}^{(j)}(x^\rho)\chi_j(y) \right] \ . \tag{10.4.10}$$

The graviton KK modes, $h_{\mu\nu}^{(j)}(x)$, are spin-2 particles in 4D. For $j \geq 1$, they are massive and therefore they have 5 degrees of freedom (corresponding angular momenta ± 2, ± 1 and 0). Two of these degrees of freedom originate in $h_{\mu\nu}(x^\rho, y)$, while the other three are given at each KK level by the graviphoton and graviscalar, which in the unitary gauge disappear from the spectrum. General coordinate invariance requires $h_{\mu 4}$ to have odd boundary conditions, and h_{44} to have even boundary conditions.[20] As a result, the 0-mode of h_{44} remains as physical spin-0 particle, which is referred to as the radion. Its properties are analyzed in Ref. [21].

Plugging the KK decomposition (10.4.10) into the 5D kinetic terms for $h_{\mu\nu}(x^\rho, y)$ gives the following differential equation for the KK functions:[22]

$$\left[\frac{d}{dy}\left(e^{-4ky}\frac{d}{dy} \right) + m_j^2 e^{-4ky} \right] \chi_j(y) = 0 \ , \tag{10.4.11}$$

where m_j is the mass of the jth KK mode of spin-2. Solving this equation with the Neumann boundary conditions

$$\frac{d}{dy}\chi_j(y)|_{y=0} = \frac{d}{dy}\chi_j(y)|_{y=L} = 0 \tag{10.4.12}$$

leads to KK functions given in terms of the J_2 and Y_2 Bessel functions:

$$\chi_j(y) = \frac{e^{2ky}}{N_j}\left[J_2\left(\frac{m_j}{k}e^{ky} \right) + \alpha_j Y_2\left(\frac{m_j}{k}e^{ky} \right) \right] \ , \tag{10.4.13}$$

where N_j is a normalization constant, and α_j are constants determined by the boundary conditions (10.4.12). The mass of the level-1 KK modes is at the TeV scale, $m_1 \simeq 3.8ke^{-kL}$, and the higher modes have masses $m_2 \simeq 1.8m_1$, $m_3 \simeq 2.7m_1$, $m_4 \simeq 3.5m_1$, ...

The couplings of the graviton KK modes to standard model fields are identical to those of the graviton 0-mode except for an overall factor of e^{kL}. Therefore, both the masses of the graviton KK modes and the suppression of the couplings are of order the TeV scale. The collider signals of these spin-2 particles are heavy resonances that decay into neutral pairs of standard model particles. The production cross section at the Tevatron or the LHC may be computed from the following couplings to gluons

$$\frac{\sqrt{2}}{\overline{M}_{\rm Pl}}e^{kL}\left(h_\rho^{(j)\rho}\eta_{\mu\nu} - 4h_{\mu\nu}^{(j)} \right) G_\sigma^\mu G^{\nu\sigma} \ , \tag{10.4.14}$$

and to quarks

$$\frac{4}{\overline{M}_{\rm Pl}}e^{kL}\left[\left(h_\sigma^{(j)\sigma}\eta_{\mu\nu} - h_{\mu\nu}^{(j)} \right)\left((\partial^\nu \bar{q})\gamma^\mu q - \bar{q}\gamma^\mu \partial^\nu q \right) - m_q h_\sigma^{(j)\sigma}\bar{q}q \right] \ . \tag{10.4.15}$$

Exercise 10.4.3:   *Compute the parton-level cross sections $gg \to h_{\mu\nu}^{(1)}$ and $q\bar{q} \to h_{\mu\nu}^{(1)}$, in the narrow width approximation, as a function of the curvature k and the $h_{\mu\nu}^{(1)}$ mass.*

The most useful decays of the graviton KK modes are into $\gamma\gamma$, e^+e^- and $\mu^+\mu^-$, due to the small backgrounds. Figure 10.11 shows the current Tevatron limit on the RS1 model. Figure 10.12 shows the predicted cross section for $pp \to h_{\mu\nu}^{(j)} \to \ell^+\ell^-$ at the LHC.

10.4.2. *Standard model in a warped extra dimension*

It is interesting to study what are the effects of the warped metric on the propagation along the extra dimension of fields other than the graviton. Boundary conditions at $y = 0$ and $y = L$ must be specified for each field propagating in the bulk.

A gauge field propagating in a warped extra dimension, with Neumann boundary conditions, has a y-independent 0-mode.[25] This flat profile, forced by gauge invariance, is in stark contrast to the exponential profile of the graviton 0-mode. The Kaluza-Klein decomposition for the gauge field is

$$A_\mu(x^\nu, y) = \frac{1}{\sqrt{2L}} \left[A_\mu^{(0)}(x^\nu) + \sum_{j \geq 1} A_\mu^{(j)}(x^\nu) f_j(y) \right]$$

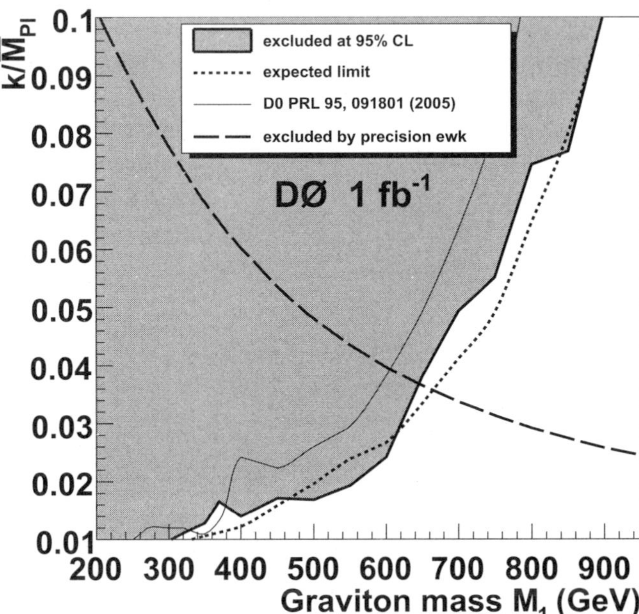

Fig. 10.11.   Exclusion region in the coupling versus mass plane for the level-1 graviton of the RS1 model, obtained by the D0 Collaboration by searching for a narrow e^+e^- or $\gamma\gamma$ resonance. This plot is reproduced from Ref. [23].

where the KK functions are given by

$$f_j(y) = \frac{e^{ky}}{N_j}\left[J_1\left(\frac{m_j}{k}e^{ky}\right) - \frac{J_0(m_j/k)}{Y_0(m_j/k)}Y_1\left(\frac{m_j}{k}e^{ky}\right)\right]$$

Here J and Y are Bessel functions, N_j is a normalization constant, M_j is the mass of the jth KK mode of spin-1. The level-1 gauge boson has a mass $M_1 = 2.5ke^{-kL}$, so at the scale associated with the standard model brane, 1.5 times lighter than the level-1 graviton. The higher spin-1 KK modes have masses $M_2 = 2.3M_1$, $M_3 = 3.6$, $M_3 = 4.8$, ... The couplings of the higher spin-1 KK modes to the fermions localized on the standard model brane are all given by the 4D gauge coupling (*i.e.*, the coupling of the 0-mode gauge boson) times a "volume" factor of $\sqrt{2kL} \approx 8.2$. Therefore, if the standard model gauge bosons propagate in the warped bulk while the quark and leptons are localized at $y = L$, then the spin-1 KK modes are strongly coupled to the fermions. The 4-fermion effective interactions induced by KK exchange are ruled out unless $M_1 > O(20)$ TeV (this is based on the assumption that single boson exchange is a good approximation of the effects induced by the rather strongly coupled KK modes).

If fermions are also propagating in the warped bulk, then their 0-modes have an exponential profile.[26] In addition, the exponential profile may be

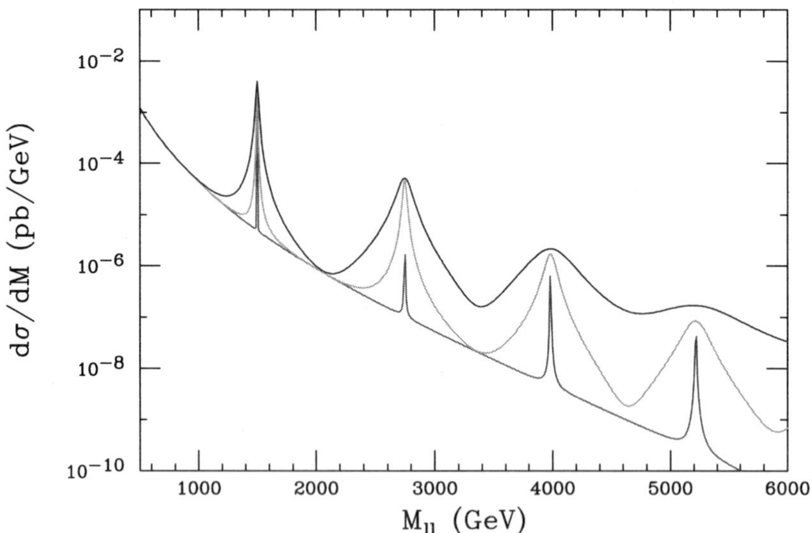

Fig. 10.12.   Cross section for $pp \to h_{\mu\nu}^{(j)} \to \ell^+\ell^-$ at the 14 TeV LHC.[24] The three curves correspond, from top to bottom, to $k/\overline{M}_{\rm Pl} = 0.1, 0.05$ and 0.01.

tuned independently for each fermion flavor, because the fermions may have bulk masses. The fermionic part of the 5D Lagrangian takes the following form:

$$\sqrt{g}\left(i\overline{\psi}\gamma^{\mu}\partial_{\mu}\psi - c_{\psi}k\overline{\psi}\psi\right) , \qquad (10.4.16)$$

where g is the determinant of the 5D metric. The ensuing y-dependence of the 0-mode is

$$f_{\psi}(y) = f_{\psi}(L)e^{-(2-c_{\psi})k(L-y)/2} . \qquad (10.4.17)$$

This result may be important for explaining the hierarchies among the standard model fermion masses.

Exercise 10.4.4: Show that the fermion bulk mass does not prevent the existence of a massless 0-mode (when the Higgs VEV is neglected).

10.5. More dimensions

Gauge theories in more than four spacetime dimensions are nonrenormalizable. This is not a problem as long as there is a range of scales where the higher-dimensional field theory is valid. For gauge couplings of order unity, as in the Standard model, the range of scales is of the order of $(4\pi)^{2/n}$, so that only low values of n are interesting. Furthermore, the low energy observables get corrections from loops with KK modes. The leading corrections are finite in the $n = 1$ case and logarithmically divergent for $n = 2$, while for $n \geq 3$ they depend quadratically or stronger on the cut-off. Therefore, the effects of the unknown physics above the cut-off scale can be kept under control only for $n = 1$ and $n = 2$.

The case of two universal extra dimensions has been analyzed less extensively compared to $n = 1$ UED. The general features of the standard model in $n = 2$ UED are presented in [28]. The hadron collider phenomenology of (1,0) modes, which are the lightest KK particles, has been explored in Ref. [29]. Cascade decays of spinless adjoints proceed through tree-level 3-body decays involving leptons as well as one-loop 2-body decays involving photons. As a result, spectacular events with as many as six charged leptons, or one photon plus four charged leptons are expected to be observed at the LHC. Unusual events with relatively large branching fractions include three leptons of same charge plus one lepton of opposite charge, or one photon plus two leptons of same charge.

The cascade decays of the (1,1) modes,[28] which are heavier than the (1,0) modes by a factor of $\sqrt{2}$, generate a series of closely-spaced narrow resonances in the $t\bar{t}$ invariant mass distribution.

10.6. Further reading

A thorough presentation of the phenomenology of various extra dimensional models can be found in the TASI lectures of G. Kribs [30]. The set of lectures [31] covers various aspects of model building in extra dimensions, while the set of lectures [35] is devoted to model building in a warped extra dimension and their holographic interpretation. A nice discussion of orbifolds, which are particular cases of boundary conditions, is given in [7]. The implications of universal extra dimensions for dark matter, which continue to be analyzed by many groups, are reviewed in [32]. The distortion of the KK spectrum in the presence of boundary kinetic terms is derived in [33]. The construction of 4D theories which are equivalent to 5D theories involving gauge fields and fermions in the bulk is detailed in [34]. Higgsless models, in which the W and Z masses arise due to the boundary conditions, are presented in [36]; 4D theories of this type are constructed in [37]. A useful review of strongly coupled theories and their relation to extra dimensions is given in [38].

The papers referenced in these notes have been chosen for their pedagogical merits rather than for their original content. Furthermore, there are many interesting papers on models with extra dimensions which have not been mentioned in these lectures. The students are encouraged to find and study them, and more importantly to explore new theories and phenomena.

Acknowledgments

I would like to thank the students who attended my lectures during the 2008 TASI summer school for their many valuable questions and comments. An earlier version of these lectures has been given in December 2006 at Harish-Chandra Research Institute. I am grateful to Tao Han and Biswarup Mukhopadhyaya for their constant encouragements without which these lecture notes would not have been completed.

References

1. *E.g.*, C. H. Hinton, "What is the fourth dimension?", Scientific Romances,

Vol. 1 (1884), pp 1.

2. O. Klein, "Quantum theory and 5-dimensional theory of relativity," Z. Phys. **37**, 895 (1926) [Surveys High Energ. Phys. **5**, 241 (1986)].

3. N. Arkani-Hamed, S. Dimopoulos and G. R. Dvali, "The hierarchy problem and new dimensions at a millimeter," Phys. Lett. B **429**, 263 (1998) [arXiv:hep-ph/9803315].

4. D. J. Kapner, T. S. Cook, E. G. Adelberger, J. H. Gundlach, B. R. Heckel, C. D. Hoyle and H. E. Swanson, "Tests of the gravitational inverse-square law below the dark-energy length scale," Phys. Rev. Lett. **98**, 021101 (2007) [arXiv:hep-ph/0611184].

5. T. Han, J. D. Lykken and R. J. Zhang, "On Kaluza-Klein states from large extra dimensions," Phys. Rev. D **59**, 105006 (1999) [arXiv:hep-ph/9811350]. G. F. Giudice, R. Rattazzi and J. D. Wells, "Quantum gravity and extra dimensions at high-energy colliders," Nucl. Phys. B **544**, 3 (1999) [arXiv:hep-ph/9811291].

6. I. Antoniadis, "A Possible new dimension at a few TeV," Phys. Lett. B **246**, 377 (1990).

7. K. R. Dienes, E. Dudas and T. Gherghetta, "Grand unification at intermediate mass scales through extra dimensions," Nucl. Phys. B **537**, 47 (1999) [arXiv:hep-ph/9806292].

8. K. m. Cheung and G. L. Landsberg, "Kaluza-Klein states of the standard model gauge bosons: Constraints from high energy experiments," Phys. Rev. D **65**, 076003 (2002) [arXiv:hep-ph/0110346].

9. T. Appelquist, H. C. Cheng and B. A. Dobrescu, "Bounds on universal extra dimensions," Phys. Rev. D **64**, 035002 (2001) [arXiv:hep-ph/0012100].

10. L.Randall and R. Sundrum, "A large mass hierarchy from a small extra dimension," Phys. Rev. Lett. **83**, 3370 (1999) [arXiv:hep-ph/9905221].

11. H. Georgi, A. K. Grant and G. Hailu, "Brane couplings from bulk loops," Phys. Lett. B **506**, 207 (2001) [arXiv:hep-ph/0012379].

12. H. C. Cheng, K. T. Matchev and M. Schmaltz, "Radiative corrections to Kaluza-Klein masses," Phys. Rev. D **66**, 036005 (2002) [hep-ph/0204342].

13. C. Macesanu, C. D. McMullen and S. Nandi, "Collider implications of universal extra dimensions," Phys. Rev. D **66**, 015009 (2002) [hep-ph/0201300].

14. J. M. Smillie and B. R. Webber, "Distinguishing spins in supersymmetric and universal extra dimension models at the Large Hadron Collider," JHEP **0510**, 069 (2005) [arXiv:hep-ph/0507170].

15. H. C. Cheng, K. T. Matchev, M. Schmaltz, "Bosonic supersymmetry? Getting fooled at the LHC," Phys. Rev. D **66**, 056006 (2002) [hep-ph/0205314].

16. C. Lin, "A search for universal extra dimensions in the multi-lepton channel from proton antiproton collisions at $\sqrt{s} = 1.8$-TeV", CDF Thesis, 2005.

17. G. Servant and T. M. P. Tait, "Is the lightest Kaluza-Klein particle a viable dark matter candidate?," Nucl. Phys. B **650**, 391 (2003) [hep-ph/0206071]. H. C. Cheng, J. L. Feng and K. T. Matchev, "Kaluza-Klein dark matter," Phys. Rev. Lett. **89**, 211301 (2002) [arXiv:hep-ph/0207125].

18. A. Datta, K. Kong and K. T. Matchev, "Discrimination of supersymmetry and universal extra dimensions at hadron colliders," Phys. Rev. D **72**, 096006

(2005) [Erratum-ibid. D **72**, 119901 (2005)] [arXiv:hep-ph/0509246].

19. J. F. Donoghue, "Introduction to the Effective Field Theory Description of Gravity," arXiv:gr-qc/9512024.

20. M. S. Carena, J. D. Lykken and M. Park, "The interval approach to braneworld gravity," Phys. Rev. D **72**, 084017 (2005) [arXiv:hep-ph/0506305].

21. W. D. Goldberger and M. B. Wise, "Modulus stabilization with bulk fields," Phys. Rev. Lett. **83**, 4922 (1999) [arXiv:hep-ph/9907447].

22. H. Davoudiasl, J. L. Hewett and T. G. Rizzo, "Phenomenology of the Randall-Sundrum Gauge Hierarchy Model," Phys. Rev. Lett. **84**, 2080 (2000) [arXiv:hep-ph/9909255].

23. V. M. Abazov *et al.* [D0 Collaboration], "Search for Randall-Sundrum gravitons with 1 fb^{-1} of data from $p\bar{p}$ collisions at $\sqrt{s} = 1.96$-TeV," Phys. Rev. Lett. **100**, 091802 (2008) [arXiv:0710.3338 [hep-ex]].

24. H. Davoudiasl, J. L. Hewett, T. G. Rizzo, "Experimental probes of localized gravity: on and off the wall," Phys. Rev. D**63**, 075004 (2001) [hep-ph/0006041].

25. H. Davoudiasl, J. L. Hewett and T. G. Rizzo, "Bulk gauge fields in the Randall-Sundrum model," Phys. Lett. B **473**, 43 (2000) [hep-ph/9911262].
A. Pomarol, "Gauge bosons in a 5-dimensional theory with localized gravity," Phys. Lett. B **486**, 153 (2000) [arXiv:hep-ph/9911294].

26. S. Chang, J. Hisano, H. Nakano, N. Okada and M. Yamaguchi, "Bulk standard model in the Randall-Sundrum background," Phys. Rev. D **62**, 084025 (2000) [arXiv:hep-ph/9912498].
T. Gherghetta and A. Pomarol, "Bulk fields and supersymmetry in a slice of AdS," Nucl. Phys. B **586**, 141 (2000) [arXiv:hep-ph/0003129].

27. D. E. Kaplan and T. M. P. Tait, "New tools for fermion masses from extra dimensions," JHEP **0111**, 051 (2001) [arXiv:hep-ph/0110126].

28. G. Burdman, B. A. Dobrescu and E. Ponton, "Resonances from Two Universal Extra Dimensions," Phys. Rev. D **74**, 075008 (2006) [hep-ph/0601186].

29. B. A. Dobrescu, K. Kong, R. Mahbubani, "Leptons and photons at the LHC: cascades through spinless adjoints", JHEP **0707**, 006 (2007) [hep-ph/0703231].

30. G. D. Kribs, "Phenomenology of extra dimensions," arXiv:hep-ph/0605325.

31. A. Perez-Lorenzana, "An introduction to extra dimensions," J. Phys. Conf. Ser. **18**, 224 (2005) [arXiv:hep-ph/0503177].

32. D. Hooper and S. Profumo, "Dark matter and collider phenomenology of universal extra dimensions," Phys. Rept. **453**, 29 (2007) [arXiv:hep-ph/0701197].

33. M. S. Carena, T. M. P. Tait and C. E. M. Wagner, "Branes and orbifolds are opaque," Acta Phys. Polon. B **33**, 2355 (2002) [arXiv:hep-ph/0207056].

34. H. C. Cheng, C. T. Hill, S. Pokorski and J. Wang, "The standard model in the latticized bulk," Phys. Rev. D **64**, 065007 (2001) [arXiv:hep-th/0104179].

35. T. Gherghetta, "Warped models and holography," arXiv:hep-ph/0601213.

36. G. Cacciapaglia, C. Csaki, G. Marandella, J. Terning, "A new custodian for a realistic Higgsless model," Phys. Rev. D**75**, 015003 (2007) [hep-ph/0607146].

37. R. S. Chivukula, B. Coleppa, S. Di Chiara, E. H. Simmons, H. J. He, M. Kurachi and M. Tanabashi, "A three site higgsless model," Phys. Rev. D **74**, 075011 (2006) [arXiv:hep-ph/0607124].
38. C. T. Hill and E. H. Simmons, "Strong dynamics and electroweak symmetry breaking," Phys. Rept. **381**, 235 (2003) [Erratum-ibid. **390**, 553 (2004)] [arXiv:hep-ph/0203079].

PART 4

Neutrino Physics, Astroparticle Physics, and Cosmology

Chapter 11

Neutrinos: Theory

M.C. Gonzalez-Garcia

C.N. Yang Institute for Theoretical Physics
State University of New York at Stony Brook
Stony Brook, NY 11794-3840, USA
and:
Institució Catalana de Recerca i Estudis Avançats (ICREA),
Departament d'Estructura i Constituents de la Matèria,
Universitat de Barcelona,
Diagonal 647, E-08028 Barcelona, Spain

These lectures are a theoretical introduction to the phenomenology of massive neutrinos. First I will present the low energy formalism for adding neutrino masses to the Standard Model and the induced leptonic mixing, and then I will describe the phenomenology associated with neutrino oscillations in vacuum and in matter. I will also comment on the laboratory probes of the absolute neutrino mass scale. Finally I will briefly describe some possible collider signatures associated to neutrino mass models.

11.1. Introduction

It is already five decades since the first neutrino was observed by Cowan and Reines[1] in 1956 in a reactor experiment, and more than seventy five years since its existence was postulated by Wolfgang Pauli,[2] in 1930, in order to reconcile the observed continuous spectrum of nuclear beta decay with energy conservation. It has been a long and winding road that has lead us from these pioneering times to the present overwhelming proof that neutrinos are massive and leptonic flavors are not symmetries of Nature. A road in which both theoretical boldness and experimental ingenuity have walked hand by hand to provide us with the first evidence of physics beyond the Standard Model. From the desperate solution of Pauli to the cathedral-size detectors built to capture and study in detail the elusive particle.

Neutrinos are copiously produced in natural sources: in the burning of the stars, in the interaction of cosmic rays... even as relics of the Big Bang. Starting from the 1960's, neutrinos produced in the sun and in the atmosphere were observed. In 1987, neutrinos from a supernova in the Large Magellanic Cloud were also detected. Indeed an important leading role in this story was played by the neutrinos produced in the sun and in the atmosphere. The experiments that measured the flux of atmospheric neutrinos found results that suggested the disappearance of muon-neutrinos when propagating over distances of order hundreds (or more) kilometers. Experiments that measured the flux of solar neutrinos found results that suggested the disappearance of electron-neutrinos while propagating within the Sun or between the Sun and the Earth.

These results called back to 1968 when Gribov and Pontecorvo[3,4] realized that flavor oscillations arise if neutrinos are massive and mixed. The disappearance of both atmospheric ν_μ's and solar ν_e's was most easily explained in terms of neutrino oscillations. The emerging picture was that at least two neutrinos were massive and mixed, unlike what it is predicted in the Standard Model.

In the last decade this picture became fully established with the upcoming of a set of precise experiments. In particular, during the last five years the results obtained with solar and atmospheric neutrinos have been confirmed in experiments using terrestrial beams in which neutrinos produced in nuclear reactors and accelerators facilities have been detected at distances of the order of hundred kilometers.

Neutrinos were introduced in the Standard Model as truly massless fermions, for which no gauge invariant renormalizable mass term can be constructed. Consequently, in the Standard Model there is neither mixing nor CP violation in the leptonic sector. Therefore, the experimental evidence for neutrino masses and mixing provided an unambiguous signal of new physics.

At present the phenomenology of massive neutrinos is in a very interesting moment. On the one hand many extensions of the Standard Model anticipated ways in which neutrinos may have small, but definitely non-vanishing masses. The better determination of the flavor structure of the leptons at low energies is of vital importance as, at present, it is our only source of positive information to pin-down the high energy dynamics implied by the neutrino masses. Needless to say that its potential will be further expanded and complemented if a positive signal on the absolute value of the mass scale is observed in kinematic searches or or in neutrinoless double beta decay as well as if the observations from a positive evidence in the precision cosmological data.

The purpose of these lectures is to provide a theoretical introduction to the present status of the phenomenology of massive neutrinos. I will present the low energy formalism for adding neutrino masses to the SM and the induced leptonic mixing, and then we describe the phenomenology associated with neutrino oscillations in vacuum and in matter. I will also describe the status of the existing probes to the absolute neutrino mass scale. I will briefly present some expected collider signatures in some of the models which provided a plausible explanation for the observed neutrino masses.

The field of neutrino phenomenology and its forward-looking perspectives is rapidly evolving and these lectures are only a partial introduction. For more details I suggest to consult the review articles, Refs. [5–15], and text books, Refs. [16–22].

11.2. Standard Model of Massless Neutrinos

The greatest success of modern particle physics has been the establishment of the connection between forces mediated by spin-1 particles and local (gauge) symmetries. Within the Standard Model, the strong, weak and electromagnetic interactions are connected to, respectively, $SU(3)$, $SU(2)$ and $U(1)$ gauge groups. The characteristics of the different interactions are explained by the symmetry to which they are related. For example, the way in which the fermions exert and experience each of the forces is determined by their representation under the corresponding symmetry group (or simply their charges in the case of Abelian gauge symmetries).

Once the gauge invariance is elevated to the level of fundamental physics principle, it must be verified by all terms in the Lagrangian, including the mass terms. This, as we will see, has important implications for the neutrino.

The Standard Model (SM) is based on the gauge group

$$G_{\mathrm{SM}} = SU(3)_{\mathrm{C}} \times SU(2)_{\mathrm{L}} \times U(1)_{\mathrm{Y}}, \tag{11.2.1}$$

with three matter fermion generations. Each generation consists of five different representations of the gauge group:

$$\left(1, 2, -\frac{1}{2}\right), \quad \left(3, 2, \frac{1}{6}\right), \quad (1, 1, -1), \quad \left(3, 1, \frac{2}{3}\right), \quad \left(3, 1, -\frac{1}{3}\right) \tag{11.2.2}$$

where the numbers in parenthesis represent the corresponding charges under the group (11.2.1). In this notation the electric charge is given by

$$Q_{\mathrm{EM}} = T_{L3} + Y. \tag{11.2.3}$$

The matter content is shown in Table 11.1, and together with the corresponding gauge fields it constitutes the full list of fields required to describe the observed elementary particle interactions. In fact, these charge assignments have been tested to better than the percent level for the light fermions.[23] The model also contains a single Higgs boson doublet, $\phi = \begin{pmatrix} \phi^+ \\ \phi_0 \end{pmatrix}$ with charges $(1, 2, 1/2)$, whose vacuum expectation value breaks the gauge symmetry,

$$\langle \phi \rangle = \begin{pmatrix} 0 \\ \frac{v}{\sqrt{2}} \end{pmatrix} \quad \Longrightarrow \quad G_{\mathrm{SM}} \to SU(3)_{\mathrm{C}} \times U(1)_{\mathrm{EM}}. \tag{11.2.4}$$

This is the only piece of the SM model which still misses experimental confirmation. Indeed, the search for the Higgs boson, remains one of the premier tasks of present and future high energy collider experiments.

As can be seen in Table 11.1 neutrinos are fermions that have neither strong nor electromagnetic interactions (see Eq. (11.2.3)), *i.e.* they are singlets of $SU(3)_{\mathrm{C}} \times U(1)_{\mathrm{EM}}$. We will refer as *active* neutrinos to neutrinos that, such as those in Table 11.1, reside in the lepton doublets, that is, that have weak interactions. Conversely *sterile* neutrinos are defined as having no SM gauge interactions (their charges are $(1, 1, 0)$), that is, they are singlets of the full SM gauge group.

Table 11.1. Matter contents of the SM.

$L_L(1,2,-\frac{1}{2})$	$Q_L(3,2,\frac{1}{6})$	$E_R(1,1,-1)$	$U_R(3,1,\frac{2}{3})$	$D_R(3,1,-\frac{1}{3})$
$\begin{pmatrix}\nu_e\\e\end{pmatrix}_L$	$\begin{pmatrix}u\\d\end{pmatrix}_L$	e_R	u_R	d_R
$\begin{pmatrix}\nu_\mu\\\mu\end{pmatrix}_L$	$\begin{pmatrix}c\\s\end{pmatrix}_L$	μ_R	c_R	s_R
$\begin{pmatrix}\nu_\tau\\\tau\end{pmatrix}_L$	$\begin{pmatrix}t\\b\end{pmatrix}_L$	τ_R	t_R	b_R

The SM has three active neutrinos accompanying the charged lepton mass eigenstates, e, μ and τ, thus there are weak charged current (CC) interactions between the neutrinos and their corresponding charged leptons given by

$$-\mathcal{L}_{\text{CC}} = \frac{g}{\sqrt{2}} \sum_\ell \bar{\nu}_{L\ell} \gamma^\mu \ell_L^- W_\mu^+ + \text{h.c.}. \qquad (11.2.5)$$

In addition, the SM neutrinos have also neutral current (NC) interactions,

$$-\mathcal{L}_{\text{NC}} = \frac{g}{2\cos\theta_W} \sum_\ell \bar{\nu}_{L\ell} \gamma^\mu \nu_{L\ell} Z_\mu^0. \qquad (11.2.6)$$

The SM as defined in Table 11.1, contains no sterile neutrinos.

Thus, within the SM, Eqs. (11.2.5) and (11.2.6) describe all the neutrino interactions. From Eq. (11.2.6) one can determine the decay width of the Z^0 boson into neutrinos which is proportional to the number of light (that is, $m_\nu \leq m_Z/2$) left-handed neutrinos. At present the measurement of the invisible Z width yields $N_\nu = 2.984 \pm 0.008$[23] which implies that whatever the extension of the SM we want to consider, it must contain three, and only three, light active neutrinos.

An important feature of the SM, which is relevant to the question of the neutrino mass, is the fact that the SM with the gauge symmetry of Eq. (11.2.1) and the particle content of Table 11.1 presents an accidental global symmetry:

$$G_{\text{SM}}^{\text{global}} = U(1)_B \times U(1)_{L_e} \times U(1)_{L_\mu} \times U(1)_{L_\tau}. \qquad (11.2.7)$$

$U(1)_B$ is the baryon number symmetry, and $U(1)_{L_e,L_\mu,L_\tau}$ are the three lepton flavor symmetries, with total lepton number given by $L = L_e + L_\mu + L_\tau$. It is an accidental symmetry because we do not impose it. It is a consequence of the gauge symmetry and the representations of the physical states.

In the SM, fermions masses arise from the Yukawa interactions which couple a right-handed fermion with its left-handed doublet and the Higgs field,

$$-\mathcal{L}_{\text{Yukawa}} = Y^d_{ij}\bar{Q}_{Li}\phi D_{Rj} + Y^u_{ij}\bar{Q}_{Li}\tilde{\phi}U_{Rj} + Y^\ell_{ij}\bar{L}_{Li}\phi E_{Rj} + \text{h.c.}, \quad (11.2.8)$$

(where $\tilde{\phi} = i\tau_2\phi^\star$) which after spontaneous symmetry breaking lead to charged fermion masses

$$m^f_{ij} = Y^f_{ij}\frac{v}{\sqrt{2}}. \quad (11.2.9)$$

However, since no right-handed neutrinos exist in the model, the Yukawa interactions of Eq. (11.2.8) leave the neutrinos massless.

In principle neutrino masses could arise from loop corrections if these corrections induced effective operators of the form

$$\frac{Z^\nu_{ij}}{v}\left(\bar{L}_{Li}\tilde{\phi}\right)\left(\tilde{\phi}^T L^C_{Lj}\right) + \text{h.c.}, \quad (11.2.10)$$

In the SM, however, this cannot happen because this operator violates the total lepton symmetry by two units. As mentioned above total lepton number is a global symmetry of the model and therefore L-violating terms cannot be induced by loop corrections. Furthermore, the $U(1)_{B-L}$ subgroup of $G^{\text{global}}_{\text{SM}}$ is non-anomalous. and therefore $B-L$-violating terms cannot be induced even by nonperturbative corrections.

It follows that the SM predicts that neutrinos are precisely massless. In order to add a mass to the neutrino the SM has to be extended.

11.3. Introducing Massive Neutrinos

As discussed above, with the fermionic content and gauge symmetry of the SM one cannot construct a renormalizable mass term for the neutrinos. So in order to introduce a neutrino mass one must either extend the particle contents of the model or abandon gauge invariance and/or renormalizability.

In what follows we illustrate the different types of neutrino mass terms by assuming that we keep the gauge symmetry and we explore the possibilities that we have to introduce a neutrino mass term if one adds to the SM an arbitrary number m of sterile neutrinos $\nu_{si}(1,1,0)$ or extends the scalar sector.

With the particle contents of the SM and the addition of an arbitrary m number of sterile neutrinos one can construct two types mass terms that arise from gauge invariant renormalizable operators:

$$-\mathcal{L}_{M_\nu} = M_{Dij}\bar{\nu}_{si}\nu_{Lj} + \frac{1}{2}M_{Nij}\bar{\nu}_{si}\nu^c_{sj} + \text{h.c.}. \tag{11.3.11}$$

Here ν^c indicates a charge conjugated field, $\nu^c = C\bar{\nu}^T$ and C is the charge conjugation matrix. M_D is a complex $m \times 3$ matrix and M_N is a symmetric matrix of dimension $m \times m$.

The first term is a Dirac mass term. It is generated after spontaneous electroweak symmetry breaking from Yukawa interactions

$$Y^\nu_{ij}\bar{\nu}_{si}\tilde{\phi}^\dagger L_{Lj} \Rightarrow M_{Dij} = Y^\nu_{ij}\frac{v}{\sqrt{2}} \tag{11.3.12}$$

similarly to the charged fermion masses. It conserves total lepton number but it breaks the lepton flavor number symmetries.

The second term in Eq. (11.3.11) is a Majorana mass term. It is different from the Dirac mass terms in many important aspects. It is a singlet of the SM gauge group. Therefore, it can appear as a bare mass term. Furthermore, since it involves two neutrino fields, it breaks lepton number by two units. More generally, such a term is allowed only if the neutrinos carry no additive conserved charge.

In general Eq. (11.3.11) can be rewritten as:

$$-\mathcal{L}_{M_\nu} = \frac{1}{2}\overline{\vec{\nu}^c}M_\nu\vec{\nu} + \text{h.c.}, \tag{11.3.13}$$

where

$$M_\nu = \begin{pmatrix} 0 & M_D^T \\ M_D & M_N \end{pmatrix}, \tag{11.3.14}$$

and $\vec{\nu} = (\vec{\nu}_L, \vec{\nu}^c_s)^T$ is a $(3+m)$-dimensional vector. The matrix M_ν is complex and symmetric. It can be diagonalized by a unitary matrix of dimension $(3+m)$, V^ν, so that

$$(V^\nu)^T M_\nu V^\nu = \text{Diag}(m_1, m_2, \ldots, m_{3+m}). \tag{11.3.15}$$

In terms of the resulting $3+m$ mass eigenstates

$$\vec{\nu}_{\text{mass}} = (V^\nu)^\dagger \vec{\nu}, \tag{11.3.16}$$

Eq. (11.3.13) can be rewritten as:

$$-\mathcal{L}_{M_\nu} = \frac{1}{2}\sum_{k=1}^{3+m} m_k \left(\bar{\nu}^c_{\text{mass},k}\nu_{\text{mass},k} + \bar{\nu}_{\text{mass},k}\nu^c_{\text{mass},k}\right) = \frac{1}{2}\sum_{k=1}^{3+m} m_k\bar{\nu}_{Mk}\nu_{Mk}, \tag{11.3.17}$$

where

$$\nu_{Mk} = \nu_{\text{mass},k} + \nu_{\text{mass},k}^c = (V^{\nu\dagger}\vec{\nu})_k + (V^{\nu\dagger}\vec{\nu})_k^c \tag{11.3.18}$$

which obey the Majorana condition

$$\nu_M = \nu_M^c \tag{11.3.19}$$

and are refereed to as Majorana neutrinos. Notice that this condition implies that there is only one field which describes both neutrino and antineutrino states. Thus a Majorana neutrino can be described by a two-component spinor unlike the charged fermions, which are Dirac particles, and are represented by four-component spinors.

From Eq. (11.3.18) we find that the weak-doublet components of the neutrino fields are:

$$\nu_{Li} = L \sum_{j=1}^{3+m} V_{ij}^\nu \nu_{Mj} \quad i = 1, 2, 3\,, \tag{11.3.20}$$

where L is the left-handed projector.

In the rest of this section we will discuss three interesting cases.

11.3.1. $M_N = 0$: *Dirac Neutrinos*

Forcing $M_N = 0$ is equivalent to imposing lepton number symmetry on the model. In this case, only the first term in Eq. (11.3.11), the Dirac mass term, is allowed. For $m = 3$ we can identify the three sterile neutrinos with the right-handed component of a four-spinor neutrino field. In this case the Dirac mass term can be diagonalized with two 3×3 unitary matrices, V^ν and V_R^ν as:

$$V_R^{\nu\dagger} M_D V^\nu = \text{Diag}(m_1, m_2, m_3)\,. \tag{11.3.21}$$

The neutrino mass term can be written as:

$$-\mathcal{L}_{M_\nu} = \sum_{k=1}^{3} m_k \bar{\nu}_{Dk} \nu_{Dk} \tag{11.3.22}$$

where

$$\nu_{Dk} = (V^{\nu\dagger}\vec{\nu}_L)_k + (V_R^{\nu\dagger}\vec{\nu}_s)_k\,, \tag{11.3.23}$$

so the weak-doublet components of the neutrino fields are

$$\nu_{Li} = L \sum_{j=1}^{3} V_{ij}^\nu \nu_{Dj}\,, \qquad i = 1, 2, 3\,. \tag{11.3.24}$$

Let us point out that in this case the SM is not even a good low-energy effective theory since both the matter content and the assumed symmetries are different. Furthermore there is no explanation to the fact that neutrino masses happen to be much lighter than the corresponding charged fermion masses as in this case all acquire their mass via the same mechanism.

11.3.2. $M_N \gg M_D$: The Type I see-saw mechanism

In this case the scale of the mass eigenvalues of M_N is much higher than the scale of electroweak symmetry breaking $\langle \phi \rangle$. The diagonalization of M_ν leads to three light, ν_l, and m heavy, N, neutrinos:

$$-\mathcal{L}_{M_\nu} = \frac{1}{2}\bar{\nu}_l M^l \nu_l + \frac{1}{2}\bar{N} M^h N \qquad (11.3.25)$$

with

$$M^l \simeq -V_l^T M_D^T M_N^{-1} M_D V_l, \qquad M^h \simeq V_h^T M_N V_h \qquad (11.3.26)$$

and

$$V^\nu \simeq \begin{bmatrix} \left(1 - \frac{1}{2}M_D^\dagger M_N^{*\,-1}M_N^{-1}M_D\right)V_l & M_D^\dagger M_N^{*\,-1}V_h \\ -M_N^{-1}M_D V_l & \left(1 - \frac{1}{2}M_N^{-1}M_D M_D^\dagger M_N^{*\,-1}\right)V_h \end{bmatrix}$$
$$(11.3.27)$$

where V_l and V_h are 3×3 and $m \times m$ unitary matrices respectively. So the heavier are the heavy states, the lighter are the light ones. This is the *Type I see-saw mechanism*.[24–28] Also as seen from Eq. (11.3.27) the heavy states are mostly right-handed while the light ones are mostly left-handed. Both the light and the heavy neutrinos are Majorana particles. Two well-known examples of extensions of the SM that lead to a Type I see-saw mechanism for neutrino masses are SO(10) GUTs[25–27] and left-right symmetry.[28]

In this case the SM is a good effective low energy theory. Indeed the Type I see-saw mechanism is a particular realization of the general case of a full theory which leads to the SM with three light Majorana neutrinos as its low energy effective realization as we discuss in Sec 11.3.5.

11.3.3. Light sterile neutrinos

This appears if the scale of some eigenvalues of M_N is not higher than the electroweak scale. As in the case with $M_N = 0$, the SM is not even a good low energy effective theory: there are more than three light neutrinos, and they are admixtures of doublet and singlet fields. Again both light and heavy neutrinos are Majorana particles.

As we will see the analysis of neutrino oscillations is the same whether the light neutrinos are of the Majorana- or Dirac-type. From the phenomenological point of view, only in the discussion of neutrinoless double beta decay the question of Majorana versus Dirac neutrinos is crucial. However, as we have tried to illustrate above, from the theoretical model building point of view, the two cases are very different.

11.3.4. *Majorana ν_L masses: Type II see-saw*

In order to be able to construct a gauge invariant neutrino mass term involving only left handed neutrinos one has to extend the Higgs sector of the Standard Model to include besides the doublet ϕ, an $SU(2)_L$ scalar triplet $\Delta \sim (1, 3, 1)$. We write the triplet in the matrix representation as

$$\Delta = \begin{pmatrix} \Delta^0 & -\Delta^+/\sqrt{2} \\ -\Delta^+/\sqrt{2} & -\Delta^{++} \end{pmatrix}. \tag{11.3.28}$$

The neutrino mass term arises from the Lagrangian:

$$\mathcal{L}_Y = -f_{\nu ij}\, \bar{L}^C_{L,i}\Delta\, L_{Lj} \; + \; \text{h.c.}, \tag{11.3.29}$$

When the neutral component of the triple acquires a vev $\langle \Delta_0 \rangle = v_\Delta/\sqrt{2}$, the 3 left handed neutrinos acquire a Majorana mass

$$M_\nu = \sqrt{2} f_\nu\, v_\Delta\,. \tag{11.3.30}$$

It is clear that v_Δ breaks L by two units. If the scalar potential preserves L the breaking is "spontaneous". In this case the model contains a massless Goldstone boson, *the triple Majoron*. Because it is part of a $SU(2)_L$ triplet, the triplet Majoron couples to the Z boson and it would contribute to its invisible decay. At present this is ruled out by the precise measurement of the Z decay width.

It could also be that the scalar potential breaks L explicitly. In this case there is no massless Goldstone boson. This explicit breaking can be induced by a triple-double mixing term in the scalar potential which would contain among others the following two terms:

$$M_\Delta^2 \; \text{Tr}(\Delta^\dagger \Delta) \; + \; \left(\mu\, \tilde{\phi}^T \, \Delta\, \tilde{\phi} \; + \; h.c. \right)$$

In this case the minimization can lead to a vev for the triple

$$v_\Delta = \frac{\mu\, v^2}{\sqrt{2}\, M_\Delta^2}. \tag{11.3.31}$$

So if $M_\Delta^2 \gg \mu v$, then $v_\Delta \ll v$ which gives an explanation to the smallness of the neutrino mass. This mechanism is labeled in the literature as Type II see-saw.[29,30]

11.3.5. *Neutrino Masses from Non-renormalizable Operators*

In general, if the SM is an effective low energy theory valid up to the scale $\Lambda_{\rm NP}$, the gauge group, the fermionic spectrum, and the pattern of spontaneous symmetry breaking of the SM are still valid ingredients to describe Nature at energies $E \ll \Lambda_{\rm NP}$. But because it is an effective theory, one must also consider non-renormalizable higher dimensional terms in the Lagrangian whose effect will be suppressed by powers $1/\Lambda_{\rm NP}^{\dim-4}$. In this approach the largest effects at low energy are expected to come from dim$= 5$ operators.

There is no reason for generic NP to respect the accidental symmetries of the SM (11.2.7). Indeed, there is a single set of dimension-five terms that is made of SM fields and is consistent with the gauge symmetry, and this set violates (11.2.7). It is given by

$$\mathcal{O}_5 = \frac{Z_{ij}^\nu}{\Lambda_{\rm NP}} \left(\bar{L}_{Li} \tilde{\phi} \right) \left(\tilde{\phi}^T L_{Lj}^C \right) + \text{h.c.}, \qquad (11.3.32)$$

which violate total lepton number by two units and leads, upon spontaneous symmetry breaking, to:

$$-\mathcal{L}_{M_\nu} = \frac{Z_{ij}^\nu}{2} \frac{v^2}{\Lambda_{\rm NP}} \bar{\nu}_{Li} \nu_{Lj}^c + \text{h.c.}. \qquad (11.3.33)$$

Comparing with Eq. (11.3.13) we see that this is a Majorana mass term built with the left-handed neutrino fields and with:

$$(M_\nu)_{ij} = Z_{ij}^\nu \frac{v^2}{\Lambda_{\rm NP}}. \qquad (11.3.34)$$

Since Eq. (11.3.34) would arise in a generic extension of the SM, we learn that neutrino masses are very likely to appear if there is NP. As mentioned above, a theory with SM plus m heavy sterile neutrinos leads to three light mass eigenstates and an effective low energy interaction of the form (11.3.32). In particular, the scale $\Lambda_{\rm NP}$ is identified with the mass scale of the heavy sterile neutrinos, that is the typical scale of the eigenvalues of M_N.

Furthermore, comparing Eq. (11.3.34) and Eq. (11.2.9), we find that the scale of neutrino masses is suppressed by $v/\Lambda_{\rm NP}$ when compared to the scale of charged fermion masses providing an explanation not only for the existence of neutrino masses but also for their smallness. Finally, Eq. (11.3.34) breaks not only total lepton number but also the lepton flavor symmetry $U(1)_e \times U(1)_\mu \times U(1)_\tau$. Therefore, as we shall see in Sec. 11.4, we should expect lepton mixing and CP violation unless additional symmetries are imposed on the coefficients Z_{ij}.

11.4. Lepton Mixing

The possibility of arbitrary mixing between two massive neutrino states was first introduced in Ref. [31]. In the general case, we denote the neutrino mass eigenstates by $(\nu_1, \nu_2, \nu_3, \ldots, \nu_n)$ and the charged lepton mass eigenstates by (e, μ, τ). The corresponding interaction eigenstates are denoted by (e^I, μ^I, τ^I) and $\vec{\nu} = (\nu_{Le}, \nu_{L\mu}, \nu_{L\tau}, \nu_{s1}, \ldots, \nu_{sm})$. In the mass basis, leptonic charged current interactions are given by

$$-\mathcal{L}_{\text{CC}} = \frac{g}{\sqrt{2}} (\bar{e}_L, \bar{\mu}_L, \bar{\tau}_L) \gamma^\mu U \begin{pmatrix} \nu_1 \\ \nu_2 \\ \nu_3 \\ \vdots \\ \nu_n \end{pmatrix} W_\mu^+ - \text{h.c..} \qquad (11.4.35)$$

Here U is a $3 \times n$ matrix[30,32,33] which verifies

$$UU^\dagger = I_{3 \times 3} \qquad (11.4.36)$$

but in general $U^\dagger U \neq I_{n \times n}$.

The charged lepton and neutrino mass terms and the neutrino mass in the interaction basis are:

$$-\mathcal{L}_M = [(\bar{e}_L^I, \bar{\mu}_L^I, \bar{\tau}_L^I) M_\ell \begin{pmatrix} e_R^I \\ \mu_R^I \\ \tau_R^I \end{pmatrix} + \text{h.c.}] - \mathcal{L}_{M_\nu} \qquad (11.4.37)$$

with $\mathcal{L}_{M_\nu}$ given in Eq. (11.3.13). One can find two 3×3 unitary diagonalizing matrices for the charge leptons, V^ℓ and V_R^ℓ, such that

$$V^{\ell\dagger} M_\ell V_R^\ell = \text{Diag}(m_e, m_\mu, m_\tau). \qquad (11.4.38)$$

The charged lepton mass term can be written as:

$$-\mathcal{L}_{M_\ell} = \sum_{k=1}^{3} m_{\ell_k} \bar{\ell}_k \ell_k \qquad (11.4.39)$$

where

$$\ell_k = (V^{\ell\,\dagger} \ell_L^I)_k + (V_R^{\ell\,\dagger} \ell_R^I)_k \qquad (11.4.40)$$

so the weak-doublet components of the charge lepton fields are

$$\ell_{Li}^I = L \sum_{j=1}^{3} V_{ij}^\ell \ell_j\,, \qquad i = 1, 2, 3 \qquad (11.4.41)$$

From Eqs. (11.3.20), (11.3.24) and (11.4.41) we find that U is:

$$U_{ij} = P_{\ell,ii}\, V_{ik}^{\ell\,\dagger}\, V_{kj}^\nu\, (P_{\nu,jj}). \qquad (11.4.42)$$

P_ℓ is a diagonal 3×3 phase matrix, that is conventionally used to reduce by three the number of phases in U. P_ν is a diagonal $n \times n$ phase matrix with additional arbitrary phases which can chosen to reduce the number of phases in U by $n - 1$ only for Dirac states. For Majorana neutrinos, this matrix is simply a unit matrix. The reason for that is that if one rotates a Majorana neutrino by a phase, this phase will appear in its mass term which will no longer be real. Thus, the number of phases that can be absorbed by redefining the mass eigenstates depends on whether the neutrinos are Dirac or Majorana particles. Altogether for Majorana [Dirac] neutrinos the U matrix contains a total of $6(n - 2)$ $[5n - 11]$ real parameters, of which $3(n - 2)$ are angles and $3(n - 2)$ $[2n - 5]$ can be interpreted as physical phases.

In particular, if there are only three Majorana neutrinos, U is a 3×3 matrix analogous to the CKM matrix for the quarks[34] but due to the Majorana nature of the neutrinos it depends on six independent parameters: three mixing angles and three phases. In this case the mixing matrix can be conveniently parametrized as:

$$U = \begin{pmatrix} 1 & 0 & 0 \\ 0 & c_{23} & s_{23} \\ 0 & -s_{23} & c_{23} \end{pmatrix} \cdot \begin{pmatrix} c_{13} & 0 & s_{13}e^{-i\delta_{\rm CP}} \\ 0 & 1 & 0 \\ -s_{13}e^{i\delta_{\rm CP}} & 0 & c_{13} \end{pmatrix} \cdot \begin{pmatrix} c_{21} & s_{12} & 0 \\ -s_{12} & c_{12} & 0 \\ 0 & 0 & 1 \end{pmatrix} \cdot \begin{pmatrix} e^{i\eta_1} & 0 & 0 \\ 0 & e^{i\eta_2} & 0 \\ 0 & 0 & 1 \end{pmatrix},$$

$$(11.4.43)$$

where $c_{ij} \equiv \cos\theta_{ij}$ and $s_{ij} \equiv \sin\theta_{ij}$. The angles θ_{ij} can be taken without loss of generality to lie in the first quadrant, $\theta_{ij} \in [0, \pi/2]$ and the phases $\delta_{\rm CP}$, $\eta_i \in [0, 2\pi]$. This is to be compared to the case of three Dirac neutrinos, where the Majorana phases, η_1 and η_2, can be absorbed in the neutrino states and therefore the number of physical phases is one (similarly to the

CKM matrix). In this case the mixing matrix U takes the form:[23]

$$U = \begin{pmatrix} c_{12}\,c_{13} & s_{12}\,c_{13} & s_{13}\,e^{-i\delta_{\rm CP}} \\ -s_{12}\,c_{23} - c_{12}\,s_{13}\,s_{23}\,e^{i\delta_{\rm CP}} & c_{12}\,c_{23} - s_{12}\,s_{13}\,s_{23}\,e^{i\delta_{\rm CP}} & c_{13}\,s_{23} \\ s_{12}\,s_{23} - c_{12}\,s_{13}\,c_{23}\,e^{i\delta_{\rm CP}} & -c_{12}\,s_{23} - s_{12}\,s_{13}\,c_{23}\,e^{i\delta_{\rm CP}} & c_{13}\,c_{23} \end{pmatrix}.$$

$$(11.4.44)$$

Note, however, that the two extra Majorana phases are very hard to measure since they are only physical if neutrino mass is non-zero and therefore the amplitude of any process involving them is suppressed a factor m_ν/E to some power where E is the energy involved in the process which is typically much larger than the neutrino mass. The most sensitive experimental probe of Majorana phases is the rate of neutrinoless $\beta\beta$ decay.

If no new interactions for the charged leptons are present we can identify their interaction eigenstates with the corresponding mass eigenstates after phase redefinitions. In this case the charged current lepton mixing matrix U is simply given by a $3 \times n$ sub-matrix of the unitary matrix V^ν.

It worth noticing that while for the case of 3 light Dirac neutrinos the procedure leads to a fully unitary U matrix for the light states, generically for three light Majorana neutrinos this is not the case when the full spectrum contains heavy neutrino states which have been integrated out as can be seen, from Eq. (11.3.27). Thus, strictly speaking, the parametrization in Eq. (11.4.43) does not hold to describe the flavor mixing of the three light Majorana neutrinos in the type I see-saw mechanism. However, as seen in Eq. (11.3.27), the unitarity violation is of the order $\mathcal{O}(M_D/M_N)$ and it is expected to be very small (at it is also severely constrained experimentally). Consequently in what follows we will ignore this effect.

11.5. Neutrino Oscillations in Vacuum

If neutrinos have masses, the weak eigenstates, ν_α, produced in a weak interaction are, in general, linear combinations of the mass eigenstates ν_i

$$|\nu_\alpha\rangle = \sum_{i=1}^{n} U_{\alpha i}^* |\nu_i\rangle \qquad (11.5.45)$$

where n is the number of light neutrino species and U is the the mixing matrix. (Implicit in our definition of the state $|\nu\rangle$ is its energy-momentum and space-time dependence). After traveling a distance L (or, equivalently for relativistic neutrinos, time t), a neutrino originally produced with a flavor α evolves as:

$$|\nu_\alpha(t)\rangle = \sum_{i=1}^{n} U_{\alpha i}^* |\nu_i(t)\rangle \,, \tag{11.5.46}$$

and it can be detected in the charged-current (CC) interaction $\nu_\alpha(t)N' \to \ell_\beta N$ with a probability

$$P_{\alpha\beta} = |\langle \nu_\beta | \nu_\alpha(t)\rangle|^2 = |\sum_{i=1}^{n}\sum_{j=1}^{n} U_{\alpha i}^* U_{\beta j} \langle \nu_j | \nu_i(t)\rangle|^2 \,, \tag{11.5.47}$$

where E_i and m_i are, respectively, the energy and the mass of the neutrino mass eigenstate ν_i.

Using the standard approximation that $|\nu\rangle$ is a plane wave $|\nu_i(t)\rangle = e^{-i E_i t}|\nu_i(0)\rangle$, that neutrinos are relativistic with $p_i \simeq p_j \equiv p \simeq E$

$$E_i = \sqrt{p_i^2 + m_i^2} \simeq p + \frac{m_i^2}{2E} \tag{11.5.48}$$

and the orthogonality relation $\langle \nu_j | \nu_i \rangle = \delta_{ij}$, we get the following transition probability

$$P_{\alpha\beta} = \delta_{\alpha\beta} - 4 \sum_{i<j}^{n} \mathrm{Re}[U_{\alpha i} U_{\beta i}^* U_{\alpha j}^* U_{\beta j}] \sin^2 X_{ij} + 2 \sum_{i<j}^{n} \mathrm{Im}[U_{\alpha i} U_{\beta i}^* U_{\alpha j}^* U_{\beta j}] \sin 2X_{ij} \,, \tag{11.5.49}$$

where

$$X_{ij} = \frac{(m_i^2 - m_j^2)L}{4E} = 1.27 \frac{\Delta m_{ij}^2}{\mathrm{eV}^2} \frac{L/E}{\mathrm{m/MeV}} \,. \tag{11.5.50}$$

Here $L = t$ is the distance between the production point of ν_α and the detection point of ν_β. The first line in Eq. (11.5.49) is CP conserving while the second one is CP violating and has opposite sign for neutrinos and antineutrinos.

The transition probability, Eq. (11.5.49), has an oscillatory behavior, with oscillation lengths

$$L_{0,ij}^{\mathrm{osc}} = \frac{4\pi E}{\Delta m_{ij}^2} \tag{11.5.51}$$

and amplitudes that are proportional to elements in the mixing matrix. Thus, in order to undergo flavor oscillations, neutrinos must have different masses ($\Delta m_{ij}^2 \neq 0$) and they must mix ($U_{\alpha i} U_{\beta i} \neq 0$). Also, as can be seen from Eq. (11.5.49), the Majorana phases cancel out in the oscillation probability as expected because flavor oscillation is a total lepton number conserving process.

A neutrino oscillation experiment is characterized by the typical neutrino energy E and by the source-detector distance L. But in general, neutrino beams are not monoenergetic and, moreover, detectors have finite energy resolution. Thus, rather than measuring $P_{\alpha\beta}$, the experiments are sensitive to the average probability

$$\langle P_{\alpha\beta}\rangle = \frac{\int dE \frac{d\Phi}{dE}\sigma_{CC}(E)P_{\alpha\beta}(E)\epsilon(E)}{\int dE \frac{d\Phi}{dE}\sigma_{CC}(E)\epsilon(E)}$$

$$= \delta_{\alpha\beta} - 4\sum_{i<j}^{n}\mathrm{Re}[U_{\alpha i}U_{\beta i}^*U_{\alpha j}^*U_{\beta j}]\langle \sin^2 X_{ij}\rangle \qquad (11.5.52)$$

$$+ 2\sum_{i<j}^{n}\mathrm{Im}[U_{\alpha i}U_{\beta i}^*U_{\alpha j}^*U_{\beta j}]\langle \sin 2X_{ij}\rangle\,,$$

where Φ is the neutrino energy spectrum, σ_{CC} is the cross section for the process in which the neutrino is detected (in general, a CC interaction), and $\epsilon(E)$ is the detection efficiency. The range of the energy integral depends on the energy resolution of the experiment.

In order to be sensitive to a given value of Δm_{ij}^2, the experiment has to be set up with $E/L \approx \Delta m_{ij}^2$ ($L \sim L_{0,ij}^{\mathrm{osc}}$). The typical values of L/E for different types of neutrino sources and experiments and the corresponding ranges of Δm^2 to which they can be most sensitive are summarized in Table 11.2.

Generically if $(E/L) \gg \Delta m_{ij}^2$ ($L \ll L_{0,ij}^{\mathrm{osc}}$), the oscillation phase does not have time to give an appreciable effect because $\sin^2 X_{ij} \ll 1$. Conversely if $L \gg L_{0,ij}^{\mathrm{osc}}$, the oscillating phase goes through many cycles before the detection and is averaged to $\langle \sin^2 X_{ij}\rangle = 1/2$. Maximum sensitivity to the oscillation phase – and correspondingly to Δm^2 – is obtained when the set up is such that:

- $E/L \approx \Delta m_{ij}^2$,
- the energy resolution is good enough, $\Delta E \ll L\Delta m_{ij}^2$,
- it is sensitive to different values of L with $\Delta L \ll E/\Delta m^2$.

For a two-neutrino case, the mixing matrix depends on a single parameter,

$$U = \begin{pmatrix} \cos\theta & \sin\theta \\ -\sin\theta & \cos\theta \end{pmatrix}\,, \qquad (11.5.53)$$

and there is a single mass-squared difference Δm^2. Then $P_{\alpha\beta}$ of

Table 11.2. Characteristic values of L and E for various neutrino sources and experiments and the corresponding ranges of Δm^2 to which they can be most sensitive.

Experiment		L (m)	E (MeV)	Δm^2 (eV^2)
Solar		10^{10}	1	10^{-10}
Atmospheric		$10^4 - 10^7$	10^2–10^5	$10^{-1} - 10^{-4}$
Reactor	SBL	$10^2 - 10^3$	1	$10^{-2} - 10^{-3}$
	LBL	$10^4 - 10^5$		$10^{-4} - 10^{-5}$
Accelerator	SBL	10^2	10^3–10^4	> 0.1
	LBL	$10^5 - 10^6$	10^4	$10^{-2} - 10^{-3}$

Eq. (11.5.49) takes the well-known form

$$P_{\alpha\beta} = \delta_{\alpha\beta} - (2\delta_{\alpha\beta} - 1)\sin^2 2\theta \sin^2 X \,. \tag{11.5.54}$$

The physical parameter space is covered with $\Delta m^2 \geq 0$ and $0 \leq \theta \leq \frac{\pi}{2}$ (or, alternatively, $0 \leq \theta \leq \frac{\pi}{4}$ and either sign for Δm^2).

Changing the sign of the mass difference, $\Delta m^2 \to -\Delta m^2$, and changing the octant of the mixing angle, $\theta \to \frac{\pi}{2} - \theta$, amounts to redefining the mass eigenstates, $\nu_1 \leftrightarrow \nu_2$: $P_{\alpha\beta}$ must be invariant under such transformation. Eq. (11.5.54) reveals, however, that $P_{\alpha\beta}$ is actually invariant under each of these transformations separately. This situation implies that there is a two-fold discrete ambiguity in the interpretation of $P_{\alpha\beta}$ in terms of two-neutrino mixing: the two different sets of physical parameters, $(\Delta m^2, \theta)$ and $(\Delta m^2, \frac{\pi}{2} - \theta)$, give the same transition probability in vacuum. One cannot tell from a measurement of, say, $P_{e\mu}$ in vacuum whether the larger component of ν_e resides in the heavier or in the lighter neutrino mass eigenstate. This symmetry is lost when neutrinos travel through regions of dense matter and/or for when there are more than two neutrinos mixed in the neutrino evolution.

11.6. Propagation of Massive Neutrinos in Matter

When neutrinos propagate in dense matter, the interactions with the medium affect their properties. These effects can be either coherent or incoherent. For purely incoherent inelastic ν-p scattering, the characteristic cross section is very small:

$$\sigma \sim \frac{G_F^2 s}{\pi} \sim 10^{-43} \text{ cm}^2 \left(\frac{E}{\text{MeV}}\right)^2 \,. \tag{11.6.55}$$

On the contrary, in coherent interactions, the medium remains unchanged and it is possible to have interference of scattered and unscattered neutrino waves which enhances the effect. Coherence further allows one to decouple the evolution equation of the neutrinos from the equations of the medium. In this approximation, the effect of the medium is described by an effective potential which depends on the density and composition of the matter.[35]

Taking this into account, the evolution equation for n ultrarelativistic neutrinos propagating in matter written in the mass basis can be casted in the following form (there are several derivations in the literature of the evolution equation of a neutrino system in matter, see for instance Ref. [36–38]):

$$i\frac{d\vec{\nu}}{dx} = H\,\vec{\nu}, \qquad H = H_m + U^{\nu\dagger}\,V\,U^{\nu}, \qquad (11.6.56)$$

where $\vec{\nu} \equiv (\nu_1, \nu_2, \ldots, \nu_n)^T$, H_m is the Hamiltonian for the kinetic energy,

$$H_m = \frac{1}{2E}\mathrm{Diag}(m_1^2, m_2^2, \ldots, m_n^2), \qquad (11.6.57)$$

and V is the effective potential that describes the coherent forward interactions of the neutrinos with matter in the interaction basis. U^{ν} is the $n \times n$ submatrix of the unitary V^{ν} matrix corresponding to the n ultrarelativistic neutrino states.

Let's consider the evolution of ν_e in a medium with electrons, protons and neutrons with corresponding n_e, n_p and n_n number densities. The effective low-energy Hamiltonian describing the relevant neutrino interactions is given by

$$H_W = \frac{G_F}{\sqrt{2}}\left[J^{(+)\alpha}(x)J_\alpha^{(-)}(x) + \frac{1}{4}J^{(N)\alpha}(x)J_\alpha^{(N)}(x)\right], \qquad (11.6.58)$$

where the J_α's are the standard fermionic currents

$$J_\alpha^{(+)}(x) = \bar{\nu}_e(x)\gamma_\alpha(1 - \gamma_5)e(x)\,, \qquad (11.6.59)$$

$$J_\alpha^{(-)}(x) = \bar{e}(x)\gamma_\alpha(1 - \gamma_5)\nu_e(x)\,, \qquad (11.6.60)$$

$$\begin{aligned}
J_\alpha^{(N)}(x) = \ &\bar{\nu}_e(x)\gamma_\alpha(1 - \gamma_5)\nu_e(x) \\
&- \bar{e}(x)[\gamma_\alpha(1 - \gamma_5) - 4\sin^2\theta_W\gamma_\alpha]e(x) \\
&+ \bar{p}(x)[\gamma_\alpha(1 - g_A^{(p)}\gamma_5) - 4\sin^2\theta_W\gamma_\alpha]p(x) \\
&- \bar{n}(x)\gamma_\alpha(1 - g_A^{(n)}\gamma_5)n(x)\,,
\end{aligned} \qquad (11.6.61)$$

and $g_A^{(n,p)}$ are the axial couplings for neutrons and protons, respectively.

Consider first the effect of the charged current interactions. The effective CC Hamiltonian due to electrons in the medium is

$$H_C^{(e)} = \frac{G_F}{\sqrt{2}} \int d^3 p_e f(E_e, T)$$
$$\times \Big\langle \langle e(s, p_e) | \bar{e}(x)\gamma^\alpha(1 - \gamma_5)\nu_e(x)\bar{\nu}_e(x)\gamma_\alpha(1 - \gamma_5)e(x) | e(s, p_e) \rangle \Big\rangle$$
$$= \frac{G_F}{\sqrt{2}} \bar{\nu}_e(x)\gamma_\alpha(1 - \gamma_5)\nu_e(x)$$
$$\int d^3 p_e f(E_e, T) \Big\langle \langle e(s, p_e) | \bar{e}(x)\gamma_\alpha(1 - \gamma_5)e(x) | e(s, p_e) \rangle \Big\rangle \,,$$

$$(11.6.62)$$

where s is the electron spin and p_e its momentum. The energy distribution function of the electrons in the medium, $f(E_e, T)$, is assumed to be homogeneous and isotropic and is normalized as

$$\int d^3 p_e f(E_e, T) = 1 \,. \qquad (11.6.63)$$

By $\langle \ldots \rangle$ we denote the averaging over electron spinors and summing over all electrons in the medium. Notice that coherence implies that s, p_e are the same for initial and final electrons. To calculate the averaging we notice that the axial current reduces to the spin in the non-relativistic limit and therefore averages to zero for a background of non-relativistic electrons. The spatial components of the vector current cancel because of isotropy and therefore the only non trivial average is

$$\int d^3 p_e f(E_e, T) \Big\langle \langle e(s, p_e) | \bar{e}(x)\gamma_0 e(x) | e(s, p_e) \rangle \Big\rangle = n_e(x) \qquad (11.6.64)$$

which gives a contribution to the effective Hamiltonian

$$H_C^{(e)} = \sqrt{2} G_F n_e \bar{\nu}_{eL}(x)\gamma_0 \nu_{eL}(x) \,. \qquad (11.6.65)$$

This can be interpreted as a contribution to the ν_{eL} potential energy

$$V_C = \sqrt{2} G_F n_e \,. \qquad (11.6.66)$$

A more detailed derivation of the matter potentials can be found, for example, in Ref. [20].

For ν_μ and ν_τ, the potential due to its CC interactions is zero for most media since neither μ's nor $\tau's$ are present.

In the same fashion one can derive the effective potential for any active neutrino due to the neutral current interactions to be

$$V_{NC} = \frac{\sqrt{2}}{2} G_F \left[-n_e(1 - 4\sin^2\theta_w) + n_p(1 - 4\sin^2\theta_w) - n_n \right] \,. \qquad (11.6.67)$$

For neutral matter $n_e = n_p$ so the contribution from electrons and protons cancel each other and we are left only with the neutron contribution

$$V_{NC} = -1/\sqrt{2}G_F n_n \qquad (11.6.68)$$

Altogether we can write the evolution equation for the three SM active neutrinos with purely SM interactions in a neutral medium with electrons, protons and neutrons as Eq. (11.6.56) with $U^\nu \equiv U$, and the effective potential:

$$V = \text{Diag}\left(\pm\sqrt{2}G_F n_e(x),\, 0,\, 0\right) \equiv \text{Diag}\left(V_e,\, 0,\, 0\right). \qquad (11.6.69)$$

In Eq. (11.6.69), the sign $+$ $(-)$ refers to neutrinos (antineutrinos), and $n_e(x)$ is the electron number density in the medium, which in general changes along the neutrino trajectory and so does the potential. For example, at the Earth core $V_e \sim 10^{-13}$ eV while at the solar core $V_e \sim 10^{-12}$ eV. Notice that the neutral current potential Eq. (11.6.68) is flavor diagonal and therefore it can be eliminated from the evolution equation as it only contributes to an overall phase which is unobservable.

The instantaneous mass eigenstates in matter, ν_i^m, are the eigenstates of H for a fixed value of x, which are related to the interaction basis by

$$\vec{\nu} = \tilde{U}(x)\vec{\nu^m}, \qquad (11.6.70)$$

while $\mu_i(x)^2/(2E)$ are the corresponding instantaneous eigenvalues with $\mu_i(x)$ being the instantaneous effective neutrino masses.

For the simplest case of the evolution of a neutrino state which is an admixture of only two neutrino species $|\nu_\alpha\rangle$ and $|\nu_\beta\rangle$

$$\mu_{1,2}^2(x) = \frac{m_1^2 + m_2^2}{2} + E[V_\alpha + V_\beta] \mp \frac{1}{2}\sqrt{[\Delta m^2 \cos 2\theta - A]^2 + [\Delta m^2 \sin 2\theta]^2},$$
$$(11.6.71)$$

and $\tilde{U}(x)$ can be written as Eq. (11.5.53) with the instantaneous mixing angle in matter given by

$$\tan 2\theta_m = \frac{\Delta m^2 \sin 2\theta}{\Delta m^2 \cos 2\theta - A}. \qquad (11.6.72)$$

The quantity A is defined by

$$A \equiv 2E(V_\alpha - V_\beta). \qquad (11.6.73)$$

Notice that for a given sign of A (which depends on the composition of the medium and on the flavor composition of the neutrino state) the mixing angle in matter is larger or smaller than in vacuum depending on whether this last one lies on the first or the second octant. Thus the symmetry present in vacuum oscillations is broken by matter potentials.

Generically matter effects are important when for some of the states the corresponding potential difference factor, A, is comparable to their mass difference term $\Delta m^2 \cos 2\theta$. Most relevant, as seen in Eq. (11.6.72), the mixing angle $\tan \theta_m$ changes sign if in some point along its path the neutrino passes by some matter density region verifying the *resonance condition*

$$A_R = \Delta m^2 \cos 2\theta \,. \tag{11.6.74}$$

Thus if the neutrino is created in a region where the relevant potential verifies $A_0 > A_R$, then the effective mixing angle in matter at the production point verifies that $\mathrm{sgn}(\cos 2\theta_{m,0}) = -\,\mathrm{sgn}(\cos 2\theta)$, this is, the flavor component of the mass eigenstates is inverted as compared to their composition in vacuum. For example for $A_0 = 2A_R$ $\theta_{m,0} = \frac{\pi}{2} - \theta$. Asymptotically, for $A_0 \gg A_R$, $\theta_{m,0} \to \frac{\pi}{2}$.

In other words, if in vacuum the lightest mass eigenstate has a larger projection on the flavor α while the heaviest has it on the flavor β, once inside a matter potential with $A > A_R$ the opposite holds. Thus for a neutrino system which is traveling across a monotonically varying matter potential the dominant flavor component of a given mass eigenstate changes when crossing the region with $A = A_R$. This phenomenon is known as *level crossing*.

In the instantaneous mass basis the evolution equation reads:

$$i\frac{d\vec{\nu}^m}{dx} = \left[\frac{1}{2E} \,\mathrm{Diag}\left(\mu_1^2(x),\, \mu_2^2(x),\, \ldots,\, \mu_n^2(x) \right) - i\,\tilde{U}^\dagger(x)\,\frac{d\tilde{U}(x)}{dx} \right] \vec{\nu}^m \,. \tag{11.6.75}$$

Because of the last term, Eq. (11.6.75) constitute a system of coupled equations which implies that the instantaneous mass eigenstates, ν_i^m, mix in the evolution and are not energy eigenstates. For constant or slowly enough varying matter potential this last term can be neglected. In this case the instantaneous mass eigenstates, ν_i^m, behave approximately as energy eigenstates and they do not mix in the evolution. This is the *adiabatic* transition approximation. On the contrary, when the last term in Eq. (11.6.75) cannot be neglected, the instantaneous mass eigenstates mix along the neutrino path so there can be *level-jumping*[39,40] and the evolution is *non-adiabatic*.

The oscillation probability takes a particularly simple form for adiabatic evolution in matter and it can be cast very similarly to the vacuum oscillation expression, Eq. (11.5.49). For example, neglecting CP violation:

$$P_{\alpha\beta} = \left| \sum_i \tilde{U}_{\alpha i}(0)\tilde{U}_{\beta i}(L) \exp\left(-\frac{i}{2E} \int_0^L \mu_i^2(x')dx' \right) \right|^2 . \qquad (11.6.76)$$

In general $P_{\alpha\beta}$ has to be evaluated numerically although there exist in the literature several analytical approximations for specific profiles of the matter potential.[41]

11.6.1. *The MSW Effect for Solar Neutrinos*

As an illustration of the matter effects discussed in the previous section we describe now the propagation of a $\nu_e - \nu_X$ neutrino system in the matter density of the Sun where X is some superposition of μ and τ.

The solar density distribution decreases monotonically with the distance R to the center of the Sun. For $R < 0.9R_\odot$ it can be approximated by an exponential

$$n_e(R) = n_e(0) \exp\left(-R/r_0\right) \qquad (11.6.77)$$

with $r_0 = R_\odot/10.54 = 6.6 \times 10^7$ m $= 3.3 \times 10^{14}$ eV^{-1}. After traversing this density the dominant component of the exiting neutrino state depends on the value of the mixing angle in vacuum, and on the relative size of $\Delta m^2 \cos 2\theta$ versus $A_0 = 2\,E\,G_F\,n_{e,0}$ (at the neutrino production point) as we describe next:

- If $\Delta m^2 \cos 2\theta \gg A_0$ matter effects are negligible and the propagation occurs as in vacuum with the oscillating phase averaged out due to the large value of L. In this case the survival probability at the sunny surface of the Earth is

$$P_{ee}(\Delta m^2 \cos 2\theta \gg A_0) = 1 - \frac{1}{2}\sin^2 2\theta > \frac{1}{2} . \qquad (11.6.78)$$

- If $\Delta m^2 \cos 2\theta \gtrsim A_0$ the neutrino does not pass any resonance region but its mixing is affected by the solar matter. This effect is well described by an adiabatic propagation, Eq. (11.6.76). Using

$$\tilde{U}(0) = \begin{pmatrix} \cos\theta_{m,0} & \sin\theta_{m,0} \\ -\sin\theta_{m,0} & \cos\theta_{m,0} \end{pmatrix} , \qquad \tilde{U}(L) = \begin{pmatrix} \cos\theta & \sin\theta \\ -\sin\theta & \cos\theta \end{pmatrix} ,$$

$$(11.6.79)$$

(where $\theta_{m,0}$ is the mixing angle in matter at the production point) we get

$$P_{ee} = \cos^2\theta_{m,0}\cos^2\theta + \sin^2\theta_{m,0}\sin^2\theta$$

$$+ \frac{1}{2}\sin^2 2\theta_{m,0}\sin^2 2\theta \cos\left(\frac{\int_0^L \mu_2^2(x') - \mu_1^2(x')}{2E}dx'\right). \qquad (11.6.80)$$

For all practical purposes, the oscillation term in Eq. (11.6.80) is averaged out in the regime $\Delta m^2 \cos 2\theta \gtrsim A_0$ and then the resulting probability reads

$$P_{ee}(\Delta m^2 \cos 2\theta \geq A_0) = \cos^2\theta_{m,0}\cos^2\theta + \sin^2\theta_{m,0}\sin^2\theta$$

$$= \frac{1}{2}\left[1 + \cos 2\theta_{m,0}\cos 2\theta\right]. \qquad (11.6.81)$$

The physical interpretation of this expression is straightforward. An electron neutrino produced at A_0 consists of an admixture of ν_1 with fraction $P_{e1,0} = \cos^2\theta_{m,0}$ and ν_2 with fraction $P_{e2,0} = \sin^2\theta_{m,0}$. At the exit ν_1 consists of ν_e with fraction $P_{1e} = \cos^2\theta$ and ν_2 consists of ν_e with fraction $P_{2e} = \sin^2\theta$ so[42-44]

$$P_{ee} = P_{e1,0}P_{1e} + P_{e2,0}P_{2e} \qquad (11.6.82)$$

which reproduces Eq. (11.6.81). Notice that as long as $A_0 < A_R$ the resonance is not crossed and consequently $\cos 2\theta_{m,0}$ has the same sign as $\cos 2\theta$ and the corresponding survival probability is also larger than $1/2$.

- If $\Delta m^2 \cos 2\theta < A_0$ the neutrino can cross the resonance on its way out if, in the convention of positive Δm^2, $\cos 2\theta > 0$ ($\theta < \pi/4$). In this case, at the production point ν_e is a combination of ν_1^m and ν_2^m with larger ν_2^m component while outside of the Sun the opposite holds. More quantitatively for $\Delta m^2 \cos 2\theta \ll A_0$ (density at the production point much higher than the resonant density),

$$\theta_{m,0} = \frac{\pi}{2} \quad \Rightarrow \quad \cos 2\theta_{m,0} = -1. \qquad (11.6.83)$$

Depending on the particular values of Δm^2 and the mixing angle, the evolution can be adiabatic or non-adiabatic. Presently we know that the oscillation parameters are such that the transition is indeed adiabatic for all ranges of solar neutrino energies. Thus the survival probability at the sunny surface of the Earth is

$$P_{ee}(\Delta m^2 \cos 2\theta < A_0) = \frac{1}{2}\left[1 + \cos 2\theta_{m,0}\cos 2\theta\right] = \sin^2\theta \qquad (11.6.84)$$

where we have used Eq. (11.6.83). Thus in this regime P_{ee} can be much smaller than $1/2$ because $\cos 2\theta_{m,0}$ and $\cos 2\theta$ have opposite signs. This is the MSW effect[35,45] which plays a crucial role in the interpretation of the solar neutrino data.

11.7. Global 3ν Analysis of Oscillation Data

In these lectures I am not going to discuss the experimental situation on neutrino oscillation searches which will be covered in other lectures. However, for consistency, I am including in these notes the output of a recent combined global combined analysis of the present oscillation search experiments.[15,46]

The minimum joint description of solar, atmospheric, long-baseline, and reactor data, requires that all three known neutrinos take part in the oscillations. In this case, the mixing parameters are encoded in the 3×3 lepton mixing matrix[31,34] which can be conveniently parametrized in the standard form of Eq. (11.4.44), since the two Majorana phases in Eq. (11.4.43) do not affect neutrino oscillations.

The results of the global combined analysis including all dominant and subdominant oscillation effects are summarized in Fig. 11.1 and Fig. 11.2 in which we show different projections of the allowed 6-dimensional parameter space.

In Fig. 11.1 we plot the correlated bounds from the global analysis several pairs of parameters. The regions in each panel are obtained after marginalization of χ^2_{global} with respect to the three undisplayed parameters. The different contours correspond to regions defined at 90%, 95%, 99% and 3σ CL for 2 d.o.f. ($\Delta\chi^2 = 4.61$, 5.99, 9.21, 11.83) respectively. From the figure we see that the stronger correlation appears between θ_{13} and Δm^2_{31} as a reflection of the CHOOZ bound. In the lower panels we show the allowed regions in the ($\sin^2 \theta_{13}$, δ_{CP}) plane. As seen in the figure, the sensitivity to the CP phase at present is marginal but we find that the present bound on $\sin^2 \theta_{13}$ can vary by about $\sim 30\%$ depending on the exact value of δ_{CP}. This is a pure 3ν mixing effects and arises from the interference of θ_{13} and Δm^2_{21} effects in the atmospheric neutrino observables. To illustrate this point we plot in the same panel the bounds on $\sin^2 \theta_{13}$ at 90%, 95%, 99% and 3σ CL if the atmospheric data is not included in the analysis (the vertical lines).

In Fig. 11.2 we plot the individual bounds on each of the six relevant parameters derived from the global analysis (full line). To illustrate the impact of the LBL and KamLAND data we also show the corresponding

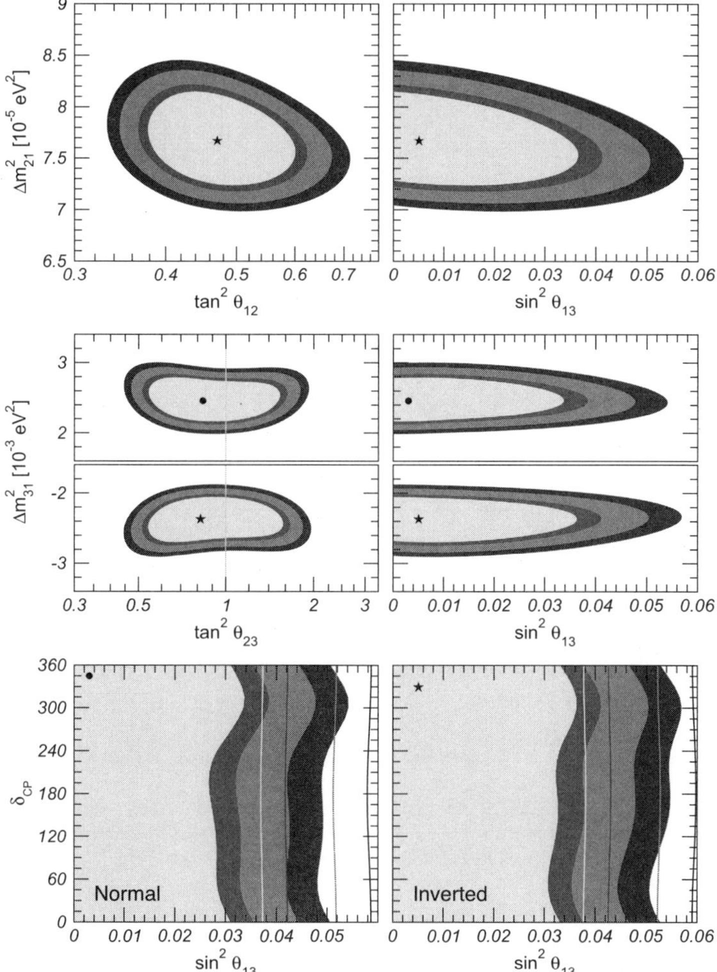

Fig. 11.1. Global 3ν oscillation analysis. Each panels shows 2-dimensional projection of the allowed 5-dimensional region after marginalization with respect to the undisplayed parameters. The different contours correspond to the two-dimensional allowed regions at 90%, 95%, 99% and 3σ CL. In the lowest panel the vertical lines correspond to the regions without inclusion of the atmospheric neutrino data.

bounds when KamLAND and the LBL data are not included in the analysis respectively. In each panel, except the lower left one, the displayed χ^2 has been marginalized with respect to the other five parameters. The lower left panel shows the χ^2 (marginalized over all parameters but θ_{13}) dependence of $\delta_{\rm CP}$ for fixed values of θ_{13}.

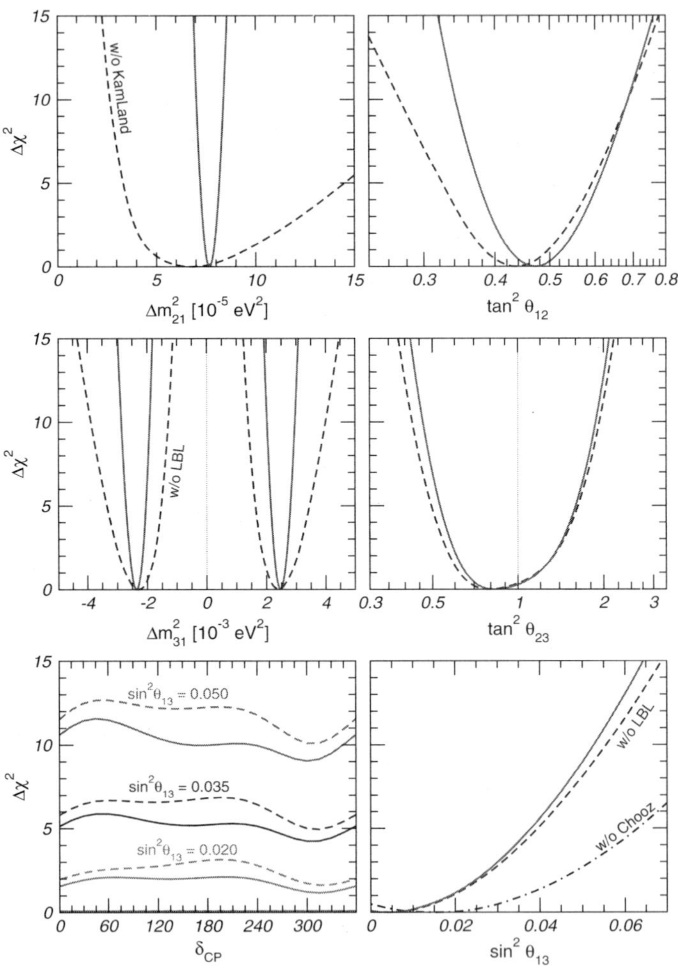

Fig. 11.2.   Global 3ν oscillation analysis. Each panels shows the dependence of $\Delta\chi^2$ on each of the parameters from the global analysis (full line) compared to the bound prior to the KamLAND data (dashed lines in the first row) and LBL data (dashed lines in the second rows). In the lowest panel we show the dependence of $\Delta\chi^2$ on θ_{13} for different sets of data as labeled in the curves. The individual 1σ (3σ) bounds in Eqs. (11.7.85) can be read from the corresponding panel with the condition $\Delta\chi^2 \leq 1$ (9).

The derived ranges at 1σ (3σ) are:

$$\Delta m_{21}^2 = 7.67\,{}^{+0.22}_{-0.21}\,\left({}^{+0.67}_{-0.61}\right) \times 10^{-5}\ \mathrm{eV}^2\,,$$

$$\Delta m_{31}^2 = \begin{cases} -2.39 \pm 0.12\,\left({}^{+0.37}_{-0.40}\right) \times 10^{-3}\ \mathrm{eV}^2 & \text{(inverted hierarchy)}\,, \\[2mm] +2.49 \pm 0.12\,\left({}^{+0.36}_{-0.39}\right) \times 10^{-3}\ \mathrm{eV}^2 & \text{(normal hierarchy)}\,, \end{cases}$$

$$\sin^2\theta_{12} = 0.321 \pm 0.023\,\left({}^{+0.08}_{-0.06}\right)\,,$$

$$\sin^2\theta_{23} = 0.47\,{}^{+0.07}_{-0.06}\,\left({}^{+0.17}_{-0.14}\right)\,,$$

$$\sin^2\theta_{13} = 0.003 \pm 0.015\,\left({}^{+.049}_{-0.0}\right)\,,$$

$$\delta_{\mathrm{CP}} \in [0,\,360]\,.$$

(11.7.85)

Figure 11.2 illustrates that the dominant effect of the inclusion of the laboratory experiments KamLAND, K2K and MINOS is the better determination of the corresponding mass differences Δm_{21}^2 and Δm_{31}^2 while the mixing angle θ_{12} is dominantly determined by the solar data and the mixing angle θ_{23} is still most precisely measured in the atmospheric neutrino experiments. The non-maximality of the best θ_{23} observed in Eq. (11.7.85) is a pure 3-ν oscillation effect associated to the inclusion of the Δm_{21}^2 effects in the atmospheric neutrino analysis.

The values on the magnitude of the elements of the leptonic mixing matrix, at 90% CL are:

$$|U|_{90\%} = \begin{pmatrix} 0.80 \to 0.84 & 0.53 \to 0.60 & 0.00 \to 0.17 \\ 0.29 \to 0.52 & 0.51 \to 0.69 & 0.61 \to 0.76 \\ 0.26 \to 0.50 & 0.46 \to 0.66 & 0.64 \to 0.79 \end{pmatrix} \qquad (11.7.86)$$

By construction these limits are obtained under the assumption that U is unitary. In other words, the ranges in the different entries of the matrix are correlated due to the fact that, in general, the result of a given experiment restricts a combination of several entries of the matrix, as well as to the constraints imposed by unitarity. As a consequence choosing a specific value for one element further restricts the range of the others. For tests of the unitarity of the lepton mixing see [47].

11.8. Direct Determination of m_ν: Kinematic Constraints

Oscillation experiments have provided us with important information on the differences between the neutrino masses-squared, Δm_{ij}^2, and on the leptonic mixing angles, U_{ij}. But they are insensitive to the absolute mass scale for the neutrinos, m_i.

Of course, the results of an oscillation experiment do provide a lower bound on the heavier mass in Δm_{ij}^2, $|m_i| \geq \sqrt{\Delta m_{ij}^2}$ for $\Delta m_{ij}^2 > 0$. But there is no upper bound on this mass. In particular, the corresponding neutrinos could be approximately degenerate at a mass scale that is much higher than $\sqrt{\Delta m_{ij}^2}$. Moreover, there is neither upper nor lower bound on the lighter mass m_j. In this section we briefly summarize the most sensitive probes of the absolute mass scale for the neutrinos.

It was Fermi who first proposed a kinematic search for the neutrino mass from the hard part of the beta spectra in ^{3}H beta decay $^3\text{H} \rightarrow {}^3\text{He} + e^- + \bar{\nu}_e$. In the absence of leptonic mixing this search provides a measurement of the electron neutrino mass.

^{3}H beta decay is a superallowed transition, which means that the nuclear matrix elements do not generate any energy dependence, so that the electron spectrum is given by the phase space alone

$$\frac{dN}{dE} = C\,p\,E\,(Q - T)\sqrt{(Q - T)^2 - m_{\nu_e}^2}\,F(E) \equiv R(E)\sqrt{(E_0 - E)^2 - m_{\nu_e}^2}.$$
$$(11.8.87)$$

where $E = T + m_e$ is the total electron energy, p its momentum, $Q \equiv E_0 - m_e$ is the maximum kinetic energy of the electron and $F(E)$ is the Fermi function which incorporates final state Coulomb interactions. In the second equality we have included in a $R(E)$ all the m_ν-independent factors.

Plotted in terms of the Curie function $K(T) \equiv \sqrt{\frac{dN}{dE}\frac{1}{pEF(E)}}$ a non-vanishing neutrino mass m_ν provokes a distortion from the straight-line T-dependence at the end point: for $m_\nu = 0 \Rightarrow T_{\max} = Q$ whereas for $m_{\nu_e} \neq 0 \Rightarrow T_{\max} = Q - m_{\nu_e}$. ^{3}H beta decay has a a very small energy release $Q = 18.6\,\text{KeV}$ which makes it particularly sensitive to this kinematic effect.

At present the most precise determination from the Mainz[48] and Troitsk[49] experiments give no indication in favor of $m_{\nu_e} \neq 0$ and one sets an upper limit

$$m_{\nu_e} < 2.2\,\text{eV} \qquad (11.8.88)$$

at 95% confidence level (CL). For the other flavors the present limits are[23]

$$m_{\nu_\mu} < 190\,\text{keV}\,(90\%\text{CL}) \qquad \text{from} \qquad \pi^- \rightarrow \mu^- + \bar{\nu}_\mu, \qquad (11.8.89)$$
$$m_{\nu_\tau} < 18.2\,\text{MeV}\,(95\%\text{CL}) \qquad \text{from} \qquad \tau^- \rightarrow n\pi + \nu_\tau. \qquad (11.8.90)$$

In the presence of mixing these limits have to be modified and in general they involve more than one flavor parameter. For neutrinos with small mass differences the distortion of the beta spectrum is given by the weighted sum of the individual spectra:[50]

$$\frac{dN}{dE} = R(E) \sum_i |U_{ei}|^2 \sqrt{(E_0 - E)^2 - m_i^2}\, \Theta(E_0 - E - m_i)\,. \qquad (11.8.91)$$

The step function, $\Theta(E_0 - E - m_i)$, reflects the fact that a given neutrino can only be produced if the available energy is larger than its mass. According to Eq. (11.8.91), there are two important effects, sensitive to the neutrino masses and mixings, on the electron energy spectrum: (i) Kinks at the electron energies $E_e^{(i)} = E \sim E_0 - m_i$ with sizes that are determined by $|U_{ei}|^2$; (ii) A shift of the end point to $E_{\rm ep} = E_0 - m_1$, where m_1 is the lightest neutrino mass. The situation is slightly more involved when the finite energy resolution of the experiment is considered.[51,52]

In general for most realistic situations the distortion of the spectrum can be effectively described by a single parameter, m_β if for all neutrino states $E_0 - E = Q_0 - T \gg m_i$. In this case one can expand Eq. (11.8.91) as:

$$\frac{dN}{dE} \simeq R(E) \sum_i |U_{ei}|^2 (E_0 - E) \left(1 - \frac{m_i^2}{2(E_0 - E)}\right)$$

$$= R(E) \sum_i |U_{ei}|^2 (E_0 - E) \left(1 - \frac{1}{2(E_0 - E)} \frac{\sum_i |U_{ei}|^2 m_i^2}{\sum_i |U_{ei}|^2}\right) \qquad (11.8.92)$$

$$\simeq R(E) \sum_i |U_{ei}|^2 \sqrt{(E_0 - E)^2 - m_\beta^2}\,,$$

with

$$m_\beta^2 = \frac{\sum_i m_i^2 |U_{ei}|^2}{\sum_i |U_{ei}|^2} = \sum_i m_i^2 |U_{ei}|^2\,, \qquad (11.8.93)$$

where the second equality holds if unitarity is assumed. So the distortion of the end point of the spectrum is described by a single parameter which is bounded to be

$$m_\beta = \sqrt{\sum_i m_i^2 |U_{ei}|^2} < 2.2 \text{ eV}\,. \qquad (11.8.94)$$

A new experimental project, KATRIN,[53] is under construction with an estimated sensitivity limit: $m_\beta \sim 0.3$ eV.

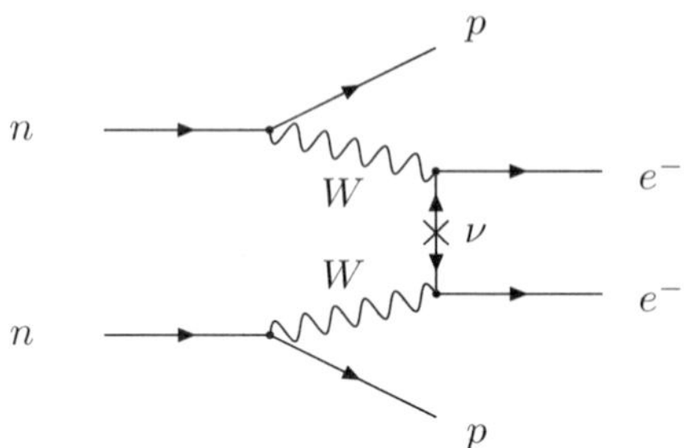

Fig. 11.3. Feynman diagram for neutrinoless double-beta decay.

11.9. Neutrinoless Double Beta Decay

Direct information on neutrino masses can also be obtained from neutrinoless double beta decay ($0\nu\beta\beta$) searches:

$$(A, Z) \rightarrow (A, Z + 2) + e^- + e^-. \tag{11.9.95}$$

Schematically in the presence of neutrino masses and mixing the process in Eq. (11.9.95) can be induced by the diagram shown in Fig. 11.3. The amplitude of this process is proportional to the product of the two leptonic currents

$$M_{\alpha\beta} \propto [\bar{e}\gamma_\alpha(1 - \gamma_5)\nu_e] \, [\bar{e}\gamma_\beta(1 - \gamma_5)\nu_e] \, . \tag{11.9.96}$$

which can only lead to a neutrino propagator from the contraction $\langle 0 \, | \, \nu_e(x)\nu_e(y)^T \, | \, 0 \rangle$. If the neutrino is a Dirac particle ν_e field annihilates a neutrino states and creates an antineutrino state which are different. Therefore the contraction $\langle 0 \, | \, \nu_e(x)\nu_e(y)^T \, | \, 0 \rangle = 0$ and $M_{\alpha\beta} = 0$. On the contrary, if ν_e is a Majorana particle, neutrino and antineutrino are the same state and $\langle 0 \, | \, \nu_e(x)\nu_e(y)^T \, | \, 0 \rangle \neq 0$.

Thus in order to induce the $0\nu\beta\beta$ decay, ν_e must be a Majorana particle. This is also obvious as the process (11.9.95) violates L by two units. The opposite also holds, if $0\nu\beta\beta$ decay is observed, neutrinos must be massive Majorana particles.[54]

However, Majorana neutrino masses are not the only mechanism which can induce neutrinoless double beta decay. In general, in models beyond the standard model there may be other sources of total lepton number violation which can induce $0\nu\beta\beta$ decay. Consequently the observation or limitation of the neutrinoless double beta decay reaction rate can only be related to a bound on the neutrino mass and mixing under some assumption about the source of total lepton number violation in the model.

For the case in which the only effective lepton number violation at low energies is induced by the Majorana mass term for the neutrinos, the rate of $0\nu\beta\beta$ decay is proportional to the *effective Majorana mass of* ν_e,

$$m_{ee} = \left| \sum_i m_i U_{ei}^2 \right| \qquad (11.9.97)$$

which, in addition to the masses and mixing parameters that affect the tritium beta decay spectrum, depends also on the leptonic CP violating phases.

Experimentally, what it is measured is the half-life of the decay. In $0\nu\beta\beta$ decay, the experimental signal is two electrons in the final state, whose energies add up to the Q-value of the nuclear transition while for the double beta decay with neutrinos ($2\nu\beta\beta$) (which constitute an intrinsic background) the energy spectrum of both electrons will be continuous as part of the Q is carried by the outgoing neutrinos. Also the decay rates for $0\nu\beta\beta$ and $2\nu\beta\beta$ have very different dependence on the available Q being the dependence much weaker for $0\nu\beta\beta$. For this reason, the sensitivity is better for isotopes with a high Q-value.

In the case that the only source of lepton number violation at low energies is induced by the Majorana neutrino mass, the decay half-life is given by:

$$(T_{1/2}^{0\nu})^{-1} = G^{0\nu} \left| M^{0\nu} \right|^2 \left(\frac{m_{ee}}{m_e} \right)^2 \qquad (11.9.98)$$

where $G^{0\nu}$ is the phase space integral and $|M^{0\nu}|$ is the nuclear matrix element of the transition.

The strongest bound from $0\nu\beta\beta$ decay was imposed by the Heidelberg-Moscow group[55] which used 11 kg of enriched Ge. After 53.9 kg yr of data taking they found no signal which allowed them to set a bound on the half-life of $T_{1/2}^{0\nu} > 1.9 \times 10^{25}$ yr (90% CL). This implies (for a given prediction of the nuclear matrix element):

$$m_{ee} < 0.26 \ (0.34) \ \text{eV} \quad \text{at } 68\% \ (90\%) \ \text{CL.} \qquad (11.9.99)$$

Taking into account the possible uncertainties in the prediction of the nuclear matrix elements, the bound may be weaken by a factor of about 3.[56]

The final result of the Heidelberg-Moscow experiment quoted above was that no positive signal was observed, but a subgroup of the collaboration found a small peak at some value of Q[57,58] which, if real, would imply a non vanishing range for the effective Majorana mass between 0.2–0.6 eV (the range might be widened by nuclear matrix uncertainties). However, the statistical analysis used, as well as the assumed background subtraction in order to establish this evidence at the claimed CL has been subject of severe criticisms[59–62] which renders the claimed signal as controversial.

Currently only two large scale experiments are running. CUORICINO at the Gran Sasso Underground Laboratory in Italy which uses uses bolometers running at very low temperature and searches for ^{120}Te decay with a Q-value of 2530 keV. The obtained half-life limit[63] $T_{1/2}^{0\nu}(^{120}\text{Te}) > 2.2 \times 10^{24}$ yr (90% CL) implies and upper bound on the effective Majorana neutrino mass $m_{ee} < 0.2$–1.1 eV. The second experiment, NEMO-3[64] in the Frejus Underground Laboratory, is built in form of time projection chambers where the double beta emitter is either the filling gas of the chamber or is included in thin foils. It has obtained a half-life bound for ^{100}Mo $T_{1/2}^{0\nu}(^{100}\text{Mo}) > 5.6 \times 10^{23}$ yr (90% CL) which result in an upper effective Majorana mass bound $m_{ee} < 0.6$–2 eV.

A series of new experiments is planned with sensitivity of up to $m_{ee} \sim 0.01$ eV. For a review of the proposed experimental techniques see Ref. [65].

11.10. Collider Signatures of ν Mass Models

The simplest and most straightforward lesson of the evidence for neutrino masses is also the most striking one: there is NP beyond the SM. This is the first experimental result that is inconsistent with the SM.

Given the relation (11.3.34), $m_\nu \sim v^2/\Lambda_{\text{NP}}$, it is straightforward to use measured neutrino masses to estimate the scale of NP that is relevant to their generation. In particular, if there is no quasi-degeneracy in the neutrino masses, the heaviest of the active neutrino masses can be estimated,

$$m_h = m_3 \sim \sqrt{\Delta m_{\text{atm}}^2} \approx 0.05 \text{ eV}. \tag{11.10.100}$$

(In the case of inverted hierarchy the implied scale is $m_h = m_2 \sim \sqrt{|\Delta m_{\text{atm}}^2|} \approx 0.05$ eV). It follows that the scale in the non-renormalizable term (11.3.32) is given by

$$\Lambda_{\text{NP}} \sim v^2/m_h \approx 10^{15} \text{ GeV}. \tag{11.10.101}$$

We should clarify two points regarding Eq. (11.10.101):

1. There could be some level of degeneracy between the neutrino masses that are relevant to the atmospheric neutrino oscillations. In such

a case Eq. (11.10.100) is modified into a lower bound and, consequently, Eq. (11.10.101) becomes an upper bound on the scale of NP.

2. It could be that the Z_{ij} couplings of Eq. (11.3.32) are much smaller than one. In such a case, again, Eq. (11.10.101) becomes an upper bound on the scale of NP.

The crucial issue is to understand the origin of this operator in a given extension of the SM in order to identify the dimensionless coupling Z and the mass scale $\Lambda_{\rm NP}$ at which the new physics enters. For illustration I will comment, what is the *standard* expectation on two simple renormalizable extensions of the Standard Model with minimal addition to generate neutrino Majorana masses:

- *Type I see-saw mechanism* discussed in Sec. 11.3.2. One can adds fermionic singlets N_i and the neutrino masses are $m_\nu \sim Y_\nu^2 v^2 / M_N$, where M_N is the right-handed neutrino mass, which sets the new physics scale $\Lambda_{\rm NP}$. From the discussion above we see that if $Y_\nu \simeq 1$ and $M_N \approx 10^{14-15}$ GeV, one obtains the natural value for the neutrino masses.

- *Type II see-saw mechanism* discussed in Sec. 11.3.4. The Higgs sector of the Standard Model is extended by adding an $SU(2)_L$ Higgs triplet Δ. The neutrino masses are $m_\nu \approx f_\nu v_\Delta$, where v_Δ is the vacuum expectation value (vev) of the neutral component of the triplet and f_ν is the Yukawa coupling. With a doublet and triplet mixing via a dimensional parameter μ, the electroweak symmetry breaking leads to a relation $v_\Delta \sim \mu v_0^2 / M_\Delta^2$, where M_Δ is the mass of the triplet. In this case the scale $\Lambda_{\rm NP} = M_\Delta^2 / \mu$, and a natural setting would be for $f_\nu \approx 1$ and $\mu \sim M_\Delta \approx 10^{14-15}$ GeV.

To directly test the above scenarios one needs to search for example for direct observations of the new heavy states responsible for the see-saw mechanism. As discussed above, most generically the characteristic mass scales of the new states are very large, rendering the new states experimentally inaccessible in the foreseeable future. However, one can think of scenarios in which this may not be necessary the case.

As an example let's take the Type II see-saw introduced in Sec. 11.3.4 (see [66] and references therein for a recent detailed study from which I have taken the following discussion). Sec. 11.3.4. The scalar potential in the Type II see-saw is given by

$$V(H, \Delta) = -m_H^2 \phi^\dagger \phi + \frac{\lambda}{4}(\phi^\dagger \phi)^2 + M_\Delta^2 Tr\Delta^\dagger \Delta + \left(\mu \, \tilde{\phi}^T \Delta \tilde{\phi} + h.c. \right) +$$

$$\lambda_1(\phi^\dagger \phi)Tr\Delta^\dagger \Delta + \lambda_2 \left(Tr\Delta^\dagger \Delta \right)^2 + \lambda_3 Tr \left(\Delta^\dagger \Delta \right)^2 + \lambda_4 \tilde{\phi}^T \Delta \Delta^\dagger \tilde{\phi}^*. \quad (11.10.102)$$

The key ingredient that makes this model testable at LHC is to assume a very small doublet-triplet mixing

$$\mu \ll M_\Delta \quad\quad\quad (11.10.103)$$

and so the Higgs triplet is heavy, typically $M_\Delta^2 > v^2/2$, but not much heavier than this bound.

Imposing the conditions of global minimum one finds that (one can neglect the contributions coming from the terms proportional to λ_1, λ_2, λ_3 and λ_4).

$$-m_H^2 + \frac{\lambda}{4}v^2 - \sqrt{2}\,\mu\,v_\Delta = 0, \quad \text{and} \quad v_\Delta = \frac{\mu\,v_0^2}{\sqrt{2}\,M_\Delta^2}, \quad (11.10.104)$$

where v and v_Δ are the vacuum expectation values of the Higgs doublet and triplet, respectively, with $v^2 + v_\Delta^2 \approx (246 \text{ GeV})^2$. Due to the simultaneous presence of the Yukawa coupling f_ν in Eq. (11.3.29), and the term proportional to the μ parameter in Eq. (11.10.102), the lepton number is explicitly broken in this theory. The present constraints form the electroweak precision data yields

$$v_\Delta \lesssim 1 \text{ GeV} \quad\quad\quad (11.10.105)$$

Once the neutral component in Δ gets the vev, v_Δ as in Eq. (11.10.104), the neutrinos acquire a Majorana mass given by the following expression:

$$M_\nu = \sqrt{2}\,f_\nu\,v_\Delta = f_\nu\,\frac{\mu\,v^2}{M_\Delta^2}, \quad\quad\quad (11.10.106)$$

We notice that in the case of small doublet-triplet mixing, Eq. (11.10.103) one can have small neutrino masses even with M_Δ as light as

$$M_\Delta \sim 110 \text{ GeV} \quad\quad\quad (11.10.107)$$

without conflicting with any existing data.

After electroweak symmetry breaking, there are seven physical massive Higgs bosons left in the spectrum:

$$
\begin{aligned}
H_1 &= \cos\theta_0 \ h^0 \ + \ \sin\theta_0 \ \delta^0, \\
H_2 &= -\sin\theta_0 \ h^0 \ + \ \cos\theta_0 \delta^0, \\
A &= -\sin\alpha \ \xi^0 \ + \ \cos\alpha \ \eta^0, \\
H^\pm &= -\sin\theta_\pm \ \phi^\pm \ + \ \cos\theta_\pm \ \Delta^\pm, \\
H^{\pm\pm} &= \Delta^{\pm\pm},
\end{aligned}
\tag{11.10.108}
$$

with

$$
\sqrt{2}\theta_\pm \approx \theta_0 \approx \alpha \approx \frac{2v_\Delta}{v}.
\tag{11.10.109}
$$

In the above equations the neutral component of the Higgs doublet and triplet are $\phi_0 = h^0 + i\eta^0$ and $\Delta_0 = \delta_0 + i\xi^0$ respectively.

Thus H_1 is SM-like (doublet) while the rest of the Higgs states are all Δ-like (triplet), and

$$
M_{H_2} \simeq M_A \simeq M_{H^\pm} \simeq M_{H^{++}} = M_\Delta.
$$

All these states are within reach of the LHC.

In the physical basis for the fermions the Yukawa interactions of the single and double charged scalars can be written as

$$
-\mathcal{L} = Y^+_{ij} \ \bar{\nu}^C_{Li} \ e_{Lj} \ H^+ \ + \ \bar{Y}^{++}_{ij} \ \bar{e}^C_{Li} \ e_{Lj} \ H^{++},
\tag{11.10.110}
$$

where

$$
Y^+ = \cos\theta_+ \ \frac{m_\nu^{diag}}{v_\Delta} \ U^\dagger_{LEP},
$$

$$
Y^{++} = U^*_{LEP} \ \frac{m_\nu^{diag}}{\sqrt{2}\ v_\Delta} \ U^\dagger_{LEP} = f_\nu \ .
\tag{11.10.111}
$$

We see that the values of the couplings Y^+ and Y^{++} are determined by the spectrum and mixing angles for the active neutrinos. Therefore, by observing lepton-number violating decays of the Higgs bosons, $H^{++} \to e_i^+ e_j^+$ and $H^+ \to e_i^+ \bar{\nu}$ ($e_i = e, \mu, \tau$) once can obtain information about neutrino masses and mixings and in particular it is possible to determine the neutrino mass spectrum.

References

1. F. Reines and C. L. Cowan, Phys. Rev. **113**, 273 (1959).
2. W. Pauli, *letter sent to the Tubingen conference*, Dec. 1930.

3. B. Pontecorvo, Sov. Phys. JETP **26**,984 (1968) [Zh. Eksp. Teor. Fiz. **53**, 1717 (1967)].

4. V. N. Gribov and B. Pontecorvo, Phys. Lett. B **28**, 493 (1969).

5. S. M. Bilenky, C. Giunti and W. Grimus, Prog. Part. Nucl. Phys. **43**, 1 (1999) [arXiv:hep-ph/9812360].

6. A. D. Dolgov, Phys. Rept. **370**, 333 (2002) [arXiv:hep-ph/0202122].

7. V. Barger, D. Marfatia and K. Whisnant, Int. J. Mod. Phys. E **12**, 569 (2003) [arXiv:hep-ph/0308123].

8. S. Pakvasa and J. W. F. Valle, Proc. Indian Natl. Sci. Acad. **70A**, 189 (2004) [arXiv:hep-ph/0301061].

9. M. Maltoni, T. Schwetz, M. A. Tortola and J. W. F. Valle, New J. Phys. **6**, 122 (2004) [arXiv:hep-ph/0405172].

10. G. L. Fogli, E. Lisi, A. Marrone and A. Palazzo, Prog. Part. Nucl. Phys. **57**, 742 (2006) [arXiv:hep-ph/0506083].

11. R. N. Mohapatra and A. Y. Smirnov, Ann. Rev. Nucl. Part. Sci. **56**, 569 (2006) [arXiv:hep-ph/0603118].

12. J. Lesgourgues and S. Pastor, Phys. Rept. **429**, 307 (2006) [arXiv:astro-ph/0603494].

13. A. Strumia and F. Vissani, arXiv:hep-ph/0606054.

14. M. C. Gonzalez-Garcia and Y. Nir, Rev. Mod. Phys. **75**, 345 (2003) [arXiv:hep-ph/0202058].

15. M. C. Gonzalez-Garcia and M. Maltoni, Phys. Rept. **460**, 1 (2008) [arXiv:0704.1800 [hep-ph]].

16. J. N. Bahcall, *Neutrino Astrophysics*, Cambridge University Press (1989).

17. F. Boehm and P. Vogel, *Physics of Massive Neutrinos*, Cambridge University Press (1987).

18. B. Kayser, F. Gibrat-Debu and F. Perrier, World Sci. Lect. Notes Phys. **25**, 1 (1989).

19. T. K. Gaisser, *Cosmic Rays And Particle Physics*, Cambridge University Press (1990).

20. C. W. Kim and A. Pevsner, Contemp. Concepts Phys. **8**, 1 (1993).

21. R. N. Mohapatra and P. B. Pal, World Sci. Lect. Notes Phys. **60**, 1 (1998) [World Sci. Lect. Notes Phys. **72**, 1 (2004)].

22. M. Fukugita and T. Yanagida, *Physics of neutrinos and applications to astrophysics*, Springer, Berlin, Germany (2003).

23. W. M. Yao *et al.* [Particle Data Group], J. Phys. G **33**, 1 (2006).

24. P. Minkowski, Phys. Lett. B **67**, 421 (1977).

25. P. Ramond, Invited talk given at Sanibel Symposium, Palm Coast, Fla., Feb 25 – Mar 2, 1979, published in *Paris 2004, Seesaw 25* 265–280 [arXiv:hep-ph/9809459].

26. M. Gell-Mann, P. Ramond and R. Slansky, in *Supergravity*, edited by P. van Nieuwenhuizen and D. Z. Freedman (North Holland).

27. T. Yanagida, *In Proceedings of the Workshop on the Baryon Number of the Universe and Unified Theories*, Tsukuba, Japan, 13–14 Feb 1979, edited by O. Sawada and A. Sugamoto (KEK).

28. R. N. Mohapatra and G. Senjanovic, Phys. Rev. Lett. **44**,912 (1980).

29. W. Konetschny and W. Kummer, "Nonconservation Of Total Lepton Number With Scalar Bosons," Phys. Lett. B **70**, 433 (1977); T. P. Cheng and L. F. Li, "Neutrino Masses, Mixings And Oscillations In SU(2) X U(1) Models Of Electroweak Interactions," Phys. Rev. D **22**, 2860 (1980); G. Lazarides, Q. Shafi and C. Wetterich, "Proton Lifetime And Fermion Masses In An SO(10) Model," Nucl. Phys. B **181**, 287 (1981); R. N. Mohapatra and G. Senjanović, "Neutrino Masses And Mixings In Gauge Models with Spontaneous Parity Violation," Phys. Rev. D **23**, 165 (1981).

30. J. Schechter and J. W. F. Valle, Phys. Rev. D **22**, 2227 (1980).

31. Z. Maki, M. Nakagawa and S. Sakata, Prog. Theor. Phys. **28**, 870 (1962).

32. J. Schechter and J. W. F. Valle, Phys. Rev. D **21**,309 (1980).

33. J. Schechter and J. W. F. Valle, Phys. Rev. D **24**, 1883 (1981) [Erratum-ibid. D **25**, 283 (1982)].

34. M. Kobayashi and T. Maskawa, Prog. Theor. Phys. **49**, 652 (1973).

35. L. Wolfenstein, Phys. Rev. D **17**, 2369 (1978).

36. A. Halprin, Phys. Rev. D **34**, 3462 (1986).

37. P. D. Mannheim, Phys. Rev. D **37**, 1935 (1988).

38. A. J. Baltz and J. Weneser, Phys. Rev. D **37**, 3364 (1988).

39. L. Landau, Phys. Z. Sov. **2**, 46 (1932).

40. C. Zener, Proc. Roy. Soc. Lond. A **137**, 696 (1932).

41. T. K. Kuo and J. T. Pantaleone, Rev. Mod. Phys. **61**, 397 (1989).

42. S. J. Parke, Phys. Rev. Lett. **57**, 1275 (1986).

43. W. C. Haxton, Phys. Rev. Lett. **57**, 1271 (1986).

44. S. T. Petcov, Phys. Lett. B **191**, 299 (1987).

45. S. P. Mikheev and A. Y. Smirnov, Sov. J. Nucl. Phys. **42**, 913 (1985) [Yad. Fiz. **42**, 1441 (1985)].

46. M. Maltoni and T. Schwetz, arXiv:0812.3161 [hep-ph].

47. S. Antusch, C. Biggio, E. Fernandez-Martinez, M. B. Gavela and J. Lopez-Pavon, JHEP **0610**, 084 (2006) [arXiv:hep-ph/0607020].

48. J. Bonn *et al.*, Nucl. Phys. Proc. Suppl. **91**, 273 (2001).

49. V. M. Lobashev *et al.*, Nucl. Phys. Proc. Suppl. **91**, 280 (2001).

50. R. E. Shrock, Phys. Lett. B **96**, 159 (1980).

51. F. Vissani, Nucl. Phys. Proc. Suppl. **100**, 273 (2001) [arXiv:hep-ph/0012018].

52. Y. Farzan, O. L. G. Peres and A. Y. Smirnov, Nucl. Phys. B **612**, 59 (2001) [arXiv:hep-ph/0105105].

53. A. Osipowicz *et al.* [KATRIN Collaboration], arXiv:hep-ex/0109033.

54. J. Schechter and J. W. F. Valle, Phys. Rev. D **25**, 2951 (1982).

55. H. V. Klapdor-Kleingrothaus *et al.*, Eur. Phys. J. A **12**, 147 (2001) [arXiv:hep-ph/0103062].

56. P. Vogel, arXiv:hep-ph/0611243.

57. H. V. Klapdor-Kleingrothaus, A. Dietz, H. L. Harney and I. V. Krivosheina, Mod. Phys. Lett. A **16**, 2409 (2001) [arXiv:hep-ph/0201231].

58. H. V. Klapdor-Kleingrothaus, I. V. Krivosheina, A. Dietz and O. Chkvorets, Phys. Lett. B **586**, 198 (2004) [arXiv:hep-ph/0404088].

59. F. Feruglio, A. Strumia and F. Vissani, Nucl. Phys. B **637**, 345 (2002) [Addendum-ibid. B **659**, 359 (2003)] [arXiv:hep-ph/0201291].

60. C. E. Aalseth *et al.*, Mod. Phys. Lett. A **17**, 1475 (2002) [arXiv:hep-ex/0202018].

61. H. L. Harney, arXiv:hep-ph/0205293.

62. Yu. G. Zdesenko, F. A. Danevich and V. I. Tretyak, Phys. Lett. B **546**, 206 (2002).

63. C. Arnaboldi *et al.*, Phys. Rev. Lett. **95**, 142501 (2005) [arXiv:hep-ex/0501034].

64. R. Arnold *et al.* [NEMO Collaboration], Nucl. Phys. A **781**, 209 (2007) [arXiv:hep-ex/0609058].

65. S. R. Elliott and J. Engel, J. Phys. G **30**, R183 (2004) [arXiv:hep-ph/0405078].

66. P. F. Perez, T. Han, G. y. Huang, T. Li and K. Wang, arXiv:0805.3536 [hep-ph].

Chapter 12

Experimentation of Neutrino Physics

Kate Scholberg

Department of Physics, Duke University,
Durham, NC 27708, USC
**E-mail: schol@phy.duke.edu*

These lectures broadly describe experimental progress in neutrino physics over the past few decades. I will describe the basic picture, how we know what we now know, and the next steps for the future.

12.1. Introduction

Neutrinos are the neutral fermion partners to the charged leptons, and their featured properties are their terribly tiny masses and their exclusively weak interactions. Until about a dozen years ago it was not known whether neutrinos have any mass at all– but we now know that at least two neutrino states have gravitational interactions as well as weak ones. The properties of neutrinos are interesting not only in that we must fit them into to the overall particle physics picture; in addition they give insight into cosmology. For the full story on largest and smallest scales, we need to understand neutrino masses and mixings, and whether neutrinos are their own antiparticles. New physics beyond the Standard Model may first manifest itself in the neutrino sector.[1] Understanding of neutrino properties also deepens our understanding of astrophysical objects, both mundane and exotic, where neutrinos are significant players in many processes.

A good deal of current activity in experimental neutrino physics is focused on measurements of mass and oscillation parameters. The purpose of this review is to provide an overview of the main recent results, and the next generation of planned experiments. In the discussion of neutrino oscillation, I will organize the review around parameter space: first, I will discuss measurements of "atmospheric" neutrino oscillation parameters, and then

565

"solar" oscillation parameters. By now, in both of these cases, oscillations have been observed for multiple kinds of neutrino sources. I then describe the LSND anomaly and its current status. I'll then summarize what we know so far, and describe the next plans to go after the unknowns in the three-flavor mixing picture with new reactor and beam experiments (and possibly supernovae, a "source of opportunity"). These unknowns are the last mixing angle θ_{13}, the nature of the mass hierarchy, and the CP-violating phase δ,

But there's more to neutrino physics than oscillations. I'll also describe progress towards a measurement of the absolute mass scale of neutrinos, via cosmology and kinematic experiments. Neutrinoless double beta decay experiments in addition may provide insight into both the absolute mass scale and the outstanding question of whether neutrinos are their own antiparticles.

For space reasons, I will omit here a number of interesting topics in neutrino physics, some of which are engaging entire communities of neutrino experimentalists. Such omissions include electromagnetic properties of neutrinos, neutrino interaction cross-sections with matter, non-standard neutrino interactions, high-energy neutrino astrophysics, geoneutrinos, and practical applications of neutrino physics.

12.1.1. *Sources of neutrinos*

To study neutrinos, we first need to have some to detect. And we need lots of them, since they reveal themselves so rarely. Fortunately, neutrinos are plentiful, if you know where to look for them.

- Wild neutrinos: neutrinos "in the wild" abound– they are made by numerous natural processes. The sources of natural neutrinos that we can aspire to catch on Earth are (in order of increasing typical energy):

  - The Big Bang: relic neutrinos at $1.95°$K infiltrate the universe; these are so cold as to be possibly non-relativistic and gravitationally clustered. Detection of these is a goal currently out of reach (see Section 12.3.4).
  - Radioactivity in the Earth: weak radioactive processes produce neutrinos of energies up to a few MeV. The Earth is gently glowing with such "geoneutrinos".
  - Cosmic rays: protons hitting the upper atmosphere create cascades of charged particles, among which are parents of neutrinos

such as pions and kaons. Atmospheric neutrinos of primarily muon and electron flavor are produced in a spherical shell around the Earth and span energies of a few tens of MeV to tens of TeV.

— The Sun: the Sun pumps out electron flavor neutrinos from a network of solar fusion processes. The highest solar neutrino energies are around 15 MeV; the bulk arise from from the "pp" process, and have energies below 1 MeV.

— Supernovae: a core collapse supernova dumps nearly all of the binding energy of the proto-neutron star into neutrinos of a few tens of MeV and all flavors.

— More exotic astrophysical objects: anywhere weak interactions play a role, you'll typically find neutrinos. Large fluxes of neutrinos of high energy, up to hundreds of TeV, are expected from exotic astrophysical objects and events, from a variety of processes. However no definitive source has yet been observed. Such potential astrophysical neutrino sources include gamma-ray bursters, hypernovae, and active galactic nuclei.

- "Tame" neutrinos: humans can create neutrinos with properties that are to some extent tunable.[*]

— Radioactive sources: sources can be created in nuclear reactors. Neutrinos and antineutrinos of electron flavor are produced in beta decay. Electron capture sources make monochromatic neutrinos. Typical energies are in the few-MeV range.[2]

— Nuclear reactors: fission reactors create large fluxes of electron antineutrinos, of few-to-several MeV energies.[3]

— Proton accelerators: proton accelerator neutrino sources can be considered "artificial cosmic ray generators". Protons are accelerated, and slammed into a target; pions and kaons are produced. These neutrino parents are focused forward with a magnetic horn, and then allowed to decay in a decay pipe. As a result of the boost of their parents, the neutrino daughters are focused forward in a beam. The beam neutrino spectrum and flavor composition can be tuned to some extent by modifying the proton beam energy, and the horn and decay pipe configurations. Detailed models and *in-situ* measurements are usually required for full understanding

---

[*]However I will note that some artificially-generated neutrinos are not necessarily all that tame: sometimes it can be nearly as difficult to understand the properties of an artificial neutrino source as to understand the properties of a natural one.

of an artificial neutrino beam.

- Stopped pion beams: in this case a proton source creates pions which are stopped in matter. Negative pions are captured by nuclei, while positive pions decay to produce a monochromatic ν_μ, via the two-body decay $\pi^+ \to \nu_\mu + \mu^+$. The muons subsequently decay to produce two more neutrinos each, of energies up to 50 MeV, according to $\mu^+ \to \bar{\nu}_\mu + \nu_e + e^+$. This decay spectrum is very well known; uncertainties arise from contamination of the beam and pions which decay in flight.

- Muon storage rings: a dream of neutrino experimentalists is a "neutrino factory": a high-intensity, clean source of many neutrino flavors, which in principle could be provided by a muon storage ring in which the muons decay. The neutrino spectrum depends on the boost of the muons, according to the well-understood muon decay spectrum. Selecting the sign of the muon charge in the storage ring allows selection of neutrinos versus antineutrinos for each of the decay product flavors.

- Beta beams: an interesting and relatively new idea for producing well-understood neutrinos is to take radioactive isotopes and circulate them in a storage ring.[4] The subsequent decays create either neutrinos or antineutrinos. The main two isotopes under consideration for a beta beam are ^{6}He and ^{18}Ne. In the same way as for a muon storage ring, the peak of the neutrino energy spectrum is tunable by selecting the Lorentz boost of the ions in the ring.

12.1.2. *Neutrino detection*

Neutrinos are aloof, but not completely unsociable. From time to time they deign to interact with matter, and the final state is observable.

12.1.2.1. *Neutrino interaction with matter*

Neutrinos interact via charged-current (CC) or neutral-current (NC) interactions. In a charged-current interaction, the neutrino exchanges a $W^\pm$ boson with a quark or lepton. In order for the interaction to occur, the incident neutrino must have enough energy to create the lepton. The outgoing lepton is of the same flavor as the incoming neutrino.

In NC interactions, the neutrino exchanges a Z boson with quarks or leptons. In this case, the flavor is irrelevant: the reaction rate is flavor-blind. Examples of NC and CC interactions are shown in Fig. 12.1.

Fig. 12.1.   Examples of CC (left) and NC (right) neutrino interactions.

What one detects in either of these cases is the outgoing electromagnetically- or strongly-interacting final state particles. For the CC case, one may see the lepton, and there may be additional accompanying particles. In the NC case, one may observe the "bumped" particles. In high-energy interactions, extra energy rattling around can end up as rest mass of other particles, too. Charged particles are detected in standard ways, by their energy loss.

12.1.2.2. *Neutrino detectors*

Detector technology is a vast topic, and experimentalists employ a diverse arsenal of technologies. When detecting neutrinos, the questions one typically wants to answer are: what flavor was the neutrino? What energy did it have? What direction was it coming from? Where did the interaction happen? What type of interaction was it? CC or NC? Detector features that matter include: energy resolution, energy threshold, time resolution, pixelization and ability to distinguish outgoing particle type. Specific capabilities depend on the details of the detection mechanism and experimental setup. However, neutrino detectors typically share some common features: first of all, they need to be *large*. Neutrino interactions are so weak that you need a lot of targets to have a measurable interaction rate, even for a large flux. Kiloton scales are typical.

Furthermore, detectors must have a low rate of background, again because event rates are terribly small. Often the entire experimental challenge has to do with reducing and understanding background. The specific nature of the background depends on what kind of neutrino interactions one is trying to measure. Very often detectors are placed underground in order to escape the swarms of cosmic rays at the surface of the Earth. Detectors of low energy neutrinos, for which radioactive background can swamp the signal, in addition usually need to be extremely clean. For some artificial neutrino sources, backgrounds can be mitigated by knowing when the neutrinos arrive and rejecting events outside of the known signal time window. "Beam-off" time may be essential to characterize background.

12.2. Neutrino Oscillations

The main recent excitement in neutrino physics has been over new understanding of neutrino oscillations. Until 1998, it was not known whether neutrinos had any mass at all; although we still do not know the absolute mass scale (see Section 12.3), we now have insight into neutrino masses via the mixing of different mass states. We assume that the N flavor interaction eigenstates $|\nu_f\rangle$ are distinct from the mass eigenstates $|\nu_i\rangle$, and can be written as a superposition of them according to

$$|\nu_f\rangle = \sum_{i=1}^{N} U_{fi}^* |\nu_i\rangle, \tag{12.2.1}$$

where U_{fi} are the elements of an $N \times N$ unitary mixing matrix. For $N = 3$ flavors (assumed since only three generations of leptons are known[5]), this mixing matrix is known as the Maki-Nakagawa-Sakata (MNS) matrix. In this framework, free neutrinos will change flavor as they propagate. As a simple example, consider the two-flavor case (see for example references [6,7] for discussions of the assumptions and implications of this simplified quantum-mechanical treatment):

$$\begin{aligned} |\nu_f\rangle &= \cos\theta|\nu_1\rangle + \sin\theta|\nu_2\rangle \\ |\nu_g\rangle &= -\sin\theta|\nu_1\rangle + \cos\theta|\nu_2\rangle \end{aligned} \tag{12.2.2}$$

The flavor states are rotated versions of the mass states, where the rotation angle θ (the "mixing angle") is a parameter of nature. Now, if you allow a neutrino produced in flavor state $|\nu_f\rangle$ by a weak interaction to propagate a distance L, $|\nu_i(t)\rangle = e^{-iE_i t}|\nu_i(0)\rangle$, in the relativistic approximation, this can be written $|\nu_i(t)\rangle \sim e^{-im_i^2 L/2p}|\nu_i(0)\rangle$, and the probability of neutrino being observed in the original flavor state is $P(\nu_f \to \nu_f) = |\langle\nu_f|\nu_f\rangle|^2$, which works out to

$$P(\nu_f \to \nu_g) = 1 - P(\nu_f \to \nu_f) = \sin^2 2\theta \sin^2(1.27\Delta m^2 L/E_\nu), \tag{12.2.3}$$

for L in km and E_ν in GeV, and mass in eV^2. This flavor transition probability depends on two parameters of nature: the mass-squared difference, $\Delta m^2 \equiv m_2^2 - m_1^2$, and the mixing angle θ. Two quantities in this expression depend on the specific parameters of the situation: the energy of the neutrino, E_ν, and the distance it travels, L. The wavelength of the oscillation is proportional to $E_\nu/\Delta m^2$; the amplitude of the flavor modulation is $\sin^2 2\theta$.

We now come to the experimental game. The basic idea of a neutrino oscillation experiment is as follows: first take some neutrinos. They can be wild or tame ones, but something about their properties, *i.e.* their energies and flavors, is known, either via model or measurement. Then the neutrinos are allowed to propagate, and their spectrum and flavor composition are measured to the extent possible. Have the flavors changed? If so, are the flavors and spectrum modified according to equation 12.2.3? If the answer to this experimental question is yes, then you have evidence for neutrino oscillation. If all neutrino states have zero mass, Δm^2 is zero, and oscillation is not possible: therefore an observation of energy-dependent flavor change according to equation 12.2.3 implies at least one non-zero mass state. One can then fit the observation to oscillation parameters: for a two-flavor oscillation, the traditional parameter space shows Δm^2 on the vertical axis and $\sin^2 2\theta$ on the horizontal axis. In this plane, an experimental setup with large L/E_ν gives access to small values of Δm^2, and small L/E_ν gives access to large values. Since a large mixing angle corresponds to large modulation of the signal, large experimental statistics are required to observe a small mixing angle. A measurement corresponds to a finite-sized region on the plot, because experimental uncertainties (either systematic or statistical) always lead to uncertainties on derived parameters. The aim of the experimentalist is to shrink the size of the allowed region down as small as possible.

Neutrino oscillation experiments can look for either disappearance or appearance of a given flavor. Since a CC interaction has an energy threshold, if a neutrino oscillates to a flavor for which the required energy for a CC interaction is below threshold, then the neutrino will effectively disappear. (Although NC interactions will still be present, the observable rate is typically small.) Therefore a loss of observed flux over the neutrino propagation distance can be interpreted as an oscillation. For example, consider ν_e at 10 MeV (such as a typical ^{8}B solar neutrino): a ν_μ must have more than 110 MeV in the lab frame to create a muon via the CC interaction with matter, *e.g.* $\nu_\mu + n \to \mu^- + p$. Similarly, a ν_τ must have at least

3.5 GeV for the corresponding CC interaction $\nu_\tau + n \to \tau^- + p$. Muon and tau neutrinos of 10 MeV just don't have the oomph to make a lepton, so if a $\nu_e \to \nu_\mu, \nu_\tau$ transition is occurring, the neutrinos will effectively disappear. One can therefore interpret an energy-dependent disappearance with respect to expectation as neutrino oscillation.

"Appearance" experiments are perhaps more satisfying, but often experimentally challenging: in this case one looks explicitly for a new flavor appearing in a neutrino flux of known flavor. An appearance experiment requires experimental capability of tagging lepton flavor in a CC interaction. For example, one might attempt to explicitly identify ν_τ-induced τ leptons in a $\nu_\mu \to \nu_\tau$ oscillation search using multi-GeV neutrinos.

More generally, for three flavors, the flavor and masses states are related by

$$\begin{pmatrix} \nu_e \\ \nu_\mu \\ \nu_\tau \end{pmatrix} = \begin{pmatrix} U_{e1}^* & U_{e2}^* & U_{e3}^* \\ U_{\mu1}^* & U_{\mu2}^* & U_{\mu3}^* \\ U_{\tau1}^* & U_{\tau2}^* & U_{\tau3}^* \end{pmatrix} \begin{pmatrix} \nu_1 \\ \nu_2 \\ \nu_3 \end{pmatrix} \qquad (12.2.4)$$

and the oscillation probabilities can be computed straightforwardly:[8]

$$P(\nu_f \to \nu_g) =$$
$$\delta_{fg} - 4 \sum_{j>i} \mathrm{Re}(U_{fi}^* U_{gi} U_{fj} U_{gj}^*) \sin^2(1.27 \Delta m_{ij}^2 L/E_\nu)$$
$$\pm 2 \sum_{j>i} \mathrm{Im}(U_{fi}^* U_{gi} U_{fj} U_{gj}^*) \sin^2(2.54 \Delta m_{ij}^2 L/E_\nu),$$

again for L in km, E_ν in GeV, and Δm_{ij}^2 in eV^2. The negative sign in front of the last term holds for neutrinos and the positive sign holds for antineutrinos.

If mass scales are well separated from each other, *i.e.* $|\Delta m_{23}^2| \gg |\Delta m_{12}^2|$, then the oscillation "decouples", *i.e.* can be described well by two-flavor oscillations with a single mass-squared difference scale. This situation holds for many real experimental situations, as will be described below.

I will now discuss the various regions of parameter space corresponding to existing signals of neutrino oscillations.

12.2.1. *Atmospheric neutrino parameter space*

The first unambiguous signal of neutrino oscillation came from study of atmospheric neutrinos. Atmospheric neutrinos are produced in collisions of cosmic rays with atoms in the upper atmosphere. The resulting cascade of hadrons includes such particles as pions and kaons, which decay to neutrinos. The neutrinos have energies ranging from about 0.1 GeV to several hundred GeV. The primary atmospheric neutrino production channel is $\pi^{\pm} \rightarrow \mu^{\pm} \, \nu_{\mu}(\bar{\nu}_{\mu}) \rightarrow e^{\pm}\nu_e(\bar{\nu}_e)\bar{\nu}_{\mu}(\nu_{\mu})$. Two muon flavor neutrinos are produced for each electron flavor neutrino. Although the absolute flux of atmospheric neutrinos is not known so precisely, the flavor ratio, resulting from this well-understood weak physics decay chain, is known to better than 5%.[9–11] Furthermore, because cosmic ray proton primaries impinge on the atmosphere isotropically (above 10 GeV or so they are almost unaffected by the geomagnetic field), neutrino daughters from the shower will observed in a detector at the surface of the Earth at equal rates above and below the horizon, for neutrinos with energy greater than about a GeV. Atmospheric neutrinos seen near the surface of the Earth will have pathlengths ranging from a few tens of km, if they originate from directly above the detector, to the diameter of the Earth, about 13,000 km, if they originate from the nadir. Atmospheric neutrinos have high enough energies to be typically above CC threshold for interactions with nucleons. The primary interaction channel is the "quasi-elastic" (QE) interaction, $\nu_l + n \rightarrow p + l^-$, or $\bar{\nu}_l + p \rightarrow n + l^+$ for $l = e, \mu$; such events have a single final state lepton. CC and NC interactions of atmospheric neutrinos which produce one or more pions or other hadrons are also possible, and tend to dominate for neutrino energies above several GeV.

If a neutrino detector can measure atmospheric neutrino direction and energy, and distinguish neutrino flavor (by determining at the flavor of the leading lepton from a CC interaction), then the flavor, pathlength and energy dependence of the neutrino flux can be inferred and the oscillation hypothesis tested.

12.2.1.1. *Super-Kamiokande*

The first hints of atmospheric neutrino oscillation came from Kamiokande II[12] in Japan and the IMB (Irvine-Michigan-Boston) experiment in the United States.[13] In 1998, Super-Kamiokande (Super-K, the successor to Kamiokande II) published definitive results.[14]

In water Cherenkov detectors, charged particles moving with speed greater than the speed of light in a vacuum create a cone of electromagnetic radiation. The less massive the particle, the lower the energy required for Cherenkov light production: electrons have a Cherenkov threshold of

0.73 MeV, and protons have a Cherenkov threshold of 1.4 GeV. The angle of the cone for a relativistic particle is $\cos\theta_C = \frac{1}{\beta n}$, where n is the index of refraction of the medium; for water, the Cherenkov angle for a relativistic particle is $42°$. Cherenkov photons are then detected by light sensors, such as photomultiplier tubes (PMTs).

Super-K, the world's largest underground water Cherenkov detector, is located in Mozumi, Japan, under a mountain providing an overburden of approximately 1 km of rock. Super-K contains a total of 50 kton of ultrapure water. Super-K ran as "Super-K I", with 11,000 photomultiplier tubes viewing an inner volume of 32 kton, from 1996 to 2001. After an accident in 2001 in which two-thirds of the PMTs were destroyed by a chain-reaction implosion, Super-K was rebuilt with 47% of the PMTs and restarted as Super-K II in 2003. After a full reconstruction which took place in 2005/2006, Super-K III ran from 2006 to 2008. It recently restarted running as Super-K IV with refurbished electronics and improved data acquisition.

The Cherenkov light detection technique allows Super-K[15] to measure properties of atmospheric neutrinos critical for oscillation studies. First, the Cherenkov cone has an intrinsic directionality. The image of the cone (typically a ring projected onto the inside surface of the detector) allows the direction of charged particles in the water to be reconstructed. Since the outgoing charged particles from a neutrino interaction are correlated with the neutrino direction, especially at high energy, one then obtains information about direction of the neutrino. Hence (under the assumption that the neutrino was produced in the atmosphere) one can infer the pathlength the neutrino traversed since its production. Furthermore, because energy loss is proportional to Cherenkov photons, one also gets a measure of the interacting neutrino energy. Finally, flavor information is available: leptons produced in neutrino interactions can be tagged by "particle ID" as muon or electron flavor. Muons just barrel through the water, losing energy uniformly: the result is a crisp, clean ("μ-like") Cherenkov ring. Electrons are lighter; they scatter and start electromagnetic showers by bremsstrahlung, yielding mushier, more diffuse (e-like) Cherenkov rings.

All of these measures of neutrino properties are imperfect. Not only does the detector have limitations of pixelization, light collection efficiency, *etc.*, but some particles produced in the interaction are often below Cherenkov threshold, so that complete information about the interaction products is lost. Also, in the absence of a magnetic field, it is very difficult to distinguish neutrinos from antineutrinos from the sign of the charged lepton produced.[16] Nevertheless significant information can be gathered overall.

"Fully-contained" events are those for which the neutrino interaction final state particles do not leave the inner part of the detector; these have their energies relatively well measured. Neutrino interactions for which the lepton is not contained in the inner detector sample higher-energy parent neutrino energy distributions: for example, in "partially-contained" events, the neutrino interacts inside the inner part of the detector but the lepton (almost always a muon, since only muons are penetrating) exits. "Upward-going muons" arise from neutrinos which interact in the rock below the detector and create muons which enter Super-K and either stop or go all the way through (entering downward-going muons cannot be distinguished from cosmic rays, which arrive from above with a rate of $\sim$3 Hz).

Super-K collects about 10 atmospheric neutrinos per day, and the current sample exceeds 25,000 events. Details of SK atmospheric neutrino samples can be found in reference [17].

Super-K's atmospheric neutrino study involves comparing angular and energy distributions of "e-like" and "μ-like" events to expectation based on knowledge of the flux. A detailed detector simulation is employed to evaluate the expected distributions under the assumption that neutrinos do not oscillate. As noted above, one expects as many neutrinos going up as going down, for each flavor. Although the angular distribution of e-like events indeed matches this expectation– about as many ν_e are upgoing as downgoing– a dramatic deficit of *up-going* muon neutrinos, which have traveled a long pathlength since their production in the atmosphere, is seen. Furthermore the deficit is energy-dependent. Figure 12.2 shows the Super-K atmospheric neutrino data. This observed pathlength- and energy-dependent loss of muon neutrinos can be interpreted in terms of $\nu_\mu \to \nu_\tau$ oscillation: given the spectrum of atmospheric neutrinos, very few ν_τ's will have sufficient energy for a CC interaction with a nucleon, so ν_μ's effectively disappear. Figure 12.3 shows the allowed parameters resulting from a two-flavor disappearance fit to equation 12.2.3, incorporating information about the event energy, direction and flavor. Best-fit parameters are $\Delta m^2 = 2.1 \times 10^{-3}$ eV^2 and $\sin^2 2\theta = 1.0$. Confirmation of this result with consistent allowed parameters came from two experiments with completely different detector technologies, Soudan II (iron calorimeter)[18] and MACRO (upward-going muon tracking).[19]

The first Super-K atmospheric neutrino measurement interpreted the disappearance as simple $\nu_\mu \to \nu_x$. In fact, current Super-K state-of-the-art analyses have much to say about which flavors are involved in the oscilla-

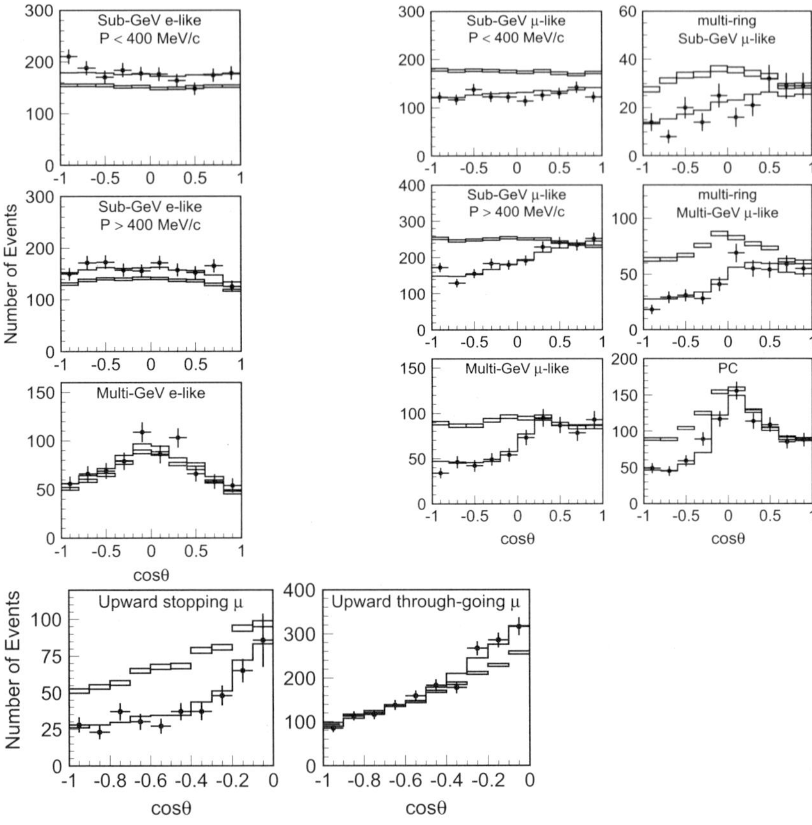

Fig. 12.2. Super-K I atmospheric neutrino zenith angle distributions for various data subsamples.[17] The boxes indicate the Monte Carlo expectation for no oscillation (where the height of the box shows the Monte Carlo statistical uncertainty). The points with errors show the data; the error bars show statistical uncertainty. The line shows the best fit of all the data to the $\nu_\mu \to \nu_\tau$ oscillation hypothesis.

tion. A pure $\nu_\mu \to \nu_e$ oscillation is clearly ruled out, because an up-going e-like excess from ν_e appearance commensurate with the ν_μ disappearance is clearly not observed. Could there oscillation of ν_μ to a sterile neutrino, ν_s of some kind, *i.e.* a neutrino which has neither CC or NC interactions (predicted in some models, *e.g.* [20,21])? Super-K employs two strategies to distinguish between $\nu_\mu \to \nu_\tau$ and $\nu_\mu \to \nu_s$. First, note that in the sterile case, the muon neutrinos would *really* disappear: they would have no NC interactions at all. A sample of NC-enriched events would exhibit an up-down asymmetry: for the $\nu_\mu \to \nu_s$ case, there would be a relative deficit of up-going events. Second, we can make use of a matter effect (see sec-

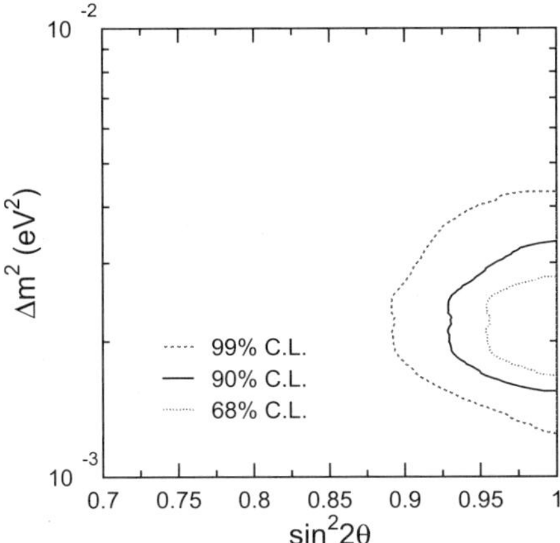

Fig. 12.3.   Allowed oscillation parameters from Super-K atmospheric neutrino data, for the Super-K I dataset.[17]

tion 12.2.2.2) arising from the fact that ν_τ's have an amplitude to exchange Z's with matter, and ν_s's do not: in the latter case, given the known density profile of the Earth, this would lead to an effective suppression of oscillation in a particular energy range.[22] The Super-K data show neither effect, and based on the absence of these signatures, the pure $\nu_\mu \to \nu_s$ hypothesis can be excluded at very high confidence,[23] although a moderate admixture is still allowed.[24]

Finally, recent work[25] has addressed the explicit appearance of "τ-like" events in the Super-K atmospheric neutrino sample.   Only a few dozen neutrino-induced τ's per year are expected at current best-fit oscillation parameters, and τ's are hard to distinguish experimentally from high energy ν_e events: in both cases, multiple particles are produced, and a large Cherenkov detector like Super-K has limited capability for making precision fits for multiple particles. Nevertheless, sophisticated selection algorithms based on event shape have shown an excess of "τ-like" events at 2.5 σ, clearly favoring the hypothesis that $\nu_\mu \to \nu_\tau$ oscillation explains the data.

Recent work has also been able more tightly constrain Δm^2 and disfavor some more exotic models for neutrino disappearance by using a high-resolution sample to resolve explicitly the "wiggle" of the oscillation probability:[26] see Fig. 12.4.

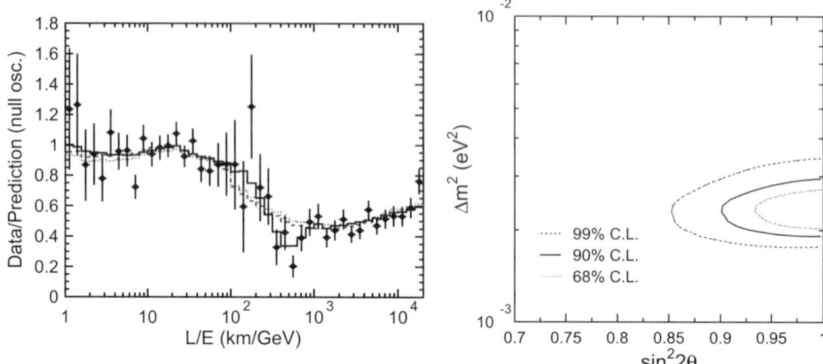

Fig. 12.4.   Left: ratio of data to predicted atmospheric neutrino events as a function of measured L/E, for a subsample of selected high-resolution events in Super-K. The solid line is the best fit to the two-flavor oscillation hypothesis. The dashed and dotted lines show the best-fit to the hypotheses of neutrino decay[27,28] and decoherence,[29,30] respectively. Right: allowed oscillation parameters for this analysis.

12.2.1.2. *Long baseline oscillation experiments*

Atmospheric neutrinos are "wild" neutrinos. Although understanding of atmospheric neutrino properties relevant for oscillation studies is reasonably good, and systematic uncertainties can be canceled by comparing upward- and downward-going fluxes, a second opinion is always welcome. For this reason, the next step after atmospheric neutrino oscillation measurements was the "long baseline experiment". Here the first goal was to provide an independent confirmation of the atmospheric neutrino oscillation hypothesis using an artificial beam of neutrinos. With $\sim$GeV neutrinos, for an L/E_ν suitable for testing the atmospheric Δm^2 parameters, baselines of hundreds of kilometers are needed. The basic method is to create an intense beam of neutrinos by making "artificial cosmic rays": protons are smashed into a target, the resulting charged hadrons are focused forward with a magnetic field, and the Lorentz-boosted pions and kaons allowed to decay to neutrinos. The neutrino beam characteristics (energy spectrum, flavor composition) are measured to the extent possible at a point near where the neutrinos are produced, before the neutrinos have a chance to oscillate. The beam then propagates hundreds of kilometers before the neutrinos are measured again. Then one plays the oscillation experiment game: have the flavors changed, and how?

K2K: K2K (KEK to Kamioka) was the first long-baseline neutrino oscillation experiment.[31] It employed a high-purity beam of ν_μ of $\sim$1 GeV energy, produced by 12 GeV protons impinging on an aluminum target. Positive pions were focused by a horn and allowed to decay by $\pi^+ \to \nu_\mu + \mu^+$ in a 200 m tunnel. One-microsecond-long bursts of neutrinos were sent 250 km through the Earth, underneath Japan, to Super-K. Beam neutrino events were tagged by comparing GPS time recorded for each beam spill to high-energy neutrino event times in Super-K. A near detector complex about 300 m from the neutrino source was used to characterize the beam, consisting of a water Cherenkov detector (same material as Super-K, to cancel systematics), fine-grained trackers, and a muon ranger. The data from the near detector were used to extrapolate to an expectation at Super-K. The presence of water in the near detector complex was important for canceling neutrino interaction cross-section uncertainties. K2K ran from 1999 to 2004, and observed a total of 112 beam-induced neutrino events in Super-K, when the no-oscillation expectation was 158.1 ± 9.[32] The observed deficit was consistent with the expectation for oscillation at Super-K atmospheric neutrino best-fit values. Furthermore, the reconstructed neutrino energy spectrum for 58 single-ring μ-like events (candidate CC quasi-elastic $\nu_\mu + n \to \mu^+ + p$ interactions) showed a spectral distortion, also consistent with atmospheric neutrino oscillation parameters. Although statistics were not huge, the K2K data disfavored the no-oscillation hypothesis at 4.3σ.

MINOS: The current state-of-the-art for long baseline disappearance oscillation is the MINOS (Main Injector Neutrino Oscillation Search) experiment in the United States. This experiment makes use of the NuMI beam,[33] which sends neutrinos over a 735 km baseline from Fermilab to the Soudan mine in Minnesota. Here again the beam is primarily ν_μ from pion decay, but energies are somewhat higher than K2K's were. The MINOS far detector is quite different from Super-K: it has 5 kton of magnetized iron instrumented with scintillating fibers used to track penetrating particles. The near detector, 1 km from the neutrino source, uses similar iron tracking detector technology. The magnetic field allows ν vs $\bar{\nu}$ selection. In a tracking detector like MINOS, neutrino flavors are tagged by event topology. CC ν_μ events have long muon tracks and hadronic activity at the vertex; NC current events are shorter and more diffuse; ν_e CC events are also short, with distinctive shower profiles. MINOS has also observed a ν_μ deficit and spectral distortion consistent with parameters describing atmospheric neutrino oscillations. Currently the best resolution on the measurement of the Δm^2 oscillation parameter comes from MINOS[34,35] (see Fig. 12.5), although best $\sin^2 2\theta$ constraint is still that from the Super-K atmospheric neutrino analysis.

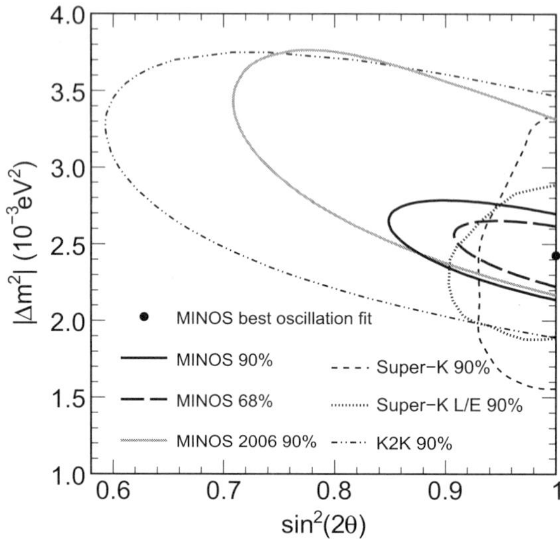

Fig. 12.5.   Allowed parameters for the MINOS ν_μ disappearance analysis.[35] Also shown are allowed regions from two Super-K analyses and from the K2K experiment.

CNGS: Another long-baseline project is CNGS (CERN Neutrinos to Gran Sasso) in Europe.[36] The baseline is (by chance) very close to that for MINOS, 730 km, but this project involves a higher energy ν_μ beam, with peak energy around 15-20 GeV, and the physics emphasis is different. For CNGS the goal is to observe τ appearance explicitly. In order to observe CC ν_τ interactions, the ν_τ must have at least 3.5 GeV of energy in the lab frame. The CNGS far detectors at Gran Sasso National Laboratory in Italy are also optimized for τ appearance: one needs a very fine-grained tracking detector in order to resolve the outgoing topology of a τ decay. The experimental signature is a kink of only millimeter scale (The first observation in 2000 of the ν_τ by the DONUT experiment[37] employed this method.) The OPERA experiment at Gran Sasso in Italy[38] is a lead/emulsion sandwich with active scintillator strip planes, along with a magnetic spectrometer. If the electronic detectors find a candidate, the appropriate brick of emulsion (a film-like detector) is removed by robot and then scanned to see if there are tracks consistent with τ decays. The active mass is 1.25 kton. OPERA observed its first long distance neutrinos in 2007.[39] In five years of running, for current best-fit atmospheric oscillation parameters OPERA expects about 10 τ-decay signal events with a background of less than one event.

The other fine-grained tracker planned for CNGS is the ICARUS experiment.[40,41] This is a liquid argon time projection chamber (TPC): this kind of detector can be considered a "digital bubble chamber". Particles create ionization charge as they lose energy. This charge is drifted through the liquid argon using an electric field and collected on cathode planes. Using the spatial pattern of the collected charge on the cathode plane, and the drift time for the third dimension, a 3D track can be reconstructed with excellent resolution. A 600 ton detector at Gran Sasso is now awaiting fill with liquid argon.

12.2.1.3. *Summary of atmospheric neutrino parameter space*

In summary for atmospheric neutrino parameter space: Super-K's atmospheric sample was the first clean signature of oscillations, and now a high-statistics sample shows good evidence that the oscillation is dominated by $\nu_\mu \to \nu_\tau$. The atmospheric oscillation was first confirmed by the K2K long-baseline experiment; now MINOS has the highest precision on Δm^2 and is continuing to improve. Coming soon are CNGS beam measurements of explicit τ appearance.

12.2.2. **Solar neutrino parameter space**

We will now consider solar neutrino oscillations. Neutrino oscillations were actually first observed in the disappearance of solar neutrinos, but it it took forty years to confirm the correct interpretation of the deficit. Solar neutrinos, of electron flavor, are produced in fusion reactions in the Sun. The spectrum is shown in Fig. 12.6. The overwhelming bulk of solar neutrinos, at energies less than 0.5 MeV, are known as the "*pp*" neutrinos since they result from the *pp* fusion chain in the Sun. Other interesting features of this spectrum are the ^{8}B neutrinos extending to $\sim$15 MeV and the monochromatic ^{7}Be neutrinos just below 1 MeV. This spectrum is well understood from weak physics.

12.2.2.1. *Radiochemical detectors*

The original celebrated observation of solar neutrinos was by Ray Davis with the chlorine detector in the Homestake mine in South Dakota. Davis used 615 tons of cleaning fluid as a neutrino target. The idea was to look for a handful of argon atoms produced by the reaction $\nu_e + {}^{37}\mathrm{Cl} \to {}^{37}\mathrm{Ar} + e^-$, which has threshold of 0.81 MeV. ^{37}Ar is radioactive with a half-life

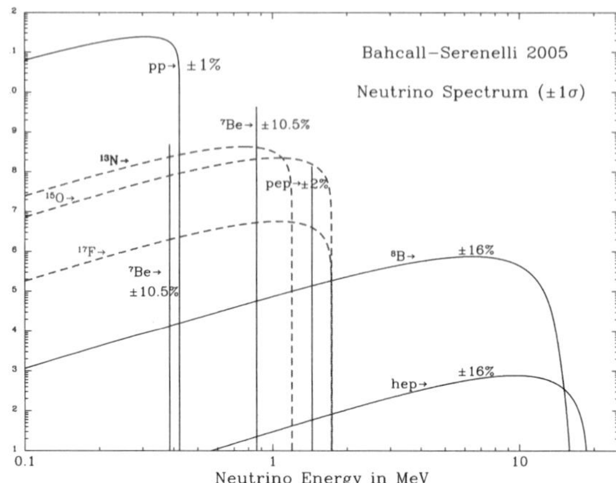

Fig. 12.6.   Predicted spectrum of solar neutrinos.[8,42]

of 35 days. Every few months the chlorine detector would be flushed to extract the argon and the decays counted. It's astounding that this worked! The monthly signal amounted to only about 15 atoms of argon produced, corresponding to about one third of the expected neutrinos from the Sun. This experiment ran from the late 1960's to the 1990's, and showed a deficit throughout this time.[43]

The next radiochemical experiments were based on gallium, and made use of the reaction $\nu_e + {}^{71}\mathrm{Ga} \to {}^{71}\mathrm{Ge} + e^-$. ${}^{71}\mathrm{Ge}$ has an 11-day half-life. The threshold for this reaction is 0.23 MeV, making gallium experiments uniquely sensitive to pp neutrinos. Two gallium experiments have taken data. One, SAGE in the Caucasus mountains in Russia, has been running since 1990.[44]   SAGE is based on 50 tons of metallic gallium kept in a liquid state. The other gallium experiment is GALLEX/Gallium Neutrino Observatory,[45] located at Gran Sasso National Laboratory in Italy, which ran over the period 1991 to2006. This experiment employed 30 tons of gallium in the form of gallium chloride. SAGE and GALLEX both reported deficits with respect to expectation for the solar ν_e flux: only about 60% of expectation was observed.

12.2.2.2. *Water Cherenkov detectors*

Water Cherenkov detection of solar neutrinos hit the scene in the early 1990's. Kamiokande-II, Super-K's 3 kton predecessor at the Kamioka mine in Japan, observed solar neutrinos via elastic scattering (ES) off electrons,

$\nu + e^- \to \nu + e^-$.[46] This process is mostly CC, but has some NC contribution (the NC cross-section is about one sixth the CC one). With a detection threshold of about 7 MeV, Kamiokande II was sensitive only to the ^{8}B component of the solar neutrino flux, but again a deficit was seen. Kamiokande II observed about 40% of the expected solar ν_e flux. This observation was significant because it was the first real-time solar neutrino measurement. Furthermore, due the directional nature of the ES interaction – neutrinos kick the electrons in the direction they're going – and the directionality of the Cherenkov cone, the neutrinos pointed back to the Sun. For the first time the source of the observed neutrinos was directly verified. Super-K has subsequently made precision observations:[47,48] see Fig. 12.7.

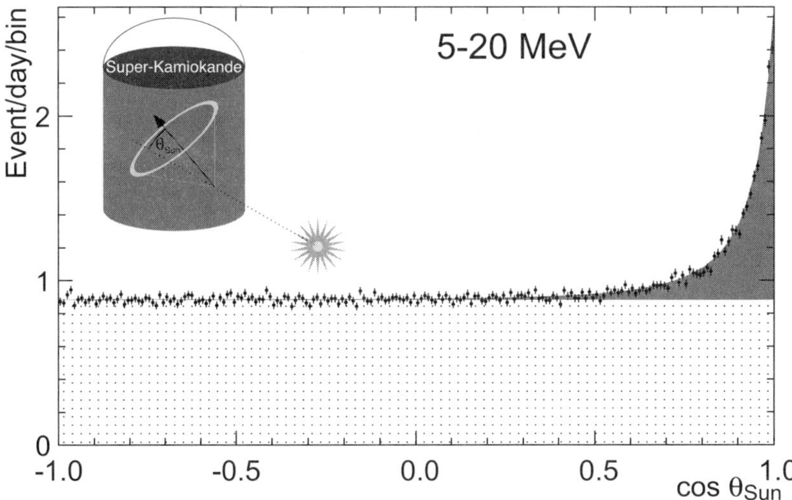

Fig. 12.7. Super-K solar neutrino angular distribution.[47] The shaded region indicates the contribution from solar neutrino elastic scattering.

At the end of the 1990's there was still a puzzle. Three different experiments– chlorine, gallium and water– had observed a deficit of solar neutrinos, and a different amount of deficit in each case. Since each detection method has a different energy threshold, the observed suppression was an *energy-dependent* one. Solar models could be tweaked somewhat to change the absolute flux, but with the spectrum shape well constrained, the data were difficult to explain from uncertainties in solar physics. However, neutrino oscillation does create an energy-dependent suppression, and for many years the prime suspect for the disappearance was $\nu_e \to \nu_{\mu,\tau}$ oscillation. At solar neutrino energies, ν_μ and ν_τ are below CC threshold for interaction with nucleons, and NC interactions with electrons have small cross-sections.

Solar neutrinos must pass through a large quantity of matter before emerging from the Sun, and the presence of matter affects the oscillation probability according to the "MSW" effect, named after Mikheyev, Smirnov and Wolfenstein.[49-51] Electron neutrinos have an CC interaction amplitude with electrons in the Sun; they may oscillate to ν_μ and ν_τ, which have no such CC amplitude with matter in the Sun ($\nu_{\mu,\tau}$ exchange only Z's with matter). Because of this difference, an extra potential term appears in the Hamiltonian, $V = \sqrt{2} G_F N_e$, where N_e is the electron density. This effect leads to a modification of the flavor transition probability, the size of which depends on the vacuum oscillation parameters and the densities of the layers traversed by the neutrinos. Either enhancement or suppression of oscillation is possible. Under some conditions, resonant conversion of neutrino flavors can occur.[†]

At the end of the 1990's, several regions of neutrino oscillation parameter space consistent with the energy-dependent ν_e disappearance from chlorine, gallium and water experiments were allowed. Δm^2 values of around 10^{-10} eV^2 were possible for "vacuum oscillations"; for these parameters MSW does not have a significant effect. Regions at both small and large mixing angle, at larger values of Δm^2 between about $10^{-6} - 10^{-4}$ eV^2, were allowed: these were known as the "large-mixing-angle" (LMA) and "small-mixing-angle" (SMA) regions. Possible experimental signatures of oscillation (at the time known as the "smoking guns") were: an explicit distortion of the electron recoil spectrum, an asymmetry between fluxes measured during the day and the night due to MSW effect in the Earth, and a seasonal variation beyond the 7% expected from ellipticity of the Earth orbit. None of these effects were observed at a significant level, and their absence constrained allowed oscillation parameters. By 2000, although none of the guns were smoking, data still showed an energy-dependent suppression. Employing all information available, large mixing was favored by Super-K combined with gallium and chlorine data.[48]

[†]In fact an MSW-induced resonant conversion was first introduced to explain how a large solar neutrino suppression could occur, given that mixing parameters should be small— clearly they would have to be small, given that quark mixings are small! The MSW effect does occur, although it turns out that the mixing angle describing solar oscillation is actually quite large.

12.2.2.3. *SNO*

Although the solar oscillation parameters was already fairly constrained in the early 2000's, the data from the Sudbury Neutrino Observatory (SNO)[52] made all the difference. SNO was a 1.0 kton heavy water detector located in Sudbury, Canada, that ran from 1999 until the end of 2006. SNO's key feature was excellent sensitivity to neutral currents via the deuteron breakup reaction $\nu_x + d \rightarrow \nu_x + p + n$, and in addition sensitivity to CC breakup $\nu_e + d \rightarrow p + p + e^-$. The NC reaction is sensitive to *all* active flavors in the flux; the CC reaction specifically measures the ν_e flux. SNO was also sensitive to elastic scattering in the 1.7 kton of light water also contained in the detector. The neutrons from deuteron breakup were detected in three ways during three successive phases. In the first phase, without any modifications to the detector, neutrons were detected by capture on d, $n + d \rightarrow t + \gamma + 6.25$ MeV; Cherenkov radiation from electrons Compton-scattered by the γ's was observed. In the second phase, which took place from 2001 to 2003, the water was spiked with NaCl. ^{35}Cl captures neutrons, $n + ^{35}$Cl $\rightarrow ^{36}$Cl$+\gamma+8.6$ MeV. Again the γ-rays were observed via Compton scattering. In the final phase, specialized low-background ^{3}He counters ("neutral current detectors" or NCDs) were deployed in the detector. These detect neutrons via $n + ^3$He $\rightarrow p + t + 0.76$ MeV, and ionization charge from energy loss of the products is recorded in a wire chamber. This third NCD phase of SNO ran from 2004 to 2006.

The first SNO CC results in 2001,[53] which combined early SNO data with Super-K's elastic scattering measurement, gave strong indications that solar neutrinos were oscillating; the ES (with both CC and NC components) combined with ν_e measurement from CC breakup showed that ν_e comprise about two-thirds of the total active flux. Subsequent results making use of SNO's NC detection strengthened the conclusions over all three phases.[54–57] See Fig. 12.8 for a summary. The observed NC event rate is consistent with the full predicted active neutrino flux from the Sun. The SNO data can be said to have definitively "solved the solar neutrino problem": SNO showed that the neutrinos are *not* disappearing, but instead are transforming into more weakly-interacting flavors. The results from SNO's three phases together have not yet been published. Preliminary results can be found in reference [58].

Overall solar neutrino fits show that the mixing is not maximal: see Fig. 12.10.

Traditionally solar oscillation parameters are shown with the horizontal

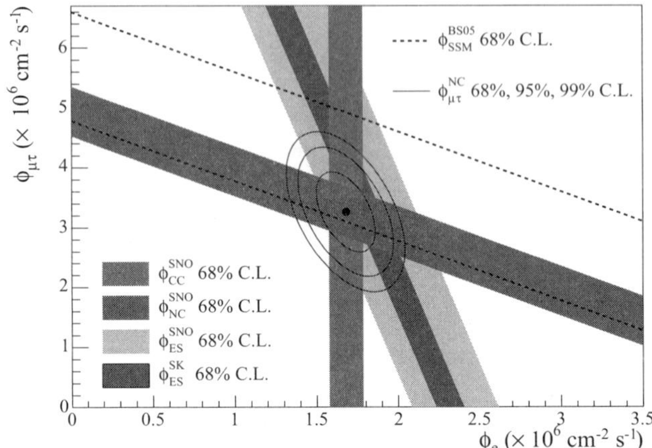

Fig. 12.8. Solar neutrino measurements including SNO data[8,57] after the salt phase: the horizontal axis represents an inferred ν_e flux and the vertical axis corresponds to $\nu_{\mu,\tau}$ flux. A CC measurement of ν_e corresponds to a vertical band on this plot, $\phi_e = \phi_{CC}$, where the width of the band indicates uncertainty. A NC measurement is sensitive to all flavors, so the measured flux is proportional to the total flux: $\phi_{NC} = \phi_e + \phi_{\mu,\tau}$. A NC measurement therefore corresponds to a band on the plot with negative slope, shown here for SNO; the dotted lines indicate the solar model prediction for the total flux. Elastic scattering has both CC and NC contributions, such that $\phi_{ES} = \phi_e + 0.15\phi_{\mu,\tau}$; and the measurement corresponds to band with steeper negative slope. Bands corresponding to both Super-K and SNO ES measurements are shown.

axis as $\tan^2 \theta$ rather than $\sin^2 2\theta$, in order to distinguish between $\tan^2 \theta < 1$ ($0 < \theta \le \frac{\pi}{4}$), the so-called "light side") and $\tan^2 \theta > 1$ ($\frac{\pi}{4} < \theta \le \frac{\pi}{2}$, the "dark side"); this selection is equivalent to a choice of sign of the Δm^2 describing the solar mixing. The MSW effect in the Sun clearly selects the light side[59] for solar neutrino oscillations.

12.2.2.4. *Scintillation detectors*

Scintillation detector technology employs hydrocarbon-based materials, typically $C_n H_{2n}$, which has a large fraction of free protons; such materials emit light when ionization energy is deposited in them by charged particles. The scintillation light is usually detected with PMTs. Scintillation light yield per unit energy loss is typically much higher than for Cherenkov light. For neutrino detection, the positron from inverse beta decay, $\bar{\nu}_e + p \to e^+ + n$, which has a threshold of 1.8 MeV, can be seen via scintillation light from its ionization energy loss. If the detector's energy threshold is low enough, Compton-scattered γ's from the positron annihilation may also be visible. Furthermore, the produced neutron is moderated and captured on a proton to produce a 2.2 MeV γ. The timescale for the neutron capture process is about 180 μs. The delayed coincidence between the positron and the neutron capture γ (and possibly the annihilation γ's) provides a powerful tag against background. Neutrino energy is also measurable, since the positron energy is related to the neutrino energy. Typically quite low energy thresholds are possible in scintillator thanks to high light yield. Effective threshold for neutrino detection usually depends on radioactive background, which tends to be troublesome in the few MeV range. The inverse beta decay reaction has only very weak directionality,[60] and furthermore scintillation light output is isotropic, so directional information from scintillation detectors is very poor.

12.2.2.5. *Reactor neutrinos and KamLAND*

The final confirmation of solar neutrino oscillation came from a reactor neutrino experiment, which also narrowed the allowed parameters. This measurement is especially significant in that the oscillation hypothesis was verified in a completely different way using an artifical neutrino source of $\bar{\nu}_e$, and a different detection reaction (and without significant matter effect). Neutrinos are produced as a by-product of nuclear reactor power generation. Nuclear reactors produce $\bar{\nu}_e$ from fission reactions, with energies in the few MeV range. The first neutrino detection by Reines and Cowan[61] observed inverse beta decay in a scintillator detector placed close to a nuclear reactor at Savannah River reactor (oscillations were not observed because the baseline was too short).

KamLAND is a 1 kton scintillator detector located in the Kamioka mine, just around the corner from Super-K. The initial goal of KamLAND was to observe reactor neutrinos produced in reactors around Japan and Korea, with baselines in the few-hundred kilometer range. For oscillation parameters in the LMA range, one expects a suppression and a spectral distortion. In 2003, the KamLAND collaboration reported exactly this effect,[62] confirming solar neutrino oscillations in a completely different channel.[‡] More recent data confirms nicely the oscillation as a function of L/E_ν:[63,64] see Fig. 12.9. This result gives considerably improved precision in Δm^2, as shown in Figs. 12.10 and 12.11. KamLAND's next goal is to observe very

---

[‡]It was the same story as for atmospheric neutrinos: "wild" hint confirmed using "tame" neutrinos.

low energy solar neutrinos by elastic scattering; this is a challenging task due to background contamination and requires a heroic scintillator purification program, currently underway.

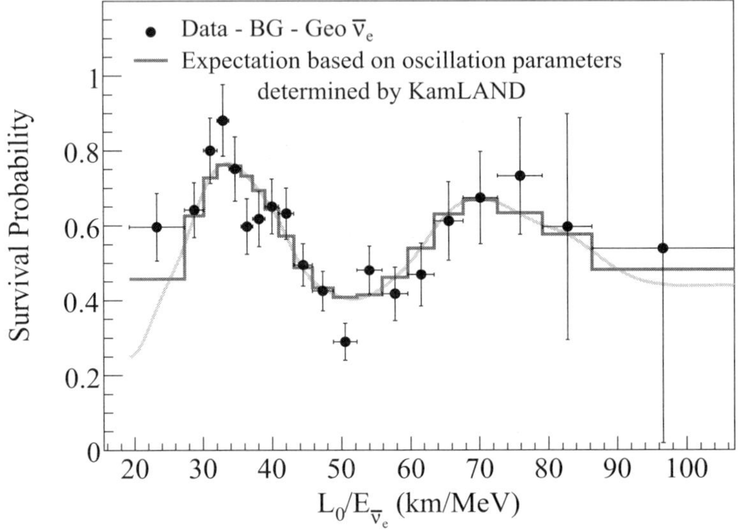

Fig. 12.9. KamLAND data:[8,64] the plot shows survival probability (ratio of measured number of neutrinos to expectation in the bin, after background subtraction) as a function of L_0/E_ν, where $L_0 = 180$ km is the average baseline and E_ν is the reconstructed energy of the antineutrino.

12.2.2.6. *Borexino*

Another experiment at Gran Sasso in Italy has made real-time detection of low-energy solar neutrinos its primary goal from the beginning. Borexino, a 0.3 kton scintillator experiment,[66] has achieved very low radioactive backgrounds for sub-MeV threshold capability. In 2007, Borexino showed a first result for the elastic scattering of solar ^{7}Be neutrinos, which appears as a shoulder in the deposited-energy distribution:[67,68] see Fig. 12.12. The measurement is consistent with expectation given the oscillation parameters from KamLAND, SNO and Super-K. Exotic models which would yield interesting effects at low energy may be constrained by these data.[69]

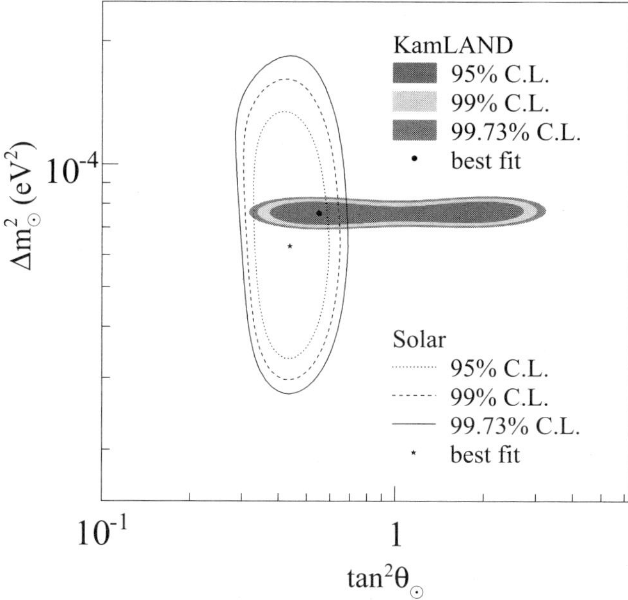

Fig. 12.10. Fit to the solar neutrino data (lines), with fit to the KamLAND data superimposed (shaded region).[8,64]

12.2.2.7. *Next-generation solar neutrino experiments*

In the next few years we can expect more and better data with reduced statistics and improved statistics from KamLAND and Borexino. A new experiment, SNO+,[70] is planned for SNOLAB in Sudbury, Canada: the acrylic vessel which previously held SNO's D_2O will be refilled with scintillator. (Because scintillator is lighter than water, it will need to be held down with ropes.) Other large scintillation detector proposals are in the works. *e.g.* LENA in Europe.[71] Such experiments will serve also for detection of neutrinos from reactors, radioactivity in the Earth, and supernovae.

Additional activity towards future solar neutrino detection lies on the low energy frontier. The *pp* neutrinos, which lie below 0.5 MeV, comprise a vast fraction of the total solar neutrino flux. So far, the *pp* neutrinos have been detected by the gallium experiments, but not in real-time and with no energy resolution. New detector technology is currently under development with the aim of observing this flux. Because the *pp* flux is so large, such detectors can be relatively small, *e.g.* 10 ton scale. The challenge is to defeat the radioactive background, which tends to be enormous at

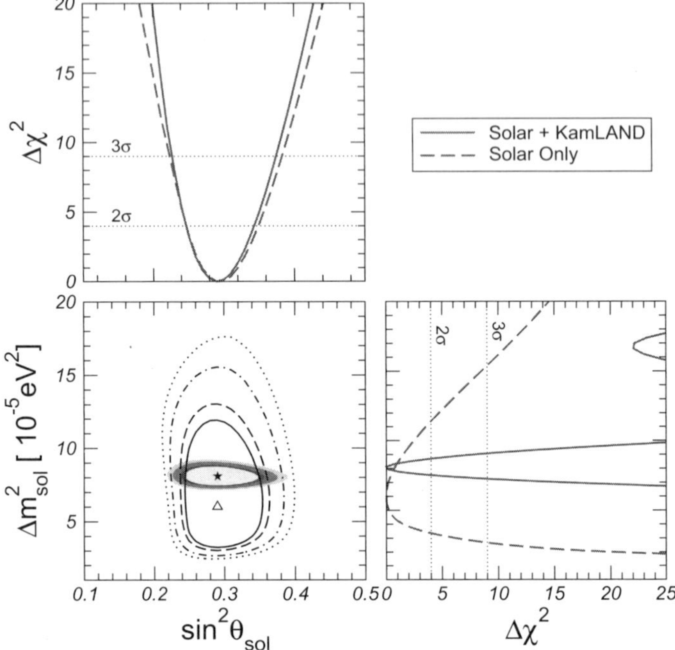

Fig. 12.11. Allowed parameters for a global fit of solar and KamLAND data from reference (shaded region); allowed parameters for the solar data only are shown by lines.[65]

energies below a few MeV. Various target materials and technologies are under development. Examples include XMASS,[72] based on liquid xenon; CLEAN/DEAP,[73] based liquid neon and argon, and LENS,[74,75] based on indium-loaded scintillator. Noble liquids are especially promising because they are easy to purify and can be scaled to relatively large sizes. Such detectors may serve also for direct dark matter searches.

12.2.3. *LSND Parameter Space*

12.2.3.1. *LSND*

I will discuss here another reported observation of neutrino oscillation which attracted considerable attention over the last 15 years. The Liquid Scintillator Neutrino Detector at Los Alamos[76] ran in the early 1990's. This was a 167 ton scintillator experiment (making use of both scintillation and Cherenkov signal in the scintillator) located 30 m away from the source of

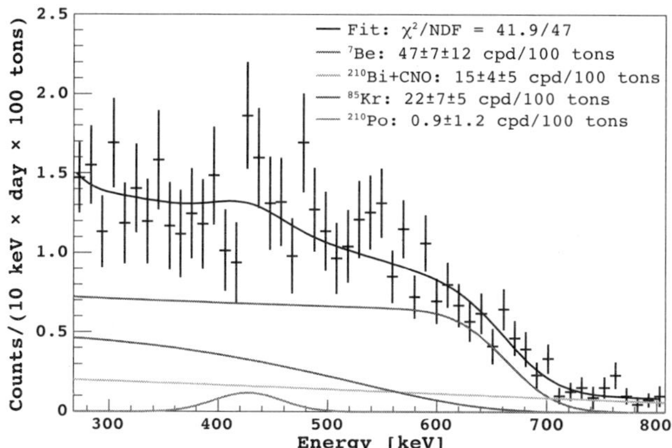

Fig. 12.12. Background-subtracted electron neutrino recoil spectrum from Borexino,[67] showing contribution from ^{7}Be solar neutrinos.

neutrinos. The neutrino source was a stopped-pion source: protons from LANSCE hit a target; these pions stopped in the target and most decayed at rest, producing a monochromatic ν_μ and a μ^+; the muon then decayed at rest to produce a $\bar{\nu}_\mu$ and a ν_e. In this decay chain, because negative π^- are captured on nuclei in the target with high probability, only a tiny fraction of $\bar{\nu}_e$ should be present in the flux. The goal of this experiment was to search for $\bar{\nu}_\mu \to \bar{\nu}_e$ oscillations, since the presence of a significant amount of $\bar{\nu}_e$ can in principle only be due to oscillations. The $\bar{\nu}_e$ can be tagged via the delayed neutrons from inverse beta decay. In 1996, LSND reported[77] an excess of $\bar{\nu}_e$ candidate events over their expected background, which they interpreted as a two-flavor oscillation with parameters. The final published analysis from 2001[78] reports an excess of $87.9 \pm 22.4 \pm 6.0$ signal events, corresponding to $\bar{\nu}_\mu \to \bar{\nu}_e$ oscillation parameters in the range Δm^2 of 0.2-10 eV^2 and $\sin^2 2\theta$ between 10^{-3} and 4×10^{-2}. Although a range of mixing angles are allowed by the data (see shaded region in Fig. 12.13), the large mixing angle part of the region is excluded by the Bugey reactor disappearance experiment.[79] In 1997, the LSND collaboration reported an excess of ν_e at higher energy, which they interpreted as $\nu_\mu \to \nu_e$ oscillation from pions decaying in flight;[80] in the final results this excess became less significant,[78] although the decay-in-flight data were not inconsistent with the fitted $\bar{\nu}_\mu \to \bar{\nu}_e$ oscillation parameters.

12.2.3.2. *KARMEN*

Another stopped-pion neutrino source experiment, KARMEN (Karlsruhe Rutherford Medium Energy Neutrino Experiment) with 56 tons of scintillator located 17.5 m from the ISIS neutrino source at Rutherford-Appleton Laboratory, ran in the 1990's. This experiment had less target mass and a shorter baseline, so its oscillation parameter sensitivity did not overlap entirely with LSND's. It did have the advantage of a pulsed beam, allowing better background rejection. KARMEN observed 15 candidate $\bar{\nu}_e$ events, with an expectation of 15.8 ± 0.5.[81] This absence of excess rules out most, but not all, of the LSND allowed oscillation space[82] (see Fig. 12.13).

12.2.3.3. *MiniBooNE*

The MiniBooNE experiment at Fermilab was designed to test the LSND oscillation hypothesis with the same L/E_ν, but higher baseline ($\sim$500 m) and energy ~ 1 GeV). The experimental setup is sufficiently different from LSND's that systematic uncertainties should be rather uncorrelated. MiniBooNE makes use of neutrinos from the FNAL Booster along with a dedicated horn and decay pipe, creating a beam of primarily ν_μ with a spectrum peaking around 700 MeV. The detector contains 1 kton of scintillator, and employs primarily Cherenkov radiation for event reconstruction. The experiment is challenging; the signature of $\nu_\mu \to \nu_e$ (for a beam consisting primarily of muon-flavor neutrinos) is an energy-dependent excess of single-ring electron-like events from quasi-elastic CC $\nu_e + n \to e^- + p$ in the scintillator. Many kinds of background (including intrinsic beam ν_e) can fake such events. In April 2007, MiniBooNE reported first oscillation results.[83] MiniBooNE observed no evidence of an energy-dependent excess of ν_e in a sample of events with reconstructed neutrino energy above 400 MeV, which is in the energy range where one would expect to see an excess if the oscillation explanation of the the LSND $\bar{\nu}_e$ excess is valid (see Fig. 12.13). The MiniBooNE results lead to exclusion of essentially all of the allowed LSND parameters: reference [84] examines compatibility. However, a new anomaly has turned up: the MiniBooNE collaboration observed an excess of e-like events at energies below 400 MeV, and this anomaly persists despite exhaustive checks of systematic effects.[85] Observation of off-axis events from the NuMI beam seen in MiniBooNE[86] are in agreement with predictions.

MiniBooNE is now running with a beam of $\bar{\nu}_\mu$, to evaluate the possibilty that neutrinos and antineutrinos do not exhibit the same behavior

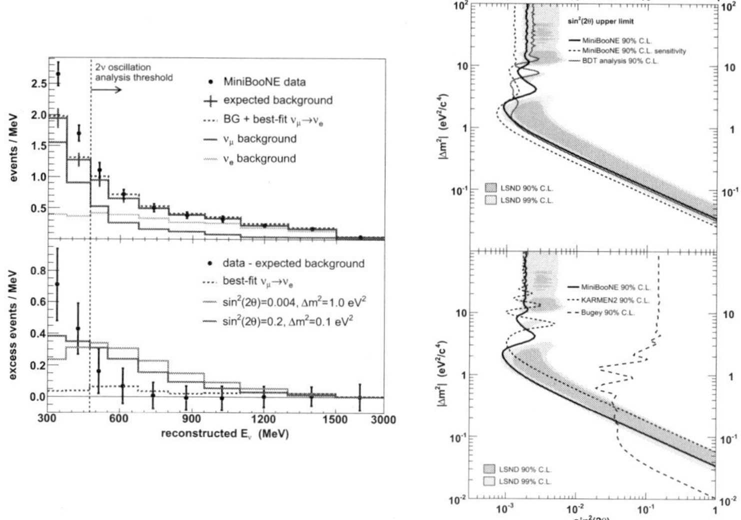

Fig. 12.13. Left: data from the MiniBooNE ν_e search analysis from reference [83]. The top plot shows the event energy distribution, with no evidence of excess above the analysis threshold of 400 MeV. The bottom plot shows excess as a function of energy. The right hand plots show excluded regions from the same analysis, with the bottom plot showing excluded regions from other experiments also.

(which would require some extraordinary, if not unthinkable, theoretical explanation). Preliminary results in December 2008[87] show neither LNSD oscillation excess nor low energy excess, although statistics so far are low. Efforts to further explore the two persistent anomalies (the LSND excess and the MiniBooNE low energy excess) include OscSNS,[88] a proposed new detector at the intense stopped-pion neutrino source available at the Spallation Neutron Source in Oak Ridge, TN; and MicroBooNE, a liquid argon TPC to explore the MiniBooNE low energy excess in the neutrino beam at FNAL.[89]

12.2.4. *Summary of two-flavor oscillation signals*

In summary, there are two firm signals of neutrino oscillation, First, oscillation described by $\nu_\mu \to \nu_\tau$ at parameters $\Delta m^2 = 2 \times 10^{-3}$ eV^2, and near-maximal $\sin^2 2\theta$, first seen in atmospheric neutrinos, was later confirmed by beam experiments. Second, solar neutrino $\nu_e \to \nu_{\mu,\tau}$ disappearance described by parameters $\Delta m^2 = 8 \times 10^{-4}$ eV^2, and large mixing angle, $\tan^2 \theta = 0.5$, has been confirmed with reactor neutrinos. We have two mass-squared differences and two mixing angles: by convention the masses in the solar mixing are m_1 and m_2, so that the solar Δm^2 is $\Delta m^2_{12} = m_1^2 - m_2^2$. The atmospheric mixing is then between mass states m_2 and m_3, and the atmospheric Δm^2 is $\Delta m^2_{23} = m_2^2 - m_3^2$ (the sign is unknown).

In both cases, we are entering an era of precision measurements. Since the LSND oscillation does not fit in to the three-flavor picture (especially in light of recent MiniBooNE results), I will ignore the LSND parameters for the rest of this review. Of course it's possible that some future experimental results will require revision of the standard three-flavor picture.

12.2.5. *What's Next*

As noted in Sec. 12.2, the large difference between the two mass-squared differences means that neutrino oscillations may "decouple", and an experimental observation may be well described by two-flavor oscillation.

However our model of neutrino mixing requires a 3×3 mixing matrix, and the oscillation is described by a total of six parameters, of which four are known. Three mixing angles and a CP-violating phase are present in the MNS matrix: only two of the mixing angles are known. The mixing matrix can be written out as a product of three "Euler-like" rotations:

$$U = \begin{pmatrix} 1 & 0 & 0 \\ 0 & c_{23} & s_{23} \\ 0 & -s_{23} & c_{23} \end{pmatrix} \begin{pmatrix} c_{13} & 0 & s_{13}e^{-i\delta} \\ 0 & 1 & 0 \\ -s_{13}e^{i\delta} & 0 & c_{13} \end{pmatrix} \begin{pmatrix} c_{12} & s_{12} & 0 \\ -s_{12} & c_{12} & 0 \\ 0 & 0 & 1 \end{pmatrix} \qquad (12.2.5)$$

where "s" represents sine of the mixing angle and "c" represents cosine. The "1-2" matrix describes solar mixing; the "2-3" matrix describes atmospheric neutrino mixing. The third angle, θ_{13}, the angle describing the "twist in the middle", is still unknown, although it is known to be small. The CP-violating phase δ is also unknown.[§]

Another unknown is the absolute mass scale, since oscillation measurement inform us only on mass *differences*. Although two of the mass-squared differences are known, we do not know how the three masses are arranged: there could be two light ones and a heavy one (the "normal" hierarchy) or two heavy ones and a light one (the "inverted" hierarchy): see Fig. 12.14.

The aim of future neutrino experiments is to hunt down these unknowns.

[§]There are also "Majorana phases", which do not affect oscillation probabilities. Although these are observable in principle in double beta decay experiments,[90] such a measurement is so far away from reality that I will leave the issue aside here.

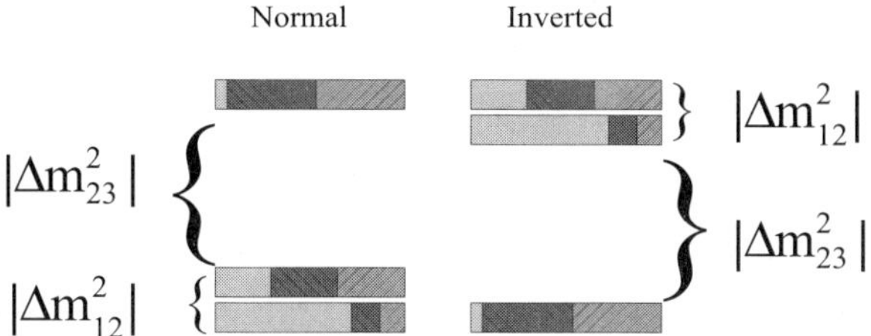

Fig. 12.14. Schematic of normal and inverted hierarchies, under the assumption of three neutrino flavors, with mass states represented by horizontal bars. The mass-squared values are indicated by the vertical dimension, and possible flavor composition is indicated by the horizontal divisions (ν_e on the left, ν_μ in the middle, and ν_τ on the right).

The next promising quarry for neutrino oscillation experiments is the unknown mixing angle θ_{13}.

12.2.5.1. *Current knowledge of* θ_{13}

The best knowledge so far about the value of θ_{13} comes from the Chooz reactor experiment, which ran in France in the late 1990's.[91] Palo Verde[92] made similar measurements, although limits were not as stringent. The Chooz detector comprised 5 tons of gadolinium-loaded scintillator and it was located ~ 1 km from two nuclear reactors. Gadolinium captures neutrons (on a 20 μs timescale) with a huge cross-section, and subsequently emits a cascade of gamma rays adding up to ~ 8 MeV; these γ's can be used to tag the neutron from inverse beta decay with very high efficiency. Chooz looked for spectral distortion of the positron spectrum from inverse beta decay. No significant distortion was observed, and hence a limit was set on the disappearance oscillation parameters for the Δm^2 greater than about 10^{-3} eV^2, range, down to $\sin^2 2\theta$ of about 0.1. Interpreted in a three-flavor context, this corresponds to a limit of about 0.05 on $\sin^2 \theta_{13}$ in the known Δm^2 range (see Fig. 12.15). Therefore we know that θ_{13} is small.

Information from atmospheric neutrinos is consistent with this limit. If θ_{13} is non-zero, one expects a small enhancement of upward-going electron neutrinos in the 5 to 10 GeV energy range, due to MSW resonant enhancement as the neutrinos traverse matter in the Earth. The Super-K three-flavor atmospheric neutrino analysis[93] is optimized for a θ_{13} search by selection of an enriched high-energy electron neutrino signal. No significant enhancement of upward-going high-energy ν_e events is seen, leading to allowed values of θ_{13} as shown in Fig. 12.15. This analysis is statistically limited.

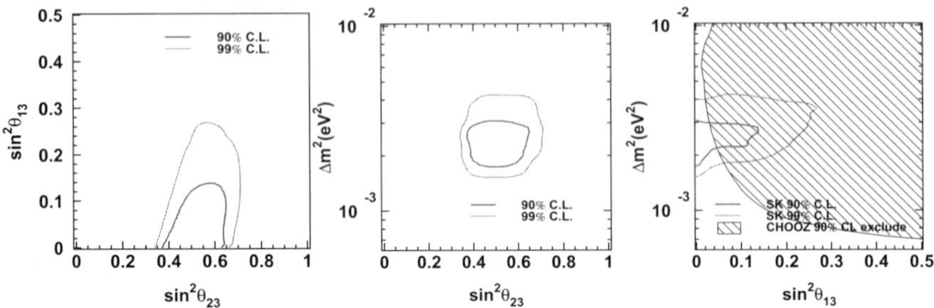

Fig. 12.15. Three-flavor allowed regions from the Super-K I atmospheric neutrino analysis.[93] The right hand plot shows also the exclusion region from the Chooz experiment.

MINOS also has recent preliminary ν_e appearance search results.[94]

There are two approaches to learning the value of θ_{13}, which are complementary. First, reactor experiments can perform disappearance experiments, *i.e.* look for spectral distortion. Second, one can look for a tiny appearance of electron neutrinos in a beam of muon neutrinos.

In both cases one is looking for a small modulation, so both high statistics and small systematic uncertainties are important to the experimental approach. Vigorous experimental efforts are proceeding for both reactor and beam approaches.

12.2.5.2. *New reactor experiments*

Three new reactor experiments are planned to begin data-taking over the next several years. The disappearance probability for $\bar{\nu}_e$ is given by $P(\bar{\nu}_e \rightarrow \nu_x) = \sin^2 2\theta_{13} \sin^2(\Delta m_{13}^2 L/4E_\nu)$. At reactor neutrino energies, the oscillation length is a few kilometers. Since the expected modulation is proportional to $\sin^2 2\theta_{13}$, which is known to be small, systematic uncertainties of less than 1% are required. This is experimentally challenging. For this reason, next-generation reactor neutrino experiments employ both near and far detectors, to cancel systematic uncertainties in the ratio between oscillated and unoscillated fluxes to the extent possible.

Double Chooz: Double Chooz,[95,96] the successor to Chooz, is located at the Chooz nuclear power station in France, and two 10 ton Gd-loaded scintillator detectors will observe neutrinos from two 4.27 GW reactor cores. The near detector has a 400 m baseline, and the far detector has a 1050 m baseline. With a planned start of data-taking in mid-2009, Double Chooz expects a sensitivity to $\sin^2 2\theta_{13}$ of about 0.06 within the first two years, with only the far detector running (the original Chooz). This sensitivity will improve as systematics decrease with the near detector in operation for another few years, down to about $\sin^2 2\theta_{13}$ of 0.03.

Daya Bay: Since Double Chooz incorporates the original Chooz site, the baseline is a bit shorter than optimal. A completely new experiment, the Daya Bay experiment in China,[97,98] plans an array of eight 10-ton detectors to be deployed in three locations around the Ling Ao nuclear reactor complex. Baselines range from a few hundred meters to 2000 m. The plan for Daya Bay is to build a railway so that detectors can be moved from place to place, to observe a baseline dependence of any disappearance signal (and cancel systematic uncertainties). This experiment is planned to start in mid-2010 and its eventual goal is a sensitivity to $\sin^2 2\theta_{13}$ of about 0.008.

RENO: Another activity is the RENO experiment at Yonggwang in Korea,[99] employing two Gd-loaded scintillation detectors. The planned start date is mid-2010, and estimated 90% CL sensitivity to $\sin^2 2\theta_{13}$ is about 0.018.

12.2.5.3. *New beam experiments*

Another approach to a θ_{13} oscillation search is to search for a small appearance of ν_e in a high-intensity beam of ν_μ. The probability of $\nu_\mu \to \nu_e$ (under the assumption that $\Delta m_{23}^2 >> \Delta m_{12}^2$ and $E_\nu \sim L\Delta m_{23}^2$) is given by $P(\nu_\mu \to \nu_e) \sim \sin^2 2\theta_{13} \sin^2 \theta_{23} \sin^2(\Delta m_{23}^2 L/4E_\nu)$. Here Δm_{23}^2 and θ_{23} are known. The overall modulation is small, so good statistics and a clean sample are important for a measurement (note that an appearance signal requires smaller statistics than a disappearance one for equivalent significance). There are two planned long-baseline beam experiments aiming for a θ_{13} measurement (and also aiming to further improve precision on the 2-3 mixing measurement).

Both of these beam experiments employ a strategy to improve systematics and background: this is the "off-axis beam". Here the idea is that due to the two-body parent pion decay kinematics, away from the beam axis the neutrino energy becomes relatively independent of the pion momentum, and the neutrino spectrum will have a narrow peak. If you site a detector a few degrees off the beam axis, there will be an overall loss of neutrino flux, but the sharp spectral peak allows tuning to maximize flux at the oscillation minimum and allows reduction of background. Overall the measurement will have improved systematic uncertainties.

T2K: the "Tokai to Kamioka" experiment[100] comprises a high intensity beam produced at the new J-PARC 50 GeV proton synchrotron, sent 295 km across Japan to Super-K IV (with freshly refurbished electronics). The total beam power in the first phase is 0.75 MW, and the neutrino intensity will be about 50 times higher than for K2K. The detector is 2.5 degrees off the beam axis; the neutrino energy spectrum will be peaked at less than 1 GeV and is optimized for oscillation physics. The beam will turn on in 2009. About 1600 neutrino events per year are expected, after oscillations. The backgrounds are neutrino interactions which can fake single-ring e-like events, which are the signal events for ν_e appearance. Backgrounds include mis-identified ν_μ, intrinsic beam ν_e, and NC π^0 production: π^0's decay electromagnetically to two γ-rays, which shower and look like electrons. To carefully characterize the backgrounds, T2K plans a near detector complex at 280 m from the target station. This detector complex includes fine-grained trackers and a detector optimized for π^0 detection. The sensitivity of T2K to $\sin^2 2\theta_{13}$ over five years of running (assuming δ of 0,π) is about ~ 0.008 (note that sensitivity is dependent on δ and hierarchy). Few-percent precision on the 2-3 mixing parameters Δm_{23}^2 and $\sin^2 2\theta_{23}$ will also be achieved.

NOνA: Another off-axis neutrino beam θ_{13} search, NOνA, is planned in the United States.[101] ¶ For this program, the Fermilab NuMI beam will be upgraded from 400 to 700 kW, and a 15 kton scintillating tracking detector at an 810 km baseline will be constructed. The neutrino spectrum for NOνA will peak at a few GeV. The NOνA detector is made of long plastic cells filled with liquid scintillator; light is collected by optical fibers and the converted to electronic signals using avalanche photodiodes. NOνA is a tracking detector, in which high energy neutrino events are categorized according to their topology (as for the Soudan and MINOS detectors; see Section 12.2.1.2). NOνA sensitivity to θ_{13} will reach about 0.01 in 6 years, and precision measurements of the 2-3 mixing parameters will also be possible.

12.2.5.4. *CP Violation in neutrino oscillation*

While the current cohort of oscillation experiments is going after a measurement of θ_{13}, the long-term goal is to observe CP violation in the lepton sector. CP violation can be parameterized by δ in the MNS mixing matrix

---

¶In Japan, the far detector exists and new beam construction was required; in the U.S., the NuMI beam exists and a new detector is required.

(see Sec. 12.2), and information may be extracted from measurements of the transition probabilities for neutrinos and antineutrinos, *e.g.* $\nu_\mu \to \nu_e$ and $\bar\nu_\mu \to \bar\nu_e$.[102] However extraction of information about δ from these observables is not completely straightforward, because transition rates depend on *all* of the MNS parameters; furthermore, matter effects come into play. For example, in the notation of reference [103]: the approximate transition probabilities for neutrinos and antineutrinos (assuming θ_{13}, $\Delta_{12}L$ and Δ_{12}/Δ_{13} are all small) are:

$$P_{\nu_e\nu_\mu(\bar\nu_e\bar\nu_\mu)} = s_{23}^2 \sin^2 2\theta_{13} \left(\frac{\Delta_{13}}{\tilde B_\mp}\right)^2 \sin^2\left(\frac{\tilde B_\mp L}{2}\right) \tag{12.2.6}$$

$$+ c_{23}^2 \sin^2 2\theta_{12} \left(\frac{\Delta_{12}}{A}\right)^2 \sin^2\left(\frac{A L}{2}\right)$$

$$+ \tilde J \frac{\Delta_{12}}{A} \frac{\Delta_{13}}{\tilde B_\mp} \sin\left(\frac{AL}{2}\right) \sin\left(\frac{\tilde B_\mp L}{2}\right) \cos\left(\pm\delta - \frac{\Delta_{13} L}{2}\right),$$

where

$$\tilde J \equiv c_{13} \sin 2\theta_{12} \sin 2\theta_{23} \sin 2\theta_{13}, \tag{12.2.7}$$

L is the baseline, $\Delta_{ij} \equiv \frac{\Delta m_{ij}^2}{2E_\nu}$, $\tilde B_\mp \equiv |A \mp \Delta_{13}|$, and A is the matter parameter $A = \sqrt{2}G_F N_e$, where G_F is the Fermi constant and N_e is the electron density of the matter traversed. The upper sign in $\pm$ refers to neutrinos and the lower to antineutrinos. The first two terms are the "non-CP" terms and the last term depends on the CP-violating phase.

Clearly one needs precision measurements of all parameters, and *multiple measurements*, if possible including both neutrinos and antineutrinos, to resolve all ambiguities (see *e.g.* [104–106]). Matter effects in particular can be considered both a help and a hindrance: they cause a matter-dependent asymmetry in transition rates between neutrinos and antineutrinos, which can mask CP violation. But the sign of the modification depends on the hierarchy: therefore one may in principle use long baseline oscillation measurements to distinguish between normal and inverted hierarchies.

With the next set of experiments, there's some hope of resolving the mass hierarchy if parameters are favorable. With its longer baseline through matter, NOνA by itself has some sensitivity, and this improves in combination with T2K.[107] Beyond NOνA and T2K, next-phase high-intensity beam programs are under serious consideration in the U.S. and Japan. Discussion

600   K. Scholberg

of future megadetectors has also been lively in Europe,[108] although the possibility there of a new high-intensity neutrino beam seems somewhat farther in the future.

T2K Phase II: For the next phase of T2K, the J-PARC beam will first be upgraded to 1.66 MW, and perhaps eventually 4 MW will be achieved. The vision for a next-generation far detector is Hyper-Kamiokande,[109] a new 0.5 Mton water detector at the Kamioka site, possibly in combination with a second detector at a 1000-1250 km baseline in Korea (the "T2KK experiment")[110,111] (see Fig. 12.16). Intermediate locations are under consideration as well: a candidate site is Okinoshima Island at 660 km from J-PARC. This program could also include different detector technology, such as liquid argon. The physics reach of such a program program (for CP violation and resolving the hierarchy) is described in references [110,112].

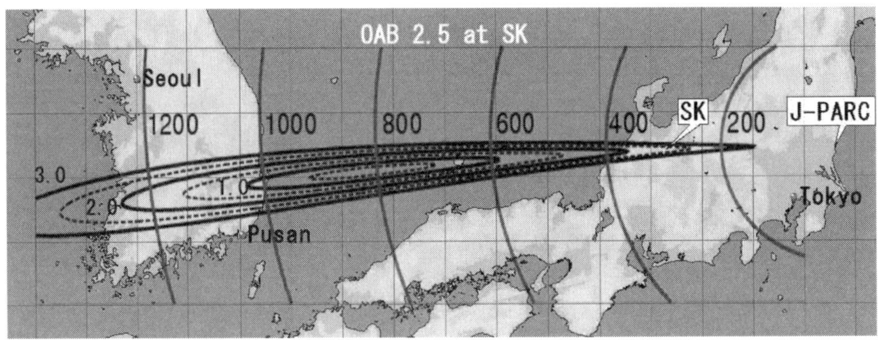

Fig. 12.16.   A possible neutrino beam from J-PARC to Korea: the solid and dashed lines indicate contours of equal angle.

Project X and DUSEL: Future "superbeam" projects are in the works in the United States as well. At Fermilab one possibility is to upgrade NuMI for a narrow-band off-axis beam. Under serious consideration at Fermilab is "Project X", which includes new high-intensity neutrino beam making use of an 8 GeV LINAC front-end, the Recycler and an upgraded Main Injector. "Project X" includes in addition beams for diverse other physics programs, as well as R&D for the International Linear Collider.[113] The prime candidate for a far site for this long baseline beam is DUSEL (the Deep Underground Science and Engineering Laboratory) proposed for Homestake, SD, at a baseline of 1300 km.

Possible future large detectors: There are two leading possibilities for the detector technology for a next-generation long-baseline neutrino detector.

First, water Cherenkov detectors like Super-K have very well-understood technology and employ very inexpensive target material. However, a disadvantage is that water Cherenkov detectors have intrinsically imperfect tracking (since final state particles from a neutrino interaction may be below Cherenkov threshold and invisible), making event reconstruction and background rejection more difficult. In contrast, liquid argon TPC technologies can be more expensive per unit target mass (although may have cost savings because needed caverns are smaller), and require more careful safety considerations. They are also unproven at very large scales. However, they offer much superior tracking capabilities, efficiency and background rejection. As a rule of thumb, 400 kt of water can be considered equivalent to 100 kt argon in physics reach.[114] In both cases, there exists also a rich physics program of non-accelerator particle physics, *e.g.* baryon number violation searches, and astrophysical (atmospheric, solar, supernova) neutrino searches. At this time, the community is entertaining both options. If resources are sufficient, both kinds of detectors will provide complementary information. A long-baseline/DUSEL program offers improvement of several orders of magnitude in sensitivity to θ_{13} and δ parameters. References [113,114] explore the physics reach of such a program.

INO: Another interesting project is the Indian Neutrino Observatory (INO),[115] to be located in southern India (6560 km from J-PARC). The ICAL tracking detector for this laboratory will consist of layers of iron target interspersed with resistive plate chambers. ICAL will be magnetized, allowing the experiment to distinguish between neutrinos and antineutrinos. Such capability has potential to improve understanding of atmospheric neutrinos.[116] INO could in principle also serve as a target for a future high intensity very long baseline neutrino factory beam.[117]

Neutrino factories: A neutrino factory is the current dream of neutrino experimentalists. The idea is to create a muon storage ring (also of potential use for even farther-future muon collider applications) and allow the stored muons to decay. The advantage is that such a factory would produce very high fluxes of very well-understood neutrinos: the spectrum of neutrinos from muon decay is very well known. A "golden channel" for oscillation searches is the "wrong sign leptons": for example, positive μ^+ stored in the ring decay via $\mu^+ \to \bar{\nu}_\mu + \nu_e + e^+$. Non-oscillated neutrinos of these $\bar{\nu}_\mu$ and ν_e flavors interact to make positive muons and electrons. If $\bar{\nu}_\mu \to \bar{\nu}_e$ or $\nu_e \to \nu_\mu$ transitions occur, then instead μ^- and e^+ will be detected. Observation of such an oscillation signature requires charge sign determination capability in the detector. A suitable boost could be chosen, and neutrino beams of energies up to 50 GeV could be achieved. A review of what could be learned from a neutrino factor can be found in references [103,118–120]. A neutrino factory presents a number of technical challenges, but R&D is currently underway along a number of fronts.[121,122] The main challenge is to learn how to "cool" the muons sufficiently, and the "ionization cooling" approach seems promising.[123]

Beta beams: An innovative idea for a future neutrino source is the "beta beam".[4] Here the idea is to store radioactive ions in a storage ring; these then decay to either neutrinos or antineutrinos, supplying an intense, clean source. The candidate isotopes are ^{6}He, which produces an antineutrinos via ^{6}He $\to^6$ Li$+\bar{\nu}_e+e^-$ (half-life 0.8 s, 2 MeV $\bar{\nu}$ produced at rest), and ^{18}Ne, which produces electron neutrinos via ^{18}Ne $\to^8$ F $+ \nu_e + e^+$ (half-life 1.7 s, 1.9 MeV ν produced at rest). Again one can design a machine to select the desired Lorentz boost for the ions in storage ring, which determines the neutrino spectrum. Such beams can be used for a variety of oscillation studies. References[124–126] review some of the possibilities. In the most ambitious scenarios with a neutrino factory or beta beams, one could reach sensitivities many orders of magnitude below current ones.

12.2.5.5. *Neutrinos from core collapse supernovae*

A perhaps somewhat less-known way that we may learn about neutrino oscillations is by gathering neutrino data when a nearby core collapse supernova occurs. When the core of a massive star at the end of its life collapses, nearly all of the binding energy of the resulting neutron star is emitted in the form of neutrinos. Because of their weak interactions, the neutrinos are able to escape on a timescale of a few tens of seconds after core collapse. The neutrinos are of all flavors, and have energies in the few tens of MeV range. An initial sharp "neutronization burst" of ν_e (representing about 1% of the total signal) is expected at the outset, from $p + e^- \to n + \nu_e$, and subsequent neutrino flux comes from NC $\nu\bar{\nu}$ pair production. Electron neutrinos have the most interactions with the proto-neutron star core; $\bar{\nu}_e$ have fewer, because neutrons dominate in the core; ν_μ and ν_τ have yet fewer, since NC interactions dominate for these. The fewer the interactions, the deeper inside the proto-neutron star the neutrinos decouple; and the deeper, the hotter. So one expects a flavor-energy hierarchy, $\langle E_{\nu_{\mu,\tau}} \rangle > \langle E_{\bar{\nu}_e} \rangle > \langle E_{\nu_e} \rangle$. This expectation may not or may not be robust (*e.g.* reference [127]), since scattering may degrade this hierarchy of energies.

Supernova neutrino detection techniques are reviewed in reference [128]. Most of the detector types described here so far in different contexts are sensitive to supernova neutrinos, *e.g.* water Cherenkov, scintillator, and noble liquid technologies. The primary flavor sensitivity of water and hydrocarbon detectors is to the $\bar{\nu}_e$ component of the flux, via inverse beta decay interactions on free protons.

A burst of neutrinos was observed from the famous supernova 1987A in the Large Magellanic Cloud (55 kpc away from Earth, just outside the Milky Way): a total of 19 neutrinos were seen in the Kamiokande II detector in Japan (the predecessor of Super-K) and the IMB detector in the United States.[129,130] Additional events were reported in scintillator detectors.[131,132] In spite of the small number of events, this observation was sufficient to confirm the baseline model of core collapse.

An observation of a neutrino burst from a Galactic supernova will bring a tremendous amount of information. We will learn about the astrophysics of the core collapse, the proto-neutron star, the possible formation of a black hole, and perhaps even nucleosynthesis, for which neutrino fluxes during a supernova may be important. A core collapse neutrino burst will in addition bring information about neutrinos themselves: we can learn about mass and oscillations, and potentially test other exotic particle physics (see *e.g.* reference [133]). The more we know about neutrinos, the more astrophysics we will learn, and vice versa: the more refined the astrophysical models, the better our ability to constrain neutrino properties.

In principle, one can learn about neutrino absolute mass from the time-of-flight delay of the neutrinos: if neutrinos have non-zero mass, a difference in energy will lead to a relative time of flight delay, $\Delta t(E) = 0.515(m_\nu/E)^2 D$ (for m_ν in units of eV, E in MeV and D in units of 10 kpc). So a potential signature of non-zero m_ν is an energy-dependent time spread of neutrinos, or a flavor-dependent delay (since average energies of the different emitted flavors are different). However, the delay for a typical supernova neutrino is less than the expected emission time spread of the burst, which limits the usefulness of the technique. Expected limits from observation of a Galactic burst[134] will not improve on current laboratory limits (see Sec. 12.3.2). However if the signal shows unusual sharp features, such as collapse to a black hole, competitive limits may be possible.[135,136] The best limit from SN1987A was about 20 eV.[137]

Supernova neutrinos may also yield information about neutrino oscillation parameters. Flavor transitions in the supernova itself may leave an imprint: since the collapsing star material has a density gradient, there may be MSW-like resonances driving flavor transitions, dependent on oscillation parameters. Hence, one may gain information about the oscillation parameters by measuring spectra of different flavors.[138,139] There has been a good deal of recent work exploring interesting effects such as neutrino-neutrino interactions (see reference [140] for a review of recent work). For example, an anomalously hot ν_e signal may be the signature of non-zero θ_{13}. Conclusions drawn from studies of the energy, flavor and time profile of the neutrinos will not be completely supernova-model-independent, however. A possibility for mitigating dependence on understanding of the supernova is to compare spectra and flavors at different locations around the Earth; matter effects may again leave an oscillation-parameter-dependent imprint on the flavors and spectra. Clearly, sensitivity to different flavors, and ability to tag different flavors is vital for disentangling information from the detection of a burst of supernova neutrinos. Therefore detector locations around the world are also desirable.[141]

Unfortunately, we can expect only a few Galactic core collapses per century, and beyond the edge of the Milky Way (at 20 kpc) the next nearest large concentration of stars is the Andromeda galaxy, 770 kpc away. We expect only about one event in Super-K for a core collapse at that distance. Nevertheless, with next-generation megadetectors, there is some hope of going beyond our Local Group: possible future experiments with such sensitivity include Hyper-K and large detectors at DUSEL mentioned previously as potential long-baseline targets (see Sec. 12.2.5.4), TITAND,[142] and MEMPHYS, a proposed European large water Cherenkov detector.[143] Large liquid argon and scintillator detectors have also been proposed[71,108]. With a Mton-scale detector, a few dozen events may be expected from Andromeda, and one may even conceive of looking beyond. Reference [144] points out that with a Mton-scale detector, one may expect a core collapse signal event per few years, and over a long timescale a large detector may collect a reasonable number of neutrinos. In this regime, tagging signal over background becomes the main challenge, and coincidence with optical signals (or perhaps gravitational waves) may be needed.

It's now even conceivable to aim to detect the "relic supernova neutrinos", or "diffuse supernova neutrino background" (DSNB): these are neutrinos from ancient core collapses which have taken place over the history of the Universe. A measurement of the DSNB flux could yield information about star formation rate, which can in turn constrain cosmological models (*e.g.* references [145,146]). Here the challenge is detector background, since the events come in singly and there is no observable supernova to associate with them. However, there is an energy regime between about 20-50 MeV in which the DSNB flux dominates other neutrino fluxes. The best limit so far comes from Super-K.[147] Since the dominant signal will be $\bar{\nu}_e$, if one can tag these, the DSNB may be observable. Large scintillator detectors,[148] or possibly water spiked with Gd for neutron tagging are possibilities currently being explored.[149]

12.3. Non-Oscillation Neutrino Physics

Understanding neutrino oscillations is clearly a high priority in neutrino physics. However other interesting neutrino physics and astrophysics questions engage considerable activity.

I will now cover a few questions in neutrino physics of great current interest. As mentioned in Sec. 12.2.4, from neutrino oscillation experiments so far, we have learned that at least two neutrino states have non-zero mass. However, neutrino oscillation experiments say nothing about the absolute mass scale. Another vital question is whether neutrinos can be described by two or four states. In the latter case, neutrinos are distinct from their antiparticles: a left-handed neutrino can be turned into a right-handed neutrino via a Lorentz boost, but it will not be an antineutrino. Such neutrinos are known as Dirac particles. In the former case, neutrinos are the same as their antiparticles: a right-handed neutrino behaves as an antineutrino. Such neutrinos are known as Majorana particles. Each of these cases is described by a different Lagrangian, and the answer to the question of the Majorana-vs-Dirac nature of the neutrino, which is an experimental one, is essential for model-building. For instance, if the proposed "see-saw" mechanism for neutrino mass generation[150] holds, neutrinos must be Majorana.

12.3.1. *Cosmology*

One way of addressing the question of absolute neutrino masses is to look on cosmological scales: the field of observational cosmology now has a wealth of data. Non-zero neutrino mass affects galaxy formation, and overall there are a host of other effects on cosmological observables. Global fits to the data– large scale structure, high Z supernovae, cosmic microwave background, and Lyman α forest measurements – yield limits on the sum of the three neutrino masses of less than about 0.3-0.6 eV, although specific results depend on assumptions. Future cosmological measurements will further constrain the absolute mass scale. References [151,152] are recent reviews.

12.3.2. *Kinematic experiments*

The most direct way of probing the question of absolute neutrino mass scale is from kinematic experiments. In nuclear beta decay, $n \to p + \bar{\nu}_e + e^-$, the total energy in the three-body final state must include the rest-mass energy of the neutrino. The endpoint energy of the electron spectrum will therefore be reduced by this energy, and the electron spectrum near the endpoint will be distorted with respect to the expectation for a zero-mass neutrino. Experiments have been attempting to observe such distortion for many years; very high resolution energy measurement is required given the tiny mass of the neutrino. There are two main experimental approaches: spectrometers and bolometers. In the spectrometer approach, the β source (typically tritium, which has an 18.6 keV endpoint) is distinct from the detector and the electron energies are measured by various techniques. In the bolometer approach, the source *is* the detector material, and electron energies are measured via temperature increase of the detector. Existing bolometer designs are based on $^{187}\text{Re} \to ^{187}\text{Os} + e^- + \bar{\nu}_e$, which has a 2.5 keV endpoint.[153] Examples are MIBETA[154] and MANU.[155]

The current best direct kinematic limits on the absolute neutrino mass scale of about 2 eV/c^2 come from Mainz[156] and Troitsk spectrometers,[157] neither of which has observed a distortion consistent with a finite-mass neutrino, and the results are now limited by systematic uncertainties. The current best bolometer limits are in the range of 15 eV.[158]

The next-generation kinematic neutrino mass search, aiming for sub-eV sensitivity, is KATRIN, a spectrometer located at Karslruhe in Germany.[159,160] The next-generation bolometer is MARE.[161] Both collaborations aim for eventual sensitivity of about 0.2 eV/c^2.

12.3.3. *Neutrinoless double beta decay*

A possible way to learn about absolute neutrino mass and at the same time answer the question of the Majorana/Dirac nature of the neutrino is via a search for neutrinoless double beta decay. The process $2n \to p + 2e^-$ is only possible (barring exotic scenarios) if neutrinos are their own antiparticles. Figure 12.17 shows a diagram of this process.

A number of nuclides exist for which the double beta decay process is energetically possible, and for which the nucleus cannot decay by α or single β decay (so that the nucleus is stable). Since two-neutrino double beta decay, $2n \to p + 2e^- + 2\nu$, is also allowed for all nuclides for which neutrinoless double beta is possible, the experimental strategy for identification of

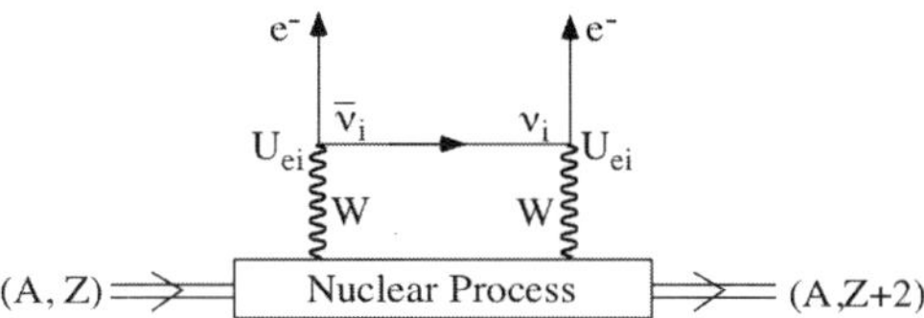

Fig. 12.17.   Diagram for neutrinoless double beta decay.[8]

neutrinoless double beta decay is a kinematic one. One expects a smooth spectrum for the two-neutrino beta decay, for which neutrinos share the available final state energy with the electrons. In contrast, neutrinoless beta decay results in a a sharp electron spectrum, since the electrons take almost all of the energy (nuclear recoil component is small) and the final state is effectively a two-body one. Therefore experiments searching for the neutrinoless double beta decay process must have excellent energy resolution to be able to distinguish the 0ν peak from the 2ν spectrum. Typically a very clean environment is also needed, to avoid background contamination in the region of interest. Since only particular isotopes of a given element are of interest, and natural abundance of the double-beta decay candidates may be small, isotopic enrichment is usually beneficial.

The observed half-life can be related to an effective mass according to $(T_{1/2,0\nu\beta\beta})^{-1} = G_{0\nu}|M_{0\nu}|^2\langle m_{\beta\beta}\rangle^2$, where the effective mass which has contributions from all mass states, is defined as: $\langle m_{\beta\beta}\rangle^2 = |\sum_i U_{ei}^2 m_i|^2$. $G_{0\nu}$ is a phase space factor, m_i is the mass of mass state i, and $M_{0\nu}$ is the matrix element. The matrix element has significant nuclear theoretical uncertainties, dependent on the nuclide under consideration.

There are a large number of current neutrinoless double beta decay search efforts, and I cannot do them all justice here: a comprehensive recent reviews are given in references [162–164]. The best limits on the effective mass are now in the sub-eV range. There is one controversial claim from a subset of collaborators of the Heidelberg-Moscow experiment, who claim a measurement of the process in ^{76}Ge, with 70 kg-years of data.[165] These authors interpret the observation as giving an $\langle m_{\beta\beta}\rangle$ of 440 meV. The result is controversial[166,167] and further confirmation from next-generation germanium experiments (*e.g.* Majorana, GERDA[168,169]) is needed.

The current situation and future view is summarized in Fig. 12.18. The left hand plot shows a parameter space of effective mass versus lightest neutrino mass in eV. The central plot's horizontal axis indicates the sum of the neutrino masses, which is the observable from cosmological measurements. The right hand plot's horizontal axis indicates the average mass $\langle m_\beta \rangle = (\sum_i |U_{ei}|^2 m_i^2)^{1/2}$, which is the observable for beta decay kinematic experiments. Currently allowed values are shown for the scenarios of normal and inverted mass patterns; these merge in the upper right of each plot for a "degenerate" scenario, for which the difference in masses is smaller than the lowest mass. The width of the bands incorporates current experimental uncertainties in the MNS matrix parameters (outer region), and the effect of the unknown Majorana phases (inner shaded region). Current neutrinoless double beta decay limits (ignoring the result of reference [165]) rule out the region above a few hundred meV (*e.g.* Cuoricino[170]). Cosmological limits encroach on the right. The plot shows that if the hierarchy is known to be inverted (from beam or atmospheric experiments), and if the neutrinoless double beta decay limits reach the 10^{-2} eV level, then one can rule out the hypothesis of a Dirac neutrino. An observation of neutrinoless double beta decay at small enough $\langle m_{\beta\beta} \rangle$ can by itself rule out the inverted hierarchy.

I will highlight only a couple of the next-generation experiments here, although many clever efforts are ongoing.[162] The EXO[160,171] experiment makes use of ^{136}Xe, which double beta decays as ^{136}Xe $\rightarrow ^{136}$Ba^{++}+e$^-$+e$^-$. The first version of EXO, EXO-200, is to run at the Waste Isolation Pilot Plant in New Mexico. It makes use of liquid xenon, in which scintillation light from ionization energy loss of the electrons is detected. The EXO collaboration's novel idea is the use of barium tagging: the idea is to reduce background by identifying the resulting nucleus by laser spectroscopy. This ambitious plan– to tag a single ion as much as ten tons of xenon– is currently under development, and there are several schemes under investigation, including gaseous versions of EXO. The expected sensitivity of EXO-200 to the neutrino effective mass is in the 100-200 meV range, and the expected sensitivity of a 10 ton version is less than 10 meV.

Another ambitious idea for a double beta decay experiment is SNO+,[70,172] already mentioned in Sec. 12.2.2.7. SNO+ is an experiment at SNOlab in Canada which plans to refill the acrylic vessel of SNO with liquid scintillator. This experiment would provide a rich physics program of solar neutrino, geoneutrino and supernova neutrino physics, as discussed in Sec. 12.2.2.7. It may also be possible to add 0.1% Nd (possibly enriched with ^{150}Nd, the $0\nu\beta\beta$ decay isotope of interest) to the scintillator. Although typically the energy resolution of such a large detector would not be expected to meet the stringent standards of a neutrinoless double beta decay search, the quantity of dissolved Nd would be so large that the neutrinoless

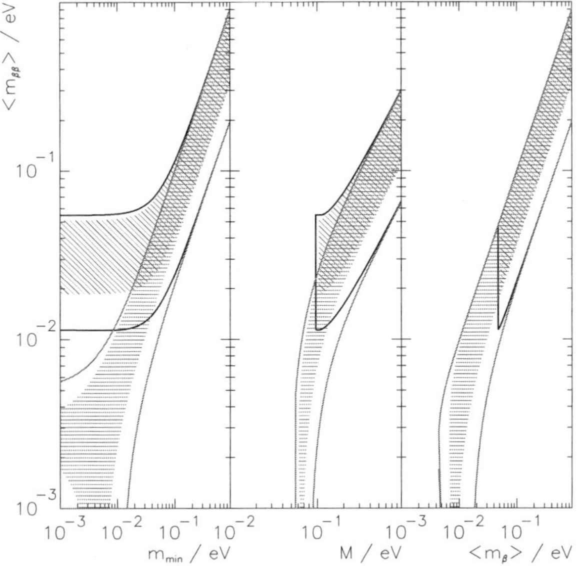

Fig. 12.18.   Neutrinoless double beta decay parameter plots.[163] The vertical axis gives effective mass $\langle m_{\beta\beta} \rangle$ measured in neutrinoless double beta decay. The left hand plot's horizontal axis gives m_{min}, the mass of neutrino state with the smallest mass; the central plot's horizontal axis is the sum of the masses; the right hand plot's horizontal axis indicates average mass. The regions shown are allowed for inverted (upper region) and normal hierarchies. The regions' widths incorporate uncertainties in oscillation parameters; the inner shaded regions include uncertainty due to unknown Majorana phases.

signal would be visible as a clear feature in the spectrum. Sensitivity of SNO+ to $\langle m_{\beta\beta} \rangle$ is predicted to be 100 meV for 3 years of running.

In general, the sensitivity of next-generation neutrinoless double beta decay experiments is in the few hundred meV range. At the ton scale, sensitivity will reach $\sim$20 meV.

12.3.4. *Big Bang relics*

A final topic is the detection of the "relic Big Bang" neutrinos, or "Cosmic Neutrino Background" (CNB). These are neutrinos which froze out of thermal equilibrium when the temperature of the early Universe was about 2 TeV. They should pervade the entire Universe today with a temperature of 1.95 keV and a number density is 113 per cm^3 per flavor (where "flavor" includes both ν and $\bar{\nu}$). They may also be gravitationally clustered.[173] The CNB neutrino energies are so low, and interaction cross-sections are so tiny that detection of the CNB presents a tremendous challenge to experimentalists. Only a few tenuous ideas exist for their detection, and all are problematic. One possibility is coherent elastic scattering off a macroscopic target, using a sensitive Cavendish-style torsion balance. Another suggestion is detection via "Z-bursts" in which ultra-high energy astrophysical neutrinos interact with the CNB at the Z pole, releasing a high-energy cascade of particles, which could conceivably be seen in cosmic ray telescopes, or as an absorption feature in a measured cosmic neutrino spectrum. There could also be interaction signatures using ultra-high energy accelerator beams. In all of these cases, required experimental parameters are currently out of reach.[174,175] Recently the idea of "zero threshold reactions"[176–179] has attracted some interest. In this case a CNB neutrino has a CC interaction with a nucleus which is unstable against beta decay, ($\nu_e + A_Z \to e^- + A^*_{Z+1}$ or $\bar{\nu}_e + A_Z \to e^- + A^*_{Z-1}$), where energy for the interaction is available from the radioactive decay. Interactions with the CNB would be distinguished from natural beta decays kinematically. There are however numerous practical problems with such an experiment; in particular, required energy resolution is not currently achievable.

12.4. Summary

Overall, the field of experimental neutrino physics has experienced tremendous progress over the last dozen years. Beyond a doubt, neutrinos have mass and mix. In the oscillation sector, we know:

- Atmospheric neutrino oscillation is confirmed by beams, and is known to be primarily $\nu_\mu \to \nu_\tau$. Parameters are measured to be in the ranges $2.1 \times 10^{-3} < |\Delta m^2_{23}| < 2.7 \times 10^{-3}$ eV^2, and $\sin^2 2\theta_{23} > 0.96$.
- The solar neutrino disappearance is clearly dominated by oscillations to active neutrinos, $\nu_e \to \nu_{\mu,\tau}$. The solar neutrino disappearance is well explained by oscillations, and the hypothesis confirmed by a reactor experiment. Best-fit parameters are pinpointed to the ranges $7 \times 10^{-5} < \Delta m^2_{12} < 9 \times 10^{-5}$ eV^2, and $0.22 < \sin^2 \theta_{12} < 0.38$.
- The two-flavor oscillation interpretation of LSND has now been strongly disfavored by MiniBooNE. The LSND anomaly is outstanding, and a low-energy excess in MiniBooNE is still unexplained.

The overall oscillation picture is summarized in Fig. 12.19 from reference [8].

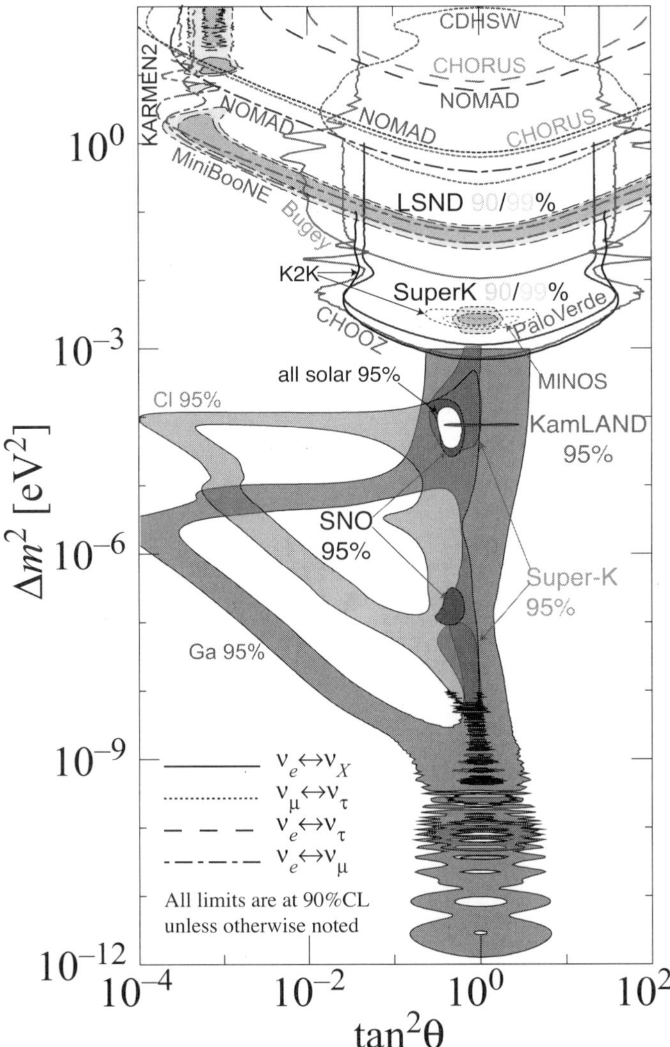

Fig. 12.19. Summary of allowed and excluded two-flavor parameters from various experiments, from reference [8].

The next quest on the horizon is first to measure a non-zero value of the third mixing angle, θ_{13}. Following that, from oscillation experiments we may learn about the mass hierarchy and CP violation in the lepton sector. New beam and reactor experiments to accomplish these goals will soon be running, and future phases are in the planning stages. We may also

K. Scholberg

be lucky, and derive answers from a supernova neutrino burst. There are good prospects for answering some other leading non-oscillation neutrino questions: kinematic experiments will push down sensitivity to absolute neutrino mass to sub-eV levels, and information from cosmological observations will also improve. An array of neutrinoless double beta decay search experiments is poised to answer the question of the Dirac-or-Majorana nature of the neutrino. The next decade will surely bring exciting answers to many of these questions.

References

1. R. N. Mohapatra et al., Theory of neutrinos: A white paper, *Rept. Prog. Phys.* **70**, 1757–1867, (2007). doi: 10.1088/0034-4885/70/11/R02.
2. W. Hampel et al., Final results of the Cr-51 neutrino source experiments in GALLEX, *Phys. Lett.* **B420**, 114–126, (1998). doi: 10.1016/S0370-2693(97)01562-1.
3. P. Vogel, G. K. Schenter, F. M. Mann, and R. E. Schenter, REACTOR ANTI-NEUTRINO SPECTRA AND THEIR APPLICATION TO ANTI-NEUTRINO INDUCED REACTIONS. 2, *Phys. Rev.* **C24**, 1543–1553, (1981). doi: 10.1103/PhysRevC.24.1543.
4. P. Zucchelli, A novel concept for a anti-nu/e / nu/e neutrino factory: The beta beam, *Phys. Lett.* **B532**, 166–172, (2002). doi: 10.1016/S0370-2693(02)01576-9.
5. C. Amsler et al., Review of particle physics, *Phys. Lett.* **B667**, 523, (2008). doi: 10.1016/j.physletb.2008.07.018.
6. A. G. Cohen, S. L. Glashow, and Z. Ligeti. (2008). arXiv:0810.4602.
7. B. Kayser, On the Quantum Mechanics of Neutrino Oscillation, *Phys. Rev.* **D24**, 110, (1981). doi: 10.1103/PhysRevD.24.110.
8. C. Amsler et al., Review of particle physics, *Phys. Lett.* **B667**, (2008). doi: 10.1016/j.physletb.2008.07.018.
9. M. Honda, T. Kajita, K. Kasahara, S. Midorikawa, and T. Sanuki, Calculation of atmospheric neutrino flux using the interaction model calibrated with atmospheric muon data, *Phys. Rev.* **D75**, 043006, (2007). doi: 10.1103/PhysRevD.75.043006.
10. T. Kajita and Y. Totsuka, Observation of atmospheric neutrinos, *Rev. Mod. Phys.* **73**, 85–118, (2001). doi: 10.1103/RevModPhys.73.85.
11. G. D. Barr, T. K. Gaisser, S. Robbins, and T. Stanev, Uncertainties in atmospheric neutrino fluxes, *Phys. Rev.* **D74**, 094009, (2006). doi: 10.1103/PhysRevD.74.094009.
12. Y. Fukuda et al., Atmospheric muon-neutrino / electron-neutrino ratio in the multiGeV energy range, *Phys. Lett.* **B335**, 237–245, (1994). doi: 10.1016/0370-2693(94)91420-6.
13. R. Becker-Szendy et al., The Electron-neutrino and muon-neutrino content

of the atmospheric flux, *Phys. Rev.* **D46**, 3720–3724, (1992). doi: 10.1103/PhysRevD.46.3720.

14. Y. Fukuda et al., Evidence for oscillation of atmospheric neutrinos, *Phys. Rev. Lett.* **81**, 1562–1567, (1998). doi: 10.1103/PhysRevLett.81.1562.

15. Y. Fukuda et al., The Super-Kamiokande detector, *Nucl. Instrum. Meth.* **A501**, 418–462, (2003). doi: 10.1016/S0168-9002(03)00425-X.

16. M. Fechner et al. (2009). arXiv:0901.1645.

17. Y. Ashie et al., A Measurement of Atmospheric Neutrino Oscillation Parameters by Super-Kamiokande I, *Phys. Rev.* **D71**, 112005, (2005). doi: 10.1103/PhysRevD.71.112005.

18. W. W. M. Allison et al., Neutrino oscillation effects in Soudan-2 upward-stopping muons, *Phys. Rev.* **D72**, 052005, (2005). doi: 10.1103/PhysRevD.72.052005.

19. M. Ambrosio et al., Measurements of atmospheric muon neutrino oscillations, global analysis of the data collected with MACRO detector, *Eur. Phys. J.* **C36**, 323–339, (2004). doi: 10.1140/epjc/s2004-01947-5.

20. B. Brahmachari and R. N. Mohapatra, Grand unification of the sterile neutrino, *Phys. Lett.* **B437**, 100–106, (1998). doi: 10.1016/S0370-2693(98)00888-0.

21. M. Bando and K. Yoshioka, Sterile neutrinos in a grand unified model, *Prog. Theor. Phys.* **100**, 1239–1250, (1998). doi: 10.1143/PTP.100.1239.

22. P. Lipari and M. Lusignoli, Comparison of numu to nutau and numu to nus oscillations as solutions of the atmospheric neutrino problem, *Phys. Rev.* **D58**, 073005, (1998). doi: 10.1103/PhysRevD.58.073005.

23. S. Fukuda et al., Tau neutrinos favored over sterile neutrinos in atmospheric muon neutrino oscillations, *Phys. Rev. Lett.* **85**, 3999–4003, (2000). doi: 10.1103/PhysRevLett.85.3999.

24. C. W. Walter, Distinguishing nu/mu → nu/tau oscillations and exotic oscillation / decay hypotheses using Super-K atmospheric neutrino data, *Nucl. Instrum. Meth.* **A503**, 110–113, (2003). doi: 10.1016/S0168-9002(03)00649-1.

25. K. Abe et al., A Measurement of Atmospheric Neutrino Flux Consistent with Tau Neutrino Appearance, *Phys. Rev. Lett.* **97**, 171801, (2006). doi: 10.1103/PhysRevLett.97.171801.

26. Y. Ashie et al., Evidence for an oscillatory signature in atmospheric neutrino oscillation, *Phys. Rev. Lett.* **93**, 101801, (2004). doi: 10.1103/PhysRevLett.93.101801.

27. V. D. Barger, J. G. Learned, S. Pakvasa, and T. J. Weiler, Neutrino decay as an explanation of atmospheric neutrino observations, *Phys. Rev. Lett.* **82**, 2640–2643, (1999). doi: 10.1103/PhysRevLett.82.2640.

28. V. D. Barger et al., Neutrino decay and atmospheric neutrinos, *Phys. Lett.* **B462**, 109–114, (1999). doi: 10.1016/S0370-2693(99)00887-4.

29. Y. Grossman and M. P. Worah, Atmospheric nu/mu deficit from decoherence. (1998).

30. E. Lisi, A. Marrone, and D. Montanino, Probing possible decoherence effects in atmospheric neutrino oscillations, *Phys. Rev. Lett.* **85**, 1166–1169, (2000).

doi: 10.1103/PhysRevLett.85.1166.

31. S. H. Ahn et al., Detection of Accelerator-Produced Neutrinos at a Distance of 250 km, *Phys. Lett.* **B511**, 178–184, (2001). doi: 10.1016/S0370-2693(01)00647-5.

32. M. H. Ahn et al., Measurement of Neutrino Oscillation by the K2K Experiment, *Phys. Rev.* **D74**, 072003, (2006). doi: 10.1103/PhysRevD.74.072003.

33. K. Anderson et al., The NuMI Facility Technical Design Report. FERMILAB-DESIGN-1998-01.

34. D. G. Michael et al., Observation of muon neutrino disappearance with the MINOS detectors and the NuMI neutrino beam, *Phys. Rev. Lett.* **97**, 191801, (2006). doi: 10.1103/PhysRevLett.97.191801.

35. P. Adamson et al., Measurement of Neutrino Oscillations with the MINOS Detectors in the NuMI Beam, *Phys. Rev. Lett.* **101**, 131802, (2008). doi: 10.1103/PhysRevLett.101.131802.

36. G. Sirri, The CNGS neutrino beam, *Nucl. Phys. Proc. Suppl.* **172**, 149–151, (2007). doi: 10.1016/j.nuclphysbps.2007.08.145.

37. K. Kodama et al., Observation of tau-neutrino interactions, *Phys. Lett.* **B504**, 218–224, (2001). doi: 10.1016/S0370-2693(01)00307-0.

38. G. Rosa, OPERA: Setting the scene for nu/tau appearance at CNGS, *J. Phys. Conf. Ser.* **136**, 022015, (2008). doi: 10.1088/1742-6596/136/2/022015.

39. R. Acquafredda et al., First events from the CNGS neutrino beam detected in the OPERA experiment, *New J. Phys.* **8**, 303, (2006). doi: 10.1088/1367-2630/8/12/303.

40. S. Amerio et al., Design, construction and tests of the ICARUS T600 detector, *Nucl. Instrum. Meth.* **A527**, 329–410, (2004). doi: 10.1016/j.nima.2004.02.044.

41. P. Cennini et al., Performance of a 3-ton liquid argon time projection chamber, *Nucl. Instrum. Meth.* **A345**, 230–243, (1994). doi: 10.1016/0168-9002(94)90996-2.

42. J. N. Bahcall, A. M. Serenelli, and S. Basu, New solar opacities, abundances, helioseismology, and neutrino fluxes, *Astrophys. J.* **621**, L85–L88, (2005).

43. B. T. Cleveland et al., Measurement of the solar electron neutrino flux with the Homestake chlorine detector, *Astrophys. J.* **496**, 505–526, (1998). doi: 10.1086/305343.

44. J. N. Abdurashitov et al., Measurement of the solar neutrino capture rate by the Russian-American gallium solar neutrino experiment during one half of the 22-year cycle of solar activity, *J. Exp. Theor. Phys.* **95**, 181–193, (2002). doi: 10.1134/1.1506424.

45. M. Altmann et al., Complete results for five years of GNO solar neutrino observations, *Phys. Lett.* **B616**, 174–190, (2005). doi: 10.1016/j.physletb.2005.04.068.

46. Y. Fukuda et al., Solar neutrino data covering solar cycle 22, *Phys. Rev. Lett.* **77**, 1683–1686, (1996). doi: 10.1103/PhysRevLett.77.1683.

47. J. Hosaka et al., Solar neutrino measurements in Super-Kamiokande-I, *Phys. Rev.* **D73**, 112001, (2006). doi: 10.1103/PhysRevD.73.112001.

48. J. P. Cravens et al., Solar neutrino measurements in Super-Kamiokande-II, *Phys. Rev.* **D78**, 032002, (2008). doi: 10.1103/PhysRevD.78.032002.

49. L. Wolfenstein, Neutrino oscillations in matter, *Phys. Rev.* **D17**, 2369–2374, (1978). doi: 10.1103/PhysRevD.17.2369.

50. S. P. Mikheev and A. Y. Smirnov, Resonant amplification of neutrino oscillations in matter and solar neutrino spectroscopy, *Nuovo Cim.* **C9**, 17–26, (1986). doi: 10.1007/BF02508049.

51. S. P. Mikheev and A. Y. Smirnov, Resonance enhancement of oscillations in matter and solar neutrino spectroscopy, *Sov. J. Nucl. Phys.* **42**, 913–917, (1985).

52. J. Boger et al., The Sudbury Neutrino Observatory, *Nucl. Instrum. Meth.* **A449**, 172–207, (2000). doi: 10.1016/S0168-9002(99)01469-2.

53. Q. R. Ahmad et al., Measurement of the charged current interactions produced by B-8 solar neutrinos at the Sudbury Neutrino Observatory, *Phys. Rev. Lett.* **87**, 071301, (2001). doi: 10.1103/PhysRevLett.87.071301.

54. Q. R. Ahmad et al., Direct evidence for neutrino flavor transformation from neutral-current interactions in the Sudbury Neutrino Observatory, *Phys. Rev. Lett.* **89**, 011301, (2002). doi: 10.1103/PhysRevLett.89.011301.

55. Q. R. Ahmad et al., Measurement of day and night neutrino energy spectra at SNO and constraints on neutrino mixing parameters, *Phys. Rev. Lett.* **89**, 011302, (2002). doi: 10.1103/PhysRevLett.89.011302.

56. S. N. Ahmed et al., Measurement of the total active B-8 solar neutrino flux at the Sudbury Neutrino Observatory with enhanced neutral current sensitivity, *Phys. Rev. Lett.* **92**, 181301, (2004). doi: 10.1103/PhysRevLett.92.181301.

57. B. Aharmim et al., Electron energy spectra, fluxes, and day-night asymmetries of B-8 solar neutrinos from the 391-day salt phase SNO data set, *Phys. Rev.* **C72**, 055502, (2005). doi: 10.1103/PhysRevC.72.055502.

58. R. G. H. Robertson, Report on the third and final phase of SNO, *J. Phys. Conf. Ser.* **136**, 022002, (2008). doi: 10.1088/1742-6596/136/2/022002.

59. A. de Gouvea, A. Friedland, and H. Murayama, The dark side of the solar neutrino parameter space, *Phys. Lett.* **B490**, 125–130, (2000). doi: 10.1016/S0370-2693(00)00989-8.

60. P. Vogel and J. F. Beacom, The angular distribution of the neutron inverse beta decay, anti-nu/e + p -¿ e+ + n, *Phys. Rev.* **D60**, 053003, (1999). doi: 10.1103/PhysRevD.60.053003.

61. C. L. Cowan, F. Reines, F. B. Harrison, H. W. Kruse, and A. D. McGuire, Detection of the free neutrino: A Confirmation, *Science.* **124**, 103–104, (1956). doi: 10.1126/science.124.3212.103.

62. K. Eguchi et al., First results from KamLAND: Evidence for reactor antineutrino disappearance, *Phys. Rev. Lett.* **90**, 021802, (2003). doi: 10.1103/PhysRevLett.90.021802.

63. T. Araki et al., Measurement of neutrino oscillation with KamLAND: Evidence of spectral distortion, *Phys. Rev. Lett.* **94**, 081801, (2005). doi: 10.1103/PhysRevLett.94.081801.

64. S. Abe et al., Precision Measurement of Neutrino Oscillation Parameters

with KamLAND, *Phys. Rev. Lett.* **100**, 221803, (2008). doi: 10.1103/PhysRevLett.100.221803.

65. M. Maltoni, T. Schwetz, M. A. Tortola, and J. W. F. Valle, Status of global fits to neutrino oscillations, *New J. Phys.* **6**, 122, (2004). doi: 10.1088/1367-2630/6/1/122.

66. G. Alimonti et al., Ultra-low background measurements in a large volume underground detector, *Astropart. Phys.* **8**, 141–157, (1998). doi: 10.1016/S0927-6505(97)00050-9.

67. C. Arpesella et al., First real time detection of Be7 solar neutrinos by Borexino, *Phys. Lett.* **B658**, 101–108, (2008). doi: 10.1016/j.physletb.2007.09.054.

68. C. Arpesella et al., Direct Measurement of the Be-7 Solar Neutrino Flux with 192 Days of Borexino Data, *Phys. Rev. Lett.* **101**, 091302, (2008). doi: 10.1103/PhysRevLett.101.091302.

69. C. Galbiati et al., New results on solar neutrino fluxes from 192 days of Borexino data, *J. Phys. Conf. Ser.* **136**, 022001, (2008). doi: 10.1088/1742-6596/136/2/022001.

70. M. C. Chen, The SNO liquid scintillator project, *Nucl. Phys. Proc. Suppl.* **145**, 65–68, (2005). doi: 10.1016/j.nuclphysbps.2005.03.037.

71. T. Marrodan Undagoitia et al., LENA: A multipurpose detector for low energy neutrino astronomy and proton decay, *J. Phys. Conf. Ser.* **120**, 052018, (2008). doi: 10.1088/1742-6596/120/5/052018.

72. Y. D. Kim, The status of XMASS experiment, *Phys. Atom. Nucl.* **69**, 1970–1974, (2006). doi: 10.1134/S106377880611024X.

73. D. N. McKinsey and K. J. Coakley, Neutrino detection with CLEAN, *Astropart. Phys.* **22**, 355–368, (2005). doi: 10.1016/j.astropartphys.2004.10.003.

74. R. S. Raghavan, Inverse beta decay of ^{115}In $\to$ $^{115}Sn^*$: a new possibility for detecting solar neutrinos from the proton-proton reaction, *Phys. Rev. Lett.* **37**, 259–262, (1976). doi: 10.1103/PhysRevLett.37.259.

75. R. S. Raghavan, LENS, MiniLENS: Status and outlook, *J. Phys. Conf. Ser.* **120**, 052014, (2008). doi: 10.1088/1742-6596/120/5/052014.

76. C. Athanassopoulos et al., The Liquid Scintillator Neutrino Detector and LAMPF Neutrino Source, *Nucl. Instrum. Meth.* **A388**, 149–172, (1997). doi: 10.1016/S0168-9002(96)01155-2.

77. C. Athanassopoulos et al., Evidence for anti-nu/mu –¿ anti-nu/e oscillation from the LSND experiment at the Los Alamos Meson Physics Facility, *Phys. Rev. Lett.* **77**, 3082–3085, (1996). doi: 10.1103/PhysRevLett.77.3082.

78. A. Aguilar et al., Evidence for neutrino oscillations from the observation of anti-nu/e appearance in a anti-nu/mu beam, *Phys. Rev.* **D64**, 112007, (2001). doi: 10.1103/PhysRevD.64.112007.

79. Y. Declais et al., Search for neutrino oscillations at 15-meters, 40-meters, and 95-meters from a nuclear power reactor at Bugey, *Nucl. Phys.* **B434**, 503–534, (1995). doi: 10.1016/0550-3213(94)00513-E.

80. C. Athanassopoulos et al., Evidence for nu/mu –¿ nu/e oscillations from pion decay in flight neutrinos, *Phys. Rev.* **C58**, 2489–2511, (1998). doi:

10.1103/PhysRevC.58.2489.

81. B. Armbruster et al., Upper limits for neutrino oscillations muon-antineutrino to electron-antineutrino from muon decay at rest, *Phys. Rev.* **D65**, 112001, (2002). doi: 10.1103/PhysRevD.65.112001.

82. E. D. Church, K. Eitel, G. B. Mills, and M. Steidl, Statistical analysis of different anti-nu/mu $-\xi$ anti-nu/e searches, *Phys. Rev.* **D66**, 013001, (2002). doi: 10.1103/PhysRevD.66.013001.

83. A. A. Aguilar-Arevalo et al., A Search for electron neutrino appearance at the $\Delta m^2 \sim 1 \text{eV}^2$ scale, *Phys. Rev. Lett.* **98**, 231801, (2007). doi: 10.1103/PhysRevLett.98.231801.

84. A. A. Aguilar-Arevalo et al., Compatibility of high-Δm^2 ν_e and $\bar{\nu}_e$ neutrino oscillation searches, *Phys. Rev.* **D78**, 012007, (2008). doi: 10.1103/PhysRevD.78.012007.

85. A. A. Aguilar-Arevalo et al. (2008). arXiv:0812.2243.

86. P. Adamson et al. (2008). arXiv:0809:2447.

87. G. Karagiorgi, (2008). http://theory.fnal.gov/jetp/talks/karagiorgi.pdf.

88. H. Ray, OscSNS: Precision Neutrino Measurements at the Spallation Neutron Source, *J. Phys. Conf. Ser.* **136**, 022029, (2008). doi: 10.1088/1742-6596/136/2/022029.

89. http://www-microboone.fnal.gov/.

90. M. Doi, T. Kotani, H. Nishiura, K. Okuda, and E. Takasugi, NEUTRINO MASSES AND THE DOUBLE BETA DECAY, *Phys. Lett.* **B103**, 219, (1981). doi: 10.1016/0370-2693(81)90746-2.

91. M. Apollonio et al., Limits on Neutrino Oscillations from the CHOOZ Experiment, *Phys. Lett.* **B466**, 415–430, (1999). doi: 10.1016/S0370-2693(99)01072-2.

92. F. Boehm et al., Final results from the Palo Verde Neutrino Oscillation Experiment, *Phys. Rev.* **D64**, 112001, (2001). doi: 10.1103/PhysRevD.64.112001.

93. J. Hosaka et al., Three flavor neutrino oscillation analysis of atmospheric neutrinos in Super-Kamiokande, *Phys. Rev.* **D74**, 032002, (2006). doi: 10.1103/PhysRevD.74.032002.

94. M. Sanchez, (2009). http://theory.fnal.gov/jetp/talks/Sanchez.pdf.

95. F. Ardellier et al. (2004). arXiv:hep-ex/0405032.

96. F. Ardellier et al. (2006). arXiv:hep-ex/0606025.

97. J. Cao, Daya Bay neutrino experiment, *Nucl. Phys. Proc. Suppl.* **155**, 229–230, (2006). doi: 10.1016/j.nuclphysbps.2006.02.057.

98. M. C. Chu, Precise measurement of θ_{13} at Daya Bay. (2008). arXiv:0810.0807.

99. S.-B. Kim, RENO: Reactor experiment for neutrino oscillation at Yonggwang, *AIP Conf. Proc.* **981**, 205–207, (2008). doi: 10.1063/1.2898934.

100. Y. Itow et al. (2001). arXiv:hep-ex/0106019.

101. D. S. Ayres et al. (2004). arXiv:hep-ex/0503053.

102. H. Minakata and H. Nunokawa, How to measure CP violation in neutrino oscillation experiments?, *Phys. Lett.* **B413**, 369–377, (1997). doi: 10.1016/

S0370-2693(97)01108-8.

103. A. Cervera et al., Golden measurements at a neutrino factory, *Nucl. Phys.* **B579**, 17–55, (2000). doi: 10.1016/S0550-3213(00)00221-2.

104. H. Minakata and H. Nunokawa, CP violation vs. matter effect in long-baseline neutrino oscillation experiments, *Phys. Rev.* **D57**, 4403–4417, (1998). doi: 10.1103/PhysRevD.57.4403.

105. H. Minakata, H. Nunokawa, and S. J. Parke, CP and T trajectory diagrams for a unified graphical representation of neutrino oscillations, *Phys. Lett.* **B537**, 249–255, (2002). doi: 10.1016/S0370-2693(02)01946-9.

106. J. Burguet-Castell, M. B. Gavela, J. J. Gomez-Cadenas, P. Hernandez, and O. Mena, On the measurement of leptonic CP violation, *Nucl. Phys.* **B608**, 301–318, (2001). doi: 10.1016/S0550-3213(01)00248-6.

107. O. Mena, H. Nunokawa, and S. J. Parke, NOvA and T2K: The race for the neutrino mass hierarchy, *Phys. Rev.* **D75**, 033002, (2007). doi: 10.1103/PhysRevD.75.033002.

108. D. Autiero et al., Large underground, liquid based detectors for astro- particle physics in Europe: scientific case and prospects, *JCAP.* **0711**, 011, (2007). doi: 10.1088/1475-7516/2007/11/011.

109. K. Nakamura, Hyper-Kamiokande: A next generation water Cherenkov detector, *Int. J. Mod. Phys.* **A18**, 4053–4063, (2003). doi: 10.1142/S0217751X03017361.

110. T. Kajita, Future possibilities with the J-PARC neutrino beam, *J. Phys. Conf. Ser.* **136**, 022020, (2008). doi: 10.1088/1742-6596/136/2/022020.

111. T. Kajita, S. B. Kim, and A. Rubbia. (2008). arXiv:0808.0650.

112. M. Ishitsuka, T. Kajita, H. Minakata, and H. Nunokawa, Resolving Neutrino Mass Hierarchy and CP Degeneracy by Two Identical Detectors with Different Baselines, *Phys. Rev.* **D72**, 033003, (2005). doi: 10.1103/PhysRevD. 72.033003.

113. J. Appel et al. Physics with a high intensity proton source at fermilab, (2008). `http://www.fnal.gov/directorate/Longrange/Steering_Public/P5/GoldenBook-2008-02-03.pdf`.

114. N. Saoulidou, Future possibilities with Fermilab neutrino beams, *J. Phys. Conf. Ser.* **136**, 022021, (2008). doi: 10.1088/1742-6596/136/2/022021.

115. V. M. Datar, Status report on INO, *J. Phys. Conf. Ser.* **136**, 022016, (2008). doi: 10.1088/1742-6596/136/2/022016.

116. A. Samanta. (2006). arXiv:hep-ph/0610196.

117. S. K. Agarwalla, S. Choubey, and A. Raychaudhuri, Unraveling neutrino parameters with a magical beta-beam experiment at INO, *Nucl. Phys.* **B798**, 124–145, (2008). doi: 10.1016/j.nuclphysb.2008.01.031.

118. C. H. Albright et al. (2004). arXiv:physics/0411123.

119. A. Donini, D. Meloni, and P. Migliozzi, The silver channel at the Neutrino Factory, *Nucl. Phys.* **B646**, 321–349, (2002). doi: 10.1016/S0550-3213(02) 00894-5.

120. A. Bandyopadhyay et al., Physics at a future Neutrino Factory and super-beam facility. (2007). arXiv:0710.4947.

121. C. R. Prior, J. S. Berg, M. Meddahi, and Y. Mori, The International Design

Study for a Neutrino Factory. EPAC'08, 11th European Particle Accelerator Conference, 23- 27 June 2008, Genoa, Italy.

122. J. S. Berg, R&D topics for neutrino factory acceleration, *AIP Conf. Proc.* **981**, 327–329, (2008). doi: 10.1063/1.2898977.

123. G. Gregoire et al., Proposal to the Rutherford Appleton Laboratory: an international muon ionization cooling experiment (MICE). MICE-NOTE-21.

124. M. Mezzetto, Physics reach of the beta beam, *J. Phys.* **G29**, 1771–1776, (2003). doi: 10.1088/0954-3899/29/8/346.

125. J. Burguet-Castell, D. Casper, J. J. Gomez-Cadenas, P. Hernandez, and F. Sanchez, Neutrino oscillation physics with a higher gamma beta- beam, *Nucl. Phys.* **B695**, 217–240, (2004). doi: 10.1016/j.nuclphysb.2004.06.021.

126. J. Burguet-Castell, D. Casper, E. Couce, J. J. Gomez-Cadenas, and P. Hernandez, Optimal beta-beam at the CERN-SPS, *Nucl. Phys.* **B725**, 306–326, (2005). doi: 10.1016/j.nuclphysb.2005.06.037.

127. M. T. Keil, G. G. Raffelt, and H.-T. Janka, Monte Carlo study of supernova neutrino spectra formation, *Astrophys. J.* **590**, 971–991, (2003). doi: 10. 1086/375130.

128. K. Scholberg. (2007). arXiv:astro-ph/0701081.

129. R. M. Bionta et al., Observation of a neutrino burst in coincidence with supernova sn1987a in the large magellanic cloud, *Phys. Rev. Lett.* **58**, 1494, (1987).

130. K. Hirata et al., Observation of a neutrino burst from the supernova sn1987a, *Phys. Rev. Lett.* **58**, 1490–1493, (1987).

131. E. N. Alekseev, L. N. Alekseeva, V. I. Volchenko, and I. V. Krivosheina, Possible detection of a neutrino signal on 23 february 1987 at the baksan underground scintillation telescope of the institute of nuclear research, *JETP Lett.* **45**, 589–592, (1987).

132. M. Aglietta et al., On the event observed in the Mont Blanc Underground Neutrino observatory during the occurrence of Supernova 1987a, *Europhys. Lett.* **3**, 1315–1320, (1987).

133. G. Raffelt and D. Seckel, Bounds on Exotic Particle Interactions from SN 1987a, *Phys. Rev. Lett.* **60**, 1793, (1988). doi: 10.1103/PhysRevLett.60. 1793.

134. J. F. Beacom and P. Vogel, Mass signature of supernova nu/mu and nu/tau neutrinos in SuperKamiokande, *Phys. Rev.* **D58**, 053010, (1998). doi: 10. 1103/PhysRevD.58.053010.

135. J. F. Beacom, R. N. Boyd, and A. Mezzacappa, Black hole formation in core collapse supernovae and time- of-flight measurements of the neutrino masses, *Phys. Rev.* **D63**, 073011, (2001). doi: 10.1103/PhysRevD.63. 073011.

136. J. F. Beacom, R. N. Boyd, and A. Mezzacappa, Technique for direct-eV scale measurements of the mu and tau neutrino masses using supernova neutrinos, *Phys. Rev. Lett.* **85**, 3568–3571, (2000). doi: 10.1103/PhysRevLett.85.3568.

137. J. N. Bahcall and S. L. Glashow, UPPER LIMIT ON THE MASS OF THE ELECTRON-NEUTRINO, *Nature.* **326**, 476, (1987). doi: 10.1038/

326476a0.

138. A. S. Dighe and A. Y. Smirnov, Identifying the neutrino mass spectrum from the neutrino burst from a supernova, *Phys. Rev.* **D62**, 033007, (2000).

139. G. M. Fuller, W. C. Haxton, and G. C. McLaughlin, Prospects for detecting supernova neutrino flavor oscillations, *Phys. Rev.* **D59**, 085005, (1999).

140. A. Dighe, Physics potential of future supernova neutrino observations, *J. Phys. Conf. Ser.* **136**, 022041, (2008). doi: 10.1088/1742-6596/136/2/022041.

141. A. Mirizzi, G. G. Raffelt, and P. D. Serpico, Earth matter effects in supernova neutrinos: Optimal detector locations, *JCAP.* **0605**, 012, (2006).

142. M. D. Kistler, H. Yuksel, S. Ando, J. F. Beacom, and Y. Suzuki. (2008). arXiv:0810.1959.

143. A. de Bellefon et al. (2006). arXiv:hep-ex/0607026.

144. S. Ando, J. F. Beacom, and H. Yuksel, Detection of neutrinos from supernovae in nearby galaxies, *Phys. Rev. Lett.* **95**, 171101, (2005). doi: 10.1103/PhysRevLett.95.171101.

145. C. Lunardini, The diffuse supernova neutrino flux, star formation rate and SN1987A, *Astropart. Phys.* **26**, 190–201, (2006). doi: 10.1016/j.astropartphys.2006.06.008.

146. L. E. Strigari, J. F. Beacom, T. P. Walker, and P. Zhang, The concordance cosmic star formation rate: Implications from and for the supernova neutrino and gamma ray backgrounds, *JCAP.* **0504**, 017, (2005). doi: 10.1088/1475-7516/2005/04/017.

147. M. Malek et al., Search for supernova relic neutrinos at Super- Kamiokande, *Phys. Rev. Lett.* **90**, 061101, (2003). doi: 10.1103/PhysRevLett.90.061101.

148. M. Wurm et al., Detection potential for the diffuse supernova neutrino background in the large liquid-scintillator detector LENA, *Phys. Rev.* **D75**, 023007, (2007). doi: 10.1103/PhysRevD.75.023007.

149. J. F. Beacom and M. R. Vagins, GADZOOKS! Antineutrino spectroscopy with large water Cherenkov detectors, *Phys. Rev. Lett.* **93**, 171101, (2004). doi: 10.1103/PhysRevLett.93.171101.

150. T. Yanagida, Horizontal gauge symmetry and masses of neutrinos. In Proceedings of the Workshop on the Baryon Number of the Universe and Unified Theories, Tsukuba, Japan, 13-14 Feb 1979.

151. J. Lesgourgues and S. Pastor, Massive neutrinos and cosmology, *Phys. Rept.* **429**, 307–379, (2006). doi: 10.1016/j.physrep.2006.04.001.

152. P. Adshead and R. Easther, Neutrinos and Future Concordance Cosmologies, *J. Phys. Conf. Ser.* **136**, 022044, (2008). doi: 10.1088/1742-6596/136/2/022044.

153. C. Arnaboldi et al., Bolometric bounds on the anti-neutrino mass, *Phys. Rev. Lett.* **91**, 161802, (2003). doi: 10.1103/PhysRevLett.91.161802.

154. S. Pirro et al., The Milano electron antineutrino mass experiment, *Nucl. Phys. Proc. Suppl.* **138**, 340–342, (2005). doi: 10.1016/j.nuclphysbps.2004.11.078.

155. D. Pergolesi et al., MANU-2: A second generation experiment for calorimetric neutrino mass determination with superconducting Re, *Nucl. Instrum.*

Meth. **A559**, 349–351, (2006). doi: 10.1016/j.nima.2005.12.170.

156. J. Bonn et al., The Mainz neutrino mass experiment, *Nucl. Phys. Proc. Suppl.* **91**, 273–279, (2001). doi: 10.1016/S0920-5632(00)00951-8.

157. V. M. Lobashev et al., Direct search for neutrino mass and anomaly in the tritium beta-spectrum: Status of 'Troitsk neutrino mass' experiment, *Nucl. Phys. Proc. Suppl.* **91**, 280–286, (2001). doi: 10.1016/S0920-5632(00)00952-X.

158. P. Doe, Direct Neutrino Mass Measurements. Talk at Neutrino 2006, Santa Fe, NM.

159. A. Osipowicz et al. (2001). arXiv:hep-ex/0109033.

160. D. Akimov et al., EXO: An advanced Enriched Xenon double-beta decay Observatory, *Nucl. Phys. Proc. Suppl.* **138**, 224–226, (2005). doi: 10.1016/j.nuclphysbps.2004.11.054.

161. E. Andreotti et al., MARE, Microcalorimeter Arrays for a Rhenium Experiment: A detector overview, *Nucl. Instrum. Meth.* **A572**, 208–210, (2007). doi: 10.1016/j.nima.2006.10.198.

162. F. T. Avignone, S. R. Elliott, and J. Engel, Double Beta Decay, Majorana Neutrinos, and Neutrino Mass, *Rev. Mod. Phys.* **80**, 481–516, (2008). doi: 10.1103/RevModPhys.80.481.

163. C. Amsler et al., Review of particle physics, *Phys. Lett.* **B667**, 525, (2008). doi: 10.1016/j.physletb.2008.07.018.

164. http://www.sc.doe.gov/np/nsac/docs/NuSAG_report_final_version.pdf.

165. H. V. Klapdor-Kleingrothaus, A. Dietz, H. L. Harney, and I. V. Krivosheina, Evidence for Neutrinoless Double Beta Decay, *Mod. Phys. Lett.* **A16**, 2409–2420, (2001). doi: 10.1142/S0217732301005825.

166. L. Baudis et al., Limits on the Majorana neutrino mass in the 0.1 eV range, *Phys. Rev. Lett.* **83**, 41–44, (1999). doi: 10.1103/PhysRevLett.83.41.

167. C. E. Aalseth et al., Comment on 'Evidence for Neutrinoless Double Beta Decay', *Mod. Phys. Lett.* **A17**, 1475–1478, (2002). doi: 10.1142/S0217732302007715.

168. V. M. Gehman, The Majorana Project: A next-generation double-beta decay experiment, *AIP Conf. Proc.* **842**, 840–842, (2006). doi: 10.1063/1.2220397.

169. C. M. Cattadori, Double beta decay searches by semiconductor detectors, *J. Phys. Conf. Ser.* **136**, 022033, (2008). doi: 10.1088/1742-6596/136/2/022033.

170. C. Arnaboldi et al., A New Limit on the Neutrinoless DBD of 130Te, *Phys. Rev. Lett.* **95**, 142501, (2005). doi: 10.1103/PhysRevLett.95.142501.

171. M. Danilov et al., Detection of very small neutrino masses in double-beta decay using laser tagging, *Phys. Lett.* **B480**, 12–18, (2000). doi: 10.1016/S0370-2693(00)00404-4.

172. M. C. Chen, SNO and SNO+, *AIP Conf. Proc.* **944**, 25–30, (2007). doi: 10.1063/1.2818544.

173. A. Ringwald and Y. Y. Y. Wong, Gravitational clustering of relic neutrinos and implications for their detection, *JCAP.* **0412**, 005, (2004). doi: 10.

1088/1475-7516/2004/12/005.

174. A. Ringwald. (2005). arXiv:hep-ph/0505024.

175. G. B. Gelmini, Prospect for relic neutrino searches, *Phys. Scripta.* **T121**, 131–136, (2005).

176. A. G. Cocco, G. Mangano, and M. Messina, Capturing Relic Neutrinos with beta-decaying nuclei, *J. Phys. Conf. Ser.* **120**, 022005, (2008). doi: 10.1088/1742-6596/120/2/022005.

177. R. Lazauskas, P. Vogel, and C. Volpe, Charged current cross section for massive cosmological neutrinos impinging on radioactive nuclei, *J. Phys.* **G35**, 025001, (2008). doi: 10.1088/0954-3899/35/2/025001.

178. M. Blennow, Prospects for cosmic neutrino detection in tritium experiments in the case of hierarchical neutrino masses, *Phys. Rev.* **D77**, 113014, (2008). doi: 10.1103/PhysRevD.77.113014.

179. B. McElrath, Laboratory tests for the cosmic neutrino background using beta-decaying nuclei. (2009). arXiv:0901.3491.

Chapter 13

Inflationary Cosmology

William H. Kinney

Department of Physics, University at Buffalo, SUNY,
Buffalo, NY 14260-1500, USA
**E-mail: whkinney@buffalo.edu*

This series of lectures gives a pedagogical review of the subject of cosmological inflation. I discuss Friedmann-Robertson-Walker cosmology and the horizon and flatness problems of the standard hot Big Bang, and introduce inflation as a solution to those problems, focusing on the simple scenario of inflation from a single scalar field. I discuss quantum modes in inflation and the generation of primordial tensor and scalar fluctuations. Finally, I provide comparison of inflationary models to the WMAP satellite measurement of the Cosmic Microwave Background, and briefly discuss future directions for inflationary physics. The majority of the lectures should be accessible to advanced undergraduates or beginning graduate students with only a background in Special Relativity, although familiarity with General Relativity and quantum field theory will be helpful for the more technical sections.

13.1. Introduction

Cosmology today is a vibrant scientific enterprise. New precision measurements are revealing a universe with surprising and unexpected properties, in particular the Dark Matter and Dark Energy which are now believed to be the dominant components of the cosmos. Galaxy surveys such as the Sloan Digital Sky Survey are making the first large-scale maps of the universe, and satellites such as WMAP are making exquisitely precise measurements of the Cosmic Microwave Background (CMB), the haze of relic photons left over from the Big Bang. In turn, these measurements are giving us clues which are helping to unravel one of the oldest and most profound questions people have ever asked: Where did the universe come from? In these lectures, I discuss what is currently the best motivated and most completely developed physical model for the first moments of the universe: cosmological inflation.[1-3][a] Inflation naturally explains how the universe came to be so large, so old, and so flat, and provides a compellingly elegant and predictive mechanism for generating the primordial perturbations which gave rise to the rich structure we see in the universe today.[7-13] Inflation provides a link between the Outer Space of astrophysics and the Inner Space of particle physics, and gives us a window to physics at energy scales far beyond the reach of particle accelerators. Furthermore, inflation makes *testable* predictions, which have so far proven to be an excellent match to the data.

The lectures are organized as follows:

- Section 13.1 provides an introduction and a brief overview of General Relativity.
- Section 13.2 discusses the Friedmann-Robertson-Walker spacetime and the standard hot Big Bang picture of cosmology, including the Cosmic Microwave Background.
- Section 13.3 explains unresolved issues in the standard cosmology, in particular the horizon and flatness problems.
- Section 13.4 introduces inflation in scalar field theories.
- Section 13.5 discusses quantum fluctuations in inflation and the generation of cosmological perturbations.
- Section 13.6 discusses the observational predictions of inflation, and current constraints from Cosmic Microwave Background measurements.
- Section 13.7 discusses conclusions and the future outlook for inflationary physics.
- Appendix A describes in detail the generation of density perturbations during inflation.

The lectures are at an advanced undergraduate or beginning graduate student level. Most of the the lectures should be accessible with only a background in Special Relativity. A working knowledge of General Relativity and quantum field theory are helpful for Secs. 13.4 and 13.5 and for Appendix A. Where possible, I reference review articles for further reading on related topics. For other reviews on inflation, see Refs. [6,14-20].

---

[a]Inflation in its current form was introduced by Guth, but similar ideas had been discussed before.[4,5] A short history of the early development of inflation can be found in Ref. [6].

13.1.1. *The Metric*

The fundamental object in General Relativity is the *metric*, which encodes the shape of the spacetime. A metric is a symmetric, bilinear form which defines distances on a manifold. For example, we can express Pythagoras' theorem in a Euclidean three-dimensional space,

$$\ell^2 = x^2 + y^2 + z^2, \tag{13.1.1}$$

as a matrix product over the identity matrix $\delta_{ij} = \mathrm{diag}\,(1,1,1)$,

$$\ell^2 = \sum_{i,j=1,3} \delta_{ij} x^i x^j. \tag{13.1.2}$$

Therefore the identity matrix δ_{ij} can be identified as the metric for the Euclidean space: if we wish to describe a non-Euclidean manifold, we replace δ_{ij} with a more complicated matrix g_{ij}, which in general can depend on the coordinates x^i. For an arbitrary path through the space, we express distances on the manifold in differential form,

$$d\ell^2 = \sum_{i,j} g_{ij} dx^i dx^j. \tag{13.1.3}$$

The distance along any path in the spacetime, or *world line*, is then given by integrating $d\ell$ along that path. A familiar example of a non-Euclidean space frequently used in physics is the Minkowski Space describing spacetime in Special Relativity. Distances along a world line in Minkowski Space are measured by the *proper time*, which is the time as measured by an observer traveling on that world line. The proper time s along a world line is given by the relation

$$\begin{aligned} ds^2 &= dt^2 - d\mathbf{x}^2 \\ &= \sum_{\mu,\nu=0,3} \eta_{\mu\nu} dx^\mu dx^\nu, \end{aligned} \tag{13.1.4}$$

where we take the speed of light $c = 1$. We express four-vectors as $\tilde{x} = (t, x, y, z) = (x^0, x^1, x^2, x^3)$, and $d\mathbf{x}^2 = dx^2 + dy^2 + dz^2$ is the Euclidean distance along a spatial interval. The metric $\eta_{\mu\nu}$ for Minkowski Space is given by

$$\eta_{\mu\nu} = \begin{pmatrix} 1 & & & \\ & -1 & & \\ & & -1 & \\ & & & -1 \end{pmatrix}. \tag{13.1.5}$$

Anything traveling the speed of light has velocity $d\left|\mathbf{x}\right|/dt = 1$. Photons therefore always travel along world lines of zero proper time, $ds^2 = dt^2 - d\mathbf{x}^2 = 0$, called *null geodesics*. Massive particles travel along world lines with real proper time, $ds^2 > 0$, called *timelike geodesics*. Causally disconnected regions of spacetime are separated by *spacelike* intervals, with $ds^2 < 0$. The set of all null geodesics passing through a given point (or *event*) in spacetime is called the *light cone* (Fig. 13.1) The interior of the light cone, consisting of all null and timelike geodesics, defines the region of spacetime causally related to that event.

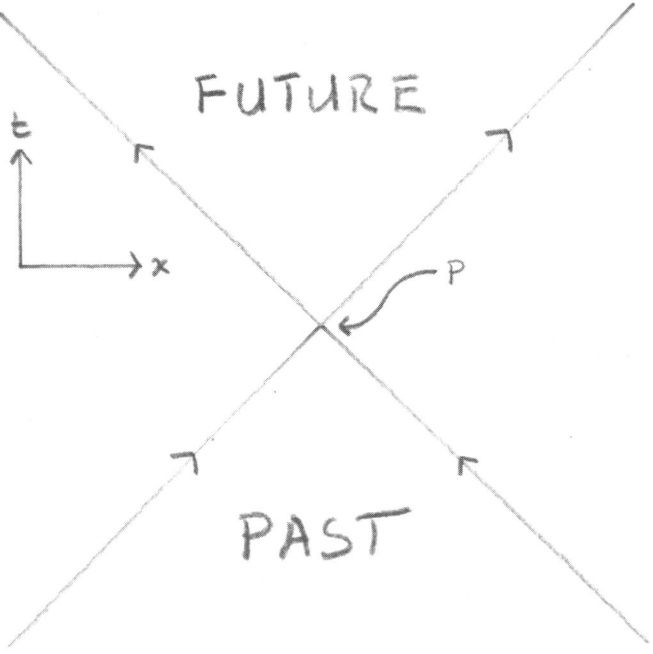

Fig. 13.1. Light cones in Minkowski Space. The past light cone defines the causal past of the event P, and the future light cone defines the causal future of P.

13.1.2. *General Relativity and the Einstein Field Equation*

The Minkowski metric $\eta_{\mu\nu}$ of Special Relativity describes a Euclidean spacetime which is static, empty, and infinite in space and time. The addition of gravity to the picture requires General Relativity, which describes gravitational fields as curvature in the spacetime. The fundamental object in General Relativity is the metric $g_{\mu\nu}(\tilde{x})$, which describes the shape of the spacetime and in general depends on the spacetime coordinate $\tilde{x}$. As in Minkowski Space, lengths in curved spacetime are measured by the proper time s, with the proper time along a world line determined by the metric

$$ds^2 = \sum_{\mu,\nu=0,3} g_{\mu\nu}(\tilde{x})\, dx^\mu dx^\nu. \tag{13.1.6}$$

As in Special Relativity, photons travel along null geodesics, with $ds^2 = 0$, and massive particles travel along timelike geodesics, with $ds^2 > 0$. However, unlike Special Relativity, null geodesics need not always be $45°$ lines defining light cones, but can be curved by gravity.

In General Relativity, the distribution of mass/energy in the spacetime determines the shape of the metric, and the metric in turn determines the evolution of the mass/energy. Electromagnetism provides a convenient analogy: in electromagnetism, the distribution of charges and currents determines the electromagnetic field, and the electromagnetic field in turn determines the evolution of the charges and currents. Given a current four-vector J^μ, Maxwell's Equations are a set of linear, first-order partial differential equations that allow us to calculate the resulting electromagnetic field

$$\partial_\nu F^{\mu\nu} \equiv \sum_{\nu=0,3} \partial_\nu F^{\mu\nu} = \frac{4\pi}{c} J^\mu. \tag{13.1.7}$$

Here we have explicitly included the speed of light c to highlight its role as an electromagnetic coupling constant. We also adopt the typical summation convention for relativity: repeated indices are implicitly summed over. In General Relativity, we describe the distribution of mass/energy in a covariant way by specifying a symmetric rank-2 *stress-energy* tensor $T_{\mu\nu}$, which acts as a source for the gravitational field similar to the way the current four-vector J^μ sources electromagnetism. The analog of Maxwell's Equations is the Einstein Field Equation, which can be written in the deceptively simple form

$$G_{\mu\nu} = 8\pi G T_{\mu\nu}, \tag{13.1.8}$$

where the coupling constant is Newton's gravitational constant G. The tensor $G_{\mu\nu}$, called the Einstein Tensor, is a symmetric 4×4 tensor consisting of the metric $g_{\mu\nu}$ and its first and second derivatives. The Einstein Field Equation therefore represents a set of ten coupled, nonlinear, second-order partial differential equations of ten free functions, which are the elements of the metric tensor $g_{\mu\nu}$. However, only six of these equations are actually indpendent, leaving four degrees of freedom. The physics of gravity is independent of coordinate system, and the additional degrees of freedom correspond to a choice of a coordinate system, or *gauge* on the four-dimensional space. Gravity is *much* more complicated than electromagnetism! As with any intractably complicated problem, we simplify the job by introducing a symmetry. In General Relativity there are a number of symmetries which allow either exact or perturbative solution to the Einstein Field Equations:

- Vacuum: $T_{\mu\nu} = 0$. If we evaluate the Einstein Field Equations for small perturbations about an empty Minkowski Space, we find that they reduce at lowest order to a wave equation, and therefore General Relativity predicts the existence of gravity waves.
- Spherical Symmetry. If we assume a spherically symmetric spacetime (also empty of matter, $T_{\mu\nu} = 0$) the Einstein Field Equation can be solved exactly, resulting in the Schwarzschild solution for black holes.
- Homogeneity and Isotropy. If we assume that the stress-energy is distributed in a fashion which is homogeneous and isotropic, this is called a *Friedmann-Robertson-Walker* (FRW) space, and is the case of interest for cosmology. Since the homogeneity and isotropy remove all spatial dependence, the Einstein Field Equations reduce from a set of partial differential equations to a set of nonlinear ordinary differential equations in time. For particular types of homogeneous, isotropic matter, these equations can be solved exactly, and perturbations about those exact solutions can be handled self-consistently.

Continuing the analogy with electromagnetism, the equivalent of charge conservation,

$$\partial_\mu J^\mu = \frac{\partial \rho}{\partial t} + \nabla \cdot \mathbf{j} = 0, \tag{13.1.9}$$

in General Relativity is stress-energy conservation

$$D_\nu T^{\mu\nu} = 0, \tag{13.1.10}$$

where D_μ represents a covariant derivative, which is a generalization of the partial derivative to a curved manifold. We will also denote covariant derivatives with a semicolon, for example $T^{\mu\nu}{}_{;\nu} = 0$. Likewise, simple partial derivatives are denoted with a comma, $\partial f/\partial x^\mu \equiv \partial_\mu f \equiv f_{,\mu}$. As in the case of electromagnetism, where the charge conservation equation is not independent, but is instead a consequence of Maxwell's Equations, stress-energy conservation in General Relativity is a consequence of the Einstein

Field Equations and does not independently constrain the solutions. In the next section, we discuss FRW spaces and their application to cosmology in more detail.

13.2. Friedmann-Robertson-Walker Spacetimes

13.2.1. *The Friedmann Equation*

A *homogeneous* space is one which is translationally invariant, or the same at every point. An *isotropic* space is one which is rotationally invariant, or the same in every direction. The two are not the same: a space which is everywhere isotropic is necessarily homogeneous, but a space which is homogeneous is not necessarily isotropic. (Consider, for example a space with a uniform electric field: it is translationally invariant but not rotationally invariant.) It is possible to show[21] that the most general metric consistent with homogeneity and isotropy is obtained by multiplying a static spatial geometry with a time-dependent *scale factor* $a(t)$:

$$ds^2 = dt^2 - a^2(t)\,d\mathbf{x}^2$$
$$= dt^2 - a^2(t)\left[\frac{dr^2}{1 - kr^2} + r^2 d\Omega^2\right], \qquad (13.2.11)$$

where we have expressed the spatial line element in terms of spherical coordinates r,θ,ϕ, and the solid angle is given by the usual $d\Omega^2 = \sin\theta d\theta d\phi$. The constant k defines the curvature of the spacetime, with $k = 0$ corresponding to flat (Euclidean) spatial sections, and $k = \pm 1$ corresponding to positive and negative curvatures, respectively. A spacetime of this general form is called a *Friedmann-Robertson-Walker* (FRW) spacetime. Likewise, the most general homogeneous, isotropic stress-energy is diagonal, with all of its spatial components identical,

$$T^\mu{}_\nu = \begin{pmatrix} \rho(t) & & & \\ & -p(t) & & \\ & & -p(t) & \\ & & & -p(t) \end{pmatrix}, \qquad (13.2.12)$$

where we identify the energy density ρ and the pressure p from the continuity equation arising from stress-energy conservation,

$$T^{\mu\nu}{}_{;\nu} = \dot{\rho} + 3\left(\frac{\dot{a}}{a}\right)(\rho + p) = 0. \qquad (13.2.13)$$

The Einstein field equations then reduce to a set of two coupled, non-linear ordinary differential equations,

$$\left(\frac{\dot{a}}{a}\right)^2 + \frac{k}{a^2} = \frac{8\pi}{3m_{\text{Pl}}^2}\rho,$$
$$\left(\frac{\ddot{a}}{a}\right) = -\frac{4\pi}{3m_{\text{Pl}}^2}\left(\rho + 3p\right). \tag{13.2.14}$$

The first is called the *Friedmann Equation*, and the second is called the *Raychaudhuri Equation*. Note that the equations for the evolution of the scale factor depend not only on the energy density ρ, but also the pressure p: pressure gravitates! The continuity equation (13.2.13) is *not* independent of the Einstein Field Equations (13.2.14), but can be derived directly from the Friedmann and Raychaudhuri Equations. The expansion rate $\dot{a}/a$ is called the *Hubble parameter H*:

$$H \equiv \frac{\dot{a}}{a}, \tag{13.2.15}$$

and has units of inverse time. A positive Hubble parameter $H > 0$ corresponds to an expanding universe, and a negative Hubble parameter $H < 0$ corresponds to a collapsing universe. (Since our actual universe is expanding, we will specialize to that case.) Minkowski Space can be recovered by assuming a flat geometry $k = 0$, and no expansion, $\dot{a} = 0$. The Hubble parameter sets the fundamental scale of the spacetime, *i.e.* a characteristic time is the *Hubble time $t \sim H^{-1}$*, and likewise the *Hubble length* is $d \sim H^{-1}$. We will see later that the Hubble time sets the scale for the age of the universe, and the Hubble length sets the scale for the size of the observable universe.

The coordinate system $(t, \mathbf{x})$ is called a *comoving* coordinate system, because observers with constant comoving coordinates are at rest relative to the expansion, *i.e.* two observers with constant separation in comoving coordinates $\Delta\mathbf{x}$ have a physical, or *proper*, separation which increases in proportion to the scale factor

$$\Delta\mathbf{x}_{\text{prop}} = a\left(t\right)\Delta\mathbf{x}_{\text{com}}. \tag{13.2.16}$$

An important kinematic effect of cosmological expansion is the phenomenon of *cosmological redshift*: we will see later that solutions to the wave equation in an FRW space have constant wavelength in *comoving* coordinates, so that the proper wavelength of (for example) a photon increases in time in proportion to the scale factor

$$\lambda \propto a\left(t\right). \tag{13.2.17}$$

For a photon emitted at time $t_{\rm em}$ and detected at time t_0, the redshift z is defined by:

$$(1 + z) \equiv \frac{\lambda_0}{\lambda_{\rm em}} = \frac{a(t_0)}{a(t_{\rm em})}. \tag{13.2.18}$$

(Here we introduce the convention used frequently in cosmology that a subscript 0 refers to the *current* time, not an initial time.) Note that the cosmological redshift is *not* a Doppler shift caused by the relative velocity of the source and detector, but is an expansion effect: the wavelength of a photon traveling through the spacetime increases because the underlying spacetime is expanding. Another way to look at this is that a photon traveling through an FRW spacetime loses momentum with time,

$$p = h\nu \propto a^{-1}(t). \tag{13.2.19}$$

By the equivalence principle, this momentum loss must apply to massive particles as well as photons: *any* particle moving in an expanding FRW spacetime will lose momentum as $p \propto a^{-1}$. For massless particles like photons, this is manifest as a redshift in the wavelength, but it means that a massive particle will asymptotically come to rest relative to the comoving coordinate system. Thus, comoving coordinates represent a preferred reference frame reminiscent of Aristotelian physics: any free body with a "peculiar" velocity relative to the comoving frame will eventually come to rest in that frame.

There are three possibilities for the curvature of the universe: flat ($k = 0$), positively curved ($k = +1$), or negatively curved ($k = -1$). The current value of the Hubble parameter is (from the Hubble Space Telescope Key Project [22]),

$$H_0 = 72 \pm 8 \ {\rm km/s/Mpc}. \tag{13.2.20}$$

Therefore, we can see from the Friedmann Equation (13.2.14) that, given the expansion rate H, the curvature is determined by the density:

$$k = a^2 \left(\frac{8\pi}{3m_{\rm Pl}^2} \rho - H^2 \right). \tag{13.2.21}$$

Note that only the *sign* of k is physically important, since any rescaling of k is equivalent to a rescaling of the scale factor a. We define the *critical density* as the density for which $k = 0$, corresponding to a geometrically flat universe,

$$\rho_c \equiv \frac{3m_{\rm Pl}^2}{8\pi} H^2 \ \Rightarrow \ k = 0. \tag{13.2.22}$$

For $\rho > \rho_c$, the universe is positively curved and *closed*, with finite volume, and for $\rho < \rho_c$, the universe is negatively curved and *open*, with infinite volume. We express the ratio of the actual density ρ to the critical density ρ_c as the parameter Ω:

$$\Omega \equiv \left(\frac{\rho}{\rho_c}\right) = \frac{8\pi}{3m_{\mathrm{Pl}}^2} \frac{\rho}{H^2}. \qquad (13.2.23)$$

(Do not confuse the density parameter Ω with the solid angle $d\Omega$ in Eq. (13.2.11)!) Table 13.1 summarizes the relation between density, curvature, and geometry. The density parameter Ω is not in general constant

Table 13.1. Cosmological density, curvature, and geometry.

density	curvature	geometry
$\Omega = 1$	$k = 0$	flat
$\Omega > 1$	$k = 1$	closed
$\Omega < 1$	$k = 0$	open

in time, and we can re-write the Friedmann Equation as

$$\Omega(t) = 1 + \frac{k}{(aH)^2}. \qquad (13.2.24)$$

Since the Hubble parameter is proportional to the inverse time $H \propto t^{-1}$, we see that the time-dependence of Ω is determined by the time dependence of the scale factor $a\,(t)$. In the next section, we tackle the problem of solving for $a\,(t)$.

13.2.2. *Solving the Friedmann Equation*

In the previous section, we considered the form and kinematics of FRW spaces, but not the *dynamics*, that is, how does the stress-energy of the universe determine the expansion history? The answer to this question depends on what kind of matter dominates the cosmological stress-energy. In this section, we consider three basic types of cosmological stress-energy: matter, radiation, and vacuum.

The simplest kind of cosmological stress-energy is generically referred to as *matter*. Imagine a comoving box with sides of length L. By *comoving* box, we mean a box whose corners are at rest in a comoving coordinate system, and whose proper dimension is therefore increasing proportional to the scale factor, $L_{\mathrm{prop}} \propto a$. That is, the box is growing with the expansion of the universe. Now imagine the box filled with N particles of mass m, also at rest in the comoving reference frame (Fig. 13.2). In units where

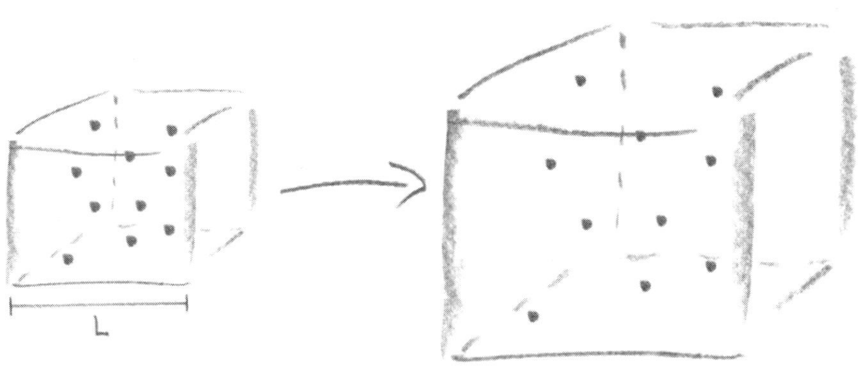

Fig. 13.2. A comoving box full of matter. The energy density in matter scales inversely with the volume of the box.

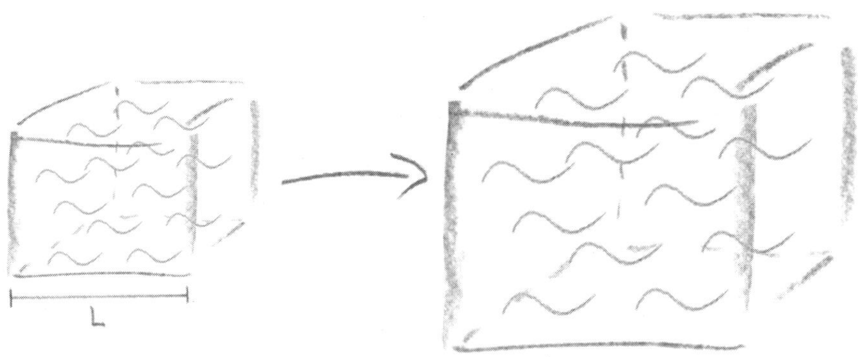

Fig. 13.3. A comoving box full of radiation. The number density of photons scales inversely with the volume of the box, but the photons also increase in wavelength.

$c = 1$, the relativistic energy density of such a system of particles is given by

$$\rho_{\mathrm{m}} = \frac{MN}{V}, \tag{13.2.25}$$

where V is the *proper* volume of the box, $V = L_{\mathrm{prop}}^3 \propto a^3$. Since neither M

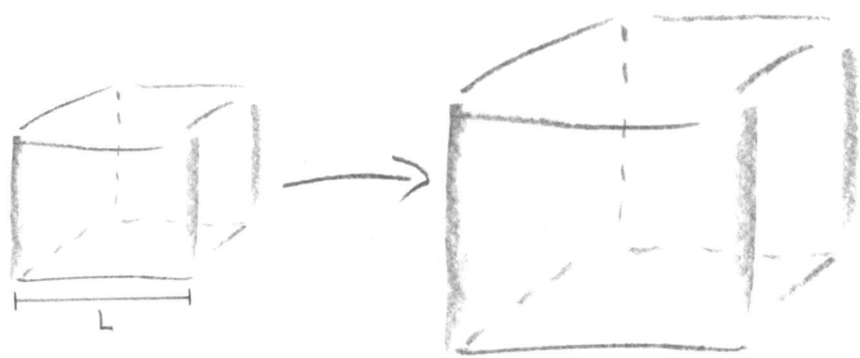

Fig. 13.4.   A comoving box full of vacuum. The energy density of vacuum does not scale at all!

nor N change with expansion, we have immediately that

$$\rho_\mathrm{m} = \frac{MN}{L^3 a^3} \propto a^{-3}, \qquad (13.2.26)$$

where L is the comoving size of the box. So the proper energy density of massive particles at rest in a comoving volume evolves as the inverse cube of the scale factor. Now imagine the same box filled with N photons with frequency ν (Fig. 13.3). The energy per photon is $h\nu$, so that the energy density in the box is then

$$\rho_\gamma = \frac{Nh\nu}{V}. \qquad (13.2.27)$$

As in the case of massive particles, the number density of photons in the box redshifts inversely with the proper volume of the box $n = N/V \propto a^{-3}$. But each photon also loses energy through cosmological redshift, $\nu \propto a^{-1}$ (13.2.19), so that the total energy density in photons or other massless degrees of freedom, which we generically refer to as *radiation*, redshifts as

$$\rho_\gamma \propto a^{-4}. \qquad (13.2.28)$$

Note also that cosmological redshift immediately gives us a rule for the behavior of a black-body spectrum of radiation with temperature T. Since all photons redshift at exactly the same rate, a system which starts out as a black-body *stays* a black-body, with a temperature that decreases with expansion,

$$T_\gamma \propto a^{-1}. \qquad (13.2.29)$$

The third type of stress-energy which is important in cosmology dates back to Einstein's introduction of a "cosmological constant" to his field equations. If we take the stress-energy $T_{\mu\nu}$ and add a term proportional to the metric, the identity $D_\nu g^{\mu\nu} = 0$ means the stress-energy conservation equation (13.1.10) is unchanged:

$$D_\nu T^{\mu\nu} \to D_\nu \left(T^{\mu\nu} + \Lambda g^{\mu\nu} \right) = 0. \tag{13.2.30}$$

In our analogy with electromagnetism, this is like adding a constant to the electromagnetic potential, $V'(x) = V(x) + \Lambda$. The constant Λ does not affect local dynamics in any way, but it does affect the cosmology. From Eq. (13.2.12), stress-energy of the form $T^{\mu\nu} = \Lambda g^{\mu\nu}$ corresponds to an equation of state

$$p_\Lambda = -\rho_\Lambda. \tag{13.2.31}$$

The continuity equation (13.2.13) then reduces to

$$\dot{\rho} + 3 \left(\frac{\dot{a}}{a} \right) (\rho + p) = \dot{\rho} = 0, \tag{13.2.32}$$

so that vacuum has a constant energy density, $\rho_\Lambda = \text{const}$. A cosmological constant is also frequently referred to as *vacuum energy*, since it is as if we are assigning an energy density to empty space. With this interpretation, a comoving box full of vacuum contains a total amount of energy which *grows* with the expansion of the universe (Fig. 13.4). This highlights the curious property of General Relativity that, while energy is conserved in a local sense, it is *not* conserved globally. We are creating energy out of nothing!

It is straightforward to solve the Einstein Field Equations for the three basic types of stress-energy. Consider first a matter-dominated universe. We can write the time derivative of the energy density as:

$$\rho_{\mathrm{m}} \propto a^{-3} \;\Rightarrow\; \dot{\rho}_{\mathrm{m}} = -3 \left(\frac{\dot{a}}{a} \right) \rho. \tag{13.2.33}$$

From the continuity equation (13.2.13), we have

$$\dot{\rho} + 3 \left(\frac{\dot{a}}{a} \right) (\rho + p) = 3 \left(\frac{\dot{a}}{a} \right) p = 0. \tag{13.2.34}$$

We then have that the pressure of matter vanishes, $p_{\mathrm{m}} = 0$. The matter-dominated Friedmann Equation becomes

$$\left(\frac{\dot{a}}{a} \right)^2 + \frac{k}{a^2} = \frac{8\pi}{3 m_{\mathrm{Pl}}^2} \rho \propto a^{-3}. \tag{13.2.35}$$

In the case of a flat universe, $k = 0$, the solution is especially simple:

$$\left(\frac{\dot{a}}{a}\right)^2 \propto a^{-3} \ \Rightarrow a(t) \propto t^{2/3}. \tag{13.2.36}$$

Similarly, for a radiation dominated universe, the continuity equation implies that

$$\rho_\gamma \propto a^{-4} \ \Rightarrow p_\gamma = \rho_\gamma/3. \tag{13.2.37}$$

Again assuming a flat geometry,

$$\left(\frac{\dot{a}}{a}\right)^2 \propto a^{-4} \ \Rightarrow a(t) \propto t^{1/2}. \tag{13.2.38}$$

Finally, solving the Friedmann Equation for the vacuum case gives

$$\left(\frac{\dot{a}}{a}\right)^2 \propto \rho_\Lambda = \text{const.} \ \Rightarrow a(t) \propto e^{Ht}, \tag{13.2.39}$$

so that the universe expands exponentially quickly, with a time constant given by the Hubble parameter

$$H = \sqrt{\frac{8\pi}{3m_{\text{Pl}}^2} \rho_\Lambda} = \text{const.} \tag{13.2.40}$$

Such a spacetime is called *de Sitter space*.

Note in particular that the energy density in radiation redshifts away more quickly than the energy density in matter, and vacuum energy does not redshift at all, so that a universe with a mix of radiation, matter and vacuum will be radiation-dominated at early times, matter-dominated at later times, and eventually vacuum-dominated (Fig. 13.5). Note also that for either matter- or radiation-domination, the universe is singular as $t \to 0$: the universe has finite age! Since the scale factor vanishes at $t = 0$, and the density scales as an inverse power of a, the initial singularity consists of infinite density. Likewise, since temperature also scales inversely with a, the initial singularity is also a point of infinite temperature. We therefore arrive at the standard hot Big Bang picture of the universe: a cosmological singularity at finite time in the past, followed by a hot, radiation-dominated expansion, during which the universe gradually cools as $T \propto a^{-1}$ and the radiation dilutes, followed by a period of matter-dominated expansion during which galaxies and stars and planets form. Finally, if the vacuum energy is nonzero, it will inevitably dominate, and the universe will enter a state of exponential expansion. Current evidence indicates that the real universe

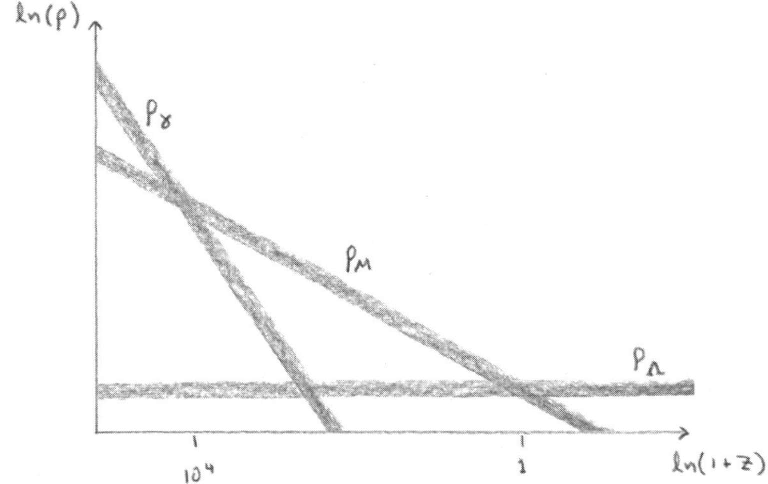

Fig. 13.5.   Schematic diagram of how the three types of stress-energy scale with redshift $1+z \propto a$: at early time, radiation dominates, followed by matter, and finally the universe is dominated by vacuum energy.

made a transition from matter-domination to vacuum-domination at a redshift of around $z = 1$, or about a billion years ago, so that the densities of the three types of matter today are of order

$$\begin{aligned}
\Omega_\Lambda &\simeq 0.7, \\
\Omega_{\mathrm{m}} &\simeq 0.3, \\
\Omega_\gamma &\simeq 10^{-4}.
\end{aligned}$$

(13.2.41)

In the next section, we discuss one important prediction of the hot Big Bang: the presence of a background of relic photons from the early universe, called the *Cosmic Microwave Background*.

13.2.3. *The Hot Big Bang and the Cosmic Microwave Background*

The basic picture of an expanding, cooling universe leads to a number of startling predictions: the formation of nuclei and the resulting primordial abundances of elements, and the later formation of neutral atoms and the consequent presence of a cosmic background of photons, the Cosmic Microwave Background (CMB).[23–27] A rough history of the universe can be given as a time line of increasing time and decreasing temperature:[28]

- $T = \infty$, $t = 0$: Big Bang.
- $T \sim 10^{15}$ K, $t \sim 10^{-12}$ sec: Primordial soup of fundamental particles.
- $T \sim 10^{13}$ K, $t \sim 10^{-6}$ sec: Protons and neutrons form.
- $T \sim 10^{10}$ K, $t \sim 3$ min: Nucleosynthesis: nuclei form.
- $T \sim 3000$ K, $t \sim 300,000$ years: Atoms form.
- $T \sim 10$ K, $t \sim 10^9$ years: Galaxies form.
- $T \sim 3$ K, $t \sim 10^{10}$ years: Today.

The epoch at which atoms form, when the universe was at an age of 300,000 years and at a temperature of around 3000 K is oxymoronically referred to as "recombination", despite the fact that electrons and nuclei had never before "combined" into atoms. The physics is simple: at a temperature of greater than about 3000 K, the universe consisted of an ionized plasma of mostly protons, electrons, and photons, with a few helium nuclei and a tiny trace of lithium. The important characteristic of this plasma is that it was *opaque*, or, more precisely, the mean free path of a photon was a great deal smaller than the Hubble length. As the universe cooled and expanded, the plasma "recombined" into neutral atoms, first the helium, then a little later the hydrogen.

If we consider hydrogen alone, the process of recombination can be described by the Saha equation for the equilibrium ionization fraction X_{e} of the hydrogen:[28]

$$\frac{1 - X_{\mathrm{e}}}{X_{\mathrm{e}}^2} = \frac{4\sqrt{2}\zeta(3)}{\sqrt{\pi}}\eta \left(\frac{T}{m_{\mathrm{e}}}\right)^{3/2} \exp\left(\frac{13.6 \text{ eV}}{T}\right). \tag{13.2.42}$$

Here m_{e} is the electron mass and 13.6 eV is the ionization energy of hydrogen. The physically important parameter affecting recombination is the density of protons and electrons compared to photons. This is determined by the *baryon asymmetry*,[b] which is described as the ratio of baryons to photons:

$$\eta \equiv \frac{n_{\mathrm{b}} - n_{\bar{\mathrm{b}}}}{n_\gamma} = 2.68 \times 10^{-8} \left(\Omega_{\mathrm{b}} h^2\right). \tag{13.2.43}$$

---

[b]If there were no excess of baryons over antibaryons, there would be no protons and electrons to recombine, and the universe would be just a gas of photons and neutrinos!

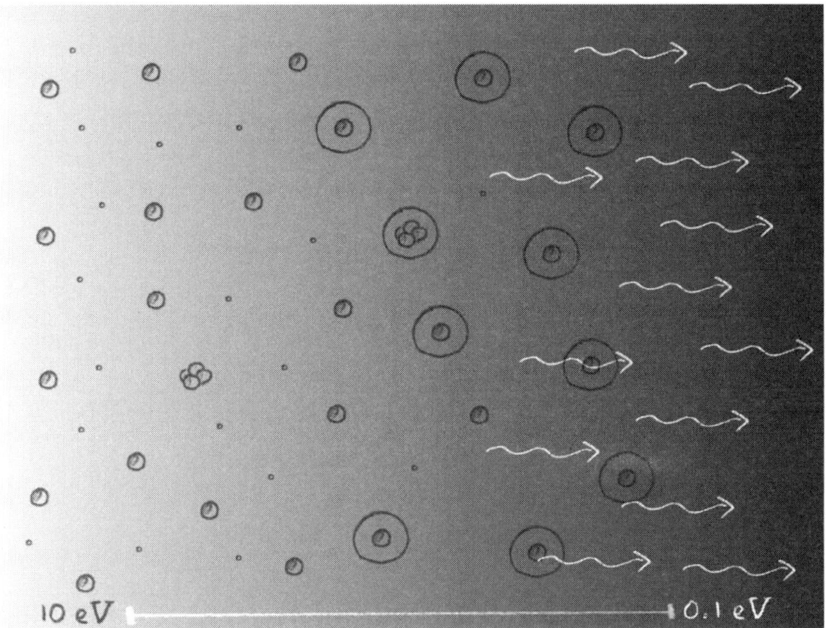

Fig. 13.6. Schematic diagram of recombination. At early time, the temperature of the universe is above the ionization energy of hydrogen and helium, so that the universe is full of an ionized plasma, and the mean free path for photons is short compared to the Hubble length. At late time, the temperature drops and the nuclei capture the electrons and form neutral atoms. Once this happens, the universe becomes transparent to photons, which free stream from the surface of last scattering.

Here $\Omega_{\rm b}$ is the baryon density and h is the Hubble constant in units of 100 km/s/Mpc,

$$h \equiv H_0/(100 \text{ km/s/Mpc}). \qquad (13.2.44)$$

The most recent result from the WMAP satellite gives $\Omega_{\rm b} h^2 = 0.02273 \pm 0.00062$.[29] Recombination happens quickly (i.e., in much less than a Hubble time $t \sim H^{-1}$), but it is not instantaneous. The universe goes from a completely ionized state to a neutral state over a range of redshifts $\Delta z \sim 200$. If we define recombination as an ionization fraction $X_{\rm e} = 0.1$, we have that the temperature at recombination $T_{\rm R} = 0.3$ eV.

What happens to the photons after recombination? Once the gas in the universe is in a neutral state, the mean free path for a photon becomes much larger than the Hubble length. The universe is then full of a background of freely propagating photons with a blackbody distribution of frequencies.

At the time of recombination, the background radiation has a temperature of $T = T_{\rm R} = 3000$ K, and as the universe expands the photons redshift, so that the temperature of the photons drops with the increase of the scale factor, $T \propto a(t)^{-1}$. We can detect these photons today. Looking at the sky, this background of photons comes to us evenly from all directions, with an observed temperature of $T_0 \simeq 2.73$ K. This allows us to determine the redshift of recombination,

$$1 + z_{\rm R} = \frac{a\left(t_0\right)}{a\left(t_{\rm R}\right)} = \frac{T_{\rm R}}{T_0} \simeq 1100. \qquad (13.2.45)$$

This is the cosmic microwave background. Since by looking at higher and higher redshift objects, we are looking further and further back in time, we can view the observation of CMB photons as imaging a uniform "surface of last scattering" at a redshift of 1100 (Fig. 13.7).

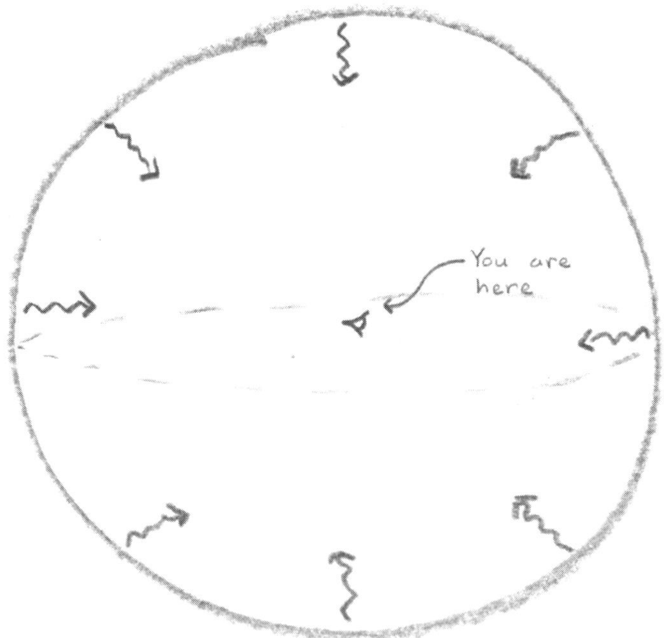

Fig. 13.7. Cartoon of the last scattering surface. From earth, we see blackbody radiation emitted uniformly from all directions, forming a "sphere" at redshift $z = 1100$.

To the extent that recombination happens at the same time and in the same way everywhere, the CMB will be of precisely uniform temperature.

While the observed CMB is highly isotropic, it is not perfectly so. The largest contribution to the anisotropy of the CMB as seen from earth is simply Doppler shift due to the earth's motion through space. (Put more technically, the motion is the earth's motion relative to a comoving cosmological reference frame.) CMB photons are slightly blueshifted in the direction of our motion and slightly redshifted opposite the direction of our motion. This blueshift/redshift shifts the temperature of the CMB so the effect has the characteristic form of a "dipole" temperature anisotropy (Fig. 13.8). The dipole anisotropy, however, is a *local* phenomenon. Any intrinsic, or primordial, anisotropy of the CMB is potentially of much greater cosmological interest. To describe the anisotropy of the CMB, we remember that the surface of last scattering appears to us as a spherical surface at a redshift of 1100. Therefore the natural parameters to use to describe the anisotropy of the CMB sky is as an expansion in spherical harmonics $Y_{\ell m}$:

$$\frac{\Delta T}{T} = \sum_{\ell=1}^{\infty} \sum_{m=-\ell}^{\ell} a_{\ell m} Y_{\ell m}(\theta, \phi). \tag{13.2.46}$$

If we assume isotropy, there is no preferred direction in the universe, and we expect the physics to be independent of the index m. We can then define

$$C_\ell \equiv \frac{1}{2\ell + 1} \sum_m |a_{\ell m}|^2. \tag{13.2.47}$$

The $\ell = 1$ contribution is just the dipole anisotropy,

$$\left(\frac{\Delta T}{T}\right)_{\ell=1} \sim 10^{-3}. \tag{13.2.48}$$

The dipole was first measured in the 1970's by several groups.[30–32] It was not until more than a decade after the discovery of the dipole anisotropy that the first observation was made of anisotropy for $\ell \geq 2$, by the differential microwave radiometer aboard the Cosmic Background Explorer (COBE) satellite,[33] launched in in 1990. COBE observed that the anisotropy at the quadrupole and higher ℓ was two orders of magnitude smaller than the dipole:

$$\left(\frac{\Delta T}{T}\right)_{\ell>1} \simeq 10^{-5}. \tag{13.2.49}$$

Fig. 13.8 shows the dipole and higher-order CMB anisotropy as measured by COBE. This anisotropy represents intrinsic fluctuations in the CMB itself,

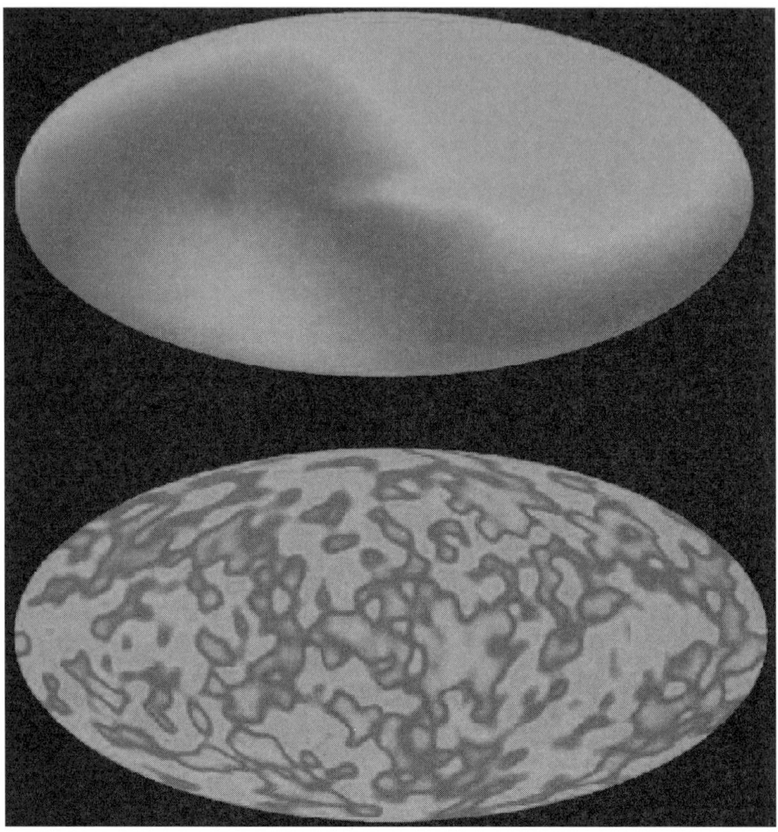

Fig. 13.8.   The COBE measurement of the CMB anisotropy.[33] The top oval is a map of the sky showing the dipole anisotropy $\Delta T/T \sim 10^{-3}$. The bottom oval is a similar map with the dipole contribution and emission from our own galaxy subtracted, showing the anisotropy for $\ell > 1$, $\Delta T/T \sim 10^{-5}$. (Figure courtesy of the COBE Science Working Group.)

due to the presence of tiny primordial density fluctuations in the cosmological matter present at the time of recombination. These density fluctuations are of great physical interest, since these are the fluctuations which later collapsed to form all of the structure in the universe, from superclusters to planets to graduate students. While the physics of recombination in the homogeneous case is quite simple, the presence of inhomogeneities in the universe makes the situation much more complicated. I describe some of the major effects in a qualitative way here, and refer the reader to the liter-

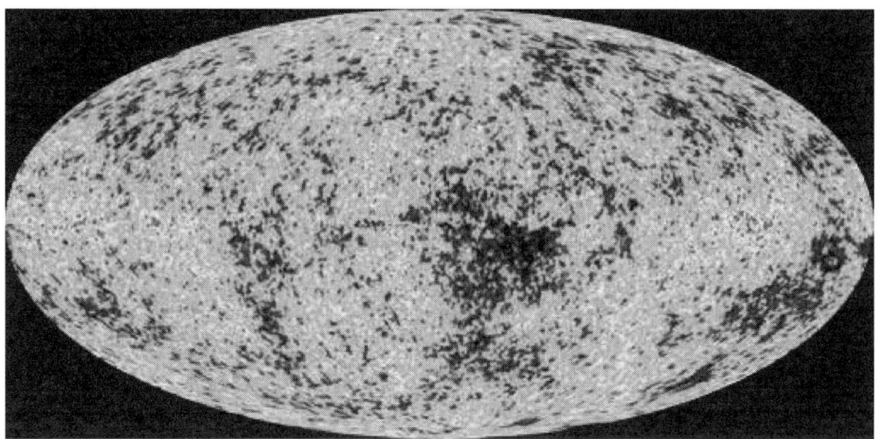

Fig. 13.9. The WMAP measurement of the CMB anisotropy.[34] (Figure courtesy of the WMAP Science Working Group.) WMAP measured the anisotropy with much higher sensitivity and resolution than COBE.

ature for a more detailed technical explanation of the relevant physics.[23–27] In these lectures, I primarily focus on the current status of the CMB as a probe of inflation, but there is much more to the story.

The simplest contribution to the CMB anisotropy from density fluctuations is just a gravitational redshift, known as the *Sachs-Wolfe effect*.[35] A photon coming from a region which is slightly denser than the average will have a slightly larger redshift due to the deeper gravitational well at the surface of last scattering. Conversely, a photon coming from an underdense region will have a slightly smaller redshift. Thus we can calculate the CMB temperature anisotropy due to the slightly varying Newtonian potential Φ from density fluctuations at the surface of last scattering:

$$\frac{\delta T}{T} = \frac{1}{3}\left[\Phi_{\rm em} - \Phi_{\rm obs}\right], \qquad (13.2.50)$$

where $\Phi_{\rm em}$ is the potential at the point the photon was emitted on the surface of last scattering, and $\Phi_{\rm obs}$ is the potential at the point of observation, which can be treated as a constant. The factor 1/3 is a General Relativistic correction. This simple kinematic contribution to the CMB anisotropy is dominant on large angular scales, corresponding to multipoles $\ell < 100$. However, the amount of information we can gain from these multipoles is limited by an intrinsic source of error called *cosmic variance*. Cosmic variance is a result of the statistical nature of the primordial power spectra:

since we have only one universe to measure, we have only one realization of the random field of density perturbations, and therefore there is an inescapable $1/\sqrt{N}$ uncertainty in our ability to reconstruct the primordial power spectrum, where N is the number of independent wave modes which will fit inside the horizon of the universe! On very large angular scales, this problem becomes acute, and we can write the cosmic variance error on any given C_ℓ as

$$\frac{\Delta C_\ell}{C_\ell} = \frac{1}{\sqrt{2\ell + 1}}, \tag{13.2.51}$$

which comes from the fact that any C_ℓ is represented by $2\ell + 1$ independent amplitudes $a_{\ell m}$. Even a perfect observation of the CMB can only approximately measure the true power spectrum — the errors in the WMAP data, for example, are dominated by cosmic variance out to $\ell \sim 400$ (Fig. 13.10).

For fluctuation modes on smaller angular scales, more complicated physics comes into play. The dominant process that occurs on short wavelengths is *acoustic oscillations* in the baryon/photon plasma. The idea is simple: matter tends to collapse due to gravity onto regions where the density is higher than average, so the baryons "fall" into overdense regions. However, since the baryons and the photons are still strongly coupled, the photons tend to resist this collapse and push the baryons outward. The result is "ringing", or oscillatory modes of compression and rarefaction in the gas due to density fluctuations. The gas heats as it compresses and cools as it expands, which creates fluctuations in the temperature of the CMB. This manifests itself in the C_ℓ spectrum as a series of peaks and valleys (Fig. 13.10). The specific shape and location of the acoustic peaks is created by complicated but well-understood physics, involving a large number of cosmological parameters. The presence of acoustic peaks in the CMB was first suggested by Sakharov,[36] and later calculated by Sunyaev and Zel'dovich[37,38] and Peebles and Yu.[39] The complete linear theory of CMB fluctuations was worked out by Ma and Bertschinger in 1995.[40] The shape of the CMB multipole spectrum depends, for example, on the baryon density Ω_b, the Hubble constant H_0, the densities of matter Ω_m and cosmological constant Ω_Λ, the amplitude of primordial gravitational waves, and the redshift z_{ri} at which the first generation of stars ionized the intergalactic medium. This makes interpretation of the spectrum something of a complex undertaking, but it also makes it a sensitive probe of cosmological models.

In addition to anisotropy in the temperature of the CMB, the photons coming from the surface of last scattering are expected to be weakly polarized due to the presence of perturbations.[41,42] This polarization is much less well measured than the temperature anisotropy, but it has been detected by WMAP and by a number of ground- and balloon-based measurements.[43-47] Measurement of polarization promises to greatly increase the amount of information it is possible to extract from the CMB. Of particular interest is the odd-parity, or *B-mode* component of the polarization, the only primordial source of which is gravitational waves, and thus provides a clean signal for detection of these perturbations. The B-mode has yet to be detected by any measurement.

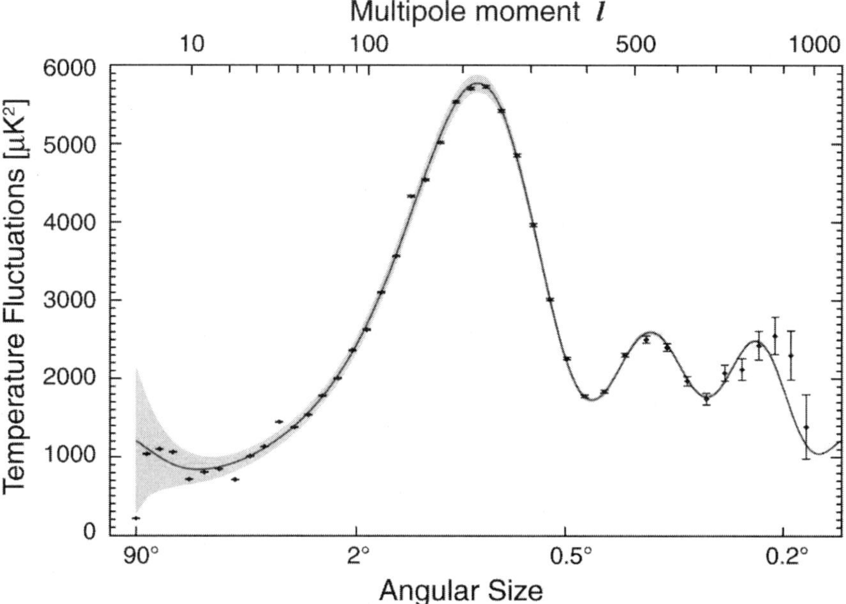

Fig. 13.10. The C_ℓ spectrum for the CMB as measured by WMAP, showing the peaks characteristic of acoustic oscillations. The gray shaded region represents the uncertainty due to cosmic variance. (Figure courtesy of the WMAP Science Working Group.)

13.3. The Flatness and Horizon Problems

We have so far considered two types of cosmological mass-energy – matter and radiation – and solved the Friedmann Equation for the simple case of a flat universe. What about the more general case? In this section, we consider non-flat universes with general contents. We introduce two related questions which are not explained by the standard Big Bang cosmology: why is the universe so close to flat today, and why is it so large?

We can describe a general homogeneous, isotropic mass-energy by its equation of state

$$p = w\rho, \tag{13.3.52}$$

so that pressureless matter corresponds to $w = 0$, and radiation corresponds to $w = 1/3$. We will consider only the case of constant equation of state, $w = \text{const}$. From the continuity equation, we have

$$\dot{\rho} + 3\left(1 + w\right)\frac{\dot{a}}{a}\rho = 0, \tag{13.3.53}$$

with solution

$$\rho \propto a^{-3(1+w)}. \tag{13.3.54}$$

The Friedmann Equation for a flat universe is then

$$\left(\frac{\dot{a}}{a}\right)^2 \propto a^{-3(1+w)}, \tag{13.3.55}$$

so that the scale factor increases as a power-law in time,

$$a\left(t\right) \propto t^{2/3(1+w)}. \tag{13.3.56}$$

What about the evolution of a non-flat universe? Analytic solutions for $a(t)$ in the $k \neq 0$ case can be found in cosmology textbooks. For our purposes, it is sufficient to consider the time-dependence of the density parameter Ω. From Eqs. (13.2.14, 13.2.23, 13.2.24) it is not too difficult to show that the density parameter evolves with the scale factor a as:

$$\frac{d\Omega}{d\ln a} = (1 + 3w)\,\Omega\,(\Omega - 1)\,. \tag{13.3.57}$$

Proof is left as an exercise for the reader. Note that a flat universe, $\Omega = 1$ remains flat at all times, but in a non-flat universe, the density parameter Ω is a time-dependent quantity, with the evolution determined by the equation of state parameter w. For matter ($w = 0$) or radiation ($w = 1/3$), the prefactor in Eq. (13.3.57) is positive,

$$1 + 3w > 0, \tag{13.3.58}$$

which means a flat universe is an *unstable* fixed point:

$$\frac{d\,|\Omega - 1|}{d\ln a} > 0, \quad (1 + 3w) > 0. \tag{13.3.59}$$

Any deviation from a flat geometry is amplified by the subsequent cosmological expansion, so a nearly flat universe today is a highly fine-tuned situation. The WMAP5 CMB measurement tells us the universe is flat to within a few percent, $|\Omega_0 - 1| < 0.02$.[29,48] If we are very conservative and take a limit on the density today as $\Omega_0 = 1 \pm 0.05$, that means that at recombination, when the CMB was emitted, $\Omega_{\mathrm{rec}} = 1 \pm 0.0004$, and at the time of primordial nucleosynthesis, $\Omega_{\mathrm{nuc}} = 1 \pm 10^{-12}$. Why did the universe start out so incredibly close to flat? The standard Big Bang cosmology provides no answer to this question, which we call the *flatness problem.*

There is a second, related problem with the standard Big Bang picture, arising from the finite age of the universe. Because the universe has a finite age, photons can only have traveled a finite distance in the time since the Big Bang. Therefore, the universe has a *horizon*: the further out in space we look, the further back in time we see. If we could look far enough out in any direction, past the surface of last scattering, we would be able to see the Big Bang itself, and beyond that we can see no further. Every observer in an FRW spacetime sees herself at the center of a spherical horizon which defines her observable universe. To calculate the size of our horizon, we use the fact that photons travel on paths of zero proper length:

$$ds^2 = dt^2 - a^2(t)\,|d\mathbf{x}|^2 = 0, \tag{13.3.60}$$

so that the comoving distance $|d\mathbf{x}|$ traversed by a photon in time dt is

$$|d\mathbf{x}| = \frac{dt}{a(t)}. \tag{13.3.61}$$

Therefore, the size of the cosmological horizon at time t after the Big Bang is

$$d_{\mathrm{H}}(t) = \int_0^t \frac{dt'}{a(t')}. \tag{13.3.62}$$

To convert comoving length to proper length, we just multiply by $a(t)$, so that the proper horizon size is

$$d_{\mathrm{H}}^{\mathrm{prop}}(t) = a(t)\, d_{\mathrm{H}}^{\mathrm{com}}(t). \tag{13.3.63}$$

Normalizing $a\left(t_0\right) = 1$, the horizon size of a 14-billion year-old flat, matter-dominated universe is $d_{\mathrm{H}} = 3t_0 \sim 13$ Gpc.

To see why the presence of a horizon is a problem for the standard Big Bang, we examine the causal structure of an FRW universe. Take the FRW metric

$$ds^2 = dt^2 - a^2\left(t\right)\left|d\mathbf{x}\right|^2 , \tag{13.3.64}$$

and re-write it in terms of a redefined clock, the *conformal time* τ:

$$ds^2 = a^2\left(\tau\right)\left[d\tau^2 - \left|d\mathbf{x}\right|^2\right] . \tag{13.3.65}$$

Conformal time is a "clock" which slows down with the expansion of the universe,

$$d\tau = \frac{dt}{a\left(t\right)}, \tag{13.3.66}$$

so that the comoving horizon size is just the age of the universe in conformal time

$$d_{\mathrm{H}}\left(t\right) = \int_0^t \frac{dt'}{a\left(t'\right)} = \int_0^\tau d\tau' = \tau. \tag{13.3.67}$$

The conformal metric is useful because the expansion of the spacetime is factored into a static metric multiplied by a time-dependent conformal factor $a\left(\tau\right)$, so that photon geodesics are simply described by $d\left|\mathbf{x}\right| = d\tau$. In a diagram of τ versus $\left|\mathbf{x}\right|$, photons travel on $45°$ angles. (Note that this is true even for curved spacetimes!) We can draw light cones and infer causal relationships with the expansion factored out, in a manner identical to the usual case of Minkowski Space.

There is one major difference between FRW and Minkowski: an FRW spacetime has a *finite age*. Therefore, unlike the case of Minkowski Space, which has an infinite past, an FRW spacetime is "chopped off" at some finite past time $\tau = 0$ (Fig. 13.11). The initial singularity is a surface of constant conformal time, and it is easy to see from Eq. (13.3.67) that our horizon size is the width of our past light cone projected on the surface defined by the initial singularity. This is a very different picture from the notion many people (even scientists) have of the Big Bang, which is something akin to an explosion, with the universe initially a cosmic "egg" of zero size. On the contrary, in the case of a flat or open universe, the universe is spatially infinite an infinitesimal amount of time after the initial singularity: the Big Bang happens everywhere at once in an infinite space! Our *observable* universe is finite because we can only see a small patch of the much larger

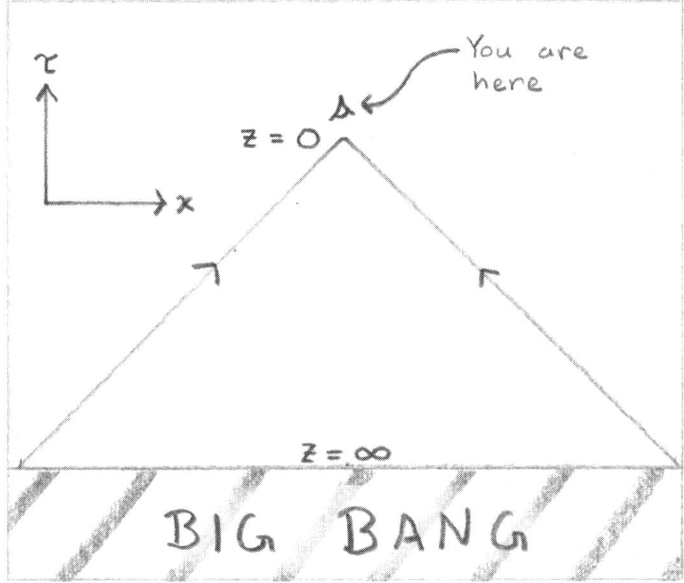

Fig. 13.11. A conformal diagram of a Friedmann-Robertson-Walker space. The FRW space is causally identical to Minkowski Space, except that it is not past-infinite, so that past light cones are "cut off" at the Big Bang, which is a spatially infinite surface at time $t = 0$.

cosmos.[c] Closed universes are spatially finite, but are still much larger in extent than our observable patch. The key point is that two events on the conformal spacetime diagram are causally connected only if they share a causal past: that is, if their past light cones overlap.

Consider two points on the CMB sky 180° degrees apart (Fig. 13.12). Their past light cones do not overlap, and the two points are causally *disconnected*. Those two points on the surface of last scattering occupy completely separate, disconnected observable universes. How did these points reach the observed thermal equilibrium to a few parts in 10^5 if they never shared a causal past? This apparent paradox is called the *horizon problem*: the universe somehow reached nearly perfect equilibrium on scales much larger than the size of any local horizon. From the Friedmann Equation, it is easy to show that the horizon problem and the flatness problem are

---

[c]Of course, this is an idealization, and the actual universe could well have a nontrivial global topology, even if it is locally flat, as long as the scale of the overall manifold is much larger than our horizon size.[49,50]

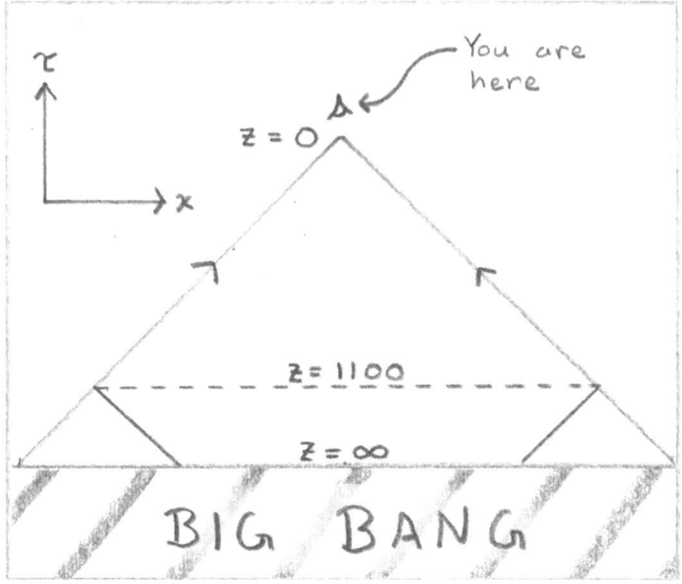

Fig. 13.12.   A conformal diagram of the Cosmic Microwave Background. Two points on opposite sides of the sky are causally separate, since their past light cones do not intersect.

related: consider a comoving length scale λ. It is is easy to show that for $w = \mathrm{const.}$, the ratio of λ to the horizon size d_{H} is related to the curvature by a conservation law

$$\left(\frac{\lambda}{d_{\mathrm{H}}}\right)^2 |\Omega - 1| = \mathrm{const.} \tag{13.3.68}$$

Proof is left as an exercise for the reader. Therefore, for a universe evolving away from flatness,

$$\frac{d\,|\Omega - 1|}{d\ln a} > 0, \tag{13.3.69}$$

the horizon size gets bigger in comoving units

$$\frac{d}{d\ln a}\left(\frac{\lambda}{d_{\mathrm{H}}}\right) < 0. \tag{13.3.70}$$

That is, more and more space "falls into" the horizon, or becomes causally connected, at late times.

What would be required to have a universe which evolves *toward* flatness, rather than away from it? From Eq. (13.3.57), we see that having

$1 + 3w$ negative will do the trick,

$$\frac{d\,|\Omega - 1|}{d\ln a} < 0, \ (1 + 3w) < 0. \tag{13.3.71}$$

Therefore, if the energy density of the universe is dominated not by matter or radiation, but by something with sufficiently negative pressure, $p < -\rho/3$, a curved universe will become flatter with time. From the Raychaudhuri Equation (13.2.14), we see that the case of $p < -\rho/3$ is exactly equivalent to an accelerating expansion:

$$\frac{\ddot{a}}{a} \propto -(1 + 3w) > 0, \ (1 + 3w) < 0. \tag{13.3.72}$$

If the expansion of the universe is slowing down, as is the case for matter- or radiation-domination, the curvature evolves away from flatness. But if the expansion is speeding up, the universe gets flatter. From Eq. (13.3.68), we see that this negative pressure solution also solves the horizon problem, since accelerating expansion means that the horizon size is shrinking in comoving units:

$$\frac{d}{d\ln a} \left(\frac{\lambda}{d_{\mathrm{H}}} \right) > 0, \ (1 + 3w) < 0. \tag{13.3.73}$$

When the expansion accelerates, distances initially smaller than the horizon size are "redshifted" to scales larger than the horizon at late times. Accelerating cosmological expansion is called *inflation*.

The simplest example of an accelerating expansion from a negative pressure fluid is the case of vacuum energy we considered in Sec. (13.2.2), for which the scale factor increases exponentially,

$$a \propto e^{Ht}. \tag{13.3.74}$$

For such expansion, the universe is driven exponentially toward a flat geometry,

$$\frac{d\ln \Omega}{d\ln a} = 2\,(1 - \Omega)\,. \tag{13.3.75}$$

We can see that the horizon problem is also solved by looking at the conformal time:

$$d\tau = \frac{dt}{a\,(t)} = e^{-Ht}dt, \tag{13.3.76}$$

so that

$$\tau = -\frac{1}{H}e^{-Ht} = -\frac{1}{aH}. \tag{13.3.77}$$

The conformal time during the inflationary period is *negative*, tending toward zero at late time. Therefore, if we have a period of inflationary expansion prior to the early epoch of radiation-dominated expansion, inflation takes place in negative conformal time, and conformal time $\tau = 0$ represents not the initial singularity but the transition from the inflationary expansion to radiation domination. The initial singularity is pushed back into negative conformal time, and can be pushed arbitrarily far depending on the duration of inflation. Figure 13.13 shows the causal structure of an inflationary spacetime. The past light cones of two points on the CMB sky do not intersect at $\tau = 0$, but inflation provides a "sea" of negative conformal time, which allows those points to share a causal past. In this way, inflation solves the horizon problem.

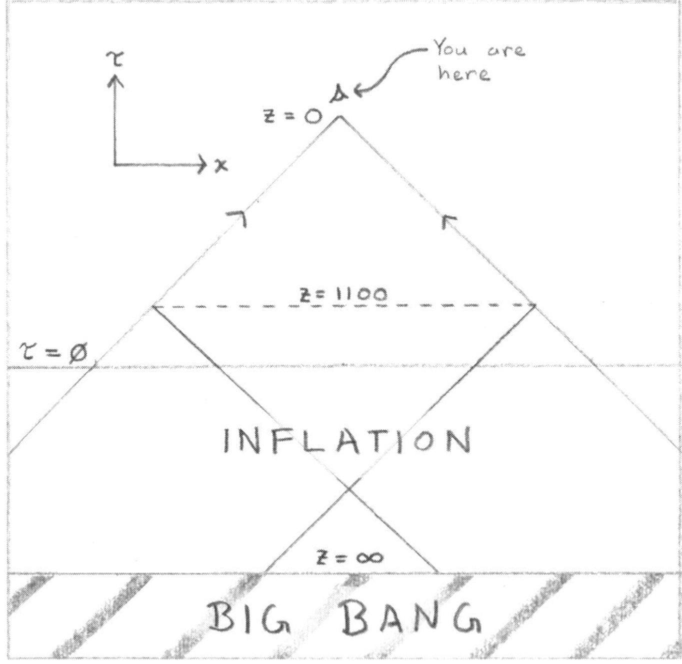

Fig. 13.13. A conformal diagram of light cones in an inflationary universe. Inflation ends in reheating at conformal time $\tau = 0$, which is the onset of the radiation-dominated expansion of the hot Big Bang. However, inflation provides a "sea" of negative conformal time, which allows the past light cones of events at the last scattering surface to overlap.

In more realistic models of inflation in the early universe, the energy density is approximately, but not exactly, constant, and the expansion is approximately, but not exactly, exponential. In such quasi-de Sitter spaces, the qualitative picture above still holds, and inflation provides a clean and compelling explanation for the peculiar boundary conditions for our universe. In the next section, we discuss how to construct more detailed models of inflation in field theory.

13.4. Inflation from Scalar Fields

The example of de Sitter evolution we considered in Sec. 13.3 gives a good qualitative picture of how inflation, or accelerated expansion, solves the horizon and flatness problems of the standard Big Bang cosmology. However, this leaves open the question: what physics is responsible for the accelerated expansion at early times? It cannot be Einstein's cosmological constant, simply because a universe dominated by vacuum energy *stays* dominated by vacuum energy for the infinite future, since in a de Sitter background matter ($\rho \propto a^{-3}$) and radiation ($\rho \propto a^{-4}$) are diluted exponentially quickly. Therefore, we will never reach a radiation-dominated phase, and we will never see a hot Big Bang. In order to transition from an inflating phase to a thermal equilibrium, radiation-dominated phase, the vacuum-like energy during inflation must be time-dependent. We model this dynamics with a scalar field ϕ, for which we assume the following action:

$$S = \int d^4x \sqrt{-g}\mathcal{L}_\phi, \tag{13.4.78}$$

where $g \equiv Det\,(g_{\mu\nu})$ is the determinant of the metric and the Lagrangian for the field ϕ is

$$\mathcal{L}_\phi = \frac{1}{2}g^{\mu\nu}\partial_\mu\phi\partial_\nu\phi - V(\phi). \tag{13.4.79}$$

Comparing the action (13.4.78) and the Lagrangian (13.4.79) with their Minkowski counterparts illustrates how we generalize a classical field theory to curved spacetime:

$$S_{\text{Minkowski}} = \int d^4x \left[\frac{1}{2}\eta^{\mu\nu}\partial_\mu\phi\partial_\nu\phi - V(\phi)\right]. \tag{13.4.80}$$

The metric appears in two places in the curved-spacetime action: First, it appears in the measure of volume in the four-space, d^4x, where the determinant of the metric takes the role of the Jacobian for arbitrary coordinate transformations, $x \to x'$. Second, the metric appears in the kinetic term for the scalar field, where we replace the Minkowski metric $\eta^{\mu\nu}$ with the general metric $g^{\mu\nu}$.

The action (13.4.78) is not the most general assumption we could make, as we can see by writing the full action including gravity,

$$S_{\rm tot} = \int d^4x \sqrt{-g} \left[\frac{m_{\rm Pl}^2}{16\pi} R + \mathcal{L}_\phi \right]. \tag{13.4.81}$$

Here R is the Ricci Scalar, composed of the metric and its derivatives. Variation of the first term in the action results in the Einstein Field Equation (13.1.8). Such a *minimally coupled* theory assumes that there is no direct coupling between the field and the metric, which would be represented in a more general action by terms which mix R and ϕ. In practice, many such non-minimally coupled theories can be transformed to a minimally coupled form by a field redefinition. We could also write a more general theory by modifying the scalar field Lagrangian (13.4.79) to contain non-canonical kinetic terms,

$$\mathcal{L}_\phi = F \left(\phi, g^{\mu\nu} \partial_\mu \phi \partial_\nu \phi \right) - V \left(\phi \right). \tag{13.4.82}$$

where $F()$ is some function of the field and its derivatives. Such Lagrangians appear frequently in models of inflation based on string theory, and are a topic of considerable current research interest. We could also complicate the gravitational sector by replacing the Ricci scalar R with a more complicated function $f(R)$. An example of such a model is the inflation model of Starobinsky,[51] which can be reduced to the form (13.4.78) through a field redefinition. We could also introduce multiple scalar fields.

Here we will confine ourselves for simplicity to a canonical Lagrangian (13.4.79) of a single scalar field, for which the only adjustable quantity is the choice of potential $V(\phi)$. For simplicity, we assume a flat spacetime,

$$g_{\mu\nu} = \begin{pmatrix} 1 & & & \\ & -a^2(t) & & \\ & & -a^2(t) & \\ & & & -a^2(t) \end{pmatrix}, \tag{13.4.83}$$

and the equation of motion for the field ϕ with a Lagrangian given by Eq. (13.4.79) is:

$$\ddot{\phi} + 3H\dot{\phi} - \nabla^2\phi + \frac{\delta V}{\delta \phi} = 0, \tag{13.4.84}$$

where an overdot indicates a derivative with respect to the coordinate time t, and $H = \dot{a}/a$ is the Hubble parameter. We will be particularly interested in the homogeneous mode of the field, for which the gradient term vanishes, $\nabla \phi = 0$, so that the the functional derivative $\delta V / \delta \phi$ simplifies to an ordinary derivative, and the equation of motion simplifies to[d]

$$\ddot{\phi} + 3H\dot{\phi} + V'(\phi) = 0. \tag{13.4.85}$$

The stress-energy for a scalar field is given by

$$T_{\mu\nu} = \partial_\mu \phi \partial_\nu \phi - g_{\mu\nu} \mathcal{L}_\phi, \tag{13.4.86}$$

and, for a homogeneous field, it takes the form of a perfect fluid with energy density ρ and pressure p, with

$$\rho = \frac{1}{2}\dot{\phi}^2 + V(\phi),$$
$$p = \frac{1}{2}\dot{\phi}^2 - V(\phi). \tag{13.4.87}$$

We see that the de Sitter limit, $p \simeq -\rho$, is just the limit in which the potential energy of the field dominates the kinetic energy, $\dot{\phi}^2 \ll V(\phi)$. This limit is referred to as *slow roll*, and under such conditions the universe expands quasi-exponentially,

$$a(t) \propto \exp\left(\int H dt\right) \equiv e^{-N}, \tag{13.4.88}$$

where it is conventional to define the number of e-folds N with the sign convention

$$dN \equiv -H dt, \tag{13.4.89}$$

so that N is large in the far past and decreases as we go forward in time and as the scale factor a increases.

This can be made quantitative by plugging the energy and pressure (13.4.87) into the Friedmann Equation

$$H^2 = \left(\frac{\dot{a}}{a}\right)^2 = \frac{8\pi}{3m_{\text{Pl}}^2}\left[\frac{1}{2}\dot{\phi}^2 + V(\phi)\right], \tag{13.4.90}$$

and the Raychaudhuri Equation, which we write in the convenient form

$$\left(\frac{\ddot{a}}{a}\right) = -\frac{4\pi}{3m_{\text{Pl}}^2}(\rho + 3p) = H^2(1 - \epsilon). \tag{13.4.91}$$

---

[d]The astute reader may well ask: if we are claiming inflation is a solution to the problems of flatness and homogeneity in the universe, why are we assuming flatness and homogeneity from the outset? The answer is that, as long as inflation gets started *somehow* and goes on for long enough, the late-time behavior of the field ϕ will always be described by Eq. (13.4.85). We will see later that we only have observational access to the *end* of the inflationary period, and therefore a consistent theory of initial conditions is not required for investigating the observational consequences of inflation.

Here H^2 is given in terms of ϕ by the Friedmann Equation (13.4.90), and the parameter ϵ specifies the equation of state,

$$\epsilon \equiv \frac{3}{2} \left(\frac{p}{\rho} + 1 \right) = \frac{4\pi}{m_{\text{Pl}}^2} \left(\frac{\dot{\phi}}{H} \right)^2. \tag{13.4.92}$$

It is a straightforward exercise to show that ϵ is related to the evolution of the Hubble parameter by

$$\epsilon = -\frac{d \ln H}{d \ln a} = \frac{1}{H} \frac{dH}{dN}, \tag{13.4.93}$$

where N is the number of e-folds (13.4.89). This is a useful parameterization because the condition for accelerated expansion $\ddot{a} > 0$ is simply equivalent to $\epsilon < 1$. The de Sitter limit $p \to -\rho$ is equivalent to $\epsilon \to 0$, so that the potential $V(\phi)$ dominates the energy density, and

$$H^2 \simeq \frac{8\pi}{3 m_{\text{Pl}}^2} V(\phi). \tag{13.4.94}$$

We make the additional approximation that the friction term in the equation of motion (13.4.85) dominates,

$$\ddot{\phi} \ll 3H\dot{\phi}, \tag{13.4.95}$$

so that the equation of motion for the scalar field is approximately

$$3H\dot{\phi} + V'(\phi) \simeq 0. \tag{13.4.96}$$

Equation (13.4.96) together with the Friedmann Equation (13.4.94) are together referred to as the *slow roll approximation*. The condition (13.4.95) can be expressed in terms of a second dimensionless parameter, conventionally defined as

$$\eta \equiv -\frac{\ddot{\phi}}{H\dot{\phi}} = \epsilon + \frac{1}{2\epsilon} \frac{d\epsilon}{dN}. \tag{13.4.97}$$

The parameters ϵ and η are referred to as *slow roll parameters*, and the slow roll approximation is valid as long as both are small, $\epsilon,\ |\eta| \ll 1$. It is not obvious that this will be a valid approximation for situations of physical interest: η need *not* be small for inflation to take place. Inflation takes place when $\epsilon < 1$, regardless of the value of η. We later demonstrate explicitly that slow roll does in fact hold for interesting choices of inflationary potential. In the limit of slow roll, we can use Eqs. (13.4.94, 13.4.96) to write the parameter ϵ approximately as

$$\epsilon = \frac{4\pi}{m_{\text{Pl}}^2} \left(\frac{\dot{\phi}}{H} \right)^2 \simeq \frac{m_{\text{Pl}}^2}{16\pi} \left(\frac{V'(\phi)}{V(\phi)} \right)^2. \tag{13.4.98}$$

The inflationary limit, $\epsilon \ll 1$ is then just equivalent to a field evolving on a flat potential, $V'(\phi) \ll V(\phi)$. The second slow roll parameter η can likewise be written approximately as:

$$\eta = -\frac{\ddot{\phi}}{H\dot{\phi}}$$

$$\simeq \frac{m_{\mathrm{Pl}}^2}{8\pi}\left[\frac{V''(\phi)}{V(\phi)} - \frac{1}{2}\left(\frac{V'(\phi)}{V(\phi)}\right)^2\right], \qquad (13.4.99)$$

so that the curvature V'' of the potential must also be small for slow roll to be a valid approximation. Similarly, we can write number of e-folds as a function $N(\phi)$ of the field as:

$$N = -\int H\,dt = -\int \frac{H}{\dot{\phi}}\,d\phi = \frac{2\sqrt{\pi}}{m_{\mathrm{Pl}}}\int \frac{d\phi}{\sqrt{\epsilon}}$$

$$\simeq \frac{8\pi}{m_{\mathrm{Pl}}^2}\int_{\phi_e}^{\phi} \frac{V(\phi)}{V'(\phi)}\,d\phi, \qquad (13.4.100)$$

The limits on the last integral are defined such that ϕ_e is a fixed field value, which we will later take to be the end of inflation, and N increases as we go *backward* in time, representing the number of e-folds of expansion which take place between field value ϕ and ϕ_e.

The qualitative picture of scalar field-driven inflation is that of a phase transition with order parameter given by the field ϕ. At early times, the energy density of the universe is dominated by the field ϕ which is slowly evolving on a nearly constant potential, so that it approximates a cosmological constant (Fig. 13.14). During this period, the universe is exponentially driven toward flatness and homogeneity. Inflation ends as the potential steepens and the field begins to oscillate about its vacuum state at the minimum of the potential. At this point, we have an effectively zero-temperature scalar in a state of coherent oscillation about the minimum of the potential, and the universe is a huge Bose-Einstein condensate: hardly a hot Big Bang! In order to transition to a radiation-dominated hot Big Bang cosmology, the energy in the inflaton field must decay into Standard Model particles, a process generically termed *reheating*. This process is model-dependent, but it typically happens very rapidly. Note that the field ϕ need not be a fundamental field like a Higgs boson (although it could in fact be fundamental). *Any* order parameter for a phase transition will do, as long as it has the quantum numbers of vacuum, and the effective potential has the correct properties. The inflaton ϕ could well be a scalar composite of more fundamental degrees of freedom, the coordinate of a brane in a

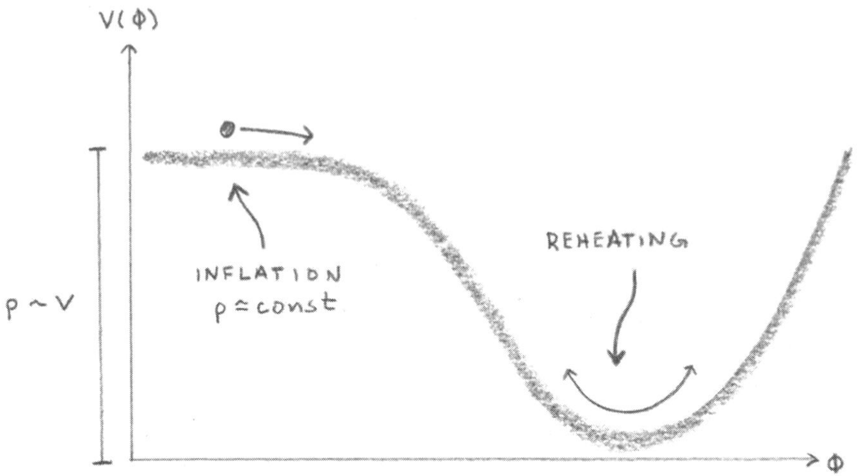

Fig. 13.14. A schematic of the potential for inflation. Inflation takes place on the region of the potential which is sufficiently "flat", and reheating takes place near the true vacuum for the field.

higher-dimensional compactification from string theory, a supersymmetric modulus, or something even more exotic. The simple single-field picture we discuss here is therefore an effective representation of a large variety of underlying fundamental theories. All of the physics important to inflation is contained in the shape of the potential $V(\phi)$. (The details of the underlying theory *are* important for understanding the epoch of reheating, since the reheating process depends crucially on the specific couplings of the inflaton to the other degrees of freedom in the theory.)

How long does inflation need to go on in order to solve the flatness and horizon problems? We use a thermodynamic argument, which rests on a simple fact about cosmological expansion: as long as there are no decays or annihilations of massive particles, all other interactions conserve photon number, so that the number of photons in a comoving volume is *constant*. Since the entropy of photons is proportional to the number density, that means the entropy per comoving volume is also constant. Therefore, the total entropy in the Cosmic Microwave Background (or, equivalently, the total number of photons) is a convenient measure of spatial volume in the universe. Since the entropy per photon s is (up to a few constants) given by the cube of the temperature,

$$s \sim T^3,$$

$$\tag{13.4.101}$$

the total photon entropy S in our current horizon volume is of order

$$S_{\text{hor}} \sim T_{\text{CMB}}^3 d_{\text{H}}^3 \sim \left(\frac{T_{\text{CMB}}}{H_0}\right)^3 \sim 10^{88}, \qquad (13.4.102)$$

where we have taken the CMB temperature to be 2.7 K and the current Hubble parameter H_0 to be 70 km/s/MpC. (The interesting unit conversion from km/s/MpC to Kelvin is left as an exercise for the reader.)

Let us consider a highly over-simplified picture of the universe, in which no particle decays or annihilations occur between the end of inflation and today. In that case, the only time when the photon number (and therefore the entropy) in the universe changes is during the reheating process itself, when the inflation ϕ decays into radiation and sets the initial state for the hot Big Bang. Therefore, we must *at minimum* create an entropy of 10^{88} during reheating. Let us say that the energy density during inflation is

$$\rho \sim V(\phi) \sim \Lambda^4, \qquad (13.4.103)$$

where Λ is some energy scale. Therefore, the horizon size during inflation is then

$$d_{\text{H}} \sim H^{-1} \sim \frac{m_{\text{Pl}}}{\Lambda^2}, \qquad (13.4.104)$$

so that the initial volume of the inflationary "patch" which undergoes exponential expansion is

$$V_i \sim d_{\text{H}}^3 \sim \frac{m_{\text{Pl}}^3}{\Lambda^6}. \qquad (13.4.105)$$

Suppose inflation continues for N e-folds of expansion, so that the scale factor a increases by a factor of e^N during inflation. The *proper* volume of the initial inflationary patch increases by the cube of the scale factor

$$V_f \sim e^{3N} d_{\text{H}}^3 \sim e^{3N} \frac{m_{\text{Pl}}^3}{\Lambda^6}. \qquad (13.4.106)$$

Inflation takes a tiny patch of the universe and blows it up exponentially large, but in such a way that the energy *density* remains approximately constant: we have created an exponential amount of energy out of nothing! During reheating, this huge store of energy in the coherently oscillating field ϕ decays into radiation and the temperature and entropy of the universe undergo an explosive increase. If reheating is highly efficient, then all or most of the energy stored in the inflaton field will be transformed into radiation, and the temperature of the universe after reheating will be of order the energy density of the inflaton field,

$$T_{\text{RH}} \sim \Lambda. \qquad (13.4.107)$$

The entropy per comoving volume after reheating will then be $s_{\rm RH} \sim T_{\rm RH}^3 \sim$ Λ^3, and the *total* entropy in our inflating patch will be

$$S_{\rm RH} \sim V_f T_{\rm RH}^3 \sim e^{3N} \frac{m_{\rm Pl}^3}{\Lambda^3}. \qquad (13.4.108)$$

Since this is our only source of entropy in our toy-model universe, this entropy must be at least as large as the entropy in our current horizon volume, $S_{\rm RH} \geq 10^{88}$. The only adjustable parameter is the number of e-folds of inflation. Taking the logarithm of both sides gives a lower bound on N,

$$N \geq 68 + \ln\left(\frac{\Lambda}{m_{\rm Pl}}\right). \qquad (13.4.109)$$

We will see later that the amplitude of primordial density fluctuations $\delta\rho/\rho \sim 10^{-5}$ typically constrains the inflationary energy scale to be of order $\Lambda \sim 10^{-4} m_{\rm Pl}$, so that we have a lower limit on the number of e-folds of inflation of

$$N > N_{\rm min} \sim 60. \qquad (13.4.110)$$

Most inflation models hugely oversaturate this bound, with $N_{\rm tot} \gg N_{\rm min}$. There is in fact no *upper* bound on the number of e-folds of inflation, an idea which is central to Linde's idea of "eternal" inflation,[52–55] in which inflation, once initiated, never completely ends, with reheating occurring only in isolated patches of the cosmos. Furthermore, it is easy to see that our oversimplified toy model of the universe gives a remarkably accurate estimate of $N_{\rm min}$. In the real universe, all sorts of particle decays and annihilations happen between the end of inflation and today, which create additional entropy. However, our lower bound (13.4.109) is only logarithmically sensitive to these processes. The dominant uncertainty is in the reheat temperature: it is possible that the energy scale of inflation is very low, or that the reheating process is very inefficient, and there are very few *observational* bounds on these scales. We do know that the universe has to be radiation dominated and in equilibrium by the time primordial nucleosynthesis happens at temperatures of order MeV. Furthermore, the baryon asymmetry of the universe is at least a good hint that the Big Bang was hot to at least the scale of electroweak unification. A typical assumption is that the reheat temperature is something between 1 TeV and 10^{16} GeV, which translates into a range for $N_{\rm min}$ of order[56,57]

$$N_{\rm min} \simeq [46, 60]. \qquad (13.4.111)$$

13.4.1. *Example: the $\lambda\phi^4$ potential*

We are now in a position to apply this to a specific case. We use the simple case of a quartic potential,

$$V\left(\phi\right) = \lambda\phi^4. \tag{13.4.112}$$

The slow roll equations (13.4.96, 13.4.94) imply that the field evolves as:

$$\dot{\phi} = -\frac{V'\left(\phi\right)}{3H} = -\sqrt{\frac{m_{\mathrm{Pl}}^2}{24\pi}}\frac{V'\left(\phi\right)}{\sqrt{V\left(\phi\right)}} \propto \phi. \tag{13.4.113}$$

Note that this potential does not much qualitatively resemble the schematic in Fig. 13.13: the "flatness" of the potential arises because the energy density $V\left(\phi\right) \propto \phi^4$ rises much more quickly than the kinetic energy, $\dot{\phi}^2 \propto \phi^2$, so that if the field is far enough out on the potential, the slow roll approximation is self-consistent. The field rolls down to the potential toward the vacuum at the origin, and the equation of state is determined by the parameter ϵ,

$$\epsilon\left(\phi\right) \simeq \frac{m_{\mathrm{Pl}}^2}{16\pi}\left(\frac{V'\left(\phi\right)}{V\left(\phi\right)}\right)^2 = \frac{1}{\pi}\left(\frac{m_{\mathrm{Pl}}}{\phi}\right)^2. \tag{13.4.114}$$

The field value ϕ_e at the end of inflation is when $\epsilon\left(\phi_e\right) = 1$, or

$$\phi_e = \frac{m_{\mathrm{Pl}}}{\sqrt{\pi}}. \tag{13.4.115}$$

For $\phi > \phi_e$, $\epsilon < 1$ and the universe is inflating, and for $\phi < \phi_e$, $\epsilon > 1$ and the expansion enters a decelerating phase. Therefore, even this simple potential has the necessary characteristics to support a period of early-universe inflation followed by reheating and a hot Big Bang cosmology. What about the requirement that the universe inflate for at least 60 e-folds? Using Eq. (13.4.98), we can express the number of e-folds before the end of inflation (13.4.100) as

$$N = \frac{2\sqrt{\pi}}{m_{\mathrm{Pl}}}\int_{\phi_e}^{\phi}\frac{dx}{\sqrt{\epsilon\left(x\right)}} = \pi\left(\frac{\phi}{m_{\mathrm{Pl}}}\right)^2 - 1, \tag{13.4.116}$$

where we integrate *backward* from ϕ_e to ϕ to be consistent with the sign convention (13.4.89). Therefore the field value N e-folds before the end of inflation is

$$\phi_N = m_{\mathrm{Pl}}\sqrt{\frac{N+1}{\pi}}, \tag{13.4.117}$$

so that

$$\phi_{60} = 4.4 m_{\rm Pl}. \qquad (13.4.118)$$

We obtain sufficient inflation, but at a price: the field must be a long way (several times the Planck scale) out on the potential. However, we do *not* necessarily have to invoke quantum gravity, since for small enough coupling λ, the energy density in the field can much less than the Planck density, and the energy density is the physically important quantity.

In this section, we have seen that the basic picture of an early epoch in the universe dominated by vacuum-like energy, leading to nearly exponential expansion, can be realized within the context of a simple scalar field theory. The equation of state for the field approximates a cosmological constant $p = -\rho$ when the energy density is dominated by the field potential $V(\phi)$, and inflation ends when the potential becomes steep enough that the kinetic energy $\dot{\phi}^2/2$ dominates over the potential. To solve the horizon and flatness problems and create a universe consistent with observation, we must have *at least* 60 or so e-folds of inflation, although in principle inflation could continue for much longer than this minimum amount. This dynamical explanation for the flatness and homogeneity of the universe is an interesting, but hardly compelling scenario. It could be that the universe started out homogeneous and flat because of initial conditions, either through some symmetry we do not yet understand, or because there are many universes, and we just happen to find ourselves in a highly unlikely realization which is homogeneous and geometrically flat. In the absence of any other observational handles on the physics of the very early universe, it is impossible to tell. However, flatness and homogeneity are not the whole story: inflation provides an elegant mechanism for explaining the *inhomogeneity* of the universe as well, which we discuss in Sec. 13.5.

13.5. Perturbations in Inflation

The universe we live in today is homogeneous, but only when averaged over very large scales. On small scales, the size of people or solar systems or galaxies or even clusters of galaxies, the universe we see is highly inhomogeneous. Our world is full of complex structure, created by gravitational instability acting on tiny "seed" perturbations in the early universe. If we look as far back in time as the epoch of recombination, the universe on all scales was homogeneous to a high degree of precision, a few parts in 10^5. Recent observational efforts such as the WMAP satellite have made exquisitely precise measurements of the first tiny inhomogeneities in the universe, which later collapsed to form the structure we see today. (We discuss the WMAP observation in more detail in Sec. 13.6.) Therefore, another mystery of Big Bang cosmology is: what created the primordial perturbations? This mystery is compounded by the fact that the perturbations we observe in the CMB exhibit correlations on scales much larger than the horizon size at the time of recombination, which corresponds to an angular multipole of $\ell \simeq 100$, or about $1°$ as observed on the sky today. This is another version of the horizon problem: not only is the universe homogeneous on scales larger than the horizon, but whatever created the primordial perturbations must also have been capable of generating fluctuations on scales larger than the horizon. Inflation provides just such a mechanism.[7–13]

Consider a perturbation in the cosmological fluid with wavelength λ. Since the proper wavelength redshifts with expansion, $\lambda_{\mathrm{prop}} \propto a\,(t)$, the *comoving* wavelength of the perturbation is a constant, $\lambda_{\mathrm{com}} = \mathrm{const}$. This is true of photons or density perturbations or gravitational waves or any other wave propagating in the cosmological background. Now consider this wavelength relative to the size of the horizon: We have seen that in general the horizon as measured in comoving units is proportional to the conformal time, $d_H \propto \tau$. Therefore, for matter- or radiation-dominated expansion, the horizon size *grows* in comoving units, so that a comoving length which is larger than the horizon at early times is smaller than the horizon at late times: modes "fall into" the horizon. The opposite is true during inflation, where the conformal time is negative and evolving toward zero: the comoving horizon size is still proportional to τ, but it now *shrinks* with cosmological expansion, and comoving perturbations which are initially smaller than the horizon are "redshifted" to scales larger than the horizon at late times (Fig. 13.15).

If the universe is inflating at early times, and radiation- or matter-dominated at late times, perturbations in the density of the universe which are initially smaller than the horizon are redshifted during inflation to superhorizon scales. Later, as the horizon begins to grow in comoving coordinates, the perturbations fall back into the horizon, where they act as a source for structure formation. In this way inflation explains the observed properties of perturbations in the universe, which exist at both super- and sub-horizon scales at the time of recombination. Furthermore, an important consequence of this process is that the last perturbations to exit the horizon are the *first* to fall back in. Therefore, the shortest wavelength

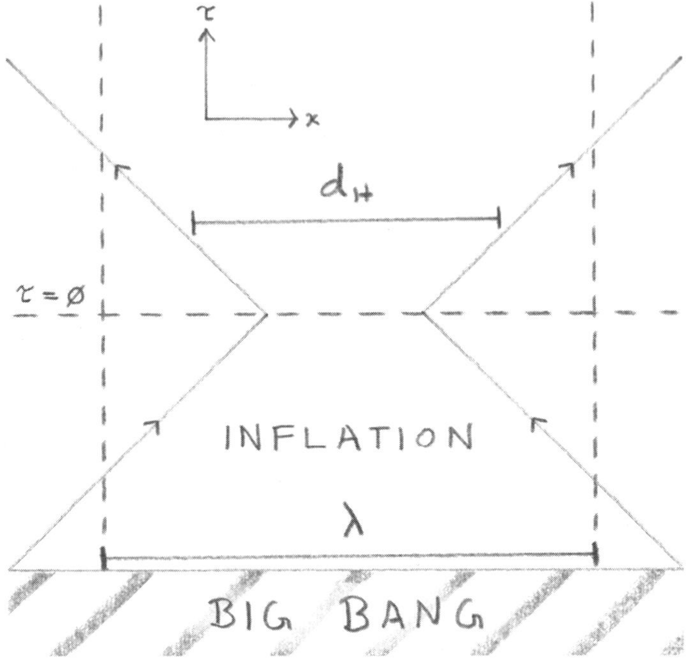

Fig. 13.15.　A conformal diagram of the horizon in an inflationary universe. The co-moving horizon shrinks during inflation, and grows during the radiation- and matter-dominated expansion, while the comoving wavelengths of perturbations remain constant. This drives comoving perturbations to "superhorizon" scales.

perturbations are the ones which exited the horizon just at the end of inflation, $N = 0$, and longer wavelength perturbations exited the horizon earlier. Perturbations about the same size as our horizon today exited the horizon during inflation at around $N = 60$. Perturbations which exited the horizon earlier than that, $N > 60$, are still larger than our horizon today. Therefore, it is only possible to place observational constraints on the *end* of inflation, about the last 60 e-folds. Everything that happened before that, including anything that might tell us about the initial conditions which led to inflation, is most probably inaccessible to us.

This kinematic picture, however, does not itself explain the physical origin of the perturbations. Inflation driven by a scalar field provides a natural explanation for this as well. The inflaton field ϕ evolving on the potential $V(\phi)$ will not evolve completely classically, but will also be subject to small quantum fluctuations about its classical trajectory, which will in general be *inhomogeneous*. Since the energy density of the universe during inflation is dominated by the inflaton field, quantum fluctuations in ϕ couple to the spacetime curvature and result in fluctuations in the density of the universe. Therefore, in the same way that the classical behavior of the field ϕ provides a description of the background evolution of the universe, the quantum behavior of ϕ provides a description of the inhomogeneous perturbations about that background. We defer a full treatment of inflaton perturbations to Appendix A, and in the next section focus on the much simpler case of quantizing a *decoupled* scalar φ in an inflationary spacetime. In addition to its relative simplicity, this case has direct relevance to the generation of gravitational waves in inflation.

13.5.1. *The Klein-Gordon Equation in Curved Spacetime*

Consider an arbitrary free scalar field, which we denote φ to distinguish it from the inflaton field ϕ. The Lagrangian for the field is

$$\mathcal{L} = \frac{1}{2} g^{\mu\nu} \partial_\mu \varphi \partial_\nu \varphi, \tag{13.5.119}$$

and varying the action (13.5.119) gives the Euler-Lagrange equation of motion

$$\frac{1}{\sqrt{-g}} \partial_\nu \left(g^{\mu\nu} \sqrt{-g} \partial_\mu \varphi \right) = 0. \tag{13.5.120}$$

It will prove convenient to express the background FRW metric in conformal coordinates

$$g_{\mu\nu} = a^2 \left(\tau \right) \eta_{\mu\nu} \tag{13.5.121}$$

instead of the coordinate-time metric (13.4.83) we used in Sec. (13.4). Here τ is the conformal time and $\eta_{\mu\nu} = \mathrm{diag.} \left(1, -1, -1, -1 \right)$ is the Minkowski metric. In conformal coordinates, the free scalar equation of motion (13.5.120) is

$$\varphi'' + 2 \left(\frac{a'}{a} \right) \varphi' - \nabla^2 \varphi = 0, \tag{13.5.122}$$

where $' = d/d\tau$ is a derivative with respect to *conformal* time. Note that unlike the case of the inflaton ϕ, we are solving for perturbations and therefore retain the gradient term $\nabla^2 \varphi$. The field φ is a decoupled spectator field evolving in a *fixed* cosmological background, and does not effect the time evolution of the scale factor $a \left(\tau \right)$. An example of such a field is gravitational waves. If we express the spacetime metric as an FRW background

$g_{\mu\nu}^{\text{FRW}}$ plus perturbation $\delta g_{\mu\nu}$, we can express the tensorial portion of the perturbation in general as a sum of two scalar degrees of freedom

$$\delta g_{0i} = \delta g_{i0} = 0$$
$$\delta g_{ij} = \frac{32\pi}{m_{\text{Pl}}} \left(\varphi_+ \hat{e}_{ij}^+ + \varphi_\times \hat{e}_{ij}^\times \right), \tag{13.5.123}$$

where $i, j = 1, 2, 3$, and $\hat{e}_{ij}^{+,\times}$ are longitudinal and transverse polarization tensors, respectively. It is left as an exercise for the reader to show that the scalars $\varphi_{+,\times}$ behave to linear order as free scalars, with equation of motion (13.5.122).

To solve the equation of motion (13.5.122), we first Fourier expand the field into momentum states φ_k,

$$\varphi(\tau, \mathbf{x}) = \int \frac{d^3 k}{(2\pi)^{3/2}} \left[\varphi_{\mathbf{k}}(\tau) b_{\mathbf{k}} e^{i\mathbf{k}\cdot\mathbf{x}} + \varphi_{\mathbf{k}}^*(\tau) b_{\mathbf{k}}^* e^{-i\mathbf{k}\cdot\mathbf{x}} \right]. \tag{13.5.124}$$

Note that the coordinates $\mathbf{x}$ are comoving coordinates, and the wavevector $\mathbf{k}$ is a comoving wavevector, which does not redshift with expansion. The proper wavevector is

$$\mathbf{k}_{\text{prop}} = \mathbf{k}/a(\tau). \tag{13.5.125}$$

Therefore, the comoving wavenumber $\mathbf{k}$ is not itself dynamical, but is just a set of constants labeling a particular Fourier component. The equation of motion for a single mode $\varphi_{\mathbf{k}}$ is

$$\varphi_{\mathbf{k}}'' + 2\left(\frac{a'}{a}\right) \varphi_{\mathbf{k}}' + k^2 \varphi_{\mathbf{k}} = 0. \tag{13.5.126}$$

It is convenient to introduce a field redefinition

$$u_k \equiv a(\tau) \varphi_{\mathbf{k}}(\tau), \tag{13.5.127}$$

and the mode function u_k obeys a generalization of the Klein-Gordon equation to an expanding spacetime,

$$u_k'' + \left[k^2 - \frac{a''}{a} \right] u_k = 0. \tag{13.5.128}$$

(We have dropped the vector notation $\mathbf{k}$ on the subscript, since the Klein-Gordon equation depends only on the magnitude of k.)

Any mode with a fixed comoving wavenumber k redshifts with time, so that early time corresponds to short wavelength (ultraviolet) and late time corresponds to long wavelength (infrared). The solutions to the mode equation show qualitatively different behaviors in the ultraviolet and infrared limits:

- *Short wavelength limit, $k \gg a''/a$.* In this case, the equation of motion is that for a conformally Minkowski Klein-Gordon field,

$$u_k'' + k^2 u_k = 0, \qquad (13.5.129)$$

with solution

$$u_k(\tau) = \frac{1}{\sqrt{2k}} \left(A_k e^{-ik\tau} + B_k e^{ik\tau} \right). \qquad (13.5.130)$$

Note that this is in terms of *conformal* time and *comoving* wavenumber, and can only be identified with an exactly Minkowski spacetime in the ultraviolet limit.

- *Long wavelength limit, $k \ll a''/a$.* In the infrared limit, the mode equation becomes

$$a'' u_k = a u_k'', \qquad (13.5.131)$$

with the trivial solution

$$u_k \propto a \;\Rightarrow\; \varphi_k = \text{const.} \qquad (13.5.132)$$

This illustrates the phenomenon of *mode freezing*: field modes φ_k with wavelength longer than the horizon size cease to be dynamical, and asymptote to a constant, *nonzero* amplitude.[e] This is a quantitative expression of our earlier qualitative notion of particle creation at the cosmological horizon. The amplitude of the field at long wavelength is determined by the boundary condition on the mode, *i.e.* the integration constants A_k and B_k.

Therefore, all of the physics boils down to the question of how we set the boundary condition on field perturbations in the ultraviolet limit. This is fortunate, since in that limit the field theory describing the modes becomes approximately Minkowskian, and we know how to quantize fields in Minkowski Space. Once the integration constants are fixed, the behavior of the mode function u_k is completely determined, and the long-wavelength amplitude of the perturbation can then be calculated without ambiguity. We next discuss quantization.

13.5.2. *Quantization*

We have seen that the equation of motion for field perturbations approaches the usual Minkowski Space Klein-Gordon equation in the ultraviolet limit,

---

[e]The second solution to this equation is a decaying mode, which is always subdominant in the infrared limit.

which corresponds to the limit of early time for a mode redshifting with expansion. We determine the boundary conditions for the mode function via canonical quantization. To quantize the field φ_k, we promote the Fourier coefficients in the classical mode expansion (13.5.124) to annihilation and creation operators

$$b_{\mathbf{k}} \to \hat{b}_{\mathbf{k}}, \quad b_{\mathbf{k}}^* \to \hat{b}_{\mathbf{k}}^\dagger, \tag{13.5.133}$$

with commutation relation

$$\left[\hat{b}_{\mathbf{k}}, \hat{b}_{\mathbf{k}'}^\dagger\right] \equiv \delta^3\left(\mathbf{k} - \mathbf{k}'\right). \tag{13.5.134}$$

Note that the commutator in an FRW background is given in terms of *comoving* wavenumber, and holds whether we are in the short wavelength limit or not. In the short wavelength limit, this becomes equivalent to a Minkowski Space commutator. The quantum field φ is then given by the usual expansion in operators $\hat{b}_{\mathbf{k}}$, $\hat{b}_{\mathbf{k}}^\dagger$

$$\varphi(\tau, \mathbf{x}) = \int \frac{d^3k}{(2\pi)^{3/2}} \left[\varphi_{\mathbf{k}}(\tau) b_{\mathbf{k}} e^{i\mathbf{k}\cdot\mathbf{x}} + \mathrm{H.C.}\right] \tag{13.5.135}$$

The corresponding canonical momentum is

$$\Pi(\tau, \mathbf{x}) \equiv \frac{\delta \mathcal{L}}{\delta(\partial_0 \varphi)} = a^2(\tau) \frac{\partial \varphi}{\partial \tau}. \tag{13.5.136}$$

It is left as an exercise for the reader to show that the canonical commutation relation

$$[\varphi(\tau, \mathbf{x}), \Pi(\tau, \mathbf{x}')] = i\delta^3(\mathbf{x} - \mathbf{x}') \tag{13.5.137}$$

corresponds to a Wronskian condition on the mode u_k,

$$u_k \frac{\partial u_k^*}{\partial \tau} - u_k^* \frac{\partial u_k}{\partial \tau} = i, \tag{13.5.138}$$

which for the ultraviolet mode function (13.5.130) results in a condition on the integration constants

$$|A_k|^2 - |B_k|^2 = 1. \tag{13.5.139}$$

This quantization condition corresponds to one of the two boundary conditions which are necessary to completely determine the solution. The second boundary condition comes from vacuum selection, *i.e.* our definition of which state corresponds to a zero-particle state for the system. In the next section, we discuss the issue of vacuum selection in detail.

13.5.3. *Vacuum Selection*

Consider a quantum field in Minkowski Space. The state space for a quantum field theory is a set of states $|n(\mathbf{k}_1), \ldots, n(\mathbf{k}_i)\rangle$ representing the number of particles with momenta $\mathbf{k}_1, \ldots, \mathbf{k}_i$. The creation and annihilation operators $\hat{a}_{\mathbf{k}}^{\dagger}$ and $\hat{a}_{\mathbf{k}}$ act on these states by adding or subtracting a particle from the state:

$$\hat{a}_{\mathbf{k}}^{\dagger} |n(\mathbf{k})\rangle = \sqrt{n+1}\, |n(\mathbf{k}) + 1\rangle$$
$$\hat{a}_{\mathbf{k}} |n(\mathbf{k})\rangle = \sqrt{n}\, |n(\mathbf{k}) - 1\rangle. \tag{13.5.140}$$

The ground state, or vacuum state of the space, is just the zero particle state:

$$\hat{a}_{\mathbf{k}} |0\rangle = 0. \tag{13.5.141}$$

Note in particular that the vacuum state $|0\rangle$ is *not* equivalent to zero. The vacuum is not nothing:

$$|0\rangle \neq 0. \tag{13.5.142}$$

To construct a quantum field, we look at the familiar classical wave equation for a scalar field,

$$\frac{\partial^2 \phi}{\partial t^2} - \nabla^2 \phi = 0. \tag{13.5.143}$$

To solve this equation, we decompose into Fourier modes $u_{\mathbf{k}}$,

$$\phi = \int d^3 k \left[a_{\mathbf{k}} u_{\mathbf{k}}(t) e^{i\mathbf{k}\cdot\mathbf{x}} + a_{\mathbf{k}}^* u_{\mathbf{k}}^*(t) e^{-i\mathbf{k}\cdot\mathbf{x}} \right], \tag{13.5.144}$$

where the mode functions $u_{\mathbf{k}}(t)$ satisfy the ordinary differential equation

$$\ddot{u}_{\mathbf{k}} + k^2 u_{\mathbf{k}} = 0. \tag{13.5.145}$$

This is a classical wave equation with a classical solution, and the Fourier coefficients $a_{\mathbf{k}}$ are just complex numbers. The solution for the mode function is

$$u_{\mathbf{k}} \propto e^{-i\omega_k t}, \tag{13.5.146}$$

where ω_k satisfies the dispersion relation

$$\omega_k^2 - \mathbf{k}^2 = 0. \tag{13.5.147}$$

To turn this into a quantum field, we identify the Fourier coefficients with creation and annihilation operators

$$a_{\mathbf{k}} \to \hat{a}_{\mathbf{k}}, \quad a_{\mathbf{k}}^* \to \hat{a}_{\mathbf{k}}^{\dagger}, \tag{13.5.148}$$

and enforce the commutation relations

$$\left[\hat{a}_{\mathbf{k}}, \hat{a}_{\mathbf{k}'}^{\dagger}\right] = \delta^3\left(\mathbf{k} - \mathbf{k}'\right). \tag{13.5.149}$$

This is the standard quantization of a scalar field in Minkowski Space, which should be familiar. But what probably is not familiar is that this solution has an interesting symmetry. Suppose we define a new mode function $u_{\mathbf{k}}$ which is a rotation of the solution (13.5.146):

$$u_{\mathbf{k}} = A(k)e^{-i\omega t + i\mathbf{k}\cdot\mathbf{x}} + B(k)e^{i\omega t - i\mathbf{k}\cdot\mathbf{x}}. \tag{13.5.150}$$

This is *also* a perfectly valid solution to the original wave equation (13.5.143), since it is just a superposition of the Fourier modes. But we can then re-write the quantum field in terms of our original Fourier modes and new *operators* $\hat{b}_{\mathbf{k}}$ and $\hat{b}_{\mathbf{k}}^{\dagger}$ and the original Fourier modes $e^{i\mathbf{k}\cdot\mathbf{x}}$ as:

$$\phi = \int d^3k \left[\hat{b}_{\mathbf{k}}e^{-i\omega t + i\mathbf{k}\cdot\mathbf{x}} + \hat{b}_{\mathbf{k}}^{\dagger}e^{+i\omega t - i\mathbf{k}\cdot\mathbf{x}}\right], \tag{13.5.151}$$

where the new operators $\hat{b}_{\mathbf{k}}$ are given in terms of the old operators $\hat{a}_{\mathbf{k}}$ by

$$\hat{b}_{\mathbf{k}} = A(k)\hat{a}_{\mathbf{k}} + B^*(k)\hat{a}_{\mathbf{k}}^{\dagger}. \tag{13.5.152}$$

This is completely equivalent to our original solution (13.5.144) as long as the new operators satisfy the same commutation relation as the original operators,

$$\left[\hat{b}_{\mathbf{k}}, \hat{b}_{\mathbf{k}'}^{\dagger}\right] = \delta^3\left(\mathbf{k} - \mathbf{k}'\right). \tag{13.5.153}$$

This can be shown to place a condition on the coefficients A and B,

$$|A|^2 - |B|^2 = 1. \tag{13.5.154}$$

Otherwise, we are free to choose A and B as we please.

This is just a standard property of linear differential equations: any linear combination of solutions is itself a solution. But what does it mean physically? In one case, we have an annihilation operator $\hat{a}_{\mathbf{k}}$ which gives zero when acting on a particular state which we call the vacuum state:

$$\hat{a}_{\mathbf{k}}\left|0_a\right\rangle = 0. \tag{13.5.155}$$

Similarly, our rotated operator $\hat{b}_{\mathbf{k}}$ gives zero when acting on some state

$$\hat{b}_{\mathbf{k}}\left|0_b\right\rangle = 0. \tag{13.5.156}$$

The point is that the two "vacuum" states are not the same

$$\left|0_a\right\rangle \neq \left|0_b\right\rangle. \tag{13.5.157}$$

From this point of view, we can define any state we wish to be the "vacuum" and build a completely consistent quantum field theory based on this assumption. From another equally valid point of view this state will contain particles. How do we tell which is the *physical* vacuum state? To define the real vacuum, we have to consider the spacetime the field is living in. For example, in regular special relativistic quantum field theory, the "true" vacuum is the zero-particle state as seen by an inertial observer. Another more formal way to state this is that we require the vacuum to be Lorentz symmetric. This fixes our choice of vacuum $|0\rangle$ and defines unambiguously our set of creation and annihilation operators $\hat{a}$ and $\hat{a}^\dagger$. A consequence of this is that an *accelerated* observer in the Minkowski vacuum will think that the space is full of particles, a phenomenon known as the Unruh effect.[58] The zero-particle state for an accelerated observer is different than for an inertial observer. The case of an FRW spacetime is exactly analogous, except that the FRW equivalent of an inertial observer is an observer at rest in comoving coordinates. Since an FRW spacetime is asymptotically Minkowski in the ultraviolet limit, we choose the vacuum field which corresponds to the usual Minkowski vacuum in that limit,

$$u_k\left(\tau\right) \propto e^{-ik\tau} \;\;\Rightarrow\; A_k = 1, \; B_k = 0. \qquad (13.5.158)$$

This is known as the *Bunch-Davies* vacuum. This is not the only possible choice, although it is widely believed to be the most natural. The issue of vacuum ambiguity of inflationary perturbations is a subject which is extensively discussed in the literature, and is still the subject of controversy. It is known that the choice of vacuum is potentially sensitive to quantum-gravitational physics,[59–61] a subject which is referred to as *Trans-Planckian* physics.[18,62,63] For the remainder of our discussion, we will assume a Bunch-Davies vacuum.

The key point is that quantization and vacuum selection together *completely* specify the mode function, up to an overall phase. This means that the amplitude of the mode once it has redshifted to long wavelength and frozen out is similarly determined. In the next section, we solve the mode equation at long wavelength for an inflationary background.

13.5.4. *Exact Solutions and the Primordial Power Spectrum*

The exact form of the solution to Eq. (13.5.128) depends on the evolution of the background spacetime, as encoded in $a\left(\tau\right)$, which in turn depends on the equation of state of the field driving inflation. We will consider the case where the equation of state is constant, which will *not* be the case in general for scalar field-driven inflation, but will nonetheless turn out to be a good approximation in the limit of a slowly rolling field. Generalizing Eq. (13.3.77) to the case of arbitrary equation of state parameter $\epsilon = \text{const.}$, the conformal time can be written

$$\tau = -\left(\frac{1}{aH}\right)\left(\frac{1}{1-\epsilon}\right), \tag{13.5.159}$$

and the Friedmann and Raychaudhuri Equations (13.2.14) give

$$\frac{a''}{a} = a^2 H^2 \left(2 - \epsilon\right), \tag{13.5.160}$$

where a prime denotes a derivative with respect to conformal time. The conformal time, as in the case of de Sitter space, is negative and tending toward zero during inflation. (Proof of these relations is left as an exercise for the reader.) We can then write the mode equation (13.5.128) as

$$u_k'' + \left[k^2 - a^2 H^2 \left(2 - \epsilon\right)\right] u_k = 0. \tag{13.5.161}$$

Using Eq. (13.5.159) to write aH in terms of the conformal time τ, the equation of motion becomes

$$\tau^2 \left(1 - \epsilon\right)^2 u_k'' + \left[\left(k\tau\right)^2 \left(1 - \epsilon\right)^2 - \left(2 - \epsilon\right)\right] u_k = 0. \tag{13.5.162}$$

This is a Bessel equation, with solution

$$u_k \propto \sqrt{-k\tau}\left[J_\nu\left(-k\tau\right) \pm iY_\nu\left(-k\tau\right)\right], \tag{13.5.163}$$

where the index ν is given by:

$$\nu = \frac{3 - \epsilon}{2\left(1 - \epsilon\right)}. \tag{13.5.164}$$

The quantity $-k\tau$ has special physical significance, since from Eq. (13.5.159) we can write

$$\left(-k\tau\right)\left(1 - \epsilon\right) = \frac{k}{aH}, \tag{13.5.165}$$

where the quantity (k/aH) expresses the wavenumber k in units of the comoving horizon size $d_{\mathrm{H}} \sim (aH)^{-1}$. Therefore, the short wavelength limit is $-k\tau \to -\infty$, or $(k/aH) \gg 1$. The long-wavelength limit is $-k\tau \to 0$, or $(k/aH) \ll 1$.

The simple case of de Sitter space ($p = -\rho$) corresponds to the limit $\epsilon = 0$, so that the Bessel index is $\nu = 3/2$ and the mode function (13.5.163) simplifies to

$$u_k \propto \left(\frac{k\tau - i}{k\tau} \right) e^{\pm ik\tau}. \tag{13.5.166}$$

In the short wavelength limit, $(-k\tau) \to -\infty$, the mode function is given, as expected, by

$$u_k \propto e^{\pm ik\tau}. \tag{13.5.167}$$

Selecting the Bunch-Davies vacuum gives $u_k \propto e^{ik\tau}$, and canonical quantization fixes the normalization,

$$u_k = \frac{1}{\sqrt{2k}} e^{-ik\tau}. \tag{13.5.168}$$

Therefore, the fully normalized exact solution is

$$u_k = \frac{1}{\sqrt{2k}} \left(\frac{k\tau - i}{k\tau} \right) e^{-ik\tau}. \tag{13.5.169}$$

This solution has no free parameters aside from an overall phase, and is valid at *all* wavelengths, including after the mode has been redshifted outside of the horizon and becomes non-dynamical, or "frozen". In the long wavelength limit, $-k\tau \to 0$, the mode function (13.5.169) becomes

$$u_k \to \frac{1}{\sqrt{2k}} \left(\frac{i}{(-k\tau)} \right) = \frac{i}{2k} \left(\frac{aH}{k} \right) \propto a, \tag{13.5.170}$$

consistent with the qualitative result (13.5.132). Therefore the field amplitude φ_k is given by

$$|\varphi_k| = \left| \frac{u_k}{a} \right| \to \frac{H}{\sqrt{2}k^{3/2}} = \text{const.} \tag{13.5.171}$$

The quantum mode therefore displays the freezeout behavior we noted qualitatively above (Fig. 13.16). The amplitude of quantum fluctuations is conventionally expressed in terms of the two-point correlation function of the field φ. It is left as an exercise for the reader to show that the vacuum two-point correlation function is given by

$$\begin{aligned}
\langle 0 | \varphi(\tau, \mathbf{x}) \varphi(\tau, \mathbf{x}') | 0 \rangle &= \int \frac{d^3k}{(2\pi)^3} \left| \frac{u_k}{a} \right|^2 e^{i\mathbf{k} \cdot (\mathbf{x} - \mathbf{x}')} \\
&= \int \frac{dk}{k} P(k) e^{i\mathbf{k} \cdot (\mathbf{x} - \mathbf{x}')}, \tag{13.5.172}
\end{aligned}$$

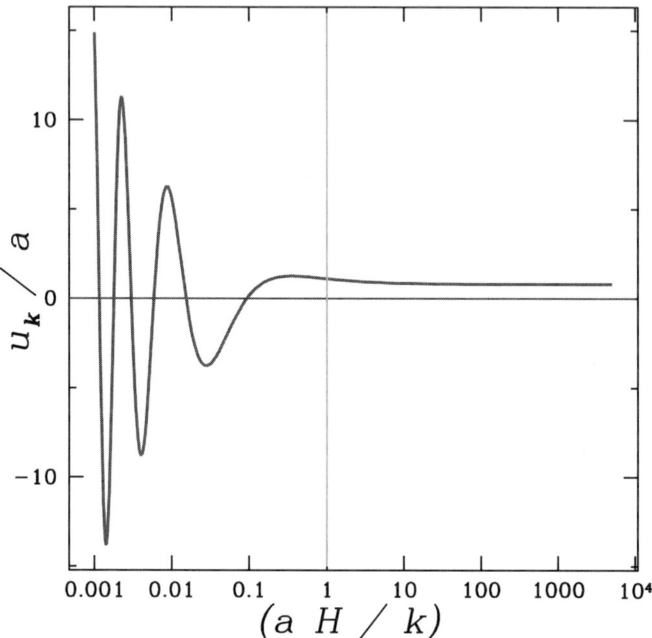

Fig. 13.16.   The normalized mode function in de Sitter space, showing oscillatory behavior on subhorizon scales $k/aH > 1$, and mode freezing on superhorizon scales, $k/aH < 1$.

where the *power spectrum* $P(k)$ is defined as

$$P(k) \equiv \left(\frac{k^3}{2\pi^2}\right)\left|\frac{u_k}{a}\right|^2 \longrightarrow \left(\frac{H}{2\pi}\right)^2, \quad -k\tau \to 0. \qquad (13.5.173)$$

The power per logarithmic interval k in the field fluctuation is then given in the long wavelength limit by the Hubble parameter $H = \text{const}$. This property of scale invariance is exact in the de Sitter limit.

In a more general model, the spacetime is only *approximately* de Sitter, and we expect that the power spectrum of field fluctuations will only be approximately scale invariant. It is convenient to express this dynamics in terms of the equation of state parameter ϵ,

$$\epsilon = \frac{1}{H}\frac{dH}{dN}. \qquad (13.5.174)$$

We must have $\epsilon < 1$ for inflation, and for a slowly rolling field $|\eta| \ll 1$ means that ϵ will also be slowly varying, $\epsilon \simeq \text{const}$. It is straightforward to show that for $\epsilon = \text{const.} \neq 0$ that:

- The Bunch-Davies vacuum corresponds to the positive mode of Eq. (13.5.163),

$$u_k \propto \sqrt{-k\tau}\left[J_\nu\left(-k\tau\right) + iY_\nu\left(-k\tau\right)\right]. \qquad (13.5.175)$$

- Quantization fixes the normalization as

$$u_k = \frac{1}{2}\sqrt{\frac{\pi}{k}}\sqrt{-k\tau}\left[J_\nu\left(-k\tau\right) + iY_\nu\left(-k\tau\right)\right]. \qquad (13.5.176)$$

- The power spectrum in the long-wavelength limit $k/aH \to 0$ is a power law in k:

$$[P\left(k\right)]^{1/2} \longrightarrow 2^{\nu-3/2}\frac{\Gamma\left(\nu\right)}{\Gamma\left(3/2\right)}\left(1-\epsilon\right)\left(\frac{H}{2\pi}\right)\left(\frac{k}{aH\left(1-\epsilon\right)}\right)^{3/2-\nu}, \qquad (13.5.177)$$

where $\Gamma\left(\nu\right)$ is a gamma function, and

$$\nu = \frac{3-\epsilon}{2\left(1-\epsilon\right)}. \qquad (13.5.178)$$

Proof is left as an exercise for the reader.[f] Note that in the case $\epsilon = \text{const.}$, both the background and perturbation equations are *exactly* solvable.

We can use these solutions as approximate solutions in the more general slow roll case, where $\epsilon \ll 1 \simeq \text{const.}$, so that the dependence of the power spectrum on k is approximately a power-law,

$$P\left(k\right) \propto k^n, \qquad (13.5.179)$$

with spectral index

$$n = 3 - 2\nu = 3 - \frac{3-\epsilon}{1-\epsilon} \simeq -2\epsilon. \qquad (13.5.180)$$

Equation (13.5.177) is curious, however, because it does not obviously exhibit complete mode freezing at long wavelength, since a and H both depend on time. We can show that $P\left(k\right)$ does in fact approach a time-dependent value at long wavelength by evaluating

$$\frac{d}{dN}\left[H\left(\frac{k}{aH}\right)^{3/2-\nu}\right] = \frac{d}{dN}\left[H\left(\frac{k}{aH}\right)^{-\epsilon/(1-\epsilon)}\right]$$

---

[f]Note that the quantization condition (13.5.137) can be applied to the solution (13.5.163) exactly, resulting in the normalization condition (13.5.139), without approximating the solution in the short-wavelength limit!

$$= H\epsilon \left(\frac{k}{aH}\right)^{-\epsilon/(1-\epsilon)} - \frac{\epsilon}{1-\epsilon}\left(\frac{k}{aH}\right)^{-\epsilon/(1-\epsilon)-1}\left(\frac{k}{aH} - \frac{\epsilon k}{aH}\right)$$
$$= 0, \tag{13.5.181}$$

which can be easily shown using $a \propto e^{-N}$ and $H \propto e^{\epsilon N}$. That is, the time-dependent quantities a and H in Eq. (13.5.177) are combined in such a way as to form an *exactly* conserved quantity. Since it is conserved, we are free to evaluate it at any time (or value of aH) that we wish. It is conventional to evaluate the power spectrum at *horizon crossing*, or at $aH = k$, so that

$$P^{1/2}(k) \simeq \left(\frac{H}{2\pi}\right)_{k=aH}, \tag{13.5.182}$$

where we have approximated the ν-dependent multiplicative factor as order unity.[g]

It is straightforward to calculate the spectral index (13.5.180) directly from the horizon crossing expression (13.5.182) by using

$$a \propto e^{-N}, \quad H \propto e^{\epsilon N}, \tag{13.5.183}$$

so that we can write derivatives in k at horizon crossing as derivatives in the number of e-folds N,

$$d\ln k|_{k=aH} = d\ln(aH) = \frac{1}{aH}\frac{d(aH)}{dN}dN = (\epsilon - 1)\,dN. \tag{13.5.184}$$

The spectral index is then, to lowest order in slow roll

$$\begin{aligned}
n = \frac{d\ln P(k)}{d\ln k} &= \frac{k}{H^2}\frac{dH^2}{dk}\bigg|_{k=aH}\\
&= \frac{1}{H^2(\epsilon-1)}\frac{dH^2}{dN}\\
&= \frac{2\epsilon}{(\epsilon-1)}\\
&\simeq -2\epsilon, \tag{13.5.185}
\end{aligned}$$

in agreement with (13.5.180). Note that we are rather freely changing variables from the wavenumber k to the comoving horizon size $(aH)^{-1}$ to the number of e-folds N. As long as the cosmological evolution is monotonic,

---

[g]This is *not* the value of the scalar field power spectrum at the moment the mode is physically crossing outside the horizon, as is often stated in the literature: it is the value of the power spectrum in the asymptotic long-wavelength limit. It is easy to show from the exact solution (13.5.176) that the mode function is still evolving with time as it crosses the horizon at $k = aH$, and the asymptotic amplitude differs from the amplitude at horizon crossing by about a factor of two. See Ref. [64] for a more detailed discussion of this point.

these are all different ways of measuring time: the time when a mode with wavenumber k exits the horizon, the time at which the horizon is a particular size, the number of e-folds N and the field value ϕ are all effectively just different choices of a clock, and we can switch from one to another as is convenient. For example, in the slow roll approximation, the Hubble parameter H is just a function of ϕ, $H \propto \sqrt{V(\phi)}$. Because of this, it is convenient to define $N(k)$ to be the number of e-folds (13.4.100) when a mode with wavenumber k crosses outside the horizon, and $\phi_N(k)$ to be the field value $N(k)$ e-folds before the end of inflation. Then the power spectrum can be written equivalently as *either* a function of k or of ϕ:

$$P^{1/2}(k) = \left(\frac{H}{2\pi}\right)_{k=aH} = \left(\frac{H}{2\pi}\right)_{\phi=\phi_N(k)} \simeq \sqrt{\frac{2V(\phi_N)}{3\pi m_{\rm Pl}^2}}. \qquad (13.5.186)$$

Wavenumbers k are conventionally normalized in units of $h{\rm Mpc}^{-1}$ as measured in the *current* universe. We can relate N to scales in the current universe by recalling that modes which are of order the horizon size in the universe today, $k \sim a_0 H_0$, exited the horizon during inflation when $N = [46, 60]$, so that we can calculate the amplitude of perturbations at the scale of the CMB quadrupole today by evaluating the power spectrum for field values between ϕ_{46} and ϕ_{60}.

One example of a free scalar in inflation is gravitational wave modes, where the transverse and longitudinal polarization states of the gravity waves evolve as independent scalar fields. Using Eq. (13.5.123), we can then calculate the power spectrum in gravity waves (or *tensors*) as the sum of the two-point correlation functions for the separate polarizations:

$$P_T = \langle \delta g_{ij}^2 \rangle = 2 \times \frac{32}{m_{\rm Pl}^2} \langle \varphi^2 \rangle = \frac{16 H^2}{\pi m_{\rm Pl}^2} \propto k^{n_T}, \qquad (13.5.187)$$

with spectral index

$$n_T = -2\epsilon. \qquad (13.5.188)$$

If the amplitude is large enough, such a spectrum of primordial gravity waves will be observable in the cosmic microwave background anisotropy and polarization, or be directly detectable by proposed experiments such as Big Bang Observer.[65,66]

The second type of perturbation generated during inflation is perturbations in the density of the universe, which are the dominant component of the CMB anisotropy $\delta T/T \sim \delta\rho/\rho \sim 10^{-5}$, and are responsible for structure formation. Density, or *scalar* perturbations are more complicated than tensor perturbations because they are generated by quantum fluctuations in the inflaton field itself: since the background energy density is dominated by the inflaton, fluctuations of the inflaton up or down the potential generate perturbations in the density. The full calculation requires self-consistent General Relativistic perturbation theory, and is presented in Appendix A. Here we simply state the result: Perturbations in the inflaton field $\delta\phi \simeq H/2\pi$ generate density perturbations with power spectrum

$$P_{\mathcal{R}}\left(k\right) = \left(\frac{\delta N}{\delta\phi}\delta\phi\right)^2 = \left.\frac{H^2}{\pi m_{\mathrm{Pl}}^2 \epsilon}\right|_{k=aH} \propto k^{n_S - 1}, \qquad (13.5.189)$$

where N is the number of e-folds. Scalar perturbations are therefore enhanced relative to tensor perturbations by a factor of $1/\epsilon$. The scalar power spectrum is also an approximate power-law, with spectral index

$$n_S - 1 = \frac{\epsilon}{H^2\left(\epsilon - 1\right)}\frac{d}{dN}\left(\frac{H^2}{\epsilon}\right) \simeq -4\epsilon + 2\eta, \qquad (13.5.190)$$

where η is the second slow roll parameter (13.4.97). Therefore, for any particular choice of inflationary potential, we have four measurable quantities: the amplitudes P_T and $P_{\mathcal{R}}$ of the tensor and scalar power spectra, and their spectral indices n_T and n_S. However, not all of these parameters are independent. In particular, the ratio r between the scalar and tensor amplitudes is given by the parameter ϵ, as is the tensor spectral index n_T:

$$r \equiv \frac{P_T}{P_S} = 16\epsilon = -8n_T. \qquad (13.5.191)$$

This relation is known as the *consistency condition* for single-field slow roll inflation, and is in principle testable by a sufficiently accurate measurement of the primordial perturbation spectra.

In the next section, we apply these results to our example $\lambda\phi^4$ potential and calculate the inflationary power spectra.

13.5.5. *Example:* $\lambda\phi^4$

For the case of our example model with $V\left(\phi\right) = \lambda\phi^4$, it is now straightforward to calculate the scalar and tensor perturbation spectra. We express the normalization of the power spectra as a function of the number of e-folds N by

$$P_{\mathcal{R}}^{1/2} = \left.\frac{H}{m_{\mathrm{Pl}}\sqrt{\pi\epsilon}}\right|_{\phi=\phi_N}$$

$$\begin{aligned}
&= \frac{4\sqrt{24\pi}}{3m_{\mathrm{Pl}}^3} \frac{[V(\phi_N)]^{3/2}}{V'(\phi_N)}\\
&= \frac{24\pi}{3}\left(\frac{N+1}{\pi}\right)\lambda^{1/2} \sim 10^{-5},
\end{aligned} \qquad (13.5.192)$$

where we have used the slow roll expressions for H (13.4.94) and ϵ (13.4.98) and Eq. (13.4.117) for ϕ_N. For perturbations about the current size of our horizon, $N = 60$, and CMB normalization forces the self-coupling to be very small,

$$\lambda \sim 10^{-15}. \qquad (13.5.193)$$

The presence of an extremely small parameter is not peculiar to the $\lambda\phi^4$ model, but is generic, and is referred to as the *fine tuning* problem for inflation.

We can similarly calculate the tensor amplitude

$$P_T^{1/2} = \frac{4H}{m_{\mathrm{Pl}}\sqrt{\pi}}, \qquad (13.5.194)$$

which is usually expressed in terms of the tensor/scalar ratio

$$\begin{aligned}
r = 16\epsilon(\phi_N) &= \frac{m_{\mathrm{Pl}}}{\pi}\left(\frac{V'(\phi_N)}{V(\phi_N)}\right)^2\\
&= \frac{16}{\pi}\left(\frac{m_{\mathrm{Pl}}}{\phi_N}\right)^2 = \frac{16}{N+1} \simeq 0.26,
\end{aligned} \qquad (13.5.195)$$

where we have again taken $N = 60$. For this particular model, the power in gravitational waves is large, about a quarter of the power in scalar perturbations. This is *not* generic, but is quite model-dependent. Some choices of potential predict large tensor contributions (where "large" means of order 10%), and other choices of potential predict very tiny tensor contributions, well below 1%.

The tensor spectral index n_T is fixed by the consistency condition (13.5.191), but the scalar spectral index n_S is an independent parameter because of its dependence on η:

$$n = 1 - 4\epsilon(\phi_N) + 2\eta(\phi_N), \qquad (13.5.196)$$

where

$$\epsilon(\phi_N) = \frac{1}{N+1}, \qquad (13.5.197)$$

and

$$\eta(\phi_N) = \frac{m_{\rm Pl}^2}{8\pi}\left[\frac{V''(\phi_N)}{V(\phi_N)} - \frac{1}{2}\left(\frac{V'(\phi_N)}{V(\phi_N)}\right)^2\right]$$

$$= \frac{m_{\rm Pl}^2}{8\pi}\left[\frac{12}{\phi_N^2} - \frac{8}{\phi_N^2}\right]$$

$$= \frac{1}{2\pi}\left(\frac{m_{\rm Pl}}{\phi_N}\right)^2 = \frac{1}{2(N+1)}. \qquad (13.5.198)$$

The spectral index is then

$$n = 1 - \frac{3}{N+1} \simeq 0.95. \qquad (13.5.199)$$

Note that we have assumed slow roll from the beginning in the calculation without *a priori* knowing that it is a good approximation for this choice of potential. However, at the end of the day it is clear that the slow roll ansatz was a good one, since ϵ and η are both of order 0.01.

Finally, we note that the energy density during inflation is characterized by a mass scale

$$\rho^{1/4} \sim \Lambda \sim \lambda^{1/4} m_{\rm Pl} \sim 10^{15}\ {\rm GeV}, \qquad (13.5.200)$$

about the scale for which we expect Grand Unification to be important. This interesting coincidence suggests that the physics of inflation may be found in Grand Unified Theories (GUTs). Different choices of potential $V(\phi)$ will give different values for the amplitudes and shapes of the primordial power spectra. Since the normalization is fixed by the CMB to be $P_\mathcal{R} \sim 10^{-5}$, the most useful observables for distinguishing among different potentials are the scalar/tensor ratio r and the scalar spectral index n_S. In single-field inflationary models, the tensor spectral index is fixed by the consistency condition (13.5.191), and is therefore not an independent parameter. The consistency condition can therefore be taken to be a *prediction* of single-field inflation, which is in principle verifiable by observation. In practice, this is very difficult, since it involves measuring not just the amplitude of the gravitational wave power spectrum, but also its *shape*. We will see in Sec. 13.6 that current data place only a rough upper bound on the tensor/scalar ratio r, and it is highly unlikely that any near-future measurement of primordial gravitational waves will be accurate enough to constrain n_T well enough to test the consistency condition. In the next section, we discuss current observational constraints on the form of the inflationary potential.

13.6. Observational Constraints

Our simple picture of inflation generated by single, minimally coupled scalar field makes a set of very definite predictions for the form of primordial cosmological fluctuations:

- Gaussianity: since the two-point correlation function of a free scalar field $\langle \varphi^2 \rangle$ is Gaussian, cosmological perturbations generated in a single-field inflation model will by necessity also form a Gaussian random distribution.
- Adiabaticity: since there is only one order parameter ϕ governing the generation of density perturbations, we expect the perturbations in all the components of the cosmological fluid (baryons, dark matter, neutrinos) to be *in phase* with each other. Such a case is called *adiabatic*. If one or more components fluctuates out of phase with others, these are referred to as *isocurvature* modes. Single-field inflation predicts an absence of isocurvature fluctuations.
- Scale invariance: In the limit of de Sitter space, fluctuations in any quantum field are exactly scale invariant, $n = 1$, as a result of the fact that the Hubble parameter is exactly constant. Since slow-roll inflation is quasi-de Sitter, we expect the perturbation spectra to be nearly, but not exactly scale invariant, with $|n_S - 1| = |2\eta - 4\epsilon| \ll 1$.
- Scalar perturbations dominate over tensor perturbations, $r = 16\epsilon$.

Furthermore, given a potential $V(\phi)$, we have a "recipe" for calculating the form of the primordial power spectra generated during inflation:

(1) Calculate the field value at the end of inflation ϕ_e from

$$\epsilon(\phi_e) = \frac{m_{\rm Pl}^2}{16\pi} \left(\frac{V'(\phi_e)}{V(\phi_e)} \right)^2 = 1. \tag{13.6.201}$$

(2) Calculate the field value N e-folds before the end of inflation ϕ_N by integrating backward on the potential from $\phi = \phi_e$,

$$N = \frac{2\sqrt{\pi}}{m_{\rm Pl}} \int_{\phi_e}^{\phi_N} \frac{d\phi'}{\sqrt{\epsilon(\phi')}}. \tag{13.6.202}$$

(3) Calculate the normalization of the scalar power spectrum by

$$P_{\mathcal{R}}^{1/2} = \left. \frac{H}{m_{\rm Pl}\sqrt{\pi\epsilon}} \right|_{\phi=\phi_N} \sim 10^{-5}, \tag{13.6.203}$$

where the CMB quadrupole corresponds to roughly $N = 60$. A more accurate calculation includes the uncertainty in the reheat temperature, which gives a range $N \simeq [46, 60]$, and a corresponding uncertainty in the observable parameters.

(4) Calculate the tensor/scalar ratio r and scalar spectral index n_S at $N = [46, 60]$ by

$$r = 16\epsilon\left(\phi_N\right), \qquad (13.6.204)$$

and

$$n_s = 1 - 4\epsilon\left(\phi_N\right) + 2\eta\left(\phi_N\right), \qquad (13.6.205)$$

where the second slow roll parameter η is given by:

$$\eta\left(\phi_N\right) = \frac{m_{\rm Pl}^2}{8\pi}\left[\frac{V''\left(\phi_N\right)}{V\left(\phi_N\right)} - \frac{1}{2}\left(\frac{V'\left(\phi_N\right)}{V\left(\phi_N\right)}\right)^2\right]. \qquad (13.6.206)$$

The key point is that the scalar power spectrum $P_\mathcal{R}$ and the tensor power spectrum P_T are both completely determined by the choice of potential $V\left(\phi\right)$.[h] Therefore, if we measure the primordial perturbations in the universe accurately enough, we can in principle constrain the form of the inflationary potential. This is extremely exciting, because it gives us a very rare window into physics at extremely high energy, perhaps as high as the GUT scale or higher, far beyond the reach of accelerator experiments such as the Large Hadron Collider.

It is convenient to divide the set of possible single-field potentials into a few basic types:[69]

- *Large-field potentials* (Fig. 13.17). These are the simplest potentials one might imagine, with potentials of the form $V\left(\phi\right) = m^2\phi^2$, or our example case, $V\left(\phi\right) = \lambda\phi^4$. Another widely-noted example of this type of model is inflation on an exponential potential, $V\left(\phi\right) = \Lambda^4 \exp\left(\phi/\mu\right)$, which has the useful property that both the background evolution and the perturbation equations are exactly solvable. In the large-field case, the field is displaced from the vacuum at the origin by an amount of order $\phi \sim m_{\rm Pl}$ and rolls down the potential toward the origin. Large-field models are typically characterized by a "red" spectral index $n_S < 1$, and a substantial gravitational wave contribution, $r \sim 0.1$.

---

[h]Strictly speaking, this is true only for scalar fields with a canonical kinetic term, where the speed of sound of perturbations is equal to the speed of light. More complicated scenarios such as DBI inflation[67] require specification of an extra free function, the speed of sound $c_S\left(\phi\right)$, to calculate the power spectra. For constraints on this more general class of models, see Ref. [68].

- *Small-field potentials* (Fig. 13.18). These are potentials characteristic of spontaneous symmetry breaking phase transitions, where the field rolls off an unstable equilibrium with $V'(\phi) = 0$ toward a displaced vacuum. Examples of small-field inflation include a simple quadratic potential, $V(\phi) = \lambda \left(\phi^2 - \mu^2\right)^2$, inflation from a pseudo-Nambu-Goldstone boson or a shift symmetry in string theory (called *Natural Inflation*) with a potential typically of the form $V(\phi) = \Lambda^4 \left[1 + \cos\left(\phi/\mu\right)\right]$, or Coleman-Weinberg potentials, $V(\phi) = \lambda \phi^4 \ln(\phi)$. Small-field models are characterized by a red spectral index $n < 1$, and a small tensor/scalar ratio, $r \leq 0.01$.

- *Hybrid potentials* (Fig. 13.19). A third class of models are potentials for which there is a residual vacuum energy when the field is at the minimum of the potential, for example a potential like $V(\phi) = \lambda \left(\phi^2 + \mu^2\right)^2$. In this case, inflation will continue *forever*, so additional physics is required to end inflation and initiate reheating. The *hybrid* mechanism, introduced by Linde,[70] solves this problem by adding a second field coupled to the inflaton which is stable for ϕ large, but becomes unstable at a critical field value ϕ_c near the minimum of $V(\phi)$. During inflation, however, only ϕ is dynamical, and these models are effectively single-field. Typical models of this type predict negligible tensor modes, $r \ll 0.01$ and a "blue" spectrum, $n_S > 1$, which is disfavored by the data, and we will not discuss them in more detail here. (Ref. [48] contains a good discussion of current limits on general hybrid models.) Note also that such potentials will also support large-field inflation if the field is displaced far enough from its minimum.

An important feature of all of these models is that each is characterized by two basic parameters, the "height" of the potential Λ^4, which governs the energy density during inflation, and the "width" of the potential μ. (Hybrid models have a third free parameter ϕ_c which sets the end of inflation.) In order to have a flat potential and a slowly rolling field, there must be a hierarchy of scales such that the width is larger than the height, $\Lambda \ll \mu$. As we saw in the case of the $\lambda \phi^4$ large-field model, typical inflationary potentials have widths of order the Planck scale $\mu \sim m_{\mathrm{Pl}}$ and heights of order the scale of Grand Unification $\Lambda \sim M_{\mathrm{GUT}} \sim 10^{15}$ GeV, although models can be constructed for which inflation happens at a much lower scale.[70–72]

The quantities we are interested in for constraining models of inflation are the primordial power spectra $P_{\mathcal{R}}$ and P_T, which are the underlying

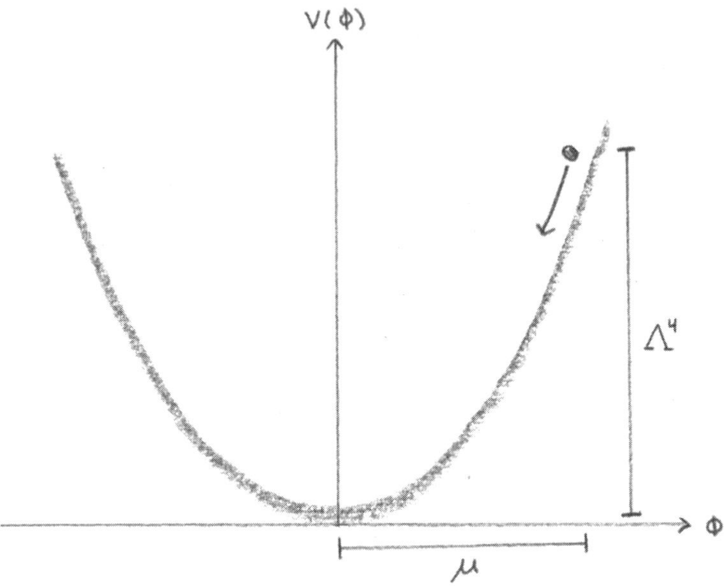

Fig. 13.17.   A schematic of a large-field potential.

source of the CMB temperature anisotropy and polarization. However, the observed CMB anisotropies depend on a handful of unrelated cosmological parameters, since the primordial fluctuations are processed through the complicated physics of acoustic oscillations. This creates uncertainties due to parameter degeneracies: our best-fit values for r and n_S will depend on what values we choose for the other cosmological parameters such as the baryon density $\Omega_{\rm b}$ and the redshift of reionization $z_{\rm ri}$. To accurately estimate the errors on r and n_S, we must fit all the relevant parameters *simultaneously*, a process which is computationally intensive, and is typically approached using Bayesian Monte Carlo Markov Chain techniques.[73] Here we simply show the results: Figure 13.20 shows the regions of the r, n_S parameter space allowed by the WMAP 5-year data set.[74,75] We have fit over the parameters $\Omega_{\rm CDM}$, $\Omega_{\rm b}$, $\Omega_{\rm Lambda}$, H_0, $P_{\mathcal{R}}$, $z_{\rm ri}$, r, and n_s, with a constraint that the universe must be flat, as predicted by inflation, $\Omega_{\rm b} + \Omega_{\rm CDM} + \Omega_{\rm Lambda} = 1$. We see that the data favor a red spectrum, $n_S < 1$, although the scale-invariant limit $n_S = 1$ is still within the 95%-confidence region. Our example inflation model $V(\phi) = \lambda\phi^4$ is convincingly ruled out by WMAP, but the simple potential $V(\phi) = m^2\phi^2$ is nicely con-

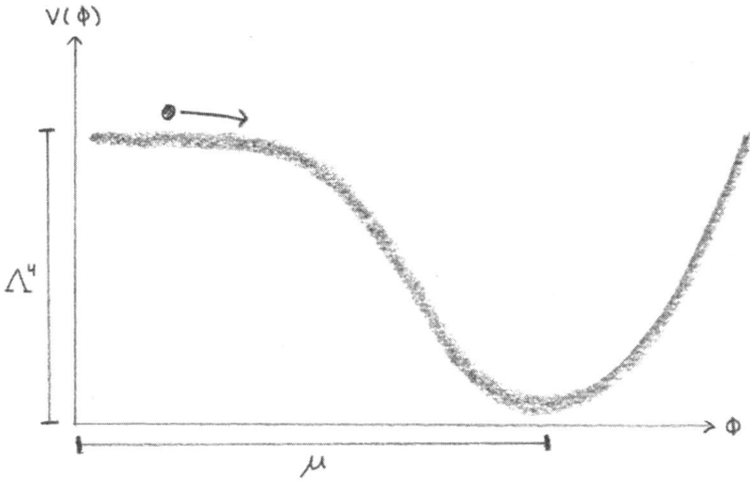

Fig. 13.18.   A schematic of a small-field potential.

sistent with the data.[i] Figure 13.21 shows the WMAP constraint with r on a logarithmic scale, with the prediction of several small-field models for reference. There is no evidence in the WMAP data for a nonzero tensor/scalar ratio r, with a 95%-confidence upper limit of $r < 0.5$. It is possible to improve these constraints somewhat by adding other data sets, for example the ACBAR high-resolution CMB anisotropy measurement[76] or the Sloan Digital Sky Survey,[77,78] which improve the upper limit on the tensor/scalar ratio to $r < 0.3$ or so. Current data are completely consistent with Gaussianity and adiabaticity, as expected from simple single-field inflation models. In the next section, we discuss the outlook for future observation.

13.7. Outlook and Conclusion

The basic hot Big Bang scenario, in which the universe arises out of a hot, dense, smooth initial state and cools through expansion, is now supported by a compelling set of observations, including the existence of the Cosmic Microwave Background, the primordial abundances of the elements,

---

[i]Liddle and Leach point out that $\lambda\phi^4$ models are special because of their reheating properties, and should be more accurately evaluated at $N = 64$.[56] However, this assumes that the potential has no other terms which might become dominant during reheating, and in any case is also ruled out by WMAP5.

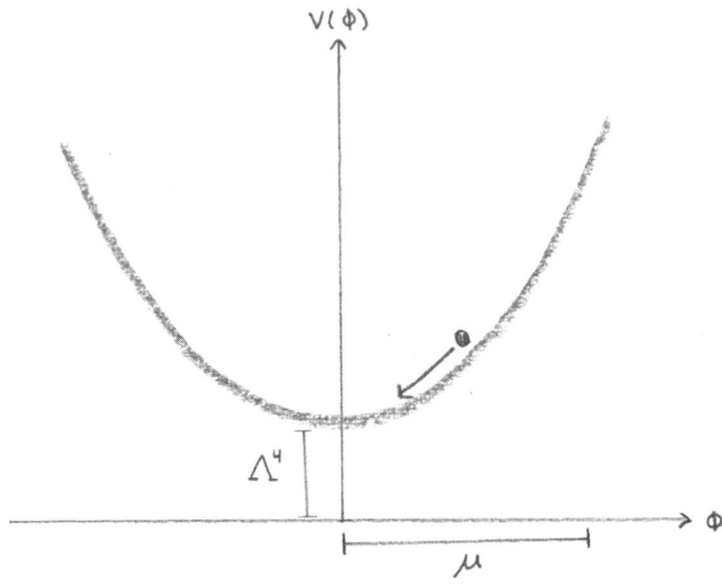

Fig. 13.19.   A schematic of a hybrid potential.

and the evolution of structure in the universe, all of which are being measured with unprecedented precision. However, this scenario leaves questions unanswered: Why is the universe so big and so old? Why is the universe so close to geometrically flat? What created the initial perturbations which later collapsed to form structure in the universe? The last of these questions is particularly interesting, because recent observations of the CMB, in particular the all-sky anisotropy map made by the landmark WMAP satellite, have directly measured the form of these primordial perturbations. A striking property of these observed primordial perturbations is that they are correlated on scales larger than the cosmological horizon at the time of last scattering. Such apparently *acausal* correlations can only be produced in a few ways:[83]

- Inflation.
- Extra dimensions.[84]
- A universe much older than H_0^{-1}.[85,86]
- A varying speed of light.[87]

WMAP 5 Limits on Inflation

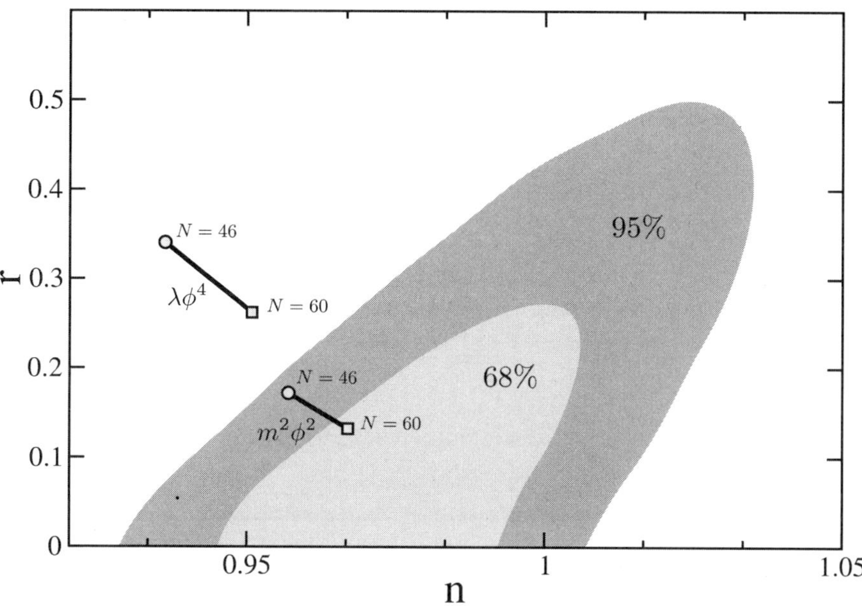

Fig. 13.20. Constraints on the r, n plane from Cosmic Microwave Background measurements. Shaded regions are the regions allowed by the WMAP5 measurement to 68% and 95% confidence. Models plotted are "large-field" potentials $V(\phi) \propto \phi^2$ and $V(\phi) \propto \phi^4$.

In addition, the WMAP data contain spectacular confirmation of the basic predictions of the inflationary paradigm: a geometrically flat universe with Gaussian, adiabatic, nearly scale-invariant perturbations. No other model explains these properties of the universe with such simplicity and economy, and much attention has been devoted to the implications of WMAP for inflation.[29,48,74,75,88-98] Inflation also makes predictions which have not been well tested by current data but *can* be by future experiments, most notably a deviation from a scale-invariant spectrum and the production of primordial gravitational waves. A non-scale-invariant spectrum is weakly favored by the existing data, but constraints on primordial gravity waves are still quite poor. The outlook for improved data is promising: over the next five to ten years, there will be a continuous stream of increasingly high-precision data made available which will allow constraint of cosmological parameters relevant for understanding the early universe. The

WMAP 5 Limits on Inflation

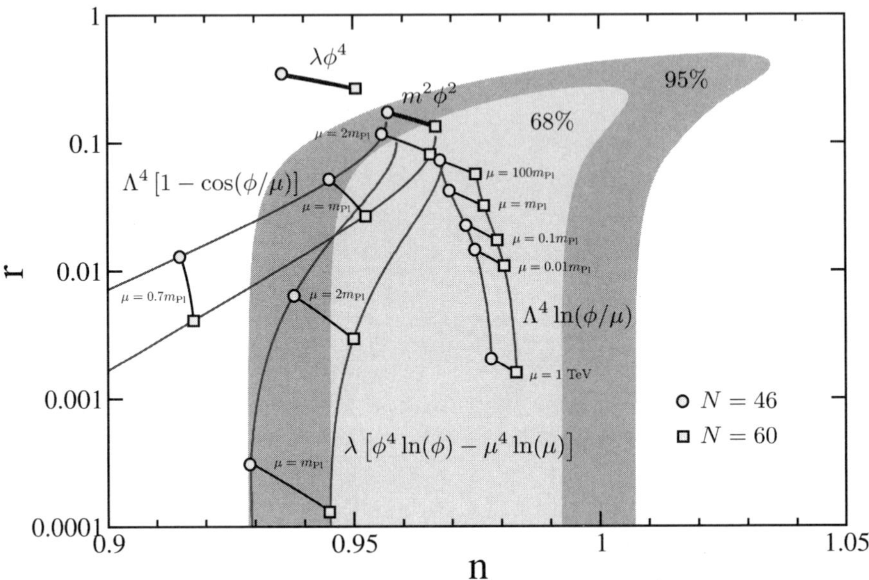

Fig. 13.21. Constraints on the r, n plane from Cosmic Microwave Background measurements, with the tensor/scalar ratio plotted on a log scale. In addition to the large-field models shown in Fig. 13.20, three small-field models are plotted against the data: "Natural Inflation" from a pseudo-Nambu-Goldstone boson,[79] with potential $V(\phi)=\Lambda^4\left[1-\cos(\phi/\mu)\right]$, a logarithmic potential $V(\phi)\propto\ln(\phi)$ typical of supersymmetric models,[80–82] and a Coleman-Weinberg potential $V(\phi)\propto\phi^4\ln(\phi)$.

most useful measurements for direct constraint of the inflationary parameter space are observations of the CMB, and current activity in this area is intense. The Planck satellite mission is scheduled to launch in 2009,[99,100] and will be complemented by ground- and balloon-based measurements using a variety of technologies and strategies.[43,44,101–106]

At the same time, cosmological parameter estimation is a well-developed field. A set of standard cosmological parameters such as the baryon density $\Omega_{\mathrm{b}}h^2$, the matter density $\Omega_{\mathrm{m}}h^2$, the expansion rate $H_0\equiv100h$ km/sec are being measured with increasing accuracy. The observable quantities most meaningful for constraining models of inflation are the ratio r of tensor to scalar fluctuation amplitudes, and the spectral index n_S of the scalar power spectrum. This kind of simple parameterization is at the moment sufficient to describe the highest-precision cosmological data sets. Furthermore, the simplest slow-roll models of inflation predict a nearly exact power-law perturbation spectrum. In this sense, a simple concordance cosmology is well-supported by both data and by theoretical expectation. It could be that the underlying universe really is that simple. However, the simplicity of concordance cosmology is at present as much a statement about the data as about the universe itself. Only a handful of parameters are required to explain existing cosmological data. Adding more parameters to the fit does no good: any small improvement in the fit of the model to the data is offset by the statistical penalty one pays for introducing extra parameters.[107–115] But the optimal parameter set is a moving target: as the data get better, we will be able to probe more parameters. It may be that a "vanilla" universe[116] of a half-dozen or so parameters will continue to be sufficient to explain observation. But it is reasonable to expect that, as measurements improve in accuracy, we will see evidence of deviation from such a lowest-order expectation. This is where the interplay between theory and experiment gains the most leverage, because we must understand: (1) what deviations from a simple universe are predicted by models, and (2) how to look for those deviations in the data. It is of course impossible to predict which of the many possible signals (if any) will be realized in the universe in which we live. I discuss below four of the best motivated possibilities, in order of the quality of current constraints. (For a more detailed treatment of these issues, the reader is referred to the very comprehensive CMBPol Mission Concept Study.[117])

Features in the density power spectrum

Current data are consistent with a purely power-law spectrum of density perturbations, $P(k) \propto k^{n_S - 1}$ with a "red" spectrum ($n_S < 1$) favored by the data at about a 90% confidence level, a figure which depends on the choice of parameter set and priors. Assuming it is supported by future data, the detection of a deviation from a scale-invariant ($n_S = 1$) spectrum is a significant milestone, and represents a confirmation of one of the basic predictions of inflation. In slow-roll inflation, this power-law scale dependence is nearly exact, and any additional scale dependence is strongly suppressed. Therefore, detection of a nonzero "running" $\alpha = dn_S/d\ln k$ of the spectral index would be an indication that slow roll is a poor approximation. There is currently no evidence for scale-dependence in the spectral index, but constraints on the overall shape of the power spectrum are likely to improve dramatically through measurements of the CMB anisotropy at small angular scales, improved polarization measurements, and better mapping of large-scale structure. Planck is expected to measure the shape of the spectrum with 2σ uncertainties of order $\Delta n \sim 0.01$ and $\Delta \alpha \sim 0.01$.[118–121] Over the longer term, measurements of 21cm radiation from neutral hydrogen promises to be a precise probe of the primordial power spectrum, and would improve these constraints significantly.[122]

Primordial Gravitational Waves

In addition to a spectrum $P_\mathcal{R}$ of scalar perturbations, inflation generically predicts a spectrum P_T of tensor perturbations. The relative amplitude of the two is determined by the equation of state of the fluid driving inflation,

$$r = 16\epsilon \tag{13.7.207}$$

Since the scalar amplitude is known from the COBE normalization to be $P_\mathcal{R} \sim H^2/\epsilon \sim 10^{-10}$, it follows that measuring the tensor/scalar ratio r determines the inflationary expansion rate H and the associated energy density ρ. Typical inflation models take place with an energy density of around $\rho \sim (10^{15} \text{ GeV})^4$, which corresponds to a tensor/scalar ratio of $r \sim 0.1$, although this figure is highly model-dependent. Single-field inflation does not make a definite prediction for the value of r: while many choices of potential generate a substantial tensor component, other choices of potential result in an unobservably small tensor/scalar ratio, and there is no particular reason to favor one scenario over another.

There is at present no observational evidence for primordial gravitational waves: the current upper limit on the tensor/scalar ratio is around $r \leq 0.3$. Detection of even a large primordial tensor signal requires extreme sensitivity. The crucial observation is detection of the odd-parity, or B-mode, component of the CMB polarization signal, which is suppressed relative to the temperature fluctuations, themselves at the 10^{-4} level, by at least another four orders of magnitude. This signal is considerably below known foreground levels,[123] severely complicating data analysis. Despite the formidable challenges, the observational community has undertaken a broad-based effort to search for the B-mode, and a detection would be a boon for inflationary cosmology. Planck will be sensitive to a tensor/scalar ratio of around $r \simeq 0.1$, and dedicated ground-based measurements can potentially reach limits of order $r \simeq 0.01$. The proposed CMBPol polarization satellite would reach r of order 10^{-3},[117,124] and direct detection experiments such as BBO could in principle detect r of order 10^{-4}.[65]

Primordial Non-Gaussianity

In addition to a power-law power spectrum, inflation predicts that the primordial perturbations will be distributed according to Gaussian statistics. Like running of the power spectrum, non-Gaussianity is suppressed in slow-roll inflation.[125] However, detection of even moderate non-Gaussianity is considerably more difficult. If the perturbations are Gaussian, the two-point correlation function completely describes the perturbations. This is not the case for non-Gaussian fluctuations: higher-order correlations contain additional information. However, higher-order correlations require more statistics and are therefore more difficult to measure, especially at large angular scales where cosmic variance errors are significant. Current limits are extremely weak,[88,126] and future high angular resolution CMB maps will still fall well short of being sensitive to a signal from slow-roll inflation or even weakly *non*-slow-roll models.[127] It will take a strong deviation from the slow-roll scenario to generate observable non-Gaussianity. However, a measurement of non-Gaussianity would in one stroke rule out virtually all slow-roll inflation models and force consideration of more exotic scenarios such as DBI inflation,[67] Warm Inflation,[128] or curvaton scenarios.[129]

Isocurvature perturbations

In a universe where the matter consists of multiple components, there are two general classes of perturbation about a homogeneous background: adiabatic, in which the perturbations in all of the fluid components are in phase, and isocurvature, in which the perturbations have independent phases. Single-field inflation predicts purely adiabatic primordial perturbations, for the simple reason that if there is a single field ϕ responsible for inflation, then there is a single order parameter governing the generation of density perturbations. This is a nontrivial prediction, and the fact that current data are consistent with adiabatic perturbations is support for the idea of quantum generation of perturbations in inflation. However, current limits on the isocurvature fraction are quite weak.[130,131] If isocurvature modes are detected, it would rule out *all* single-field models of inflation. Multi-field models, on the other hand, naturally contain multiple order parameters and can generate isocurvature modes. Multi-field models are naturally motivated by the string "landscape", which is believed to contain an enormous number of degrees of freedom. Another possible mechanism for the generation of isocurvature modes is the curvaton mechanism, in which cosmological perturbations are generated by a field other than the inflaton.[132,133]

The rich interplay between theory and observation that characterizes cosmology today is likely to continue for the foreseeable future. As measurements improve, theory will need to become more precise and complete than the simple picture of inflation that we have outlined in these lectures, and single-field inflation models could yet prove to be a poor fit to the data. However, at the moment, such models provide an elegant, compelling, and (most importantly) scientifically useful picture of the very early universe.

Acknowledgments

I would like to thank the organizers of the Theoretical Advanced Studies Institute (TASI) at Univ. of Colorado, Boulder for giving me the opportunity to return to my alma mater to lecture. Various versions of these lectures were also given at the Perimeter Institute Summer School on Particle Physics, Cosmology, and Strings in 2007, at the Second Annual Dirac Lectures at Florida State University in 2008, and at the Research Training Group at the University of Würzburg in 2008. This research is supported in part by the National Science Foundation under grants NSF-PHY-0456777 and NSF-PHY-0757693. I thank Dennis Bessada, Richard Easther, Hiranya Peiris, and Brian Powell for comments on a draft version of the manuscript.

A.1. The Curvature Perturbation in Single-Field Inflation

In this section, we discuss the generation of perturbations in the density $\delta\left(\mathbf{x}\right) \equiv \delta\rho/\rho$ generated during inflation. The process is similar to the case of a free scalar field discussed in Sec. 13.5: the inflaton field ϕ, like any other scalar, will have quantum fluctuations which are stretched to superhorizon scales and subsequently freeze out as classical perturbations. The difference is that the energy density of the universe is dominated by the inflaton potential, so that quantum fluctuations in ϕ generate perturbations in the density ρ. Dealing with such density perturbations is complicated by the fact that in General Relativity, we are free to choose any coordinate system, or *gauge*, we wish. To see why, consider the case of an FRW spacetime evolving with scale factor $a\left(t\right)$ and uniform energy density $\rho\left(t, \mathbf{x}\right) = \bar{\rho}\left(t\right)$. What we mean here by "uniform" energy density, or homogeneity, is that the density is a constant in *comoving* coordinates. But the physics is independent of coordinate system, so we could equally well work in coordinates t', $\mathbf{x}'$ for which constant-time hypersurfaces do *not* have constant density (Fig. A.1). Such a division of spacetime into a time coordinate and a set

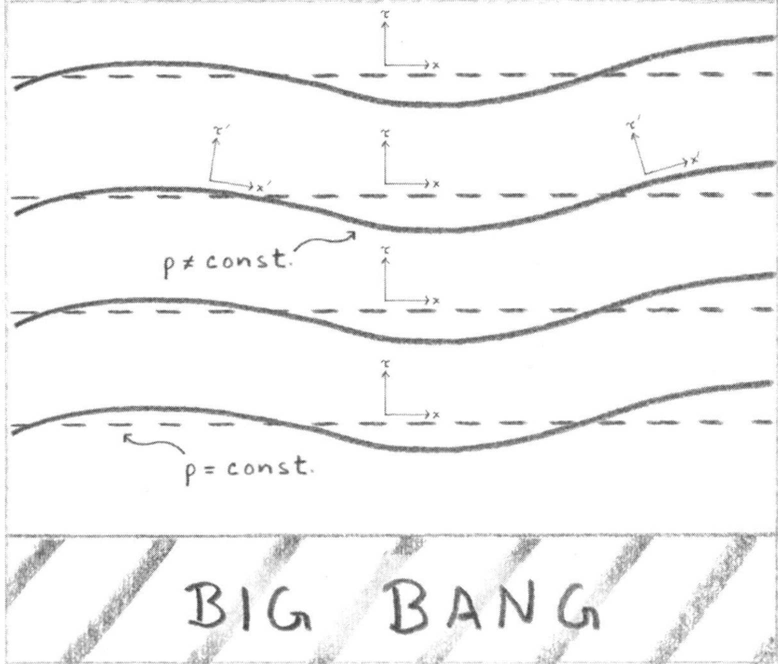

Fig. A.1.   Foliations of an FRW spacetime. Comoving hypersurfaces (dashed lines) have constant density, but another choice of gauge (solid lines) will have unphysical density fluctuations which are an artifact of the choice of gauge.

of orthogonal spacelike hypersurfaces is called a *foliation* of the spacetime, and is an arbitrary choice equivalent to a choice of coordinate system.

For an FRW spacetime, comoving coordinates correspond to a foliation of the spacetime into spatial hypersurfaces with constant density: this is the most physically intuitive description of the spacetime. Any other choice of foliation of the spacetime would result in density "perturbations" which are entirely due to the choice of coordinate system. Such unphysical perturbations are referred to as *gauge modes*. Another way to think of this is that the division between what we call "background" and what we call "perturbation" is itself gauge-dependent. For perturbations with wavelength smaller than the horizon, it is possible to define background and perturbation without ambiguity, since all observers can agree on a definition of time coordinate t and on an average density $\bar{\rho}(t)$. Not so for superhorizon modes: if we consider a perturbation mode with wavelength much larger than the horizon size, observers in different horizons will see themselves in independently evolving, homogeneous patches of the universe: a "perturbation" can be defined only by comparing causally disconnected observers, and there is an inherent gauge ambiguity in how we do this. The canonical paper on gauge issues in General Relativistic perturbation theory is by Bardeen.[134] A good pedagogical treatment with a focus on inflationary perturbations can be found in Ref. [135].

In practice, instead of the density perturbation δ, the quantity most directly relevant to CMB physics is the Newtonian potential Φ on the surface of last scattering. For example, this is the quantity that directly appears in Eq. (13.2.50) for the Sachs-Wolfe Effect. The Newtonian potential is related to the density perturbation δ through the Poisson Equation:

$$\nabla^2 \Phi = 4\pi G \bar{\rho} a^2 \delta, \qquad (A.1)$$

where the factor of a^2 comes from defining the gradient ∇ relative to comoving coordinates. Like δ, the Newtonian potential Φ is a gauge-dependent quantity: its value depends on how we foliate the spacetime. For example, we are free to choose spatial hypersurfaces such that the density is constant, and the Newtonian potential vanishes everywhere: $\Phi(t, \mathbf{x}) = 0$. This foliation of the spacetime is equivalent to the qualitative picture above of different horizon volumes as independently evolving homogeneous universes. Observers in different horizons use the density ρ to synchronize their clocks with one another. Such a foliation is not very useful for computing the Sachs-Wolfe effect, however! Instead, we need to define a gauge which corresponds to the Newtonian limit in the present universe. To accomplish this, we describe the evolution of a scalar field dominated cosmology using the useful fluid flow approach.[136–140] (An alternate strategy involves the construction of gauge-invariant variables: see Refs. [141,142] for reviews.)

Consider a scalar field ϕ in an arbitrary background $g_{\mu\nu}$. The stress-energy tensor of the scalar field may be written

$$T_{\mu\nu} = \phi_{,\mu}\phi_{,\nu} - g_{\mu\nu}\left[\frac{1}{2}g^{\alpha\beta}\phi_{,\alpha}\phi_{,\beta} - V(\phi)\right]. \qquad (A.2)$$

Note that we have not yet made any assumptions about the metric $g_{\mu\nu}$ or about the scalar field ϕ. Equation (A.2) is a completely general expression. We can define a fluid four-velocity for the scalar field by

$$u_\mu \equiv \frac{\phi_{,\mu}}{\sqrt{g^{\alpha\beta}\phi_{,\alpha}\phi_{,\beta}}}. \qquad (A.3)$$

It is not immediately obvious why this should be considered a four-velocity. Consider any perfect fluid filling spacetime. Each element of the fluid has four-velocity $u^\mu(x)$ at every point in spacetime which is everywhere timelike,

$$u^\mu(x)\, u_\mu(x) = 1\ \forall x. \tag{A.4}$$

Such a collection of four-vectors is called a *timelike congruence*. We can draw the congruence defined by the fluid four-velocity as a set of flow lines in spacetime (Fig. A.2). Each event P in spacetime has one and only one flow line passing through it. The fluid four-velocity is then a set of unit-normalized tangent vectors to the flow lines, $u^\mu u_\mu = 1$. For a scalar field,

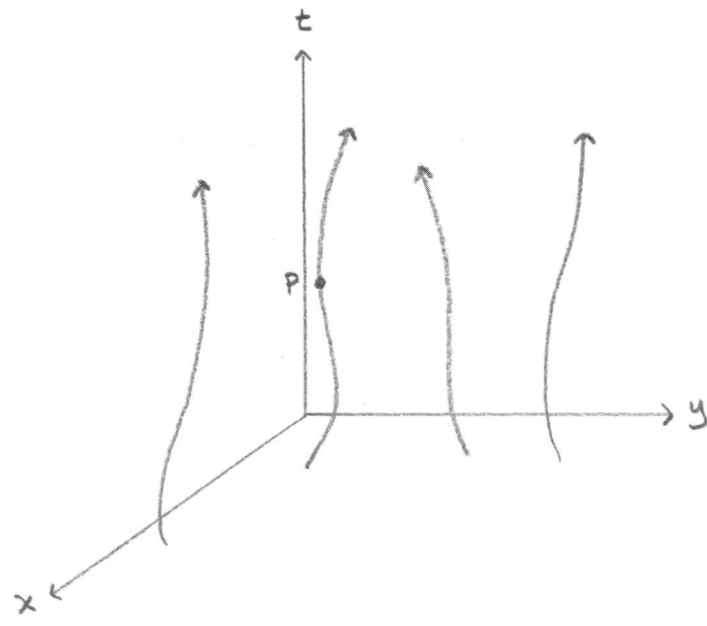

Fig. A.2.   A timelike congruence in spacetime. Each event P is intersected by exactly one world line in the congruence.

we construct a timelike congruence by Eq. (A.3), which is by construction unit normalized:

$$u^\mu u_\mu = \frac{g^{\mu\nu}\phi_{,\mu}\phi_{,\nu}}{g^{\alpha\beta}\phi_{,\alpha}\phi_{,\beta}} = 1. \tag{A.5}$$

We then define the "time" derivative of any scalar quantity $f(x)$ by the projection of the derivative along the fluid four-velocity:

$$\dot{f} \equiv u^{\mu} f_{,\mu}. \tag{A.6}$$

In particular, the time derivative of the scalar field itself is

$$\dot{\phi} \equiv u^{\mu} \phi_{,\mu} = \sqrt{g^{\alpha\beta} \phi_{,\alpha} \phi_{,\beta}}. \tag{A.7}$$

Note that in the homogeneous case, we recover the usual time derivative,

$$\nabla \phi = 0 \Rightarrow \dot{\phi} = \sqrt{g^{00} \phi_{,0} \phi_{,0}} = \frac{d\phi}{dt}. \tag{A.8}$$

The stress-energy tensor (A.2) in terms of $\dot{\phi}$ takes the form

$$T_{\mu\nu} = \left[\frac{1}{2} \dot{\phi}^2 + V(\phi) \right] u_{\mu} u_{\nu} + \left[\frac{1}{2} \dot{\phi}^2 - V(\phi) \right] (u_{\mu} u_{\nu} - g_{\mu\nu}). \tag{A.9}$$

We can then define a generalized density ρ and and pressure p by

$$\begin{aligned} \rho &\equiv \frac{1}{2} \dot{\phi}^2 + V(\phi), \\ p &\equiv \frac{1}{2} \dot{\phi}^2 - V(\phi). \end{aligned} \tag{A.10}$$

Note that despite the familiar form of these expressions, they are defined without any assumption of homogeneity of the scalar field or even the imposition of a particular metric.

In terms of the generalized density and pressure, the stress-energy (A.2) is

$$T_{\mu\nu} = \rho u_{\mu} u_{\nu} + p h_{\mu\nu}, \tag{A.11}$$

where the tensor $h_{\mu\nu}$ is defined as:

$$h_{\mu\nu} \equiv u_{\mu} u_{\nu} - g_{\mu\nu}. \tag{A.12}$$

The tensor $h_{\mu\nu}$ can be easily seen to be a projection operator onto hypersurfaces orthogonal to the four-velocity u^{μ}. For any vector field A^{μ}, the product $h_{\mu\nu} A^{\nu}$ is identically orthogonal to the four-velocity:

$$(h_{\mu\nu} A^{\nu}) u^{\mu} = A^{\nu} (h_{\mu\nu} u^{\mu}) = 0. \tag{A.13}$$

Therefore, as in the case of the time derivative, we can define gradients by projecting the derivative onto surfaces orthogonal to the four-velocity

$$(\nabla f)^{\mu} \equiv h^{\mu\nu} f_{,\nu}. \tag{A.14}$$

In the case of a scalar field fluid with four-velocity given by Eq. (A.3), the gradient of the field identically vanishes,

$$(\nabla\phi)^\mu = 0. \tag{A.15}$$

Note that despite its relation to a "spatial" gradient, ∇f is a covariant quantity, *i.e.* a four-vector.

Our fully covariant definitions of "time" derivatives and "spatial" gradients suggest a natural foliation of the spacetime into spacelike hypersurfaces, with time coordinate orthogonal to those hypersurfaces. We can define spatial hypersurfaces to be everywhere orthogonal to the fluid flow (Fig. A.3). This is equivalent to choosing a coordinate system for which $u^i = 0$ everywhere. Such a gauge choice is called *comoving* gauge. In the case of a scalar field, we can equivalently define comoving gauge as a coordinate system in which spatial gradients of the scalar field $\phi_{,i}$ are defined to vanish. Therefore the time derivative (A.6) is just the derivative with respect to the coordinate time in comoving gauge

$$\dot{\phi} = \left(\frac{\partial\phi}{\partial t}\right)_{\rm c}. \tag{A.16}$$

Similarly, the generalized density and pressure (A.10) are just defined to be those quantities as measured in comoving gauge.

The equations of motion for the fluid can be derived from stress-energy conservation,

$$T^{\mu\nu}{}_{;\nu} = 0 = \dot{\rho}u^\mu + (\nabla p)^\mu + (\rho + p)\left(\dot{u}^\mu + u^\mu\Theta\right), \tag{A.17}$$

where the quantity Θ is defined as the divergence of the four-velocity,

$$\Theta \equiv u^\mu{}_{;\mu}. \tag{A.18}$$

We can group the terms multiplied by u^μ separately, resulting in familiar-looking equations for the generalized density and pressure

$$\dot{\rho} + \Theta\left(\rho + p\right) = 0,$$
$$(\nabla p)^\mu + (\rho + p)\,\dot{u}^\mu = 0. \tag{A.19}$$

The first of these equations, similar to the usual continuity equation in the homogeneous case, can be rewritten using the definitions of the generalized density and pressure (A.10) in terms of the field as

$$\ddot{\phi} + \Theta\dot{\phi} + V'(\phi) = 0. \tag{A.20}$$

This suggests identifying the divergence Θ as a generalization of the Hubble parameter H in the homogeneous case. In fact, if we take $g_{\mu\nu}$ to be a

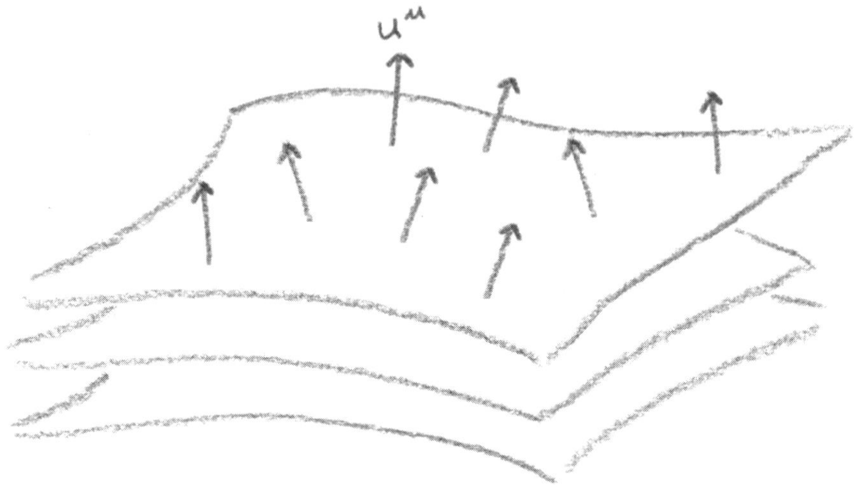

Fig. A.3. A comoving foliation of spacetime. Spatial hypersurfaces are everywhere orthogonal to the fluid four-velocity u^μ.

flat Friedmann-Robertson-Walker (FRW) metric and take comoving gauge, $u^\mu = (1, 0, 0, 0)$, we have

$$u^\mu{}_{;\mu} = 3H, \tag{A.21}$$

and the generalized equation of motion (A.20) becomes the familiar equation of motion for a homogeneous scalar,

$$\ddot{\phi} + 3H\dot{\phi} + V'(\phi) = 0. \tag{A.22}$$

Now consider perturbations $\delta g_{\mu\nu}$ about a flat FRW metric,

$$g_{\mu\nu} = a^2(\tau)\left[\eta_{\mu\nu} + \delta g_{\mu\nu}\right], \tag{A.23}$$

where τ is the conformal time and η is the Minkowski metric $\eta = \text{diag}(1, -1, -1, -1)$. A general metric perturbation $\delta g_{\mu\nu}$ can be separated into components which transform independently under coordinate transformations,[134]

$$\delta g_{\mu\nu} = \delta g_{\mu\nu}^{\text{scalar}} + \delta g_{\mu\nu}^{\text{vector}} + \delta g_{\mu\nu}^{\text{tensor}}. \tag{A.24}$$

The tensor component is just the transverse-traceless gravitational wave perturbation, discussed in Sec. 13.5, and vector perturbations are not sourced by single-field inflation. We therefore specialize to the case of scalar perturbations, for which the metric perturbations can be written generally in terms of four scalar functions of space and time A, B, $\mathcal{R}$, and H_T:

$$
\begin{aligned}
\delta g_{00} &= 2A \\
\delta g_{0i} &= \partial_i B \\
\delta g_{ij} &= 2\left[\mathcal{R}\delta_{ij} + \partial_i\partial_j H_T\right].
\end{aligned}
\tag{A.25}
$$

We are interested in calculating $\mathcal{R}$. Recall that in the Newtonian limit of General Relativity, we can write perturbations about the Minkowski metric in terms of the Newtonian potential Φ as:

$$
ds^2 = (1 + 2\Phi)\,dt^2 - (1 - 2\Phi)\,\delta_{ij}dx^i dx^j.
\tag{A.26}
$$

Similarly, we can write Newtonian perturbations about a flat FRW metric as

$$
ds^2 = a^2\left(\tau\right)\left[(1 + 2\Phi)\,d\tau^2 - (1 - 2\Phi)\,\delta_{ij}dx^i dx^j\right].
\tag{A.27}
$$

We therefore expect $\Phi \propto \mathcal{R}$ in the Newtonian limit. A careful calculation[138,141] gives

$$
\Phi = -\frac{3\left(1 + w\right)}{5 + 3w}\mathcal{R},
\tag{A.28}
$$

so that in a matter-dominated universe,

$$
\Phi = -\frac{3}{5}\mathcal{R}.
\tag{A.29}
$$

In these expressions, $\mathcal{R}$ is the curvature perturbation measured on comoving hypersurfaces. To see qualitatively why comoving gauge corresponds correctly to the Newtonian limit in the current universe, consider the end of inflation. Since inflation ends at a particular field value $\phi = \phi_e$, comoving gauge corresponds to a foliation for which inflation ends at *constant time* at all points in space: all observers synchronize their clocks to $\tau = 0$ at the end of inflation. This means that the background, or unperturbed universe is exactly the homogeneous case diagrammed in Fig. 13.13, and the comoving curvature perturbation $\mathcal{R}$ is the Newtonian potential measured relative to that background.

To calculate $\mathcal{R}$, we start by calculating the four-velocity u^μ in terms of the perturbed metric.[j] If we specialize to comoving gauge, $u^i \equiv 0$, the norm of the four-velocity can be written

$$
u^\mu u_\mu = a^2\left(1 + 2A\right)\left(u^0\right)^2 = 1,
\tag{A.30}
$$

---

[j]This treatment closely follows that of Sasaki and Stewart.[139]

and the timelike component of the four-velocity is, to linear order,

$$u^0 = \frac{1}{a}\left(1 - A\right)$$
$$u_0 = a\left(1 + A\right). \tag{A.31}$$

The velocity divergence Θ is then

$$\begin{aligned}
\Theta &= u^\mu{}_{;\mu} = u^0{}_{,0} + \Gamma^\alpha{}_{\alpha 0} u^0 \\
&= 3H\left[1 - A - \frac{1}{aH}\left(\frac{\partial \mathcal{R}}{\partial \tau} + \frac{1}{3}\partial_i \partial_i \frac{\partial H_T}{\partial \tau}\right)\right],
\end{aligned} \tag{A.32}$$

where the unperturbed Hubble parameter is defined as

$$H \equiv \frac{1}{a^2}\frac{\partial a}{\partial \tau}. \tag{A.33}$$

Fourier expanding H_T,

$$\partial_i \partial_i H_T = k^2 H_T, \tag{A.34}$$

we see that for long-wavelength modes $k \ll aH$, the last term in Eq. (A.32) can be ignored, and the velocity divergence is

$$\Theta \simeq 3H\left[1 - A - \frac{1}{aH}\frac{\partial \mathcal{R}}{\partial \tau}\right]. \tag{A.35}$$

Remembering the definition of the number of e-folds in the unperturbed case,

$$N \equiv \int H\,dt. \tag{A.36}$$

we can define a generalized number of e-folds as the integral of the velocity divergence along comoving world lines:

$$\mathcal{N} \equiv \frac{1}{3}\int \Theta\,ds = \frac{1}{3}\int \Theta\left[a\left(1 + A\right)d\tau\right]. \tag{A.37}$$

Using Eq. (A.35) for Θ and evaluating to linear order in the metric perturbation results in

$$\mathcal{N} = \int H\,dt - \mathcal{R}, \tag{A.38}$$

and we have a simple expression for the curvature perturbation,

$$\mathcal{R} = N - \mathcal{N}. \tag{A.39}$$

This requires a little physical interpretation: we defined comoving hypersurfaces such that the field has no spatial variation,

$$(\nabla\phi)^\mu = 0 \;\Rightarrow\; \phi = \text{const.} \tag{A.40}$$

Then $\mathcal{N}$ is the number of e-folds measured on comoving hypersurfaces. But we can equivalently foliate the spacetime such that spatial hypersurfaces are flat, and the field exhibits spatial fluctuations:

$$A = \mathcal{R} = 0 \;\Rightarrow\; \phi \neq \text{const.} \tag{A.41}$$

On flat hypersurfaces, the field varies, but the curvature does not, so that the metric on these hypersurfaces is exactly of the FRW form (13.2.11) with $k = 0$. We then see immediately that

$$N = \int Hdt = \text{const.} \tag{A.42}$$

is the number of e-folds measured on flat hypersurfaces, and has no spatial variation. The curvature perturbation $\mathcal{R}$ is the difference in the number of e-folds between the two sets of hypersurfaces (Fig. A.4). This can be expressed to linear order in terms of the field variation $\delta\phi$ on flat hypersurfaces as

$$\mathcal{R} = N - \mathcal{N} = \frac{\delta N}{\delta\phi}\delta\phi \tag{A.43}$$

where $\mathcal{R}$ is measured on comoving hypersurfaces, and $\delta N/\delta\phi$ and $\delta\phi$ are measured on flat hypersurfaces. We can express N as a function of the field

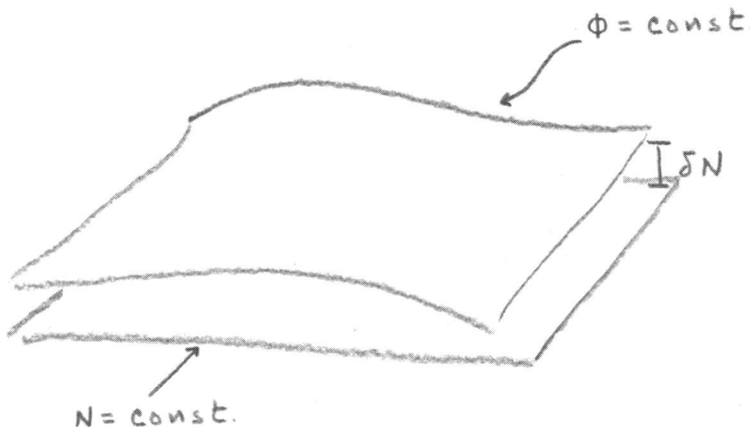

Fig. A.4. Flat and comoving hypersurfaces.

ϕ:

$$N = \int H \, dt = \int \frac{H}{\dot\phi} \, d\phi. \tag{A.44}$$

For monotonic field evolution, we can express $\dot\phi$ as a function of ϕ, so that

$$\frac{\delta N}{\delta \phi} = \frac{H}{\dot\phi}, \tag{A.45}$$

and the curvature perturbation is given by

$$\mathcal{R} = N - \mathcal{N} = \frac{\delta N}{\delta \phi} \delta \phi = \frac{H}{\dot\phi} \delta \phi. \tag{A.46}$$

Note that this is an expression for the metric perturbation $\mathcal{R}$ on comoving hypersurfaces, calculated in terms of quantities defined on *flat* hypersurfaces. For $\delta\phi$ produced by quantum fluctuations in inflation, the two-point correlation function is

$$\sqrt{\langle \delta \phi^2 \rangle} = \frac{H}{2\pi}, \tag{A.47}$$

and the two-point correlation function for curvature perturbations is

$$\sqrt{\langle \mathcal{R}^2 \rangle} = \frac{H^2}{2\pi\dot\phi} = \frac{H}{m_{\mathrm{Pl}}\sqrt{\pi\epsilon}}, \tag{A.48}$$

which is the needed result.

References

1. A. H. Guth, Phys. Rev. D **23**, 347 (1981).
2. A. D. Linde, Phys. Lett. B **108**, 389 (1982).
3. A. Albrecht and P. J. Steinhardt, Phys. Rev. Lett. **48**, 1220 (1982).
4. E. B. Gliner, Sov. Phys.-JETP **22**, 378 (1966).
5. E. B. Gliner and I. G. Dymnikova, Sov. Astron. Lett. **1**, 93 (1975).
6. A. Linde, Lect. Notes Phys. **738**, 1 (2008) [arXiv:0705.0164 [hep-th]].
7. A. A. Starobinsky, JETP Lett. **30** (1979) 682 [Pisma Zh. Eksp. Teor. Fiz. **30** (1979) 719].
8. V. F. Mukhanov and G. V. Chibisov, JETP Lett. **33** (1981) 532 [Pisma Zh. Eksp. Teor. Fiz. **33** (1981) 549].
9. S. W. Hawking, Phys. Lett. B **115**, 295 (1982).
10. S. W. Hawking and I. G. Moss, Nucl. Phys. B **224**, 180 (1983).
11. A. A. Starobinsky, Phys. Lett. B **117** (1982) 175.
12. A. H. Guth and S. Y. Pi, Phys. Rev. Lett. **49**, 1110 (1982).
13. J. M. Bardeen, P. J. Steinhardt and M. S. Turner, Phys. Rev. D **28**, 679 (1983).
14. D. H. Lyth and A. Riotto, Phys. Rept. **314**, 1 (1999) [arXiv:hep-ph/9807278].

15. G. S. Watson, arXiv:astro-ph/0005003.
16. A. Riotto, arXiv:hep-ph/0210162.
17. C. H. Lineweaver, arXiv:astro-ph/0305179.
18. W. H. Kinney, arXiv:astro-ph/0301448.
19. M. Trodden and S. M. Carroll, arXiv:astro-ph/0401547.
20. D. Baumann and H. V. Peiris, arXiv:0810.3022 [astro-ph].
21. S. Weinberg, "Gravitation and Cosmology: Principles and Applications of the General Theory of Relativity," John Wiley & Sons (1972) Ch. 13.
22. W. L. Freedman *et al.* [HST Collaboration], Astrophys. J. **553**, 47 (2001) [arXiv:astro-ph/0012376].
23. M. J. White, D. Scott and J. Silk, Ann. Rev. Astron. Astrophys. **32**, 319 (1994).
24. W. Hu and S. Dodelson, Ann. Rev. Astron. Astrophys. **40**, 171 (2002) [arXiv:astro-ph/0110414].
25. A. Kosowsky, arXiv:astro-ph/0102402.
26. D. Samtleben, S. Staggs and B. Winstein, Ann. Rev. Nucl. Part. Sci. **57**, 245 (2007) [arXiv:0803.0834 [astro-ph]].
27. W. Hu, arXiv:0802.3688 [astro-ph].
28. E. W. Kolb and M. S. Turner, "The Early Universe," Addison-Wesley (1990), Ch. 3.
29. J. Dunkley *et al.* [WMAP Collaboration], arXiv:0803.0586 [astro-ph].
30. P. S. Henry, Nature, **231**, 516 (1971).
31. B. E. Corey and D. T. Wilkinson, Bull. Amer. Astron. Soc,. **8**, 351 (1976).
32. G. F. Smoot, M. V. Gorenstein, and R. A. Muller, Phys. Rev. Lett **39**, 898 (1977).
33. C. L. Bennett *et al.*, Astrophys. J. **464**, L1 (1996) [arXiv:astro-ph/9601067].
34. G. Hinshaw *et al.* [WMAP Collaboration], arXiv:0803.0732 [astro-ph].
35. R. K. Sachs and A. M. Wolfe, Astrophys. J. **147**, 73 (1967).
36. A. D. Sakharov, JETP 49, 345 (1965).
37. Y. B. Zeldovich and R. A. Sunyaev, Astrophys. Space Sci. **4**, 301 (1969).
38. R. A. Sunyaev and Y. B. Zeldovich, Astrophys. Space Sci. **7**, 3 (1970).
39. P. J. E. Peebles and J. T. Yu, Astrophys. J. **162**, 815 (1970).
40. C. P. Ma and E. Bertschinger, Astrophys. J. **455**, 7 (1995) [arXiv:astro-ph/9506072].
41. A. Kosowsky, New Astron. Rev. **43**, 157 (1999) [arXiv:astro-ph/9904102].
42. M. Zaldarriaga, arXiv:astro-ph/0305272.
43. E. M. Leitch, J. M. Kovac, N. W. Halverson, J. E. Carlstrom, C. Pryke and M. W. E. Smith, Astrophys. J. **624**, 10 (2005) [arXiv:astro-ph/0409357].
44. J. L. Sievers *et al.*, arXiv:astro-ph/0509203.
45. T. E. Montroy *et al.*, Astrophys. J. **647**, 813 (2006) [arXiv:astro-ph/0507514].
46. J. H. Wu *et al.*, arXiv:astro-ph/0611392.
47. M. R. Nolta *et al.* [WMAP Collaboration], arXiv:0803.0593 [astro-ph].
48. E. Komatsu *et al.* [WMAP Collaboration], arXiv:0803.0547 [astro-ph].
49. M. Kamionkowski, Science **280**, 1397 (1998) [arXiv:astro-ph/9806347].
50. J. J. Levin, Phys. Rept. **365**, 251 (2002) [arXiv:gr-qc/0108043].

51. A. A. Starobinsky, Phys. Lett. B **91** (1980) 99.
52. A. D. Linde, Mod. Phys. Lett. A **1**, 81 (1986).
53. A. H. Guth, Phys. Rept. **333**, 555 (2000) [arXiv:astro-ph/0002156].
54. A. Aguirre, arXiv:0712.0571 [hep-th].
55. S. Winitzki, "Eternal inflation," *Hackensack, USA: World Scientific (2008)*.
56. A. R. Liddle and S. M. Leach, Phys. Rev. D **68**, 103503 (2003) [arXiv:astro-ph/0305263].
57. W. H. Kinney and A. Riotto, JCAP **0603**, 011 (2006) [arXiv:astro-ph/0511127].
58. W. G. Unruh, Phys. Rev. D **14**, 870 (1976).
59. L. Hui and W. H. Kinney, Phys. Rev. D **65**, 103507 (2002) [arXiv:astro-ph/0109107].
60. U. H. Danielsson, Phys. Rev. D **66**, 023511 (2002) [arXiv:hep-th/0203198].
61. R. Easther, B. R. Greene, W. H. Kinney and G. Shiu, Phys. Rev. D **66**, 023518 (2002) [arXiv:hep-th/0204129].
62. J. Martin and R. H. Brandenberger, Phys. Rev. D **63**, 123501 (2001) [arXiv:hep-th/0005209].
63. J. C. Niemeyer, Phys. Rev. D **63**, 123502 (2001) [arXiv:astro-ph/0005533].
64. W. H. Kinney, Phys. Rev. D **72**, 023515 (2005) [arXiv:gr-qc/0503017].
65. T. L. Smith, M. Kamionkowski and A. Cooray, Phys. Rev. D **73**, 023504 (2006) [arXiv:astro-ph/0506422].
66. B. C. Friedman, A. Cooray and A. Melchiorri, Phys. Rev. D **74**, 123509 (2006) [arXiv:astro-ph/0610220].
67. E. Silverstein and D. Tong, Phys. Rev. D **70**, 103505 (2004) [arXiv:hep-th/0310221].
68. N. Agarwal and R. Bean, Phys. Rev. D **79**, 023503 (2009) [arXiv:0809.2798 [astro-ph]].
69. S. Dodelson, W. H. Kinney and E. W. Kolb, Phys. Rev. D **56**, 3207 (1997) [arXiv:astro-ph/9702166].
70. A. D. Linde, Phys. Rev. D **49**, 748 (1994) [arXiv:astro-ph/9307002].
71. L. Knox and M. S. Turner, Phys. Rev. Lett. **70**, 371 (1993) [arXiv:astro-ph/9209006].
72. W. H. Kinney and K. T. Mahanthappa, Phys. Rev. D **53**, 5455 (1996) [arXiv:hep-ph/9512241].
73. A. Lewis and S. Bridle, Phys. Rev. D **66**, 103511 (2002) [arXiv:astro-ph/0205436].
74. W. H. Kinney, E. W. Kolb, A. Melchiorri and A. Riotto, Phys. Rev. D **74**, 023502 (2006) [arXiv:astro-ph/0605338].
75. W. H. Kinney, E. W. Kolb, A. Melchiorri and A. Riotto, Phys. Rev. D **78**, 087302 (2008) [arXiv:0805.2966 [astro-ph]].
76. C. L. Reichardt *et al.*, arXiv:0801.1491 [astro-ph].
77. J. Loveday [the SDSS Collaboration], arXiv:astro-ph/0207189.
78. K. N. Abazajian *et al.* [SDSS Collaboration], arXiv:0812.0649 [astro-ph].
79. K. Freese, J. A. Frieman and A. V. Olinto, Phys. Rev. Lett. **65**, 3233 (1990).
80. G. R. Dvali, Q. Shafi and R. K. Schaefer, Phys. Rev. Lett. **73**, 1886 (1994) [arXiv:hep-ph/9406319].

81. E. D. Stewart, Phys. Rev. D **51**, 6847 (1995) [arXiv:hep-ph/9405389].
82. J. D. Barrow and P. Parsons, Phys. Rev. D **52**, 5576 (1995) [arXiv:astro-ph/9506049].
83. D. N. Spergel and M. Zaldarriaga, Phys. Rev. Lett. **79**, 2180 (1997) [arXiv:astro-ph/9705182].
84. J. Khoury, B. A. Ovrut, P. J. Steinhardt and N. Turok, Phys. Rev. D **64**, 123522 (2001) [arXiv:hep-th/0103239].
85. J. Khoury, P. J. Steinhardt and N. Turok, Phys. Rev. Lett. **92**, 031302 (2004) [arXiv:hep-th/0307132].
86. R. Brandenberger and N. Shuhmaher, JHEP **0601**, 074 (2006) [arXiv:hep-th/0511299].
87. A. Albrecht and J. Magueijo, Phys. Rev. D **59**, 043516 (1999) [arXiv:astro-ph/9811018].
88. D. N. Spergel *et al.* [WMAP Collaboration], Astrophys. J. Suppl. **170**, 377 (2007) [arXiv:astro-ph/0603449].
89. L. Alabidi and D. H. Lyth, JCAP **0608**, 013 (2006) [arXiv:astro-ph/0603539].
90. U. Seljak, A. Slosar and P. McDonald, JCAP **0610**, 014 (2006) [arXiv:astro-ph/0604335].
91. J. Martin and C. Ringeval, JCAP **0608**, 009 (2006) [arXiv:astro-ph/0605367].
92. J. Lesgourgues, A. A. Starobinsky and W. Valkenburg, JCAP **0801**, 010 (2008) [arXiv:0710.1630 [astro-ph]].
93. H. V. Peiris and R. Easther, JCAP **0807**, 024 (2008) [arXiv:0805.2154 [astro-ph]].
94. L. Alabidi and J. E. Lidsey, arXiv:0807.2181 [astro-ph].
95. J. Q. Xia, H. Li, G. B. Zhao and X. Zhang, Phys. Rev. D **78**, 083524 (2008) [arXiv:0807.3878 [astro-ph]].
96. J. Hamann, J. Lesgourgues and W. Valkenburg, JCAP **0804**, 016 (2008) [arXiv:0802.0505 [astro-ph]].
97. T. L. Smith, M. Kamionkowski and A. Cooray, Phys. Rev. D **78**, 083525 (2008) [arXiv:0802.1530 [astro-ph]].
98. H. Li *et al.*, arXiv:0812.1672 [astro-ph].
99. [Planck Collaboration], arXiv:astro-ph/0604069.
100. F. R. Bouchet [Planck Collaboration], Mod. Phys. Lett. A **22**, 1857 (2007).
101. C. L. Kuo *et al.*, arXiv:astro-ph/0611198.
102. J. E. Ruhl *et al.* [The SPT Collaboration], arXiv:astro-ph/0411122.
103. K. W. Yoon *et al.*, large angular scale CMB polarimeter," arXiv:astro-ph/0606278.
104. A. C. Taylor [the Clover Collaboration], New Astron. Rev. **50**, 993 (2006) [arXiv:astro-ph/0610716].
105. D. Samtleben and f. t. Q. collaboration, arXiv:0806.4334 [astro-ph].
106. B. P. Crill *et al.*, arXiv:0807.1548 [astro-ph].
107. R. Trotta, Mon. Not. Roy. Astron. Soc. **378**, 72 (2007) [arXiv:astro-ph/0504022].
108. J. Magueijo and R. D. Sorkin, Mon. Not. Roy. Astron. Soc. Lett. **377**, L39

(2007) [arXiv:astro-ph/0604410].

109. D. Parkinson, P. Mukherjee and A. R. Liddle, Phys. Rev. D **73**, 123523 (2006) [arXiv:astro-ph/0605003].

110. A. R. Liddle, P. Mukherjee and D. Parkinson, Astron. Geophys. **47**, 4.30-4.33 (2006) [arXiv:astro-ph/0608184].

111. A. R. Liddle, Mon. Not. Roy. Astron. Soc. Lett. **377**, L74 (2007) [arXiv:astro-ph/0701113].

112. C. Pahud, A. R. Liddle, P. Mukherjee and D. Parkinson, Mon. Not. Roy. Astron. Soc. **381**, 489 (2007) [arXiv:astro-ph/0701481].

113. E. V. Linder and R. Miquel, arXiv:astro-ph/0702542.

114. A. R. Liddle, P. S. Corasaniti, M. Kunz, P. Mukherjee, D. Parkinson and R. Trotta, arXiv:astro-ph/0703285.

115. G. Efstathiou, arXiv:0802.3185 [astro-ph].

116. R. Easther, AIP Conf. Proc. **698**, 64 (2004) [arXiv:astro-ph/0308160].

117. D. Baumann *et al.*, arXiv:0811.3919 [astro-ph].

118. W. H. Kinney, Phys. Rev. D **58**, 123506 (1998) [arXiv:astro-ph/9806259].

119. E. J. Copeland, I. J. Grivell and A. R. Liddle, Mon. Not. Roy. Astron. Soc. **298**, 1233 (1998) [arXiv:astro-ph/9712028].

120. L. P. L. Colombo, E. Pierpaoli and J. R. Pritchard, arXiv:0811.2622 [astro-ph].

121. P. Adshead and R. Easther, JCAP **0810**, 047 (2008) [arXiv:0802.3898 [astro-ph]].

122. V. Barger, Y. Gao, Y. Mao and D. Marfatia, arXiv:0810.3337 [astro-ph].

123. A. Kogut *et al.*, arXiv:0704.3991 [astro-ph].

124. J. Dunkley *et al.*, arXiv:0811.3915 [astro-ph].

125. J. M. Maldacena, JHEP **0305**, 013 (2003) [arXiv:astro-ph/0210603].

126. P. Creminelli, A. Nicolis, L. Senatore, M. Tegmark and M. Zaldarriaga, JCAP **0605**, 004 (2006) [arXiv:astro-ph/0509029].

127. M. Liguori, A. Yadav, F. K. Hansen, E. Komatsu, S. Matarrese and B. Wandelt, Phys. Rev. D **76**, 105016 (2007) [Erratum-ibid. D **77**, 029902 (2008)] [arXiv:0708.3786 [astro-ph]].

128. I. G. Moss and C. Xiong, JCAP **0704**, 007 (2007) [arXiv:astro-ph/0701302].

129. M. Sasaki, J. Valiviita and D. Wands, Phys. Rev. D **74**, 103003 (2006) [arXiv:astro-ph/0607627].

130. K. Moodley, M. Bucher, J. Dunkley, P. G. Ferreira and C. Skordis, [arXiv:astro-ph/0407304].

131. R. Bean, J. Dunkley and E. Pierpaoli, Phys. Rev. D **74**, 063503 (2006) [arXiv:astro-ph/0606685].

132. D. H. Lyth, C. Ungarelli and D. Wands, Phys. Rev. D **67**, 023503 (2003) [arXiv:astro-ph/0208055].

133. D. H. Lyth and D. Wands, Phys. Rev. D **68**, 103516 (2003) [arXiv:astro-ph/0306500].

134. J. M. Bardeen, Phys. Rev. D **22** (1980) 1882.

135. E. Komatsu, arXiv:astro-ph/0206039.

136. S. W. Hawking, Astrophys. J. **145**, 544 (1966).

137. G. F. R. Ellis and M. Bruni, Phys. Rev. D **40**, 1804 (1989).

138. A. R. Liddle and D. H. Lyth, Phys. Rept. **231**, 1 (1993) [arXiv:astro-ph/9303019].
139. M. Sasaki and E. D. Stewart, Prog. Theor. Phys. **95**, 71 (1996) [arXiv:astro-ph/9507001].
140. A. Challinor and A. Lasenby, Astrophys. J. **513**, 1 (1999) [arXiv:astro-ph/9804301].
141. H. Kodama and M. Sasaki, Prog. Theor. Phys. Suppl. **78**, 1 (1984).
142. V. F. Mukhanov, H. A. Feldman and R. H. Brandenberger, Phys. Rept. **215**, 203 (1992).

Chapter 14

Particle Dark Matter

Dan Hooper

*Theoretical Astrophysics Group, Fermi National Accelerator Laboratory,
Department of Astronomy and Astrophysics, University of Chicago*

`dhooper@fnal.gov`

Based on lectures given at the 2008 Theoretical Advanced Study Institute (TASI), I review here some aspects of the phenomenology of particle dark matter, including the process of thermal freeze-out in the early universe, and the direct and indirect detection of WIMPs. I also describe some of the most popular particle candidates for dark matter and summarize the current status of the quest to discover dark matter's particle identity.

14.1. Evidence For Dark Matter

A wide variety of evidence has accumulated in support of dark matter's existence. At galactic and sub-galactic scales, this evidence includes galactic rotation curves,[1] the weak gravitational lensing of distant galaxies by foreground structure,[2] and the weak modulation of strong lensing around individual massive elliptical galaxies.[3] Furthermore, velocity dispersions of stars in some dwarf galaxies imply that they contain as much as $\sim 10^3$ times more mass than can be attributed to their luminosity. On the scale of galaxy clusters, observations (of radial velocities, weak lensing, and X-ray emission) indicate a total cosmological matter density of $\Omega_M \approx 0.2 - 0.3$,[4] which is much larger than the corresponding density in baryons. In fact, it was measurements of velocity dispersions in the Coma cluster which led Fritz Zwicky to claim for the first time in 1933 that large quantities of non-luminous matter are required to be present.[5] On cosmological scales, observations of the anisotropies in the cosmic microwave background have lead to a determination of the total matter density of $\Omega_M h^2 = 0.1326 \pm 0.0063$, where h is the Hubble parameter in units of 100 km/sec per Mpc (this improves to $\Omega_M h^2 = 0.1358^{+0.0037}_{-0.0036}$ if distance measurements from baryon acoustic oscillations and type Ia supernovae are included).[6] In contrast, this information combined with measurements of the light chemical element abundances leads to an estimate of the baryonic density given by $\Omega_B h^2 = 0.02273 \pm 0.00062$ ($\Omega_B h^2 = 0.02267^{+0.00058}_{-0.00059}$ if BAO and SN are included).[6,7] Taken together, these observations strongly lead us to the conclusion that 80-85% of the matter in the universe (by mass) consists of non-luminous and non-baryonic material.

The process of the formation of large scale structure through the gravitational clustering of collisionless dark matter particles can be studied using N-body simulations. When the observed structure in our universe[8] is compared to the results of cold (non-relativistic at time of structure formation) dark matter simulations, good agreement has been found. The large scale structure predicted for hot dark matter, in contrast, is in strong disagreement with observations.

Although there are many pieces of evidence in favor of dark matter, it is worth noting that they each infer dark matter's presence uniquely through its gravitational influence. In other words, we currently have no conclusive evidence for dark matter's electroweak or other non-gravitational interactions. Given this, it is natural to contemplate whether, rather than being indications of dark matter's existence, these observations might instead be revealing departures from the laws of gravity as described by general relativity.

Since first proposed by Milgrom in 1983,[9] efforts have been made to explain the observed galactic rotation curves without dark matter within the context of a phenomenological model known as modified Newtonian dynamics, or MOND. The basic idea of MOND is that Newton's second law, $F = ma$, is modified to $F = ma \times \mu(a)$, where μ is very closely approximated by unity except in the case of very small accelerations, for which μ behaves as $\mu = a/a_0$. Applying the modified form of Newton's second law to the gravitational force acting on a star outside of a galaxy of mass M leads us to

$$F = \frac{GMm}{r^2} = ma\mu, \qquad (14.1.1)$$

which in the low acceleration limit (large r, $a \ll a_0$) yields

$$a = \frac{\sqrt{GMa_0}}{r}. \qquad (14.1.2)$$

Equating this with the centrifugal acceleration associated with a circular orbit, we arrive at

$$\frac{\sqrt{GMa_0}}{r} = \frac{v^2}{r} \quad \Longrightarrow \quad v = (GMa_0)^{1/4}. \qquad (14.1.3)$$

In other words, MOND yields the prediction that galactic rotation curves should become flat (independent of r) for sufficiently large orbits. This result is in good agreement with galaxy-scale observations for a value of $a_0 \sim 1.2 \times 10^{-10}$ m/s^2, even without the introduction of dark matter. For this value of a_0, the effects of MOND are imperceptible in laboratory or Solar System scale experiments.

MOND is not as successful in explaining the other evidence for dark matter, however. In particular, MOND fails to successfully describe the observed features of galaxy clusters. Other evidence, such as the cosmic microwave background anisotropies and large scale structure, are not generally able to be addressed by MOND, as MOND represents a phenomenological modification of Newtonian dynamics and thus is not applicable to questions addressed by general relativity, such as the expansion history of the universe. Efforts to develop a viable, relativistically covariant theory which yields the behavior of MOND in the non-relativistic, weak-field limit have mostly been unsuccessful. A notable exception to this is Tensor-Vector-Scalar gravity, or TeVeS.[10] TeVeS, however, fails to explain cluster-scale observations without the introduction of dark matter.[11] This problem has been further exacerbated by recent observations of two merging clusters, known collectively as the bullet cluster. In the bullet cluster, the locations of the baryonic material and gravitational potential (as determined using X-ray observations and weak lensing, respectively) are clearly spatially separated, strongly favoring the dark matter hypothesis over modifications of general relativity.[12]

14.2. The Production of Dark Matter in the Early Universe

The nucleons, electrons and neutrinos that inhabit our universe can each trace their origin back to the first fraction of a second following the Big Bang. Although we do not know for certain how the dark matter came to be formed, a sizable relic abundance of weakly interacting massive particles (WIMPs) is generally expected to be produced as a byproduct of our universe's hot youth. In this section, I discuss this process and the determination of the relic abundance of a WIMP.[13–15]

Consider a stable particle, X, which interacts with Standard Model particles, Y, through some process $X\bar{X} \leftrightarrow Y\bar{Y}$ (or $XX \leftrightarrow Y\bar{Y}$ if X is its own antiparticle). In the very early universe, when the temperature was much higher than m_X, the processes of $X\bar{X}$ creation and annihilation were equally efficient, leading X to be present in large quantities alongside the various particle species of the Standard Model. As the temperature of the universe dropped below m_X, however, the process of $X\bar{X}$ creation became exponentially suppressed, while $X\bar{X}$ annihilation continued unabated. In thermal equilibrium, the number density of such particles is given by

$$n_{X,\,\mathrm{eq}} = g_X \left(\frac{m_X T}{2\pi} \right)^{3/2} e^{-m_X/T}, \qquad (14.2.4)$$

where g_X is the number of internal degrees of freedom of X.

If these particles were to remain in thermal equilibrium indefinitely, their number density would become increasingly suppressed as the universe cooled, quickly becoming cosmologically irrelevant. There are ways that a particle species might hope to avoid this fate, however. For example, baryons are present in the universe today because of a small asymmetry which initially existed between the number of baryons and antibaryons; when all of the antibaryons had annihilated with baryons, a small residual of baryons remained. The baryon-antibaryon asymmetry prevented the complete annihilation of these particles from taking place.

While it is possible that a particle-antiparticle asymmetry is also behind the existence of dark matter, there is an even simpler mechanism which can lead to the survival of a sizable relic density of weakly interacting particles. In particular, the self-annihilation of weakly interacting species can be contained by the competing effect of Hubble expansion. As the expansion and corresponding dilution of WIMPs increasingly dominates over the annihilation rate, the number density of X particles becomes sufficiently small that they cease to interact with each other, and thus survive to the present day. Quantitatively, the competing effects of expansion and annihilation are described by the Boltzmann equation:

$$\frac{dn_X}{dt} + 3H n_X = -\langle \sigma_{X\bar{X}} |v| \rangle (n_X^2 - n_{X,\,\mathrm{eq}}^2), \qquad (14.2.5)$$

where n_X is the number density of WIMPs, $H \equiv \dot{R}/R = (8\pi^3 \rho / 3 M_{\mathrm{Pl}})^{1/2}$ is the expansion rate of the universe, and $\langle \sigma_{X\bar{X}} |v| \rangle$ is the thermally averaged $X\bar{X}$ annihilation cross section (multiplied by their relative velocity).

From Eq. (14.2.5), we can identify two clear limits. As I said before, at very high temperatures ($T \gg m_X$) the density of WIMPs is given by the

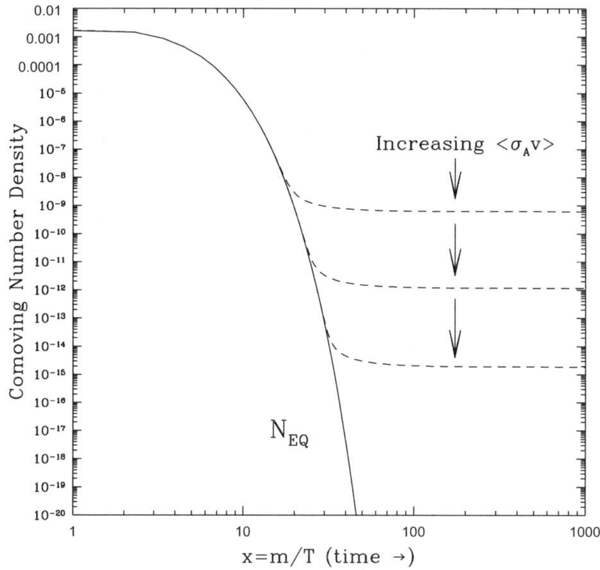

Fig. 14.1. A schematic of the comoving number density of a stable species as it evolves through the process of thermal freeze-out.

equilibrium value, $n_{X,\text{eq}}$. In the opposite limit ($T \ll m_X$), the equilibrium density is very small, leaving the terms $3Hn_X$ and $\langle \sigma_{X\bar{X}}|v|\rangle n_X^2$ to each further deplete the number density. For sufficiently small values of n_X, the annihilation term becomes insignificant compared to the dilution due to Hubble expansion. When this takes place, the comoving number density of WIMPs becomes fixed — thermal freeze-out has occurred.

The temperature at which the number density of the species X departs from equilibrium and freezes out is found by numerically solving the Boltzmann equation. Introducing the variable $x \equiv m_X/T$, the temperature at which freeze-out occurs is approximately given by

$$x_{\text{FO}} \equiv \frac{m_X}{T_{\text{FO}}} \approx \ln\left[c(c+2)\sqrt{\frac{45}{8}}\frac{g_X}{2\pi^3}\frac{m_X M_{\text{Pl}}(a + 6b/x_{\text{FO}})}{g_\star^{1/2} x_{\text{FO}}^{1/2}}\right]. \qquad (14.2.6)$$

Here, $c \sim 0.5$ is a numerically determined quantity, $g_\star$ is the number of external degrees of freedom available (in the Standard Model, $g_\star \sim 120$ at $T \sim 1$ TeV and $g_\star \sim 65$ at $T \sim 1$ GeV), and a and b are terms in the non-relativistic expansion, $\langle \sigma_{X\bar{X}}|v|\rangle = a + b\langle v^2 \rangle + \mathcal{O}(v^4)$. The resulting density of WIMPs remaining in the universe today is approximately given by

$$\Omega_X h^2 \approx \frac{1.04 \times 10^9 \, \mathrm{GeV}^{-1}}{M_{\mathrm{Pl}}} \frac{x_{\mathrm{FO}}}{g_\star^{1/2}(a + 3b/x_{\mathrm{FO}})}. \tag{14.2.7}$$

If X has a GeV-TeV scale mass and a roughly weak-scale annihilation cross section, freeze-out occurs at $x_{\mathrm{FO}} \approx 20 - 30$, resulting in a relic abundance of

$$\Omega_X h^2 \approx 0.1 \left(\frac{x_{\mathrm{FO}}}{20}\right)\left(\frac{g_\star}{80}\right)^{-1/2}\left(\frac{a + 3b/x_{\mathrm{FO}}}{3 \times 10^{-26} \mathrm{cm}^3/\mathrm{s}}\right)^{-1}. \tag{14.2.8}$$

In other words, if a GeV-TeV scale particle is to be thermally produced with an abundance similar to the measured density of dark matter, it must have a thermally averaged annihilation cross section on the order of 3×10^{-26} cm^3/s. Remarkably, this is very similar to the numerical value arrived at for a generic weak-scale interaction. In particular, $\alpha^2/(100 \, \mathrm{GeV})^2 \sim$ pb, which in our choice of units (and including a factor of velocity) is $\sim 3 \times 10^{-26}$ cm^3/s. The similarly between this result and the value required to generate the observed quantity of dark matter has been dubbed the "WIMP miracle". While far from constituting a proof, this argument has lead many to conclude that dark matter is likely to consist of particles with weak-scale masses and interactions and certainly provides us with motivation for exploring an electroweak origin of our universe's missing matter.

14.2.1. *Case Example – The Thermal Abundance of a Light or Heavy neutrino*

At first glance, the Standard Model itself appears to contain a plausible candidate for dark matter in the form of neutrinos. Being stable and weakly interacting, neutrinos are a natural place to start in our hunt for dark matter's identity.

In the case of a Standard Model neutrino species, the relatively small annihilation cross section ($\langle \sigma|v|\rangle \sim 10^{-32}$ cm^3/s) and light mass leads to an overall freeze-out temperature on the order of $T_{\mathrm{FO}} \sim$ MeV, and a relic density of

$$\Omega_{\nu+\bar{\nu}} h^2 \approx 0.1 \left(\frac{m_\nu}{9 \, \mathrm{eV}}\right). \tag{14.2.9}$$

As constraints on Standard Model neutrino masses require m_ν to be well below 9 eV, we are forced to conclude that only a small fraction of the dark matter could possibly consist of Standard Model neutrinos. Furthermore, even if these constraints did not exist, such light neutrinos would be highly relativistic at the time of freeze-out ($T_{\rm FO}/m_\nu \sim {\rm MeV}/m_\nu \gg 1$) and thus would constitute hot dark matter, in conflict with observations of large scale structure.

Moving beyond the Standard Model, we could instead consider a heavy 4th generation Dirac neutrino. In this case, the annihilation cross section can be much larger, growing with the square of the neutrino's mass up to the Z pole, $m_\nu \sim m_Z/2$, and declining with m_ν^{-2} above $m_\nu \sim m_Z/2$. For a GeV-TeV mass neutrino, the process of freeze-out yields a cold relic ($T_{\rm FO}/m_\nu \sim \mathcal{O}(0.1)$), with an abundance approximately given by

$$\Omega_{\nu+\bar\nu}h^2 \approx 0.1 \left(\frac{5.5\,{\rm GeV}}{m_\nu}\right)^2, \qquad {\rm MeV} \ll m_\nu \ll m_Z/2 \quad (14.2.10)$$

$$\Omega_{\nu+\bar\nu}h^2 \sim 0.1 \left(\frac{m_\nu}{400\,{\rm GeV}}\right)^2, \qquad m_Z/2 \ll m_\nu. \quad (14.2.11)$$

Thus, if the relic density of a heavy neutrino species is to constitute the bulk of the observed dark matter abundance, we find that it must have a mass of approximately 5 GeV,[16] or several hundred GeV. The former case is excluded by LEP's measurement of the invisible width of the Z, however, which rules out the possibility of a fourth active neutrino species lighter than half of the Z mass.[17] The later case, although consistent with the bounds of LEP and other accelerator experiments, is excluded by the limits placed by direct dark matter detection experiments (which I will discuss in Sec. 14.4).

14.2.2. *A Detour Into Coannihilations*

In some cases, particles other than the WIMP itself can play an important role in the freeze-out process.[18] Before such a particle can significantly impact the relic density of a WIMP, however, it must first manage to be present at the temperature of freeze-out.

The relative abundances of two species at freeze-out can be very roughly estimated by

$$\frac{n_Y}{n_X} \sim \frac{e^{-m_Y/T_{\rm FO}}}{e^{-m_X/T_{\rm FO}}}. \quad (14.2.12)$$

Considering, for example, a particle with a mass twice that of the WIMP and a typical freeze-out temperature of $m_X/T_{\rm FO} \approx 20$, there will be only $\sim e^{-40}/e^{-20} \sim 10^{-9}$ Y particles for every X at freeze-out, thus making Y

completely irrelevant. If m_Y were only 10% larger than m_X, however, we estimate $n_Y/n_X \sim e^{-22}/e^{-20} \sim 10^{-1}$. In this quasi-degenerate case, the additional particle species can potentially have a significant impact on the dark matter relic abundance.

To quantitatively account for other species in the calculation of the relic abundance of a WIMP, we make the following substitution (for both a and b) into Eqs. (14.2.6) and (14.2.7):

$$\sigma_{\mathrm{Ann}} \rightarrow \sigma_{\mathrm{Eff}}(x) = \sum_{i,j} \sigma_{i,j} \frac{g_i g_j}{g_{\mathrm{Eff}}^2(x)} (1 + \Delta_i)^{3/2} (1 + \Delta_j)^{3/2} e^{-x(\Delta_i + \Delta_j)},$$

$$(14.2.13)$$

where the double sum is over all particle species (i, j=1 denoting the WIMP itself) and $\sigma_{i,j}$ is the cross section for the coannihilation of species i and j (or self-annihilation in the case of $i = j$) into Standard Model particles. As the effective annihilation cross section has a strong dependence on x, we must integrate Eqs. (14.2.6) and (14.2.7) over x (or T). The quantities $\Delta_i = (m_i - m_1)/m_1$ denote the fractional mass splittings between the species i and the WIMP. The effective number of degrees of freedom, $g_{\mathrm{Eff}}(x)$, is given by:

$$g_{\mathrm{Eff}}(x) = \sum_i g_i (1 + \Delta_i)^{3/2} e^{-x\Delta_i}. \qquad (14.2.14)$$

To better understand how the introduction of particles other than the WIMP can effect the process of freeze-out, lets consider a few simple cases. First, consider one additional particle with a mass only slightly above that of the WIMPs ($\Delta_2 \ll 1$), and with a comparatively large coannihilation cross section, such that $\sigma_{1,2} \gg \sigma_{1,1}$. In this case, $g_{\mathrm{Eff}} \approx g_1 + g_2$, and $\sigma_{\mathrm{Eff}} \approx \sigma_{1,2}\, g_1 g_2/(g_1 + g_2)^2$. Since σ_{Eff} is much larger than the WIMP's self-annihilation cross section, the relic density of WIMPs will be sharply suppressed. This is the case that is usually meant by the term "coannihilation".

Alternatively, consider the opposite case in which the WIMP and the additional quasi-degenerate particle do not coannihilate efficiently ($\sigma_{1,2} \ll \sigma_{1,1}, \sigma_{2,2}$). Here, $\sigma_{\mathrm{Eff}} \approx \sigma_{1,1}\, g_1^2/(g_1+g_2)^2 + \sigma_{2,2}\, g_2^2/(g_1+g_2)^2$, which in some cases can actually be *smaller* than that for the process of self-annihilation alone, leading to an *enhanced* relic abundance. Physically speaking, what is going on here is that the two species are each freezing out independently of each other, after which the heavier species decays, producing additional WIMPs as a byproduct.

As an extreme version of this second case, consider a scenario in which the lightest state is not a WIMP, but is instead a purely gravitationally interacting particle. A slightly heavier particle with weak interactions will self-annihilate much more efficiently than it will coannihilate with the lightest particle ($\sigma_{1,2}$ is negligible), leading the two states to freeze-out independently. The gravitationally interacting particle, however, never reaches thermal equilibrium, so could potentially have not been produced in any significant quantities up until this point. Well after freezing out, the heavier particles will eventually decay, producing the stable gravitationally interacting lightest state. Although the resulting particles are not WIMPs (they do not have weak interactions), they are naturally produced with approximately the measured dark matter abundance because of the WIMP-like properties of the heavier state. In other words, this case – known as the "superWIMP" scenario[19] – makes use of the coincidence between the electroweak scale and the measured dark matter abundance without the dark matter actually consisting of WIMPs. Because gravitationally interacting particles and other much less than weakly interacting particles are almost impossible to detect astrophysically, superWIMPs are among the dark matter hunter's worst nightmares.

14.3. Beyond The Standard Model Candidates For Dark Matter

There has been no shortage of dark matter candidates proposed over the years. A huge variety of beyond the Standard Model physics models have been constructed which include a stable, electrically neutral, and colorless particle, many of which could serve as a phenomenologically viable candidate for dark matter. I could not possibly list, must less review, all of the proposed candidates here. Finding the "WIMP miracle" (as discussed in Sec. 14.2) to be fairly compelling (along with the hierarchy problem, which strongly suggests the existence of new particles at or around the electroweak scale), I chose to focus my attention on dark matter in the form of weak-scale particles. So although the dark matter of our universe could plausibly consist of particles ranging from 10^{-6} eV axions to 10^{16} GeV WIMPzillas, I will ignore everything but those particle physics frameworks which predict the existence of a stable, weakly interacting particle with a mass in the few GeV to few TeV range.

14.3.1. *Supersymmetry*

For a number of reasons, supersymmetry is considered by many to be among the most attractive extensions of the Standard Model. In particular, weak-scale supersymmetry provides us with an elegant solution to the hierarchy problem,[20] and enables grand unification by causing the gauge couplings of the Standard Model to evolve to a common scale.[21] From the standpoint of dark matter, the lightest superpartner is naturally stable in models that conserve R-parity. R-parity is defined as $R = (-1)^{3B+L+2S}$ (B, L and S denoting baryon number, lepton number and spin), and thus is assigned as $R = +1$ for all Standard Model particles and $R = -1$ for all superpartners. R-parity conservation, therefore, requires superpartners to be created or destroyed in pairs, leading the lightest supersymmetric particle (LSP) to be stable, even over cosmological timescales.

The identity of the LSP depends on the hierarchy of the supersymmetric spectrum, which in turn is determined by the details of how supersymmetry is broken. The list of potential LSPs which could constitute a plausible dark matter candidate is somewhat short, however. The only electrically neutral and colorless superparnters in the minimal supersymmetric standard model (MSSM) are the four neutralinos (superpartners of the neutral gauge and Higgs bosons), three sneutrinos, and the gravitino. The lightest neutralino, in particular, is a very attractive and throughly studied candidate for dark matter.[22]

Before discussing the phenomenology of neutralino dark matter, lets briefly contemplate the possibility that sneutrinos might make up the dark matter of our universe. In many respects, sneutrino dark matter would behave very similarly to a heavy 4th generation neutrino, as discussed in Sec. 14.2.1. In particular, like neutrinos, sneutrinos are predicted to annihilate to Standard Model fermions efficiently through the s-channel exchange of a Z boson (as well as through other diagrams). As a result, sneutrinos lighter than about 500-1000 GeV would be under produced in the early universe (a $\sim$10 GeV sneutrino would also be produced with approximately the measured dark matter abundance, but is ruled out by LEP's invisible Z measurement).

The Feynman diagram corresponding to sneutrino annihilation into quarks through a s-channel Z exchange can be turned on its side to produce an elastic scattering diagram with quarks in nuclei. When the elastic scattering cross section of a $\sim$100-1000 GeV sneutrino is calculated, we find that it is several orders of magnitude larger than current experimental constraints.[23] We are thus forced to abandon MSSM sneutrinos as candidates for dark matter.

In the MSSM, the superpartners of the four Standard Model neutral bosons (the bino, wino and two neutral higgsinos) mix into four physical states known as neutralinos. Often times, the lightest of these four states is simply referred to as "the neutralino". The neutralino mass matrix can be used to determine the masses and mixings of these four states. In the $\widetilde{B}$-$\widetilde{W}^3$-$\widetilde{H}_1$-$\widetilde{H}_2$ basis, this matrix is given by

$$\mathcal{M}_{\chi^0} =$$

$$(14.3.15)$$

$$\begin{pmatrix} M_1 & 0 & -m_Z \cos\beta \sin\theta_W & m_Z \sin\beta \sin\theta_W \\ 0 & M_2 & m_Z \cos\beta \cos\theta_W & -m_Z \sin\beta \cos\theta_W \\ -m_Z \cos\beta \sin\theta_W & m_Z \cos\beta \cos\theta_W & 0 & -\mu \\ m_Z \sin\beta \sin\theta_W & -m_Z \sin\beta \cos\theta_W & -\mu & 0 \end{pmatrix},$$

where M_1 and M_2 are the bino and wino masses, μ is the higgsino mass parameter, θ_W is the Weinberg angle, and $\tan\beta \equiv v_2/v_1$ is the ratio of the vacuum expectation values of the Higgs doublets. This matrix can be diagonalized into mass eigenstates by the unitary matrix N,

$$\mathcal{M}_{\chi^0}^{\text{diag}} = N^* M_{\chi^0} N^{-1}. \qquad (14.3.16)$$

In terms of the elements of the matrix, N, the lightest neutralino is given by the following mixture of gaugino and higgsino components:

$$\chi^0 = N_{11}\widetilde{B} + N_{12}\widetilde{W}^3 + N_{13}\widetilde{H}_1 + N_{14}\widetilde{H}_2. \qquad (14.3.17)$$

The quantities $|N_{11}|^2 + |N_{12}|^2$ and $|N_{13}|^2 + |N_{14}|^2$ are often referred to as the gaugino fraction and higgsino fraction of the lightest neutralino, respectively.

The lightest neutralino can annihilate through a wide variety of Feynman diagrams. In Fig. 14.2, we show some of the most important of these; although it is far from an inclusive list. Which of these diagrams dominate the process of thermal freeze-out in the early universe depends on the composition of the lightest neutralino, and on the masses and mixings of the exchanged particles.[24] Since so many different diagrams can potentially contribute to neutralino annihilation (not to mention the many possible coannihilation processes[25–28]), the resulting relic density depends on a large number of supersymmetric parameters and is not trivial to calculate accurately. Publicly available tools such as DarkSUSY[29] and MicroOmegas[30] are often used for this purpose.

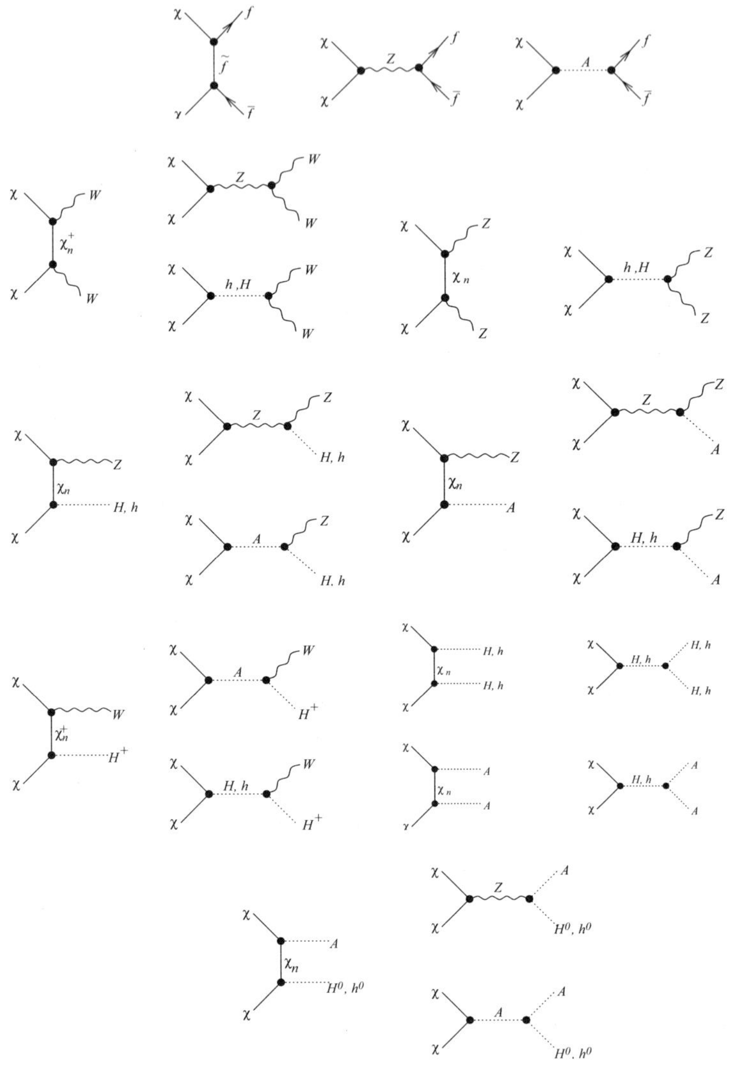

Fig. 14.2. Some of the most important Feynman diagrams for neutralino annihilation.

The mass and composition of the lightest neutralino is a function of four supersymmetric parameters: M_1, M_2, μ and $\tan\beta$. This becomes further simplified if the gaugino masses are assumed to evolve to a single value at the GUT scale, yielding a ratio at the electroweak scale of $M_1 = \frac{5}{3}\tan^2\theta_W M_2 \approx 0.5 M_2$. In this case, the lightest neutralino has only a small wino fraction and is largely bino-like (higgsino-like) for $M_1 \ll |\mu|$ $(M_1 \gg |\mu|)$.

Over much or most of the supersymmetric parameter space, the relic abundance of neutralinos is predicted to be in excess of the observed dark matter density. To avoid this, we are forced to consider the regions of parameter space which lead to especially efficient neutralino annihilation in the early universe. In particular, the following scenarios are among those which can lead to a phenomenologically viable density of neutralino dark matter:

- If the lightest neutralino has a significant higgsino or wino fraction, it can have fairly large couplings and, as a result, annihilate very efficiently.
- If the mass of the lightest neutralino is near a resonance, such as the CP-odd Higgs pole, it can annihilate efficiently, even with relatively small couplings.
- If the lightest neutralino is only slightly lighter than another superpartner, such as the lightest stau, coannihilations between these two states can very efficiently deplete the dark matter abundance.

In Fig. 14.3, we illustrate these regions within the context of a specific subset of the MSSM known as the CMSSM (C stands for constrained). In this framework, all of the scalar masses are set to a common value m_0 at the GUT scale, from which the electroweak scale values are determined by RGE evolution. Similarly the three gaugino masses are each set to $m_{1/2}$ at the GUT scale. In each frame, the narrow blue regions denote the parameter space in which neutralino dark matter is predicted to be generated with the desired abundance ($0.0913 < \Omega_{\chi^0} h^2 < 0.1285$). In the corridor along side of the LEP chargino bound ($m_{\chi^\pm} > 104$ GeV), μ and M_1 are comparable in magnitude, leading to a mixed bino-higgsino LSP with large couplings. Within the context of the CMSSM, this is often called the "focus point" region. In the bottom portion of each frame, the lightest stau ($\tilde{\tau}$) is the LSP, and thus does not provide a viable dark matter candidate. Just outside of this region, however, the stau is slightly heavier than the lightest neutralino, leading to a neutralino LSP which efficiently coannihilates with the nearly degenerate stau. In the lower right frame, a viable region also appears along the CP-odd Higgs resonance ($m_{\chi^0} \approx m_A/2$). This is often called the A-funnel region.

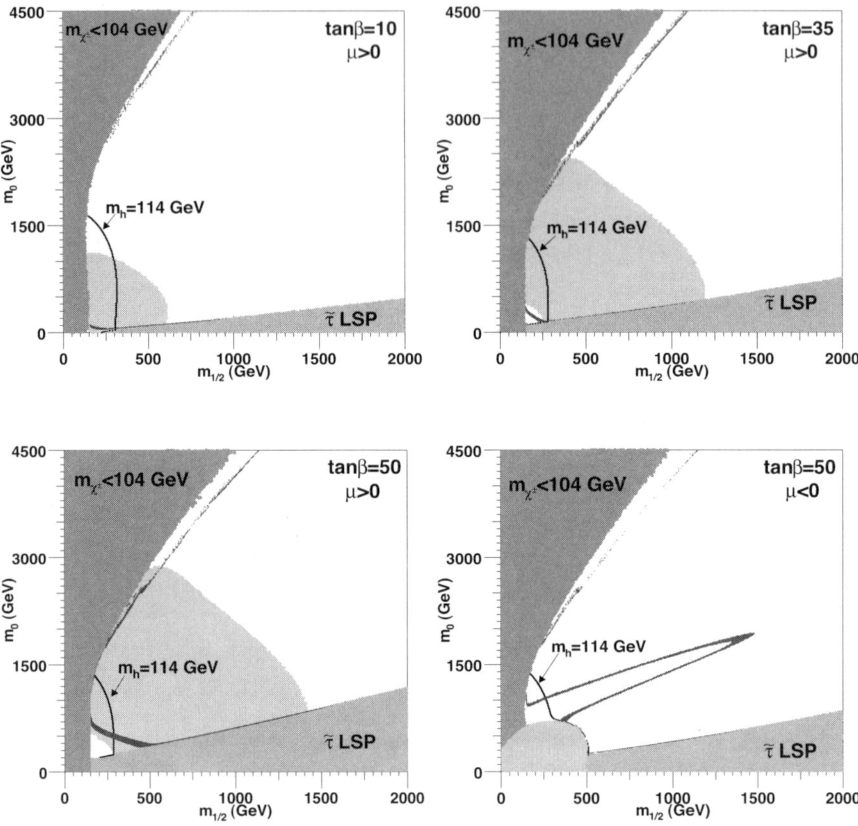

Fig. 14.3. Representative regions of the CMSSM parameter space. The blue regions predict a neutralino density consistent with the measured dark matter abundance. The shaded regions to the upper left and lower right are disfavored by the LEP chargino bound and as a result of containing a stau LSP, respectively. The LEP bound on the light Higgs mass is shown as a solid line ($m_h = 114$ GeV). The region favored by measurements of the muon's magnetic moment are shown as a light shaded region (at the 3σ confidence level).[31] In each frame, we have used $A_0 = 0$ and $\mu > 0$. (Figures generated by Gabriel Caceres, using the package DarkSUSY.[29])

14.3.2. *Kaluza-Klein Dark Matter in Models With Universal Extra Dimensions*

Supersymmetry is not by any means the only particle physics framework from which viable a dark matter candidate can arise. As an alternative, I will discuss in this section the possibility of Kaluza-Klein dark matter in models with a universal extra dimension.

In recent years, interest in theories with extra spatial dimensions has surged. In particular, a colossal amount of attention has been given to two classes of extra dimensional theories over the past decade: scenarios featuring one or more large (millimeter-scale), flat extra dimensions (ADD),[32,33] and the Randall-Sundrum scenario, which introduces an additional small dimension with a large degree of spatial curvature.[34] A somewhat less studied but interesting class of extra-dimensional models, which goes by the name of universal extra dimensions (UED), postulates the existence of a flat extra dimension (or dimensions) through which all of the Standard Model (SM) fields are free to propagate (rather than being confined to a brane as some or all of them are in the ADD and Randall-Sundrum models).[35]

For a number of reasons, extra dimensions of size $R \sim \text{TeV}^{-1}$ are particularly well motivated within the context of UED[36] (much smaller than those found in the ADD model). Among these reasons is the fact that a TeV-scale Kaluza-Klein (KK) state, if stable, colorless, and electrically neutral, could potentially serve as a viable candidate for dark matter.[37,38]

Standard Model fields with momentum in an extra dimension appear as heavy particles, called KK states. This leads to a tower of KK states for each Standard Model field, with tree-level masses given by:

$$m^2_{X^{(n)}} = \frac{n^2}{R^2} + m^2_{X^{(0)}}, \qquad (14.3.18)$$

where $X^{(n)}$ is the nth Kaluza-Klein excitation of the Standard Model field, X, and $R \sim \text{TeV}^{-1}$ is the size of the extra dimension. $X^{(0)}$ denotes the ordinary Standard Model particle (known as the zero mode).

If the extra dimensions were simply wrapped (compactified) around a circle or torus, then extra dimensional momentum conservation would ensure the conservation of KK number (n) and make the lightest first level KK state stable. Realistic models, however, require an orbifold to be introduced, which leads to the violation of KK number conservation. A remnant of KK number conservation called KK-parity, however, can remain and lead to the stability of the lightest KK particle (LKP) in much the same way that R-parity conservation prohibits the decay of the lightest supersymmetric particle.

In order for the LKP to be a viable dark matter candidate, it must be electrically neutral and colorless. Possibilities for such a state include the first KK excitation of the photon, Z, neutrinos, Higgs boson, or graviton. Assuming that R^{-1} is considerably larger than any of the Standard

Model zero mode masses, Eq. (14.3.18) leads us to expect a highly degenerate spectrum of Kaluza-Klein states at each level (although this picture is somewhat modified when radiative corrections and boundary terms are included). Of our possible choices for the LKP, the relatively large zero-mode mass of the Higgs make its first level KK excitation an unlikely candidate. Furthermore, KK neutrinos are excluded by direct detection experiments, just as sneutrinos or heavy 4th generation Dirac neutrinos are. For these reasons, we focus on the mixtures of the KK photon and KK Z as our dark matter candidate (note that, unlike higgsino and gauginos, the KK Higgs has a different spin than the KK photon and KK Z, and thus does not mix with these states).

The mass eigenstates of the KK photon and KK Z are very nearly identical to their gauge eigenstates, $B^{(n)}$ and $W^{3(n)}$. The reason for this can be seen from their mass matrix:

$$\begin{pmatrix} \frac{n^2}{R^2} + \delta m^2_{B^{(n)}} + \frac{1}{4}g_1^2 v^2 & \frac{1}{4}g_1 g_2 v^2 \\ \frac{1}{4}g_1 g_2 v^2 & \frac{n^2}{R^2} + \delta m^2_{W^{(n)}} + \frac{1}{4}g_2^2 v^2 \end{pmatrix}. \tag{14.3.19}$$

Here $v \approx 174$ GeV is the Higgs vacuum expectation value. In the well known zero-mode case ($n = 0$), there is significant mixing between $B^{(0)}$ and $W^{3(0)}$ ($\sin^2 \theta_W \approx 0.23$). In the absence of radiative corrections ($\delta m^2_{B^{(n)}} = \delta m^2_{W^{(n)}} = 0$), the same mixing angle is found at the first KK level as well. If the difference between $\delta m^2_{B^{(n)}}$ and $\delta m^2_{W^{(n)}}$ is larger than the (rather small) off diagonal terms, however, the mixing angle between these two KK states is driven toward zero. Using typical estimates of these radiative corrections, the effective first KK level Weinberg angle is found to be approximately $\sin^2 \theta_{W,1} \sim 10^{-3}$. Thus the mass eigenstate often called the "KK photon" is not particularly photon-like, but instead is nearly identical to the state $B^{(1)}$.

The KK state $B^{(1)}$ annihilates largely to Standard Model (zero-mode) fermions through the t-channel exchange of KK fermions, with a cross section given by

$$\sigma v(B^{(1)} B^{(1)} \to f\bar{f}) = \frac{95}{32,256} \sum_f \frac{N_c(Y_{f_L}^4 + Y_{f_R}^4)g_1^4}{\pi\, m^2_{B^{(1)}}}. \tag{14.3.20}$$

As the $B^{(1)}$ couples to the fermions' hypercharge, the cross section scales with $Y_{f_L}^4 + Y_{f_R}^4$ and most of its annihilations proceed to charged lepton pairs.

If Fig. 14.4, the thermal relic abundance of KK dark matter is shown as a function of its mass. Because the first level KK spectrum is expected

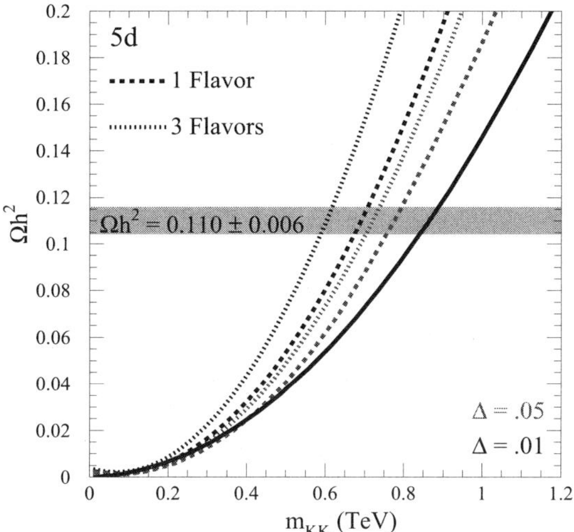

Fig. 14.4.   The thermal relic abundance of KK dark matter ($B^{(1)}$) without the effects of any other KK species (solid line) and including the effects of KK leptons 5% and 1% heavier than the LKP (dashed and dotted lines).   Shown as a horizontal band is the measured dark matter abundance.[6] Adapted from Ref. [37].

to be quasi-degenerate, coannihilations are likely to play an important role. In the figure, results are shown ignoring the effects of other KK states (solid line) and including the effects of KK leptons 5% or 1% heavier than the LKP (dashed and dotted lines, respectively). Note that the KK leptons lead to a larger relic abundance, due to the fact that they freeze-out quasi-independently from the LKP and then increase the number of LKPs through their decays. Depending on the details of the KK spectrum, many other states could potentially effect the relic density of KK dark matter as well. LKP masses from approximately 500 GeV to several TeV can potentially lead to a relic density consistent with the measured dark matter abundance.[39,40]

14.3.3. *A Note On Other Possibilities For TeV-Scale Dark Matter*

At this time, I would like to make a general comment about some of the many other possibilities for the particle identity of dark matter. It is interesting to note that a wide range of solutions to the gauge hierarchy problem also introduce a candidate for dark matter. In particular, in order to be consistent with electroweak precision measurements, new physics at the TeV scale typically must possess a discrete symmetry. Supersymmetry, for example, accomplishes this through R-parity conservation. Similarly, Little Higgs models can avoid problems with electroweak precision tests by introducing a discrete symmetry called T-parity. And just as R-parity stabilizes the lightest superpartner, T-parity can enable the possibility of dark matter by stabilizing the lightest T-parity odd state in the theory.[41] So, thinking outside of any specific particle physics framework, we can imagine a much larger class of TeV-scale physics scenarios which address the hierarchy problem without violating electroweak precision measurements by introducing a discrete symmetry, which in turn leads to a stable dark matter candidate. In this way, many of the motivations and much of the phenomenology discussed within the context of supersymmetry can actually be applied to a more general collection of particle physics models associated with the electroweak scale.

14.4. Direct Detection

Turning our attention now to dark matter detection, we begin with those experiments which attempt to detect dark matter particles through their elastic scattering with nuclei, including CDMS,[42] XENON,[43] ZEPLIN,[44] EDELWEISS,[45] CRESST,[46] CoGeNT,[47] DAMA/LIBRA,[48] COUPP,[49] WARP,[50] and KIMS.[51] This class of techniques is collectively known as direct detection, in contrast to indirect detection efforts which attempt to observe the annihilation products of dark matter particles.

A WIMP striking a nucleus will induce a recoil of energy given by

$$E_{\text{recoil}} = \frac{|\vec{q}|^2}{2M_{\text{nucleus}}} = \frac{2\mu^2 v^2(1-\cos\theta)}{2M_{\text{nucleus}}} = \frac{m_X^2 M_{\text{nucleus}}\, v^2(1-\cos\theta)}{(m_X + M_{\text{nucleus}})^2},$$

$$(14.4.21)$$

where $\vec{q}$ is the WIMP's momentum, v is its velocity, and μ is the reduced mass. For $m_X \gg M_{\text{nucleus}}$ and a velocity of ~ 300 km/s, we expect typical recoil energies of $E_{\text{recoil}} \sim M_{\text{nucleus}}\, v^2 \sim 1\text{-}100$ keV.

WIMPs scatter with nuclei in a target at a rate given by

$$R \approx \int_{E_{\text{min}}}^{E_{\text{max}}} \int_{v_{\text{min}}}^{v_{\text{max}}} \frac{2\rho}{m_X} \frac{d\sigma}{d|\vec{q}|}\, v\, f(v)\, dv\, dE_{\text{recoil}}, \qquad (14.4.22)$$

where ρ is the dark matter density, σ is the WIMP-nuclei elastic scattering cross section, and $f(v)$ is the velocity distribution of WIMPs. The limits of integration are set by the galactic escape velocity, $v_{\text{max}} \approx 650$ km/s, and kinematically by $v_{\text{min}} = (E_{\text{recoil}} M_{\text{nucleus}}/2\mu^2)^{1/2}$. The minimum energy is set by the energy threshold of the detector, which is typically in the range of several keV to several tens of keV.

WIMPs can potentially scatter with nuclei through both spin-independent and spin-dependent interactions. The experimental sensitivity to spin-independent couplings benefits from coherent scattering, which leads to cross sections (and rates) proportional to the square of the atomic mass of the target nuclei. The cross sections for spin-dependent scattering, in contrast, are proportional to $J(J+1)$, where J is the spin of the target nucleus, and thus do not benefit from large target nuclei. As a result, the current experimental sensitivity to spin-dependent scattering is far below that of spin-independent interactions. For this reason, we consider first the case of spin-independent scattering of WIMPs with nuclei (we will return to spin-dependent scattering in Sec. 14.5.3).

The spin-independent WIMP-nucleus elastic scattering cross section is given by

$$\sigma \approx \frac{4m_{\chi^0}^2 m_{\text{nucleus}}^2}{\pi(m_{\chi^0} + m_{\text{nucleus}})^2}[Zf_p + (A - Z)f_n]^2, \qquad (14.4.23)$$

where Z and A are the atomic number and atomic mass of the nucleus. f_p and f_n are the WIMP's couplings to protons and neutrons, given by[24]

$$f_{p,n} = \sum_{q=u,d,s} f_{T_q}^{(p,n)} a_q \frac{m_{p,n}}{m_q} + \frac{2}{27} f_{TG}^{(p,n)} \sum_{q=c,b,t} a_q \frac{m_{p,n}}{m_q}, \qquad (14.4.24)$$

where a_q are the WIMP-quark couplings and $f_{T_u}^{(p)} \approx 0.020 \pm 0.004$, $f_{T_d}^{(p)} \approx 0.026 \pm 0.005$, $f_{T_s}^{(p)} \approx 0.118 \pm 0.062$, $f_{T_u}^{(n)} \approx 0.014 \pm 0.003$, $f_{T_d}^{(n)} \approx 0.036 \pm 0.008$, $f_{T_s}^{(n)} \approx 0.118 \pm 0.062$ are quantities measured in nuclear physics experiments.[52] The first term in Eq. (14.4.24) corresponds to interactions with the quarks in the target nuclei. The second term corresponds to interactions with the gluons in the target through a colored loop diagram. $f_{TG}^{(p)}$ is given by $1 - f_{T_u}^{(p)} - f_{T_d}^{(p)} - f_{T_s}^{(p)} \approx 0.84$, and analogously, $f_{TG}^{(n)} \approx 0.83$.

14.4.1. *Direct Detection of Neutralino Dark Matter*

Neutralinos can elastically scatter with quarks through either t-channel CP-even Higgs exchange, or s-channel squark exchange:

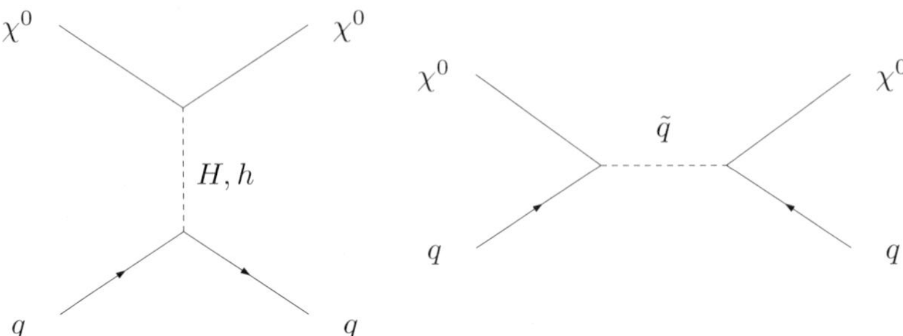

In addition to these diagrams, we can write analogous processes in which the WIMP couples to gluons in the target through a quark/squark loop. By calculating the WIMP-quark couplings, a_q, we can also implicitly include the interactions of neutralinos with gluons in the target nuclei as well (see Eq. (14.4.24)).

The neutralino-quark coupling, in which all of the supersymmetry model-dependent information is contained, is given by[53]

$$
a_q = -\frac{1}{2(m_{1i}^2 - m_\chi^2)} Re\left[(X_i)(Y_i)^*\right] - \frac{1}{2(m_{2i}^2 - m_\chi^2)} Re\left[(W_i)(V_i)^*\right]
$$

$$
- \frac{g_2 m_q}{4 m_W B}\left[Re\left(\delta_1[g_2 N_{12} - g_1 N_{11}]\right) DC \left(-\frac{1}{m_H^2} + \frac{1}{m_h^2}\right)\right.
$$

$$
\left. + Re\left(\delta_2[g_2 N_{12} - g_1 N_{11}]\right)\left(\frac{D^2}{m_h^2} + \frac{C^2}{m_H^2}\right)\right],
\tag{14.4.25}
$$

where

$$
X_i \equiv \eta_{11}^* \frac{g_2 m_q N_{1,5-i}^*}{2 m_W B} - \eta_{12}^* e_i g_1 N_{11}^*,
$$

$$
Y_i \equiv \eta_{11}^* \left(\frac{y_i}{2} g_1 N_{11} + g_2 T_{3i} N_{12}\right) + \eta_{12}^* \frac{g_2 m_q N_{1,5-i}}{2 m_W B},
$$

$$
W_i \equiv \eta_{21}^* \frac{g_2 m_q N_{1,5-i}^*}{2 m_W B} - \eta_{22}^* e_i g_1 N_{11}^*,
$$

$$
V_i \equiv \eta_{22}^* \frac{g_2 m_q N_{1,5-i}}{2 m_W B} + \eta_{21}^* \left(\frac{y_i}{2} g_1 N_{11}, + g_2 T_{3i} N_{12}\right),
\tag{14.4.26}
$$

where throughout $i = 1$ for up-type quarks and $i = 2$ for down type quarks. m_{1i}, m_{2i} denote the squark mass eigenvalues and η is the matrix which diagonalizes the squark mass matrices, $diag(m_1^2, m_2^2) = \eta M^2 \eta^{-1}$. y_i, T_{3i}

and e_i denote hypercharge, isospin and electric charge of the quarks. For scattering off of up-type quarks

$$\delta_1 = N_{13}, \quad \delta_2 = N_{14}, \quad B = \sin\beta, \quad C = \sin\alpha, \quad D = \cos\alpha, \quad (14.4.27)$$

whereas for down-type quarks

$$\delta_1 = N_{14}, \quad \delta_2 = -N_{13}, \quad B = \cos\beta, \quad C = \cos\alpha, \quad D = -\sin\alpha. (14.4.28)$$

The quantity α is the angle that diagonalizes the CP-even Higgs mass matrix.

The first two terms in Eq. (14.4.25) correspond to interactions through the exchange of a squark, while the final term is generated through Higgs exchange. To help develop some intuition for what size neutralino-nucleon cross sections we might expect, lets consider a few simple limits. First, consider the case in which the scattering is dominated by heavy Higgs (H) exchange through its couplings to strange and bottom quarks. This behavior is often found when the squarks are heavy and $\cos\alpha \approx 1$ (which implies moderate to large $\tan\beta$ and $m_A \sim m_H \sim m_{H^\pm}$). In this case, the leading contribution to the neutralino-nucleon cross section is

$$\sigma_{\chi N} \sim \frac{g_1^2 g_2^2 |N_{11}|^2 |N_{13}|^2 \, m_N^4}{4\pi m_W^2 \cos^2\beta \, m_H^4} \left(f_{T_s} + \frac{2}{27} f_{TG} \right)^2, \quad (m_{\tilde{q}} \text{ large}, \cos\alpha \approx 1).$$

$$(14.4.29)$$

Here, the cross section scales with m_H^{-4} and with $\tan^2\beta$, a leads to the possibility of very large rates. For a $\sim$100 GeV neutralino, for example, a 200 GeV heavy Higgs mass leads to cross sections with nucleons on the order of 10^{-5} to 10^{-7} pb for $|\mu| \sim 200$ GeV, or 10^{-7} to 10^{-9} pb for $|\mu| \sim 1$ TeV.

Second, we can consider the case in which the cross section is dominated by light Higgs boson (h) exchange through its couplings to up-type quarks. This is often found in the case of heavy squarks and heavy to moderate H. In this limiting case

$$\sigma_{\chi N} \sim \frac{g_1^2 g_2^2 |N_{11}|^2 |N_{14}|^2 \, m_N^4}{4\pi m_W^2 \, m_h^4} \left(f_{T_u} + \frac{4}{27} f_{TG} \right)^2, \quad (m_{\tilde{q}}, m_H \text{ large}, \cos\alpha \approx 1).$$

$$(14.4.30)$$

If the heavy Higgs (H) is heavier than about $\sim$500 GeV, exchange of the light Higgs generally dominates, leading to cross sections of around 10^{-8} to 10^{-10} pb for $|\mu|$ in the range of 200 GeV to 1 TeV.

Third, consider the case in which the elastic scattering cross section is dominated by the exchange of squarks through their couplings to strange and bottom quarks. This is found for large to moderate $\tan\beta$ and squarks with masses well below 1 TeV. In this limiting case, and with approximately diagonal squark mass matrices,

$$\sigma_{\chi N} \sim \frac{g_1^2 g_2^2 |N_{11}|^2 |N_{13}|^2 \, m_N^4}{4\pi m_W^2 \cos^2\beta \, m_{\tilde{q}}^4} \left(f_{T_s} + \frac{2}{27} f_{TG} \right)^2, \quad (\tilde{q} \text{ dominated}, \tan\beta \gg 1).$$

$$(14.4.31)$$

For squarks lighter than $\sim$1 TeV, squark exchange can potentially provide the dominant contribution to neutralino-nuclei elastic scattering.

Using Eq. (14.4.22), we can crudely estimate the minimum target mass required to potentially detect neutralino dark matter. A detector made up of Germanium targets (such as CDMS or Edelweiss, for example) would expect a WIMP with a nucleon-level cross section of 10^{-6} pb (10^{-42} cm^2) to yield approximately 1 elastic scattering event per kilogram-day of exposure. Such a target mass could thus be potentially sensitive to strongly mixed gaugino-higgsino neutralinos with light m_H and large $\tan\beta$. The strongest current limits on spin-independent scattering have been obtained using $\sim$10^2 kilogram-days of exposure. In contrast, reaching sensitivities near the 10^{-10} pb level will require ton-scale detectors capable of operating for weeks, months or longer with very low backgrounds.

In Fig. 14.5, we show the current constraints on the WIMP-nucleon spin-independent elastic scattering cross section, as a function of the WIMP's mass. The most stringent constraints currently come from the CDMS and XENON-10 collaborations, which each have obtained limits at around the 10^{-7} pb level. The CDMS and XENON-100 collaborations are each anticipated to place limits a factor of several times more stringent within the next year (unless a positive detection is made, that is). Although the more distant future is more difficult to project, it is generally expected that experiments approaching the ton-scale will reach sensitivies near 10^{-9} or 10^{-10} pb within the next few to several years.

14.4.2. *Direct Detection of Kaluza-Klein Dark Matter*

In the case of Kaluza-Klein dark matter (as described in Sec. 14.3.2), the WIMP-quark coupling a_q receives contributions from the s-channel exchange of KK quarks and the t-channel Higgs boson exchange.[38,54] This leads to a contribution to the cross section from Higgs exchange which is proportional to $1/(m_{B^{(1)}}^2 \, m_h^4)$ and a contribution from KK-quark exchange which is approximately proportional to $1/(m_{B^{(1)}}^6 \, \Delta^4)$, where $\Delta = (m_{q^{(1)}} - m_{B^{(1)}})/m_{B^{(1)}}$ is the fractional mass splitting of the KK quarks and

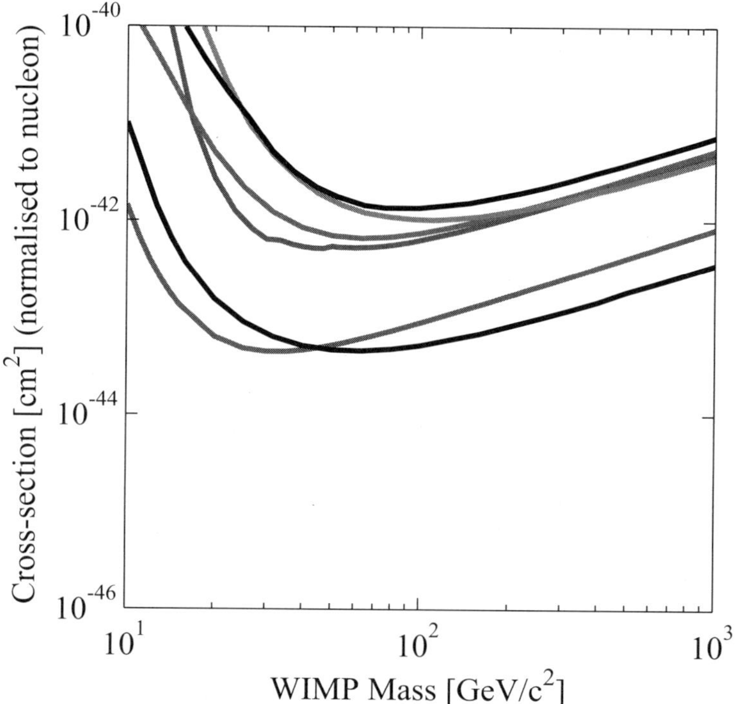

Fig. 14.5.   Current constraints on the WIMP-nucleon, spin-independent elastic scattering cross section. From bottom-to-top on the right side of the figure, the lines correspond to the limits from the CDMS,[42] XENON-10,[43] WARP,[50] CRESST,[46] ZEPLIN,[44] and Edelweiss[45] experiments. This figure was generated using the Dark Matter Limit Plotter.[55]

the $B^{(1)}$. The WIMP-quark coupling in this case is given by

$$a_q = \frac{m_q\, g_1^2\, (Y_{q_R}^2 + Y_{q_L}^2)\,(m_{B^{(1)}}^2 + m_{q^{(1)}}^2)}{4 m_{B^{(1)}}(m_{B^{(1)}}^2 - m_{q^{(1)}}^2)^2} + \frac{m_q\, g_1^2}{8 m_{B^{(1)}}\, m_h^2}. \quad (14.4.32)$$

Numerically, the $B^{(1)}$-nucleon cross section is approximately given by

$$\sigma_{B^{(1)}n,\mathrm{SI}} \approx 1.2 \times 10^{-10}\,\mathrm{pb} \left(\frac{1\,\mathrm{TeV}}{m_{B^{(1)}}}\right)^2 \left[\left(\frac{100\,\mathrm{GeV}}{m_h}\right)^2 + 0.09 \left(\frac{1\,\mathrm{TeV}}{m_{B^{(1)}}}\right)^2 \left(\frac{0.1}{\Delta}\right)^2\right]^2.$$

$$(14.4.33)$$

With such small cross sections, we will most likely have to wait for ton-scale detectors before this model will be tested by direct detection experiments.

14.4.3. *Some Model Independent Comments Regarding Direct Detection*

In the cases of the two dark matter candidates discussed thus far in this section, we are lead to expect a rather small elastic scattering cross sections between WIMPs and nuclei – typically below or well below current experimental constraints. This is *not* a universal prediction for a generic WIMP, however. In fact, both neutralinos and Kaluza-Klein dark matter represent somewhat special cases in which direct detection rates are found to be particularly low.

To illustrate this point, consider a Dirac fermion or a scalar WIMP which annihilates in the early universe to fermions with roughly equal couplings to each species – a heavy 4th generation neutrino or sneutrino, for example. We can take the Feynman diagram for the process of this WIMP annihilating to quarks and turn it on its side, and then calculate the resulting elastic scattering cross section. What we find is that, if the interaction is of scalar or vector form, such a WIMP will scatter with nuclei several orders of magnitude more often than is allowed by the limits of CDMS, XENON and other direct detection experiments. Similar conclusions are reached for many otherwise acceptable WIMP candidates.[56] A warning well worth keeping in mind for any WIMP model builder is, "Beware the crossing symmetry!".

So what is it about neutralinos or Kaluza-Klein dark matter than enable them to evade these constraints? In the case of neutralinos, the single most important feature is the suppression of its couplings to light fermions. Being a Majorana fermion, a neutralino's annihilation cross section to fermion pairs (at low velocity) scales with $\sigma v \propto m_f^2/m_{\chi^0}^2$. As a result, neutralinos annihilate preferentially to heavy fermions (top quarks, bottom quarks, and taus) or gauge/Higgs bosons. As heavy fermions (and gauge/Higgs bosons) are largely absent from nuclei, the potentially dangerous crossing symmetry does not apply. More generally speaking, current direct detection constraints can be fairly easily evaded for any WIMP which interacts with quarks through Higgs exchange, as the Yukawa couplings scale with the fermion's mass.

Alternatively, if the WIMP's couplings are simply very small, direct detection constraints can also be evaded. Small couplings, however, leave us in need of a mechanism for efficiently depleting the WIMP in the early universe. But even with very small couplings, a WIMP might efficiently coannihilate in the early universe, or annihilate through a resonance, leading to an acceptable relic abundance. In this way, coannihilations and resonances can considerably suppress the rates expected in direct detection experiments.

14.5. Indirect Detection

Direct detection experiments are not the only technique being pursued in the hope of identifying the particle nature of dark matter. Another major class of dark matter searches are those which attempt to detect the products of WIMP annihilations, including gamma rays, neutrinos, positrons, electrons, and antiprotons. These methods are collectively known as indirect detection and are the topic of this section.

14.5.1. *Gamma Rays From WIMP Annihilations*

Searches for the photons generated in dark matter annihilations have a key advantage over other indirect detection techniques in that these particles travel essentially unimpeded. In particular, unlike charged particles, gamma rays are not deflected by magnetic fields, and thus can potentially provide valuable angular information. For example, point-like sources of dark matter annihilation radiation might appear from high density regions such as the Galactic Center or dwarf spheroidal galaxies. Furthermore, over galactic distance scales, gamma rays are not attenuated, and thus retain their spectral information. In other words, the spectrum that is measured is the same as the spectrum generated in the dark matter annihilations.

The spectrum of photons produced in dark matter annihilations depends on the details of the WIMP being considered. Neutralinos, for example, typically annihilate to final states consisting of heavy fermions ($b\bar{b}$, $t\bar{t}$, $\tau^+\tau^-$) or gauge and/or Higgs bosons (ZZ, W^+W^-, HA, hA, ZH, Zh, ZA, $W^\pm H^\pm$, where H, h, A and $H^\pm$ are the Higgs bosons of the minimal supersymmetric standard model).[24] With the exception of the $\tau^+\tau^-$ channel, each of these annihilation modes result in a very similar spectrum of gamma rays. The gamma ray spectrum from a WIMP which annihilates to leptons can be quite different, however. This can be particularly important in the case of Kaluza-Klein dark matter in models with one universal extra dimension, for example, in which dark matter particles annihilate significantly to e^+e^-, $\mu^+\mu^-$ and $\tau^+\tau^-$.[37,38] In Fig. 14.6, we show the predicted gamma ray spectrum, per annihilation, for several possible WIMP annihilation modes.

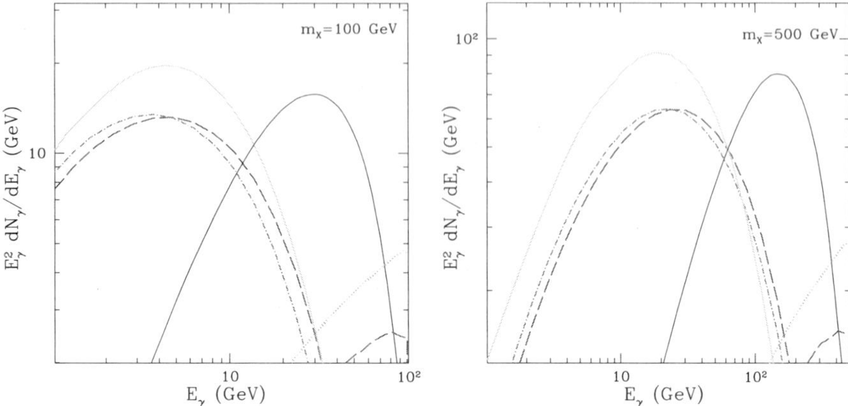

Fig. 14.6.   The gamma ray spectrum per WIMP annihilation for a 100 GeV (left) and 500 GeV (right) WIMP. Each curve denotes a different choice of the dominant annihilation mode: $b\bar{b}$ (solid cyan), ZZ (magenta dot-dashed), W^+W^- (blue dashed), $\tau^+\tau^-$ (black solid), e^+e^- (green dotted) and $\mu^+\mu^-$ (red dashed).

In addition to generating continuum gamma rays through the decays of quarks, leptons, Higgs bosons or gauge bosons, dark matter particles can produce gamma rays directly, leading to monoenergetic spectral signatures. If a gamma ray line could be identified, it would constitute a "smoking gun" for dark matter annihilations. By definition, however, WIMPs do not annihilate through tree level processes to final states containing photons (if they did, they would be EMIMPs rather than WIMPs). On the other hand, they may be able to produce final states such as $\gamma\gamma$, γZ or γh through loop diagrams. Neutralinos, for example, can annihilate directly to $\gamma\gamma$[57] or γZ[58] through a variety of charged loops. These final states lead to gamma ray lines with energies of $E_\gamma = m_{\rm dm}$ and $E_\gamma = m_{\rm dm}(1 - m_Z^2/4m_{\rm dm}^2)$, respectively. Such photons are produced in only a very small fraction of neutralino annihilations, however. The largest neutralino annihilation cross sections to $\gamma\gamma$ and γZ are about 10^{-28} cm^3/s, and even smaller values are more typical.[59]

The Galactic Center has long been considered to be one of the most promising regions of the sky in which to search for gamma rays from dark matter annihilations.[59,60] The prospects for this depend, however, on a number of factors including the nature of the WIMP, the distribution of dark matter in the region around the Galactic Center, and our ability to understand the astrophysical backgrounds present.

The gamma ray flux from dark matter annihilations is given by

$$\Phi_\gamma(E_\gamma, \psi) = \frac{1}{2}\langle\sigma_{XX}|v|\rangle \frac{dN_\gamma}{dE_\gamma} \frac{1}{4\pi m_X^2} \int_{\text{los}} \rho^2(r)dl(\psi)d\psi. \tag{14.5.34}$$

Here, $\langle\sigma_{XX}|v|\rangle$ is the WIMP's annihilation cross section (times relative velocity), ψ is the angle observed relative to the direction of the Galactic Center, $\rho(r)$ is the dark matter density as a function of distance to the Galactic Center, and the integral is performed over the line-of-sight. dN_γ/dE_γ is the gamma ray spectrum generated per WIMP annihilation. Averaging over a solid angle centered around a direction, ψ, we arrive at

$$\Phi_\gamma(E_\gamma) \approx 2.8 \times 10^{-12}\,\text{cm}^{-2}\,\text{s}^{-1}\,\frac{dN_\gamma}{dE_\gamma}\left(\frac{\langle\sigma_{XX}|v|\rangle}{3\times 10^{-26}\,\text{cm}^3/\text{s}}\right)\left(\frac{1\,\text{TeV}}{m_X}\right)^2$$
$$\times J(\Delta\Omega, \psi)\Delta\Omega, \tag{14.5.35}$$

where $\Delta\Omega$ is the solid angle observed. The quantity $J(\Delta\Omega, \psi)$ depends only on the dark matter distribution, and is the average over the observed solid angle of the quantity

$$J(\psi) = \frac{1}{8.5\,\text{kpc}}\left(\frac{1}{0.3\,\text{GeV}/\text{cm}^3}\right)^2 \int_{\text{los}} \rho^2(r(l, \psi))dl. \tag{14.5.36}$$

$J(\psi)$ is normalized such that a completely flat halo profile, with a density equal to the value at the solar circle, integrated along the line-of-sight to the Galactic Center would yield a value of one. In dark matter distributions favored by N-body simulations, however, this value is much larger. A commonly used parameterization of halo profiles is given by

$$\rho(r) = \frac{\rho_0}{(r/R)^\gamma[1 + (r/R)^\alpha]^{(\beta-\gamma)/\alpha}}, \tag{14.5.37}$$

where $R \sim 20$ kpc is the scale radius and ρ_0 is fixed by imposing that the dark matter density at the distance of the Sun from the Galactic Center is equal to the value inferred by observations (~ 0.3 GeV/cm^3). Among the most frequently used parameterizations is the Navarro-Frenk-White (NFW) profile, which is described by $\alpha = 1$, $\beta = 3$ and $\gamma = 1$.[61] When considering the region of the Galactic Center, the most important feature of the halo profile is the inner slope, γ. In some halo profiles, this slope can be considerably steeper than the value used in the NFW parameterization. For example, the Moore *et al.* profile is described by $\alpha = 1.5$, $\beta = 3$, $\gamma = 1.5$.[62] Note that for $\gamma \geq 1.5$, the integral in Eq. (14.5.36) diverges. To avoid this behavior, one must truncate the profile within very small galactic radii.

The Narvarro-Frenk-White and Moore *et al.* profiles lead to values of $J(\Delta\Omega = 10^{-5}\,\text{sr}, \psi = 0) \sim 10^5$ and $\sim 10^8$, respectively. And although these profiles serve as useful benchmarks, they certainly do not exhaust the range of possibilities. For a number of reasons, it is very difficult to predict the dark matter distribution in the highly important inner parsecs of the Galaxy. Firstly, the resolution of N-body simulations is limited to scales of approximately $\sim 10^2$ parsecs or so. Furthermore, the gravitational potential in the inner region of the Milky Way is dominated not by dark matter, but by baryons, whose effects are not generally included in such simulations. The precise impact of baryons on the dark matter distribution is difficult to predict, although an enhancement in the dark matter annihilation rate due to adiabatic compression is generally expected.[63] The adiabatic accretion of dark matter onto the central supermassive black hole may also lead to the formation of a density spike in the dark matter distribution. If present, such a spike would result in a very high dark matter annihilation rate.[64]

The recent observation of a bright, very high energy gamma ray source in the region of the Galactic Center by HESS and other ground based Atmospheric Cerenkov Telescopes[65] has made efforts to identify gamma rays from dark matter annihilations more difficult. This source appears to be coincident with the dynamical center of the Milky Way (Sgr A*) and has no detectable angular extension (less than 1.2 arcminutes). Its spectrum is well described by a power-law, $dN_\gamma/dE_\gamma \propto E_\gamma^{-\alpha}$, where $\alpha \approx 2.25$ over the range of 160 GeV to 20 TeV, and thus appears to be inconsistent with a dark matter interpretation. Although this gamma ray source represents a formidable background for experiments searching for dark matter annihilation radiation,[66] it may be possible to reduce the impact of this and other backgrounds by studying the angular distribution of gamma rays from this region of the sky.[67]

Telescopes potentially capable of detecting gamma rays from dark matter annihilations include the satellite based Fermi gamma ray space telescope (formerly known as GLAST), and a number ground based Atmospheric Cerenkov Telescopes, including HESS, MAGIC and VERITAS. These two classes of experiments play complementary roles in the search for dark matter. On one hand, Fermi will continuously observe a large fraction of the sky, but with an effective area far smaller than possessed by ground based telescopes. Ground based telescopes, in contrast, study the emission from a small angular field, but with far greater exposure. Furthermore, while ground based telescopes can only study gamma rays with energy greater than $\sim$100 GeV, Fermi will be able to directly study gamma

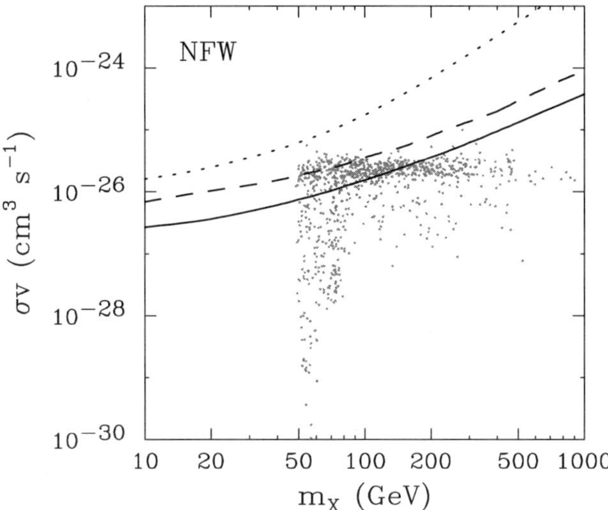

Fig. 14.7. The projected exclusion limits at the 95% confidence level from five years of observation with the Fermi gamma ray space telescope (formerly known as GLAST) on the WIMP annihilation cross section as a function of its mass, for the case of an NFW halo profile. The region above the dotted line is already excluded by EGRET.[68] The dashed and solid lines show the projections for Fermi for an assumed isotropic diffuse background and in the limiting case in which the astrophysical background has the same angular distribution as the dark matter signal, respectively. Also shown are points representing a random scan of supersymmetric models. Figure from Ref. [67].

rays with energies over the range of 100 MeV to 300 GeV.

In Fig. 14.7, the sensitivity of Fermi to dark matter annihilations in the Galactic Center region is shown for the case of an NFW halo profile and a WIMP annihilating to W^+W^-. For this halo profile, WIMPs with an annihilation cross sections of $\langle \sigma_{XX}|v| \rangle \sim 3 \times 10^{-26}$ cm^3/s are near the threshold for detection by Fermi. For WIMPs with approximately this cross section, halo profiles more cuspy than NFW are thus likely to be observable by Fermi, whereas less dense profiles are unlikely to lead to an identifiable signal.

If the dark matter density in the inner parsecs of the Milky Way is not particularly high, or if the astrophysical backgrounds turn out to be particularly foreboding, the prospects for identifying dark matter annihilation radiation from the Galactic Center may be quite unfavorable. In this case, regions of the sky away from the Galactic Center may be more advantageous for dark matter searches. In particular, dwarf spheroidal galaxies within and near the Milky Way provide an opportunity to search for dark matter annihilation radiation with considerably less contamination from astrophysical backgrounds. The flux of gamma rays from dark matter annihilations in such objects, however, is also expected to be lower than from a cusp in the center of the Milky Way.[69] As a result, planned experiments are likely to observe dark matter annihilation radiation from dwarf galaxies only in the most favorable range of particle physics models. Alternatively, Fermi may also be sensitive to dark matter annihilations taking place throughout the halo of the Milky Way, or throughout the cosmological distribution of dark matter.[70]

14.5.2. *Charged Cosmic Rays From WIMP Annihilations*

In addition to gamma rays, WIMP annihilations throughout the galactic halo are expected to create charged cosmic rays, including electrons, positrons, protons and antiprotons. Unlike gamma rays, which travel along straight lines, charged particles move under the influence of the Galactic Magnetic Field, diffusing and steadily losing energy, resulting in a diffuse spectrum at Earth. By studying the spectrum of these particles, it may be possible to identify signatures of dark matter annihilations. In fact, multiple experiments have recently announced results which have been interpreted as possible products of WIMPs.

The PAMELA experiment, which began its three-year satellite mission in June of 2006, recently reported an anomalous rise in the cosmic ray positron fraction (the positron to positron-plus-electron ratio) above 10 GeV,[71] confirming earlier indications from HEAT[72] and AMS-01.[73] Additionally, the ATIC balloon experiment has recently published data revealing a feature in the cosmic ray electron (plus positron) spectrum between approximately 300 and 800 GeV, peaking at around 600 GeV.[74] These measurements are each shown in Fig. 14.8, compared to the standard astrophysical predictions. These observations suggest the presence of a relatively local (within $\sim$1 kpc) source or sources of energetic cosmic ray electrons and positrons. Furthermore, in addition to the observations of PAMELA and ATIC, the WMAP experiment has revealed an excess of microwave emission from the central region of the Milky Way which has been interpreted as synchrotron emission from a population of electrons/positrons with a hard spectral index.[75,76] Taken together, these observations suggest that energetic electrons and positrons are surprisingly ubiquitous throughout our galaxy.

Although the origin of these electrons and positrons is not currently

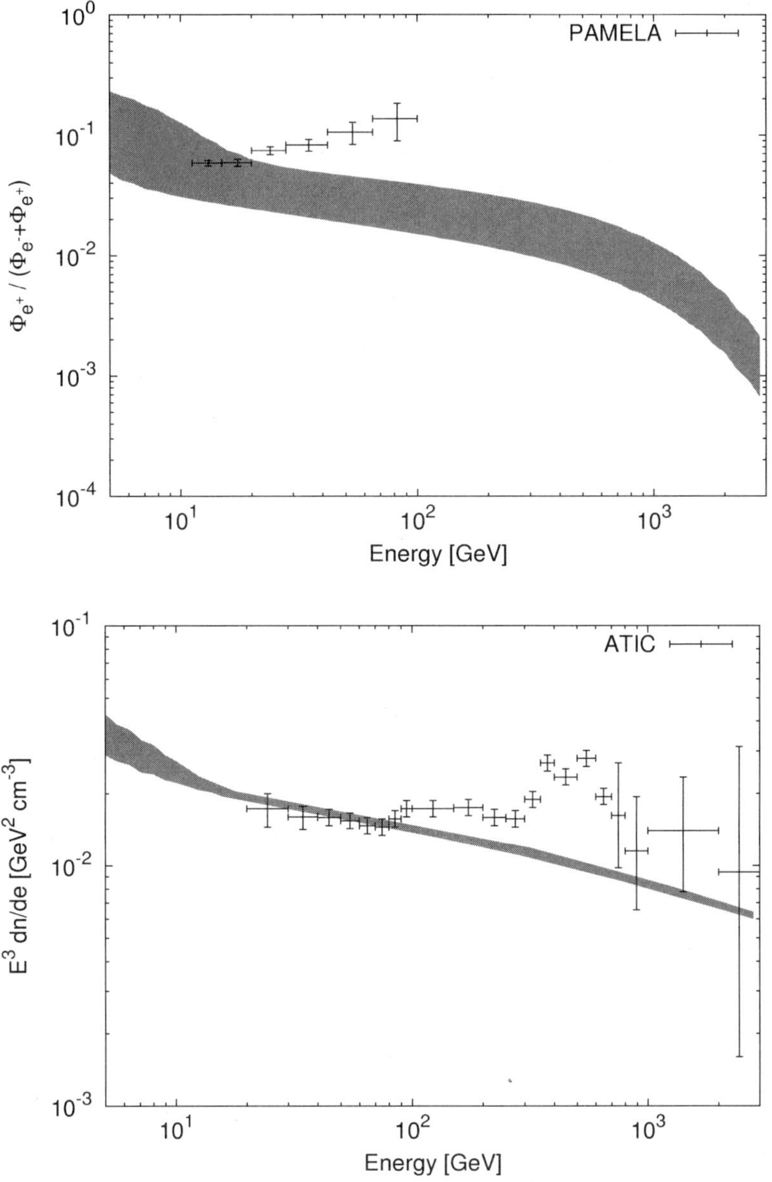

Fig. 14.8. The positron fraction above 10 GeV as measured by the PAMELA experiment (top)[71] and the electron-plus-positron spectrum as measured by ATIC (bottom).[74] In each frame, the range of standard astrophysical expectations is shown for comparison. These observations clearly require an additional source of energetic electrons and positrons. Figures generated by Melanie Simet.

known, interpretations of the observations have focused two possibilities: emission from pulsars,[77] and dark matter annihilations.[78,79] In order for dark matter annihilations throughout in the Milky Way halo to produce a spectrum with a shape similar to that observed by PAMELA and ATIC, however, a large fraction of the annihilations must proceed to electron-positron pairs, or possibly to $\mu^+\mu^-$ or $\tau^+\tau^-$.[78] Furthermore, WIMPs annihilating to other final states typically exceed the observed flux of cosmic ray antiprotons if normalized to generate the PAMELA and ATIC signals.[80]

Dark matter particles which annihilate directly to e^+e^- are predicted to generate a distinctive feature in the cosmic ray electron spectrum: an edge that drops off suddenly at $E_e = m_X$. In contrast, pulsars and other astrophysical sources of cosmic ray electrons are expected to produce spectra which fall off more gradually. Although the current data from ATIC is not detailed enough to discriminate between a feature with a sudden edge (dark matter-like) or graduate cutoff (pulsar-like), such a discrimination could become possible if the electron spectrum were measured with greater precision. Interestingly, such a measurement should be possible for ground based gamma ray telescopes such as HESS or VERITAS.[81]

Once electrons and positrons are injected into the local halo through dark matter annihilations (or from pulsars), they propagate under the influence of the Galactic Magnetic Field, gradually losing energy through synchrotron emission and through inverse Compton scattering with radiation fields. At energies of a few GeV and higher, the resulting spectrum at Earth can be calculated by solving the diffusion-loss equation:[82]

$$\frac{\partial}{\partial t}\frac{dn_e}{dE_e} = \vec{\nabla} \cdot \left[K(E_e,\vec{x})\vec{\nabla}\frac{dn_e}{dE_e}\right] + \frac{\partial}{\partial E_e}\left[b(E_e,\vec{x})\frac{dn_e}{dE_e}\right] + Q(E_e,\vec{x}),$$

$$(14.5.38)$$

where dn_e/dE_e is the number density of positrons/electrons per unit energy, $K(E_e,\vec{x})$ is the diffusion constant, $b(E_e,\vec{x})$ is the rate of energy loss, and $Q(E_e,\vec{x})$ is the source term, which contains all of the information regarding the dark matter annihilation modes, cross section, and spatial distribution. As we expect the cosmic ray distribution from dark matter annihilations to be in or near the steady state limit, the left hand side of the equation is generally set to zero. The energy dependance of the diffusion constant is typically parameterized by $K(E_e) = K_0 E_e^\alpha$.

In the relativistic limit, the energy loss rate resulting from inverse Compton scattering and synchrotron emission is given by

$$b(E_e) = \frac{4}{3}\sigma_T \rho_{\rm rad}\left(\frac{E_e}{m_e}\right)^2 + \frac{4}{3}\sigma_T \rho_{\rm mag}\left(\frac{E_e}{m_e}\right)^2$$

$$\approx 1.02 \times 10^{-16}\,{\rm GeV/s}\left(\frac{\rho_{\rm rad} + \rho_{\rm mag}}{\rm eV/cm^3}\right) \times \left(\frac{E_e}{\rm GeV}\right)^2, \quad (14.5.39)$$

where σ_T is the Thompson cross section and $\rho_{\rm rad}$ and $\rho_{\rm mag}$ are the energy densities of radiation (including starlight, cosmic microwave background and emission from dust) and the galactic magnetic field, respectively.

To find a solution to the diffusion-loss equation, a set of boundary conditions must also be adopted. For example, under the assumption of cylindrical symmetry, the distance from the galactic plane at which electrons and positrons freely escape, L, must be selected.

The diffusion parameters K_0, α and L can be constrained by studying the spectra of various species of cosmic ray nuclei. In particular, by studying the ratios of cosmic ray secondaries-to-primaries (most importantly boron-to-carbon) and unstable primaries-to-stable secondaries (Be^{10}-to-Be^9), information can be inferred regarding the size of the diffusion zone and the length of time that cosmic rays are confined to the galaxy. Current cosmic ray data favor values of approximately $K_0 \sim 10^{28}$ cm^2/s, $\alpha \approx 0.4 - 0.5$ and $L \sim 1 - 10$ kpc. For an excellent review of this subject, see Ref. [83].

The spectral shape of the source term, Q, depends on the leading annihilation modes of the WIMP in the low velocity limit. Neutralinos and other WIMPs which annihilate primarily to combinations of heavy fermions and Higgs or gauge bosons generally produce somewhat soft spectra. In contrast, Kaluza-Klein dark matter and other WIMPs which annihilate significantly to electrons and muons are predicted to generate a considerably harder spectrum.[84]

The normalization of the source term is determined by the annihilation cross section of the WIMP and the spatial distribution of the dark matter in the galactic halo. In particular, small scale inhomogeneities in the dark matter distribution can boost the average annihilation rate. The factor by which inhomogeneities enhance the resulting cosmic ray flux is called the "boost factor", and is defined as

$$\text{boost factor} = \int_V \frac{\langle \rho^2 \rangle}{\langle \rho \rangle^2} \frac{dV}{V}, \quad (14.5.40)$$

where the integral is performed over the volume that contributes to the observed spectrum. The results of N-body simulations lead us to expect that this quantity could be as large as 5 to 10. Although considerably larger boost factors are not impossible, they would be somewhat surprising.

In order to generate the signals observed by PAMELA and ATIC with a WIMP annihilating with an annihilation cross section of $\sigma v \approx 3 \times 10^{-26}$ cm^3/s (as required to thermally produce the observed dark matter abundance with a fixed value of σv) an annihilation rate many hundred times larger than is naively expected would be required. This could potentially be accommodated in several ways. Firstly, the boost factor may be larger than expected. Secondly, the dark matter particles may have been generated through a non-thermal mechanism, allowing for considerably larger annihilation cross sections. Thirdly, WIMPs interacting through the exchange of very light particles can annihilate through non-perturbative processes such that $\sigma v \propto v^{-1}$. In this case, the annihilation cross section in the present galaxy (with $v \sim 10^{-3} c$) would be much larger than the corresponding cross section at the time of thermal freeze-out (when $v \sim 10^{-1} c$).[85]

14.5.3. *Neutrinos From WIMP Annihilations in the Sun*

As the Solar System moves through the halo of the Milky Way, WIMPs become swept up by the Sun. Although dark matter particles interact only weakly, they occasionally scatter elastically with nuclei in the Sun and lose enough momentum to become gravitationally bound. Over the lifetime of the Sun, a sufficient density of WIMPs can accumulate in its center so that an equilibrium is established between their capture and annihilation rates. The annihilation products of these WIMPs include neutrinos, which escape the Sun with minimal absorption, and thus potentially constitute an indirect signature of dark matter. Such neutrinos can be generated through the decays of heavy quarks, gauge bosons, and other products of WIMP annihilation, and then proceed to travel to Earth where can be efficiently identified using large volume neutrino detectors.

Beginning with a simple estimate, we expect WIMPs to be captured in the Sun at a rate approximately given by:

$$C^{\odot} \sim \phi_X (M_{\odot}/m_p)\, \sigma_{Xp}, \qquad (14.5.41)$$

where ϕ_X is the flux of WIMPs in the Solar System, $M_{\odot}$ is the mass of the Sun, and σ_{Xp} is the WIMP-proton elastic scattering cross section. Reasonable estimates of the local distribution of WIMPs leads to a capture rate of $C^{\odot} \sim 10^{20}\,\text{sec}^{-1} \times (100\,\text{GeV}/m_X)\,(\sigma_{Xp}/10^{-6}\,\text{pb})$. This neglects, however, a number of potentially important factors, including the gravitational focusing of the WIMP flux toward the Sun, and the fact that not every scattered WIMP will ultimately be captured. Taking these and other effects into account leads us to a solar capture rate of:[86]

$$C^{\odot} \approx 1.3 \times 10^{21} \, \text{sec}^{-1} \left(\frac{\rho_{\text{local}}}{0.3 \, \text{GeV/cm}^3} \right) \left(\frac{270 \, \text{km/s}}{\bar{v}_{\text{local}}} \right)$$

$$\times \left(\frac{100 \, \text{GeV}}{m_X} \right) \sum_i \left(\frac{A_i \left(\sigma_{Xi,\text{SD}} + \sigma_{Xi,\text{SI}} \right) S(\frac{m_X}{m_i})}{10^{-6} \, \text{pb}} \right), \quad (14.5.42)$$

where ρ_{local} is the local dark matter density and $\bar{v}_{\text{local}}$ is the local rms velocity of halo dark matter particles. $\sigma_{Xi,\text{SD}}$ and $\sigma_{Xi,\text{SI}}$ are the spin-dependent and spin-independent elastic scattering cross sections of the WIMP with nuclei species i, and A_i is a factor denoting the relative abundance and form factor for each species. In the case of the Sun, $A_{\text{H}} \approx 1.0$, $A_{\text{He}} \approx 0.07$, and $A_{\text{O}} \approx 0.0005$. The quantity S contains dynamical information and is given by

$$S(x) = \left[\frac{F(x)^{3/2}}{1 + F(x)^{3/2}} \right]^{2/3} \quad (14.5.43)$$

where

$$F(x) = \frac{3}{2} \frac{x}{(x-1)^2} \left(\frac{v_{\text{esc}}}{\bar{v}_{\text{local}}} \right)^2, \quad (14.5.44)$$

and $v_{\text{esc}} \approx 1156 \, \text{km/s}$ is the escape velocity of the Sun. Notice that for WIMPs much heavier than the target nuclei $S \propto 1/m_X$, leading the capture rate to be suppressed by two factors of the WIMP mass. In this case ($m_X \gtrsim 30$ GeV), we can write the capture rate as:

$$C^{\odot} \approx 3.35 \times 10^{20} \, \text{sec}^{-1} \left(\frac{\rho_{\text{local}}}{0.3 \, \text{GeV/cm}^3} \right) \left(\frac{270 \, \text{km/s}}{\bar{v}_{\text{local}}} \right)^3 \left(\frac{100 \, \text{GeV}}{m_X} \right)^2$$

$$\times \left(\frac{\sigma_{X\text{H,SD}} + \sigma_{X\text{H,SI}} + 0.07 \, \sigma_{X\text{He,SI}} + 0.0005 \, S(\frac{m_X}{m_O}) \sigma_{X\text{O,SI}}}{10^{-6} \, \text{pb}} \right). \, (14.5.45)$$

If the capture rate and annihilation cross sections are sufficiently large, equilibrium will be reached between these processes. For $N(t)$ WIMPs in the Sun, the rate of change of this quantity is given by

$$\dot{N}(t) = C^{\odot} - A^{\odot} N(t)^2 - E^{\odot} N, \quad (14.5.46)$$

where $C^{\odot}$ is the capture rate described above and $A^{\odot}$ is the annihilation cross section times the relative WIMP velocity per volume. $E^{\odot}$ is the inverse time for a WIMP to escape the Sun via evaporation. Evaporation is highly suppressed for WIMPs heavier than a few GeV.[87,88] $A^{\odot}$ can be approximated by

$$A^{\odot} = \frac{\langle \sigma v \rangle}{V_{\text{eff}}}, \qquad (14.5.47)$$

where V_{eff} is the effective volume of the core of the Sun determined roughly by matching the core temperature with the gravitational potential energy of a single WIMP at the core radius and is given by[87,88]

$$V_{\text{eff}} = 5.7 \times 10^{27} \, \text{cm}^3 \left(\frac{100 \, \text{GeV}}{m_X} \right)^{3/2}. \qquad (14.5.48)$$

Neglecting evaporation, the present WIMP annihilation rate is given by

$$\Gamma = \frac{1}{2} A^{\odot} N(t_{\odot})^2 = \frac{1}{2} C^{\odot} \tanh^2 \left(\sqrt{C^{\odot} A^{\odot}} \, t_{\odot} \right), \qquad (14.5.49)$$

where $t_{\odot} \approx 4.5$ billion years is the age of the solar system. The annihilation rate is maximized when it reaches equilibrium with the capture rate. This occurs when

$$\sqrt{C^{\odot} A^{\odot}} t_{\odot} \gg 1. \qquad (14.5.50)$$

If this condition is met, the final annihilation rate is determined entirely by the capture rate and has no further dependence on the dark matter particle's annihilation cross section.

Through their annihilations, WIMPs can generate neutrinos through a variety of channels. Annihilations to heavy quarks, tau leptons, gauge bosons and/or Higgs bosons can each generate energetic neutrinos in their subsequent decays.[89] In some models, WIMPs can also annihilate directly to neutrino-antineutrino pairs. Annihilations to light quarks or muons, however, do not contribute to the high energy neutrino spectrum, as these particles come to rest in the solar medium before decaying.

Neglecting the effects of oscillations and interactions with the solar medium, the spectrum of neutrinos from WIMP annihilations to a final state, $Y\bar{Y}$, is given by:

$$\frac{dN_\nu}{dE_\nu} = \frac{1}{2} \int_{E_\nu/\gamma(1+\beta)}^{E_\nu/\gamma(1-\beta)} \frac{1}{\gamma\beta} \frac{dE'}{E'} \left(\frac{dN_\nu}{dE_\nu} \right)_{Y\bar{Y}}^{\text{rest}}, \qquad (14.5.51)$$

where $\gamma = m_X/m_Y$, $\beta = \sqrt{1 - \gamma^{-2}}$, and $(dN_\nu/dE_\nu)_{Y\bar{Y}}^{\text{rest}}$ is the spectrum of neutrinos produced in the decay of an Y at rest.

Gauge bosons produced in WIMP annihilations produce the most energetic neutrinos through their decays, $W \to l\nu$, $Z \to \nu\bar{\nu}$, but also produce neutrinos through the subsequent decays of muons, taus and other particles. Tau leptons produce neutrinos through a variety of channels, including through the semi-leptonic decays $\tau \to \mu\nu\nu$, $e\nu\nu$, and the hadronic decays $\tau \to \pi\nu$, $K\nu$, $\pi\pi\nu$, and $\pi\pi\pi\nu$. Top quarks decay to a $W^{\pm}$ and a bottom quark essentially 100% of the time, each of which can generate neutrinos in their subsequent decay. For bottom and charm quarks, only the semileptonic decays contribute to the neutrino spectrum (with the exception of the neutrinos resulting from taus and c-quarks produced in decays of b-quarks). In the case of b and c quark decays, the process of hadronization reduces the fraction of energy that is transferred to the resulting neutrinos and other decay products.

Once produced in the Sun's core, neutrinos then propagate through the solar medium and to the Earth. Over this journey, they can potentially be absorbed, lose energy, and/or change flavor. In particular, charged current interactions of electron and muon neutrinos in the Sun lead to their absorption. The probability of absorption taking place can be estimated by $1 - \exp(-E_\nu/E_{\mathrm{abs}})$, where E_{abs} is approximately 130 GeV for electron or muon neutrinos and 200 GeV for electron or muon antineutrinos. Absorption, therefore, only plays an important role for relatively heavy WIMPs.

The effect of charged current interactions on tau neutrinos in the Sun is somewhat more complicated. The tau leptons produced in such interactions quickly decay and thus regenerate the absorbed tau neutrino, albeit with a reduced energy. Neutral current interactions of all three neutrino flavors similarly reduce the neutrinos' energy without depleting their number.

Vacuum oscillations lead to the full mixing of muon and tau neutrinos over their propagation to the Earth, making the observed muon neutrino spectrum (which is the flux relevant for detection in neutrino telescopes) effectively the average of the muon and tau flavors prior to mixing. Electron neutrinos can also oscillate into muon flavor through matter effects in the Sun (the MSW effect). Electron antineutrinos can generally be neglected, as their oscillations to muon or tau flavors are highly suppressed.[90]

Program such as DarkSUSY,[29] which include effects such as hadronization, absorption, regeneration, and oscillations, are very useful in making detailed predictions for the neutrino spectrum resulting from WIMP annihilations in the Sun.

Once they reach Earth, neutrinos can potentially be detected in large volume neutrino telescopes. Neglecting oscillations and solar absorption, the muon neutrino spectrum at the Earth resulting from WIMP annihilations in the Sun is given by

$$\frac{dN_{\nu_\mu}}{dE_{\nu_\mu}} = \frac{C_\odot F_{\mathrm{Eq}}}{4\pi D_{\mathrm{ES}}^2} \left(\frac{dN_{\nu_\mu}}{dE_{\nu_\mu}}\right)^{\mathrm{Inj}}, \tag{14.5.52}$$

where $C_\odot$ is the WIMP capture rate in the Sun, F_{Eq} is the non-equilibrium suppression factor ($F_{\mathrm{Eq}} = 1$ for capture-annihilation equilibrium), D_{ES} is the Earth-Sun distance and $(\frac{dN_{\nu_\mu}}{dE_{\nu_\mu}})^{\mathrm{Inj}}$ is the muon neutrino spectrum from the Sun per WIMP annihilating.

Muon neutrinos can produce muons through charged current interactions with ice or water nuclei inside or near the detector volume of a high energy neutrino telescope. The rate of neutrino-induced muons observed in a high energy neutrino telescope is estimated by

$$N_{\mathrm{events}} \approx \int \int \frac{dN_{\nu_\mu}}{dE_{\nu_\mu}} \frac{d\sigma_\nu}{dy}(E_{\nu_\mu}, y)\left[R_\mu(E_\mu) + L\right] A_{\mathrm{eff}}\, dE_{\nu_\mu}\, dy$$

$$+ \int \int \frac{dN_{\bar{\nu}_\mu}}{dE_{\bar{\nu}_\mu}} \frac{d\sigma_{\bar{\nu}}}{dy}(E_{\bar{\nu}_\mu}, y)\left[R_\mu(E_\mu) + L\right] A_{\mathrm{eff}}\, dE_{\bar{\nu}_\mu}\, dy, \tag{14.5.53}$$

where σ_ν ($\sigma_{\bar{\nu}}$) is the neutrino-nucleon (antineutrino-nucleon) charged current interaction cross section, $(1-y)$ is the fraction of neutrino/antineutrino energy which goes into the muon, A_{eff} is the effective area of the detector, $R_\mu(E_\mu)$ is the distance a muon of energy, $(1-y)\,E_\nu$, travels before falling below the energy threshold of the experiment (ranging from approximately 1 to 100 GeV), called the muon range, and L is the depth of the detector volume. The muon range in water/ice is approximately given by

$$R_\mu(E_\mu) \approx 2.4\,\mathrm{km}\, \times\, \ln\left[\frac{2.0 + 0.0042\, E_\mu(\mathrm{GeV})}{2.0 + 0.0042\, E_\mu^{\mathrm{thr}}(\mathrm{GeV})}\right], \tag{14.5.54}$$

where , E_μ^{thr} is the threshold of the experiment.

When completed, the IceCube experiment will possess a full square kilometer of effective area and kilometer depth, and will be sensitive to muons above approximately 50 GeV.[91] The Deep Core extension of Icecube will be sensitive down to 10 GeV. The Super-Kamiokande detector, in contrast, has 10^{-3} times the effective area of IceCube and a depth of only 36.2 meters.[92] For low mass WIMPs, however, Super-Kamiokande benefits over large volume detector such as IceCube by being sensitive to muons with as little energy as ~ 1 GeV.

The spectrum and flux of neutrinos generated in WIMP annihilations depends on the annihilation modes which dominate, and thus are model dependent. As long as the majority of annihilations proceed to final states

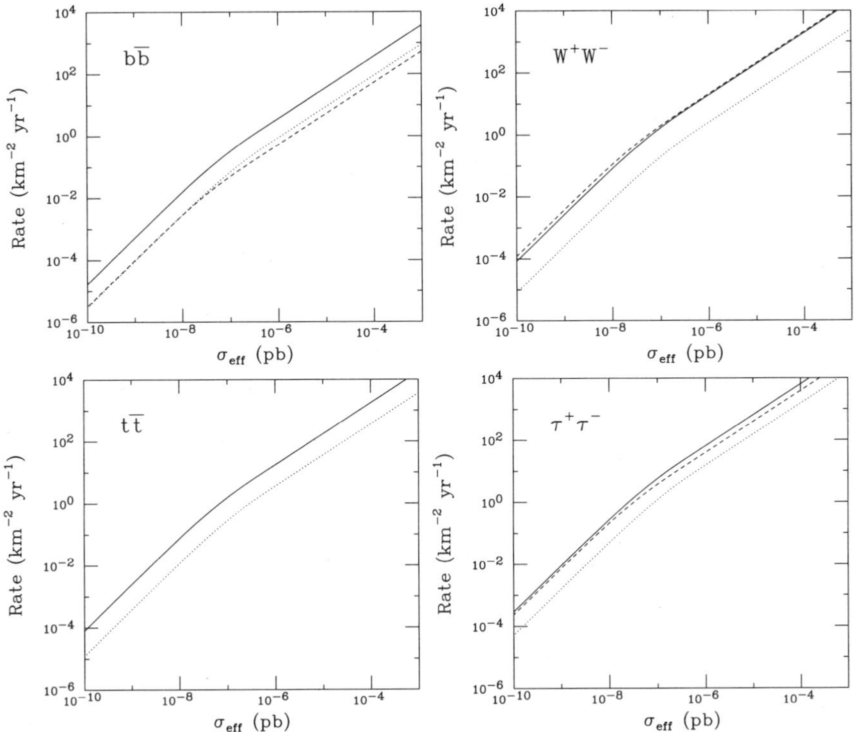

Fig. 14.9.   The event rate in a kilometer-scale neutrino telescope such as IceCube as a function of the WIMP's effective elastic scattering cross section in the Sun for a variety of annihilation modes. The effective elastic scattering cross section is defined as $\sigma_{\rm eff} = \sigma_{\rm XH,SD} + \sigma_{\rm XH,SI} + 0.07\,\sigma_{\rm XHe,SI} + 0.0005\,S(m_X/m_0)\,\sigma_{\rm X0,SI}$ (see Eq. (14.5.45)). The dashed, solid and dotted lines correspond to WIMPs of mass 100, 300 and 1000 GeV, respectively. A 50 GeV muon energy threshold and an annihilation cross section of 3×10^{-26} cm^3 s^{-1} have been adopted.

such as $b\bar{b}$, $t\bar{t}$, $\tau^+\tau^-$, W^+W^-, ZZ, or some combination of Higgs and gauge bosons, however, the variation between different final states is not dramatic. In Fig. 14.9, we plot the event rate in a kilometer-scale neutrino telescope such as IceCube as a function of the WIMP's effective elastic scattering cross section for four possible annihilation modes. The effective elastic scattering cross section used here is defined as $\sigma_{\rm eff} = \sigma_{\rm XH,SD} + \sigma_{\rm XH,SI} + 0.07\,\sigma_{\rm XHe,SI} + 0.0005\,S(m_X/m_0)\,\sigma_{\rm X0,SI}$ (see Eq. (14.5.45)).

The elastic scattering cross section of a WIMP is constrained by the absence of a positive signal in direct detection experiments. As described in Sec. 14.4, the strongest limits on the WIMP-nucleon spin-independent elastic scattering cross section have been made by the CDMS[42] and XENON[43] experiments. These results exclude spin-independent cross sections larger than approximately 5×10^{-8} pb for a 25-100 GeV WIMP or 2×10^{-7} pb $\times (m_X/500$ GeV) for a heavier WIMP.

With these results in mind, consider as an example a 300 GeV WIMP with an elastic scattering cross section with nucleons which is largely spin-independent. With a cross section near the CDMS bound, say 1×10^{-7} pb, we can determine from Fig. 14.9 the corresponding rates in a kilometer-scale neutrino telescope, such as IceCube. Sadly, we find that this cross section yields less than 1 event per year for annihilations to $b\bar{b}$, about 2 events per year for annihilations to W^+W^- or $t\bar{t}$ and about 8 per year for annihilations to $\tau^+\tau^-$, none of which are likely to be sufficient to be identified over the background of atmospheric neutrinos by IceCube or other planned experiments. Clearly, WIMPs that scatter with nucleons mostly through spin-independent interactions are not likely to be detected with IceCube or other planned neutrino telescopes.

The same conclusion is not reached for the case of spin-dependent scattering, however. The strongest bounds on the WIMP-proton spin-dependent cross section have been placed by the COUPP[49] and KIMS[51] collaborations. These constraints are approximately 7 orders of magnitude less stringent than those corresponding to spin-independent couplings, how-ever. As a result, a WIMP with a largely spin-dependent scattering cross section with protons may be capable of generating large event rates in high energy neutrino telescopes. Again considering a 300 GeV WIMP with a cross section near the experimental limit, Fig. 14.9 suggests that rates as high as $\sim 10^6$ per year could be generated if purely spin-dependent scattering contributes to the capture rate of WIMPs in the Sun.

Spin-dependent, axial-vector, scattering of neutralinos with the quarks or gluons within a nucleon is made possible through the t-channel exchange of a Z-boson, or the s-channel exchange of a squark. Although the cross sections for these processes can vary dramatically depending on the details of the supersymmetric model under consideration, it is often the case that the neutralino's spin-dependent cross section is considerably larger than its corresponding spin-independent interaction. In particular, very large spin-dependent cross sections ($\sigma_{\rm SD} \gtrsim 10^{-3}$pb) are possible even in models with very small spin-independent scattering rates. Such a model would go easily undetected in all planned direct detection experiments, while still generating on the order of ~ 1000 events per year at IceCube.

In Fig. 14.10, we demonstrate this by plotting the rate in a kilometer-

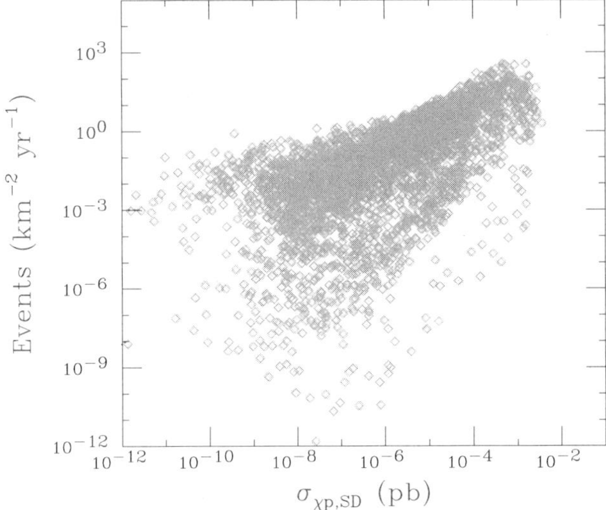

Fig. 14.10.   The rate of events at a kilometer-scale neutrino telescope such as IceCube from neutralino annihilations in the Sun, as a function of the neutralino's spin-dependent elastic scattering cross section with protons. Each point shown is beyond the reach of current direct detection experiments.

scale neutrino telescope from WIMP annihilation in the Sun verses the WIMP's spin-dependent cross section with protons. In this figure, each point shown is beyond the reach of present direct detection experiments. We thus conclude that neutralinos may be observable by IceCube while remaining unobserved by current and near future direct detection experiments.

A neutralino which has a large spin-dependent cross section generally has a sizable coupling to the Z, and thus has a large higgsino component. In particular, the spin-dependent scattering cross section through the exchange of a Z is proportional to the square of the quantity $|N_{13}|^2 - |N_{14}|^2$. As a result, neutralinos with a few percent higgsino fraction or more are likely to be within the reach of IceCube.[93,94] This makes the focus point region of supersymmetric parameter space especially promising. In this region, the lightest neutralino is typically a strong mixture of bino and higgsino components, often leading to the prediction of hundreds of events per year at IceCube.

Considering Kaluza-Klein dark matter, the range of elastic scattering cross sections predicted are quite challenging to reach with direct detection experiments, but are more favorable for detection using neutrino telescopes. The spin-independent $B^{(1)}$-nucleon cross section, which is generated through the exchange of KK quarks and the Higgs boson, is rather small and well beyond the sensitivity of current or upcoming direct detection experiments. The spin-dependent scattering cross section for the $B^{(1)}$ with a proton, however, is considerably larger and is given by[54]

$$\sigma_{H,SD} = \frac{g_1^4 \, m_p^2}{648\pi m_{B^{(1)}}^4 r_q^2} (4S_u^p + S_d^p + S_s^p)^2$$

$$\approx 4.4 \times 10^{-6}\, \text{pb} \left(\frac{800\,\text{GeV}}{\text{m}_{B^{(1)}}}\right)^4 \left(\frac{0.1}{\Delta_q}\right)^2, \qquad (14.5.55)$$

where $\Delta_q \equiv (m_{q^{(1)}} - m_{B^{(1)}})/m_{B^{(1)}}$ is fractional shift of the KK quark masses over the $B^{(1)}$ mass, which is expected to be on the order of 10%. The S's parameterize the fraction of spin carried by each variety of quark within the proton.

In addition to this somewhat large spin-dependent scattering cross section, the annihilation products of the $B^{(1)}$ are very favorable for the purposes of generating observable neutrinos. In sharp contrast to neutralinos, approximately 60% of $B^{(1)}$ annihilations generate a pair of charged leptons (20% to each type). Although most the remaining 40% of annihilations produce up-type quarks, about 4% generate neutrino pairs directly. The neutrino and tau lepton final states each contribute substantially to the event rate in a neutrino telescope.

Taken together, this leads to the prediction of a fairly high neutrino event rate from Kaluza-Klein dark matter annihilating in the Sun. In particular, for masses in the 500-1000 GeV range and a 10% mass degeneracy with the Kaluza-Klein quarks, we expect $\sim$10-1000 events per year in a kilometer-scale neutrino telescope such as IceCube.[95]

Currently, the strongest constraints on the neutrino flux from WIMPs annihilating in the Sun have been placed by the IceCube (for $m_X \gtrsim 200$ GeV) and Super-Kamiokande ($m_X \lesssim 200$ GeV) collaborations. The current IceCube limit constrains the neutrino-induced muon rate from dark matter to be less than $\sim$400-500 per square kilometer per year. Ultimately, IceCube is expected to reach a sensitivity about an order of magnitude below this current level.

14.6. Signals, Hints, and... Otherwise

Over the past several years, a number of observations have been interpreted as possible products of dark matter annihilations. In this section, I take the opportunity to summarize and discuss some of these observations.

14.6.1. *The PAMELA and ATIC Excesses*

As I have already discussed the recent observations of the ATIC[74] and PAMELA[71] experiments in Sec. 14.5.2, in this section I will only briefly summarize some of the arguments in favor of and against these signals being likely products of dark matter annihilations.

Although the observations of ATIC and PAMELA support a very compelling case that a powerful source of energetic positrons and electrons is present within $\sim$1 kpc of the Solar System, the nature of this source or sources is not yet clear. In order to produce the PAMELA and ATIC signals through the annihilations of dark matter distributed throughout the galactic halo, the WIMPs must annihilate dominantly to charged leptons (to produce a sufficiently hard spectrum, and to avoid the overproduction of cosmic ray antiprotons).[78–80] Furthermore, a very large annihilation rate is also required – hundreds or thousands of times higher than is naively expected for a thermal relic. To accommodate such a high annihilation rate, we must require that either the WIMPs possess a very large annihilation cross section (which, in turn, requires a non-thermal mechanism for their production in the early universe), or that local inhomogeneities in the dark matter distribution boost the annihilation rate far more efficiently than N-body simulations would lead us to expect. Large annihilation cross sections might also result from non-perturbative processes in some models.[85] Alternatively, in the relatively unlikely event that a large and dense clump of dark matter happened to reside within $\sim$1 kpc of the Solar System, a sufficiently high annihilation rate, hard electron/positron spectrum, and low antiproton flux could plausibly be generated.[96]

Taken together, these considerations lead me to conclude that although the ATIC and PAMELA signals could potentially be explained by dark matter annihilations, such a scenario would require WIMPs which possess rather special properties, or that are distributed in a somewhat unlikely way. The leading astrophysical alternative for the origin of these signals is a nearby pulsar (or pulsars).[77] Although pulsars are known to be sites of electron-positron pair production, it is not possible to reliably predict the spectrum of or total power injected from these objects. To accommodate the PAMELA and ATIC observations, a nearby (within $\sim$1 kpc) and somewhat young (10^5-10^6 years) pulsar or pulsars must have deposited a few percent or more of their total energy output in the form of a very hard spectrum of electron-positron pairs. While this appears to be a larger fraction of the total energy than was generally expected, it is certainly not implausible.

In the relatively near future, measurements by the Fermi gamma ray space telescopes, as well as ground based telescopes,[81] should clarify this situation considerably. The origin of the PAMELA and ATIC signals will likely not remain a mystery for long.

14.6.2. *The WMAP Haze*

In addition to its measurements of the cosmic microwave background, data from the Wilkinson Microwave Anisotropy Probe (WMAP) has been used to provide the best measurements to date of the standard interstellar medium emission mechanisms, including thermal dust, spinning dust, ionized gas, and synchrotron. In addition to these expected foregrounds, the observations have revealed an excess of microwave emission in the inner $20°$ around the center of the Milky Way, distributed with approximate radial symmetry. This excess is known as the "WMAP Haze".[75]

Although the WMAP Haze was initially thought likely to be thermal bremsstrahlung (free-free emission) from hot gas ($10^4\,\mathrm{K} \gg \mathrm{T} \gg 10^6\,\mathrm{K}$), this interpretation can be ruled out by the relative absence of the Hα recombination line and X-ray emission. Other possible origins for this signal, such as thermal dust, spinning dust, and Galactic synchrotron as traced by low-frequency surveys, also seem unlikely. Alternatively, it has been suggested that the WMAP Haze could be a product of dark matter annihilations.[76] In particular, annihilating dark matter particles produce relativistic electrons and positrons which travel under the influence of the Galactic magnetic field. As they do, they will emit synchrotron photons, which naturally fall within the frequency range measured by WMAP.

The angular distribution of the Haze can be used to constrain the shape of the required dark matter halo profile. In particular, the morphology of the Haze is consistent with originating from dark matter distributed as $\rho(r) \propto r^{-1.2}$ within the inner kiloparsecs of our galaxy.[76] This slope falls between those predicted by the NFW, $\rho(r) \propto r^{-1}$, and Moore *et al.*, $\rho(r) \propto r^{-1.5}$, halo profiles. The annihilation cross section required of a $\sim$100-1000 GeV WIMP to produce the observed intensity of the WMAP Haze is of the same order magnitude as the value predicted for a simple thermal relic. No large boost factors are needed to generate this signal.

It is also interesting to note that the dark matter halo profile and annihilation cross section required to generate the WMAP Haze with dark matter imply a flux of prompt gamma rays from the Galactic Center region that is within the reach of the upcoming Fermi gamma ray space telescope.[97] Additionally, the upcoming Planck satellite will provide substantially improved measurements of the spectrum and angular distribution of the Haze.

14.6.3. *DAMA's Annual Modulation*

The direct detection experiment DAMA has reported evidence for an annual modulation in its rate of nuclear recoil events.[48] It has been claimed that this signal is the result WIMP interactions, which result from variations in the relative velocity of the Earth with respect to the dark matter halo as the Earth orbits the Sun. This effect is predicted to lead to a variation in the flux of dark matter particles and their velocity distribution, with expected extrema occurring at June 2 and December 2. The DAMA experiment observes a maximum rate at low nuclear recoil energies on May 24, plus or minus 8 days, and have accumulated enough data to put the significance of the observed modulation at approximately 8σ. The collaboration has not been able to identify any other systematic effects capable of producing this signal. The claim that this signal is the result of dark matter interactions has been controversial, in part because a number of other experiments appear to be in direct conflict with the DAMA result.

Several studies have attempted to reconcile the DAMA modulation signal with the null results of other direct-detection experiments.[98] In particular, an elastically scattering WIMP with a mass in the several GeV range can satisfy the results of DAMA while remaining marginally consistent with the null results of CDMS,[42] CRESST,[46] CoGeNT,[47] and XENON.[43] The allowed parameter region depends crucially on the occurrence of an effect known as channeling in the NaI crystals of the DAMA apparatus.[99]

Another possibility to have been proposed is that the DAMA signal might arise from a WIMP which does not scatter with nuclei elastically, but instead scatters inelastically, leaving the interaction in the form of a slightly heavier state (with a mass splitting on the order of 100 keV).[100] For kinematic reasons, this scenario allows for the efficient scattering of the WIMPs with iodine nuclei in DAMA, while suppressing the scattering rate off of germanium and other comparatively light nuclei used in other experiments.

14.6.4. *The INTEGRAL 511 keV Line*

In 2003, the SPI spectrometer onboard the INTEGRAL satellite confirmed the very bright emission of 511 keV photons from the region of the Galactic Bulge, corresponding to an injection rate of approximately 3×10^{42} positrons per second in the inner Galaxy.[101] This is orders of magnitude larger than the expected rate from pair creation via cosmic ray interactions with the interstellar medium. The signal appears to be approximately (but not perfectly[102]) spherically symmetric (with a full-width-half-maximum of approximately $6°$), with little of the emission tracing the Galactic Disk. A stellar origin of this signal, such as type Ia supernovae, hypernovae or gamma ray bursts, would therefore also require a mechanism (such as substantial coherent magnetic fields) by which the positrons could be transported from the disk to throughout the volume of the Bulge.[103] Furthermore, type Ia supernovae do not inject enough positrons to generate the observed intensity of this signal.[104] It is possible, however, that hypernovae[105] or gamma ray bursts[105,106] might be capable of injecting positrons at a sufficient rate. A large population of several thousand X-ray binaries has also been proposed as a possible source of these photons.[107]

Given the challenges involved with generating the observed 511 keV emission with astrophysical sources, it was suggested that this signal could potentially be the product of dark matter annihilations.[108] In order for dark matter particles to generate the observed spectral line width of this signal, however, their annihilations must inject positrons with energies below a few MeV.[109] This, in turn, implies that the dark matter's mass be near the 1-3 MeV range – much lighter than annihilating dark matter particles in most theoretically attractive models (for an interesting exception, see Ref. [110]).

Although weakly interacting particles with masses smaller than a few GeV (but larger than ~ 1 MeV) tend to be overproduced in the early universe relative to the measured dark matter abundance,[16] this can be avoided if a new light mediator is introduced which makes dark matter annihilations more efficient.[111] For example, although neutralinos within the MSSM are required by relic abundance considerations to be heavier than ~ 20 GeV,[112] they can be much lighter in extended supersymmetric models in which light Higgs bosons can mediate neutralino annihilations.[113]

For dark matter particles with MeV-scale masses to generate the measured dark matter abundance, they must annihilate during the freeze-out epoch with a cross section of $\sigma v \sim 3 \times 10^{-26}$ cm^3/s. To inject the flux of positrons needed to generate the signal observed by SPI/INTEGRAL, however, an annihilation cross section four to five orders of magnitude smaller is required. Together, these requirements force us to consider dark matter particles with annihilation cross sections of the form, $\sigma v \propto v^2$. Such behavior can be found, for example, in the case of fermionic or scalar dark matter particles annihilating through a vector mediator. For such a dark matter particle to not be overabundant today, the mediating boson also must be quite light.[114]

As an alternative explanation for the 511 keV line, it has been proposed that ~ 500 GeV dark matter particles could be collisionally excited to states which are ~ 1 MeV heavier, which produce electron-positron pairs in their subsequent de-excitations to the ground state.[115]

14.6.5. *EGRET's Diffuse Galactic Spectrum*

The satellite-based gamma ray detector, EGRET, has measured the diffuse spectrum of gamma rays over the entire sky. When compared to conventional galactic models, these measurements appear to contain an excess at energies above approximately 1 GeV. This has been interpreted as evidence for dark matter annihilations in the halo of the Milky Way.[116]

Among the most intriguing features of the observed EGRET excess is its similar spectral shape over all regions of the sky. Furthermore, this shape is consistent with that predicted from the annihilations of a 50-100 GeV WIMP. There are, however, some substantial challenges involved with interpreting the EGRET excess as a product of dark matter annihilations. In particular, to accommodate the required normalization for the annihilation rate in various regions of the Galaxy, the distribution of dark matter has to depart substantially from the predictions of standard dark matter halo profiles. In particular, Ref. [116] adopts a distribution which includes two very massive ($\sim 10^{10}\, M_\odot$) toroidal rings of dark matter near or within the Galactic Plane, at distances of approximately 4 and 14 kiloparsecs from the Galactic Center. The authors motivate the presence of these rings by observed features in the Galactic rotation curve, and suggest that they may be remnants of very massive dwarf galaxies which have been tidally disrupted.

The other difficulty involved with the interpretation of the EGRET excess as dark matter annihilation radiation is the large flux of antiprotons which is expected to be generated in such a scenario.[117] In particular, the flux of cosmic antiprotons produced is expected to exceed the measured flux by more than an order of magnitude. To avoid this conclusion, one is forced to consider significant departures from standard galactic diffusion models. In particular, an anisotropic diffusion model featuring strong convection away from the Galactic Disk and a large degree of inhomogeneities in the local environment could reduce the cosmic antiproton flux to acceptable levels.[118]

The dark matter interpretation of EGRET's measurement of the galactic diffuse spectrum has also been challenged on the grounds that the data could plausibly be explained without the addition of an exotic component, from dark matter or otherwise. In particular, uncertainties in the cosmic ray propagation and diffusion model lead to considerable variations in the predicted diffuse gamma ray backgrounds.[119] More recently, it has also been suggested that the observed excess could also be the result of systematic errors in EGRET's calibration.[120]

The Fermi gamma ray space telescope should be able to quickly clarify the origin of the EGRET diffuse emission.

14.6.6. *EGRET's Diffuse Extragalactic Spectrum*

In addition to galactic emission, it has also been proposed that dark matter annihilation radiation might constitute a significant fraction of the extragalactic (isotropic) diffuse gamma ray flux as measured by EGRET.[121]

The origin of EGRET's diffuse extragalactic gamma ray background is currently unknown. Although all or most of the observed spectrum could be the product of astrophysical source such as blazars, much of the flux observed in the 1-20 GeV range could also plausibly be the result of dark matter annihilations taking place throughout the universe.[70] In particular, the observed spectrum fits reasonably well the predictions for a WIMP with a mass of roughly 500 GeV.

The dark matter annihilation rate needed to normalize to the diffuse flux measured by EGRET is, however, quite high and requires either a very large dark matter annihilation cross section or dark matter halos which are very cuspy. In particular, if an NFW profile[61] is adopted for all halos throughout the universe, then an annihilation cross section 10^2 to 10^3 times larger than is predicted for a thermal relic is required to generate the observed gamma ray flux. If the dark matter distribution in the Milky Way is similar to that found in halos throughout the universe, however, then the gamma ray flux from the center of our galaxy would far exceed that which is observed.[122] To generate the isotropic diffuse flux observed by EGRET without conflicting with observations of the Galactic Center, therefore, requires extremely cusped halo profiles in most or at least many of the galaxies throughout the universe, and a far less dense cusp in our own Milky Way.

14.7. Summary and Outlook

The next few years will be a very exciting time for research in particle dark matter. As direct detection experiments move closer to the ton-scale, many of the most attractive models of dark matter will become within their reach. The lack of a detection in these experiments would be capable of severely constraining the nature of supersymmetry or other TeV-scale physics containing a WIMP candidate. A confirmed detection of a WIMP would open the door to the age of precision dark matter studies, in which measurements of the WIMP's mass, interactions, and distribution would begin to be made.

Indirect detection efforts are currently developing very rapidly. In particular, much of what I have written here about electron, positron and gamma ray signals for dark matter will be hopelessly out of date even a year from now. But rapid progress makes for exciting times! With new data from the Fermi gamma ray space telescope on the horizon, the near future will likely hold many exciting results for indirect detection.

The single most remarkable aspect of these lecture notes is that I have managed to write almost 50 pages about TeV-scale dark matter without mentioning the Large Hadron Collider. This should not be taken as an indication that the LHC is not important in the hunt for dark matter's identity. In contrast, I fully expect the next twenty years of particle physics (including particle-astrophysics) to be largely defined by what this incredible machine reveals to us. And while I remain largely agnostic regarding what the LHC is likely to discover, it will almost certainly give us insights into the nature of dark matter (even if that insight is that dark matter is not made up of WIMPs).

Colliders and astrophysical experiments each provide very different and complementary types of information regarding dark matter. It is very unlikely that any single experiment or class of experiments will be sufficient to conclusively identify the particle nature of dark matter. The direct or indirect detection of the dark matter particles making up our galaxy's halo will probably not be able to provide enough information to reveal the underlying physics (supersymmetry, etc.) behind these particles. In contrast, collider experiments may identify a long-lived, weakly interacting particle, but will not be able to test its cosmological stability or abundance. Only by combining the information provided by many different experimental approaches is the mystery of dark matter's particle nature likely to be solved. Although a confirmed detection of dark matter in any one search channel would constitute a discovery of the utmost importance, it would almost certainly leave many important questions unanswered as well.

Acknowledgments

I would like to thank Tao Han and Robin Erbacher for organizing TASI 2008, and K. T. Mahanthappa for his hospitality. This work was supported in part by the Fermi Research Alliance, LLC under Contract No. DE-AC02-07CH11359 with the US Department of Energy and by NASA grant NNX08AH34G.

References

1. A. Borriello and P. Salucci, Mon. Not. Roy. Astron. Soc. **323**, 285 (2001) [arXiv:astro-ph/0001082].
2. H. Hoekstra, H. Yee and M. Gladders, New Astron. Rev. **46**, 767 (2002) [arXiv:astro-ph/0205205].
3. R. B. Metcalf, L. A. Moustakas, A. J. Bunker and I. R. Parry, arXiv:astro-ph/0309738; L. A. Moustakas and R. B. Metcalf, Mon. Not. Roy. Astron. Soc. **339**, 607 (2003) [arXiv:astro-ph/0206176].
4. N. Bahcall and X. Fan, Astrophys. J. **504** (1998) 1; A. Kashlinsky, Phys. Rep. **307** (1998) 67; R. G. Carlberg et al., Astrophys. J. **516** (1999) 552; J. A. Tyson, G. P. Kochanski and I. P. Dell'Antonio, Astrophys. J. **498**, L107 (1998) [arXiv:astro-ph/9801193]. H. Dahle, arXiv:astro-ph/0701598.
5. F. Zwicky, Helv. Phys. Acta 6 (1933) 110.
6. E. Komatsu *et al.* [WMAP Collaboration], arXiv:0803.0547 [astro-ph].
7. K. A. Olive, G. Steigman and T. P. Walker, Phys. Rept. **333**, 389 (2000) [arXiv:astro-ph/9905320].
8. M. Tegmark *et al.* [SDSS Collaboration], Astrophys. J. **606**, 702 (2004) [arXiv:astro-ph/0310725].
9. M. Milgrom, Astrophys. J. **270**, 365 (1983).
10. J. D. Bekenstein, Phys. Rev. D **70**, 083509 (2004) [Erratum-ibid. D **71**, 069901 (2005)] [arXiv:astro-ph/0403694].
11. C. Skordis, D. F. Mota, P. G. Ferreira and C. Boehm, Phys. Rev. Lett. **96**, 011301 (2006) [arXiv:astro-ph/0505519].
12. D. Clowe, M. Bradac, A. H. Gonzalez, M. Markevitch, S. W. Randall, C. Jones and D. Zaritsky, arXiv:astro-ph/0608407.
13. M. Srednicki, R. Watkins and K. A. Olive, Nucl. Phys. B **310**, 693 (1988).
14. P. Gondolo and G. Gelmini, Nucl. Phys. B **360** (1991) 145.

15. E. W. Kolb and M. S. Turner, "The Early Universe".
16. B. W. Lee and S. Weinberg, Phys. Rev. Lett. **39**, 165 (1977).
17. C. Amsler *et al.* (Particle Data Group), Phys. Lett. B **667**, 1 (2008).
18. K. Griest and D. Seckel, Phys. Rev. D **43** (1991) 3191.
19. J. L. Feng, A. Rajaraman and F. Takayama, Phys. Rev. Lett. **91** (2003) 011302 [arXiv:hep-ph/0302215].
20. For a review of supersymmetry phenomenology, see: S. P. Martin, arXiv:hep-ph/9709356.
21. J. R. Ellis, S. Kelley and D. V. Nanopoulos, Phys. Lett. B **260**, 131 (1991).
22. H. Goldberg, Phys. Rev. Lett. **50**, 1419 (1983); J. R. Ellis, J. S. Hagelin, D. V. Nanopoulos, K. A. Olive and M. Srednicki, Nucl. Phys. B **238**, 453 (1984).
23. T. Falk, K. A. Olive and M. Srednicki, Phys. Lett. B **339**, 248 (1994) [arXiv:hep-ph/9409270].
24. G. Jungman, M. Kamionkowski and K. Griest, Phys. Rept. **267**, 195 (1996).
25. J. Edsjo and P. Gondolo, Phys. Rev. D **56** (1997) 1879 [arXiv:hep-ph/9704361].
26. J. R. Ellis, T. Falk, K. A. Olive and M. Srednicki, Astropart. Phys. **13**, 181 (2000) [Erratum-ibid. **15**, 413 (2001)] [arXiv:hep-ph/9905481].
27. J. R. Ellis, K. A. Olive and Y. Santoso, Astropart. Phys. **18**, 395 (2003) [arXiv:hep-ph/0112113].
28. J. Edsjo, M. Schelke, P. Ullio and P. Gondolo, JCAP **0304** (2003) 001 [arXiv:hep-ph/0301106].
29. P. Gondolo, J. Edsjo, P. Ullio, L. Bergstrom, M. Schelke and E. A. Baltz, JCAP **0407**, 008 (2004) [arXiv:astro-ph/0406204].
30. G. Belanger, F. Boudjema, A. Pukhov and A. Semenov, Comput. Phys. Commun. **177**, 894 (2007).
31. K. Hagiwara, A.D. Martin, D. Nomura, and T. Teubner, (2006) [arXiv:hep-ph/0611102].
32. N. Arkani-Hamed, S. Dimopoulos and G. R. Dvali, Phys. Lett. B **429**, 263 (1998) [arXiv:hep-ph/9803315].
33. N. Arkani-Hamed, S. Dimopoulos and G. R. Dvali, Phys. Rev. D **59**, 086004 (1999) [arXiv:hep-ph/9807344].
34. L. Randall and R. Sundrum, Phys. Rev. Lett. **83**, 3370 (1999) [arXiv:hep-ph/9905221].
35. T. Appelquist, H. C. Cheng and B. A. Dobrescu, Phys. Rev. D **64**, 035002 (2001) [arXiv:hep-ph/0012100].
36. D. Hooper and S. Profumo, Phys. Rept., in press, arXiv:hep-ph/0701197.
37. G. Servant and T. M. Tait, Nucl. Phys. B **650** (2003) 391 [arXiv:hep-ph/0206071].
38. H. C. Cheng, J. L. Feng and K. T. Matchev, Phys. Rev. Lett. **89**, 211301 (2002) [arXiv:hep-ph/0207125].
39. F. Burnell and G. D. Kribs, Phys. Rev. D **73**, 015001 (2006) [arXiv:hep-ph/0509118].
40. K. Kong and K. T. Matchev, JHEP **0601**, 038 (2006) [arXiv:hep-ph/0509119].

41. H. C. Cheng and I. Low, JHEP **0309**, 051 (2003) [arXiv:hep-ph/0308199].

42. Z. Ahmed *et al.* [CDMS Collaboration], arXiv:0802.3530 [astro-ph].

43. J. Angle *et al.* [XENON Collaboration], Phys. Rev. Lett. **100**, 021303 (2008) [arXiv:0706.0039 [astro-ph]].

44. G. J. Alner *et al.*, Astropart. Phys. **28**, 287 (2007) [arXiv:astro-ph/0701858]; G. J. Alner *et al.* [UK Dark Matter Collaboration], Astropart. Phys. **23**, 444 (2005).

45. V. Sanglard *et al.* [The EDELWEISS Collaboration], Phys. Rev. D **71**, 122002 (2005) [arXiv:astro-ph/0503265].

46. G. Angloher *et al.*, Astropart. Phys. **23**, 325 (2005) [arXiv:astro-ph/0408006].

47. C. E. Aalseth *et al.*, arXiv:0807.0879 [astro-ph].

48. R. Bernabei *et al.* [DAMA Collaboration], arXiv:0804.2741 [astro-ph]; R. Bernabei *et al.* [DAMA Collaboration], arXiv:0804.2741 [astro-ph].

49. E. Behnke *et al.* [COUPP Collaboration], Science **319**, 933 (2008) [arXiv:0804.2886 [astro-ph]].

50. P. Benetti *et al.*, arXiv:astro-ph/0701286; R. Brunetti *et al.*, New Astron. Rev. **49**, 265 (2005) [arXiv:astro-ph/0405342].

51. H. S. Lee. *et al.* [KIMS Collaboration], Phys. Rev. Lett. **99**, 091301 (2007) [arXiv:0704.0423 [astro-ph]].

52. A. Bottino, F. Donato, N. Fornengo and S. Scopel, Astropart. Phys. **18**, 205 (2002) [arXiv:hep-ph/0111229]; Astropart. Phys. **13**, 215 (2000) [arXiv:hep-ph/9909228]; J. R. Ellis, K. A. Olive, Y. Santoso and V. C. Spanos, Phys. Rev. D **71**, 095007 (2005) [arXiv:hep-ph/0502001].

53. G. B. Gelmini, P. Gondolo and E. Roulet, Nucl. Phys. B **351**, 623 (1991); M. Srednicki and R. Watkins, Phys. Lett. B **225**, 140 (1989); M. Drees and M. Nojiri, Phys. Rev. D **48**, 3483 (1993) [arXiv:hep-ph/9307208]; M. Drees and M. M. Nojiri, Phys. Rev. D **47**, 4226 (1993) [arXiv:hep-ph/9210272]; J. R. Ellis, A. Ferstl and K. A. Olive, Phys. Lett. B 481, (2000) 304, [arXiv:hep-ph/0001005].

54. G. Servant and T. M. Tait, New J. Phys. **4**, 99 (2002) [arXiv:hep-ph/0209262].

55. http://dendera.berkeley.edu/plotter/entryform.html

56. M. Beltran, D. Hooper, E. W. Kolb and Z. C. Krusberg, arXiv:0808.3384 [hep-ph].

57. L. Bergstrom and P. Ullio, Nucl. Phys. B **504**, 27 (1997) [arXiv:hep-ph/9706232].

58. P. Ullio and L. Bergstrom, Phys. Rev. D **57**, 1962 (1998) [arXiv:hep-ph/9707333].

59. L. Bergstrom, P. Ullio and J. H. Buckley, Astropart. Phys. **9**, 137 (1998) [arXiv:astro-ph/9712318].

60. V. Berezinsky, A. Bottino and G. Mignola, Phys. Lett. B **325**, 136 (1994) [arXiv:hep-ph/9402215].

61. J. F. Navarro, C. S. Frenk and S. D. M. White, Astrophys. J. **462**, 563 (1996) [arXiv:astro-ph/9508025]; J. F. Navarro, C. S. Frenk and S. D. M. White, Astrophys. J. **490**, 493 (1997).

62. B. Moore, S. Ghigna, F. Governato, G. Lake, T. Quinn, J. Stadel and P. Tozzi, Astrophys. J. **524**, L19 (1999).

63. F. Prada, A. Klypin, J. Flix, M. Martinez and E. Simonneau, arXiv:astro-ph/0401512; G. Bertone and D. Merritt, Mod. Phys. Lett. A **20**, 1021 (2005) [arXiv:astro-ph/0504422]; G. Bertone and D. Merritt, Phys. Rev. D **72**, 103502 (2005) [arXiv:astro-ph/0501555].

64. P. Gondolo and J. Silk, Phys. Rev. Lett. **83**, 1719 (1999) [arXiv:astro-ph/9906391]; P. Ullio, H. Zhao and M. Kamionkowski, Phys. Rev. D **64**, 043504 (2001) [arXiv:astro-ph/0101481]; G. Bertone, G. Sigl and J. Silk, Mon. Not. Roy. Astron. Soc. **337**, 98 (2002) [arXiv:astro-ph/0203488].

65. F. Aharonian *et al.* [The HESS Collaboration], arXiv:astro-ph/0408145; J. Albert *et al.* [MAGIC Collaboration], Astrophys. J. **638**, L101 (2006) [arXiv:astro-ph/0512469].

66. G. Zaharijas and D. Hooper, Phys. Rev. D **73**, 103501 (2006). [arXiv:astro-ph/0603540].

67. S. Dodelson, D. Hooper and P. D. Serpico, arXiv:0711.4621 [astro-ph].

68. D. Hooper and B. L. Dingus, Phys. Rev. D **70**, 113007 (2004) [arXiv:astro-ph/0210617].

69. N. W. Evans, F. Ferrer and S. Sarkar, Phys. Rev. D **69**, 123501 (2004) [arXiv:astro-ph/0311145]; L. Bergstrom and D. Hooper, Phys. Rev. D **73**, 063510 (2006) [arXiv:hep-ph/0512317]; L. E. Strigari, S. M. Koushiappas, J. S. Bullock, M. Kaplinghat, J. D. Simon, M. Geha and B. Willman, arXiv:0709.1510 [astro-ph].

70. P. Ullio, L. Bergstrom, J. Edsjo and C. G. Lacey, Phys. Rev. D **66**, 123502 (2002) [arXiv:astro-ph/0207125]; D. Elsaesser and K. Mannheim, Astropart. Phys. **22**, 65 (2004) [arXiv:astro-ph/0405347].

71. O. Adriani *et al.*, arXiv:0810.4995 [astro-ph].

72. S. W. Barwick *et al.* [HEAT Collaboration], Astrophys. J. **482**, L191 (1997) [arXiv:astro-ph/9703192]; S. Coutu *et al.* [HEAT-pbar Collaboration], in Proceedings of 27th ICRC (2001).

73. Olzem, Jan [AMS Collaboration], Talk given at the 7th UCLA Symposium on Sources and Detection of Dark Matter and Dark Energy in the Universe, Marina del Ray, CA, Feb 22-24, 2006.

74. J. Chang *et al.* [ATIC Collaboration], Nature **456**, 362 (2008).

75. G. Dobler and D. P. Finkbeiner, Astrophys. J. **680**, 1222 (2008) [arXiv:0712.1038 [astro-ph]].

76. D. P. Finkbeiner, arXiv:astro-ph/0409027; D. Hooper, D. P. Finkbeiner and G. Dobler, Phys. Rev. D **76**, 083012 (2007) [arXiv:0705.3655 [astro-ph]].

77. D. Hooper, P. Blasi and P. D. Serpico, arXiv:0810.1527 [astro-ph]; S. Profumo, arXiv:0812.4457 [astro-ph]; F. A Aharonian, A. M. Atoyan and H. J. Volk, Astron. Astrophys. **294**, L41-L44 (1995); L. Zhang and K. S. Cheng, Astron. Astrophys. **368**, 1063-1070 (2001); I. Buesching, O. C. de Jager, M. S. Potgieter and C. Venter, arXiv:0804.0220 [astro-ph].

78. I. Cholis, L. Goodenough, D. Hooper, M. Simet and N. Weiner, arXiv:0809.1683 [hep-ph].

79. L. Bergstrom, T. Bringmann and J. Edsjo, arXiv:0808.3725 [astro-ph];

M. Cirelli and A. Strumia, arXiv:0808.3867 [astro-ph]; V. Barger, W. Y. Keung, D. Marfatia and G. Shaughnessy, arXiv:0809.0162 [hep-ph]; N. Arkani-Hamed, D. P. Finkbeiner, T. Slatyer and N. Weiner, arXiv:0810.0713 [hep-ph]; I. Cholis, G. Dobler, D. P. Finkbeiner, L. Goodenough and N. Weiner, arXiv:0811.3641 [astro-ph]; P. J. Fox and E. Poppitz, arXiv:0811.0399 [hep-ph]; K. M. Zurek, arXiv:0811.4429 [hep-ph].

80. F. Donato, D. Maurin, P. Brun, T. Delahaye and P. Salati, arXiv:0810.5292 [astro-ph]; M. Cirelli, M. Kadastik, M. Raidal and A. Strumia, arXiv:0809.2409 [hep-ph].

81. J. Hall and D. Hooper, arXiv:0811.3362 [astro-ph]; E. A. Baltz and D. Hooper, JCAP **0507**, 001 (2005) [arXiv:hep-ph/0411053].

82. W. R. Webber, M. A. Lee and M. Gupta, Astrophys. J.**390** (1992) 96; I. V. Moskalenko, A. W. Strong, S. G. Mashnik and J. F. Ormes, Astrophys. J. **586**, 1050 (2003) [arXiv:astro-ph/0210480]; I. V. Moskalenko and A. W. Strong, Phys. Rev. D **60**, 063003 (1999) [arXiv:astro-ph/9905283]; E. A. Baltz and J. Edsjo, Phys. Rev. D **59** (1999) 023511 [arXiv:astro-ph/9808243]; D. Hooper and J. Silk, Phys. Rev. D **71**, 083503 (2005) [arXiv:hep-ph/0409104].

83. A. W. Strong, I .V. Moskalenko and V. S. Ptuskin, ARNPS, 57, 285 (2007).

84. D. Hooper and G. D. Kribs, Phys. Rev. D **70**, 115004 (2004) [arXiv:hep-ph/0406026].

85. M. Cirelli and A. Strumia, arXiv:0808.3867 [astro-ph]; I. Cholis, G. Dobler, D. P. Finkbeiner, L. Goodenough and N. Weiner, arXiv:0811.3641 [astro-ph]; N. Arkani-Hamed, D. P. Finkbeiner, T. Slatyer and N. Weiner, arXiv:0810.0713 [hep-ph].

86. A. Gould, Astrophys. J. **388**, 338 (1991).

87. A. Gould, Astrophys. J. **321**, 571 (1987).

88. K. Griest and D. Seckel, Nucl. Phys. B **283**, 681 (1987) [Erratum-ibid. B **296**, 1034 (1988)].

89. G. Jungman and M. Kamionkowski, Phys. Rev. D **51** (1995) 328 [arXiv:hep-ph/9407351]; For a more recent calculation see: M. Cirelli, N. Fornengo, T. Montaruli, I. Sokalski, A. Strumia and F. Vissani, arXiv:hep-ph/0506298.

90. R. Lehnert and T. J. Weiler, Phys. Rev. D **77** (2008) 125004 [arXiv:0708.1035].

91. T. DeYoung [IceCube Collaboration], Int. J. Mod. Phys. A **20**, 3160 (2005); J. Ahrens *et al.* [The IceCube Collaboration], Nucl. Phys. Proc. Suppl. **118**, 388 (2003) [arXiv:astro-ph/0209556].

92. S. Desai *et al.* [Super-Kamiokande Collaboration], Phys. Rev. D **70**, 083523 (2004) [Erratum-ibid. D **70**, 109901 (2004)] [arXiv:hep-ex/0404025].

93. L. Bergstrom, J. Edsjo and P. Gondolo, Phys. Rev. D **55**, 1765 (1997) [arXiv:hep-ph/9607237]; Phys. Rev. D **58**, 103519 (1998) [arXiv:hep-ph/9806293]; V. D. Barger, F. Halzen, D. Hooper and C. Kao, Phys. Rev. D **65**, 075022 (2002) [arXiv:hep-ph/0105182].

94. F. Halzen and D. Hooper, Phys. Rev. D **73**, 123507 (2006) [arXiv:hep-ph/0510048].

95. D. Hooper and G. D. Kribs, Phys. Rev. D **67**, 055003 (2003) [arXiv:hep-

ph/0208261].

96. D. Hooper, A. Stebbins and K. M. Zurek, arXiv:0812.3202 [hep-ph].

97. D. Hooper, G. Zaharijas, D. P. Finkbeiner and G. Dobler, Phys. Rev. D **77**, 043511 (2008) [arXiv:0709.3114 [astro-ph]].

98. A. Bottino, F. Donato, N. Fornengo and S. Scopel, Phys. Rev. D **77**, 015002 (2008) [arXiv:0710.0553 [hep-ph]]; R. Foot, arXiv:0804.4518 [hep-ph]; J. L. Feng, J. Kumar and L. E. Strigari, arXiv:0806.3746 [hep-ph]; F. Petriello and K. M. Zurek, arXiv:0806.3989 [hep-ph]; A. Bottino, F. Donato, N. Fornengo and S. Scopel, arXiv:0806.4099 [hep-ph].

99. E. M. Drobyshevski, arXiv:0706.3095 [physics.ins-det]; R. Bernabei *et al.*, Eur. Phys. J. C **53**, 205 (2008) [arXiv:0710.0288 [astro-ph]].

100. S. Chang, G. D. Kribs, D. Tucker-Smith and N. Weiner, arXiv:0807.2250 [hep-ph]; D. R. Smith and N. Weiner, Phys. Rev. D **64**, 043502 (2001) [arXiv:hep-ph/0101138]; D. Tucker-Smith and N. Weiner, arXiv:hep-ph/0402065.

101. P. Jean *et al.*, Astron. Astrophys. **407**, L55 (2003) [arXiv:astro-ph/0309484].

102. G. Weidenspointner *et al.*, Nature **451**, 159 (2008).

103. N. Prantzos, Astron. Astrophys. **449**, 869 (2006) [arXiv:astro-ph/0511190].

104. E. Kalemci, S. E. Boggs, P. A. Milne and S. P. Reynolds, Astrophys. J. **640**, L55 (2006) [arXiv:astro-ph/0602233].

105. M. Casse, B. Cordier, J. Paul and S. Schanne, Astrophys. J. **602**, L17 (2004) [arXiv:astro-ph/0309824].

106. G. Bertone, A. Kusenko, S. Palomares-Ruiz, S. Pascoli and D. Semikoz, Phys. Lett. B **636**, 20 (2006) [arXiv:astro-ph/0405005]; E. Parizot, M. Casse, R. Lehoucq and J. Paul, arXiv:astro-ph/0411656.

107. R. M. Bandyopadhyay, J. Silk, J. E. Taylor and T. J. Maccarone, arXiv:0810.3674 [astro-ph].

108. C. Boehm, D. Hooper, J. Silk, M. Casse and J. Paul, Phys. Rev. Lett. **92**, 101301 (2004) [arXiv:astro-ph/0309686].

109. J. F. Beacom and H. Yuksel, Phys. Rev. Lett. **97**, 071102 (2006) [arXiv:astro-ph/0512411]; J. F. Beacom, N. F. Bell and G. Bertone, Phys. Rev. Lett. **94**, 171301 (2005) [arXiv:astro-ph/0409403].

110. D. Hooper and K. M. Zurek, Phys. Rev. D **77**, 087302 (2008) [arXiv:0801.3686 [hep-ph]].

111. C. Boehm, T. A. Ensslin and J. Silk, J. Phys. G **30**, 279 (2004) [arXiv:astro-ph/0208458].

112. D. Hooper and T. Plehn, Phys. Lett. B **562**, 18 (2003) [arXiv:hep-ph/0212226]; A. Bottino, F. Donato, N. Fornengo and S. Scopel, Phys. Rev. D **68**, 043506 (2003) [arXiv:hep-ph/0304080].

113. J. F. Gunion, D. Hooper and B. McElrath, Phys. Rev. D **73**, 015011 (2006) [arXiv:hep-ph/0509024].

114. C. Boehm and P. Fayet, Nucl. Phys. B **683**, 219 (2004) [arXiv:hep-ph/0305261]; P. Fayet, Phys. Rev. D **70**, 023514 (2004) [arXiv:hep-ph/0403226].

115. D. P. Finkbeiner and N. Weiner, arXiv:astro-ph/0702587.

116. W. de Boer, M. Herold, C. Sander, V. Zhukov, A. V. Gladyshev and

D. I. Kazakov, arXiv:astro-ph/0408272; W. de Boer, C. Sander, V. Zhukov, A. V. Gladyshev and D. I. Kazakov, Astron. Astrophys. **444**, 51 (2005) [arXiv:astro-ph/0508617]; W. de Boer, C. Sander, V. Zhukov, A. V. Gladyshev and D. I. Kazakov, Phys. Lett. B **636**, 13 (2006) [arXiv:hep-ph/0511154]; W. de Boer, C. Sander, V. Zhukov, A. V. Gladyshev and D. I. Kazakov, Phys. Rev. Lett. **95**, 209001 (2005) [arXiv:astro-ph/0602325].

117. L. Bergstrom, J. Edsjo, M. Gustafsson and P. Salati, JCAP **0605**, 006 (2006) [arXiv:astro-ph/0602632].

118. W. de Boer, I. Gebauer, C. Sander, M. Weber and V. Zhukov, AIP Conf. Proc. **903**, 607 (2007) [arXiv:astro-ph/0612462]; See also: W. de Boer and V. Zhukov, arXiv:0709.4576 [astro-ph].

119. I. V. Moskalenko, S. W. Digel, T. A. Porter, O. Reimer and A. W. Strong, arXiv:astro-ph/0609768.

120. F. W. Stecker, S. D. Hunter and D. A. Kniffen, arXiv:0705.4311 [astro-ph].

121. D. Elsaesser and K. Mannheim, Phys. Rev. Lett. **94**, 171302 (2005) [arXiv:astro-ph/0405235].

122. S. Ando, Phys. Rev. Lett. **94**, 171303 (2005) [arXiv:astro-ph/0503006].

TASI 2008 Student Talk Schedule
Talks will be 30 minutes, with 5 minutes for questions

Thursday 6/5 (starting at 7)

Randy Kelley "Effective field theories"
Andrew Hornig "Basics of HQET and SCET"
Jui-Yu Chiu "An example using SCET"

Friday 6/6 (starting at 3:45)

Claudia Frugiuele "Small x resummation (pQCD)"
Rob Putnam "The factorization scale in pQCD"

Monday 6/9 (starting at 7)

Abram Krislock "Supercritical string cosmology at the LHC"
Brian Glover "Galaxy cluster surveys for constraining new physics"

Tuesday 6/10 (starting 7)

Jamie Gainer "General Features of the MSSM at the ILC"
Sebastian Grab "Phenomenology of R-parity violating mSUGRA models"
Ricky Fok "Baryogenesis with SUSY and 4th generation"

Thursday 6/12 (starting at 7)

Brian Dudley "SUSY flavor charging sum rules and $b \to s$ gamma"
Saba Zuberi "Higher order perturbative corrections in extracting $|V_u b|$"
Johannes Heinonen "Little Higgs, T-parity, and anomalies"

Friday 6/13 (starting at 3:45)

Christian Spethmann "Anomaly free UV completion of the Littlest Higgs with T-Parity"
Andy Spray "Indirect detection of little Higgs dark matter"

Monday 6/16 (starting at 7)

Phill Grajek "Identifying multi-top events at LHC"
Haiying Cai "A spin-1 top partner"
Seth Quakenbush "Measuring Z' couplings at the LHC"

Tuesday 6/17 (starting at 7)

Mathew McCaskey "LHC phenomenology of a complex singlet extended SM"
Won Sang Cho "Measuring superparticle masses at hadron colliders using transverse mass kink"

Wednesday 6/18 (starting at 7)

Travis Martin & Ken Moats "WHiZard Monte Carlo event generator"
Willie Klemm "Discriminating spin through quantum interference"
Chee Sheng Fong "Soft leptogenesis: flavored and unflavored"

Monday 6/23 (starting at 3:45)

Nausheen Shah "Gauge-Higgs unification in warped extra dimensions"
Ryan Gavin "Color octet scalar phenomenology"